Graph Neural Networks

Scrivener Publishing
100 Cummings Center, Suite 541J
Beverly, MA 01915-6106

Leading-Edge Breakthroughs in Artificial Intelligence

Series Editor: Dr. S. Balamurugan (sbnbala@gmail.com)

The world is on the cusp of several breakthroughs in artificial intelligence (AI), rapidly transforming industries and the research field. The "Leading-Edge Breakthroughs in Artificial Intelligence" book series is dedicated to exploring the most pioneering developments in artificial intelligence. The series aims to highlight both theoretical and practical advancements, offering insights into several cutting-edge advancements in AI and how the same could be deployed to solve complex problems in fields such as healthcare, finance, manufacturing, robotics, marketing, energy management, agriculture, autonomous systems, and many more. It also addresses critical issues like AI governance, ethics, trust, security, transparency, and societal impacts, making it an essential resource for researchers, engineers, and industry leaders looking to leverage AI for future breakthroughs.

Publishers at Scrivener
Martin Scrivener (martin@scrivenerpublishing.com)
Phillip Carmical (pcarmical@scrivenerpublishing.com)

Graph Neural Networks

Concepts and Applications

Edited by

J. Ramkumar
Department of Computer Science, School of Quantum Science, Computing and AI, Rathinam Global (Deemed to be University), Coimbatore, Tamil Nadu, India

Sridaran Rajagopal
Department of Computer Applications, Ganpat University, Gujarat, India

B. Suchitra
Department of Information Technology and Cognitive Systems, Sri Krishna Arts and Science College, Coimbatore, Tamil Nadu, India

and

S. Balamurugan
Intelligent Research Consultancy Services (iRCS), Coimbatore, Tamil Nadu, India

Scrivener Publishing

WILEY

This edition first published 2026 by John Wiley & Sons, Inc., 111 River Street, Hoboken, NJ 07030, USA and Scrivener Publishing LLC, 100 Cummings Center, Suite 541J, Beverly, MA 01915, USA

For more information about Scrivener publications please visit www.scrivenerpublishing.com.

Wiley Global Headquarters
111 River Street, Hoboken, NJ 07030, USA

For details of our global editorial offices, customer services, and more information about Wiley products visit us at www.wiley.com.

The manufacturer's authorized representative according to the EU General Product Safety Regulation is Wiley-VCH GmbH, Boschstr. 12, 69469 Weinheim, Germany, e-mail: Product_Safety@wiley.com.

Library of Congress Cataloging-in-Publication Data

ISBN 978-1-394-42273-9

Cover image: Generated with AI using Adobe Firefly
Cover design by Russell Richardson

Set in size of 11pt and Minion Pro by Manila Typesetting Company, Makati, Philippines

Printed in the USA

10 9 8 7 6 5 4 3 2 1

Contents

Series Preface

This book provides a comprehensive introduction to the principles, architectures, and real-world use cases of graph neural networks, highlighting their impact on complex data representations and relational learning. This significant volume explains one of the more powerful and fast-moving areas of deep learning, which is capable of effectively modeling complex and even non-Euclidean data structures such as social networks, biological processes, or knowledge graphs. Because graph neural networks (GNN) can allow machines to learn from structured relationships between interrelated entities, GNNs introduces novel possibilities in areas such as the recommendation system, drug discovery, fraud detection, natural language understanding, etc. Included in the series *Leading Edge Breakthroughs in Artificial Intelligence*, it explores a cutting-edge paradigm that fundamentally augments standard AI models and offers potentially innovative solutions in an area where relational intelligence and context-aware inference are highly relevant.

Dr. S. Balamurugan
Director, Intelligent Research Consultancy Services, Coimbatore, Tamil Nadu, India

Preface

This book combines scholarly contributions that examine the theoretical foundations, architectural innovations, and real-world applications of Graph Neural Networks (GNNs). Authored by domain experts, the chapters span fundamental principles, model design, scalability challenges, and specialized use cases across science, healthcare, cybersecurity, and industry, offering readers a coherent and progressive understanding of how GNNs operate in diverse contexts.

The first chapter explains why traditional neural networks struggle with non-Euclidean data and how GNNs bridge this gap through neighborhood aggregation and iterative learning, with applications spanning social networks to molecular biology. The second chapter provides the structural and mathematical backbone—nodes, edges, adjacency matrices, and isomorphism—linking pure graph theory to GNN frameworks and preparing readers for more advanced architectures.

Chapter 3 details message passing as the core mechanism of GNNs, comparing synchronous and asynchronous schemes, and surveying models such as GCNs, GATs, and MPNNs. Chapter 4 expands to a survey of architectures, exploring how components like layer depth, attention, pooling, and skip connections shape performance, while introducing newer hybrids such as Graph Transformers and GNNs integrated with reinforcement learning.

The fifth chapter examines fairness, bias, privacy, and explainability, emphasizing the risks of incomplete or manipulated graphs and mitigation strategies like adversarial defenses and fairness-aware protocols. The sixth explores scalability, addressing computational bottlenecks, neighbor explosion, and distributed training, with benchmarks and strategies for making GNNs more efficient at industrial scale.

Chapter 7 analyzes the effect of topological indices—such as centrality, connectivity, and symmetry—on GNN expressiveness, particularly in chemistry and biology. Chapter 8 turns to deployment at scale, describing infrastructures used by leading firms to apply GNNs in recommendation, fraud detection, and network routing, highlighting graph databases, parallel processing, and real-world system trade-offs.

The ninth chapter presents federated learning with GNNs to preserve privacy while enabling collaboration across decentralized data, key for healthcare, finance, and IoT. Chapter 10 turns to scientific computing, showing how GNNs aid in physics and chemistry simulations, molecule prediction, and surrogate modeling, often surpassing traditional solvers.

Chapter 11 addresses zero-day attack detection, demonstrating how stratified sampling combined with GNNs improves anomaly detection in communication networks. Chapter 12 introduces an enhanced spatio-temporal GNN for urban traffic prediction, integrating spatial and temporal data to improve forecasting accuracy and efficiency for smart cities.

Chapter 13 surveys graph-based approaches in cybersecurity and critical infrastructure, outlining supervised and contrastive methods for detecting anomalies in large datasets. Chapter 14 applies these strategies specifically to fintech, focusing on fraud detection, transaction auditing, and regulatory compliance, with temporal and heterogeneous graphs capturing hidden transactional patterns.

Chapter 15 presents crowd analytics, showing how relational modeling enhances people counting and behavior analysis in smart retail, safety, and event monitoring. Chapter 16 provides a bibliometric analysis of India's GNN research, charting trends, institutions, and collaborations to position India in the global graph AI landscape.

Chapter 17 presents GNN-based methods for early osteoporosis prediction, modeling patient records as relational graphs to capture latent health trajectories. Chapter 18 surveys broader medical applications—from genomics and radiology to drug discovery and EHRs—while addressing adoption challenges such as privacy, validation, and interpretability.

Chapter 19 explores autism prediction, mapping behavioral screening data into graph structures that capture relational patterns, improving diagnostic reliability. Chapter 20 examines GNNs in combinatorial optimization, applying them to supply chains, chip design, and transportation, with reinforcement learning and hybrid models enabling scalable, adaptive solutions.

Chapter 21 demonstrates how GNNs improve software quality assurance and defect prediction by modeling code interactions as graphs, providing insights for integration into software pipelines. Chapter 22 identifies emerging themes—lifelong learning, dynamic graphs, quantum models, privacy, and ethics—laying out a forward-looking roadmap for research and cross-disciplinary applications.

In closing, we extend our thanks to Martin Scrivener and the team at Scrivener Publishing for their effort and guidance in the publication of this volume.

Part I

CONCEPTUAL FOUNDATIONS AND LEARNING FRAMEWORKS

1

Introduction to Graph Neural Networks

K. Sangeetha and P. Solairani*

Department of Mathematics, RVS College of Arts and Science, Coimbatore, Tamilnadu, India

Abstract

We can naturally express the connections between the set of objects as patterns of graphs. A graph is a mathematical structure consists of the vertices and the links, where each node can have multiple nodes and edges connected to it. Mathematical graphs are often represented as arrays of elements, such as adjacency and incidence matrices. Graph theory provides a crucial framework for handling complicated, wide data easily in various fields, which includes network modeling, routing algorithms, and more. Graph theory helps to design and analyze complicated networks, such as the ecological networks, social networks and transportation systems, and also provides efficient algorithms for finding shortest paths and optimal routes in networks. Graph theory can be used to produce predictive models like the graph neural networks which help to learn about patterns and connections between the set of data. These graph networks are one of the neural networks to handle the graphical structured data in which, the nodes represent objects such as persons, organizations, locations and the edges represent the connection and the interactions between the objects, such as friendships, collaborations and the transactions. One of the emerging technologies is the machine learning algorithm which follows the concept of a human neuron. A neuron collects input from multiple neurons and transmits signals to multiple neurons. Artificial neural networks consist of a three-stage process, which includes the input layer, iteration or hidden layer and output layer. An input layer observes the input from the system or environment. A hidden or iteration layer may process and transform the input data through multiple layers of interconnected neurons. The output layer determines the output or prediction based on processed input data. Graph neural networks play some significant roles in FinTech and fraud detection, like transaction monitoring, digital payments, identity verification, biometric authentication,

**Corresponding author*: prsolairani@gmail.com

J. Ramkumar, Sridaran Rajagopal, B. Suchitra and S. Balamurugan (eds.) Graph Neural Networks: Concepts and Applications, (3–42) © 2026 Scrivener Publishing LLC

anomaly detection, predictive modeling, etc. The notable benefits of neural networks in FinTech and fraud detection are reduced losses, regulatory compliance, and increased efficiency.

Keywords: Predictive models, graph neural networks, architecture, Fintech, fraud detection

1.1 Introduction

The graph neural networks (GNNs) are the fundamental tool for analyzing the complicated relationships in the graphical data. By incorporate the mechanism of the neural networks within graph; the graph neural networks incorporate the links in between the vertices and the in data. They are particularly useful in various applications such as the social media networks, bio molecular structures and the money transactions. In social network analysis, the spammer detection methods are of two types such as the feature based and the propagation based approaches. The feature based methods are used in the matrix factorization with text features and the social graph and the propagation based approaches are used in random fields to monitor the user relationship patterns but often overlook valuable text features. The recommendation systems utilize machine learning techniques to suggest products based on the user behavior, preferences and the interactions [2]. These systems are valuable for the content providers, driving consumers to the various products and the services tailored to their interests. The GNNs can improve the existing regulatory requirements and increase the efficiency of fraud detection [15], and using in many tasks such as transaction monitoring and risk assessment. The graph neural networks are the crucial tool to make a huge change in molecular biology by modeling molecules as graphs. Also, they are using in the traffic forecasting [29] by modeling the traffic networks as graphs. In this chapter, we discussed about the fundamentals of GNNs with several types and their advancements in different fields including FinTech [10] and fraud detection [1, 8].

1.2 Graph Structure

The graph structure contains the set of nodes and the links, denoted by G. The vertex is also called the node. Generally, the nodes are used to denote the objects or entities and the edges are used to denote the connections between the nodes. For example, the set of edges and vertices of

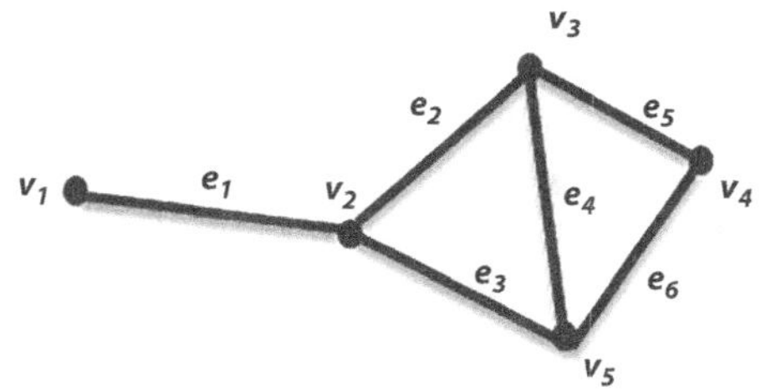

Figure 1.1 Graph structure.

G are given in Figure 1.1 is as follows. That is, $V = \{v_1, v_2, v_3, v_4, v_5\}$ and $E = \{e_1, e_2, e_3, e_4, e_5, e_6\}$.

1.3 Types of Graphs

The graphs, which are using in machine learning techniques can be classified as the directed and undirected, weighted and unweighted, the cyclic and acyclic based on their several characteristics including the direction, weights and the presence of cycles, etc. Also, the graphs can be categorized as the homogeneous and the heterogeneous. In the homogeneous graph, the vertices and the links are of the same type. In the heterogeneous graph, the vertices and the links are of different types and allowing for the diverse representations and analyses.

1.3.1 Directed Graph

A graph in which all the edges are directed is said to be the directed graph. The directed graph is also noted as the digraph. They represent the one way relationship between its vertices. An example of a directed graph is shown in Figure 1.2.

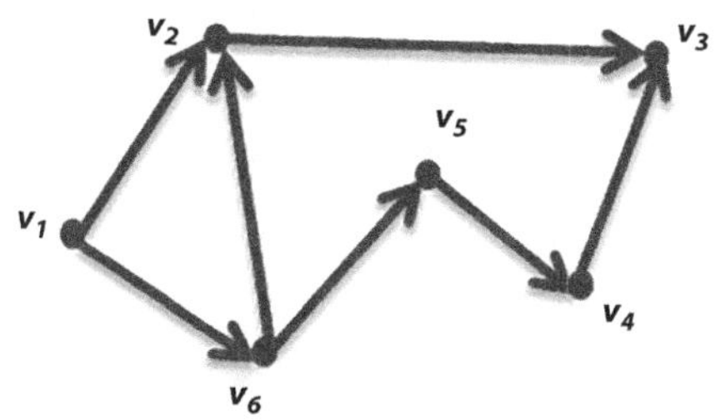

Figure 1.2 Directed graph (one-way).

1.3.2 Undirected Graph

A graph in which, there is no connections with the direction is said to be an undirected graph. It is also called bidirectional and represents the two-way relationship between its vertices. An example of an Undirected graph is shown in Figure 1.3.

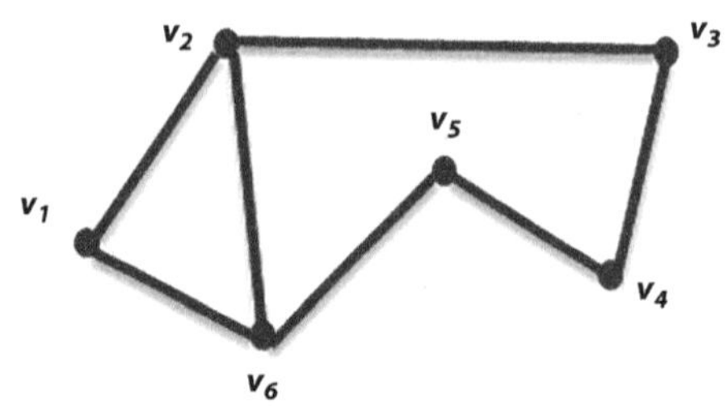

Figure 1.3 Undirected graph (two-way).

1.3.3 Weighted Graph

A graph in which, all the links are allocated by some quantity is said to be a weighted graph. They quantity may be corresponding to any one of the characteristics such as time, cost, distance and capacity. Examples of an undirected weighted graph and a directed weighted graph are shown in the following Figure 1.4 and Figure 1.5.

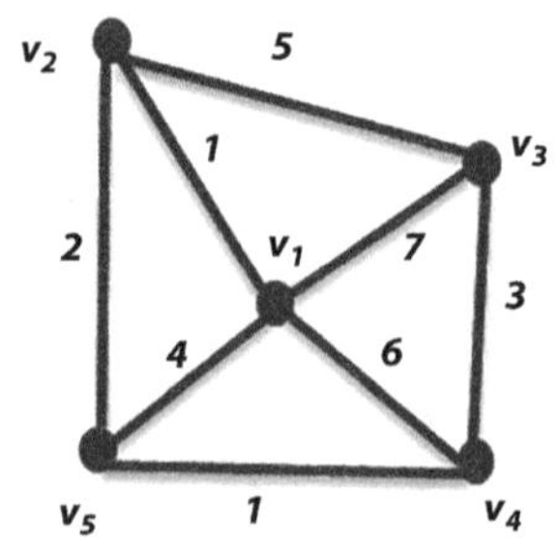

Figure 1.4 Undirected weighted graph.

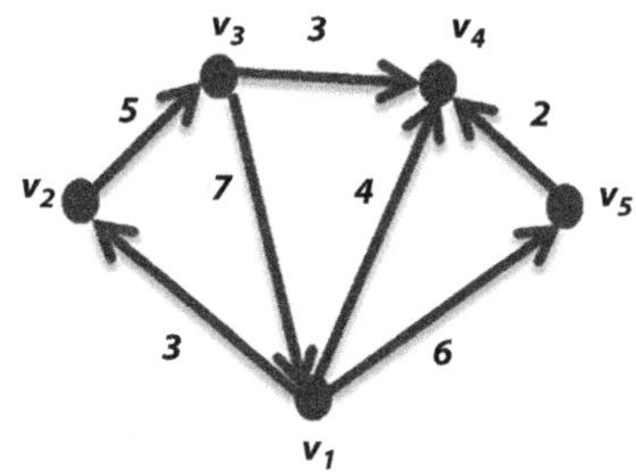

Figure 1.5 Directed weighted graph.

1.3.4 Unweighted Graph

A graph which is not a weighted graph is known as an unweighted graph. That is, there is no value is assigned to any edge of the graph. Examples of a directed unweighted graph and an undirected unweighted graph are shown in the following Figure 1.6 and Figure 1.7.

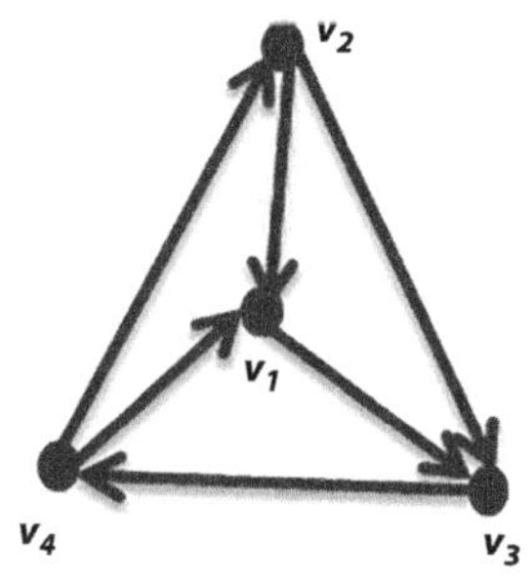

Figure 1.6 Directed unweighted graph.

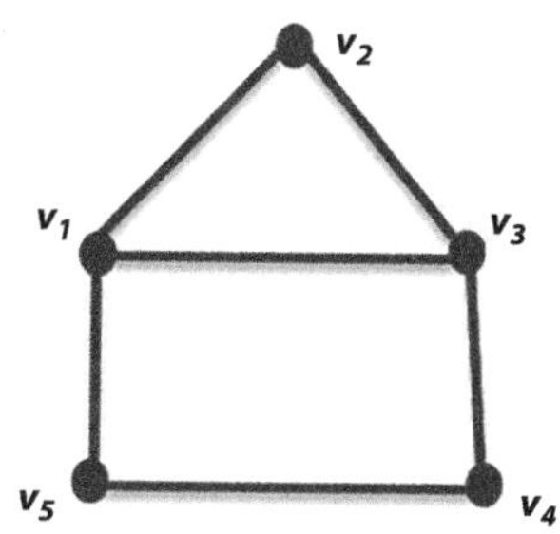

Figure 1.7 Undirected unweighted graph.

1.3.5 Cyclic Graph

In a graph, if there exist at least one cycle forming a closed loop then it is known as a cyclic graph. Examples of cyclic graphs without self loop and with self loops are shown in Figure 1.8 and Figure 1.9.

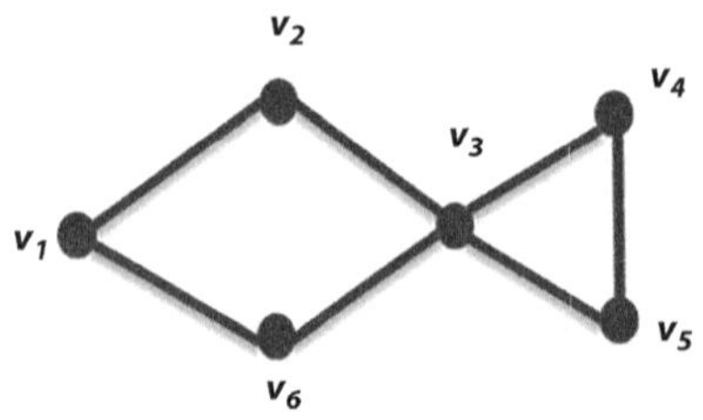

Figure 1.8 Cyclic graph without self-loops.

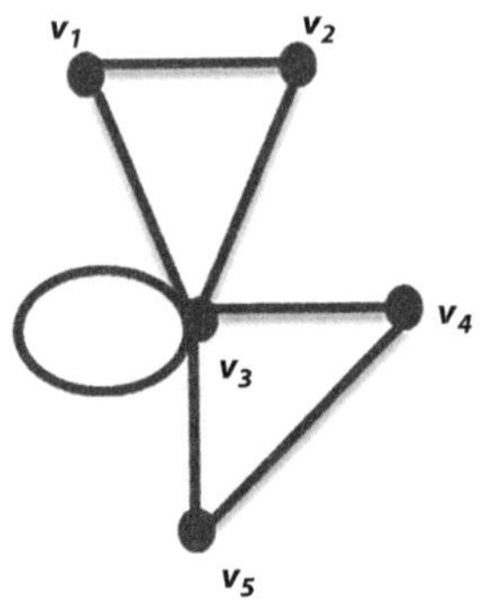

Figure 1.9 Cyclic graph with self-loops.

1.3.6 Acyclic Graph

A graph without loops is known as an acyclic graph. An example of an acyclic graph is shown in Figure 1.10.

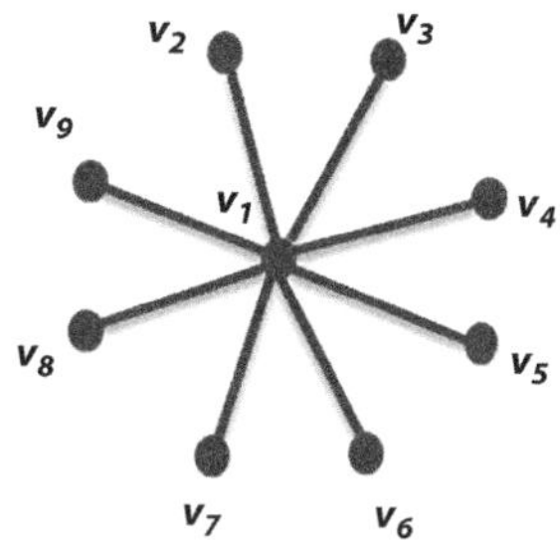

Figure 1.10 Acyclic graph.

1.3.7 Homogeneous Graph

A graph in which, all vertices and links are of the same type is said to be a homogeneous graph. This approach is used to ensure that there is a uniform relationship in the graph structure. An example of a homogeneous graph is shown in Figure 1.11.

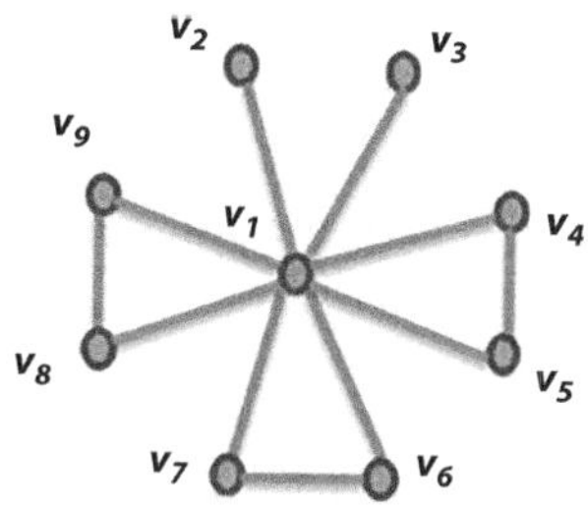

Figure 1.11 Homogeneous graph.

1.3.8 Heterogeneous Graph

In a heterogeneous graph, the nodes or the edges or both can have different types. This is the example for the representation of diverse relationships and complex systems such as the knowledge graphs, recommendation systems and social networks. An example of a heterogeneous graph is shown in Figure 1.12.

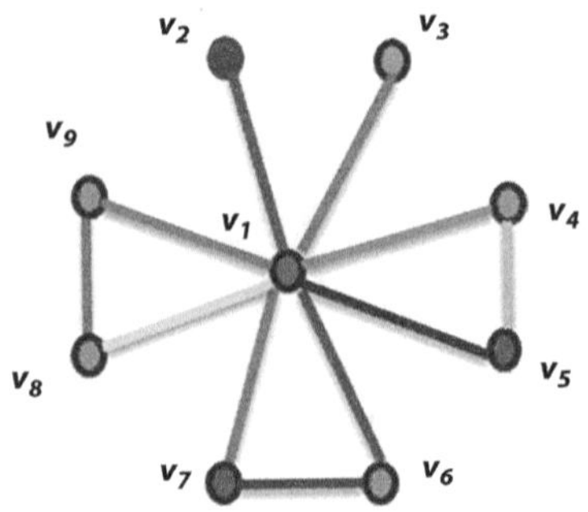

Figure 1.12 Heterogeneous graph.

1.3.9 Static Graph

A graphical representation of the data points in a coordinate system is known as static graph, in which the data is fixed. Its visual appearance can be customized using elements like graph lines, grid lines and labels.

1.3.10 Dynamic Graph

In a dynamic graph, the vertices or the links or both can change over time, reflecting evolving relationships or the updates in the data. This process includes the additions, deletions or modifications to the graph structure.

1.3.11 Transductive Graph

A transductive graph is an approach, where the entire graph, including both labeled and the unlabelled nodes are provided as input, and the aim is to identify the labels for unlabelled nodes, leveraging the structure and the relationships within the graph.

1.3.12 Inductive Graph

An Inductive graph is a process, where a model is trained on a sample of graph data, often with masked labels, to learn patterns and make predictions for unseen or valid data, generalizing beyond the training set.

1.4 Network Structure

The network structure refers to the system of connections and relationships within a network. It defines how the nodes interact and how the information and the resources flow through the network. In the network, the vertices are representing the elements and entities as well as the edges or links represent the connections or the relationships. These connections can be one-way or two-way and can represent various connections like communications, friendship, collaboration and data flow. The understanding of the network structure helps to analyze information flow, identify key nodes, and optimize the network performance. The open and enclosed network structures are shown in Figure 1.13 and Figure 1.14.

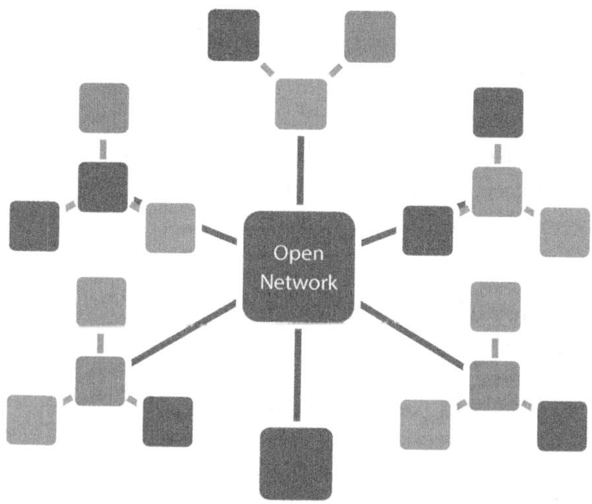

Figure 1.13 Open network structure.

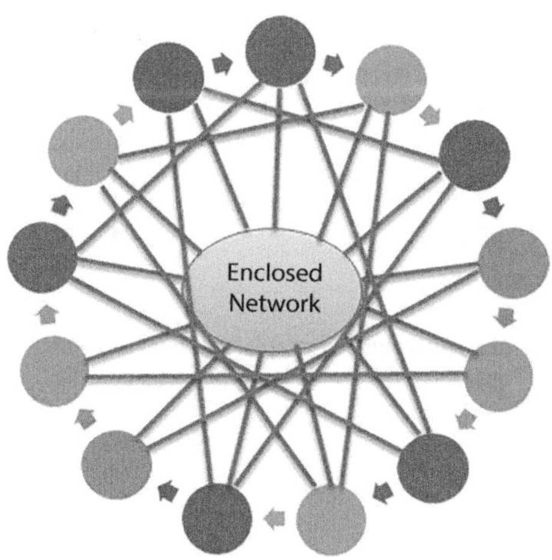

Figure 1.14 Enclosed network structure.

1.5 Graph Data Representation

There are some different ways to describe the graph data in graph theory and each way is appropriate for a particular set of circumstances and uses. This allows the flexibility and efficiency in graph-based solutions.

1.5.1 Adjacency Matrix

In an adjacency matrix, each cell indicates whether there is a link connecting the corresponding vertices or not. It is a square matrix. In dense networks, when the majority of nodes are related to one another, this form is especially helpful. It is a method of expressing the graph as Boolean matrix of 0s and 1s. Make an $n \, x \, n$ two-dimensional (2D) array for a graph having n vertices, where $adjM[i][j]=1$ indicates there is a connection between the vertices i and j; $adjM[i][j]=0$ means that there is no connection between the vertices i and j.

1.5.1.1 Adjacency Matrix Representation for Undirected and Unweighted Graph

Let G be an undirected unweighted graph having 5 nodes and 6 links. To represent this G as adjacency matrix, let us start with all elements of the entire matrix to be initialized as 0. If there is a connection between vertex i to vertex j, we insert 1 in both the following cases, such as $adjM[i][j]$ and $adjM[j][i]$, because we can go either way. An example of the adjacency matrix of an undirected weighted graph is shown in Figure 1.15.

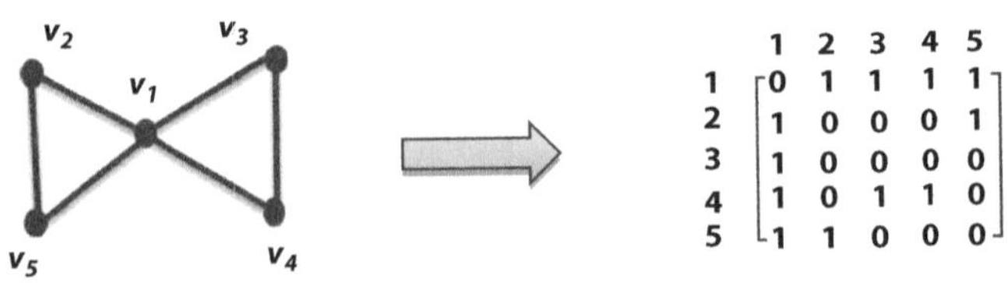

Figure 1.15 Undirected and unweighted graph ⇒ adjacency matrix.

1.5.1.2 *Adjacency Matrix Representation for Undirected and Weighted Graph*

Let G be an undirected weighted graph which contains 5 nodes and 8 links. For the graph G, the entries $adjM[i][j]$ and $adjM[j][i]$ in the adjacency matrix are represented by the value of 'w' if there is a connection between the vertices i and j having weight 'w': otherwise, mark it as 0. An example of the adjacency matrix of an undirected weighted graph is shown in Figure 1.16.

The adjacency matrix is always symmetric for the undirected graph. That is, $adjM[i][j] = adjM[j][i]$ for all i and j.

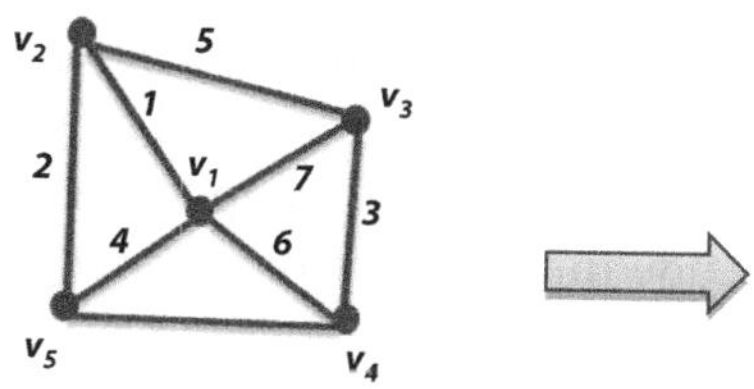

	1	2	3	4	5
1	0	1	7	6	4
2	1	0	5	0	2
3	7	5	0	3	0
4	6	0	3	0	1
5	4	2	0	1	0

Figure 1.16 Undirected and weighted graph ⇒ adjacency matrix.

1.5.1.3 *Adjacency Matrix Representation for Directed and Unweighted Graph*

Figure 1.17 shows a directed and unweighted graph as an adjacency matrix. Initially, the entries of the entire matrix are initialized to 0. If there is an edge from the vertex i to j, we insert 1 for that particular $adjM[i][j]$.

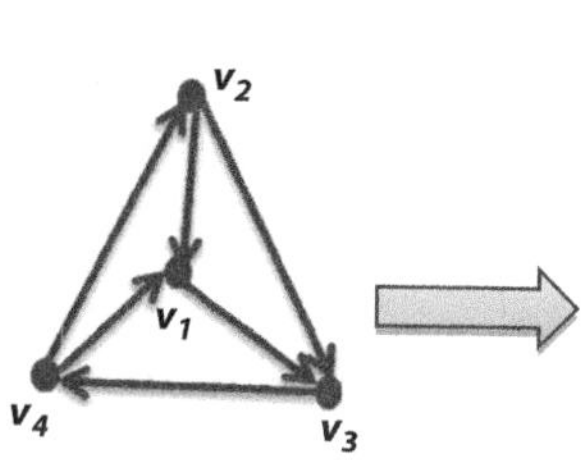

	1	2	3	4
1	0	0	1	0
2	1	0	1	0
3	0	0	0	1
4	1	1	0	0

Figure 1.17 Directed and unweighted graph ⇒ adjacency matrix.

1.5.1.4 *Adjacency Matrix Representation for Directed and Weighted Graph*

Consider a directed weighted graph G having 5 nodes and 7 links. For the graph G, the entries $adjM[i][j]$ in the adjacency matrix are represented by the value of 'w' if there is a connection between vertex i and j having weight 'w'; otherwise, mark it as 0. The graph G is represented as adjacency matrix in Figure 1.18.

The diagonal entries of an adjacency matrix $adjM[i][i]$ are usually set as 0, it denotes that the graph has no self-connections. However, in some cases a graph may allow the self-loop in which case the diagonal entries would be set as 1 to indicate the presence of an edge from a vertex to itself. This modification can be crucial for certain applications such as representing states in a Markov chain analysis. The adjacency matrix representation has various applications in the graph representation and analysis. One of its important uses is that representing a graph computationally, providing a straightforward way to encode the presence or absence of the nodes between the pair of nodes.

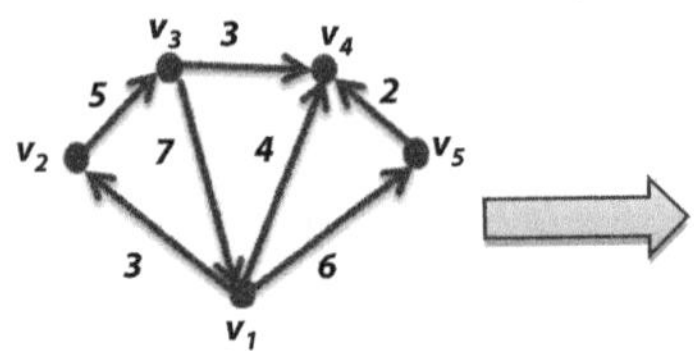

	1	2	3	4	5
1	0	3	0	4	6
2	0	0	5	0	0
3	7	0	0	3	0
4	0	0	0	0	0
5	0	0	0	2	0

Figure 1.18 Directed and weighted graph ⇒ adjacency matrix.

1.5.2 Adjacency List

An adjacency list consists of an array of vertices and the collection of connected lists representing the edges. The most effective access to the nearby nodes is made possible by the association of the each vertex with the list of its neighboring vertices and the edges.

1.5.2.1 *Adjacency List Representation for Undirected and Unweighted Graph*

Let G be an unweighted undirected graph containing 5 elements in both vertex and the edge set. The adjacency list for the graph G is to compute as follows.

- Count the nodes placed in the given graph G
- Create an array of lists, with each index representing a vertex.
- Initialize each list as empty.
- Add all the edges given in the graph to the corresponding lists.

For example, if the edges are $(v_1, v_2), (v_1, v_4), (v_2, v_4), (v_3, v_4)$ *and* (v_4, v_5), the adjacency list representation as shown in Figure 1.19.

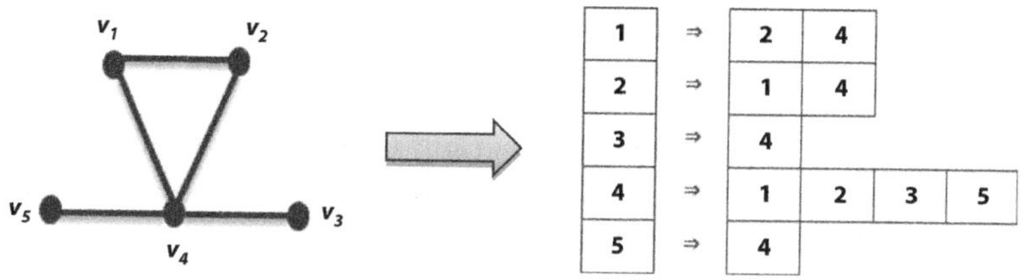

Figure 1.19 Undirected and unweighted graph ⇒ adjacency list.

1.5.2.2 *Adjacency List Representation for Undirected and Weighted Graph*

Let us consider an undirected weighted graph G having 5 nodes and 7 edges. The adjacency list for G is to compute by the steps given below.

- Find out how many vertices there are (5 in this instance).
- Construct an array of lists, with a vertex represented by each index.
- Set each list's initial value to zero.
- To the respective lists, append each edge of the graph with its weight 'w' enclosed in parenthesis.

As an illustration, if the edges have the weights of 5, 7, 4, 3, 8, 2, 4 and 4 respectively and if the edges are $(v_1, v_2), (v_1, v_4), (v_1, v_5)$, $(v_2, v_3), (v_2, v_4), (v_3, v_4)$, and (v_4, v_5) respectivelythen the adjacency list representation as shown in Figure 1.20.

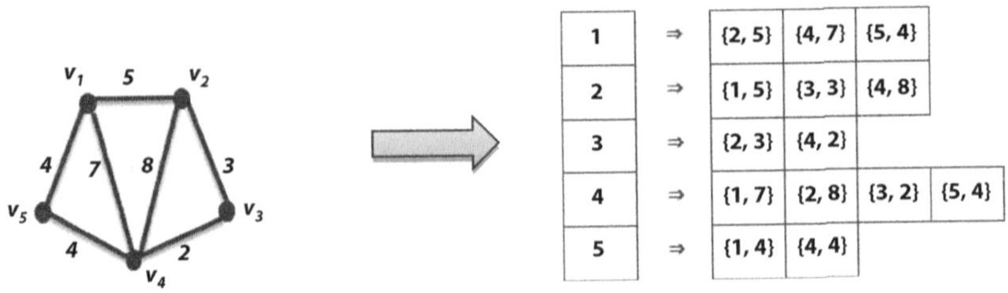

Figure 1.20 Undirected and weighted graph ⇒ adjacency list.

1.5.2.3 *Adjacency List Representation for Directed and Unweighted Graph*

Let us consider a directed unweighted graph G with five vertices and seven edges. Use the procedures listed below to generate an adjacency list for the provided directed and unweighted graph G.

- Determine the number of vertices (5 in this case).
- Create an array of lists, with each index representing a vertex.
- Initialize each list as empty.
- Add all the directed edges (v_i, v_j), given in the graph to the corresponding lists.

For example, if the edges are $(v_1, v_2), (v_1, v_5), (v_2, v_3), (v_2, v_4), (v_3, v_4)$, (v_4, v_5) and (v_5, v_2), the adjacency list representation as shown in Figure 1.21.

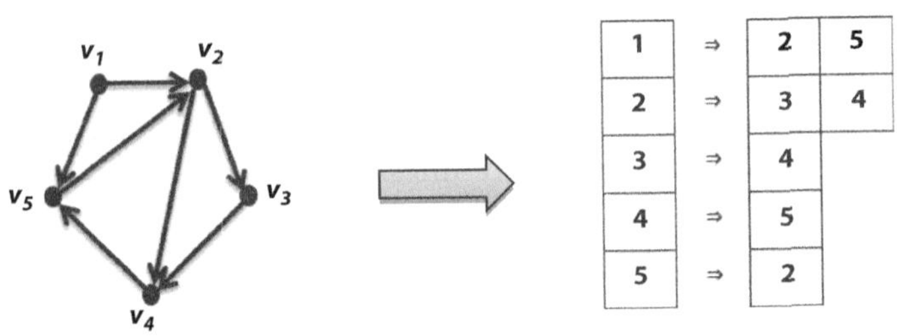

Figure 1.21 Directed and unweighted graph ⇒ adjacency list.

1.5.2.4 *Adjacency List Representation for Directed and Weighted Graph*

Consider the graph G, which contains five nodes and seven edges and is weighted and directed.

To generate the adjacency list for the specified directed and weighted graph G, use the steps given below.

- Count the nodes in G.
- Create an array of lists with each index representing a vertex.
- Initialize each list as empty.
- Add all the directed edges (v_i, v_j), given in the graph G with their weight 'w' within the parenthesis, to the corresponding lists.

For example, if the edges are $(v_1, v_2), (v_1, v_4), (v_1, v_5), (v_2, v_3), (v_3, v_1)$, (v_3, v_4), *and* (v_5, v_4) having weight 5, 7, 4, 3, 8, 2, 4 respectively, then the representation of adjacency list as shown in Figure 1.22.

Compared to adjacency matrices, the adjacency lists are take up less space and are especially effective for sparse graphs. Additionally, the adjacency lists make it simple to add vertices to a graph and provide easy traversal of all of its edges. Adjacency lists are also used to speed up the implementation of certain graph algorithms. The algorithms such as Prim's and Dijkstra's are more efficient when the adjacency lists are used. However, the adjacency lists have some drawbacks. One significant disadvantage is that if two vertices have an edge can be costly because it requires searching through the adjacency list. Also, adjacency lists are not a good fit for thick graphs where the number of links is almost equal to the maximum. In some circumstances, alternative representations like adjacency matrices might be more suitable.

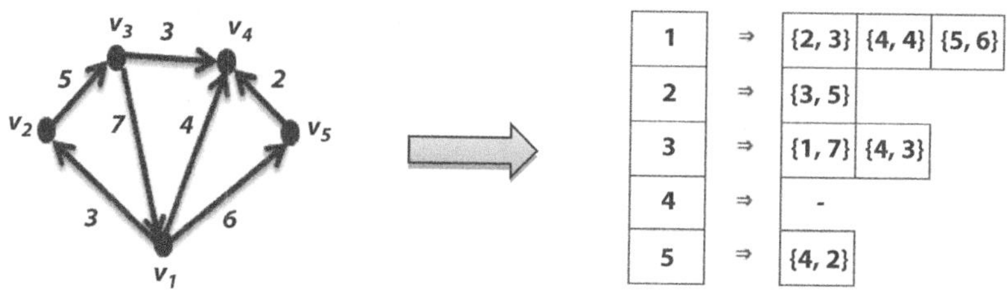

Figure 1.22 Directed and weighted graph ⇒ adjacency list.

1.6 Basics of Graph Neural Networks

The neural networks are made to handle the complex graphical structured data, uncovering the patterns and the connections between the connected objects. Inspired by the CNNs [4] and the graph embedding [11], graph neural networks improve in applications like molecular studies, social network analysis, and more. The GNNs are tailored to process intricate structures with arbitrary sizes and complexities, made them static for the tasks like the node, edge and the graph based predictions [25]. The graphs offer the flexibility and adaptability, making them suitable for various disciplines. While the convolutional neural networks [23] are well suited for the image classification [3]; the RNNs [21] are perfectly adopted for the string data like text and specially designed for the graph structured data. These features allow GNNs to succeed in dealing the limitations of networks in handling complex interdependencies and links found in graph data. They have established themselves as highly effective at handling difficult problems, often surpassing convolutional neural networks in this regard.

The major key differences between the traditional neural networks and the graph neural networks are tabulated as follows.

1.7 Types of Graph Neural Networks

There are several types of the GNNs such as the graph convolutional networks [16], global attention networks [29], GraphSAGE networks [10], graph auto-encoder networks [20] and the recurrent graph neural networks [3]. The working flow and types of these networks are discussed below. The differences between traditional neural networks and graph neural networks are listed below in Table 1.1.

Table 1.1 Difference between traditional neural networks and GNNs.

Traditional neural networks	Graph neural networks
This kind of networks is designed to handle the grid like data and the sequences.	This kind of networks is particularly designed to handle the data structured as graphs.
This is mainly focused on the individual data points.	This is to capture the relationships between the data points.
CNNs, RNNs are some example for the traditional neural networks.	GCNs, GATs, GraphSAGE are few notable examples for the GNNs.

1.7.1 Graph Convolutional Networks (GCNs)

The GCNs are deep learning techniques which are able to learn the node properties, identify the intricate patterns and the connections in graph data. The graph convolutional neural networks are created to handle the normal; grid like data directly [11]. The graph neural networks can capture and spread information *via* the nodes of the networks by analyzing the neighboring nodes and learning features using the convolutional operations on graph data [3]. The graph based semi supervised learning methodologies with multiple layers were introduced in graph convolutional neural networks [24] to improve the standard of the output. Within the multiple layers, everyone performing the convolutional and the aggregation steps, the graph neural networks can refine the node consideration and capture the complex dependencies within the graph data. This is one of the powerful tools for the huge range of applications like the image classification, text classification traffic prediction and forecasting and the visual question answering.

1.7.1.1 Architecture of Graph Convolutional Networks

The GCNs are detecting the node connections to extract insights from the graph data, facilitating accurate analysis and the forecasting through the information aggregation and the dissemination. An example of a graph convolutional neural network model is shown in Figure 1.23. The working flow of graph convolutional networks is as in Figure 1.24.

The graph convolutional networks work with following steps

- ❖ **Initialization:** Every node in the graph is linked to the set of attributes or features, which can vary depending on the specific use case. For instance, in a social network, nodes representing users might be associated with profile details, interests, or behavioral data.
- ❖ **Convolution Process:** The central mechanism of the graph convolutional networks is graph convolutional operation,

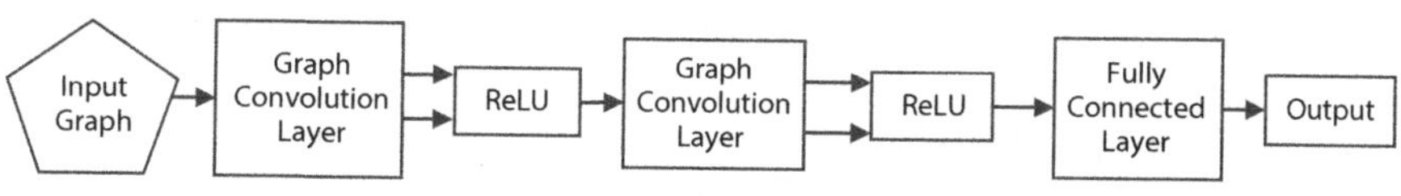

Figure 1.23 Graph convolutional neural network model.

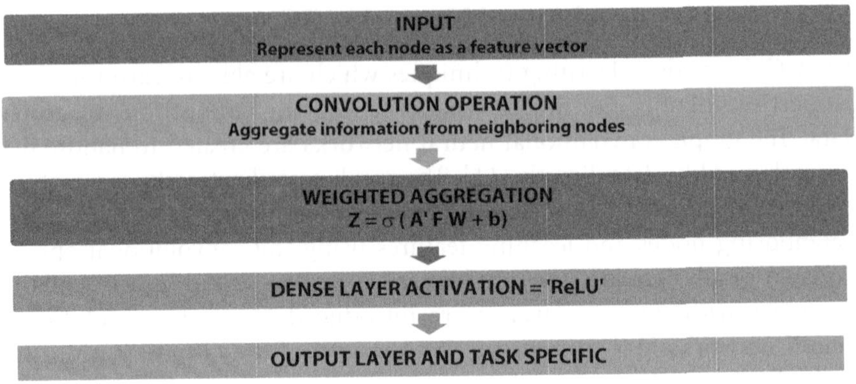

Figure 1.24 Architecture of graph convolutional networks.

which draws inspiration from the convolutional neural networks. They collects and combines the message from the adjacent nodes by calculating the weighted sum of their feature vectors, with weights derived from the graph's adjacency matrix, resulting in updated node representations.

- **Weighted Aggregation:** The adjacency matrix that assigns weights to the neighboring nodes during aggregation. The neighboring features for each point are noted by adjacency matrix and then concatenated, allowing for efficient information propagation.
- **Activation Function and Weight:** The collected node features are subsequently sent *via* a nonlinear activation function such that ReLU, to enhance the model expressiveness. Meanwhile, the weight matrix is optimized during training, allowing the graph convolutional networks to adjust its aggregation process to the unique graph architecture and job requirements.
- **Stacking Layers:** The graph convolutional networks are often composed with many layers, allowing the design to predict complicated patterns and abstract elements with the graph. The result of each layer serves as an input for successive steps, iteratively refining node representations over a fixed number of levels.
- **Task-Specific Output:** The outputs of the graph convolutional networks can be utilized for the variety of graph-related events, which includes categorizing nodes,

predicting connections and classifying entire graphs, depending on the use case. This architecture allows the graph convolutional networks to efficiently process the graphical structured data for a variety of applications.

1.7.1.2 *Equation of Graph Convolutional Network*

The GCN layers can be mathematically formulated as

$$Z = \sigma\left(A_j' F' W' + b \right) \tag{1.1}$$

The components used in the equation of graph convolutional networks are defined as follows. The adjacency matrix A_j' reflects the structure, feature matrix F' indicates node attributes, the trainable weighted matrix W' and b, the bias vector are optimized during training and σ refers the nonlinear activation function such as ReLU, introducing non-linearity to the model. Equation 1.1 enables node feature smoothing based on their neighboring vertices, allowing the GCNs to predict the complicated node representations through the multiple layers.

1.7.1.3 *Types of Graph Convolution Networks*

There are two major types of GCNs, which are as follows.

- **Spectral-based GCNs:** The spectral based graph convolutional networks are work in spectrum domain, using the Fourier transform to produce the convolution. The Chebyshev polynomial and the first order approximations are used in the key models such as ChebNet and graph convolutional networks to improve the speed of computation. Based on graph signal processing theory, the spectral convolutional networks transport information across nodes using mathematical approaches such as the Eigen decomposition and the polynomial approximations, which are inspired by signal processing concepts.
- **Spatial-based GCNs:** This network can conduct the convolution directly in the spatial field by aggregating data from the neighboring vertices, making them more understandable and simpler to implement. The key models include GraphSAGE, which combines data from the neighbors

using mean, LSTM, pooling and GAT, which employs attention processes to determine the value of neighbors. The spatial graph convolutional networks [29] function similarly to graph convolutional networks; merging neighboring node attributes into a central node and is frequently more scalable due to localized filters, though implementing local invariance for nodes with different neighbors can be difficult.

1.7.2 Graph Attention Networks (GATs)

The GATs are a graph neural network design that uses hidden self-attention layers to cope with graph structured input. It includes an attention mechanism that distributes values to the neighboring points based on the quality, allowing the design to reflect complicated interactions and varied degrees of influence between the nodes. This adaptive focus improves performance on tasks where certain connections are more significant than others. The graph attention networks consist of input layers, hidden layers with graph convolutional layers, activation functions, and optional pooling layers, generating final node embedding or predictions for effective graph-based learning.

1.7.2.1 Architecture of Graph Attention Networks

The GATs are the neural network model that improves the graph convolutional neural networks by embedding the attention mechanisms, allowing for adaptive weighting of the neighboring nodes when analyzing the graph structured data [17]. The working flow of graph attention network is as shown in Figure 1.25.

The GATs works with the steps as follows.

- ❖ **Initialization:** In graph attention networks, each and every point in the graph is denoted by the feature vector, capturing its distinct attributes and the properties.
- ❖ **Self-Attention Mechanism:** The self-attention techniques in the graph attention networks allow vertices to selectively concentrate on the neighbors during the message aggregation. It computes the attention scores for each neighbor based on the central features of the vertices and its neighbors and then uses these scores to the weight contributions. This enables for the variable and the context-dependent aggregation [17].

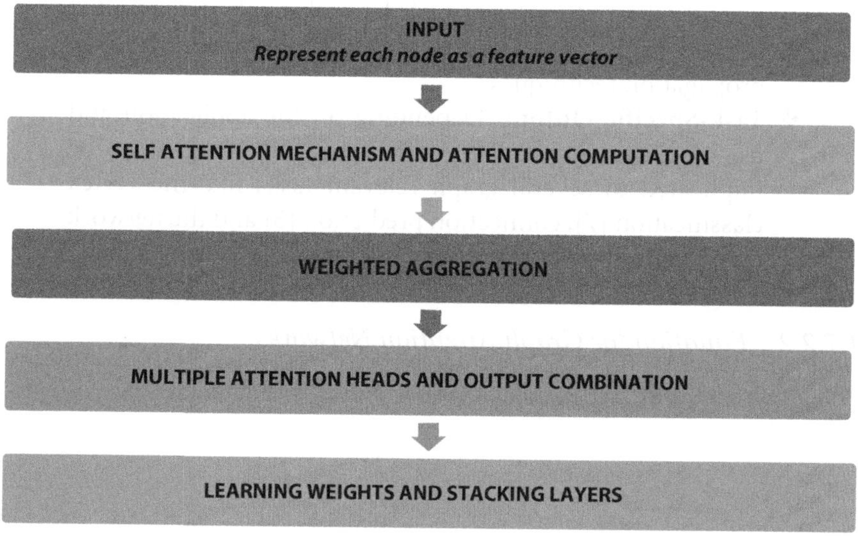

Figure 1.25 Architecture of graph attention networks.

- **Weighted Aggregation:** The attention values determine how the features of each neighbor contribute to the aggregation of the center vertex. The neighbors with better attention scores have a greater influence, while those with lower scores contribute less. The resulting weighted aggregate generates an updated vector for the core vertex.
- **Multi- Head Attention:** The multi-head method improves the expressiveness of the design and the capacity to predict critical graph representations. GAT networks use the multiple attention heads in everywhere equidistant , each of which records different relationships and patterns in the graph, and the results from all of these attention controls are merged by concatenation and averaging, to generate the complete feature vector for every vector [17].
- **Learning Weights and Stacking Layers:** The graph attention networks acquire complete knowledge about the weight parameters during training, just like GCNs, to maximize the feature of attention mechanisms for the specific graphs and tasks. By stacking numerous GAT layers, the model is able to represent complex linkages and higher-level features. The subsequent layer, which enhances the node representations,

receives the output from each layer through the repeated process. Learning is aided by the optimization and the back propagation techniques.

- **Task-Specific Output:** Depending on the application and use case, the output of a graph attention networks can be implicated to several graph relevant tasks like the vertex classification [7], connection prediction [5] and the network classification.

1.7.2.2 Equation for Graph Attention Networks

The graph attention network layer equation is formulated as

$$H_i^{(l+1)} = \sigma(\sum_{j \in N(i)} \frac{1}{C_{i,j}} \omega^{(l)} H_j^{(l)}) \tag{1.2}$$

where $H_i^{(l+1)}$ - output of the current vertex, σ - ReLU non-linearity function, $C_{i,j}$ - normalized attention coefficient, $\omega^{(l)}$ - weight matrix, $H_j^{(l)}$ - feature vector of neighboring node j, $N(i)$ - One-hop neighbors of node i.

1.7.2.3 Types of Graph Attention Networks

We identify that there are six major categories of graph attention networks based on their methodological approaches, attention mechanisms, feature capture and techniques for addressing graph learning challenges. This taxonomy gives the comprehensive framework for learning the diverse range of graph attention network methodologies, complemented by a table summarizing influential studies in the field [3].

- **Global Attention Networks:** By adding the attention mechanisms that capture the local and the global surrounding knowledge within the graphical structured data, the global attention graph neural network mechanism improves on conventional graph neural networks. Applications of the graph attention networks include the semi supervised vertex classification (HGAT), stock prediction and suggestion (RA-AGAT), visual position recognition in changing environments, high accuracy road marking recognition (GATSCNet) and metabolic pathway prediction for drug discovery (HFGAT). A global-based attention mechanism is used by the global attention networks such as the holistic graph neural networks to train and produce node embedding. By combining the local and the global properties, holistic graph neural networks allow the nodes to grasp the larger graph context. They variants investigate the various pooling approaches and hierarchical global feature extraction. These models exhibit adaptability and effectiveness in a range of applications, attaining cutting edge outcomes in medication development, sentiment classification, recommendation systems and other fields.

- **Multi-Layer GATs:** Several attention layers are stacked in multilayer graph attention networks in order to produce intricate, high level features from graph structured data. To create the new node features that are passed into later layers, each layer compiles the data from nearby nodes. The multilayer GATs can learn abstract representations by integrating data from an ever larger neighborhood of nodes thanks to this hierarchical architecture. The adaptability and resilience of multilayer graph attention networks have been demonstrated through their application in a variety of fields. For stock recommendations, the FinGAT model uses the multi-layer GATs, which record the price trends over time. The efficiency of the graph attention networks in locating suitable edges, identifying hacker groups and forecasting the agent trajectories is shown by the other models such as SGATs, HGHAN and HEA. Further innovations include DuGa-DIT, a dual gated graph attention model for entity matching and RRL-GAT, a robust representation learning model for the multilayer image characterization. Additional applications span predicting miRNA disease associations (HGANMDA), travel recommendations (MV-GAN) and anomaly detection in the complex equipment. According to the studies, the multilayer graph attention networks can handle graph structured data effectively, extract valuable characteristics and perform better than the conventional methods. We may expect more developments and the uses of graph attention networks in various domains as research progresses, which will improve the model performance and increase its applicability.
- **Graph Embedding GATs:** The graph embedding graph attention networks learn the latent vertex representations

that capture the graph structure by using graph embedding approaches with attention techniques. They are effectively denoted and handle the graphical structured data, solving several issues across domains by converting the network into the smaller dimensional space while taking neighbor importance into consideration. The graph embedding graph attention networks combine attention mechanisms with graph embedding techniques to learn expressive vertex notations, catching both local and the global graph structures. The successful applications include MS-GAT, which analyses traffic systems by capturing subtle spatial temporal interactions and a heterogeneous graph neural network framework that manages multi relational data with hierarchical attention, both achieving the results. These mechanisms show how the graph embedding graph attention networks can be used to handle the complex graph data and drive graph deep learning further. The notable applications of graph attention networks across a range of fields demonstrate their adaptability and efficiency. Predicting the lncRNA-disease associations, mash up tagging (CAMT), fault recognition (SGANM), trust assessment in online social networks (GATrust), java method summarization (GSCS), drug to drug interaction prediction (DGAT-DDI), web based service classification (Graph4Web), semi supervised node classification, explainable recommendations, the drug target interactions (DTIHNC), Parkinson's disease prediction using multi modal features, program classification and the defect prediction in source code, identifying potential treatments for SARS-CoV-2 through drug virus association analysis and road marking recognition (GAT-SCNet) are some of the valuable applications. Furthermore, the graph attention networks have been applied to recommendation systems including trip recommendation (MV-GAN) and the computer vision tasks like monocular depth estimation and semantic segmentation (AG-CRF). These examples show how the graph attention networks can promote innovation and progress in several fields.

- **Spatial GATs:** The spatial graph attention networks emphasize the proximity and the dependency, focusing on the geographical connections between the nodes. They use attention processes to rate the surrounding nodes based on their spatial significance in order to record the spatial patterns and local interactions. They are therefore appropriate for uses such as traffic flow forecasting and urban planning. In order to attain the results, models such as STGA capture spatial and temporal linkages in video anomaly detection, one of the many uses for spatial GATs. The spatial GATs are also utilized in the graph classification. For instance, SA-GAT introduces the substructure interaction attention, which enhances the existing techniques. They are also employed in multi domain discussion systems and land change simulation, where they perform exceptionally well and highlight the importance of topological order modification. The integration of the dense graphs and the multi relational knowledge graphs was improved with the proposal of a censor-based graph attention network for the knowledge graphs of smart corporate management systems. On smart EMS datasets, MR-GAT exhibits improved representation learning performance for knowledge fusion by utilizing a relation attention method and a combined objects and connection alignment framework. The two prominent models are HFGAT and gGATLDA. The gGATLDA model is effective in detecting lncRNAs linked to cancer because it integrates the graph-level attention networks with graph neural networks to predict lncRNA-disease connections. Using spatial graph attention networks for drug discovery, HFGAT is one of the hybrid frameworks that predict the metabolic pathways by fusing the global and local properties of the substances. Both models perform better than the conventional approaches.

- **Variation GATs:** The variation graph attention networks (VGATs) efficiently handle the heterogeneous or multi-modal data by combining the variation inference and the graph attention methods to capture intricate node interactions. VGAT has several applications, including in power systems, where it differentiates between the temporary rotor angle instability and the short term voltage instability. Additionally, VGAT is used in drug-target interactions to integrate the heterogeneous networks and the cross-modal similarities. In medical image processing, VGAT achieves high accuracy in distinguishing complex pneumonia related PPE from simple cases. Furthermore, a variant called

QPGAT combines the graph attention mechanisms and the quantum probability for the natural language processing tasks like the sentiment classification, emotion cause pair extraction and joint dialogue.

- **Hybrid GATs:** The multiple technologies are combined by the hybrid graph attention networks (GATs) to provide optimal performance in a variety of jobs. These networks have been used in a number of domains, such as the finding of linkages between miRNA and diseases, where GATMDA employs the graph attention networks and the multi-source data to find the connections. DQ-GAT provides a superior safety-efficiency balance and higher success rates in multi-agent dynamic traffic control. MRGAT demonstrates superior performance in knowledge graph completion. Hybrid GATs are also used in character representation, session-based recommendation, anomaly detection of Internet routing traffic, biomarker-disease association discovery, RNA-seq data analysis, COVID-19 transmission network representation, pedestrian trajectory prediction, fake news detection, IoT data analysis, crowd flow prediction, food safety risk detection, genetic variation analysis, autonomous driving safety systems, domain adaptation, anomaly detection in complex equipment, robot learning from demonstration, protein-protein interaction site prediction, direction of arrival estimation, object recognition and relationship categorization, manufacturing process quality prediction, aspect-based sentiment analysis, multi-agent reinforcement learning, chemical screening, autism spectrum disorder identification, hyper spectral image classification, brain cortical surface parcellation, cloud classification, single-cell dropout imputation, anti-TB drug resistance prediction, sentiment analysis, resource management in cellular networks, circRNA-disease association prediction, automated synthesis planning, malware detection, semi-supervised fault diagnosis, and robot team scheduling.

1.7.3 GraphSAGE Model

GraphSAGE is an inductive learning system that uses two primary processes, such as aggregation and sampling. It considers the fact that, its observations are linked to the edges of the training set during the training.

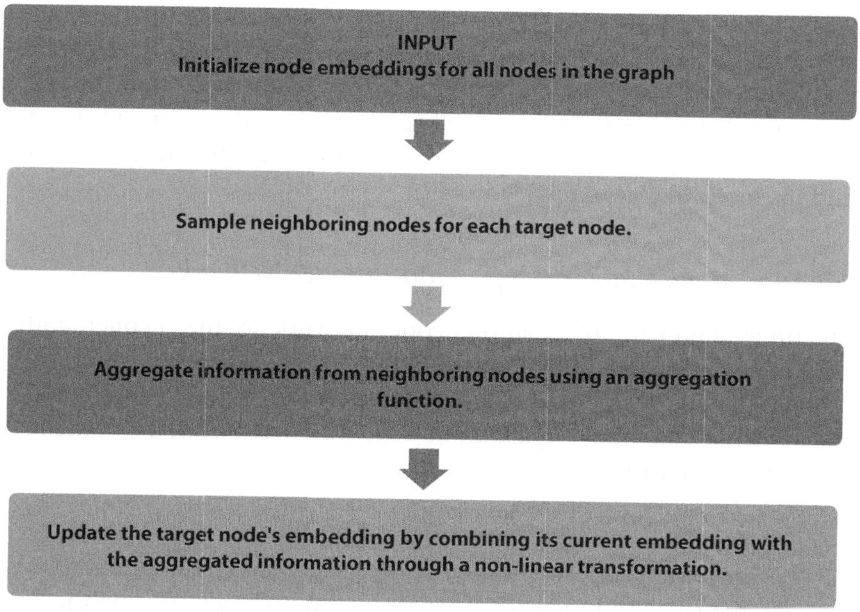

Figure 1.26 Architecture of GraphSAGE network model.

The vertex representation vector is coupled with the aggregated vector and then traverses the entire layer that is connected with the nonlinear activation function. Since every network layer has the similar weight and aggregator matrix, the significance of layer count in connection to aggregator count is evident. A regularized step applied for the output layer to complete the procedure.

Architecture of GraphSAGE Model

The mechanism of the inductive notations learning on graphs, the GraphSAGE (SAmple and aggreGatE) creates smaller dimensional vector notations for the vertices. The applications such as the vertex classification, connection prediction and the graph categorization are made possible by its exceptional efficiency in the graphs with rich vertex attribute information. The GraphSAGE model functions as shown in Figure 1.26.

1.7.4 Graph Auto-Encoder Networks (GAEs)

The type of graph neural network that can develop effective data representations for applications like feature extraction and dimensionality reduction is the graph auto-encoder [29]. The graphs auto-encoders are particularly managing the graph data [20], documenting the topological structure [12] and the node content. The vertex classification, connection prediction and the graph clustering are some of the tasks that the graph auto-encoders [20] can perform well for the various developments like the variational GAEs and their practical applications in the image clustering and the molecular research [5].

1.7.4.1 Architecture of Graph Auto-Encoder Networks

The three main parts of an auto-encoder network are the encoder, bottleneck and the decoder. They are used to develop the efficient data representations. The size of the input data is reduced by the encoder, represented by the bottleneck and the original input is reconstructed from the bottleneck, which is possible by decoder.

- **Encoder:** This minimizes the dimension and preserves the important information by compressing the input data into the smaller dimensional representation when capturing the important features.
  - **Input Layer:** The original data such that the image collection features enters the network through the input layer.
  - **Hidden Layers:** The encoders create the hidden layers from the input data. The major motivation of these encoders is to minimize the dimension of the data and take out the key features.
  - **Output of Encoder:** The encoder can reduce the dimensionality and extract relevant features by altering the input data through the hidden layers.
- **Bottleneck (Latent Space):** The bottleneck is the smallest layer which represents the data in its most compressed form. Sometimes it is called latent space. The network is able to analyze the essential data from incoming input by capturing the most important patterns and structures by retaining only the most important properties.
- **Decoder:** The goal of the decoder is to precisely restore the structure and the properties of the input data by reconstructing the original data from the compressed form of data in latent space.
  - **Hidden Layers**: For the purpose of recreating the original data, the decoder performs a sequence of layers to

gradually up sample the compressed data and expands it back to the original dimension of the input.

- **Output Layer**: This output layer produces the reconstructed data, with the goal of closely matching the original input data, effectively reversing the compression process.

In order to minimize the loss by reconstruction, the auto-encoder tries to find the key features in the latent space by optimizing its weights. The cost function guides the optimization process to increase the reconstruction accuracy by evaluating the performance of auto-encoder [13]. For continuous data, the mean squared error is a common technique. Calculations are made to determine the average squared discrepancies between the input and replicated data. Binary cross entropy approach can be applied to the data in binary format, which consists of values that are either 0 or 1. It calculates the likelihood that the original and reconstructed outputs will differ. The optimization mechanism such as, the gradient descent or the Adam optimizer, which modify the weights of the encoder as well as decoder networks to lower error and enhance the auto-encoder's efficiency, are used to minimize the remodeling error.

1.7.4.2 *Types of Auto-Encoders*

There are several types of auto-encoders, each with unique properties and intended for particular applications.

- **Denoising Auto-Encoder:** A denoising auto-encoder reconstructs the original clean data by learning to eliminate noise from the distorted input. By doing this, it promotes robust representation learning by capturing the key characteristics and the patterns instead of memorization of the input.
- **Sparse Auto-Encoder:** Even though, there are usually more hidden units than inputs, a sparse auto-encoder ensures sparsity by allowing only a few hidden units to be active at one. This can be accomplished through methods such as hand zeroing, modifying the activation function, or including a loss term that generates sparsity and results in a significant and concise representation of the data.
- **Variational Auto-Encoder:** A variational auto-encoder (VAE) assumes the probability distribution over the data and learns an approximation of it. It uses the stochastic gradient descent to optimize and learn the distribution of latent

variables, enabling the generative capabilities and the probabilistic modeling.

- **Convolutional Auto-Encoder:** The CNNs are used for the image classification and processing by a convolutional auto-encoder [13]. The decoder successfully captures the spatial hierarchies and the patterns in data from images by reconstructing the image by deconvolution [19], while the encoder uses the convolutional layers to extract the features.

1.7.5 Recurrent Graph Neural Networks (RGNNs)

The RGNNs are perfect for the applications like traffic predictions on road networks, social network analysis, and modeling time evolving knowledge graphs because they excel at managing dynamic graphs with changing the structures and the node properties [21]. In the complex systems, the recurrent graph neural networks [29] provide the accurate predictions and the valuable insights by effectively capturing the time fluctuations and deep linkages. Their capacity for the natural language processing techniques [3] such as text production and multimedia analysis, which includes image description and video tagging, demonstrates their adaptability in managing the range of the dynamic graph-based data.

1.7.5.1 Architecture of Recurrent Graph Neural Networks

The RGNNs are integrating RNNs to capture the temporal dependencies in dynamic graphs [3]. In order to discover the complex connections and the patterns, the recurrent graph neural networks enhance the connections by capturing local and global network structures through iterative information flow between the pairs of nodes.

The constructions of the RGNNs are uses the method of the banach fixed point theorem. Consider the entire metric space (X, d) and let the mapping $\mathrm{T}: \mathrm{X} \rightarrow \mathrm{X}$ be the contraction mapping. Once T reaches its one and only fixed point x^*, then sequence $\mathrm{T}\{\mathrm{x}_\mathrm{n}\}$ for $\mathrm{n} \rightarrow \infty$ converges to $\{\mathrm{x}\}$. Accordingly, if we apply the mapping T on x k times, x^k should be almost equivalent to $x^{(k-1)}$.

1.7.5.2 Examples of Recurrent Graph Neural Networks

The two popular types of RGNN designs are the diffusion convolutional RNNs and the graph convolutional recurrent networks. This enables accurate predictions and insights in the range of applications. In this section, we discussed about few categories of RGNN techniques.

- **Gated Graph Neural Networks (GGNNs):** The long-term dependency tasks are better performed by the **GGNNs** than by RGNNs. They enhance RGNNs by managing long term dependencies through the use of node, edge and the temporal gates. As with the gated recurrent units, these gates enable the model to selectively remember and discard the data across many states.
- **Diffusion Convolutional Recurrent Neural Networks (DCRNNs):** This type of networks is effectively capturing periodic and spatial correlations in the graph structured data by combining the diffusion convolutions with the gated recurrent units in encoder-decoder architecture. These kinds of networks are suitable for the time series analysis and the traffic forecasting.
- **Long Short-Term Memory (LSTM) networks:** These types are used to capture time dependent relationships in the graph structured data.

1.8 Key Tasks in Graph Neural Networks

The GNN is an emerging mechanism for analyzing and producing the complex relational data because they are highly adaptable models that can handle several graph-based functions such as the vertex classification, connection prediction, graph embedding, etc., are given below.

1.8.1 Graph Classification

Applications like molecular classifications and social network analysis are made possible by the implementation of the graph based neural networks, which separate the entire graphs into discrete groups according to their structure and the link properties.

1.8.2 Node Classification

The vertex classification challenge involves predicting unknown or missing labels for the nodes inside the network by using the data gathered from the neighboring node labels and accurately classifying the nodes based on the vertex interactions and the graph structure [7].

1.8.3 Link Prediction

Predicting the possible links between the node pairs in network with incomplete adjacency matrix is known as the link prediction. This approach plays a vital role in numerous applications like the social networks; recommendation systems and the biological networks where, spotting missing or upcoming links provide valuable information.

1.8.4 Community Detection

By clustering the nodes according to the edge weights, edge structure and node similarities, the graph clustering identify the underlying patterns and the tightly connected communities. This method assists in revealing the relationships and the hidden structures in intricate networks.

1.8.5 Graph Embedding

The graph embedding involves in mapping the entire graphs or the nodes into vector representations that preserve the crucial information about the nodes, edges and the structural properties. With the help of this technology, the complicated graph data may be efficiently processed for the later tasks such as the categorization, clustering and the similarity search by being transferred to some format, which is acceptable for building the machine learning models.

1.8.6 Graph Generation

The graph generation involves in training the models on existing graph structures to generate the new graphs that exhibit similar properties, patterns and distributions. This capability is valuable in applications such as molecule generation, network synthesis and simulation where creating the realistic graph structured data is essential.

1.9 Operations in Graph Neural Networks

The major operations in the GNNs are shown in Figure 1.27.

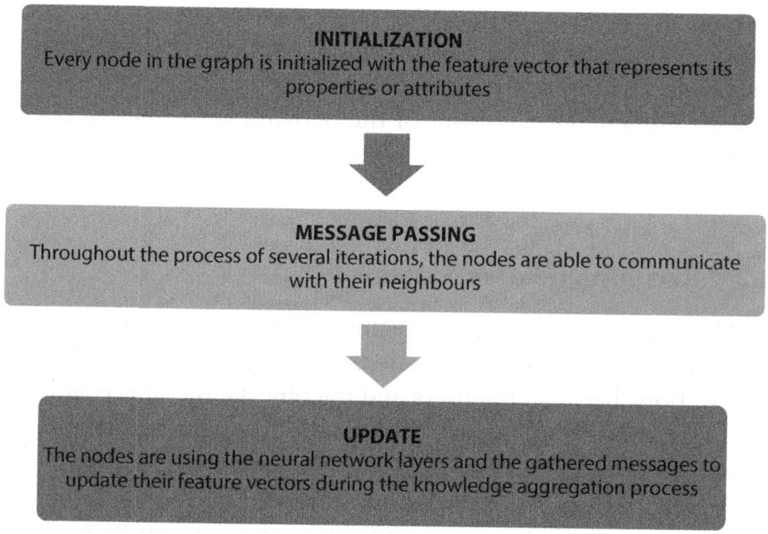

Figure 1.27 Operations in graph neural networks.

1.10 The Applications of Graph Neural Networks in Different Fields

The most important graph neural network applications in various domains are discussed in this section.

1.10.1 Social Network Analysis

The graph neural networks are able to identify user clusters, predict future connections, and classify the individuals by analyzing the links between the points in their domain. They can also spot the possible fraudulent activities, track the movement of the activities, the effect of information inside the social networks and provide the information about the social trends and the behaviors [14]. Moreover, the graph neural networks can develop the customized recommender systems that provide the recommendations for friends, goods and the relevant content by using the social connections [3]. They use the unique message passing technique to unravel the complex data and allowing each vertex in the network to learn about hidden patterns within network by exchanging the information within its nearest neighbors. This approach has greatly benefited from the social media network analysis and the other graph data structures. The graph neural networks have various advancements in the social media analysis [14], including redefining community detection with higher accuracy, predicting the new connections through the link prediction and transforming the recommendations by analyzing the connections, interests and network activity. By enhancing the power of the network connections, GNNs are transforming the social media networking and improve new technologies for the advancement in the field.

1.10.2 Recommendation Systems

The GNNs are significantly enhancing personalized recommendations by modeling the user item interactions as the graph, effectively leveraging connections between the users and the objects. By capturing the complicated relationships, the graph neural networks achieve better accuracy and performance compared to the traditional models [2]. They can manage huge, complicated datasets and use the information from the current nodes to solve the cold start issue. The graph neural networks are the effective tools for the real-world applications since they can be used in several recommendation scenarios such as social, session based and product recommendation systems. The graph neural networks are enabling to understand user and object representations from their relationships in the graph and they generate suggestions by estimating the probability that the user will interact with the given item. The recommendation systems are categorized into four main types. The popularity-based filtering makes the recommendations for the products based on their general level of popularity, presuming that many consumers will find popular products appealing. By examining the characteristics like actors or genre, the content-based filtering engine recommends products that are comparable to previous choices of the users. Furthermore, by identifying the patterns in the user item interactions, the collaborative filtering system produces the recommendations based on the similar actions of the users. Additionally, the hybrid recommender systems integrate the range of strategies including the collaborative and the content-based filtering is to overcome the new ideas of the individual techniques and generate recommendations [26] that are accurate and more customized.

1.10.3 Chemistry and Biology

The GNNs are used to forecast the properties and their interactions between the molecules [4]. This approach highly improves the medication development, protein to protein interaction prediction and the material design. They are used to predicting the molecular properties, generating novel chemical compounds; developing drugs and examining the protein interaction networks. These networks provide the valuable insights into disease mechanisms, personalized treatment and the biological processes by analyzing the complicated biological data. In these molecular networks, the atoms are denoted by the vertices and the chemical bonds are denoted by the links. The computational biology provides the crucial biological information and insights from the protein structures, genetic data and high throughput screening which enhances graph neural networks. The researchers can create new chemicals and the materials, improve prediction accuracy and understand the biological systems more thoroughly when the GNNs are embed in computational biology [4]. Biomolecules and biological interactions can now be denoted as graphs according to the GNNs which are able to recognize the complex connections inside the chemical structures.

1.10.4 Knowledge Graph

The determination of this graph is the crucial step in applying graph neural networks for the knowledge graph processing. Depending on the nature of the data, two scenarios arise that the homogeneous or the heterogeneous graphs. The images, classes and the values for the heterogeneous graphs can be considered as nodes, with attributes serving as edges that connect them. However, this approach has few limitations, particularly in utilizing the edge weights which motivates the use of homogeneous graphs. The weighted knowledge graph can be created in the homogeneous graphs by using the nodes that represent attribute or the value properties, one node for each class and one node for each picture instance. This structure improves the efficiency of the graph neural networks by allowing efficient use of the edge weights and the node interactions. By describing the nodes as well as the edges as vectors that define the graph structure and semantics, the graph neural networks can extract messages from the neighboring vertices and the edges to update the vertex representations and identify the patterns. By enhancing the entity representation and the relation prediction, the graph neural networks significantly enhance the knowledge graph processing.

1.10.5 Traffic and Transportation

The graph neural networks are the valuable tools which are used in the transportation and the traffic systems for predicting demand, congestion and the traffic flow. They are used to model the road networks as graphs.

The graph neural networks are enabling the accurate traffic forecasting, demand prediction and the autonomous vehicle control by capturing the spatial temporal dependencies. These networks are able to anticipate the accident hotspots, detect possible congestion hotspots and enhance the transportation safety. They are ideally suited for the deployment in smart cities due to their capacity to handle the extensive road networks, adjust to changing the traffic patterns and process the data instantly. These kinds of networks employ the graph model of the traffic network, while the vertices are representing the traffic sensors and the edges indicate separations between them to capture the intricate temporal and the geographical patterns. The dynamic input attributes of every vertex including the average traffic speed over time, provide precise estimates of the traffic density, volume and speed. By utilizing the graph structure, this method improves the forecasting performance and gives some important information to help the transportation authorities to optimize their traffic management plans. Consequently, the graph neural networks empower the transportation authorities to make the informed decisions by providing deep insights into the traffic mechanisms, predicting traffic conditions, identifying hotspots and optimizing the traffic flow, thereby creating more efficient and safe transportation systems.

1.10.6 Computer Vision

The graph neural networks are revolutionizing the computer vision by making it possible to process the structured data such as graphs [16]. By simulating the spatial relationships between the objects and extracting features that capture the underlying structure of the picture, they are essential to object recognition, image segmentation and 3D reconstruction [16]. The graph neural networks offer the flexible tool for processing the structured data with applications including the object detection, semantic labeling and image generation. There are several primary forms of graph neural networks, each with unique strengths. Also, the graph convolutional networks use the convolutional operations to propagate the message between the nodes while the graph attention networks employ attention techniques to allocate the numerical values to the edges based on the vertex relationships. The graph recurrent networks, designed for the sequential data processing, use the recurrent neural network architecture to update node outputs based on their neighbors and the current states. The GNNs are good alternative for more challenging visual comprehension tasks, even though the conventional neural networks are good in identifying the objects in the pictures and the videos. For example, the graph neural networks computing the pixel into the semantic graph objects and their connections while generating the scene graphs.

1.10.7 Natural Language Processing (NLP)

The graph representation and transformer designs have been used to spot the long-range connections in several neural network types including the transformer based on GNNs, GCNs and GATs. The graphs are increasingly being used in NLP to supplement more conventional sequential models such as the LSTMs and the RNNs. The machine translation, text categorization, user spot, relation extraction and the question answering are few of the NLP tasks for which the graph neural networks have gained the popularity. Generalization to the unseen nodes is made possible through some notable techniques such as GraphSAGE which broadens the use of graph neural networks in NLP brings new life to continue the problems like question answering. The graph neural networks provide the new possibilities for the knowledge representation and the retrieval by representing the items as vertices and their links as edges. This enhances the performance on the natural language processing tasks that need an understanding of the complex relationships.

1.10.8 FinTech and Fraud Detection

A combination of the two strings "financial" and "technology," "fintech" describes the use of technology to provide the financial services, frequently in novel ways. FinTech businesses use the digital platforms, sophisticated software and the algorithms to revolutionize the financial services and increase their accessibility and convenience. There are two types of credit card fraud [6] such as the inner card fraud and the exterior card fraud [22]. While exterior card fraudsters steal credit cards and PINs, inner credit card fraud happens when the cardholder presents false identification [9]. Conventional approaches use expert-based rules, which may not always be sufficient [27]. The neural networks are having more efficient method for analyzing very large amount of transaction data and uncovering the hidden patterns. The applications of the graph neural networks play a crucial role in FinTech [8], specifically in the areas such as the fraud prevention and the detection [1]. They are used in detecting and spotting the fraudulent conduct through the examination of the intricate connections in the financial transactions. The GCNs, graph transformers and the GAT networks are some of the specialized designs which are used in credit card fraud detection [6, 18, 27, 28]. They are successfully detecting suspicious behaviors including the money laundering, account compromise and the payment schemes between the different companies. In the financial applications, the financial data are represented by the graphs in which the entities such as the accounts, people, transactions and the devices are represented by the nodes and the relationships such as the shared qualities, transactions and the interactions are represented by the edges [10]. By utilizing the knowledge of context related and camouflage related behaviors, the researchers can create more potent graph neural network-based fraud detection techniques.

1.11 Conclusion

The GNNs have revolutionized graph data mining by offering the powerful tool for spotting the complicated connections and the patterns in graph structured data. The applications in a huge range of fields including the recommendation systems, chemistry, biology, social network analysis, traffic forecasting and the natural language processing have been effective due to their adaptability, expressive capacity and ease of visualization. The node label prediction, the graph categorization and the graph creation are among the activities that the graph neural networks excel at [10]. The GNNs are used in financial industry to capture the potential fraud and reduce the risk in the transaction data [15]. The specific architectures of graph neural networks such as the graph transformers, GCNs and GATs are made possible to enhance the detection skills [8]. Finally, we conclude that GNNs have demonstrated a great deal of promise in processing complicated graph structure through their successful advancements in various fields.

AI disclosure statement: The author confirms that no artificial intelligence (AI)-generated content or AI-assisted writing tools were used in the preparation of this chapter (including summaries and the abstract). All text, analysis, and conclusions are the result of the author's original work.

References

1. Chouiekh, A. and Hassane, E.L., ConvNets for Fraud Detection analysis. *Procedia Comput. Sci.*, 127, 133–138, 2018, doi: https://doi.org/10.1016/j.procs.2018.01.107.
2. Pucci, A., *et al.*, Investigations into the application of graph neural networks to large-scale recommender systems. *Int. J. Syst. Sci.*, 32, 4, 17–26, 2006.

3. Khemani, B., Patil, S., Kotecha, K., *et al.*, A review of graph neural networks: concepts, architectures, techniques, challenges, datasets, applications, and future directions. *J. Big Data*, 11, 18, Jan 2024, https://doi.org/10.1186/s40537-023-00876-4.
4. Choubisa, D., An overview of graph neural networks for molecular biology: Challenges and solutions. *Int. J. Sci. Res. Arch.*, 13, 01, 2670–2673, Oct 2024, doi: https://doi.org/10.30574/ijsra.2024.13.1.1985.
5. Chang, D.T., Deep Learning for Molecular Graphs with Tiered Graph Autoencoders and Graph Prediction. *Mach. Learn.*, July 2021, http://arxiv.org/abs/1910.11390v2.
6. Ileberi, E., Sun, Y., Wang, Z., A machine learning based credit card fraud detection using the ga algorithm for feature selection. *J. Big Data*, 9, 24, 1–17, Feb 2022, https://doi.org/10.1186/s40537-022-00573-8.
7. Khan, I., *et al.*, Distance driven graph neural network for advanced node classification through feature augmentation. *Discov. Comput.*, 28, 70, May 2025, doi: https://doi.org/10.1007/s10791-025-09569-3.
8. Jafri, J.A., *et al.*, A systematic literature review of the role of trust and security on Fintech adoption in banking. *Heliyon*, 10, 1, e22980, Jan 2024, doi: https://doi.org/10.1016/j.heliyon.2023.e22980.
9. Leach, J. and Thayasivam, U., Optimizing data evaluation metrics for fraud detection using machine learning, *Int. J. Comput. Math. Sci.*, 16, 8, 52–59, Aug 2022.
10. Wang, J., *et al.*, A Review on Graph Neural Network Methods in Financial Applications. *J. Data Sci.*, 1–32, July 2020, https://arxiv.org/pdf/2111.15367.
11. Zhoua, J., *et al.*, Graph neural networks: A review of methods and applications. *AI Open 2021*, 1, 57–81, 2020, https://arxiv.org/abs/1812.08434.
12. Doaud, M. and Mayo, M., Using Swarm Optimization To Enhance Autoencoders Images. *14th International Conference on Modeling Decisions for Artificial Intelligence*, Kitakyushu, Japan, Oct 2017, pp. 118–131, https://arxiv.org/abs/1807.03346.
13. Ma, M., *et al.*, AEGCN: An Autoencoder-Constrained Graph Convolutional Network. *Neurocomputing 2021*, Feb 2021, http://arxiv.org/abs/2007.03424v3.
14. Kulkarni, M.M., *et al.*, Enhancing Social Network Analysis using Graph Neural Networks. *Adv. Nonlinear Var. Inequal.*, ISSN: 1092-910X, 27, 3, 213–230, Jun 2024, doi: https://doi.org/10.52783/anvi.v27.1502.
15. Turaba, M.Y., *et al.*, Fraud Detection During Financial Transactions Using Machine Learning and Deep Learning Techniques. *2022 International Conference on Communications, Computing, Cybersecurity, and Informatics (CCCI)*, Dalian, China, Oct 2022, pp. 1–8, doi: 10.1109/CCCI55352.2022.9926503.
16. Cao, P., *et al.*, Applications of graph convolutional networks in computer vision. *Neural Comput. Appl.*, 34, 13387–13405, Aug 2022, doi:10.1007/s00521-022-07368-1.

17. Velickovic, P., *et al.*, Graph Attention Networks, ICLR 2018, Feb 2018, https://doi.org/10.48550/arXiv.1710.10903.
18. Asha, R.B. and Kumar, K.R.S., Credit card fraud detection using artificial neural network. *Global Transitions Proc.*, 2, 1, 35–41, Jun 2021, doi: https://doi.org/10.1016/j.gltp.2021.01.006.
19. Kearnes, S., *et al.*, Decoding Molecular Graph Embeddings with Reinforcement Learning. *ICML 2019 Workshop on Learning and Reasoning with Graph-Structured Data*, Jun 2019, http://arxiv.org/abs/1904.08915v2.
20. Pan, S., *et al.*, Adversarially Regularized Graph Autoencoder for Graph Embedding. *Mach. Learn.*, Jan 2019, http://arxiv.org/abs/1802.04407v2.
21. Banu, S.R., *et al.*, Financial Fraud Detection Using Hybrid Convolutional and Recurrent Neural Networks: An Analysis of Unstructured Data in Banking. *2024 10th International Conference on Communication and Signal Processing (ICCSP)*, June 2024, IEEE, doi:10.1109/ICCSP60870.2024.10543545.
22. Singh, S. and Grag, M., Credit card fraud detection system. *Int. J. Creat. Res. Thoughts (IJCRT)*, 9, 6, b312–b316, June 2021, ISSN: 2320-2882, https://ijcrt.org/papers/IJCRT2106161.pdf.
23. Berhane, T., *et al.*, A Hybrid Convolutional Neural Network and Support Vector Machine-Based Credit Card Fraud Detection Model. Hindawi, *Math. Probl. Eng.*, 2023, 8134627, 1–10, June 2023, doi: https://doi.org/10.1155/2023/8134627.
24. Kipf, T.N. and Welling, M., Semi-Supervised Classification with Graph Convolutional Networks. *ICLR 2017 Conference Proceedings*, Feb 2017, pp. 1–14, doi:https://arxiv.org/pdf/1609.02907.
25. Cai, Y. and Jelovica, J., Efficient graph representation in graph neural networks for stress predictions in Stiffened panels. *Thin Walled Struct.*, 203, 112157, 2–3, July 2024, doi: 10.1016/j.tws.2024.112157.
26. Lavryk, Y. and Kryvenchuk, Y., Product Recommendation System Using Graph Neural Network. *5th International Workshop on Modern Machine Learning Technologies and Data Science, CEUR Workshop Proceedings (CEUR-WS.org)*, Jun 2023, https://ceur-ws.org/Vol-3426/paper15.pdf.
27. Murugan, Y., et al., Credit Card Fraud Detection Using CNN, Internet of Things and Connected Technologies. *LNSS*, 340, 194–204, Springer, Cham, Jan 2022, doi: https://doi.org/10.1007/978-3-030-94507-7_19.
28. Wu, Y., *et al.*, Graph convolutional networks with markov random field reasoning for social spammer detection. *Proc. AAAI Conf. Artif. Intell.*, 34, 01, 1054–1061, Apr 2020, doi: https://doi.org/10.1609/aaai.v34i01.5455.
29. Wu, Z., *et al.*, A Comprehensive Survey on Graph Neural Networks. *IEEE Trans. Neural Netw. Learn. Syst.*, 32, 1, 4–24, Jan. 2021, doi: 10.1109/TNNLS.2020.2978386.

2

Graph Theory Foundations for Neural Network Models

Nidhi Asthana*, Divya Gautam and Gaurav Paliwal

School of Technology Management and Engineering, SVKM's Narsee Monjee Institute of Management Studies (NMIMS), Indore, MP, India

Abstract

This chapter provides a concise introduction to graph-theoretic models of neural networks with an emphasis on their mathematical underpinnings, computational structure, richness of architecture, and utility in applications to real-world problems in various areas. We begin with the foundations of graph theory. With this context, the chapter proceeds to spectral graph theory, the foundation upon which most graph-based machine learning algorithms are built. Graph Laplacian, being a crucial operator in this context, is provided both combinatorically and normalized form, and how their use towards structural relations bound in eigenvalues and eigenvectors. Graph Fourier transform idea is introduced through which filtering as well as convolution becomes feasible within the spectral space. The chapter also redefines traditional neural network architectures in terms of graph theory. A close discussion of Graph Neural Networks (GNNs) and their extensions forms the basis of the chapter. We also cover classical models such as Graph Convolutional Networks (GCNs), which utilize first-order neighborhood propagation to describe spectral convolutions.

For social networks, GNNs enable node classification, community detection, and link prediction. Recommender systems use graph-based models of user-item interactions, while GNNs on KG are used to reasoning, relation extraction, and to entities' disambiguation. Not just application-wise but also as having critically examined existing constraints and continuing research challenges. The chapter ends with the future directions for future graph-based learning development. The future directions are universal graph learning models which can learn to generalize task and domain, combination of graph learning with foundation models and

**Corresponding author*: drnidhi80@gmail.com

J. Ramkumar, Sridaran Rajagopal, B. Suchitra and S. Balamurugan (eds.) Graph Neural Networks: Concepts and Applications, (43–100) © 2026 Scrivener Publishing LLC

transformers, and construction of neuromyotonic systems combining statistical learning and symbolic reasoning.

Keywords: Graph theory, neural networks, graph neural networks, spectral graph theory, deep learning, network topology, artificial intelligence

2.1 Introduction

2.1.1 Background and Motivation

The rampant explosion of machine learning (ML) and artificial intelligence (AI) over the past decade was overwhelmingly fueled by progress in the neural network architecture. These artificial computer networks, inspired by their biological counterparts in the human brain, have enabled machines to perform high-level tasks such as image recognition, speech recognition, natural language processing, autonomous driving, and strategic games with previously unattained accuracy [1]. The underlying theme that cannot be seen across this revolution is the increasing capability of neural networks to learn complicated patterns from well-organized data by optimizing multi-layered computational graphs.

Conventional deep learning architectures like Convolutional Neural Networks (CNNs) and Recurrent Neural Networks (RNNs) rely on the Euclidean assumption, i.e., data is organized in Euclidean spaces, e.g., a grid for images and visual data or a sequential list for time-series and text data. These data structures are regular in nature with fixed neighborhood structures and uniform dimensionalities. But more and more real-life scenarios have to deal with non-Euclidean, sparse, and irrationally networked data—making the Euclidean assumption outdated and even misleading [2]. Examples include:

- Social networks, where nodes are tied together by social relations of various types (edges), e.g., friendship, professional acquaintance, or follower-followed.
- Biological systems, where molecular structure or protein-protein interaction networks are graph-like in connectively.
- Financial networks, where transactions, customers, and banks create complicated webs of dependency and risk components.
- Transit networks, as in road networks in a town or flight networks, with non-grid like topologies that are dynamic.

- Knowledge graphs, used in natural language processing and semantic web, where entities are related by complicated relationships.

In all such domains, graphs are a more natural and expressive data type. A graph can be rigorously defined as G=(V,E), with V being a set of nodes and $E \subseteq V \times V$ being the set of edges between sets of nodes. Graphs inherently capture the node-local properties and the topological relationships between nodes, and therefore are naturally well-suited to represent real-world systems [3].

To apply the potential of graphs in artificial intelligence tasks, it is essential to generalize the typical neural network models to represent graph-structured data. The requirement has generated a new model category named Graph Neural Networks (GNNs). The models generalize and transfer deep learning techniques to graph spaces so that models can reason, classify, regress, and generate over graph-represented information [4]. Graph theory application to neural network models allows more advanced feature extraction, more solid contextual learning, and better comprehension of data structure and meaning. Graph-based learning not only removes the shortcomings of Euclidean models but also holds the promise of being able to overcome challenges in domains that had been stubbornly resistant to pattern recognition and automation.

2.1.2 Theoretical Significance of Graph Theory in AI

Integration of graphs into neural networks is neither structural change for the sake of conceptual accommodation nor conceptual accommodation also—mappings of relations and their use in guiding the learning process both count as conceptual change. The majority of ML models were conventionally constructed in such a way that the input sample of data was identically and independently distributed. Graph information, however, does reflect interdependence, in the sense that the value of the node is not just a function of the node's properties but its neighbors and their positions [5].

As an illustration, within a citation network, a research paper will be semantically linked to the papers that it references or which reference it. Nodes (papers) and edges (papers referring) form an interdependent pattern of relations to be read together, and not in isolation. Similarly, chemical bonds between atoms in molecules in molecular chemistry also create graph structures which govern molecular properties. Simulation of such interaction renders such activity like drug design, virtual screening, and prediction of properties with awe-inspiring accuracy [6]. Graph theory is used almost to provide tools like adjacency matrices, incidence matrices, Laplacian matrices, and spectral decompositions to facilitate the study of complex structures. They play a very critical role in converting raw graph data to a machine learning system into a comprehensible form. Graph Laplacian, for instance, simplifies obtaining an idea of the overall picture of a graph and facilitates operations like spectral clustering, manifold learning, and graph convolution [7].

Also, spectral graph theory, which studies the behavior of the graph in relation to the eigenvalues and eigenvectors of their Laplacian matrices, is the basis for the majority of graph-based learning methods. With the aid of graph Fourier transforms, scientists are able to treat graph signals as it is treated for ordinary domains, thus taking operations such as convolution, filtering, and frequency analysis into irregular domains [8]. Graph-theoretic perspectives allow for new architecture design of neural networks. Neural networks are a directed acyclic graph (DAG) in feedforward networks, and recurrent ones include cycles to hold temporal dependency. Even for convolutional networks, receptive field and weight-sharing are their spatial graph analogs. Optimization, compression, and interpretability of neural models are facilitated by graph algorithms based on such analogies [9].

2.1.3 The Rise of Graph Neural Networks (GNNs)

GNNs are a milestone in deep learning. The models can be trained from random, sparse, and dynamic graphs, thereby unleashing the potential of non-Euclidean spaces for machine learning. Message passing mechanism is one of the greatest successes of GNNs wherein each node at every step accumulates information from neighboring nodes to update its feature representation. This is done for a certain number of iterations (or layers), and the data thus propagates over the graph [10].

Graph Convolutional Networks (GCNs), or perhaps one of the most widely used GNN variants as illustrated in the Figure 2.1, are computationally efficient spectral convolution approximations. GCNs give each node an opportunity to make individual contributions to its new representation in aggregate *via* re-normalization of adjacency matrices and addition of self-loops. Such a setup has been found to be beneficial for semi-supervised node classification, link prediction, and clustering [11]. Another strong model is the Graph Attention Network (GAT) that applies attention to learn dynamic weights for neighboring nodes adaptively in order of relevance. This enables models to focus more on informative neighbors while

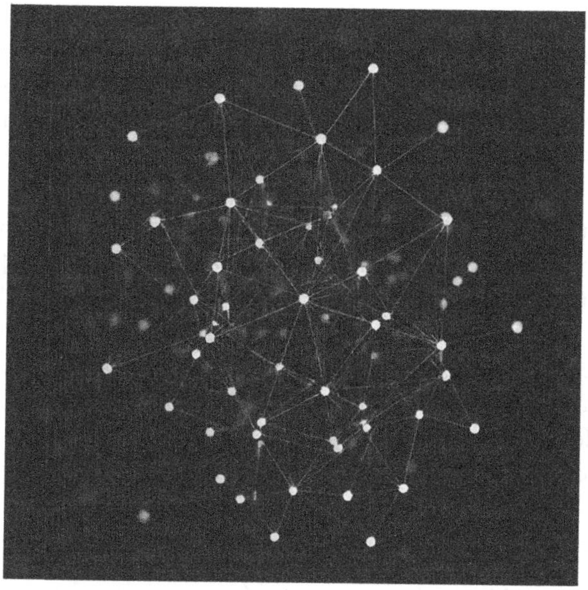

Figure 2.1 Graph neural networks (GNNs).

aggregating features and makes GATs particularly resilient in noisy or heterogeneous graphs [12].

MPNN abstraction provides a general formulation for decently different GNN architecture. MPNN separates learning into two separate processes: message computation (from node state and edge attribute) and state update (from messages received). This makes it modular in design, where task-specific types like quantum chemistry, natural language processing, and graph generation are byproducts [13]. These advances placed GNNs at the forefront of modern AI technologies. Not only do they hold out the promise of performance improvement on standard tasks but they also open up new frontiers to attack to apply to fields ranging from bioinformatics, recommender systems, to cyber security and autonomous systems.

2.1.4 Practical Motivation Across Disciplines

Following are real-world, practical examples showing how graph theory is applied directly in models of neural networks, i.e., Graph Neural Networks (GNNs), in various domains:

a) Drug Discovery and Molecular Chemistry
Graph Theory Role: Molecules are graphs naturally — atoms as vertices and chemical bonds as edges. Neural Network Model: Message Passing Neural Networks (MPNNs) process molecular graphs to learn about molecular properties. Example Application:

- Solubility, toxicity, or drug-target interaction prediction.
- For example, GNN models like DeepChem are helpful in identifying possible antiviral drugs against COVID-19.

b) Social Network Analysis
Role of Graph Theory: Users are vertices; friendships, likes, or retweets are edges. Neural Network Model: Graph Convolutional Networks (GCNs) and GraphSAGE. Example Application:

- Prediction of user attributes (e.g., interests, age).
- Friend suggestion or community detection.
- GNN-based models are used by Facebook and LinkedIn for friend recommendation and ranking of content.

c) Recommender Systems
Graph Theory Role: Products and users form a bipartite graph. Neural Network Model: PinSage, a GNN model for Pinterest content recommendation. Example Application:

- Learning item-user embeddings from interaction data.
- Pin, product, or video recommendation from user behavior.

d) Biomedical Knowledge Graphs
Graph Theory Role: Diseases, symptoms, and drugs form a heterogeneous graph. Neural Network Model: HetGNN (Heterogeneous Graph Neural Networks) and HAN (Heterogeneous Attention Networks). Example Application:

- Gene-disease association or personalized treatment workflow identification.
- Used in Electronic Health Records (EHR) for treatment suggestion.

e) Knowledge Graph Completion (e.g., Google, Wikidata)
Graph Theory Role: Objects (e.g., people, places) are nodes, and relations (e.g., "is married to") are edges. Neural Network Model: Relational GCNs (R-GCNs), CompGCN. Example Application:

- Missing fact completion like "X is capital of Y."
- Used in semantic search, question answering, and chatbots.

f) Cybersecurity and Fraud Detection
Graph Theory Role: Transaction networks or access logs are graphs with the accounts/users as nodes and transactions as edges. Neural Network Model: Dynamic GNNs or Graph-based anomaly detection models. Example Application:

- Credit card fraud detection, collusive fraud rings, or anomalous login patterns.
- Used by banks and cybersecurity firms for risk scoring and anomaly detection.

g) Traffic Prediction and Smart Cities
Graph Theory Role: Roads and intersections are traffic graph nodes and edges. Neural Network Model: Spatio-temporal GNNs, e.g., ST-GCN. Example Application:

- Real-time traffic flow prediction, congestion recognition, and route planning.
- Integrated in smart city apps and platforms like Google Maps.

h) Robotics and Multi-Agent Systems
Graph Theory Role: Robots or agents are nodes; proximity or communication form edges. Neural Network Model: Graph-based coordination networks. Example Application:

- Cooperative control of swarms of drones or warehouse robots (e.g., Amazon Robotics).
- GNNs encapsulate message-passing between agents for successful cooperation.

i) Computer Vision (Scene Graphs)
Graph Theory Role: Objects in an image form a graph based on spatial or semantic relationships. Neural Network Model: Scene Graph GNNs. Example Application:

- Object detection with relation awareness (e.g., "person sitting on chair").
- Used in autonomous cars, surveillance, and robot vision.

j) Document and Text Analysis
Graph Theory Role: Sentences, words, or entities are encoded as a graph (e.g., word co-occurrence, syntactic parse trees). Neural Network Model: TextGCN, GraphBERT. Example Application:

- Text classification, sentiment analysis, and co-reference resolution.
- Enhances document understanding in legal tech, academic indexing, and chatbots.

Graph-based neural model application extends beyond experimental studies—its use is increasing in industries and research areas.

- Computer vision graphs represent spatial relationships between objects or super pixels viewed. Scene graphs allow machines not just to identify what objects an image contains, but also how objects are linked together—i.e., "man on the bike" or "cup on table" [14].
- NLP's co-reference resolution, dependency parsing, and knowledge graph completion are all based on the relational text structure. GNNs have now taken over from the traditional RNN and Transformer-based architectures for reasoning-over-relations operations between entities and concepts [15].
- Gene-disease associations, patient similarity, and protein-protein interaction networks in medicine are graph-structured. Predictive graph-based models enable early diagnosis, targeted treatment, and drug discovery [16].
- In recommenders, the user-item interactions are represented as graphs. GNNs outperform the standard collaborative filtering because they can learn more complex relations, i.e.,

users with similar patterns of behavior or items with similar characteristics [17].
- Discovery of financial fraud enables graph models, which will detect atypical patterns of transactions or groupings of accounts outside the usual. Transaction graphs reveal concealed patterns typical of collusive fraud or money laundering operation [18].

The ubiquity of such applications testifies to growing need for saleable and high-throughput graph models that will be capable of satisfying a gigantic amount of application instances.

2.1.5 Research Challenges and Future Directions

Graph-based learning is full of benefits but possesses a few drawbacks as well. One of them is the scalability of GNNs to very large graphs with billions or millions of nodes and edges. Multi-message passing or whole-graph computation in standard GNNs might be computationally expensive and entail massive memory usage.

In response, there have been some solutions proposed:

- Graph mini-batching and sampling: Techniques such as GraphSAGE [19] sample the neighboring objects at training time in an attempt to reduce computational costs.
- Graph scarification: Edge removal in a manner that retains significant structural properties.
- Asynchronous and distributed GNNs: Parallelization so that one can handle large graphs efficiently.

Another issue area is interpretability and explainability of GNN models. Because training is being done through mass manipulation of information from various sources, it is becoming increasingly difficult to understand why a specific model was approximating something in some specific way. Saliency map, attention view, and subgraph attribution form techniques are being developed with the aim of making the process transparent and to hold [20] accountable.

Temporal and dynamic graphs add even higher complexity in which nodes and edges evolve over time. Temporal relationships must be addressed through such hybrid models, which combine the use of GNNs along with sequence models such as RNNs or attention mechanisms.

They also take place in more expressive node and edge-type heterogeneous graph models. HetGNNs abstract message passing with node/edge types to achieve more fine-grained learning in more complicated domains like research networks, ecological networks, and multi-modal artificial intelligence systems [21].

Finally, ethics of fairness, bias, and privacy are rising in graph models. Because graphs represent relationships in social, demographics, and behaviors, the graph can multiply a biased input or it's an invasion of privacy of users. Differential privacy, fair-aware learning, and de-biasing graphs should be investigated in order to ethically use AI [22].

2.2 Fundamentals of Graph Theory

Graph theory, which emerged in discrete mathematics, is strong enough as a formalism to capture object-to-object relations. Graph theory pervades most computer science disciplines, from algorithms and network science to machine learning. Due to the creation of Graph Neural Networks (GNNs), graph-theoretic primitives came to be familiar enough to establish, study, and utilize learning algorithms for processing structured data. Graphs are the most intuitive data structure to encode pairwise relations between objects. In contrast to vector encoding, where point independency is generally assumed, graph-based structures can encode relational dependencies of a higher order. For this reason, graphs are very convenient

Table 2.1 Comparison of graph types.

Graph type	Edge direction	Edge weights	Self-loops allowed	Typical applications
Undirected Graph	No	Optional	Sometimes	Social networks, biological systems
Directed Graph (Digraph)	Yes	Optional	Often	Web links, citation graphs
Weighted Graph	Yes/No	Yes	Optional	Road networks, recommendation systems
Multigraph	Yes/No	Yes/No	Yes	Transport systems, semantic networks
Heterogeneous Graph	Yes	Yes	Yes	Knowledge graphs, EHR systems

in applications where connectivity or interaction among items is equally as important as item attributes themselves—e.g., social networks, protein-protein interactions, recommendation systems, and traffic flow models [1]. As presented in Table 2.1, a comparison of various graphs and their applications is provided.

2.2.1 Basic Concepts

A graph G=(V,E) is a non-empty, finite set of edges $E \subseteq V \times V$ and vertices V (or nodes) inducing pairwise relations between the vertices. Having $(u,v) \in E$ implies an edge from vertex v to vertex u. Graphs can be categorized in multiple ways depending upon the edges and vertices as depicted in the Figure 2.2.

- Undirected Graphs: An undirected graph is a graph whose statement that $(u,v) \in E$ is the same as saying that, the relation being symmetric. Undirected graphs come in handy to represent mutual relations such as friendships or bidirectional communication channels as demonstrated in the Figure 2.3.
- Directed Graphs (Digraphs): Direction is indicated by directed edges of directed graphs. Directed edge (u,v) is a directed edge from source node u to target node v as indicated in the Figure 2.4. Directed graphs surround us in web linkages, citation networks, and flow networks [2].

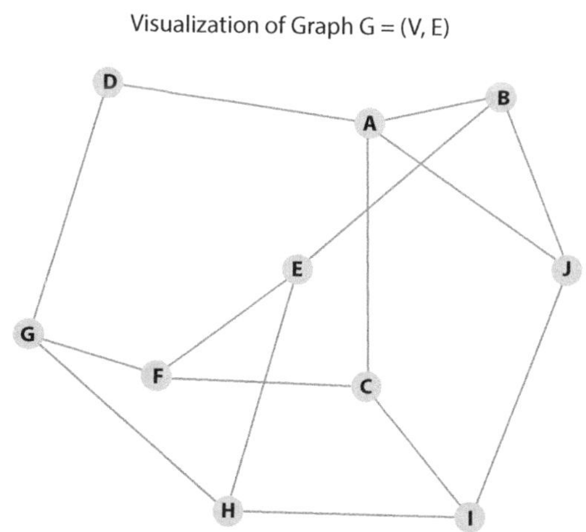

Figure 2.2 Graph G=(V,E) with 10 vertices and 15 edges.

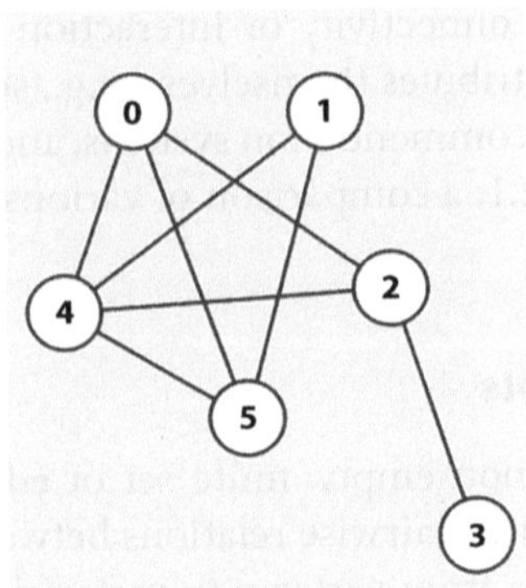

Figure 2.3 Undirected graph.

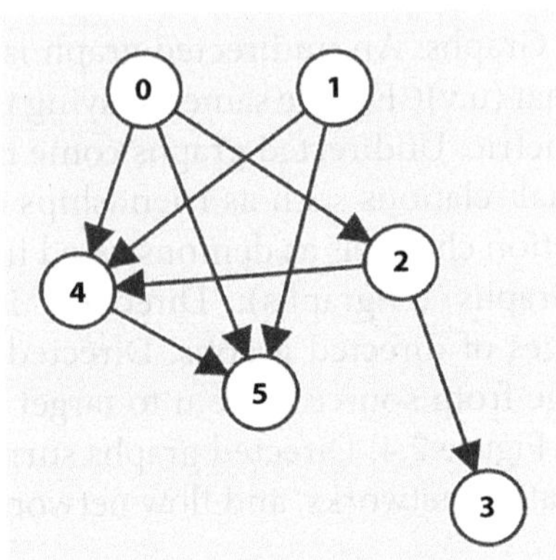

Figure 2.4 Directed graph.

- Weighted Graphs: Each edge of a weighted as shown in Figure 2.5 has an associated numeric value (weight), w(u,v), along with it, such as cost, capacity, similarity, or any other value depending on the particular problem being modeled. They find widespread use in route finding, neural networks, and physics simulation.
- Unweighted Graphs: Unweighted edges of a graph are not independent weights and are usually represented in binary form (no relationship = 0, related = 1) as presented in the Figure 2.6. Such graphs arise fairly frequently in conception system drawings where presence or absence will suffice.

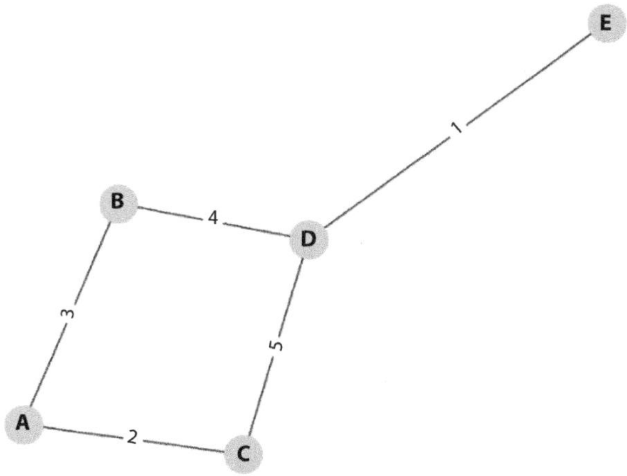

Figure 2.5 Simple weighted undirected graph.

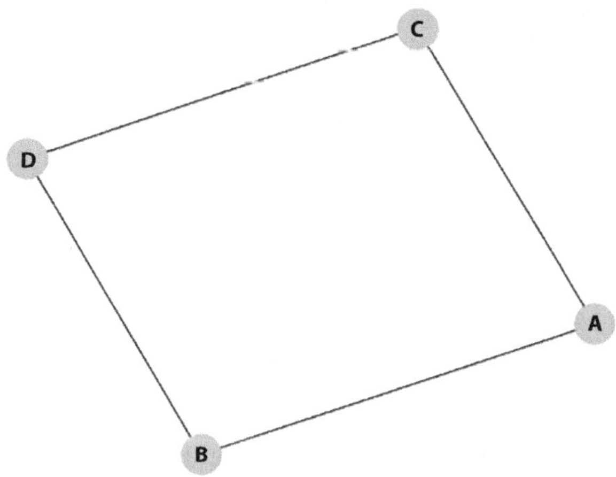

Figure 2.6 Simple unweighted undirected graph with four vertices.

- Basic Graphs: A basic graph has no loops (edges between vertices by the same vertex) and more than one edge between identical vertices.
- Multigraphs: Any two nodes may have multiple edges in a multigraph. Multigraphs may be used to define parallel or alternative modes of relationship as seen in the Figure 2.7.

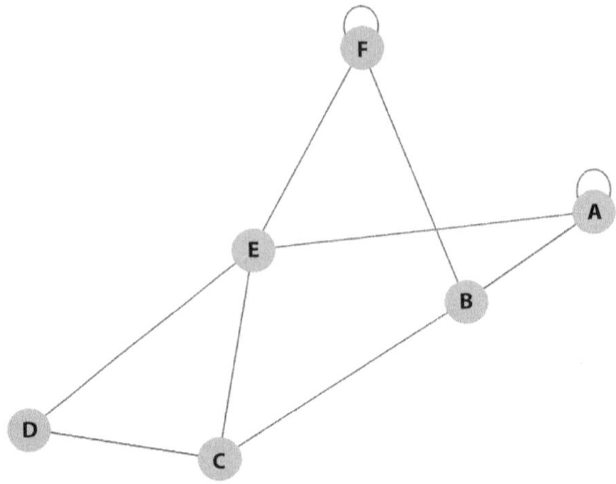

Figure 2.7 Complex multigraph with six vertices and loops.

Machine learning graph is typically enriched by node features and edge features in such a way that for each vertex $v_i \in V$ there is a bound feature vector $x_i \in R$, and each edge $(u,v) \in E$ may have feature data like weights, orientation, or semantic labels. The topological flexibility of graph structure makes graphs ideally suited to represent non-Euclidean, irregular data. It is the realization that graphs can represent domain-specific data both feature-wise and topologically that has driven most of the key advances in GNNs [3].

2.2.2 Graph Representations

Graph representation is graph structure description to be computed. Representation impacts time and space complexity of algorithms. Four primitive representation schemes are: adjacency matrices, incidence matrices, edge lists, and adjacency lists.

- Adjacency Matrix

The $n=|V|$ adjacency matrix $A \in R^{n \times n}$, graphs represent edges. For an unweighted graph:

$$A(i,j) = \begin{cases} 1 & \text{if } (v_i, v_j) \in E \\ 0 & \text{otherwise} \end{cases}$$

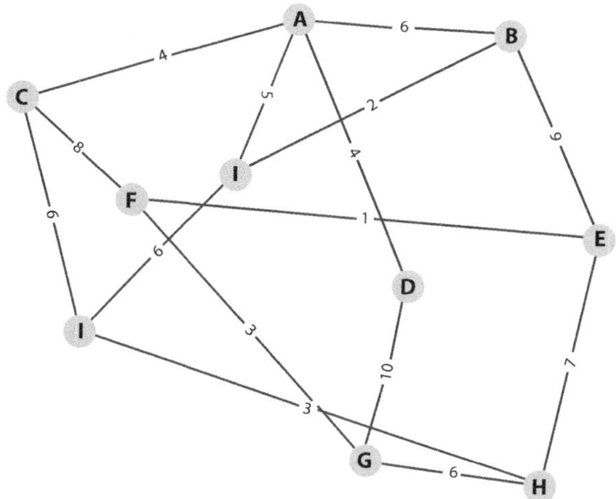

Figure 2.8 Weighted graph where each edge has been assigned a weight between 1 and 10.

For a weighted graph, A(i, j) holds the edge weight w(vi,vj) instead of a binary value. A is symmetric in case of undirected graphs as represented in the Figure 2.8.

Even though adjacency matrices provide constant time in checking for the existence of an edge or not, they are space inefficient with O(n2) space complexity when handling sparse graphs. They are easy to implement in spectral graph theory and matrix operation like Eigen decomposition and Laplacian computation [4].

- **Incidence Matrix**

The m=|E| dimension edge-node incidence matrix $M \in R^{n \times m}$ preserves the edge-node incidence relation. There is one row for every vertex and one column for each edge as evidenced by the Figure 2.9. A matrix value M(i,j) in an undirected graph will likely be 1 if the vertex vi is incident with the edge e_j, or 0 otherwise. A matrix value M(i,j) in a directed graph might use one standard notation:

$$M(i,j) = \begin{cases} +1 & \text{if } v_i \text{ is the source of } e_j \\ -1 & \text{if } v_i \text{ is the target of } e_j \\ 0 & \text{otherwise} \end{cases}$$

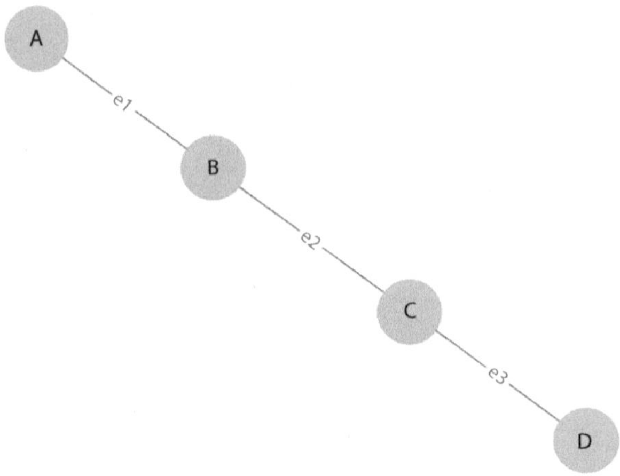

Figure 2.9 Undirected graph with labeled edges.

Incidence matrices are employed in circuit analysis and graph flow when edge-node interaction is significant [5].

Suppose we have a graph with 4 vertices and 3 edges:

Vertices: V = {A, B, C, D}

Edges: e_1: A — B, e_2: B — C and e_3: C — D

Edges: e_1 connects v_1 and v_2, e_2 connects v_2 and v_3 and e_3 connects v_3 and v_4

Incident Matrix is given by:

$$\begin{bmatrix} & e_1 & e_2 & e_3 \\ v_1(A) & 1 & 0 & 0 \\ v_{21}(B) & 1 & 1 & 0 \\ v_3(C) & 0 & 1 & 1 \\ v_4(D) & 0 & 0 & 1 \end{bmatrix}$$

- **Edge List**

Adjacency list is the list or dictionary representation of the graph, and for each node vi, a list of its neighbors. E={(v_i, v_j),(v_k, v_l),… } as outlined in the Figure 2.10. In weighted graphs, the list holds pairs (v_j, w_{ij}).

edges = [(0, 1), (0, 2), (0, 4), (1, 3), (2, 3), (3, 4)]

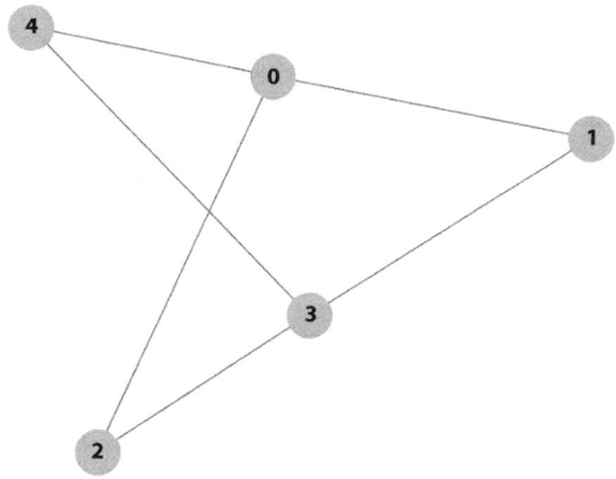

Figure 2.10 Graph with five vertices.

Adjacency lists save space more effectively for sparse graphs and support quicker traversal, and hence are the representation of choice in GNN implementation and traversal algorithms such as BFS or DFS [7].

- **Adjacency List**

Adjacency list is the list or dictionary representation of the graph, and for each node vi, a list of its neighbors. $Adj[v_i]=\{v_j:(v_i,v_j)\in E\}$ as can be observed in the Figure 2.11. In weighted graphs, the list holds pairs (v_j, w_{ij}).

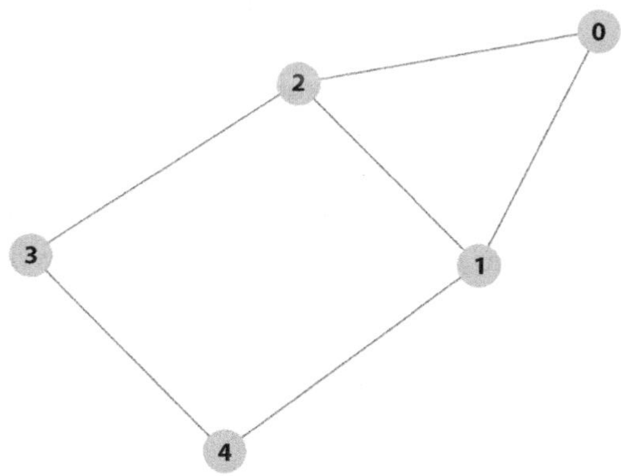

Figure 2.11 Graph visualization.

Table 2.2 Comparison of graph representations.

Representation	Space complexity	Edge existence check	Neighbor lookup	Suitable for
Adjacency Matrix	O(n2)	O(1)	O(n)	Dense graphs, spectral operations
Incidence Matrix	O(n,m)	O(m)	O(m)	Flow problems, circuit analysis
Edge List	O(m)	O(m)	O(m)	Edge-centric algorithms
Adjacency List	O(n+m)	O(d)	O(d)	Sparse graphs, graph traversal

For the above graph Adjacency list is given by: As illustrated in the Table 2.2.

Adjacency List of the Graph:
0: [1, 2]
1: [0, 2, 4]
2: [0, 1, 3]
3: [2, 4]
4: [1, 3]

Adjacency lists save space more effectively for sparse graphs and support quicker traversal, and hence are the representation of choice in GNN implementation and traversal algorithms such as BFS or DFS [7].

Most used is adjacency matrix and adjacency list in graph neural networks. Adjacency matrix becomes a component when Laplacian-based operations are used and adjacency lists are used in spatial GNNs where message passing is limited to the neighbors.

Representations like these must be interpretable to the one who's doing good graph computation. Aside from execution time complexity and memory consumption, representability influences interpretability and scalability of downstream machine learning models as well.

2.3 Spectral Graph Theory and Signal Processing

As machine learning moved into non-standard data structures, learning from non-Euclidean spaces meant new emphasis on spectral graph theory—a mathematical field that deals with the study of graphs *via* spectral (eigenvalue) theory of matrices such as the adjacency matrix and the Laplacian. Spectral graph theory serves as the basis for graph-structured data generalization of deep learning since it enables one to conduct operations analogous to convolution, filtering, and signal processing on graph data. Development of GNN has been spurred to a significant extent by advances in spectral graph theory, the mathematics whereby one formulates and studies functions acting on non-Euclidean space. A glimpse of what spectral graph theory has in store is a description of the graph Laplacian, a description of the graph Fourier transform, and a description of why the above tools enable deep models to process signals on graphs.

2.3.1 The Graph Laplacian: Capturing Graph Structure

One of the earliest, and in some sense, simplest operators of spectral graph theory is the graph Laplacian that maintains the connectivity between the vertices and the degree of the graph. One of the primary objects to define the structure of the graph, as well as one of the essential building blocks in most of the spectral methods like graph partitioning, spectral clustering, and graph signal processing, is the graph Laplacian.

For an unweighted, undirected graph G=(V,E) with n vertices, node-to-node edges can be described using the adjacency matrix A∈ $R^{n\times n}$: Aij=1 if i→j edge exists, otherwise 0. The degree matrix D is diagonal, where each diagonal term Di has the vertex degree of i.

The combinatorial Laplacian L is: L=D−A. This matrix is the value of every node to the neighborhood and is discrete for the continuous differential equation Laplacian operator. Normalized graph Laplacian finds most common usage in spectral analysis. There are two forms used most widely:

Symmetric normalized Laplacian: I−D−1A

Both of these equations include scale and degree variation of nodes and hence non-regular graph analysis is more effective. Symmetry and positive semi-definiteness of the Laplacian matrix guarantee that all its eigenvalues are real and non-negative. Its application to spectral filtering and Fourier-like transformations is based on this spectral property.

2.3.2 Eigenvalues and the Spectrum of a Graph

The spectrum of a graph consists of eigenvalues of the Laplacian matrix together with its associated eigenvectors. Its spectral values possess profound structural information about the graph. Let

$$L\mathbf{u}_i = \lambda_i \quad \text{for } i = 0, \dots, n-1$$

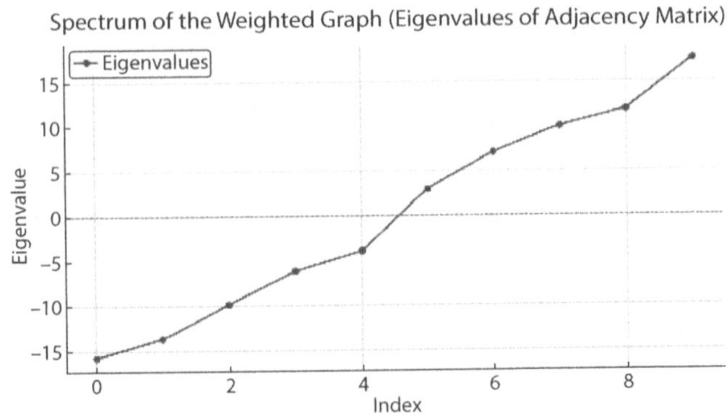

Figure 2.12 Spectrum of the weighted graph.

where λ_i are eigenvalues and u_i are associated eigenvectors. The eigenvectors form an orthonormal basis as well and therefore graph signals can be written in the same form as the conventional Fourier transform as visualized in the Figure 2.12.

L's minimum eigenvalue is always zero. Multiplicity of the minimum eigenvalue is the number of connected components of the graph. The second smallest eigenvalue, or algebraic connectivity, or Fiedler value, carries the connectivity of the graph—a larger value carries more connectivity [1].

2.3.3 Graph Signal Processing and the Graph Fourier Transform

Graph Signal Processing (GSP) signals live on the nodes of a graph—i.e., the sensor temperatures in a sensor network or the user features in a social network. A graph signal $x \in R^n$ can be modeled as a linear combination of Laplacian eigenvectors:

$$\mathbf{x} = \sum_{i=0}^{n-1} \hat{x}_i \mathbf{u}_i$$

Here in the spectral domain, filtering is equivalent to element-wise multiplications reducing computational cost by gargantuan orders of magnitude. The low-frequency elements are piecewise constant signals in the graph (i.e., same value at next-to nodes), and high-frequency elements are catching discontinuities. Such spatial to spectral transform ability enables GNNs to utilize localized convolution operators, for instance, in imaging, without loss of global graph structure.

2.3.4 Spectral Filtering and Convolution on Graphs

Convolution in standard deep learning is a local operation that produces feature extraction in the vicinity of local neighboring pixels or time steps. To extend this to graphs, introduce convolution in the frequency or spectral domain.

$$g_\theta * \mathbf{x} = \mathbf{U} g_\theta(\Lambda) \mathbf{U}^T \mathbf{x}$$

Where g_θ is a diagonal matrix with each filter value applied to every eigenvalue λi. However, computing the full Eigen decomposition of L is costly on large graphs. Chebyshev polynomial approximations are hence used, which approximate $g_\theta(\lambda)$ without ever actually computing eigenvectors:

$$g_\theta(\mathbf{L})\mathbf{x} \approx \sum_{k=0}^{K} \theta_k T_k(\tilde{\mathbf{L}})\mathbf{x}$$

Where Tk is the k-th Chebyshev polynomial and L is a scaled L to avoid numerical instability [2]. This type of approach, as defined in ChebNet, is filter-friendly in a local manner and receptive field size is governed by K. It is also at the foundation of Graph Convolutional Networks (GCNs), which even simplify the process to first-order approximation.

2.3.5 Laplacian Eigenmaps and Spectral Embedding

Laplacian Eigenmaps is spectral dimensionality reduction technique preserving local neighborhood structure. Smallest non-zero Laplacian eigenvalues are used for projecting data to low dimension in point similarity graph. Objective of optimization is to minimize below cost:

$$\min_{\mathbf{Y}} \sum_{i,j} \|\mathbf{y}_i - \mathbf{y}_j\|^2 A_{ij}$$

Subject to constraints against trivial solutions. This definition ensures that two well-connected nodes yi and yj will remain close to one another. Laplacian Eigenmaps have inspired GNN pre-processing transformations and positional embeddings in graph transformers.

2.3.6 Role of GNNs

2.3.6.1 Spectral Methods

- Deal with global-graph structures.
- More fit for theoretical works with well-behaved graphs (citation networks, for example).

Use cases:

- Spectral clustering
- Graph signal filtering
- Early GCN-type methods

2.3.6.2 Spatial Methods

- Better for really large graphs (social, recommender systems, etc.).
- Inductive learning; in other words, it generalizes to unseen nodes/graphs.

Use cases:

- Social network graphs (Facebook)
- E-commerce graphs (Amazon, Pinterest)
- Dynamic and heterogeneous graphs

2.3.6.3 Spectral vs. Spatial Methods in GNNs

As depicted in Table 2.3, a comparison between spectral and spatial methods in GNNs is presented.

Table 2.3 Spectral vs. spatial methods in GNNs.

Aspect	**Spectral methods**	**Spatial methods**
Origin	Derived from spectral graph theory (eigenvalues/eigenvectors of Laplacians).	Inspired by neighborhood-based aggregation in traditional graph traversal.

(Continued)

Table 2.3 Spectral vs. spatial methods in GNNs. (*Continued*)

Aspect	**Spectral methods**	**Spatial methods**
Core Idea	Perform convolutions in the frequency domain using graph Laplacian Eigen basis.	Perform convolutions directly in the spatial domain by aggregating neighbors' features.
Graph Dependency	Graph-specific (e.g., trained filters tied to a graph's structure).	Graph-agnostic (generalizes well across graphs of different sizes or structures).
Mathematical Basis	Uses Graph Fourier Transform.	Aggregates feature vectors from neighboring nodes using functions like mean, sum, or max.
Computational Cost	High, due to Eigen decomposition (costly for large graphs).	Low, scalable *via* local message passing and sampling.
Localization	Initially non-local (global filters), but localization possible *via* approximation (e.g., Chebyshev polynomials).	Naturally local, working with k-hop neighborhoods.
Transferability	Poor: learned filters are not easily transferable across graphs.	Good: works across graphs with varying topology.
Examples	ChebNet (Chebyshev filters), GCN	GraphSAGE, GAT (Graph Attention Networks), GIN (Graph Isomorphism Network)

2.3.6.4 Applications in GNN Architectures

Spectral methods are not just theoretical; they also inspire actual GNN architectures. For example:

- GCNs use a low-order spectral filter with the renormalized adjacency matrix A~=A+I and obtaining the propagation rule

- Random diffusion process such as heat diffusion or random walk is described using powers of normalized Laplacian.
- Graph transformers utilize Laplacian eigenvectors for position encoding.

2.3.6.5 Limitations and Advances Beyond Spectral Methods

Despite that they work when used, spectral methods have some limitations.

- Computation of eigenvalue and eigenvector is expensive when working with large graphs.
- Transferability: Learned graph-specific spectral filters will never be transferable to graphs of varying structure.
- Localization: Spectral filters cannot be localized unless they are highly approximated with utmost care.

These are handled in spatial GNNs such as GraphSAGE [3], GAT [4], etc., which perform convolutions over the neighborhoods of nodes without using Laplacian spectra. But the spectral context remains beneficial in theoretical analysis and hybrid model design.

2.4 Neural Networks from a Graph Perspective

Neural networks, previously considered to be computational models of biological neurons, form the foundation of current artificial intelligence. Such systems have usually been understood as layer systems with more grid-like inputs like images or time-series. There exists, however, a complementary and now trendy manner of thinking in order to see neural networks as graphs. This viewpoint offers deep insights into the network flow of information, facilitates novel structural calculations, and paves the way for extending neural models to non-Euclidean spaces. Graph-theoretic understanding of neural networks not only enables structural visualization and optimization of models but also carries over naturally to a Graph Neural Network (GNN) construction approach. We can see here how feed-forward, convolutional, and recurrent types of neural networks are understood in graph terms and how graph theory can be employed in order to design, optimize, and analyze neural structures as portrayed in the Figure 2.13.

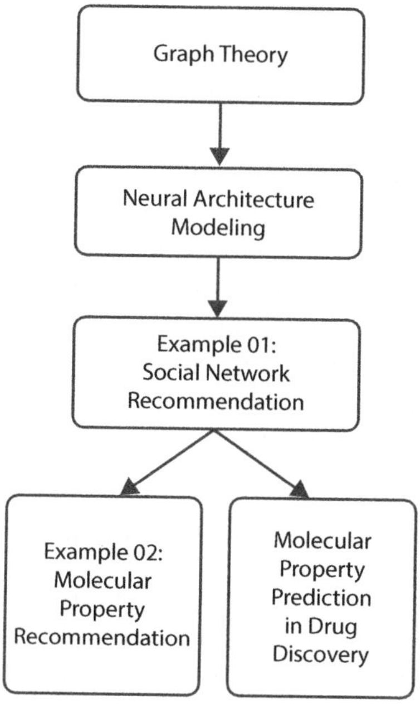

Figure 2.13 Neural networks from a graph perspective.

- **Example 01: Social Network Recommendation**
Consider a social networking website like Facebook or LinkedIn, with the users as nodes and their interactions—friendships, follows, or exchanges—as edges. Proposing a prospective new friend or professional acquaintance is essentially a link prediction task, which the fields of graph theory naturally set up. GNNs act on the graph topology by learning node representations from neighborhood aggregation. This enables the model to make new connections not only from explicit links, but from higher-order relational patterns—such as shared friends or group memberships—that are present in the graph structure. Conventional neural models would not be able to tap into these relational signals owing to their inflexible input formats, whereas GNNs are optimized for them.

- **Example 02: Molecular Property Prediction in Drug Discovery**
In molecular chemistry, a molecule may be represented as a graph with nodes for atoms and edges for chemical bonds. The effectiveness or safety of a drug typically depends on the type of atoms as well as how they are connected. By using graph-based neural models like Message Passing

Neural Networks (MPNNs), researchers are able to deduce molecular properties like toxicity, solubility, or protein binding affinity directly from the graph structure. Unlike conventional approaches that require hand-designed features, GNNs learn the important patterns from the topology and local substructures of the molecule, making predictions more accurate and generalizable in drug discovery research. So, graph theory does not simply contribute to neural network modeling—it revolutionizes it. It makes it possible for neural networks to be used in new fields of learning by embracing the native topological and relational structure provided by graphs. What begins as representational mathematical abstraction now turns, in this blending, into a highly useful tool for learning structured, dynamic, and interdependent data.

2.4.1 Graph-Based Abstraction of Neural Networks

Neural network can be naturally graphically represented as a directed graph G=(V,E) where vertices V are computation nodes (neurons) and edges E are one-way connections (synaptic weights) between neurons. Each edge e=(u,v)∈E is neuron u's output fed as input to neuron v. Neural layers here are a chain of connected subgraphs with one-way data flow from inputs to outputs. For the basic feedforward network layer, the formula can be described as:

$$y = \sigma(W\mathbf{x} + \mathbf{b})$$

Where x is input vector, W is weight matrix defining the inter-layer relation, b is bias vector, and σ is any non-linear activation function. The W matrix may be viewed as being responsible for adjacency structure of a bipartite graph between adjacent layers. This graph notation described above is not a metaphor but is actually implemented computationally by today's deep learning algorithms whose computational graphs control operations flow as well as gradients for the forward and the backward passes. Every operation of a network is under the computational graph, and automatically differentiability of it, deep learning's fundamental concept, [1] holds.

2.4.2 Connectivity and Topological Features

Connectivity between neurons constitutes the backbone of a neural network's power. Interconnections, as the edges of a graph, are the structure of the network. Depth (number of layers), width (size of layers), and connectivity (edge pattern) all determine expressiveness and learnability for the model. In graph-theory terms:

- Depth corresponds to the longest path in the network graph and dictates the number of sequential operations on data.
- Width is analogous to node degree and parallel paths, i.e., the capacity of the model to learn heterogeneous features in parallel.
- Sparsity of edges—especially of convolutional networks—reduces parameter counts and computation, analogously to sparse graphs.

These properties can also be quantitatively calculated in the form of measures such as graph diameter, centrality, and clustering coefficients. For instance, betweenness centrality can identify bottleneck neurons or layers through which information must flow and spectral properties will inform us regarding redundancy and robustness [2].

2.4.3 Feedforward Neural Networks as DAGs

Feedforward Neural Networks (FNNs) according to the Figure 2.14 can be usually described as Directed Acyclic Graphs (DAGs). In feedforward networks, data travels only in a single direction from the input layer to possibly a single or multiple hidden layers and finally to the output layer without any possibility to form cycles. Each hidden layer performs an affine transformation and then a non-linear activation. Layering allows us to have a hierarchical graph in the sense that nodes at each layer become connected only with the subsequent node.

The calculation of a multi-layer FNN may be given in terms of nested function composition:

$$f(\mathbf{x}) = f_L(f_{L-1}(\cdots f_1(\mathbf{x})))$$

Where for every f_i there exists the computation in layer i. Topological sort of the FNN's graph is topologically possible such that information is communicated without recursion and loops and thus gradient computation with backpropagation is simple.

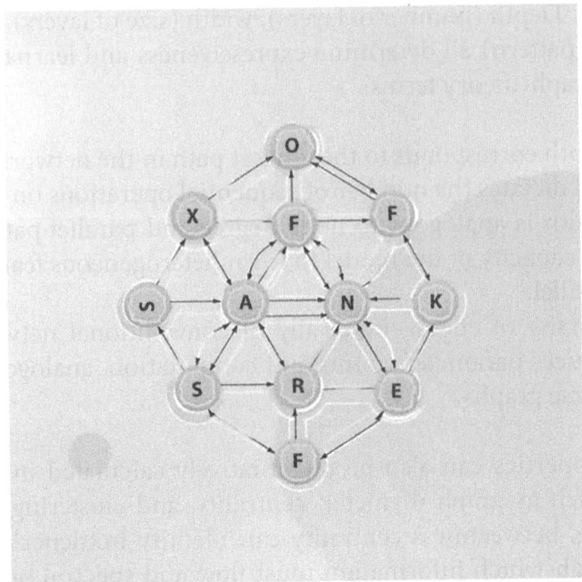

Figure 2.14 Feedforward neural networks as DAGs.

2.4.4 Convolutional Neural Networks as Grid Graphs

Convolutional Neural Networks (CNNs) are local data networks, i.e., images or video. Each convolutional layer applies a local filter (or kernel) to the local neighborhoods of the input with parameter sharing and translation invariance. In graph theory terms, CNNs operate on grid graphs, where pixels or patches are nodes with edges between them and their spatial neighbors [3]. The pattern of connectivity produced because of the convolution kernel creates a structured graph. For example, in a 2D image, it is possible to connect every node to the eight surrounding nodes, and that creates a lattice graph. Performing convolutional filtering repeatedly on the grid imitates weight-sharing on a structured graph. In addition, CNN networks are generally made up of pooling operations that down sample the graph through reducing the number of nodes (e.g., max pooling). They parallel graph coarsening operations of spectral graph theory and have impacted hierarchical pooling methods in GNNs.

2.4.5 Recurrent Neural Networks as Temporal Graphs

Recurrent Neural Networks (RNNs) achieve sequential representation of data by adding temporal recurrence to the network structure. RNNs differ

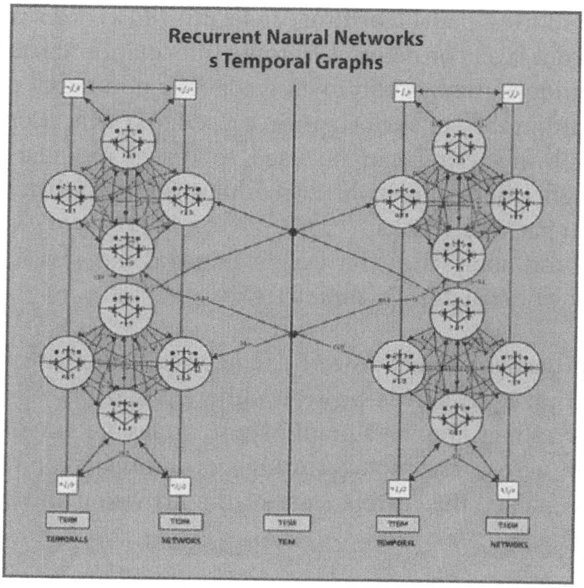

Figure 2.15 Recurrent neural networks as temporal graphs.

from FNNs in that they have loops that form cycles such that the output at neuron ttt is utilized as input for computation at time t+1. This results in a cycle graph structure as reflected in the Figure 2.15.

Mathematically, updating the hidden state in an RNN is:

$$\mathbf{h}_t = \sigma(\mathbf{W}_h \mathbf{h}_{t-1} + \mathbf{W}_x \mathbf{x}_t + \mathbf{b})$$

Propagation of this topology over time creates a linear chain DAG, node per step and data along edges. Unrolling can be very easy to use in employing backpropagation through time (BPTT) during training. LSTMs and GRU-like variants include gate mechanisms but otherwise the same identical general graph structure with extra internal edges [4].

2.4.6 Graph-Theoretic Analysis and Optimization of Neural Networks

With graphs to represent neural networks, there is encroachment by traditional graph algorithms to optimize the design, prune the networks, and explainability.

- The shortest path algorithm can be utilized to determine significant layers or neurons to route information through.
- Pruning a graph is eliminating edges or nodes that contribute minimally to performance, e.g., edge scarification.
- Graph clustering can be used to direct modular neural design, where subnetworks (modules) are specialized to perform particular tasks.
- Spectral analysis of the weight graph could reveal redundancy or structural symmetries.

These approaches are very powerful in neural architecture search (NAS), wherein optimization across possibly different settings of a network is conceived as a super graph—a graph whose nodes are graphs themselves. Optimization across such space through evolutionary or reinforcement learning has enabled the current state-of-the-art vision and language processing models [5].

2.4.7 Hybrid Architectures and Emerging Graph-Based Designs

The latest developments in deep learning are the integration of traditional neural networks with graph concepts. For instance, transformers can be considered computation over a completely connected graph where every token (node) is listening to all other tokens through attention edges. The attention mechanism itself imposes a learned, dynamic graph structure that changes with input context [6]. Similarly, Capsule Networks represent part-whole relationships in tree-structured graphs and Residual Networks (ResNets) in skip-connected graphs. Hybrid networks take advantage of graph theory expressiveness to impose more compostable and expressive graphs. In future applications such as multi-agent systems, robotics, and physics-informed learning, neural architectures increasingly compute on communication topologies that are graph-based where agents or entities exchange and receive messages with other agents or entities—a direct predecessor to GNNs.

2.5 Graph Neural Networks (GNNs)

Emerging needs to reason about complex, relational, and non-Euclidean data have motivated the introduction of Graph Neural Networks (GNNs) as a new class of machine learning models. Unlike the conventional neural

Table 2.4 Graph neural network model variants.

Model	Aggregation strategy	Handles edge features?	Attention mechanism	Example use cases
GCN	Mean (spectral-based)	No	No	Node classification, citation networks
GraphSAGE	Mean / LSTM / Pooling	Yes	No	Large-scale social graphs
GAT	Weighted sum (learned)	No	Yes	Heterogeneous graphs, NLP tasks
MPNN	Message + Update functions	Yes	Optional	Molecular property prediction
HAN	Meta-path + attention	Yes	Yes	Knowledge graphs, academic networks

networks—e.g., Convolutional Neural Networks (CNNs) and Recurrent Neural Networks (RNNs)—bound to fixed-size, Euclidean-shaped data (e.g., sequences, grids), GNNs are tailored to learn from graph-structured data [1]. As indicated in the Table 2.4.

Graph representation learning is a compact data structure for representing object interactions in domains like social networks, chemical molecules, transport graphs, and knowledge graphs. In the representation, entities and relations are represented by nodes and edges, respectively. GNNs offer representation learning on such graphs with the capability of learning the local neighborhood and global structure. The core idea relies on repeatedly optimizing node representations in terms of message passing and information aggregation from the neighbors, a method often referred to as message passing or neighborhood aggregation [2].

2.5.1 Core Principles of GNNs

The key notion of any GNN is that the state (or embedding) of a node can be manipulated by combining its current state with messages from its neighbors. This aggregation process is distributed across multiple layers in a way that all nodes have an opportunity to learn from nodes which are more remotely positioned relative to it within the graph. Node embedding preserve topological and attribute-level information such that they become appropriate for tasks like classification, regression, link prediction, and clustering [3]. The GNN model is generalized to many instances of relational learning by defining AGGREGATE and UPDATE functions, typically as neural modules. This makes GNNs end-to-end trainable and extremely flexible across domains.

2.5.2 Graph Convolutional Networks (GCNs)

Graph Convolutional Networks are probably the most prevalent type of GNNs as illustrated in the Figure 2.16. They extend the concept of convolution from Euclidean space to graphs using spectral theory. The Kipf and Welling's GCN model approximates this concept with a first-order approximation to spectral graph convolutions without the costly computation of Laplacian eigenvectors [4].

The model employs a normalization trick wherein it introduces self-loops and symmetrically normalizes the adjacency matrix, hence maintaining it stable but preserving the node's own features. The operation smooths local neighborhood feature representations, and the model thus becomes stable and efficient with sparse labels. GCNs have been shown to be extremely helpful for semi-supervised classification tasks such as citation networks [5].

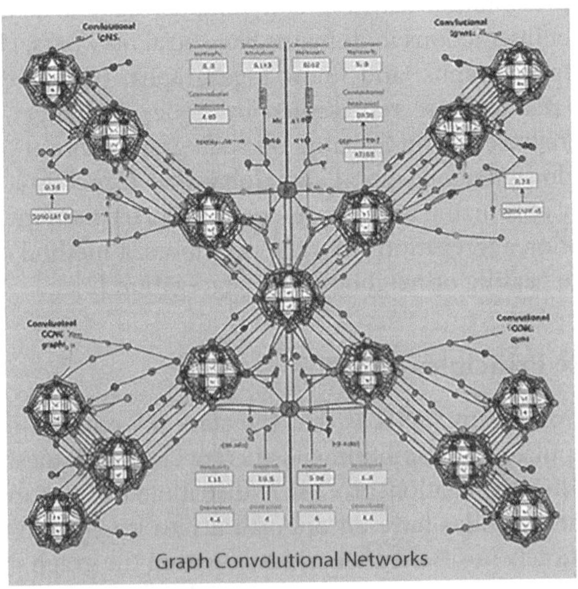

Figure 2.16 Graph convolutional networks (GCNs).

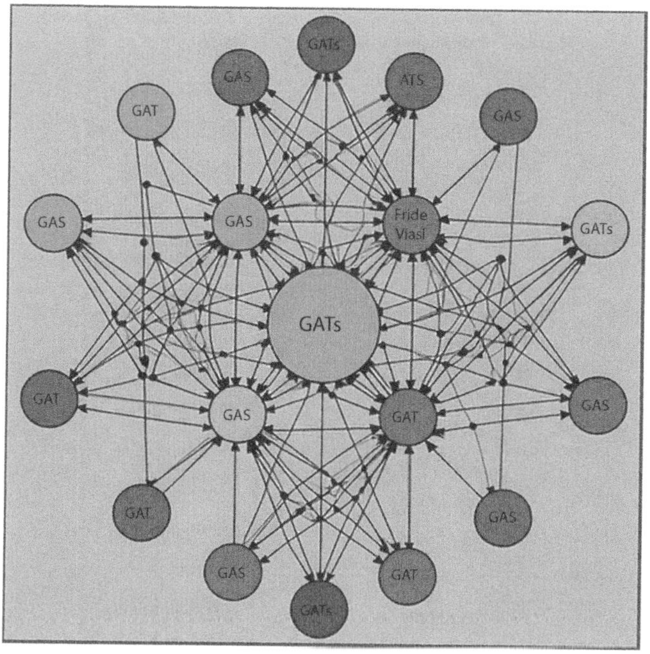

Figure 2.17 Graph attention networks (GATs).

2.5.3 Graph Attention Networks (GATs)

Whereas GCNs aggregate neighbor features equally or use node degrees, Graph Attention Networks (GATs) use attention to attach various weights to each neighbor. It is one method suggested by Veli Ković *et al.* [6], which uses self-attention to graph nodes in an attempt to diffuse features more complexly as depicted in the Figure 2.17.

GATs are particularly effective for noisy or heterogeneous graphs where nodes within the local neighborhood may not be all helpful. GATs also enhance expressiveness and stability of the model through multi-head attention. GATs learned attention weights evolve during training in modeling complex relationships.

2.5.4 Message Passing Neural Networks (MPNNs)

The Message Passing Neural Network (MPNN) paradigm [7] offers a generic abstraction for most GNN variants. Learning is split into two phases: message aggregation and state update. During the first phase, a node collects information (messages) from neighbors, possibly a function

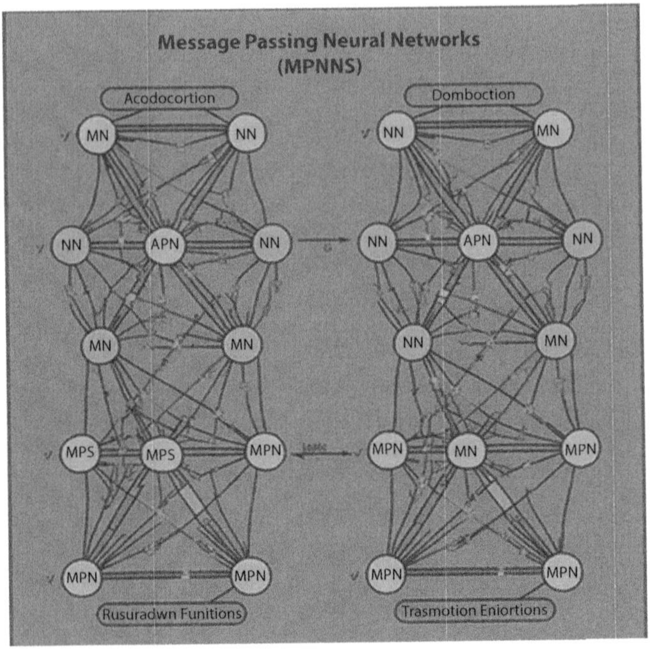

Figure 2.18 Message passing neural networks.

of edge features and feature states of both nodes. During the second phase, the node updates its embedding given the messages.

This model performs well in the regions where edge data is crucial, such as quantum chemistry as demonstrated in the Figure 2.18, where bonds have varying properties. MPNNs are highly expressive and have outperformed traditional machine learning models in molecular property types of tasks, such as QM9 [8].

2.5.5 Pooling and Readout Operations

For graph-level task like graph classification or property prediction, it is required to convert node-level embedding into a single vector that captures the whole graph. It is achieved by using readout functions such as max pooling, mean pooling, or more complex hierarchical pooling methods such as DiffPool [9]. Hierarchical pooling mimics the spatial down sampling operation in CNNs to allow the model to learn coarse-grained graph structures with global context. These approaches enhance the generalization capability of GNNs and render scaling GNNs to large graph datasets feasible.

2.5.6 Training GNNs and Loss Functions

GNNs are typically trained on standard gradient descent algorithms and loss functions for the task. Node classification is typically performed using cross-entropy loss. Contrastive learning models such as DGI (Deep Graph Infomax) maximize mutual information between global graph summaries and node representations in unsupervised settings [10]. In semi-supervised environments, structure in graphs enables propagation of labels from small labeled sets to the entire graph. GATs and GCNs have excelled in these environments, especially when they are regularized with mechanisms that preserve smoothness as well as local consistency.

2.5.7 Applications of GNNs

Applicability space for GNNs is truly immense and continues to be as indicated in the Figure 2.19.

- Social Networks: GNNs can identify communities, estimate influence, and identify opinion leaders from learning on interaction graphs [11].

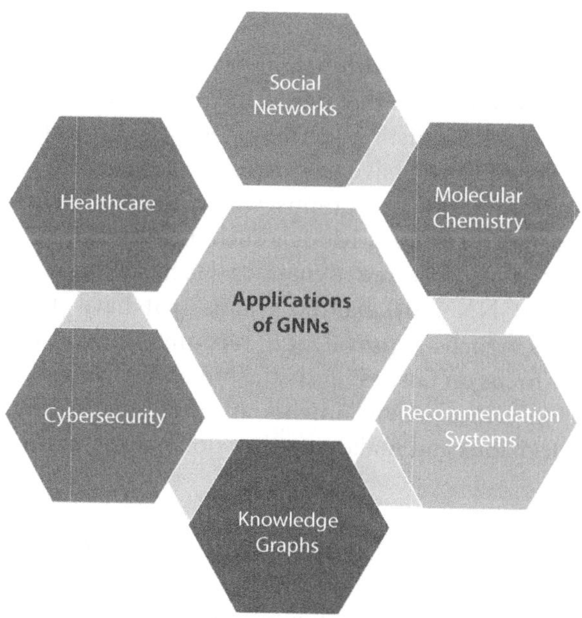

Figure 2.19 Applications of GNNs.

- Molecular Chemistry: Atom and bond representation of nodes as atoms and edges as chemical bonds enable prediction of properties, virtual screening, and molecule generation [8, 12].
- Recommendation Systems: Bipartite graph modeling of user-item interactions enables GNNs to learn latent preferences and make personalized recommendations [13].
- Knowledge Graphs: GNNs enable link prediction, entity classification, and reasoning by processing graph-embedded knowledge bases [14].
- Cybersecurity: GNNs model network traffic and user behavior to identify anomalies, intrusions, and coordinated attacks [15].
- Healthcare: Patient history, symptoms, and genomic information can be modeled as heterogeneous graphs to support personalized medicine and disease prediction [16].

2.5.8 Challenges and Future Directions

Even though extremely powerful, GNNs are plagued with a number of problems. Quite likely the most important one is scalability. It is not possible to scale the full-batch version of GNNs to large graphs like the whole Facebook social graph or an international transport network. In order to get around this bottleneck, sampling strategies like GraphSAGE [17] or mini-batch training with neighbor sampling are employed. Oversmoothing is also an issue in that deeper GNN layers draw node representations in and become indistinguishable. Residual connections, skip links, and attention-based gating have been proposed to prevent this effect [18].

The growing use of dynamic graphs, where nodes and edges evolve with time, presents new directions. Models such as EvolveGCN and Temporal GNNs attempt to understand dynamics [19]. Likewise, heterogeneous GNNs (e.g., HetGNN [20]) handle graphs that have different types of nodes and edges, which are prevalent in recommendation systems. Finally, there is greater focus on fairness and interpretability. Since GNNs are being used in mission-critical environments, it is essential to understand how GNNs make decisions and make sure they don't perpetuate structural bias.

2.6 Applications of Graph-Theoretic Neural Models

The beauty of graph-theoretic neural models, such as Graph Neural Networks (GNNs), has unlocked the area of machine learning applications

Table 2.5 Applications of graph-theoretic neural models.

Domain	Graph representation	GNN role	Representative models
Social Networks	Users and connections	Community detection, link prediction	GCN, GraphSAGE
Drug Discovery	Atoms and bonds	Molecular property prediction	MPNN, GAT
Recommender Systems	Users and items	Collaborative filtering	PinSAGE, NGCF
Knowledge Graphs	Entities and relations	Triplet reasoning, completion	R-GCN, CompGCN
Cybersecurity	Systems and access logs	Anomaly detection	Dynamic GNNs, TGNs
NLP	Words, phrases, documents	Semantic parsing, summarization	TextGCN, HAN

to a great extent. As demonstrated in the Table 2.5. GNNs offer a general and expressive framework on which to represent and reason about data whose relational structure between entities is paramount. Unlike the Euclidean domains (grids or sequences) on which classical neural networks are limited, GNNs can represent the structure and dynamics of non-Euclidean data and therefore are naturally suited to hard systems whose data is relational.

This chapter covers the extensive and rich real-world applications of graph neural models and illustrates how they facilitate breakthroughs in a wide range of fields ranging from social network analysis, bioinformatics, chemistry, computer vision, natural language processing, recommender systems, financial systems, to cyber-physical systems.

2.6.1 Social Network Analysis and Community Detection

Social networks—web-based sites such as Facebook and Twitter or non-web relational networks such as co-authorship in academia—are naturally represented as graphs. In social networks, the entities are defined by nodes and relationships such as friendship, retweeting, or co-authorship are defined by edges as presented in the Figure 2.20. Graph-theoretic neural models excel in this class by enabling operations such as:

- Node classification: Prediction of user properties (e.g., political affiliation, age) from the activity and connectivity of the user [1].
- Link prediction: Forecasting of likely links among very dissimilar users on which friend recommendation systems rely.
- Community detection: Identification of densely connected groups of users exhibiting strong similarity in tastes or behavior [2].

Inductive graph-neural GNN model, GraphSAGE [3], etc., has, for example, been employed at scale in social networks to enhance content

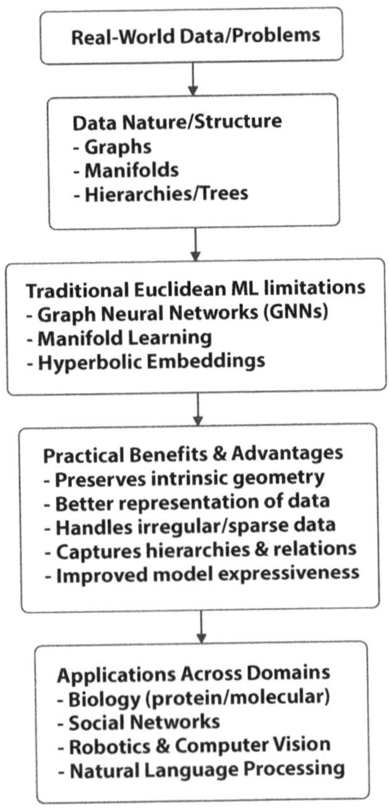

Figure 2.20 Social network analysis and community detection.

suggestion and network expansion algorithms with node attribute- and structure-induced learning of feature representations.

2.6.2 Molecular Property Prediction and Drug Discovery

Drug discovery and computational chemistry is one of the most powerful uses of GNNs. Molecules are naturally graphs, with chemical bonds as edges and atoms as nodes. Graph neural models are trained to predict molecular properties by observing both atomic data and the topological shape of the compound. Tasks are:

- Toxicity prediction: Whether a compound is safe for biological systems.
- Binding affinity and solubility: Prediction of the molecular behavior in different solvents or protein target binding [4].
- Molecule generation and optimization: Generation of new molecules with certain properties according to generative GNN models.

Message Passing Neural Network (MPNN) paradigm [5] has been found to be of state-of-the-art performance for benchmark datasets such as QM9 on which it is trained to precisely predict quantum properties. Graph models are also being combined with reinforcement learning in order to automate the design of molecules for target uses, e.g., for inhibiting virus proteins or side effect avoidance.

2.6.3 Biomedical and Healthcare Systems

Graph-theoretic neural models are also being used in biomedical informatics since proteins, genes, and diseases do not interact singly but as a network of networks. Some of them include:

- Protein-protein interaction (PPI) networks: GNNs allow the discovery of functional protein complexes by learning embeddings to represent biological plausibility of interaction [6].
- Gene-disease association: Gene, symptom, and disease interaction graphs allow models to predict therapeutics or diagnostics.

- Electronic Health Records (EHR): Patient data, symptoms, and treatment can be modeled as heterogeneous graphs to enable personalized care recommendations [7].

Graph-based models augment clinical knowledge graphs with patient-specific data to enable decision-making for treatment planning, diagnosis, and disease modeling.

2.6.4 Knowledge Graph Completion and Reasoning

Knowledge graphs (KGs) can be modeled as structured data in the format of triplets (head entity, relation, tail entity). Graph neural models operate on such knowledge graphs to enable:

- Entity classification and resolution: Prediction of missing entity type or duplicate entities resolution.
- Link prediction: Prediction of missing fact in a knowledge base.
- Multi-hop reasoning: Computation of indirect relations tracing the graph.

These models like Relational Graph Convolutional Networks (R-GCNs) [8] and CompGCN extend simple GNNs to learn edge types and directions, and can be used on reasoning tasks for KGs like Google Knowledge Graph or Wiki data. These models extensively get utilized in question answering, semantic search, and ontology alignment tasks.

2.6.5 Recommendation Systems

Recommender algorithms have evolved from collaborative filtering and matrix factorization to graph-based approaches, where users and items are bipartite graph nodes and interaction (e.g., clicks, ratings, purchases) are edges.

GNNs improve recommendations by:

- Modeling multi-hop user-item interaction.
- Modeling temporal dynamics and context.
- Learning generalizable user and item embeddings over sparse data [9].

For example, PinSage, constructed by Pinterest, combined GNNs and random walks to learn embeddings of hundreds of millions of users and pins and achieved state-of-the-art performance on content recommendation [10].

2.6.6 Financial Fraud Detection and Risk Analysis

GNNs are being used more and more by banks to detect suspicious transactions and quantify systemic risk. In transaction graphs, entities and accounts are nodes, and transactions are links. Anomaly detection in GNNs indicates well-performing patterns of fraud such as:

- Money laundering *via* chain-like patterns.
- Synthetic identity fraud *via* densely connected but deviating subgraphs.
- Credit default risk *via* analysis of borrower–lender relationships.

Because GNNs can reason about node features and relational graphs, they outperform traditional machine learning models in terms of recall and accuracy for transaction graph-based fraud detection [11].

2.6.7 Cybersecurity and Intrusion Detection

Cyber-physical systems such as enterprise networks, Internet of Things (IoT), and cloud computing are graph-structured by nature. In such systems:

- Devices and users are nodes.
- Communication links or authentication records are edges.

GNNs enable the identification of anomaly, intrusion detection, and vulnerability analysis by way of detecting unusual interaction patterns. Dynamic graph neural networks have also been employed in next-generation attack graphs to predict threat propagation [12].

2.6.8 Traffic Prediction and Smart Cities

In ITS, road intersections or vehicles are nodes and connectivity between them (roads or traffic) as edges. Graph neural models can capture both spatial relationships and temporal patterns.

Applications are:

- Route planning and optimization with predicted flows.
- Spatiotemporal GNN-based traffic congestion prediction.
- Accident detection and avoidance by edge anomaly detection [13].

These models are increasingly being integrated into smart city infrastructure to enable real-time adaptive traffic light and public transportation management.

2.6.9 Computer Vision and Scene Understanding

Besides grid-structured convolution, graphs also offer a different view in scene understanding, object interaction, and image-based reasoning. Some of the interesting applications are:

- Scene graphs, where image objects are nodes and spatial or semantic relations (e.g., "man sitting on bench") are edges.
- Human pose estimation, where body joints are a graph.
- 3D mesh analysis, where GNNs analyses surfaces for shape classification or reconstruction tasks.

Hybrid models, with the interaction of GNNs and CNNs or transformers, show improved performance in visual question answering, image captioning, and robotics perception [14].

2.6.10 Natural Language Processing and Document Modeling

The language data can be represented as words, sentences, or entities as nodes and syntactic/semantic relations as edges. GNNs are applied to Natural Language Processing (NLP) tasks like:

- Dependency parsing and semantic role labeling.
- Document-level sentiment analysis and co-reference resolution.
- Text classification through constructing keyword and phrase graphs.

For example, GNNs can be used for entity graphs or word co-occurrence graphs to have a sense of more context than sequence models. Transformers enable GNNs to handle multi-hop reasoning and robust discourse structure representation [15].

2.7 Practical Advantages of Non-Euclidean Space Handling in Learning Models

Traditional deep learning and machine learning algorithms primarily operate in Euclidean space, where data resides in flat, continuous vector spaces with typical geometric attributes (e.g., points, distances, angles). However, many real-world data and relationships invariably reside in non-Euclidean spaces such as graphs, manifolds, or hyperbolic spaces. Effective handling of these non-Euclidean geometries can yield significant practical advantages in learning models as seen in the Figure 2.21:

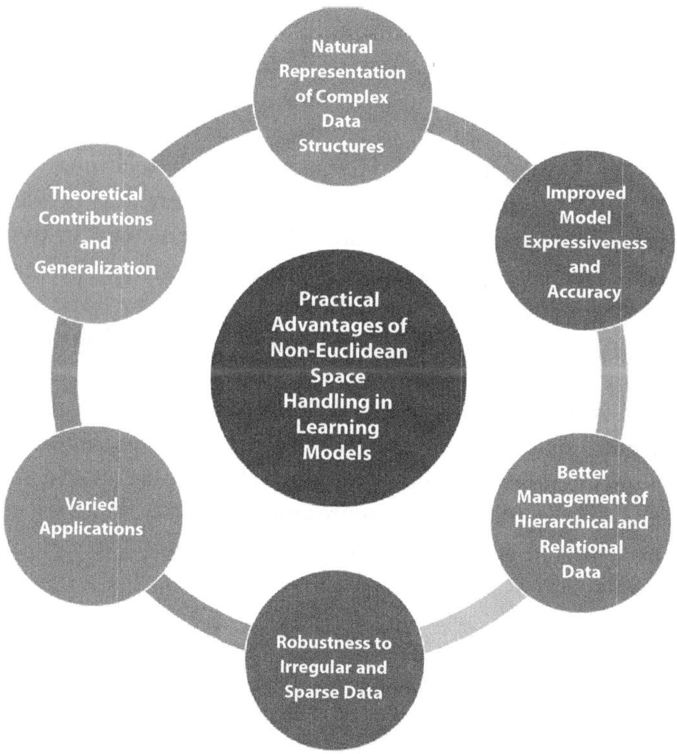

Figure 2.21 Practical advantages of non-Euclidean space.

a) Natural Representation of Complex Data Structures: The majority of data types are not Euclidean in nature:

- Graphs: Social networks, molecules, road networks, knowledge graphs.
- Manifolds: 3D geometries, human motion patterns, sensor readings on curved surfaces.
- Hyperbolic spaces: Hierarchical data, taxonomies, language structures.

Modeling directly on these structures preserves their natural geometry, enabling models to better capture relationships, connectivity, and topology without unnatural flattening or distortion.

b) Improved Model Expressiveness and Accuracy: Non-Euclidean techniques such as Graph Neural Networks (GNNs) or manifold learning can tap into relational and geometric data information. This translates to better performance on node classification, link prediction, 3D shape analysis, and natural language understanding since the models learn from the true geometry of the data.
c) Better Management of Hierarchical and Relational Data: Many systems have hierarchical structures (knowledge bases, org charts), which are inherently encoded in hyperbolic or tree geometries. Hierarchies can be preserved by non-Euclidean embeddings at low distortion, which improves recommendation, clustering, and information retrieval tasks.
d) Robustness to Irregular and Sparse Data: Euclidean representations prefer regular grid-like data (images, sequences). Non-Euclidean methods can natively support sparse, missing, or irregular data by representing entities and their relationships in flexible ways in graph or manifold spaces.
e) Varied Applications

- Biology: Protein structures and molecular interactions as graphs enable better drug discovery.
- Social Networks: Robust capture of community structure and influence diffusion.

- Robotics and Computer Vision: 3D environment navigation and interpretation and sensor inputs on curved surfaces.
- Natural Language Processing: Semantic hierarchies and relation modeling.

f) Theoretical Contributions and Generalization: Non-Euclidean geometry provides an even more mathematically expressive way to handle complex data distributions and invariances. This can lead to more generalizable models that are able to transfer knowledge more effectively between domains and tasks.

In essence, working with non-Euclidean space in learning paradigms allows the algorithms to handle hierarchical, relational, and complex data naturally and more effectively. This leads to increased precision, improved understanding, and applicability in real-world problems where the simplistic assumptions of Euclidean space do not apply.

2.8 Research Challenges and Future Directions

As shown in Table 2.6, an overview of key challenges and future research directions is provided. While graph-theoretic neural models, especially

Table 2.6 Key challenges and future directions in GNN research.

Challenge	Description	Ongoing solutions
Scalability	GNNs struggle with very large graphs	Sampling (FastGCN), Cluster-GCN, GraphSAINT
Over-smoothing	Deeper networks make node embeddings indistinguishable	Residuals, skip connections, rewiring
Dynamic Graphs	Time-evolving topologies are hard to model	EvolveGCN, TGN, temporal attention
Interpretability	Hard to understand GNN decisions	GNNExplainer, counterfactual reasoning
Heterogeneous Graphs	Nodes and edges of multiple types	HAN, HetGNN, type-aware attention mechanisms

Graph Neural Networks (GNNs), are suitable for an enormously vast range of applications, they also have to bear enormous hurdles that make it difficult for them to scale, generalize, and interpret. While there has been immense development in the building of expressive, low-complexity, and application-specific graph models, there indeed are a few research open questions. For research and for safe, responsible, and efficient deployment of GNNs in real-world systems, it is essential that these constraints are appreciated. We here present an overview of the most critical technical and conceptual issues currently being tackled by researchers and outline promising directions for future graph-based deep learning research as represented in the Figure 2.22.

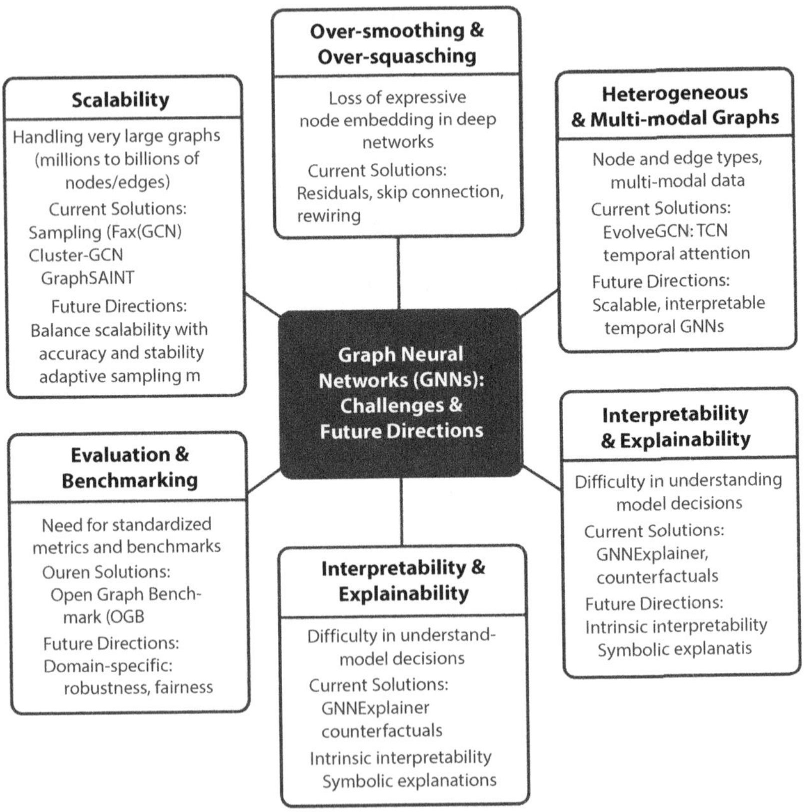

Figure 2.22 Key challenges and future directions in GNN research.

2.8.1 Scalability to Large Graphs

The majority of real graphs, e.g., web graphs, social networks, and biological interaction graphs, have millions or even billions of nodes and edges. Classical GNNs, particularly full-batch training-based or spectral methods-based GNNs, are weakly scalable because of:

- High memory demand: Storing and updating all the nodes' hidden states.
- Neighborhood explosion: Neighborhood sampling in the message passing results in computation exploding exponentially with network depth.
- Redundant computation: Redundantly computing the same messages between overlapping neighborhoods in mini-batch scenarios.

Cluster-based approaches (e.g., Cluster-GCN [3]) and sampling-based approaches (e.g., GraphSAGE [1], FastGCN [2]) have solved such problems to some extent by sampling informative regions of the graph at training time. Scalability vs. accuracy and stability is still an open research problem.

2.8.2 Over-Smoothing and Over-Squashing

The central issue of deep GNNs is over-smoothing, wherein node representations become indistinguishable following a couple of message-passing layers. The phenomenon was first empirically realized by Li *et al.* [4] through formal investigation. Over-smoothing occurs as a result of consecutive aggregation, whereby node embeddings become homogenized and lose their ability to distinguish semantic variations. Another issue is over-squashing, where more than the node representation of fixed size is crammed into it, particularly where long-range dependencies must be encoded but there are just local aggregations [5].

Mitigation methods are:

- Residual connections and skip layers to maintain node identity.
- Attention mechanisms to learn selectively weighting relevant messages.
- Jumping knowledge networks to facilitate layer-wise combination of features [6].

- Graph rewiring methods to topology optimize the learning process.

Adaptive depth mechanisms, multi-scale architectures, and gradient-preserving propagation rules can be explored in the future to offer expressivity without convergence to trivial embeddings.

2.8.3 Dynamic and Temporal Graphs

Most GNNs are based on the static graph assumption, but most real-world applications are dynamic graphs where the nodes and edges change over time—user behavior in recommendation systems, dynamic communication networks, and evolving financial relations.

Representing such graphs introduces a range of challenges:

- Temporal dependencies have to be represented between snapshots or in streaming time.
- Scalability is also limited by the necessity of processing streaming updates.
- Catastrophic forgetting is possible if earlier states of the graph are not stored properly.

Methods such as EvolveGCN [7], Temporal Graph Networks (TGNs) [8], and GCRNs try to mitigate such challenges by blurring the boundaries of GNNs with recurrent or attention-based temporal representation. General-purpose, scalable, and interpretable temporal GNNs yet need to be developed.

One of the most important directions of research is to integrate event-based modeling, time-series prediction, and graph learning into one framework, potentially by graph-based reinforcement learning or neurosymbolic systems.

2.8.4 Heterogeneous and Multi-Modal Graphs

In real graphs, nodes and edges are typically heterogeneous. For example, in a bibliographic graph:

- Nodes may be authors, papers, venues.
- Edges may be citations, authorship, or co-authorship.
- Modalities are text, graphs, and numerical attributes.

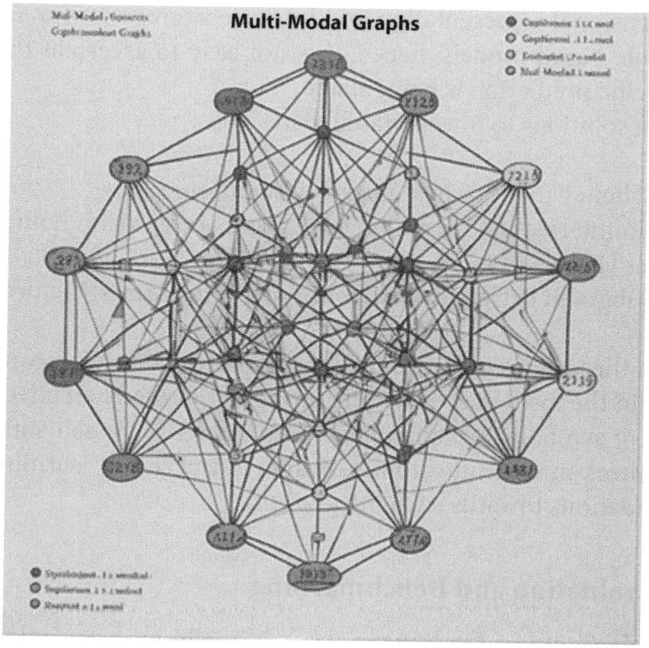

Figure 2.23 Multi-modal graphs.

- Learning from such heterogeneous graphs needs models which should be capable of Preserving type-specific meanings, Sensing cross-type interactions, Integrating multimodal data (e.g., text + structure + image).

Models such as Heterogeneous GNNs (HetGNN) [9] and HAN (Heterogeneous Attention Networks) try to do this, but usually at the cost of generality, training performance, or interpretability as evidenced by the Figure 2.23.

Future models might be aided by:

- Meta-learning over node types.
- Edge-type flexible message schemes.
- Cross-modal fusion methods, perhaps combined with transformers and attention-based architectures.

2.8.5 Interpretability and Explainability

Lack of interpretability is the greatest challenge of applying GNNs to risk-sensitive applications like healthcare, finance, or law. GNNs use high-dimensional representation and intricate aggregations over decision trees or rule-based models; hence, it is not easy to ascertain the reasons why a specific prediction is being made.

Current solutions to interpretability are:

- Saliency maps or heat maps for attention.
- Counterfactual reasoning to reason over decision boundaries [10].
- Subgraph extraction for identifying influential structures

Most of these, however, are *post-hoc* and presumably not rigorous. More is needed in the form of intrinsically interpretable GNN architectures, i.e., provision of symbolic explanations, causal inference, or semantic anchoring. Advances in neuromyotonic AI and causal graph learning provide solid foundations towards such integration.

2.8.6 Evaluation and Benchmarking

In the background of the large number of GNN models deployed, standardization is a new phenomenon in:

- Evaluation metrics.
- Dataset quality.
- Task definitions.

Some of the popular benchmarks (e.g., Citeseer, Cora) are small, homogeneous, and do not capture real-world hardness. Recent work such as Open Graph Benchmark (OGB) [11] offers large-scale, heterogeneous settings and standardized test pipelines.

Future work needs to tackle:

- Domain-specific benchmarks under real-world constraints.
- Demographic and graph type-biased fairness and bias testing.
- Adversarial attack or noisy graph robustness testing.

2.8.7 Towards Universal Graph Learning Models

The majority of existing GNNs are task- or domain-specific. Long-term ultimate goals are to develop universal graph learning systems that:

- Generalize across domains (social science, NLP, biology).
- Flexible process different types of graphs (temporal, static, heterogeneous).
- Are minimally supervised and learnable.

This may involve borrowing ideas from:

- Graph signal processing.
- Transformer architectures (e.g., Graphormer [12]).
- Graph-based foundation models were trained on gigantic corpora of structured information.

Recent advancements in pre-training graphs, contrastive learning, and self-supervised tasks can be leveraged as the building blocks for these models. Graph Neural Networks (GNNs) have been heralded as extremely powerful tools for graph learning and representing complex relational structures present in many real-world settings such as social networks, biological organisms, and recommendation systems. Despite all the potential they offer, GNNs are extremely difficult to address in a manner that will render them scalable, resilient, and deployable at scale. Scalability is one of the biggest challenges: graphs in the real world can have millions or even billions of nodes and edges, and typical GNN architectures are not able to manage the attendant memory and computational overheads that arise from neighborhood explosion and redundant message passing. While sampling and clustering-based approaches have assisted in improving scalability to some extent, finding the right trade-off among efficiency, accuracy, and stability remains an open research problem. Deep GNNs also suffer from over-smoothing and over-squashing, which weaken representation expressiveness of node embeddings by making representations close to each other or by forcing long-distance dependencies into sparse representations. Dynamic and temporal graphs introduce another level of complexity, where models must actually represent temporal dependencies and changing relationships and deal with streaming updates and avoid catastrophic forgetting. Furthermore, heterogeneous and multi-modal graphs present unique challenges resulting from node and edge type heterogeneity and the integration of two or more data modalities and need more sophisticated and type-aware architectures to preserve semantic information and capture cross-type interactions effectively.

Beyond scalability and expressiveness, interpretability of GNNs is the critical barrier to their applications in high-risk domains such as medicine, finance, and law, where predictions must be explained. Current interpretability methods are weakly rigorous and *post-hoc*, so intrinsically interpretable GNN architectures that can produce symbolic, causal, or semantically rich explanations are highly sought after. In addition, the area suffers from the absence of standardized benchmarks and testing procedures that capture the heterogeneity and richness of actual graphs, preventing equitable comparison and thorough testing of model skills, such as resistance to adversarial attacks and bias. The future direction of GNN research is to develop universal graph learning architecture capable of generalizing to different domains, supporting different types of graphs (static, temporal, heterogeneous), and operating with limited supervision. In this direction, breakthrough would be possible by building on advances in graph signal processing, transformer designs like Graphormer, and foundation models learned from large structured datasets, as well as advances in self-supervised and contrastive learning. Mitigating these interconnected challenges will be essential to achieving the ultimate potential of GNNs and implementing safe, scalable, interpretable, and adaptive graph-based AI systems capable of mitigating the richness of real-world data.

2.9 Conclusion and Summary

The graph-theoretic framework of neural study is among the most trending and impactful new frontiers of present-day machine learning. By tapping into synergy between concepts borrowed from traditional graph theory and representational power of deep neural networks, Graph Neural Networks (GNNs) and related models have offered novel avenues to learn relational, temporal, and high-dimensional data. The chapter "Graph Theory Foundations for Neural Network Models" is a significant theoretical contribution to the intersection of graph theory and deep learning, especially within GNN formulation and analysis. It presents a formal, systematic, and coherent theoretical framework that re-expresses neural networks in graph structures and pushes the boundaries of AI model representation, design, and learning.

a) Foundation Graph Theory Integration with Neural Networks: Offers a formal mathematical framework to describe neural networks as graph-structured representations—neurons as nodes, edges as edges. Reframes classical neural architectures (feedforward, convolutional, and recurrent) in terms of directed graphs or grid graphs to enable graph-theoretic optimization, visualization, and abstraction. Offers the basis for building non-Euclidean learning systems that inherently model relational, sparse, and irregular data.

b) Spectral Graph Theory as a Computational Backbone: Introduces Graph Laplacian, eigenvalues, eigenvectors, and the graph Fourier transform as computational primitives to describe computation over graphs. These spectral primitives allow for convolutional operations over graphs and yield spectral-based GNN models such as GCNs and ChebNets. Theoretical treatments illustrate how signal filtering, clustering, and embedding can be done on graphs with strong linear algebraic underpinnings.
c) Message Passing Paradigm and Representation Learning: Makes the message-passing paradigm the core computational model of GNNs. Mathematically formulates aggregation and update functions, demonstrating how node embeddings are learned iteratively from local neighborhoods. Demonstrates how the mechanism generalized classic learning to graph-structured data with preservation of topological and semantic relationships.
d) Architectural Generalization and Graph Perspective: The chapter sets up neural models as computation graphs, introducing DAG structures for feedforward networks, lattice graphs for CNNs, and cyclic graphs for RNNs. Offers quantitative features for graph-theoretic analysis of neural networks such as connectivity, centrality, diameter, and sparsity. Enables hybrid graph architecture, i.e., transformers, attention networks, and capsule networks, as graph structures.
e) Graph Representations and Data Structures: Offers comparative methodology for different graph representations: adjacency matrices, incidence matrices, edge lists, and adjacency lists. Illustrates the impact of each representation on computational efficiency, interpretability of the model, and scalability of algorithms. Emphasizes the importance of representational selection both in theory building and operational GNN deployment.
f) Graph-Based Generalization of Deep Learning: Demonstrates that traditional deep learning methods assume i.i.d. Euclidean input structures, which are incapable

of modeling interdependencies. Graph-based models generalize learning to structured, interdependent data, capturing non-local correlations, hierarchical dependencies, and multiscale interactions. Demonstrates how GNNs preserve locality and structure, enabling models to learn global graph topology and node-level fine-grained features jointly.

g) Addressing Real-World Complicated Graphs: Explains theoretical solutions to managing heterogeneous, dynamic, and temporal graphs based on hybrid models. Presents models like HetGNNs, TGNs, and EvolveGCNs and explains their theoretical foundations along with generalization to evolving topologies in time. Refers to the future in which universal GNNs can manage all modalities, time scales, and semantic layers.

h) Scalability and Optimization Challenges: Recognizes the following critical theoretical bottlenecks of over-smoothing, neighborhood explosion, and computational complexity in big graphs. Suggests residual connections, graph sampling, subgraph batching, rewiring, and message-passing reformulations as mathematically grounded solutions. Accounts for the trade-off between local aggregation and global expressiveness of deep GNNs.

i) Explainability, Interpretability, and Ethical AI: Stresses the need for intrinsically interpretable graph models, especially for sensitive domains like healthcare and finance. Examines saliency maps, subgraph extraction, and counterfactual reasoning as explanatory tools for explainable graph learning. Sees a vision of transparent AI, with graph-theoretic intuition delivering both functional performance and human-meaningful explanation.

j) Future Vision for Graph-Based AI: Foresees a trajectory to foundation models for graphs by pretraining, contrastive learning, and universal embeddings. Demands neuromyotonic integration—fusing symbolic reasoning with GNNs to build robust, generalizable, and logic-savvy learning systems. Seeks research in causal inference, fairness, privacy, and benchmarking in graph-based models. Key Takeaways are as follows:

- Graph theory is not just a structural metaphor—it's a substrate for learning and computation.
- Algebraic and frequency-based view in graph signal processing, convolution, and dimensionality reduction is offered by spectral methods.
- GNNs represent a generalization of deep learning to the structured, interdependent world based on a well-established theoretical toolkit.
- The article presents a complete mathematical framework for exploring neural architectures from a graph-based viewpoint—spanning algebra, topology, optimization, and logic.
- Future research will require holistic graph models that are explainable, scalable, and multimodal—capable of learning, reasoning, and adapting across domains in the real world.

This chapter thus gives a multi-dimensional theoretical basis to neural computation in graph spaces, with the community not only being given a methods and definitions toolkit but also being offered an ambitious research agenda founded on mathematical depth and worldly relevance. Essentially, this chapter provides a unifying theoretical framework that brings together the combinatorially elegant world of graph theory with the functionally rich world of neural computation. In doing so, it supports a reimagination of how machines can learn from structured, irregular, and dynamic information. The chapter doesn't just define current models—it proposes a conceptual vision for the future of AI: one of reasoning, structure, and learning harmoniously co-existing in graph-based neural systems. This contribution is not only opportune but foundational, planting the intellectual seeds for the construction of smart, explainable, and universal learning architectures in the era of data complexity and interconnectedness.

AI disclosure statement: The author acknowledges the use of AI-based tools for minor language editing or grammatical correction. However, all intellectual contributions, analyses, and conclusions presented in this chapter are entirely the author's own.

Bibliography

1. Alon, U. and Yahav, E., On the Bottleneck of Graph Neural Networks and its Practical Implications, in: *Proc. Int. Conf. Learn. Representations (ICLR)*, 2021.
2. Bronstein, M., Bruna, J., LeCun, Y., Szlam, A., Vandergheynst, P., Geometric Deep Learning: Going Beyond Euclidean Data. *IEEE Signal Process Mag.*, 34, 4, 18–42, 2017.
3. Chen, J., Ma, T., Xiao, C., FastGCN: Fast Learning with Graph Convolutional Networks via Importance Sampling, in: *Proc. Int. Conf. Learn. Representations (ICLR)*, 2018.
4. Chiang, J., *et al.*, Cluster-GCN: An Efficient Algorithm for Training Deep and Large Graph Convolutional Networks, in: *Proc. ACM SIGKDD Int. Conf. Knowl. Discov. Data Min. (KDD)*, 2019.
5. Chung, F., *Spectral Graph Theory*, AMS, Providence, RI, USA, 1997.
6. Connolly, A.R., *et al.*, GNN-FiLM: Graph Neural Networks with Feature-wise Linear Modulation, in: *Proc. Int. Conf. Mach. Learn. (ICML)*, 2021.
7. Cormen, T.H., Leiserson, C.E., Rivest, R.L., Stein, C., *Introduction to Algorithms*, 3rd ed., MIT Press, Cambridge, MA, USA, 2009.
8. Defferrard, M., Bresson, X., Vandergheynst, P., Convolutional Neural Networks on Graphs with Fast Localized Spectral Filtering, in: *Proc. Adv. Neural Inf. Process. Syst. (NeurIPS)*, 2016.
9. Dettmers, T., *et al.*, Convolutional 2D Knowledge Graph Embeddings, in: *Proc. AAAI Conf. Artif. Intell.*, 2018.
10. Diestel, R., *Graph Theory*, 5th ed., Springer, Berlin, Germany, 2017.
11. Elsken, T., Metzen, J.H., Hutter, F., Neural Architecture Search: A Survey. *J. Mach. Learn. Res.*, 20, 55, 1–21, 2019.
12. Fiedler, M., Algebraic Connectivity of Graphs. *Czech. Math. J.*, 23, 298–305, 1973.
13. Fortunato, S., Community Detection in Graphs. *Phys. Rep.*, 486, 3–5, 75–174, 2010.
14. Fout, A., *et al.*, Protein Interface Prediction Using Graph Convolutional Networks, in: *Proc. Adv. Neural Inf. Process. Syst. (NeurIPS)*, 2017.
15. Gilmer, J., Schoenholz, S.S., Riley, P.F., Vinyals, O., Dahl, G.E., Neural Message Passing for Quantum Chemistry, in: *Proc. Int. Conf. Mach. Learn. (ICML)*, 2017.
16. Hamilton, W., Ying, Z., Leskovec, J., Inductive Representation Learning on Large Graphs, in: *Proc. Adv. Neural Inf. Process. Syst. (NeurIPS)*, 2017.
17. Hochreiter, S. and Schmidhuber, J., Long Short-Term Memory. *Neural Comput.*, 9, 8, 1735–1780, 1997.
18. Hu, C., *et al.*, Heterogeneous Graph Neural Network, in: *Proc. ACM SIGKDD Int. Conf. Knowl. Discov. Data Min. (KDD)*, 2019.
19. Hu, W., *et al.*, Open Graph Benchmark: Datasets for Machine Learning on Graphs, in: *Proc. Adv. Neural Inf. Process. Syst. (NeurIPS)*, 2020.

20. Johnson, J., *et al.*, Image Retrieval Using Scene Graphs, in: *Proc. IEEE Conf. Comput. Vis. Pattern Recognit. (CVPR)*, 2015.
21. Joseph, A.D., *et al.*, Fairness and Abstraction in Sociotechnical Systems, in: *Proc. ACM Conf. Fairness, Accountability, and Transparency (FAT)*, 2018.
22. Kipf, T.N. and Welling, M., Semi-Supervised Classification with Graph Convolutional Networks, in: *Proc. Int. Conf. Learn. Representations (ICLR)*, 2017.
23. LeCun, Y., Bengio, Y., Hinton, G., Deep Learning. *Nature*, 521, 7553, 436–444, 2015.
24. Leskovec, J., Rajaraman, A., Ullman, J.D., *Mining of Massive Datasets*, Cambridge Univ. Press, Cambridge, U.K., 2020.
25. Li, Q., Han, Z., Wu, X., Deeper Insights into Graph Convolutional Networks for Semi-Supervised Learning, in: *Proc. AAAI Conf. Artif. Intell.*, 2018.
26. Liu, S., *et al.*, Graph Neural Networks for Financial Fraud Detection: A Survey. *ACM Comput. Surv.*, 55, 3, 1–34, 2023.
27. Pareja, H., *et al.*, EvolveGCN: Evolving Graph Convolutional Networks for Dynamic Graphs, in: *Proc. AAAI Conf. Artif. Intell.*, 2020.
28. Paszke, A., *et al.*, Automatic Differentiation in PyTorch, in: *Proc. NeurIPS Workshop Autodiff.*, 2017.
29. Rossi, E., *et al.*, Temporal Graph Networks for Deep Learning on Dynamic Graphs, in: *Proc. Int. Conf. Mach. Learn. (ICML)*, 2020.
30. Scarselli, F., *et al.*, The Graph Neural Network Model. *IEEE Trans. Neural Netw.*, 20, 1, 61–80, Jan. 2009.
31. Schlichtkrull, M., *et al.*, Modelling Relational Data with Graph Convolutional Networks, in: *Proc. Extended Semantic Web Conf. (ESWC)*, 2018.
32. Shuman, D.I., *et al.*, The Emerging Field of Signal Processing on Graphs. *IEEE Signal Process Mag.*, 30, 3, 83–98, May 2013.
33. Silver, D., *et al.*, Mastering the Game of Go with Deep Neural Networks and Tree Search. *Nature*, 529, 7587, 484–489, 2016.
34. Vaswani, A., *et al.*, Attention is All You Need, in: *Proc. Adv. Neural Inf. Process. Syst. (NeurIPS)*, 2017.
35. Veličković, P., *et al.*, Graph Attention Networks, in: *Proc. Int. Conf. Learn. Representations (ICLR)*, 2018.
36. Veličković, P., *et al.*, Deep Graph Infomax, in: *Proc. Int. Conf. Learn. Representations (ICLR)*, 2019.
37. Wang, X., *et al.*, Knowledge Graph Convolutional Networks for Recommender Systems, in: *Proc. Int. World Wide Web Conf. (WWW)*, 2019.
38. Wang, X., *et al.*, Neural Graph Collaborative Filtering, in: *Proc. ACM SIGIR Conf. Res. Develop. Inf. Retr. (SIGIR)*, 2019.
39. Wasserman, S. and Faust, K., *Social Network Analysis: Methods and Applications*, Cambridge Univ. Press, Cambridge, U.K., 1994.
40. West, D.B., *Introduction to Graph Theory*, 2nd ed., Prentice Hall, Upper Saddle River, NJ, USA, 2001.

41. Wu, F., *et al.*, Simplifying Graph Convolutional Networks, in: *Proc. Int. Conf. Mach. Learn. (ICML)*, 2019.
42. Wu, Z., *et al.*, A Comprehensive Survey on Graph Neural Networks. *IEEE Trans. Neural Netw. Learn. Syst.*, 32, 1, 4–24, Jan. 2021.
43. Xu, K., *et al.*, How Powerful are Graph Neural Networks?, in: *Proc. Int. Conf. Learn. Representations (ICLR)*, 2019.
44. Xu, K., *et al.*, Representation Learning on Graphs with Jumping Knowledge Networks, in: *Proc. Int. Conf. Mach. Learn. (ICML)*, 2018.
45. Yao, Y., *et al.*, Graph-Based Neural Multi-Document Summarization, in: *Proc. Conf. Empir. Methods Nat. Lang. Process. (EMNLP)*, 2020.
46. Ying, R., *et al.*, GNNExplainer: Generating Explanations for Graph Neural Networks, in: *Proc. Adv. Neural Inf. Process. Syst. (NeurIPS)*, 2019.
47. Ying, R., *et al.*, Graph Convolutional Neural Networks for Web-Scale Recommender Systems, in: *Proc. ACM SIGKDD Int. Conf. Knowl. Discov. Data Min. (KDD)*, 2018.
48. Ying, Z., *et al.*, Do Transformers Really Perform Bad for Graph Representation?, in: *Proc. Adv. Neural Inf. Process. Syst. (NeurIPS)*, 2021.
49. Yu, B., Yin, H., Zhu, Z., Spatio-Temporal Graph Convolutional Networks: A Deep Learning Framework for Traffic Forecasting, in: *Proc. Int. Joint Conf. Artif. Intell. (IJCAI)*, 2018.
50. Zhang, Z., Cui, P., Zhu, W., Deep learning on graphs: A survey, *arXiv preprint*, 2018 .
51. Zheng, Z., *et al.*, Graph Neural Networks for Graph-Based Anomaly Detection in Cybersecurity: A Review. *IEEE Access*, 9, 155579–155593, 2021.
52. Zhou, T., *et al.*, Graph Neural Networks for Cybersecurity: A Review. *IEEE Access*, 10, 89245–89262, 2022.
53. Zitnik, A., Agrawal, M., Leskovec, J., Modelling Polypharmacy Side Effects with Graph Convolutional Networks. *Bioinformatics*, 34, 13, i457–i466, 2018.
54. Zitnik, A. and Leskovec, J., Predicting Multicellular Function through Multilayer Tissue Networks. *Bioinformatics*, 33, 14, i190–i198, 2017.

3

Message Passing Techniques in Graph-Based Learning

Kulkarni Manjusha Manikrao* **and Savitha Hiremath**

Department of Computer Science and Engineering, School of Engineering, Dayananda Sagar University, Bengaluru, Karnataka, India

Abstract

The graph data structure consists of nodes and its connecting edges. Graph establishes relationship between nodes and exploits the dependencies among. The graphs enable various applications of the algorithm which enhances the relationship between the nodes in such a way that they can be used for optimization, improve user experience, and decision making. Graph-based learning is a machine learning paradigm based on graphical data that analyses and represents them by exploring hidden patterns. These new insights can be applied in various fields such as knowledge graphs, biological systems, recommender systems, molecular structures, network analysis, fraud detection, and information retrieval systems. Deeper understanding of data through the lens of graphical structures can provide power to organizations to make data-driven decisions. In the field of deep learning with graph data, GNNs (Graph Neural Networks) are modeled specially to work with graph data structures. They are successful, dominant, and fast-growing paradigms in artificial neural networks and deep learning. To learn complex relationships and exploit hidden patterns among nodes in a graph, GNN uses message passing technique. In this technique the representations of nodes are updated by exchanging information between them. This updated information is propagated throughout the graph. Each node aggregates data from neighboring nodes iteratively forming new relationships. This process helps in graph related tasks such as node classification, link prediction, and graph representation learning. The goal of this chapter is to provide a synthesis and overview of message passing technique in graph-based learning.

**Corresponding author*: manjushakulkarni-rs-cse@dsu.edu.in

J. Ramkumar, Sridaran Rajagopal, B. Suchitra and S. Balamurugan (eds.) Graph Neural Networks: Concepts and Applications, (101–142) © 2026 Scrivener Publishing LLC

Keywords: Graph, graph neural networks (GNNs), message passing technique, graph convolution network (GCN), knowledge representation learning

3.1 Introduction to Graph-Based Learning

Nowadays network structures are represented as graphs more often as collection of nodes and connecting edges can give simpler understanding of complex relationship. For example, if we consider a particular domain of social network, chemical bonds, transportation system, E-commerce system, entertainment field, or similar information retrieval systems, there exists a complex relation. Various machine learning methods are applied to find relation between nodes in graphs. Recent advancement shows that to improve the learning performance with exploitation of interconnected complex connections is done by graph based leaning approach more efficiently. Graph based learning is defined as a class of machine learning paradigm which uses graphs to analyze data. The interconnections between nodes are modeled by exploiting important features at each node in the graph. Hence it enhances the learning process with the help of every data point(node) and its connecting edge (link or relationship) between them.

3.1.1 Types of Graphs

To represent this node and edge structure of real-world data, various graph types are used depending on the nature of data and their relationship such as:

- Directed graph
- Undirected graph
- Weighted graph
- Hyper graph
- Dynamic graph
- Heterogeneous graph
- Temporal graph
- Tree based graph
- K-partite graph
- Multiple graphs

When we see graph in mathematical concept, we imagine only some geometric shapes or few types of charts such as bar chart, pie chart etc. But if we consider graph in computer science, they will unfold the relationships among entities. Let us see few examples of graphs in computer science way. In Figure 3.1 [1] authors have defined heterogeneous graph with authors academic information and we have Figure 3.2 denoting E-Commerce data using k-partite graph.

When looked at such graphical representation of information, it is more expressive and more meaningful, decision-making analysis is possible.

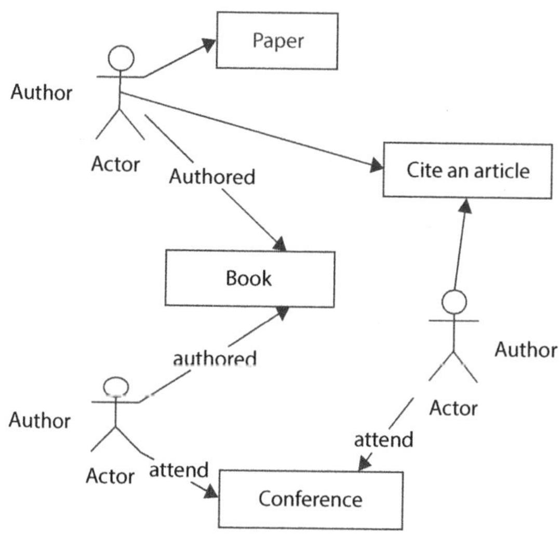

Figure 3.1 Source: [1]. Heterogeneous graph denoting academic information about author.

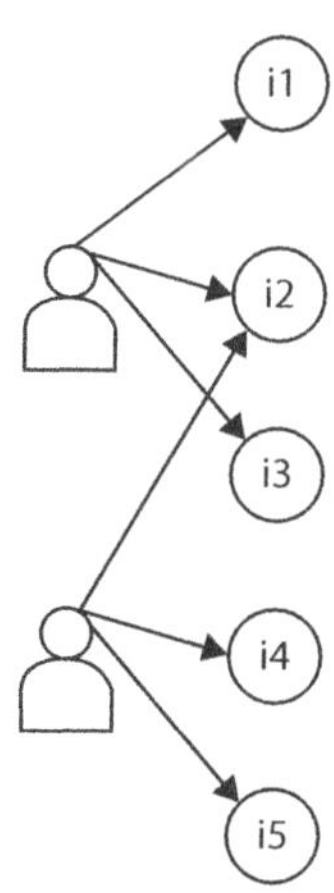

Figure 3.2 K-partite graphs showing left side users and right-side items interacted in E-commerce domain.

That is power of graph representation learning. As the number of entities are huge in real world data to find the connection/association between them becomes complex.

3.1.2 Graph-Based Tasks

Graph based tasks are divided into node-based tasks, edge based and graph-based tasks to perform tasks. This can be summarized as shown in the below figure with example.

As shown in below Figure 3.3, it is a general graph structure of points (nodes) and connecting links (edges) proposed in [2]. To understand exactly what graph level tasks this is the perfect figure as mentioned above. Before we move further about machine learning method that uses graph structured data, let us quickly see what type of graph learning based tasks [3, 7] are with reference to Figure 3.3 with examples such as node classification can be a finding a fake account of user in social network, for edge-level, finding protein interactions in biological applications, at graph-level we can take example from molecular systems to predict whether a molecule is toxic or not. The real-world raw data comes in huge number of attributes. The selection of useful and measured information is necessary. Hence, therefore nowadays graph-based learning is playing crucial role where more meaningful representations are required.

- Node-level tasks

Node level tasks are categorized into three types focusing on nodes:

✓ Node Classification- Nodes are being classified into number of classes

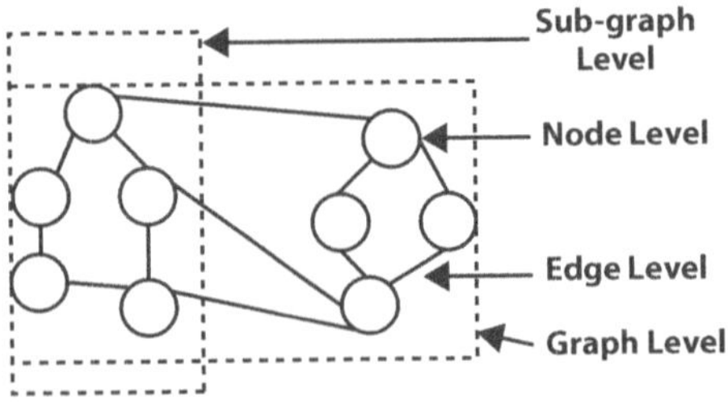

Figure 3.3 Source: [2]. Graph based tasks. Node level task – on nodes in each graph, edge level – on relationship between the nodes, subgraph level – forming smaller graph based on some criteria and Graph level – on independent graphs.

✓ Node Regression-Prediction of continuous value for every node
✓ Node Clustering- Finding number of disjoint clusters of nodes with same nodes

- Edge level tasks

This task will help to find relation between nodes. They are categorized into two types based on existence of edge/connection between nodes:

✓ Edge Prediction-Model try to classify similar edge types in the graph
✓ Link Prediction-Model try to predict the existence of link/ edge between nodes

- Graph level tasks

Based on graph representations these tasks are categorized into:

✓ Graph classification
✓ Graph Regression
✓ Graph Matching

3.1.3 Types of Graph-Based Learning

Graph-based learning is a machine learning paradigm. In the graphs depending on nature of nodes (labeled or unlabeled) different types and supervision perspective are defined as below:

A. Supervised graph learning: In this type of graph learning whole graph is labeled and used to perform the task of node classification, edge/link prediction, and graph classification. Examples: molecular graph (toxic or nontoxic graph), citation network (to predict research domain based on features of nodes).
B. Unsupervised learning on graphs: In this type of graph learning, nodes in the graph do not have labels. Graphs are used to find patterns or some relations among nodes to cluster them.
C. Semi supervised learning on graphs : In this type of graph-based learning, in the given graph some nodes will be labeled and using these labeled nodes and their connecting

edges the information will be propagated to find similar nodes to classify them in a group.

D. Graph Neural Networks: Graph data is learnt by using graph neural networks which is most popular, fast growing and most successful method in the domain of deep learning. The node representation is achieved by aggregating information from neighboring nodes in the graph. There are different types of GNNs are defined depending on type of data and task to be performed.

- GCN (Graph Convolutional Network)
- GraphSAGE
- GAT (Graph Attention Network)
- GIN (Graph Isomorphism Network)

E. Graph reinforcement learning: This type of graph learning is used to solve decision making problems using graphs or to solve dynamic graph problems.

F. Graph-based probabilistic models: In this type of graph learning probabilistic graphical models are used for example, Morkov Random Fields, or Bayesian networks.

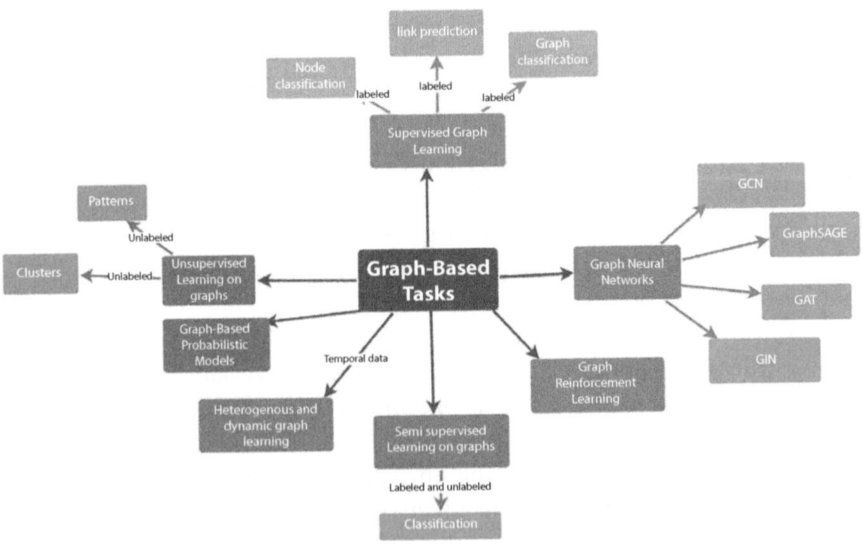

Figure 3.4 Summary of graph-based tasks.

G. Heterogeneous and dynamic graph learning: In this type of graph learning type of graphs used are heterogeneous or dynamic to solve real world problems where in the given graph nodes and edges are changing continuously with time.

Above Figure 3.4 shows pictorially graph-based learning tasks based on type of data used i.e. labeled or unlabeled.

3.1.4 Applications of Graph-Based Learning

Graphs are providing a meaningful relational shape to our real-world information. It's a wonderful experience when someone scrolling down videos on YouTube, purchasing "Anything" on online platforms like Amazon or Myntra, using social media and suddenly you will see a friend recommendation and it could be just surfing your favorite stuff on web. In everyday life almost 99% people are using maps to commute, many applications to order food and grocery that too with fasted delivery. How is it possible always that you are getting a stuff which is almost you are interested in! Well, that's because Networks. Graphs are seen as pie chart or bar chart but in computer science graphs are used to represent relational structures. Graph structures have various application areas, few of them are listed below with type of graphical representation of data used:

- Finance and fraud detection-The type of graph used here are transaction graphs where nodes and edges will be showing relationship of users, banks, and merchants.
- Recommendation system-graphs will be user to represent relation between user and items, with the edges representing purchase history, or interaction information.
- Social networks-Graphs will represent friends' connections (nodes are individuals and links exists based on features). Graphs can be used to find a community or group of similar feature nodes and to find wrong or misleading content by detecting pattern propagated by node-to-node.
- Transportation system- Graphs are constructed using destinations as nodes and roads as edges.

  These types of graphs have wide applications of traffic management, finding congestion and using spatial-temporal graph learning to apply optimal routing which we basically see in maps.

- Biological systems- Graph structures denote interaction between protein-protein.
- Natural Language processing-In this domain graphs can be knowledge graphs or built by decoding semantic information used in chatbots, question answering scenarios as well as in search engines.

From above applications it is obvious that graphs play vital role in establishing relationship between real world data and to solve complex structural problems. Graphs have become one of the frontiers in Artificial Intelligence world which is used by Graph neural networks. Graph neural networks are neural networks. These neural networks process graph data which contains relational information. These application areas will use graphs to compute the graph-based tasks such as prediction, classification, and regression by considering edge, node, and subgraphs of a graph respectively. In the following sessions investigation of each one of the tasks as well as methods to be applied can be seen.

3.2 Graph Neural Networks

The graph neural network is a general methodology used in deep learning which used graph structured data. We have seen that our agenda is to use graph to exploit hidden relationship between nodes using features of nodes and links that exists between them. This is achieved by GNNs by generating node representations from given graph structure and feature information from nodes. This node representation is modeled by exchanging and aggregating feature information among neighboring nodes in the graph. This process is called as "neural message passing." This is very crucial concept in graph neural networks.

3.2.1 Core Concepts of Message Passing Techniques

In this section the general framework of neural message passing is explained and shown that how GNNs use them as a core concept in node representation. This chapter will describe overview of basic message passing technique in graph neural network and other common variants of it. To put it in simple language we can say that in the message passing information is shared among neighboring nodes in a graph where nodes are connected by an edge to infer and learn complex relationship between them. Another terminology is "embedding", which is a core part of this process is a simple, low dimensional representation of nodes and edges or whole graph without harming original graph structure.

3.2.2 Overview of Message Passing Technique

Message passing depends on graph structure completely, that means how the nodes are represented (what entities are represented by nodes), edges show relationship between nodes, and the message passing is flow of data between nodes *via* edges. Table 3.1 describes preliminaries used to define the graph, methods, and its associated terms.

There are two arbitrable differentiable functions [5], those are two neural networks, used in message passing, update, and aggregate.

- UPDATE function

This function in message passing is defined as:

$$
\begin{aligned}
h_u^{(k+1)} &= UPDATE^{(k)}\left(h_u^{(k)}, AGGREGATE^{(k)}\left(h_v^{(k)}, \forall\, v \in N(u)\right)\right) \\
&= UPDATE^{(k)}\left(h_u^{(k)}, m_{N(u)}^{(k)}\right)
\end{aligned}
\tag{3.1}
$$

In the above equation $m_{N(u)}^{(k)}$ is information/message which is aggregated from neighbourhood *N(u)* of graph. Iterations denoted by k in message passing are sometimes referred as layers in graph neural networks.

Table 3.1 Preliminaries.

Terminology	**Description**
G	Graph
V	Set of nodes
E	Set of edges- An edge from u to v and v to u $(u, v) \in E$
h	Hidden Embedding
k	Iterations
$h_u^{(k)}$	Embedding corresponding to every node $u \in V$
N	Neighborhood of given nodes

The above two functions can be broken down in simple way as every node in a graph will compute a message for its neighboring node we can put in this way-m (node, neighbor, edge) [4, 5].

Same thing can be seen from Figure 3.5 where in [6] authors have pictorially shown message passing among nodes in graph. At target node A aggregates messages are coming from its neighboring nodes D, B, C (connecting edge available between all nodes). At the right side we can see B, C, and D are aggregating information from their neighboring nodes (B aggregating information from A and C). The figure shows a two-layer version of message passing framework. This aggregation is simple average or it can be some of information gathered from neighboring nodes. Finally, nodes will update their parameters with the aggregated information from current values. This process is happening simultaneously on all the nodes in graph, hence updating of attribute information also same time. This same theory is described by equations of UPDATE and AGGREGATE arbitrary functions. At every iteration k of graph neural network, each node aggregates information as an input coming from set of embeddings from its neighborhood nodes N(u) i.e. and generating $m_{N(u)}^{(k)}$ message. This UPDATE function combines these messages with previous iteration embedding $h_u^{(k-1)}$ of node u to update embedding in current iteration $h_u^{(k)}$ embedding.

$h_u^{(0)} = xu$ are the embeddings at k=0 set for all the nodes at initial phase k=0 $\forall u \in V$. Final embeddings can be defined at the end k iterations of every node in this message passing process as:

$$z_u = h_u^{(k)}, \forall u \in V \tag{3.2}$$

In this process of message passing in GNN, the way theses UPDATE and AGRREGATE functions are applied they are independent of node order in graph instead what matters is given graph structure and this will be

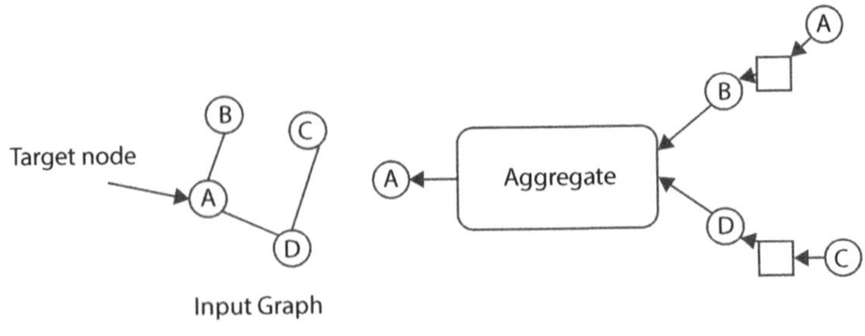

Figure 3.5 Source: [6]. Overview of update and aggregate function in message passing technique.

followed throughout the process. Due to GNNs are called as permutation equivariant. The basic message passing in GNN is defined as:

$$h_u^{(k)} = \sigma\left(W_{self}^k h_u^{(k-1)} + W_{neigh_{v \in N(u)}}^k h_v^{(k-1)} + b^k\right) \tag{3.3}$$

W_{self}^k, $W_{neigh_{v \in N(u)}}^k$ are trainable parameter matrices ϵ $R^{(d^{(k)} \times d^{(k-1)}\}}$.

σ is non-linear function (tanh or ReLU). A very strong performance is achieved by considering bias terms b^k but it is always excluded for simplicity of notations.

The message generation process starts by adding messages coming from its immediate nodes as described mathematically below:

$$m_{N(u)} = \sum_{v \in N(u)} h_v \tag{3.4}$$

which also combines neighboring information with previous embedding of node denoted by h_v with linear combination. At the end elementwise nonlinearity is applied.

- UPDATE and AGGREGATE functions in basic GNN.

$$UPDATE\left(h_u, m_{N(u)}\right) = \sigma\left(W_{self} h_u + W_{neigh} m_{N(u)}\right) \tag{3.5}$$

$$m_{N(u)} = AGGREGATE^{(k)}\left(\{h_u^{(k)}, \forall v \in N(u)\}\right) \tag{3.6}$$

This is the message notation aggregated from graph u's neighborhood.

Message passing techniques are used in various application domains such as social networks, recommendation systems, traffic management and in healthcare etc. We can take example of recommendation system. Suppose we are working on social networks, friend recommender systems, we will be defining node with users, friendships are edges and features are posts, followers of that user and connection score or engagement score (their comments or likes or views).

Let us consider predicting joining of a new group or new friend based on their behavior is a task. Using the preliminary data defines a graph using number of users, their posts, followers, and their engagement score.

Here, the feature factors h(v) is nothing but the given data. E.g. posts=10, followers=80 and so on. From here if we apply aggregate mean then given values for attributes are summed and taken from them to form new parameters as aggregated mean. Later update function will update with neighboring node feature E.g. A's own features: [0.9, 20, 100] for engagement score, posts, followers etc. then concatenation happen with the mean value and feature values of node. This process will be done for every node in the graph This is updated embedding N(v). In this way layer by layer these functions are applied and embeddings are updated.

3.2.3 Message Passing with Self-Loop

In this approach self-loops (nodes connecting itself) are added to the input graph. This enables nodes to communicate neighboring nodes as well as to itself sending message to itself and other immediate nodes. Such approach will ease the overfitting [9] and exclude explicit UPDATE step. This can be explained with equation by modified message step as:

$$m_v^{(k+1)} = \sum_{v \in N(v) \cup (v)} M^{(k)}\left(h_v^{(k)}, h_u^{(k)}, e_{vu}\right) \tag{3.7}$$

UPDATE function is modified as:

$$h_u^{(k+1)} = U^{(k)}(M_v^{(k+1)}) \tag{3.8}$$

Which excludes nodes own features. There are different variants of message passing techniques are there based on type of graph (graph structure) and how the neighboring nodes are exchanging information:

- Graph Convolutional Network (GCN)
- GraphSAGE
- Graph Attention Networks (GAT)
- Message Passing Neural Network (MPNN)
- Gated Graph Neural Network (GGNN)

Every type has its variation in number of layers available in the network structure (how far information is sent to its next connected nodes) i.e. number of hops. In simple words we can say that checking possibility of

Figure 3.6 Variants of message passing technique in GNN and their application areas in this chapter every technique is explained with its definition, equation parameters, and its applications.

available edge between nodes. These types have variation in UPDATE and AGGREGATE functions as they can be based on node and edge features. Let us see them together in summarized Figure 3.6.

3.2.4 Different Variants of Message Passing Techniques

3.2.4.1 Graph Convolution Networks

We are familiar with the word "Graph," "Network" but what does this "convolution" here in this graph convolution network? Let us understand.

While working with image processing convolutional layers are used in Convolution Neural network. To find patterns of links(edges) or other textures a small matrix is used for small region of the image pixels in the grid.

This matrix is called as a filter. In similar manner, convolution in graph structures, neighboring nodes will aggregate and transform information as these structures are not regular [10–12]. Data is processed from surrounding pixels in CNN while in GCN neighboring node information is gathered from the graph [13, 14]. We can see graph convolution network is a technique to compute node embeddings based on graph neural network which uses graph Laplacian. Node embedding is the task where a new node representation is achieved by using all available data about the graph, its node features, and the connecting edges. Graph Laplacian is a useful structural representation of a graph in a matrix form which is used in machine learning, graph theory, network analysis domain. Definition of Laplacian Matrix: [15] Graph G (V, E) is defined by V nodes and E edges with adjacency matrix

$A \in \{0,1\}^{n \times n}$, where degree matrix and adjacency matrix are as follows:

$$A_{i,j} := \{1\,0 \qquad 1 \text{ if there exist edge between } v_i \; to \; v_j \text{ else } 0$$

Degree matrix $D \in Z^{\{n \times n)}$
Where,

$$D_{i,j} := \{degree_{(v_i)}\; 0 \qquad \text{if i=j (} \; degree_{(v_i)}) \text{ else } 0$$

Laplacian matrix is defined as:

$$\text{L: = D-A}$$

It is the discrete analogue to the Laplacian operator on multivariate continuous functions. The idea is allowing greater Laplacian smoothing when node features of one-layer GCN are passed to next layer. Now let us see AGRREGATE and Combine (UPDATE) functions. GCN interprets mean aggregator [11] as:

$$a_v^l = Mean\left(h_u^{(l-1)}, u \in N(v) \cup (v)\right) \tag{3.9}$$

The aggregated node feature a_v^l for the lth layer is calculated by the mean of the neighbourhood node features along with its feature taken as input by the algorithm that is a self-loop. This is computed at every node in a graph. Because of this formulation GCN becomes an inductive graph neural network. In combine function, node feature $h_v^{(l)}$ for lth layer is defined by multiplying aggregated node features a_v^l with a learnable [16] weight matrix and non-linear activation function (ReLU).

$$h_v^{(l)} = ReLU\left(W^{(l)} a_v^{(l)}\right) \tag{3.10}$$

learnable weight matrix $W^{(l)} \in R^{d_{(l-1)}} \times d^{(l)}$ as mentioned in above equation.

3.2.4.2 Graph Attention Networks (GAT)

3.2.4.2.1 GAT Architecture

Attention is referred here in the graph as to give importance to every node [17]. How to assign/denote importance or give attention to the node? It is done by assigning weight (size of) link as shown in Figure 3.7 from [13] where authors have visualized the attention of neighboring node in the graph with specific importance.

This weightage or attention can be of 1) Learn attention weights 2) Similarity based attention 3) Attention guided walks. What makes GAT differ than GCN as both have aggregate and combine functions? [18] In GCN node degree plays crucial role while selecting neighboring node. But in GAT it will consider node degree as well as node features for generating weightage. In GCN if a node has lesser neighbors will have normalized coefficient which make it important for selection. GAT's can handle more complex data as they have control over information flow through complicated graph structure.

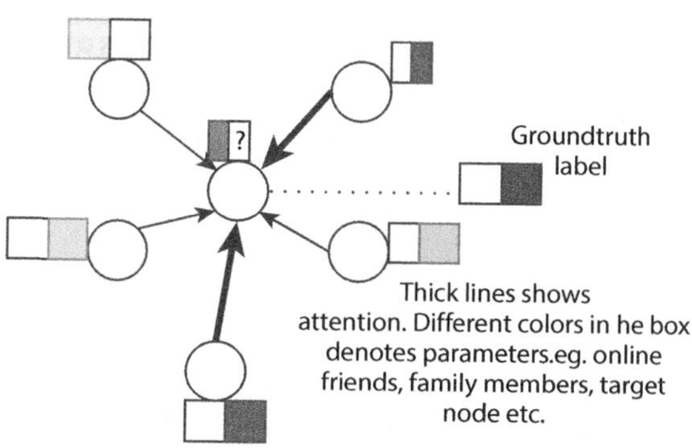

Figure 3.7 Source: [13]. Every neighboring node in the graph is assigned with an attention to specify importance. In the figure this weight is shown by thickness of link. The classmate node is focused by adding attention and the target nodes interest is predicted.

The process of graph attention layer in GAT can be summarized stepwise as:

i. Every node passes through joint linear transformation with applying weightage W is a weight parameter matrix
ii. To compute attention coefficient-This will assign importance of node through weights between nodes
iii. Concatenation of z embeddings of immediate nodes to form a combined vector. To integrate node features efficiently in attention mechanism, this combined vector used in dot product operation with learnable weight vector
iv. Application of activation function LeakyReLU at the output of dot product operation to add nonlinearity into the model
v. To normalize attention coefficient y applying softmax function to maintain consistency among nodes
vi. Aggregation phase where model combines neighbors' embeddings based on attention weight

To avoid potential instability of self- attention in GATs, the multi-head concept is employed. How it differs from self-attention? In this process multiple attention mechanisms are created. These heads or mechanisms are associated with each of different parameters. Model is more capable and stable due to its parallel operation of multiple heads. Final output is the integration of all heads output. As we know that there are intermediate layers will combine the output of each head and output layer will averaging will happen to consolidate the output by inspecting various perspectives. Attentional weights are computed implicitly by input comparison i.e. self-attention. From each head the aggregated features are combined. This aggregation produces nodes final vector representation. Vector representation of node features with multi head attentional weights is computed as:

$$h_i = \sum_{j \in N_i} a_{ij} \, Wx_j \tag{3.11}$$

a_{ij} are the attentional weights computed by network dynamically.

To define aggregate and combine functions following preliminaries are used as shown in Table 3.2:

The attention coefficient is computed by:

Table 3.2 Preliminaries.

Terminology	Description
N	No of nodes in graph G
F	Number of features in every node
$W \in R^{FxF`}$	Weight matrix
x_i	Input vector of neighbors of selected node, here A

$$e_{ij} = a\left(W\vec{h}_i, W\vec{h}_j\right) \tag{3.12}$$

It shows significance of node *j's* feature with respect to node *i.* a is shared attentional mechanism.

a:$R^{F`} x R^{F`} \rightarrow R$, h $= \{\vec{h}_1, \vec{h}_2, \vec{h}_3 \ldots . \vec{h}_N\}$ and $h =' \{\vec{h'}_1, \vec{h'}_2, \vec{h'}_3 \ldots . \vec{h'}_N\}$ are set of node features given to input layer and new node features generated as output respectively. In this structure every node will visit every other node in the graph irrespective of structural information.

Graph Attention network uses self-attention mechanism to aggregate features of neighborhood nodes. As discussed, attention provides weighted mean of neighboring node features. Here we would like to address two variations of attention mechanisms to understand single self-attention block and k masked self-attention modules. Using one self-attention block definition of aggregate and combine function: Here as attention module is aggregating features from neighboring nodes by computing weighted mean---

$$\alpha_u^{(l)} = Attention\left(\left(Wh_u^{(l-1)} \,||\, Wh_v^{(l-1)}\right), u \in Nv \cup v\right) \tag{3.13}$$

Attention(.) is a single fully connected layer with activation function LeakyReLU, $W \in R^{d_{(l-1)} \times d_{(l)}}$, $\alpha_u^{(l)}$ is the coefficient for u's neighboring nodes.

- **Combine Function**

Multiplying node feature $\alpha_u^{(l)}$ multiplied with learnable weight matrix and non-linear activation function to generate new node features $h_u^{(l)}$ at the lth layer.

$$h_u^{(l)} = ReLU\left(sum\alpha_u^{(l)} Wh_u^{(l-1)}, u \in Nv \cup v\right) \tag{3.14}$$

- **Masked Attention Mechanism**

In this operation, attention coefficient $e_{(ij)}$ computed for nodes I,j which are neighbors is then normalized by applying softmax function to make is comparable across all choices of node j neighbor.

$$\gamma_{ij} = softmax_j e_{(ij)} = \frac{exp\ e_{(ij)}}{\sum_{k\epsilon N_i} exp\ e_{(ik)}}$$

$$\gamma_{ij} = \frac{exp\left(LeakyReLU\left(\vec{a}^T\left[W\vec{h_i} \,||\, W\vec{h_j}\right]\right)\right)}{\sum_{k\epsilon N_i} exp\left(LeakyReLU\left(\vec{a}^T\left[W\vec{h_i} \,||\, W\vec{h_k}\right]\right)\right)} \tag{3.15}$$

$.T\ and\ ||$ shows transportation and concatenation operation. At the output layer normalised attention coefficients are used to compute linear combination of features of every node in graph.

3.2.4.3 Graph SAGE

Graph SAmple and AggreGatE (GraphSAGE) [19, 67] is the framework used to generate node embeddings i. e low order vector representation of nodes having large number of feature parameters. It is called as inductive learning algorithm. It generates node embeddings by sampling and aggregating neighboring nodes using multiple hops. To do this operation GraphSAGE does not require full graph structure, that is why it is inductive learning. There are hierarchies of applications associated with node embeddings. The low dimensional feature vector representation of nodes will help in graph analysis and prediction tasks as well as when fed to machine learning systems it will be applied for node classification, link prediction and clustering tasks. In inductive learning, it helps to find new sub graphs, explore complex relationships between nodes by learning unseen nodes.

In GraphSAGE, aggregators are learnt from neighboring nodes by sampling the neighbor nodes. It will not consider all the nodes in graph; it will sample them. The aggregator functions should operate on an unordered set of vectors of nodes. We are describing here these aggregators mean, LSTM and pooling.

- **Mean Aggregator**

Mean operator is one of the candidate aggregate functions. In this operation elementwise mean of vectors $h_u^{(k-1)}$ is taken ((k-1) layers).

$$h_v^k \leftarrow \sigma(\{W.MEAN(\{(h_v^{(k-1)}\} \cup \{h_u^{(k-1)}\}, \forall u \in N(v))\})$$

This aggregator is rough, linear approximation of localized spectral convolution hence called as mean-based aggregator convolutional. The performance is majorly gained with the concatenation of $h_v^{(k-1)}$ and $h_{N(v)}^{(k)}$. They are vectors of previous layer node and its neighbouring nodes making it in depth search between layers from GraphSAGE algorithm.

- **LSTM Aggregator**

LSTM aggregators are more expressive. They process input in sequential manner hence are not inherently symmetric. How LSTMs are applied on unordered set of vectors?

In [19] authors have shown that aggregated feature vector $\alpha_u^{(l)}$ from lth layer is generated by shuffling neighbourhood nodes in graph as:

$$a_u^{(l)} = LSTM\left(h_u^{(l-1)}, u \in N(v)\right) \tag{3.16}$$

While the combine function joins node features from (l-1)th layer with $a_u^{(l)}$ which is aggregated feature vector. This function will pass the output node feature vector at every layer through fully connected neural network:

$$h_u^{(l)} = ReLU\left(W^{(l)}\left[h_u^{(l-1)} a_v^l\right]\right) \tag{3.17}$$

$W^{(l)} \in R^{d_{(l-1)} \times d_{(l)}}$ is lth layer learnable weight matrix.

- **Pooling Aggregator**

Pooling aggregator is trainable and symmetric. Pooling strategy follows feeding feature vectors of each neighbor independently to fully connected neural network. So, you can imagine here "pooling" output vector from neighbor as input and producing output at every layer. To aggregate information across set of neighbors, this transformation is followed by every element go through a max-pooling operation.

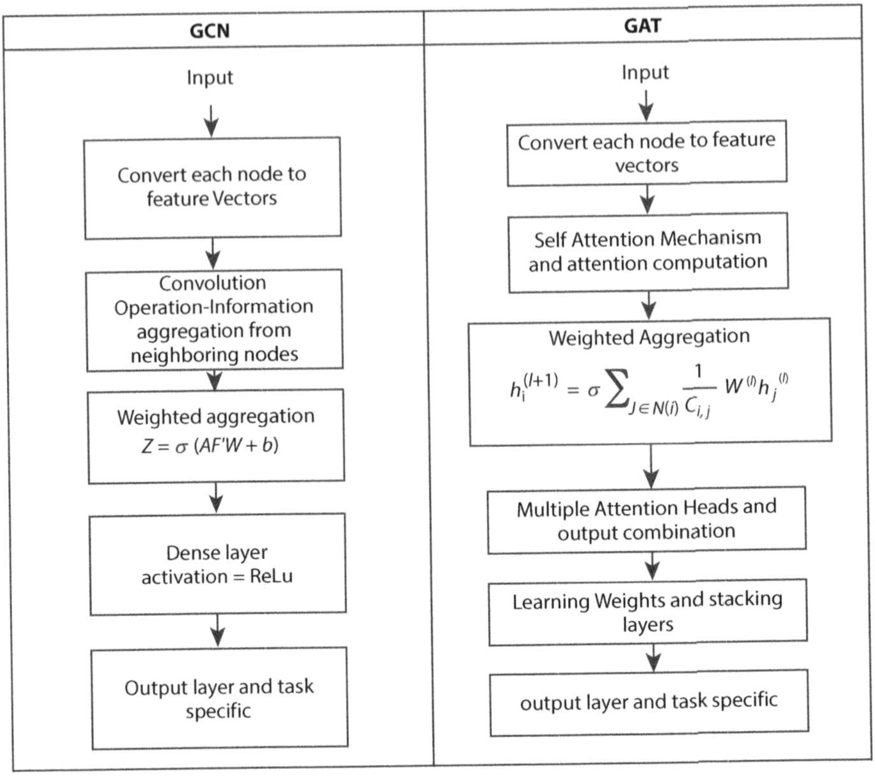

Figure 3.8 Source: [8]. Step by step approach of GCN and GAT GNN models.

$$AGGREGATE_k^{POOL} = max\left(\sigma\left(W_poolh_{ui}^l + b\right), \forall u_i \in N(v)\right) \quad (3.18)$$

σ and max are non-linear activation function and elementwise max operation respectively. Application of max pooling operation for computed features of each node will help to explore many of aspects/relations of neighboring nodes.

In study [8] authors have highlighted working of these three GNN models visualized with diagrams. Let us see diagrammatically working of these models together with Figure 3.8.

3.2.4.4 *Message Passing Neural Networks (MPNN)*

Deep neural networks are used in various application areas such as natural language, complex audio signals, feature detection from real world videos and images, molecules and many more.

Message passing neural network [4] is a supervised learning framework designed to work on graphs to exploit complex relationship, make them simpler to understand and to produce meaningful connection between parameters. Architecture definition in mathematical terms is discussed below:

Authors [2] have considered undirected graphs on which MPPN is applied. G is the undirected graph, x_v *and* e_{vw} are node features and edge features respectively. There are two phases under forward pass –

a) A message passing phase defined with M_l, it will run for T time steps represented as M_t.

During this message passing phase, every node will go through hidden state/layer(embedding) in the graph shown by h_u^t are updated based on messages M_v^{t+1} at step time T given by-

$$M_v^{t+1} = \sum_{w \in n(v)} M_t(h_v^t, h_w^t, e_{vw}) \tag{3.19}$$

b) A readout phase defined with U_l vertex update function and at T time steps represented by U_t

It will compute whole graph feature vector using R as readout function according to:

$$h_v^{t+1} = U_t(h_v^t, M_v^{t+1}) \tag{3.20}$$

In the readout phase feature vector of complete graph is also computed by readout function R as:

$$\hat{y} = R(h_v^T \mid v \in G) \tag{3.21}$$

All these functions defined above readout, update, message function are learned differentiable functions. The same functions can be applied to learn edge features by applying hidden states between edges of nodes and modifying above functions.

3.2.4.5 *Gated Graph Neural Network (GGNN)*

This form of graph neural network is used for non-sequential output. This mechanism has built with Gated Recurrent Unit [20, 21]. The recurrence is unrolled for fixed number of steps T. The gradient is by using backpropagation through time. We will briefly see what is GRU and its functionality.

3.2.4.5.1 Gated Recurrent Unit (GRU)

Gated Recurrent unit [22] is a recurrent neural design to address the challenges from traditional recurrent design such as vanishing gradient issue. In [22] authors have introduced GRU which is simpler than LSTM (Long Short-Term Memory) model. Data flow is controlled using gates hence it is efficient mechanism for sequential data over long time/period. This will store data for longer time. Retention of information is applicable to time series prediction or language modeling tasks [23].

GRU has three core components:

a. Update Gate

Whether to store past data or discard, this decision is taken by update gate. This will help to keep only relevant data over some time. Mathematically this is shown by following equation:

$$z_t = \sigma(W_z.[h_{t-1}, x_t]) \tag{3.22}$$

b. Reset Gate

Most related features are considered by GRU with the help of reset gate by forgetting historical data. This makes sure that memory is not retaining irrelevant historical information. Mathematically this is shown by following equation:

$$r_t = \sigma(W_r.[h_{t-1}, x_t]) \tag{3.23}$$

c. Candidate Activation

This component will compute possible new memory content at the current time step with the help of input state and hidden state. Candidate activation is impacted by reset gate in consideration of past state for computing candidate. Basically, it provides possible new state influenced by reset gate. Mathematically this is shown by following equation:

$$\hat{h}^t = tanh(W_h.[r_t * h_{t-1}, x_t]) \tag{3.24}$$

In the final hidden state, candidate activation and previous state is combined based on weights assigned by update gate. Now, after GRU concept description as it was a part of GGNN, let us move towards study of GGNN in depth.

GGNN mechanism consists of following three steps:

I. Node Annotation

In Gated Graph Neural Networks node labels are considered as additional inputs as in GNN. Node annotations are denoted with vector x. Initial hidden state is shown by $h_v^{(0)}$

II. Propagation Model

As this method is also a part of GNN, we can consider this step as message passing through GRU

Let us see its mathematical founded equation for every step mentioned above:

Initialization step:

In the initialization step node annotations are copied into hidden states first components and remaining annotations are padded with zeros. Below equation shows initialization step:

$$h_v^1 = [x_v^T, 0]^T \tag{3.25}$$

III. Message Passing through GRU

In this step data is transferred among nodes of graph through bidirectional edges carrying features based on type of edge. It also depends on edge direction either incoming or outgoing. This is shown with below equation:

$$a_v^t = A_v^T : \left[h_1^{(t-1)T} \dots h_{|v|}^{(t-1)T} \right]^T + b \tag{3.26}$$

a_v^t denotes both directions activations from edges. How we have defined mathematically GRU components, in similar manner GRU type updates happens for incorporation of data from other nodes to previous time which updated every node hidden state. Reset and update gates are denoted with z and r. Logistic sigmoid function $is\ \sigma(x) = \frac{1}{1+e^{-x}}$, elementwise multiplication is denoted by $\odot$.

IV. Output Model

Node selection task:

Node selection task is supported by GGNN by making $o_v = g(h_v^T, x_v)$ where $v \in V$ output node score. Later softmax function is applied over scores of nodes. Graph level representation score is defined for graph level output as given by equation:

$$h_G = tanh\left(\sum_{v \in V} \sigma\left(i\left(h_v^{(T)}, x_v\right)\right) \odot tanh\left(j\left(h_v^{(T)}, x_v\right)\right)\right) \quad (3.27)$$

Where, $\sigma\left(i\left(h_v^{(T)}, x_v\right)\right)$-is a soft attention mechanism to decide suitable nodes at current level graph task and j, i - neural networks.

These neural networks are used in concatenation of inputs $h_v^{(T)}$ and x_v and produce output of real valued vectors. In equation, activation function tanh can be replaced by identity.

3.3 Other Popular Graph Neural Networks and Message Passing Techniques

3.3.1 DimNet-Directional Message Passing Neural Network

In this message passing scheme where directional messages are transformed based on angle between nodes in graph, like belief propagation. This method is particularly used in molecular predictions to predict molecules quantum mechanical properties. To achieve better performance authors in [24] used spherical Bessel functions and spherical harmonics for to constructing theoretically well-founded, orthogonal representations. Mathematical representation of Message passing in DimNet [25] as well as description using graph with nodes is shown in Figure 3.9:

m_{ij} -is the message transferred from atom j to atom i

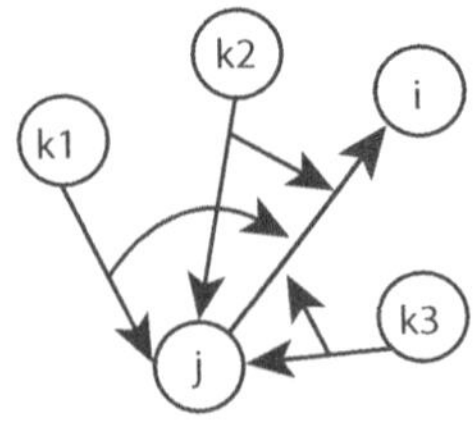

Figure 3.9 Source: [25]. Aggregation scheme for message embeddings and the message from node can be defined with $m_{k1,j}$ and so on.

Embedding every item i with set of incoming messages m_{ji} as $h_i = \sum_{j \in N_i} m_{ji}$ and message m_{ji} is updated as m_{kj} based on incoming message.

Message embedding with UPDATE and AGRREGATE function is defined as below:

$$m_{ji}^{(l+1)} = f_u pdate\left(m_{ji}^{(l)}\right) \sum_{\frac{k \in N_j}{i}} f_{int}\left(m_{kj}^{(l)}, e_R^{ji} BF, a_S^{(kj,ji)} BF\right) \tag{3.28}$$

The terminology meanings from above equation:

- Radial basis function of interatomic distance
- Angle between atom pairs or relation/bond between atoms

This definition of aggregation is having analogy towards belief propagation. These types of message embeddings are used by higher-order Weisfeiler-Lehman tests of isomorphism which are provably more powerful GNNs.

3.3.2 Edge Conditioned Convolution Neural Network (ECCNN)

Generally, convolution operators are applied on regular grids. In edge conditioned convolution applied on graph structured data to work on various size graphs and connections. How this is achieved? Edge conditioning!

In edge conditioned convolution, convolution filter modified according to edge features. Mathematical representation of convolution operation of ECCNN [26]:

$$x_{(i)}^{l} = 1/N(i) \sum_{j \in N_{(i)}} F^l(L(j,i); w^l) X_{(j)}^{(l-1)} + b^l$$

$$\begin{aligned} x_{(i)}^{l} &= 1/N(i) \textstyle\sum_{j \in N_{(i)}} \theta_{(ji)}^{l} X_{(j)}^{(l-1)} + b^l \\ &= 1/N(i) \textstyle\sum_{j \in N_{(i)}} F^l(L(j,i); w^l) X_{(j)}^{(l-1)} + b^l \end{aligned} \tag{3.29}$$

Where learnable bias $b^l \in R^{d_l}$, filter generating network F^l is parameterised w^l, w^l it is learnable network weight. b^l, w^l, are updated dynamically generated parameter $\theta^l_{(ji)}$ which is edge label in specific input graph. $l \in \{0,\ldots,l\_max$ layer index in feed forward neural network, labels are assigned by functions, $X^l: V \rightarrow R^{d_l}, L: E \rightarrow R^s$. $X^l \in R^{n x d_l}$ $L \in R^{m \times s}$ are the matrix functions, R^s assigns features to every vertex and attribute to every edge, input signal is given by X^0. $N_{(i)}$ is a neighbourhood of vertex i. It is defined as $\{j; (j,i) \in E \cup \{i\}\}$ all vertex including itself.

As proposed in [26] conditioning filtering weight on each edge label respectively can solve issue of undefined vertex ordering and variation of neighborhood sizes. Even it helps in smoothing structural information. This is a commutative approach where, $x^l_{(i)} \in R^{d_l}$ is a filtered signal at node i as a weighted sum of signals in its neighbourhood of j, $X^{(l-1)}_{(j)} R^{d_{l-1}}$.

3.3.3 GIN (Graph Isomorphism Network)

To define graph isomorphism network, let us begin with below theorem where GIN [27] satisfies conditions stated in theorem.

Theorem:
Statement: Let A denotes a GNN where $A: G \rightarrow R^d$. Any graphs $G_1 and G_2$ satisfying Weisfeiler-Lehman test of isomorphism deciding a non-isomorphism mapped by A with sufficient layers network, of other embeddings if the below conditions are satisfied:

- aggregate and update functions operated by A to generate new node features iteratively as given by-

$$h_v^{(k)} = \varnothing\left(h_v^{(k-1)}, f\left(h_u^{(k-1)}, u \in N_{(v)}\right)\right) \tag{3.30}$$

f *and* $\varnothing$ are functions operating on multiset and are injective.

- Readout function of A at graph level is operated on multiset of node feature h_v^K is injective.

In above theorem we can see one word "injective", it refers to distinctiveness on input is saved by functions or not with respect to countable sets and for uncountable it refers to continuity of node features or some other considerations are needed. GIN generalizes Weisfeiler-Lehman test of isomorphism deciding a non-isomorphism making them achieving discriminative power.

Injective multiset functions are modeled for aggregating neighbors with deep multisets. Deep multisets are parameterized universal multiset functions in neural networks.

By satisfying conditions stated in the theorem a simple architecture of GIN is modeled. To satisfy first condition 1 in theorem i.e. injective-ness, universal multiset function has been modeled as given in below as a building block stated in lemma 1.

Lemma 1:

Input feature space X is assumed countable. Function f exists where $f: X \rightarrow R^n$ such that $h(X) = \sum_{x \in X} f(x)$ is unique for every multiset $X \subset \chi$ within size limit. Multiset function denoted by g can be degenerated for some function $\varnothing$ as $g(X) = \varnothing\left(\sum_{x \in X} f(x)\right)$

Deep multisets are differed from sets in terms of the infectiveness, not all injective set functions such as popular function mean aggregator is not injective multiset function. Next, corollary provides simple and concrete foundation for multiset functions.

Corollary:

χ is assumed as countable. A function $f: X \rightarrow R^n$ exists providing infinitely may options for $\in$ that includes irrational numbers $h(c, X) = (1+\in).f(c) + \sum_{x \in X} f(x)$. These numbers are unique for every pair denoted by (c, X).

Here $c \rightarrow X \; and \; X \rightarrow \chi$ and it is a multiset with bounded size. Any function g over this pair such as (c, X) can be distributed as $g(c, X) = \varphi\left(\left((1+\in).f(c) + \sum_{x \in X} f(x)\right)\right)$ for some φ function.

To learn f and φ multilayer perceptron can be used by using universal approximation theorem [28, 29]. After learning the background theorem, lemma and corollary behind graph isomorphism networks let us see how node representation is updated in GIN. $\in$ is made fixed scalar or a learnable parameter.

$$h_v^{(k)} = MLP^{(k)}\left((1+\in)^{(k)}.h_v^{(k-1)}\right) + u \sum_{u \in N_{(v)}} h_u^{(k-1)} \tag{3.31}$$

GIN is powerful among GNNs as there exist many more.

3.3.3.1 *Graph-Level Readout Function of Graph Isomorphism Network*

Node embeddings generated by GIN can be used for various graph-based tasks such as node classification or link prediction. Let us define readout function for graph classification task by using individual node embeddings those produces whole graph. For Node representation of subtree structure with more refined and global form readout function at graph-level is an important aspect as number of iterations increases there. A good discriminative power comes with sufficient iterations. Not to lose useful/important features from past iterations and considering entire structural data of the design all the iterations are considered. Jumping knowledge architecture which similar model used to achieve discriminative power.

According to [30] graph representations multiplied at from all the iterations of graph isomorphic network layers.

$$h_G = CONCAT\left(READOUT\left(h_v^{(k)} \mid v \in G\right) \mid k = 0,1, \ldots . K\right) \quad (3.32)$$

Replacing above equation of READOUT in defined theorem and corollary in previous sections of GIN it generalizes WL test and the WL subtree kernel.

The conditions mentioned in the theorem are not satisfied by some of GNNs such as GCN and GraphSAGE but still they are powerful equal to WL test and GINs power is equal to WL test in terms of representations.

3.3.4 Signed GNN/HeteroGNNs

Let us elaborate on what are signed GNN. As the name suggest "sign" is nothing but connection between nodes with positive and negative relation [31, 32]. Learning representations in signed network will provide broader analysis of data, we can say two types of associations like opposites and synonyms in simple words. Basically, one of the objectives here is to classify link and to achieve that goal proposed a new message propagation model by improving scalability of signed network. In GIN [32] we can study weighted Multiple aggregators (WMA).WMA helps to solve issue of confusion of information when node feature aggregation happens in GIN. These aggregators are used in application of speech emotion recognition.

3.3.4.1 *Learnable Aggregators*

Another variation of message passing technique in GNN is based on learning aggregators [33]. In this type of message passing technique, general methodology of message passing is same but only difference that aggregators will be learning modules. What do you mean by aggregators? They are neighbors in all the techniques but here in this type instead of neighbor's LSTMs, pooling mechanisms or transformers will be learning aggregators. Here, aggregation function will adapt at the time of training this makes design more expressive. We have few of them in previous sections such as GAT, using attentions. Now we will discuss about LSTM/GRU aggregators and learnable pooling as well as kernel aggregators. We can find description of aggregator architecture [34]. This involves arbitrarily ordered node neighborhood feature set is applied and trained by neural network. Generally, node neighbors are not naturally ordered hence making aggregator functions are variant to permutations of its input neighbors. To ensure this thing three aggregator functions are detailed and examined here.

- **Mean Aggregator**
In this aggregator function, element-wise mean of the vector taken in the form of

$$\{h_{u\vdash}^{(k-1)}, \forall u \in N_{\vdash}(v)\}$$
$$h_{u\dashv}^{(k-1)}, \forall u \in N_{\dashv}(v)$$

- **Pooling Aggregator**
A fully connected neural network is fed with all-nodes neighbors' vector from given graph and later max pooling operation is applied elementwise as shown below:

$$AGGREGATE_k^{\vdash} = max\left(\sigma\left(W_{pool}h_{u\vdash}^{(k)} + b\right), u \in N_{\vdash}(v)\right)$$
$$AGGREGATE_k^{\dashv} = max\left(\sigma\left(W_{pool}h_{u\dashv}^{(k)} + b\right), u \in N_{\dashv}(v)\right)$$

Where, max-elementwise max operator, activation function is denoted by σ which is a nonlinear. In this way to capture various information/relationship across neighbouring nodes application of max-pooling can be done.

- **LSTM Aggregator**

These are all general message passing techniques in graph-based learning. We have given brief introduction as well as mathematical foundations for all the techniques. Long Short-Term Memory (LSTM) or Gated Recurrent Unit (GRU) are commonly used variation of recurrent neural network. If we look at the general framework of architecture of LSTM it has three components memory cell, input gate and forget gate. Retention and deletion of memory unit happens through these gates [35].

LSTMs have two types of connections as it handles sequential data:

I. Previous time step connections (output of previous units as input to next time step)
II. Previous layer connections

Memory cell unit maintains state over time. The LSTM block is the main body of LSTM unit. All the three gates have sigmoid activation functions with [0,1] restrictions. Usually, output activation function and LSTM block input is a tanh activation function. The longer/deeper range events are in sequence is enabled by LSTM core unit. They are used to aggregate variable length sequences whenever required to be in fixed size representations. LSTMs remembers over time dependencies in sequential data. They make them more expressive by assigning appropriate weightage to sequential data based on its context and order in the sequence. As it stores and uses information it can be used to avoid vanishing gradient problem [37]. We can see variation of LSTM and GRU with Bidirectional LSM/GRU with forward and backward pass. Let us see update unit with mathematical equations as below:

$$h^{(t)} = g_o^{(t)} f_h\left(s^{(t)}(t)\right)$$

$$s^{(t-1)} = g_f^{(t)} s^{(t-1)} + g_i^{(t)} f_s\left(Wh^{(t-1)}\right) + uX^{(t)} + b$$

$$g_i^{(t)} = sigmoid\left(w_i h^{(t-1)} + u_i X^{(t)} + b_i\right)$$

$$g_f^{(t)} = sigmoid\left(w_f h^{(t-1)} + u_f X^{(t)} + b_f\right)$$

$$g_o^{(t)} = sigmoid\left(w_o h^{(t-1)} + u_o X^{(t)} + b_o\right)$$

In the above equations we have f_h and f_s are activation functions, f_h is system state and f_h is internal state and uses hyperbolic tangent (tanh) function. g is the gating operation which is a feedforward network having sigmoid activation function. It manages the output values between o to 1 range. These values are interpreted as weights. Description of the subscripts from above equation are: i- input, o -output and f -forget gate [38]. While using LSTM aggregators in GNN messages from neighboring nodes are fed to LSTM network in sequence. Total number of messages are denoted by N, coming from neighbors at one computational step. LSTM run at N time steps, every time step $1 \leq t \leq N$ *with this* input to the LSTM computation of N messages using φ is given by:

$$z_t^u = LSTM\left(\varphi(x_u, x_t), z_{(t-1)}^u\right) \tag{3.32}$$

z_{t0}^u is the initial cell state of LSTM which is fixed zero vector 0. Using output of LSTM final updated embeddings at last step of time are calculated by below equation. They are final aggregated messages of N total messages:

$$x_u' = \varnothing\left(x_u, z_N^{(u)}\right) \tag{3.33}$$

As input to the LSTM is in specific order, $z_N^{(u)}$ will get affected hence it breaks the principle of permutation invariance being not invariant to message receiving order. Just to see the architectural and function difference between LSTM and GRU. Gated recurrent unit is another RNN model addresses short term memory. There is a small difference compared to LSTM is about number of gates. In GRU input and forget gate are combined into one gate adding simplicity to structure. It has 3 components, reset gate, update gate, and current memory state. To capture long term dependencies from given sequence with the help of gates in GRU. The GRU can selectively update use data from previous time step.

Update gate: $z_t = \sigma\left(W_z\left[h_{(t-1)}, x_t\right] + b_z\right)$

Reset gate: $r_t = \sigma\left(W_r\left[h_{(t-1)}, x_t\right] + b_r\right)$

Candidate activation function defined by hyperbolic tangent function: $\tilde{h}_t = tanh\left(W_r\left[h_{(t-1)}, x_t\right]\right)$

The final memory state: $h_t = (1 - z_t)h_{(t-1)} + z_t\tilde{h}_t$

The output gate: $o_t = \sigma_o(W_o h_t + b_o)$

b_o, W_o are bias vector and weight matrix of output layer. GRU operates on fewer tensor operations and faster training is allowed. Both LSTM and GRU have their pros and cons depending on nature of task. Now as we are aware of basic functionality and structure of LSTM, let us see the LSTM aggregators [19, 40, 41] with mathematical foundations in message passing method.

3.3.4.2 Integration of Message Passing with Global Graph Features

Global graph features are includes applying sum, mean, max readout functions over node attributes. These global features are defined by considering whole graph not from node neighborhood only. They can be graph-level data like diameter of graph or degree distribution globally. It can be a transformer mechanism from global attention mechanism and presenting global context using virtual nodes. Integration of message passing with global graph features is a difficult task while designing GNN. To handle this message integration various techniques are defined based on graph level task such as pooling-global or hierarchical for classifying graph and learning larger graphs and multi scale features for hidden/complex representation learning, global attention mechanism for representing larger dependencies in a graph. Let us see which models work on these mechanisms for more clarity.

- Global Pooling-GCN and Usage of Readout functions [42]
- Hierarchical Pooling-DiffPool, SAGPool and TopKPool
- Global Attention Mechanism-GIN Virtual and Graphomer [43]
- Multi-stage Features-JK-Net and GIN-JK [44]

The dynamic or temporal graphs are having evolving structures over time. Their nodes, edges as well as attributes changes timely. We have social networks where new members can be added or few members (nodes) can be deleted, in financial networks we have transaction related graphs where data changes timely manner as an example of dynamic graph. Because of this evolving nature of dynamic graph, adapting message passing with them is challenging. For maintaining performance and catching new temporal dependency in timely manner, dynamic graph needs modifications in traditional message passing techniques. These modifications should address scalability, memory, changing patterns and time dependency in dynamic or temporal graphs [45–47]. Recently, authors have proposed methods to apply message passing variations such as Snapshot-Based Method: GC-LSTM and Temporal Message Passing: TeMP. Specifically temporal GNN architecture have TGN (Temporal Graph Networks), TGAT (Temporal Graph Attention), DyRep and Evolve GCN variations. Let us put together in summary Table 3.3 comparing message-passing strategies across different GNN architectures.

Table 3.3 Summary of comparison of message-passing strategies across different GNN architectures.

GNN model	Message passing strategy	Application	Graph-based task accomplished	Type of graph used
GCN	Normalization used for aggregating neighbor features	Citation networks, social networks Recommendation system [71]	Node classification	Static, Homogeneous, Node-featured
GraphSAGE	Aggregation is learnable on sampled neighbors	Web graphs, social networks	Works on larger graphs, Node embedding Learning, Link Prediction	Static/Dynamic, Homogeneous, Node Attributed
GAT	Attention based aggregation for weighing neighbors	Citation, molecular, traffic networks	Node classification, link prediction, Knowledge Graph Reasoning	Static, Homogeneous, node-attributed
GGNN	Step updating happen using GRU	Program analysis, source code graphs	Global Graph context	Static, Directed, Edge-featured
MPNN	Update happens with user defined message	Molecule graphs, chemicals	Edge based tasks (link prediction), graph classification	Static, Homogeneous, Edge-featured
GIN	Summation function with MLP applied in aggregation	Molecular property prediction	graph isomorphism, Graph Classification	Static, Homogeneous, Node-attributed
Transformer based GNN	Considers entire graph, like global graph attention	Molecular graphs, traffic prediction	Edge based tasks (link prediction), node classification	Global, Large-scale, Optional edge/node features

3.4 Core Challenge of Optimal Depth of Message Passing in GNN

There is a stack of graph architecture with neighbor aggregation that shows receptive field size of neighbors. Each GNN layer aggregates information from neighboring nodes and creates representations that becomes its depth of graph network. Hence depth of message passing is equal to this network depth [48, 49]. As the network goes deeper the aggregated features will cover the graph and the end the nodes lose their discriminability due to similar convergence of information/features at nodes. Enumerations can be considered for optimal depth to avoid this discriminability. Learning optimal depth of network is important factor for improving performance of GNN to save time and cost. For each of these steps we have mathematical denotation and description shown below:

3.5 Recent Research Achievement in Message Passing Techniques

[50] The dependency of GNNs message passing technique on graph structure makes them complex when it comes to large scale graphs. To tackle this high computational complexity in graph neural networks many hardware-specific optimization techniques proposed and still trying improve the speed of GNNs. To handle this issue, we can see a new "dedicated algorithm-architecture co-design framework" proposed method named MePa. This accelerates GNN speed with algorithm coordination and innovations at hardware-level. This method also applied on any graph-based task at node level, edge level or channel related task. Using message passing technique by dynamically cutting task-irrelevant graph data. This pruning approach significantly decreases the overall complexity of GNN with negligible accuracy loss. This new design is structured to assist dynamic pruning and add performance improvements in GNNs speed. When it comes to show the performance improvements in terms of measure and number this new method gains average 1.95× speedups and 2.6× energy efficiency when compared to the state-of-the-art GNN accelerator AWB-GCN.

[51] This study reveals understanding of message passing in terms of power iterations. As we have mentioned basic framework of message passing, by making use of same functional dependencies some new approach is proposed. This includes, removing activation functions are partly along with reduction of GNN layer weights subspace power iteration clustering

(SPIC) is modeled. This design iteratively learns only one aggregator which process random featured network.

[52] The couple of limitations have been addressed handled here by proposing Hierarchical Message-passing Graph Neural Networks framework. One is, to encode deep tree structured data flow from deeper inside nodes in graph becomes costlier and other one is, GNN does information aggregation among observed edges in the given original graph but fails to encode parameter features in high order neighboring nodes. This work generates inter and intra level propagation through a hierarchical structure where nodes of a flat graph are reorganized into multi-level super graphs. The long-distance nodes are connected by this generated hierarchy creating shortest paths between far a node away. This makes important far away node interactions easily accessible by core technique of message passing. This makes learned node representations filled with meso- and macro-level semantics of information. This framework uses hierarchical community detection algorithm, hence named as Hierarchical Community aware Graph Neural Network (HC-GNN).

[53] One of the successful application areas of GNN is in molecules/molecular biology and advancement of message passing mechanism in heterogeneous graph. To exploit interactions at multi-molecular level and to address other issues such as larger in-depth graph data, over smoothing issue and to improve predictive performance framework called as GONNMDA. Authors have proposed an ordered gating message-passing mechanism. The mechanism works on heterogeneous graph built by applying root-tree hierarchical alignment. This message passing mechanism addresses challenge of over-smoothing problem as well as heterogeneity of graphs.

[54] We can have different scenario of message passing mechanism in this study. The concept of synchronous message passing neural network (MPNN) have introduced to address the issues of over-smoothing, over squashing and heterogeneity of larger graphs. Here authors have proposed Asyn MPNN, an asynchronous MPNN framework which can be applied on various GNN variants as we have studies in previous sections. Authors have proposed an automated asynchronous (aAsyn) layer to overcome local priority to retain the original information. They have mentioned shortest path based multi-hop message passing [55, 56].

[57] We can study more in-depth theoretical concepts about message passing mechanism in areas of chemistry, drug discovery, and materials science [36, 39]. Over-squashing and expressivity, these two limitations have known solutions according to their problem and solution approach

such as residual connections, positional encoding, skip connections, graph rewiring and attention mechanisms [58, 59].

3.6 Challenges Associated with Graph Structured Data [8, 60]

Graph structures data gives very insightful relationship among nodes. But to exploit this complex relationship between features of nodes and connecting edges in larger graphs is a challenge and a tedious job. Some of the challenges are have briefed here so that one can investigate and starts learning them deeply.

i. Irregular graph structures.
ii. Node ordering is not canonical
iii. Dynamic nature of real-world applications which gives dynamic graph topology
iv. Scalability
v. Over smoothing
vi. Managing edge features
vii. Data sparsity and noise issue
viii. Interpretability-Working with graph structured data in graph neural network is like black box as they miss explanations sometimes.
ix. Heterogeneous data-Graphs are complex and flexible in real life applications. To define their heterogeneity or real time flexible data through proper representation is crucial. Learning multiple graphs, heterogeneous graphs and dynamic graph will improve this issue.

To represent real world data in graph structures is a challenge as the topologies of graphs are flexible and complex. As mentioned earlier in the chapter we have different types of graphs and many studies we have seen those are based on dynamic and heterogeneous graphs to handle various type of data including temporal. Another noticing challenge is while using graph convolution layers. Many layers can affect performance GNN based on convolution. There are chances of performance decrease at node aggregation due to convergence of all node's representations. Scalability is another challenge because of huge graph data e.g. Twitter social network data. It has millions of nodes and edges. Also, if we think about distributing and parallelizing larger dataset what is happening with deep learning strategies but it is not that easy to perform with graph structures [61, 62]. The research shows solution to such scenarios but still are appropriate and becomes non-trivial in distributed environment graph storage. When it comes to explainability again the similar issue may arrive. As there are less graph structure related algorithms for explainability are available but the work has been consistently going on in this area [63]. Message compression or optimization techniques have been recently proposed in [64, 65]. Another challenge we would like to mention here is about high-order structures such as graphlets, simplicial complexes, and motifs. For understanding more complex networks and their characteristics such as interaction bond between protein structures or in any other biological applications they are important [68–70]. When it comes to GNN, how to apply these high-order structures in message passing, will it give more expressive power by considering them in such transmission scheme? [66] But this will make GNN more efficient by considering high order structures with more expressive capability.

3.7 Conclusion

In today's AI world graph neural networks have gain huge success due their ability to work on graph structured data. Graph structures data exploits possible and hidden relations among data points for complex systems. They are widely used in decision making and data analysis using various machine learning algorithms. In this chapter we have provided introduction to graph-based learning, types of graphs and their use in computations in computer science. Later in the chapter we moved towards brief introduction to graph neural network and its core concept message passing technique. We have discussed variants of message passing techniques, their applications and how they differ from each other.

AI disclosure statement: The author confirms that no artificial intelligence (AI)-generated content or AI-assisted writing tools were used in the preparation of this chapter (including summaries and the abstract). All text, analysis, and conclusions are the result of the author's original work.

References

1. Ma, Y. and Tang, J., *Deep learning on graphs*, Cambridge University Press, 2021.
2. Hamilton, W.L., Representation Learning on Graphs: Methods and Applications. *IEEE*, 2017.
3. Zhou, J., Cui, G., Hu, S., Zhang, Z., Yang, C., Liu, Z., Wang, L., Li, C., Sun, M., Graph neural networks: A review of methods and applications. *AI Open*, *1*, 57–81, 2020.
4. Gilmer, J., Schoenholz, S.S., Riley, P.F., Vinyals, O., Dahl, G.E., Neural message passing for quantum chemistry. *34th International Conference on Machine Learning, ICML*, pp. 2053–2070, 2017.
5. Grattarola, D., A practical introduction to GNNs - Part 2, 12 Mar 2021, https://danielegrattarola.github.io/posts/2021-03-12/gnn-lecture-part-2.html.
6. Hamilton, W.L., *Graph Representation Learning*, Morgan & Claypool publishers, 2020.
7. Faik, L., Graph Neural Networks: Merging Deep Learning With Graphs (Part I), 2022, https://blog.dataiku.com/graph-neural-networks-merging-deep-learning-with-graphs.
8. Khemani, B., Patil, S., Kotecha, K., Tanwar, S., A review of graph neural networks: concepts, architectures, techniques, challenges, datasets, applications, and future directions. *J. Big Data*, 11, 1, 2024.
9. Lampert, M. and Scholtes, I., The Self-Loop Paradox: Investigating the Impact of Self-Loops on Graph Neural Networks, in: *Second Learning on Graphs Conference (LoG 2023)*, 2023.
10. Kipf, T.N. and Welling, M., Semi-Supervised Classification with Graph Convolutional Networks, arXiv, 2017.
11. Zhang, S., Tong, H., Xu, J., Maciejewski, R., Graph convolutional networks: a comprehensive review. *Comput. Social Networks*, 6, 1, 2019.
12. Bhatti, U.A., Tang, H., Wu, G., Marjan, S., Hussain, A., Deep Learning with Graph Convolutional Networks: An Overview and Latest Applications in Computational Intelligence. *Int. J. Intell. Syst.*, 2023.
13. Duvenaud, D., *et al.*, Convolutional Networks on Graphs for Learning Molecular Fingerprints. *Advances in Neural Information Processing Systems 28*, Curran Associates, Inc., 2015.
14. Nayak, N.V., Graph neural networks-notes, Update, 2020.
15. Bernstein, M.N., The graph Laplacian, 2020, https://mbernste.github.io/.
16. Lee, J.B., Rossi, R.A., Kim, S., Ahmed, N.K., Koh, E., Attention Models in Graphs: A Survey. *Assoc. Comput. Mach.*, 13, 6, 2019.
17. Veličković, P., Cucurull, G., Casanova, A., Romero, A., Liò, P., Bengio, Y., Graph Attention Networks, in: *ICLR 2018*, arXiv, 2018.

18. Vrahatis, A. and Lazaros, K.S., Graph Attention Networks: A Comprehensive Review of of methods and applications. *Future Internet*, MDPI, 16, 318, 16, 2024.
19. Hamilton, W., Zhitao, Y., Leskovec, J., Inductive Representation Learning on Large Graphs, in: *Advances in Neural Information Processing Systems*, vol. 30, Curran Associates, Inc., 2017.
20. Li, Y., Tarlow, D., Brockschmidt, M., Zemel, R., Gated Graph Sequence Neural Networks, in: *ICLR2016*, arXiv, 2016.
21. Ruiz, L., Gama, F., Ribeiro, A., Gated Graph Recurrent Neural Networks. *IEEE Trans. Signal Process.*, 68, 6303–6318, 2020.
22. Cho, K., *et al.*, Learning Phrase Representations using RNN Encoder-Decoder for Statistical Machine Translation, in: *EMNLP2014*, arXiv, 2014.
23. Shasrma, K., Understanding Gated Recurrent Unit (GRU) in Deep Learning, 2024, https://www.pickl.ai/blog/gated-recurrent-unit-in-deep-learning.
24. Gasteiger, J., Groß, J., Günnemann, S., Directional Message Passing for Molecular Graphs, in: *ICLR2020*, arxiv, 2020.
25. Yedidia, J.S., Freeman, W.T., Weiss, Y., Understanding belief propagation and its generalizations, Exploring Artificial Intelligence in the New Millennium, Morgan Kaufmann Publishers Inc., pp. 239–269, 2003.
26. Simonovsky, M. and Komodakis, N., Dynamic Edge-Conditioned Filters in Convolutional Neural Networks on Graphs, in: *IEEE Conference on Computer Vision and Pattern Recognition (CVPR)*, Honolulu, HI, USA, pp. 29–38, 2017.
27. Xu, K., Hu, W., Leskovec, J., Jegelka, S., How Powerful are Graph Neural Networks?, arXiv, 2019.
28. Hornik, K., Stinchcombe, M., White, H., Multilayer feedforward networks are universal approximators. *Neural Networks*, 2, 5, 359–366, 1989.
29. Hornik, K., Approximation capabilities of multilayer feedforward networks. *Neural Networks*, 4, 2, 251–257, 1991.
30. Xu, K., Li, C., Tian, Y., Sonobe, T., Kawarabayashi, K., Jegelka, S., Representation Learning on Graphs with Jumping Knowledge Networks, in: *Proceedings of the 35th International Conference on Machine Learning*, vol. *80*, PMLR, pp. 5453-5462, 2018.
31. Jin, J., Bi, J., Qu, C., SGNN: A New Method for Learning Representations on Signed Networks, in: *Artificial Neural Networks and Machine Learning – ICANN 2023*, Springer Nature Switzerland, pp. 447–458, 2023.
32. Hu, Y., Tang, Y., Huang, H., He, L., A Graph Isomorphism Network with Weighted Multiple Aggregators for Speech Emotion Recognition, in: *Interspeech 2022*, ISCA, pp. 4705–4709, 2022.
33. You, J., Ying, R., Ren, X., Hamilton, W.L., Leskovec, J., GraphRNN: Generating Realistic Graphs with Deep Auto-regressive Models, in: *ICML 2018*, arXiv, 2018.
34. Xu, K., Wu, L., Wang, Z., Feng, Y., Witbrock, M., Sheinin, V., Graph2Seq: Graph to Sequence Learning with Attention-based Neural Networks, arxiv, 2018.

35. Patterson, J. and Gibson, A., *Deep Learning, A Practitioners Approach*, O'Reilly Media, Inc., 2017.
36. Greff, K., Srivastava, R.K., Koutník, J., Steunebrink, B.R., Schmidhuber, J., LSTM: A Search Space Odyssey. *IEEE Trans. Neural Networks Learn. Syst.*, 28, 10, 2222–2232, 2017.
37. Shiri, F.M., Perumal, T., Mustapha, N., Mohamed, R., A Comprehensive Overview and Comparative Analysis on Deep Learning Models. *J. Artif. Intell.*, 6, 2024.
38. Jürgen Schmidhuber, S.H., Long short-term memory. *Neural Comput.*, 9, 8, 1735–1780, 1997.
39. Gumaei, A., *et al.*, A Hybrid Deep Learning Model for Human Activity Recognition Using Multimodal Body Sensing Data. *IEEE Access, 7*, 99152–99160, 2019.
40. Smagulova, K. and James, A.P., Overview of Long Short-Term Memory Neural Networks. In *Deep Learning Classifiers with Memristive Networks. Modeling and Optimization in Science and Technologies*, A. James (ed.), vol. 14, Springer, Cham, 2020.
41. Ward, I.R., Joyner, J., Lickfold, C., Guo, Y., Bennamoun, M., A Practical Tutorial on Graph Neural Networks. *Assoc. Comput. Mach.*, 54, 10, 2022.
42. Tang, J., Song, S., Du, Q., Yao, Y., Qu, J., Graph convolutional networks with the self-attention mechanism for adaptive influence maximization in social networks. *Complex Intell. Syst.*, 10, 6, 8383–8401, 2024.
43. Fu, D., *et al.*, VCR-Graphormer: A Mini-batch Graph Transformer *via* Virtual Connections, arxiv, 2024.
44. Tao, Y., Li, Y., Wu, Z., Revisiting Attention-Based Graph Neural Networks for Graph Classification, in: *Lecture Notes in Computer Science*, vol. *13398*, Springer, Cham, 2022.
45. Zheng, Y., Yi, L., Wei, Z., A survey of dynamic graph neural networks. *Front. Comput. Sci.*, *19*, 2025.
46. Chen, J., Wang, X., Xu, X., GC-LSTM: Graph convolution embedded LSTM for dynamic network link prediction. *Appl. Intell.*, Springer, 2022.
47. Cai, *et al.*, Temporal knowledge graph completion: A survey, arxiv, 2022.
48. Rong, Y., Huang, W., Xu, T., Huang, J., Dropedge: Towards deep graph convolutional networks on node classification, arxiv, 2019.
49. Li, Q., Han, Z., Wu, X.-M., Deeper insights into graph convolutional networks for semi-supervised learning, in: *Proceedings of the AAAI conference on artificial intelligence*, 2018.
50. Zou, *et al.*, Efficient Message Passing Algorithm and Architecture Co-Design for Graph Neural Networks. *IEEE Trans. Emerging Top. Comput. Intell.*, 2024.
51. Li, X. and Cheng, Y., Understanding the message passing in graph neural networks *via* power iteration clustering. *Neural Networks*, 140, 131–135, 2021.
52. Zhong, Z., Li, C.-T., Pang, J., Hierarchical message-passing graph neural networks. *Data Min. Knowl. Discovery*, 37, 1, 381–408, 2023.

53. Zeng, S., Zhang, S., Wang, Z., Yang, C., Yuan, S., GONNMDA: A Ordered Message Passing GNN Approach for miRNA–Disease Association Prediction. *Genes*, MDPI, 16, 4, 425, 2025.
54. Chen, J., LiaoChen, T.C., Zheng, Z., Improving Message-Passing GNNs by Asynchronous Aggregation, in: *Proceedings of the 33rd ACM International Conference on Information and Knowledge Management*, Association for Computing Machinery, Boise, ID, USA, pp. 228–238, 2024.
55. Michel, G., Nikolentzos, G., Lutzeyer, J.F., Vazirgiannis, M., Path neural networks: Expressive and accurate graph neural networks, in: *International Conference on Machine Learning*, PMLR, pp. 24737–24755, 2023.
56. Abboud, R., Dimitrov, R., Ceylan, I.I., Shortest path networks for graph property prediction, in: *Learning on Graphs Conference*, PMLR, 2022.
57. Gilmer, J., Schoenholz, S.S., Riley, P.F., Vinyals, O., Dahl, G.E., Message passing neural networks, in: *Machine learning meets quantum physics*, pp. 199–214, Springer, 2020.
58. Ying, *et al.*, Do transformers really perform badly for graph representation? *Advances in neural information processing systems*, vol. 34, pp. 28877–28888, 2021.
59. Topping, J., Di Giovanni, F., Chamberlain, B.P., Dong, X., Bronstein, M.M., Understanding over-squashing and bottlenecks on graphs *via* curvature, *arXiv preprint arXiv:2111.14522,* 2021.
60. Waikhom, L. and Patgiri, R., Graph Neural Networks: Methods, Applications, and Opportunities, arxiv, 2021.
61. Zheng, *et al.*, DistDGL: Distributed graph neural network training for billion-scale graphs. *2020 IEEE/ACM 10th Workshop on Irregular Applications: Architectures and Algorithms (IA3)*, pp. 36–44, 2020.
62. Serafini, M., Scalable Graph Neural Network Training: The Case for Sampling. *ACM SIGOPS Oper. Syst. Rev., 55*, 1, 68–76, 2021.
63. Yuan, H., Yu, H., Gui, S., Ji, S., Explainability in graph neural networks: A taxonomic survey. *IEEE Trans. Pattern Anal. Mach. Intell.*, 45, 5, 5782–5799, 2022.
64. Ma, *et al.*, Eliminating Data Processing Bottlenecks in GNN Training over Large Graphs *via* Two-level Feature Compression. *Proc. VLDB Endow.*, 17, 11, 2854–2866, 2024.
65. Kurt, S.E., Yan, J., Sukumaran-Rajam, A., Pandey, P., Sadayappan, P., Communication optimization for distributed execution of graph neural networks, in: *2023 IEEE International Parallel and Distributed Processing Symposium (IPDPS)*, pp. 512–523, 2023.
66. Chen, J., Yang, L., Wang, Z., Gong, M., Higher-order GNN with Local Inflation for entity alignment. *Knowledge-Based Syst.*, 293, 111634, 2024.
67. Hamilton, W., Ying, Z., Leskovec, J., Inductive representation learning on large graphs. *Advances in neural information processing systems*, vol. 30, 2017.
68. Zaheer, *et al.*, Deep sets. *Advances in neural information processing systems*, vol. 30, 2017.

69. Li, P. and Leskovec, J., The expressive power of graph neural networks, in: *Graph Neural Networks: Foundations, Frontiers, and Applications*, pp. 63–98, Springer, 2022.
70. Gilmer, J., Schoenholz, S.S., Riley, P.F., Vinyals, O., Dahl, G.E., Neural message passing for quantum chemistry, in: *International conference on machine learning*, pp. 1263–1272, 2017.
71. Kulkarni, M.M. and Hiremath, S., A Comprehensive Survey on Deep Learning-based Recommender Systems for Market Analysis. *Int. J. Comput. Digit. Syst.*, 16.1, 1617–1632, 2024.

4

Architectures Defining Graph Neural Networks: Capitalizing on the Potential of GNN's for Real-World Solutions

A. Priyadharshini[1]* and Saju Mathew[2]

[1]Department of BCA, Shri Nehru Maha Vidyalaya College of Arts and Science, Tamil Nadu, Coimbatore, India
[2]Department of Computer Science with Data Analytics, Shri Nehru Maha Vidyalaya College of Arts and Science, India

Abstract
A fundamental construction for complex interrelationships in domains such as natural science, finance, social networking, and recommendation systems is a simple data structure called graph which is real world systems are expressed through. The management of non-Euclidean data structures, which are graph-based, pose exceptional challenges because of their irregularity compared to traditional machine learning approaches which rely on grid structured data. By means of deep learning techniques, GNNs are able to manage optimally graph-structured data efficiently. There has been a marked advancement in GNN design in recent years with regards to scaling issues and over-smoothing in addition to enhancing their capability to learn efficient representations. Learning on graphs was made possible by early GNN models like Spectral Based Graph Convolutional Networks (GCNs) and spatial based Graph SAGE and Graph Attention Networks (GATs). In the detection of fraudulent data and peer referral, Graph Neural Networks improve the effectiveness of social network analysis. During drug development, they help forecast molecular traits and probe protein interactions for the analysis. In the creation of tailored recommendations, GNNs are employed in recommendation systems, while in cyber defense, they scrutinize networks for intrusion anomalies and fraudulent activities. Furthermore, GNNs offer more complex graph-based learning, affecting disciplines such as natural language processing, traffic prediction, and financial modeling. Considering their achievements, GNNs are still facing a

**Corresponding author*: phdpriyadharshini@gmail.com

J. Ramkumar, Sridaran Rajagopal, B. Suchitra and S. Balamurugan (eds.) Graph Neural Networks: Concepts and Applications, (143–180) © 2026 Scrivener Publishing LLC

number of obstacles, such as the need for improved interpretability, computational inefficiencies, and trouble managing dynamic graphs. Resolving these problems is essential to increasing their usefulness and enhancing their functionality in a variety of applications. The development of GNN designs, their main practical uses, and the persisting difficulties are all thoroughly covered in this study. This paper explores the evolution of GNN architectures, highlighting key innovations, their limitations, applications and future research directions.

Keywords: Graph neural network, graph attention network, non-Euclidean character, machine-learning methods, convolutional network, analyze, graph convolutional networks, spatial-based GraphSAGE

4.1 Introduction

GNNs' initial motivation originates from the lengthy history of neural networks for graphs. The use of Recursive Neural Networks on directed acyclic networks dates back to the 1990s. Then, in order to address cycles, Recurrent Neural Networks and Feedforward Neural Networks are, respectively, introduced. Despite their success, these approaches' common concept of creating state transition systems on graphs and iterating until convergence limited their capacity for representation and extend ability. GNNs have been re-discoverable as a result of recent developments in deep neural networks, particularly convolutional neural networks (CNNs). The ability of CNNs to extract multi-scale localized spatial features and combine them to create highly expressive representations sparked the new era of deep learning and led to advancements in nearly every area of machine learning. CNNs are characterized by their use of several layers, shared weights, and local connections these are also crucial for resolving graph-related issues. Though these data structures can be thought of as examples of graphs, CNNs can only work with standard Euclidean data, such as texts (1D sequences) and images (2D grids). Consequently, generalizing CNNs on graphs is simple.

Graphs are one of the many common use cases for deep learning, which has grown significantly in recent years. A common data format for many machine learning and deep learning tasks, graph data offers relational information between items. Effective use of graph data requires models that are able to learn from such inputs. In order to construct graph topologies, this study finds nodes and edges within particular applications, including text, entities, and relations. Specific purposes may require different GNN models to be implemented. With GNNs, the nodes in a graph can interface and understand the dependencies of the nodes and edges.

This work examines specific models of GNNs, including graph convolution networks (GCNs), graph attention networks (GATs), and Graph SAGE, all of which are widely used in multiple domains today. In the field of machine learning and artificial intelligence, graph neural networks, or GNNs, have become a revolutionary phenomenon. The rapid advancement of GNNs is attributed to the abundant availability of networked data across numerous domains such as social networking, biology, recommendation systems, and even cybersecurity. These networks aid in addressing real-world problems which conventional machine learning models struggle to deal with due to their exceptional modeling and understanding skills of intricate interrelations.

Graphs are literally everywhere; real-world objects are often depicted in relation to other objects. A collection of entities can be naturally represented as a graph, along with the connections between those entities. For more than ten years, researchers have created neural networks known as graph neural networks, or GNNs, that function on graph data. Their expressive strength and skills have grown in recent years. Graphs are a fundamental data structure used to model complex relationships in Practical applications are beginning to appear in fields including traffic prediction, fake news detection, physics simulations, social media, transportation system networks, and biological protein–protein networks, and also antibiotic discovery are commonly represented using graphs and recommendation systems. Traditional machine learning models struggle with graph-based data due to their irregular structure. GNNs address this challenge by leveraging deep learning techniques to process and analyze graph-structured data effectively. The majority of graph data are high-dimensional and complicated, making it challenging for researchers to glean useful information. Thus, it is crucial to process graph data and convert it into a form (fixed-dimensional vectors) so that researchers can use it for various downstream activities. In order to support a variety of downstream tasks, including node classification, link prediction, community discovery, etc., graph representation learning aims to acquire vector representations of graph elements (e.g., nodes, edges, subgraphs, etc.). Consequently, graph representation learning is crucial since it has the potential to greatly improve downstream task performance.

4.1.1 Graph SAGE

GraphSAGE learns to "aggregate" data from a node's local neighborhood to generate its visualization, instead of establishing a straight reference table for every node. Therefore, it is extremely adaptable and capable of dealing with dynamic networks and unknown nodes, which is critical for applications in reality such as recommendation systems (e.g., Pinterest's PinSAGE), fraud detection, and social network analysis. GraphSAGE (Graph Sample and aggreGatE) is one of the innovations that has enabled the learning of node embedding's in large graphs with abundant node feature information. It has graphed to address a gap in the previous graph embedding approaches which were usually transductive. Transductive uses Graph DeepWals, Node2Vec which is the traditional graph embedding method. The difficulty with these methods are often transductive in the sense that they only learn embedding's for a certain set of nodes within a graph in the training phase. So, if new nodes are added to the graph or a new graph is encountered complete retraining of the model is required which is inefficient and computationally expensive for static or non-evolving graphs. To overcome the problem of transductive Graph SAGE uses inductive approach. In this approach instead of learning a distinctive embedding for each node, it opts to learn a function that can be used to produce embeddings for any node including those that are unseen. Figure 4.1 below shows Graph SAGE architecture.

4.1.2 Graph Attention Network (GAT)

The GAT or Graph Attention Network is a sophisticated form of a neural network that learns from graph-structured data. It focuses on the

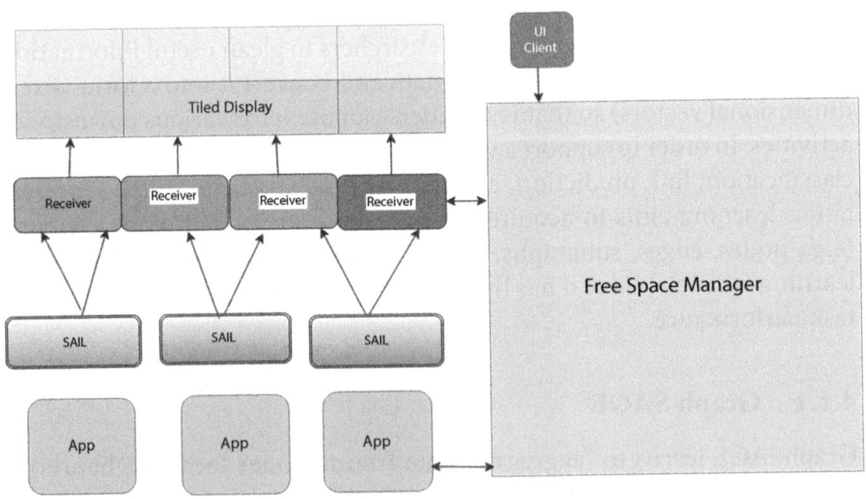

Figure 4.1 Graph SAGE architecture.

application of an attention mechanism that helps determine the weight of neighboring nodes with respect to a node for computing its new representation. Comparison with earlier GNNs like GCNs clearly show differences since GNNs like GCNs usually apply fixed and uniform weights to neighbors. Following are some of the key advantages of GAT Architecture:

a) Automatically allocates various significance: Unlike GCNs, which assign static weights, GATs may learn to allocate varied significance to various neighbors, making them more adaptable and descriptive.
b) Inductive Capability: GATs are inductive just like GraphSAGE. Since they acquire an attention function which might be assigned to any pair of nodes, they can generalize to previously unseen nodes or even completely new networks.
c) Handles Varying Neighborhood Sizes: The Softmax normalization naturally manages a variety of neighbor counts.
d) Computationally Efficient: The attention coefficients may be determined concurrently across every edges, and the aggregation can be parallelized across all nodes. It may also be done with sparse matrix operations, which reduces memory requirements.

In the below Figure 4.2, **w** represents shared weight, $\mathbf{a}_{ij}$ represents attention head (attention mechanism) which is one of the innovative idea in which GAT tries to calculate how attention needs to be given to itself and to its neighboring nodes, $\mathbf{e}_{ij}$ represents the raw attention scores which are normalized by using the Softmax functions.

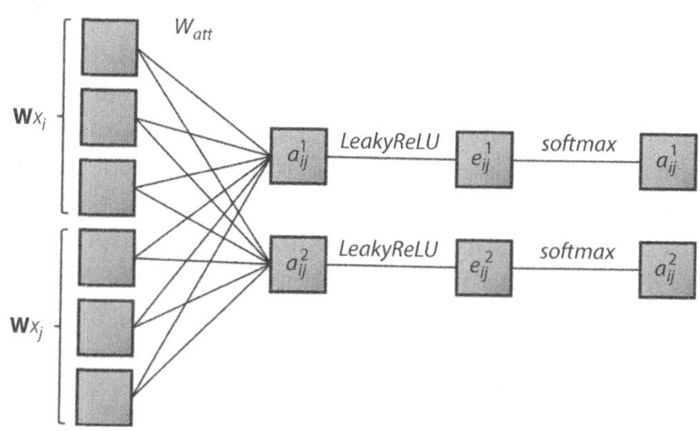

Figure 4.2 Graph attention network architecture.

GIN (Graph Isomorphism Network)

Graph Isomorphism Network (GIN) is a sophisticated architecture of Graph Neural Networks (GNN) that aims to be highly expressive with the goal of capturing the discriminative power of the "Weisfeiler-Lehman (WL) test." WL test is a known algorithm to determine whether two graphs are isomorphic or not (i.e., if they are structurally identical). It is well-known for its effectiveness in differentiating between diverse graph structures. GIN's overarching purpose is to create an aggregation function that approaches the potential of the WL test. More primitive GNNs, such as GCNs or GraphSAGE that used mean or max aggregators, were proven to be weaker than the WL test in differentiating certain graph structures. GIN overcomes this limitation with a specific form of aggregation and update. As compared to the GCN and GAT, GIN is different in the following ways;

a) Discriminative Power: GIN aimed to optimize discriminative power and showed resemblance to the WL test in some conditions. Non-isomorphic graphs that GCNs and GATs may struggle to classify are sometimes separable using the WL test (and therefore GIN).

b) Aggregate Function:
Different aggregation strategies are used by different Graph Neural Network (GNN) architectures, and each has advantages and disadvantages with regard to discriminative power and information preservation. For example, Graph Convolutional Networks (GCNs) usually aggregate neighbor information using a normalized summation or averaging. Despite being straightforward, this method may result in "oversmoothing," where node representations become excessively similar between layers, losing valuable, unique feature information within a neighborhood. This is addressed by Graph Attention Networks (GATs), which use an attention mechanism that learns weighted sums of neighbor features. This enables GATs to give different neighbors different levels of importance, but in some situations, the attention mechanism's non-uniform injectivity may limit its discriminative power in comparison to the Weisfeiler-Leman (WL) test. By the contrary, Graph Isomorphism Networks (GINs) employ a transparent sum aggregation with a learnable parameter and an MLP (Multi-Layer Perceptron). In principle, this sum aggregation is more successful at distinguishing among various feature sets, closely

resembling the core purpose of the WL test and providing more robust discriminative capabilities.

c) Update Function: GIN's update function is based on a robust Multi-Layer Perceptron (MLP), as opposed to some other GNNs' use of single-layer neural networks or straightforward linear transformations. In order to approximate the injective (one-to-one) functions necessary for the equivalence of the WL test, the MLP's non-linearity is essential.

d) Isomorphism is the Key: GIN was developed in order to create GNNs that are as strong as the Weisfeiler-Lehman test for graph isomorphism. This makes them especially effective for graph classification tasks where it's critical to discern between subtle differences in graph structures.

Figure 4.3 shows the GIN Architecture. Table 4.1 shows the summarized table for the different architecture. Table 4.2 shows the difference between the GCN and GAT.

4.2 Basic Graph Theory and Its Representation

A graph is defined as a collection of nodes and edges. A graph G = (V, E) is made up of vertices (also known as nodes) and edges (E). If an edge

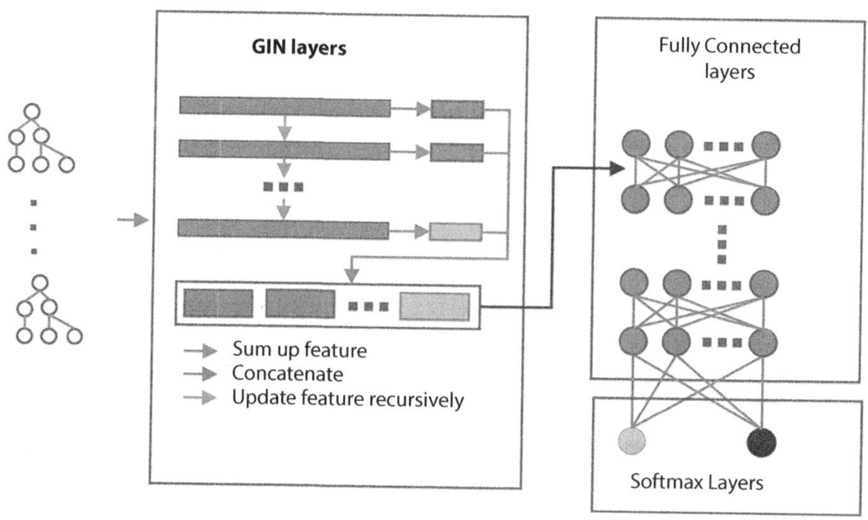

Figure 4.3 GIN architecture.

Table 4.1 Summary of different architectures.

Model	Supports edge types	Handles large graphs	Learns node importance	Good for dynamic graphs
GCN	No	Moderate	No	No
GAT	No (basic)	Less efficient	Yes	Yes
GraphSAGE	No	Yes (scalable)	Partial	Yes
GIN	No	No	No	No

Table 4.2 Difference between GCN and GAT.

Aspect	GCN (Graph Convolutional Network)	GAT (Graph Attention Network)
Aggregation Method	Uses weighted average based on normalized adjacency matrix	Uses attention mechanism to assign dynamic weights to neighbors
Neighbor Weighting	Uniform (based on degree)	Adaptive (learned importance)
Computational Cost	Lower	Higher (due to attention calculations)
Scalability	More scalable to large graphs	Less scalable without optimization
Suitability for Node Features	Works well with uniform or homogenous node importance	Better for heterogeneous or dynamic node importance

connects to a vertex, we call the vertex the edge's terminus. An edge with only one terminus is known as a loop edge. An edge connects two vertices, which are referred to as neighboring vertices. A graph can be weighted, directed, or undirected. A directed graph is one in which each edge may only be visited in one direction. Edges in a simple directed graph may be specified by an ordered pair (v_i, v_j) of the two vertices that the edge connects. We say that v_i is adjacent to v_j and v_j is adjacent from v_i. In a directed

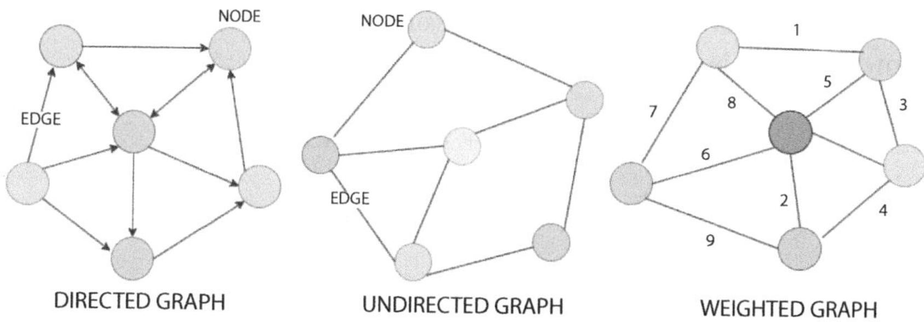

Figure 4.4 Directed, undirected and weighted graph.

graph, the in-degree of a vertex is the number of edges incident to the vertex and the out-degree of a vertex is the number of edges incident from the vertex. A weighted graph is a graph G = (V, E) along with a function w: E → R that associates a numerical weight to each edge. If G is a weighted graph, then T is a minimal spanning tree of G if it is a spanning tree and no other spanning tree of G has smaller total weight. Above Figure 4.4 illustrates directed, undirected, and weighted graph.

4.3 Neural Network

Neural networks are computerized learning programs or models that employ approaches similar to that used when real neurons collaborate to recognize events, assess options, and draw inferences. This enables them to make decisions in a manner similar to those of the brains of humans. Every neural network consists of an input layer, one or more hidden layers, an output layer, and layers of nodes or artificial neurons. Each individual node possesses its own unique weight and threshold, and it communicates with additional nodes. If any node's output surpasses the predetermined threshold value, it becomes operational and delivers data to the subsequent tier of the network. If not, no data is sent to the network's next layer. Neural networks require data for training to improve efficiency over time. They utilize successful computer science and artificial intelligence approaches for fast classifying and clustering data once it has been corrected for precision. Comparing to traditional identification by human specialists, picture or voice recognition jobs might take hours rather than minutes. Google's search algorithm is one of the best-known use of a neural network. Consider every single node as a separate linear regression model, consisting of input data, weights, a bias (or threshold), and an output. This is how the formula would appear:

$$\Sigma wixi + bias = w1x1 + w2x2 + w3x3 + bias$$

$$Output = f(x) = 1 \text{ if } \Sigma w1x1 + b >= 0; 0 \text{ if } \Sigma w1x1 + b < 0.$$

where wi is the weights and xi is the input data.

Weights are allocated once a starting layer is discovered. Higher variables contribute considerably to the outcome compared to various inputs, hence these weights help determine the relevance of any given variable. After that, every single input gets multiplied by its respective weight prior to being joined together. The output is then calculated by running it through an activation function. When the node's output exceeds a predefined threshold, it is "fired" (triggered), and data is transmitted to the next network layer. As a result, one node's output becomes the input for the node that follows it. This neural network is known as a feedforward network as it governs how data is transmitted from each successive layer over to the next. Figure 4.5 below illustrates neural network.

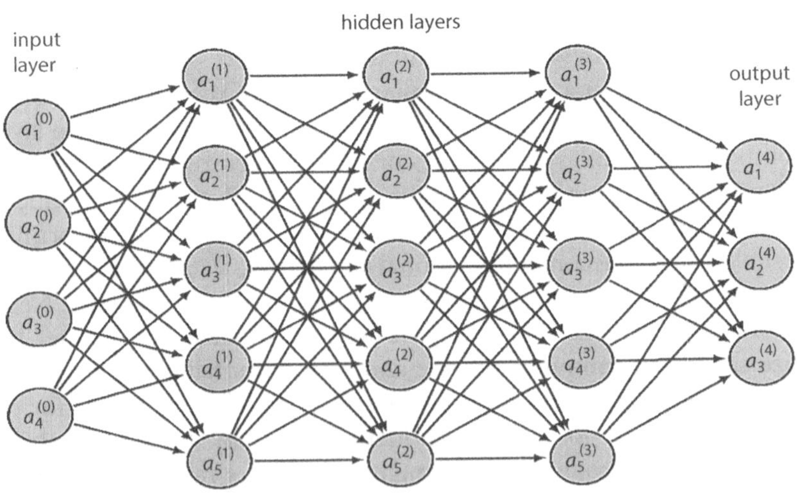

Figure 4.5 Neural network.

4.3.1 Types of Neural Network

Neural networks are computer simulations that simulate the information processing of biological neural networks found in the human brain. They are composed of multiple levels (layers) of neurons that use a sequence of mathematical operations to convert the input data into meaningful outputs.

a) **Feedforward Neural Networks:** A type of ANN which is known as a feedforward neural network allows inputs in the direction of data through nodes in the hidden level to the output nodes without creating any cycles between layers or nodes. In this type of network layers with a unidirectional data flow from the input layer passing through hidden layer and reaching the output layer which makes up its architecture. Its primary goal is to minimize prediction errors during training. It is used in many applications such as financial forecasting, natural language processing, visual and audio recognition, and recommendation systems. Figure 4.6 below illustrates the feed forward neural network.
b) **Convolutional Neural Networks (CNNs):** The construction of convolutional neural networks is centered on processing grid-type input, such as pictures and movies, by

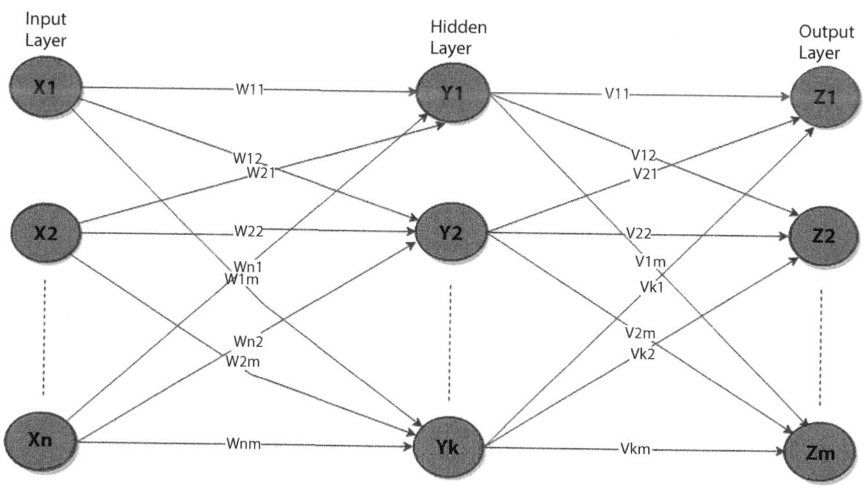

Figure 4.6 The feed forward neural network.

employing filtering layers that drive patterns and spatial hierarchies. It is fully connected nodes, pooling, and convolutional layers. It is used in several applications such as autonomous driving, medical imaging analysis, object detection, image classification, and augmented reality visualization. Figure 4.7 below illustrates the convolutional neural network.

c) **Recurrent Neural Network (RNN):** Recurrent neural networks use looping over themselves to store internal state (memory) when handling sequential data, where the current output is the outcome of earlier inputs. Its architecture is made up of recurrent connections that allows a feedback loops to process sequence. Since they restrict the mode detectors' capacity to fully record interdependence on a long scale, issues like disappearing gradients become evident. They are used in many applications which include time series prediction, one-to-one interaction, open-ended text classification, and language translation. Figure 4.8 below illustrates recurrent neural network.

d) **Long Short-Term Memory Networks (LSTMs):** One category of recurrent neural network is called a Long Short-Term Memory Network (LSTM). To cope with the vanishing incline problem and retain a wide variety of information in their memory, they display memory cells. It basically captures memory cells in the graduate greediness problem and pass information flowing. RNNs may be used in multiple domains, including time-series forecasting as well as linguistic translation for incorporating long-term memory into models.
Figure 4.9 below illustrates long short-term memory network (LSTM).

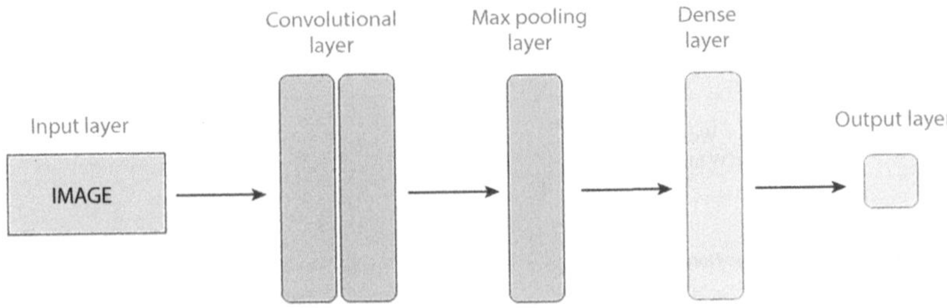

Figure 4.7 Convolutional neural network.

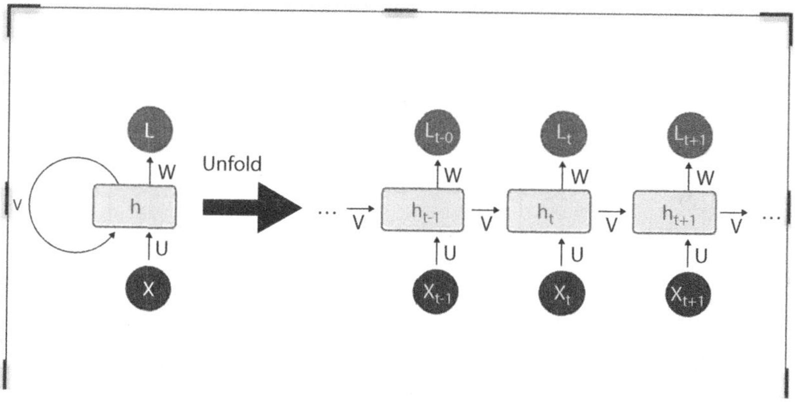

Figure 4.8 Recurrent neural network.

e) **Gated Recurrent Units (GRUs):** The second common variation of RNNs, known as Gated Recurrent Units (GRUs) which operates on a gating mechanism similar to LSTM but with fewer parameters. As compared to LSTM it is compute efficient and addresses the disappearing gradient issue. It is used in the application such as text monitoring and speech recognition. Following Figure 4.10 illustrates GRUs.

f) **Generative Adversarial Networks (GANs)** are made up of two neural networks, the generator and the discriminator, that contend with each other. The discriminator learns how distinguish between genuine and spurious data, while the generator generates phony information. As the fictitious

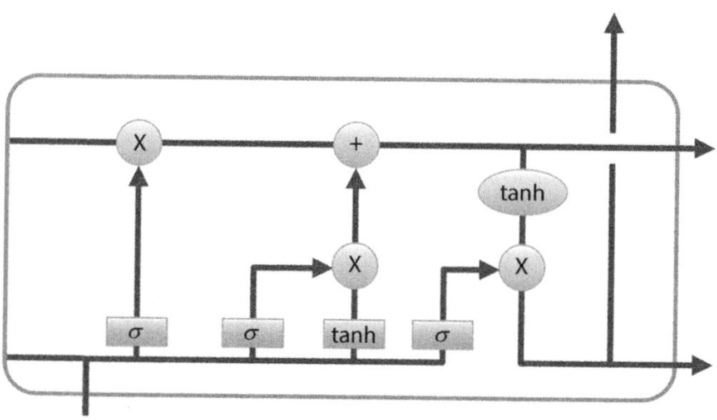

Figure 4.9 Long short-term memory network (LSTM).

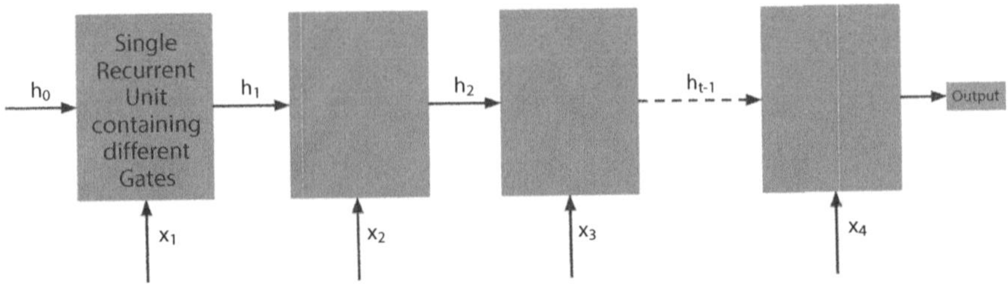

Figure 4.10 Gated recurrent units (GRUs).

data is being generated, the generator changes with each iteration. In the process of determining whether the components are produced or real, the discriminator becomes even more discriminatory. It is most used in the applications which deals with pattern formation and they have also shown promise in data augmentation, style transfer, and unsupervised learning. Following Figure 4.11 illustrates GANs.

g) **Auto-Encoders (AE):** Auto-encoders are feedforward networks which one of the type of artificial neural network, which are taught to re-code the input data in order to obtain the most useful informational displays. While the decoder decodes the space from this representation, the encoder is calibrated in the direction that will accurately translate the

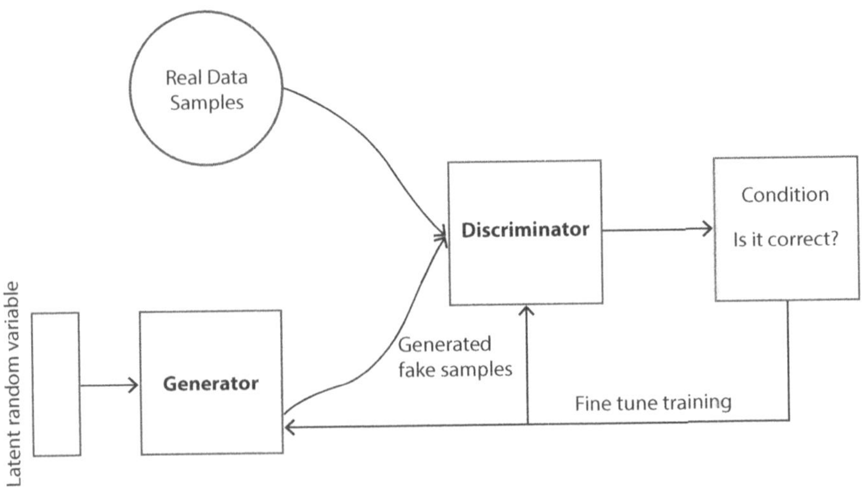

Figure 4.11 Generative adversarial networks (GANs).

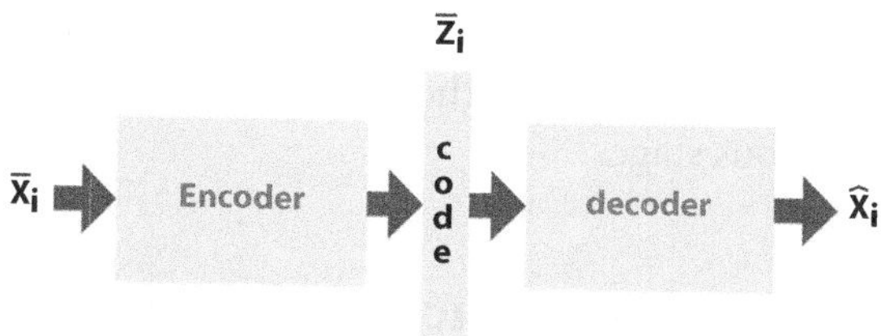

Figure 4.12 Auto-encoder (AE).

input into the legal latent space representation. The main function of auto-encoder is to assist in making the images understandable through methods like reducing the dimensions, data mining, eliminating the noise, and generative simulation. There are various types of auto-encoder such as undercomplete, overcomplete, and variational auto-encoders. Figure 4.12 above shows Auto-encoder (AE).

4.3.2 Graph Neural Network (GNN)

GCNs (Graph Convolutional Networks) bear similarities to conventional CNNs. It gathers features by looking at surrounding nodes. Non-linearity is added by means of the activation function in GNNs after the node vectors have been aggregated and sent to a dense layer. This consists of an activation function without learning, a graph convolution with a linear layer, and a non-linear component. GCNs chiefly divide into two types: Spatial Convolutional Networks and Spectral Convolutional Networks. Graph Auto-Encoder Networks learn the graph representation with the encoder and attempt reconstruction of the input graphs with the decoder afterward. Encoders and decoders are connected by a so-called bottleneck layer. These auto-encoders are common in link prediction since they manage class imbalance efficiently.

RGNNs (Recurrent graph neural networks) can handle multi-relational graphs, where just one node has several connections, and can learn the best propagation structure. This type of graph neural network employs regularization to reduce over-parameterization and improve smoothness. RGNNs produce better outcomes with less processing resources. Their applications include text production, machine translation, speech recognition, picture description generation, video tagging, and text summarization. For tasks with long-term dependencies, GGNNs (Gated Graph Neural Networks) outclass RGNNs. GGNN improves recurrent graph neural networks by including node, edge, and temporal gates based on long-term dependency. Similar to Gated Recurrent Units, the gates are used to recall and forget information in various stages.

4.4 Implementation of Euclidean Theorem-in GNN Architecture

In the literature on mathematics and machine learning, the phrase "Euclidean theorem" is not commonly used. However, when working with geometric or geographical data, Euclidean geometry concepts and principles are crucial for improving Graph Neural Networks (GNNs). Euclidean geometry supports GNN architectures in the following ways:

a) **Geometric Deep Learning and Euclidean Symmetry:** GDL (Geometric Deep Learning) extends deep learning models to non-Euclidean domains like graphs and manifolds. It is crucial to use Euclidean symmetry groups, especially the Euclidean group E(n) which consists of translations, rotations, and reflections, in GDL. GDL frameworks ensure that GNNs employ these symmetries so that invariant relative motion can be performed on the data. This enables the geospatial networks and 3D molecular activities where the arrangement of data coordinates is critical.

b) **Graph Geometric Algebra Networks (GGANs):** This new theory applying geometric algebra into GNNs is known as Graph Geometric Algebra Networks (GGANs). The integration of geometric spaces allows GNNs to learn geometric embedding's for nodes and graphs. GGANs enhance the learning of graph representations while simultaneously reducing the model's complexity and reinforcing the correlations between nodes.

c) **Constructing Graphs with Euclidean Distances:** Building graphs that rely on the Euclidean distances between the locations is a common procedure in spatial data analysis. The collection of the points in a Euclidean space is joined by the Euclidean Minimum Spanning Tree (EMST) which has the smallest total edge length. This tree can be utilized in GNNs to describe spatial arrangements efficiently and offers fundamental structure for numerous applications including clustering and network building.

d) Node feature updates are essential for GNNs to capture relationships better. They improve their representations resilience by transforming node embedding through Euclidean transformations such as translations and rotations. These transformations have guarantees which make the model spatially agnostic in changeably inconsistent orientations or positions – especially beneficial for 3D data or spatial reasoning tasks.

e) Assuming the proper design and training regime, GNNs can represent complex relationships in Euclidean spaces. In approximate neural network theory, a GNN can represent interconnections and continuous functions, given enough time and energy. In node, link, and graph classification, data is naturally situated in Euclidean domains, meaning this property is important for these tasks.

4.5 Fundamentals of Graph Neural Network Architecture

Understanding the intricacies of GNN systems requires appreciation of the underlying concepts and constituent elements. In its simplest configuration, a graph consists of nodes also termed as vertices, and edges or links that connect the nodes together. These links may be directed (one way) or undirected (two way). In addition, edges can be weighted to indicate the strength or type of connection between the nodes. Incorporation of nodes and graph modeling is fundamental within GNNs. The fundamental purpose of a GNN is typically to assign to every node in the graph a unique identifier which in this case is a low dimensional vector called embedding space. In this region, those nodes that are the same which means have the same location and relations in the graph are closer to each other. A graph can be represented in many forms such as, adjacency matrices which depict the presence and movement of edges joining nodes are incidence matrices which describe the relationship between a node and an edge and among edges. Furthermore, feature matrices are frequently used to contain associated with nodes and edges which provides additional information to the model. Message passing is the primary operating principle of GNNs. With this strategy, nodes are able to continuously update their internal representations by sharing information with adjacent nodes. Generally, this strategy consists of two major sub-processes: **aggregation**, where a node gathers information from neighboring nodes, and **combination**, where the gathered information is integrated with the node's data to form an updated representation or depiction. This step can be repeated for further layers so that information can diffuse throughout the graph with nodes acquiring representations influenced by regions more remote to the network's structure and more distant networks 'cross the structure.

Also, with graphs, one fact that stands out is that the node sequences have a level of randomness and do not have any type of structural order with respect to the network. Therefore, for GNNs, randomness invariance is a crucial consideration. Precisely, this means that the GNN's output cannot depend on the specific order in which the nodes are provided to the model. In order to achieve this, GNNs make use of order-independent summation, which also includes functions for average, maximum, and other forms of aggregating. Such considerations guarantee that irrespective of the organization of the nodes, the results with regards to node incorporation and graph-level projections would remain unchanged.

Algorithms dedicated to machine learning can manage and obtain insights from structured data due to the ability of transforming discrete graph data into continuous vector spaces through processes such as node embedding. The familiarity of vertices in the surrounding space, which reflects their shared characteristics inside the network, is an effective method of storing relational information. Furthermore, message passing, similar to information diffusion, is determined through the total number of message passing layers, which determines each node's receptive field and controls how much global context is retained in the node's embedding. However, the necessity for recombination consistency, caused by the variable sequencing of nodes, mandates the employment of order-invariant aggregating functions, which might limit the sorts of information that can be explicitly collected.

Historically, samples in Deep Learning applications such as computer vision and natural language processing have been expressed in Euclidean space. However, there has been an increase in the use of graphs to depict non-Euclidean data. To achieve this resolution, GNNs constitute a potential attempt to use deep learning principles in graphs. The word GNN often refers towards a collection of diverse procedures rather than a sole design. Many alternative designs have evolved throughout the years to offer an early look of GNN.

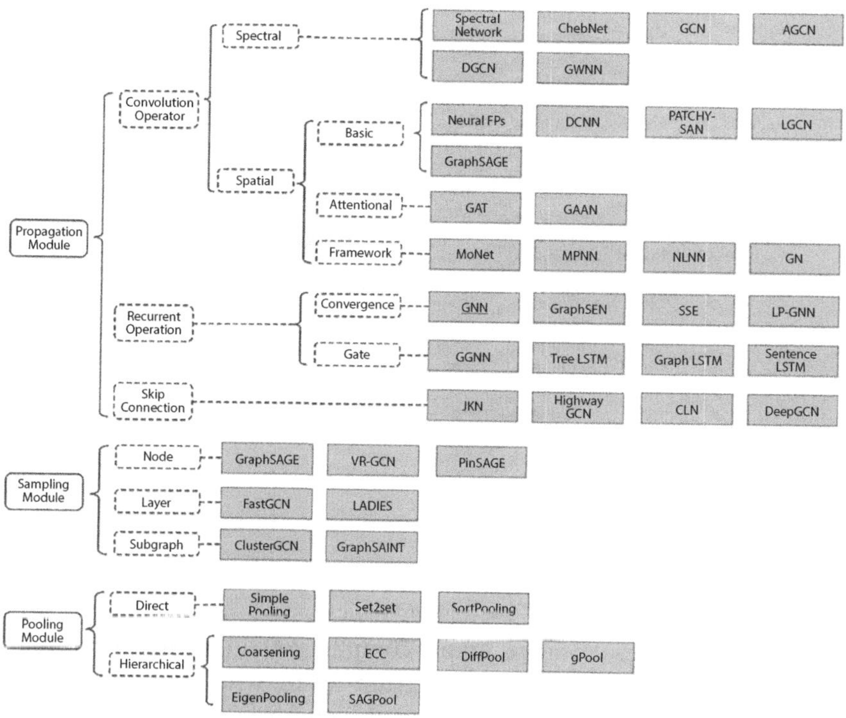

Figure 4.13 Graph neural networks. Source: Graph Neural Networks: A Review of Methods and Applications [1].

The following Figure 4.13 shows the Graph Neural Networks (GNN) which was adapted from a current assessment work on GNNs by Zhou J. *et al.*

4.6 Addressing Fundamental GNN Architectures: Spectral and Spatial Methods

GNN architectures are diverse, with several techniques aimed to address the distinct problems associated with learning from data generated by graphs. These structures can potentially be classified into three distinct types: spectral techniques, spatial methods, and sampling methods, each one having its own set of fundamental concepts and advantages. Figure 4.14 illustrates the progression of GNN architecture.

Following are the most common and important GNN architecture;

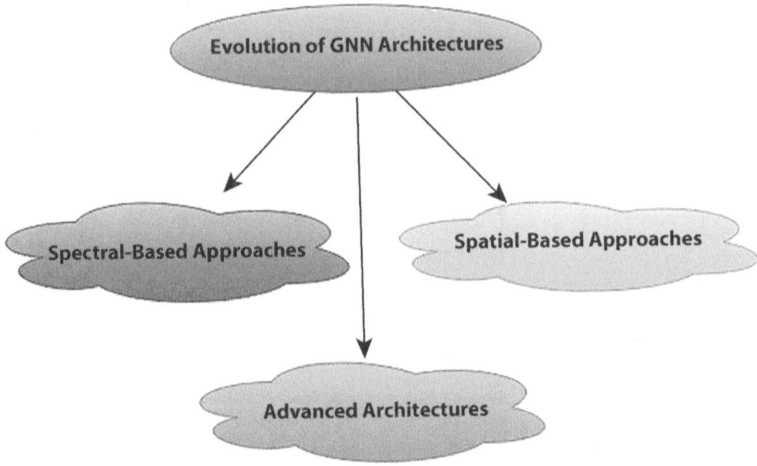

Figure 4.14 Evolution of GNN architecture.

4.6.1 Spectral Approach

Spectral techniques leverage graph computation ideas to establish the operation of convolution within spectrum realm, which is represented by the graph Fourier transform F. The graph Fourier transform F first transforms the graph computed information x into spectral domain. These approaches frequently use the eigenvectors and eigenvalues of the graph Laplacian matrix, which offers a spectral description of the graph's structure. One of the first spectrum techniques led to the creation of Spectral Networks (SCNN), which substituted the spectral convolutional kernel with an autonomous learning diagonal matrix. Sometimes the SCNNs can be computationally wasteful, especially on big networks. The Graph Convolutional Network (GCN) is a more often used spectral approach that streamlines spectral convolutions by employing a first-order approximation of spectral filters. GCNs are often made up of convolutional layers that combine data from surrounding nodes, accompanied by linear transformations and non-linear activation functions. Once the operation of the convolution is accomplished, the resultant indication (signal) is processed back using the inverse graph Fourier transform F^{-1}.

$$F(x) = U^T x$$

$$\mathrm{F}^{-1}(\mathrm{x}) = \mathrm{Ux}$$

U is a matrix defined by the eigenvectors of L, where $L = U\Lambda U^{T}$, Λ is a diagonal matrix with the eigenvalues of the graph $\Lambda = diag([\lambda 0,\ldots,\lambda 1])$.

The convolution operation is defined as:

$$g * x = F^{-1}(F(g)\cdot F(x)) = U\,(U^{T}g \cdot U^{T}x)$$

L is the normalized graph Laplacian and is constructed as depicted below:

$$L = I - D^{-1/2} A D^{-1/2}$$

Spectral approaches in GNNs use the spectral features of graphs to solve a variety of graph-related tasks. Following are some of the significant application for spectral approaches.

a. **Node Categorization:** Spectral GNNs, especially Graph Convolutional Networks (GCNs), have shown extremely effective in node categorization applications. They work by conducting spectral convolution, which allows for fast information collection from a node's vicinity in order to forecast its label. For example, GCNs may be used to categorize research publications based on their citation networks. **Graph Classification:** Spectral Networks (SCNN) were designed to train convolutional filters in the spectral domain with the objective of categorizing complete graphs.
b. **Network Embedding:** Graph Auto encoders (GAEs), that can use graph convolutional layers (a spectrum approach), serve a purpose for unsupervised learning applications like network embedding. They understand graph representations in weak dimensions by first encoding followed by decoding the graph structure.
c. **Graph Pooling:** Spectral clustering algorithms may be used with GNNs to conduct graph pooling activities. Clustering nodes according on the graph's spectral qualities is the first step, followed by aggregating the features of nodes within the same cluster.

Spectral GNNs thrive at collecting worldwide graph features as well as having extensive conceptual backing, making them ideal for jobs that need a thorough grasp of the entire graph structure and long-term relationships.

Following are the advantages of spectral approach;

a) **Worldwide Information Gather:** Spectral approaches are more successful in extracting global patterns and long-range relationships throughout graph structures than spatial methods, which are primarily concerned with local neighborhoods. By operating in the spectral domain, the full structure of the graph is taken into consideration.
b) Conceptual Foundation: The roots of spectral GNNs are based on the well-established domains of spectral graph theory and graph signal processing which give reasoned grounds for understanding their actions and attributes.
c) Distinctiveness: As proposed by some research, it seems that spectral GNNs are to some extent highly expressive and are optimal for graph signal extraction in the frequency domain.

4.6.2 Spatial Approach

Unlike spectral methods, spatial techniques perform convolution integrals directly on the network topology, working within each node and modifying and aggregating data from the local neighbors. These methods are also frequently less expensive in terms of computation when compared to their spectral counterparts, and can be applied to both directed and undirected graphs. The Message Passing Neural Network (MPNN) is a basic example of a spatial method which imposes a generic architecture for information flow in a network using messages at the nodes. Another well-known design is the Graph Attention Network (GAT) which employs an attention mechanism to capture the importance of different surrounding nodes during the aggregation step. GATs are well-known for their scalability and computational efficiency. Spatial GNNs, are capable of interpreting graph-structured information by focusing on local neighborhoods and offer an extensive number of applications spanning several areas. Following are some of the few significant applications of spatial approach

a) Discovering user qualities such as interests and demographics according to their relationships with others and neighbors' attributes.
b) Identifying probable future connections between users and groupings of heavily linked people in a network. It also

involves recognizing users with influence who may successfully disseminate knowledge.

c) To forecast future interactions, users and items can be seen representing vertices within a network, with their interactions (e.g., purchases, ratings) as edges. Using the network structure and item/user features, it also suggest things to prospective or current users.
d) Using social connections or knowledge graph data to make more personalized suggestions.
e) It can also be used for forecasting chemical and physical characteristics of molecules using their graphical structure, predicting drug-target interactions which requires recognizing possible interactions between medicines and target proteins.
f) Drug repurposing is the process of discovering new applications for existing medications by analyzing drug-drug and drug-disease interactions.
g) Traffic flow forecasting is the process of projecting upcoming traffic situations (such as speed and volume) by modeling road networks as graphing and taking into account the spatial connections between road segments.
h) It can also be used for congestion detection which involves identifying busy locations in a traffic network using real-time data and graph analysis.
i) Estimation of time can be done to get from one site to another on a road network.
j) Fraudulent actions in banking or social media platforms can be discovered.
k) The identification of anomalies involves identifying strange patterns or nodes in numerous kinds of networks.

Spectral GNNs thrive at collecting worldwide graph features as well as having extensive conceptual backing, making them ideal for jobs that need a thorough grasp of the entire graph structure and long-term relationships. Following are the advantages of spatial approach;

a) Spatial approaches are straightforward and localized since they act straight on the graph, analogous to Convolutional Neural Networks (CNNs) used to process pictures. They concentrate on collecting information from every node's local neighborhood, which is consistent with the idea that a node's attributes are frequently impacted by its direct connections.

b) Spatial techniques may automatically address networks with variable degrees of nodes and complicated, irregular layouts without the need for an established graph representation, such as the adjacency matrix in spectral approaches.

c) Many spatial GNNs, like GraphSAGE, are inductive. This implies that they may generalize to previously unknown nodes and even completely new networks without upgrading the whole dataset. This provides a considerable benefit in changeable graph circumstances.

d) Since spatial techniques focus on small neighborhoods, they might prove more computationally effective for large, small networks than spectral approaches, which might demand operations on the whole graph structure.

e) Spatial approaches enable the direct aggregate of characteristics from neighboring nodes, which is commonly done by employing a variety of aggregation algorithms. This allows more versatility in how information is integrated. It also offers a greater level of scalability.

f) Some spatial GNNs include parameters that can be learnt during the aggregation phase, enabling the network to adjust the relevance for various neighbors or attributes.

g) Spatial approaches excel in capturing local patterns and relationships within the network, which is critical for jobs in which a node's attributes are heavily impacted by its surroundings.

Table 4.3 below shows the comparison between the spectral and spatial approaches for GNN based on the following features.

4.7 Applications of Graph Neural Networks

GNNs have demonstrated their effectiveness across various domains by capturing complex relationships in graph-structured data. Some of the most prominent applications include as follows. Following Figure 4.15 shows the application of GNN.

Social Network Analysis: GNNs are emerging as a helpful technique for Social Network Analysis (SNA), outperforming older approaches in

Table 4.3 Comparison between spectral and spatial approaches.

Features	**Spectral approach**	**Spatial approach**
Domain of Operation	Spectral (Laplacian domain)	Spatial (vertex domain)
Computational Cost	Generally higher	Generally lower
Convolution	Linear operations in the spectral domain	Direct multiplication with adjacency matrix
Interpretability	Generally lower	Generally higher
Expressiveness	Potentially higher	Potentially lower
Strength	Potentially more expressive in capturing complex graph relationships. May be more suitable for graphs with intricate structures.	Can be computationally more efficient than spectral GNNs, particularly for large graphs. May be easier to interpret and understand, as they directly leverage the graph structure.
Weakness	Can be computationally expensive, especially for large graphs. May be less intuitive to interpret, as they operate in the spectral domain.	May be less flexible in capturing complex relationships compared to spectral GNNs. May struggle with graphs where node relationships are not clearly defined.

managing the intricacy and interconnection of social network data. GNNs offer an excellent foundation for analyzing social networks through generating significant representations of users and their interactions. Their capacity to record complicated network structures and carry out numerous forecasting functions presents them as an indispensable tool for analyzing social behavior, identifying important people, and developing intelligent social applications. Platforms like Facebook and LinkedIn use GNNs to predict potential connections. Fake News Detection, Analyzing propagation patterns and relationships between users and content. Some of the key responsibility of social network analysis that are enriched by GNN and which can be summarized as follows;

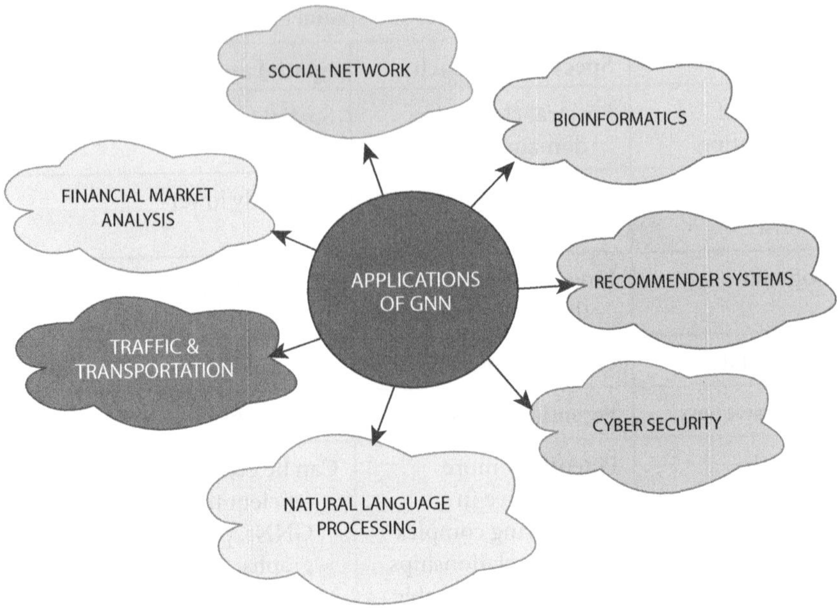

Figure 4.15 Application of graph neural network.

Graph Neural Networks (GNNs) have quickly become go-to tools in network analysis because they boost both the reach and accuracy of predictions. An obvious use is node tagging, where the model forecasts traits or labels for individual users, also called nodes. With this approach, GNNs can flag VIP members, sort people by shared hobbies, or even spot automated accounts. They do so by looking at each nodes own data and the data of its immediate neighbors, combining the two to sharpen the final class. A second key task is link prediction, the art of finding broken connections or guessing which users might buddy up next. Here, GNNs track how edges form over time and, by studying those patterns, assign a probability score to every potential link. Community detection runs along similar lines; by squeezing users features and ties into compact embedding's, the network can naturally group neighbors who think or behave alike. Beyond that, GNNs assist in influence maximization, helping planners locate the few nodes that can spread news far and fast. Each embedding carries the weight of a nodes standing and reach, so decision-makers can simply rank those weights and activate the best few champions. Lastly, GNNs are useful for emotion exploration, which involves analyzing user sentiment or opinions in a social network. To ascertain general sentiments regarding a topic or product, they can spread sentiment data across the network.

Some of the key advantages of social network analysis can be summarized as follows;

Graph Neural Networks (GNNs) shine in social-network work simply because they were built to think in graphs. Compared with older techniques, they handle the lopsided, non-Euclidean structure of friendship circles, reading edges and nodes straight to learn context. Another plus is that the networks design their own features on the fly, so users do not spend day's hand-crafting summaries of who speaks to whom. Huge graphs still strain any computer, yet tricks from models like GraphSAGE let GNNs sample neighborhoods rather than load the whole city map at once, easing the crunch. Because of these gains, GNNs routinely outscore standard machine-learning tools and decades-old statistical tests on tasks such as link prediction, community discovery, and rumor tracking.

a) Bioinformatics and Drug Discovery: GNN, as they are doing a great job for the representation and analysis of complex biological data, are changing the course of bioinformatics and drug discovery. Several range of methods based on graph convolutional neural network increasing with time, which will allow for the solution of critical questions in the field of healthcare, such as understanding the origin of diseases and developing new therapies. Additionally, GNNs find affirmation in the new representation of molecules as graphs, which provides a road map to the drug design field, as well as models directly correlating terminology with application in the case of proteins for biomedical purposes of course. The summarized positive effects of utilizing GNN in the bioinformatics and drug discovery sector can be explained as:

Graph Neural Networks (GNNs) are transforming how scientists study living systems and speeding up the search for new medicines. In protein research, these nets model each amino acid as a node and each bond as an edge, letting them guess a proteins 3-D shape with surprising accuracy. They also scan known sequences and pairings to map Protein-Protein Interactions (PPIs), a key step for grasping how a cell stays alive. Move beyond proteins, and GNNs untangle gene networks by linking mRNA, transcription factors, and other controls, illuminating both healthy tissue and tumors. In the search for drugs, GNNs quickly estimate how tightly a small molecule sticks to a target protein, letting chemists sort huge libraries in minutes. The networks can even draw up brand-new compounds that dissolve well and hit the right site, all by learning the rules of structure and effect. For medicines already on the shelf, GNNs scan drug-target and disease-gene maps to point out repurposing chances that might have been overlooked. Lastly, by modeling how multiple drugs dance together at their targets, these models flag risky combinations early, helping doctors keep patients safer.

Despite promising sparks, several bumps still slow the drive toward applying GNNs in biology and drug discovery. Graph neural networks (GNNs) still have a rewarding yet rocky road in biology and drug discovery. Chief among the roadblocks is interpretability; it is hard to see why a model reaches a decision when the biology is already tangled, so calmer minds are pushing for clearer, traceable designs. Solid models also ask for a mountain of labeled examples, and those samples-shared human records, wetlab proofs, or hard-to-find test sets-are not always piled up in a lab drawer. Scalability matters too; once networks expand to a full proteome or a compound library on steroids, the memory and time skyrocket, pushing less-hasty researchers to search for leaner blueprints. Because living things keep changing, the still-image assumption of many GNNs falls short; what is really needed are temporal layers that can track mutation, treatment, and feedback in near-real time. Last (for now), the field cries out for shared benchmarks and honest scores so every team knows whether its new trick actually helps drugs hit the bench or keeps biology on the shelf.

b) Recommender Systems: GNNs have been a potent framework for creating complex and successful recommender systems. They are a vital resource for customizing customer interactions across a range of online platforms because of their capacity to model intricate relationships and integrate a variety of data sources. E-Commerce & Streaming Services Platforms like Amazon and Netflix leverage GNNs to enhance personalized recommendations. Session-Based Recommendations improves real-time recommendations based on user-item interactions.

How GNNs can enhance the recommendation system?

Graph Neural Networks (GNNs) are quietly reshaping how recommendations are made by capturing different types of links between people and products. A standard technique builds a bipartite graph, placing users in one group, items in another, and drawing edges for every click, purchase, or star rating. Adding a knowledge graphlet, which explains products-brands, categories-and users-age, hobby-crumbles the system into a richer, more colorful network. Toss in social links-friendships, follows-and the model starts eavesdropping on what your circle likes, broadening its own view. For time-sensitive suggestions, it treats each step of a browsing trail as a mini-graph and scouts the next move you are most likely to make.

A range of graph neural networks powers modern, clever recommendation engines. At the base, Graph Convolutional Networks, or GCNs, operate by pulling in a neighbor's info and averaging it, weighted by trust. Newer setups like Neural Graph Collaborative Filtering, NGCF for short, build on that idea to untangle the many ways users and items link up. Then Graph

Attention Networks, also called GATs, add attention scores so the model can spot which neighbors matter most when the average is formed. For fast-moving sessions, designers turn to Session-based GNNs, or SR-GNNs, treating every short burst of clicks as a tiny graph and learning embedding's that trace the click order. Lastly, Knowledge Graph Convolutional Networks, KGCNs, fold in outside facts-what items mean and how they relate-boosting accuracy and letting users see why a choice was made.

Graph Neural Networks (GNNs) improve recommender systems in a number of important ways. First and foremost, GNNs are naturally adapted to deal with the graph-structured data that is common in recommendation tasks, like user-item discussions. Since GNN-based recommenders usually outperform traditional approaches by successfully capturing intricate user-item relationships and utilizing a variety of supplemental information, this architectural fit usually results in enhanced performance. The "cold-start" problem is another area in which GNNs shine because they can use graph connectivity and built-in features to make more precise recommendations for prospective clients and items. Additionally, GNNs demonstrate exceptional adaptability in incorporating a wide range of data, fusing item content, user profiles, social connections, and user-item interactions with ease. Lastly, GNNs enhance comprehensibility by examining the learnt graph structures and attention weights, which aids in providing an explanation of the reasoning behind.

c) Cybersecurity & Fraud Detection: GNNs provide a robust and diverse solution to combating sophisticated cybersecurity risks and criminal activity by using the relational structure inherent in data. As graph-based data grows more ubiquitous, GNNs are going to perform an ever more important role in safeguarding digital property and averting monetary damages. The main objective of using GNN in cybersecurity is to identifying fraudulent transactions in financial systems, identifying the intrusion detection by analyzing communication graphs in networks.

Due to their innate ability to comprehend relational data, graph neural networks (GNNs) provide a potent method for cybersecurity and fraud detection. GNNs can simulate network traffic as a graph in network security, with nodes standing for systems or IP addresses and edges for connections. GNNs can identify discrepancies suggestive of botnets, DDoS attacks, and insider threats through recognizing typical patterns of behavior. The identification of intricate, multiple levels attacks is further made possible by their capacity to record dynamic interactions between nodes. In order to recognize new malware variants or malicious behaviors,

GNNs examine the graph structure of code and how it interacts. By looking at the connections between websites and domains, they can also identify phishing attempts by spotting questionable trends in the larger web entity network. When detecting financial fraud, depicting transactions as graphs with accounts, users, merchants, or devices as nodes and transactions as edges, GNNs are excellent at detecting financial fraud. This enables them to use multi-hop link analysis and trend detection to spot complex fraud patterns that traditional systems frequently overlook, such as money laundering or collusive fraud rings.

Because of their innate capacity to manage intricate, linked data, graph neural networks (GNNs) provide notable benefits in cybersecurity and fraud detection. GNNs provide a more thorough contextual understanding of the surroundings by taking into account the entire network topology, in contrast with conventional techniques that examine individual data points separately. This is important because the majority of fraud and cybersecurity cases involve complex relationships between entities. Additionally, GNNs have the potent ability to constantly gain insight from and adapt to graph data, which allows them to identify new trends and changing fraud or threat tactics. One of their main advantages is their ability to learn features automatically, which speeds up the detection process by automatically extracting crucial node and edge attributes from the graph and doing away with the necessity for human feature engineering.

d) Roads and Publically Modes of Transportation: GNNs offer robust infrastructure for the analysis and prediction of numerous features for traffic and transportation networks. Their capacity to simulate complicated geographical and temporal connections serves as an invaluable tool for increasing transportation system effectiveness, security, and longevity. As accessibility to data and processing capacity increase, we may anticipate more usage and more complex GNN-based technologies in this vital sector. Using GNN can helps use to predict the traffic flow by modeling the road networks to predict congestion pattern. It also uses the route optimization algorithm that will enhance the navigation in logistics and smart cities.

The potential of GNNs to easily integrate a variety of data as node or edge attributes—such as weather, points of interest, or demographic data—enhances the model's comprehension of the network and constitutes one of its main advantages. Even though very large graphs can be computationally demanding, GNN architectures are becoming more flexible and scalable as a result of ongoing advancements. Crucially, in numerous traffic and transportation prediction tasks, GNN-based models have continuously surpassed conventional statistical and machine learning techniques, demonstrating their unparalleled efficacy and accuracy.

When used in traffic and transportation networks, Graph Neural Networks (GNNs) present a number of significant obstacles regardless of their benefits. The high degree of uncertainty and flexible nature of these networks is a significant obstacle. Research on creating GNNs that can accurately depict and forecast these sporadic, fast changes is still crucial. Integrating the effects of outside events, such as sporting events or concerts, which have the potential to significantly change traffic patterns but are challenging to smoothly integrate into existing models, is another major challenge. Clarifying the "why" behind GNN forecasts is essential for trust and real-world implementation, and comprehending the interpretability of these predictions in intricate transportation networks is a continuous research area. Lastly, high-quality, real-time data is necessary for GNN models to be successful in this field. Deficits or irregularities in such data can significantly impact model effectiveness and reliability.

e) Financial Market Analysis: GNNs provide an effective toolbox for financial market evaluation by allowing to perform modeling of complicated interactions and the synthesis of multiple sources of information. As research in this field advances, GNNs are likely to acquire a more significant part in a variety of financial applications such as prediction by capturing relationships between stocks to improve predictive modeling, risk management by identifying hidden dependencies among financial instruments, fraud detection, and algorithmic trading.

Graph Neural Networks (GNNs) are a powerful tool for financial market analysis because they are excellent at modeling high-order and nonlinear connections that conventional approaches frequently overlook. They effortlessly acquire valuable representations of financial institutions and their interactions, and by taking into account every aspect of the network structure, they offer an increased awareness of market trends. Additionally, GNNs can combine various data sources, including market statistics, press releases, and accounting records, into a single graph structure. As a result, in numerous financial applications, GNN-based models have performed better than conventional machine learning and statistical techniques.

f) Natural Language Processing (NLP): GNN have begun to make advances into Natural Language Processing (NLP), providing novel methods to convey and handle written information beyond classic sequential models such as RNNs and Transformers. By utilizing the intrinsic network structures inside and across text, GNNs may capture intricate linkages and dependencies that are sometimes neglected. Knowledge Graph Completion can he used to enhancing search engines and Chatbots with structured relational data. Text Classification & Sentiment Analysis using document-level graphs to improve classification accuracy.

By accurately simulating the innate relationships found in language, Graph Neural Networks (GNNs) provide notable benefits in Natural Language Processing (NLP). For in-depth analysis and context understanding, they are excellent at identifying relationships between words, entities, and documents. Because GNNs work with syntactic and semantic graphs, they can readily integrate language structure into their learning. This makes it easy to incorporate structured data, such as knowledge graphs, into NLP models. Additionally, even though text is sequential, GNNs can directly handle many basic linguistic connections that are intrinsically graph-like (such as dependencies). As a result, in many NLP tasks, especially those where relational information is crucial, GNN-based models are performing better than traditional sequential models.

4.8 Advantages and Disadvantages of GNN

GNNs tries to develop as strong tools for scrutinizing and learning through statistics represented as graphs, which capture links and dependencies between things. GNNs offers an effective framework for learning through linked data by taking use of the network structure directly. This enables them to succeed at activities that need correlations across data points, resulting in considerable advances across a variety of disciplines. When working with such data, they provide numerous significant benefits over typical machine learning approaches.

Following are some of the advantages of GNN

a) GNNs are especially built to manage the intricate linkages and interconnections that accompany graph data, as opposed to standard approaches, which frequently encounter problems with non-Euclidean structures.
b) They can use the graph's architecture (connections between nodes) and node characteristics to develop meaningful representations.
c) GNNs can handle manage graphs of varied dimensions in addition to unconventional topologies, thus rendering them suitable for an extensive array of real-life applications. They can deal with a comprehensive range of graphs, containing directed, undirected, weighted graphs etc.

d) GNNs have accomplished more effective outcomes when compared to conventional methods in multiple graph-related tasks. They have been effectively functional in varied sectors.
e) Recent GNN architectures are developed to manage massive and complicated graphs smoothly, which makes them useful to real-world datasets.
f) Some spatial GNNs can make inferences to new, undiscovered nodes or even completely new graphs without asking for retraining.

Following are some of the disadvantages of GNN

a) Gathering information from neighbor's continuously might result in node embedding by becoming excessively similar, particularly in deep GNNs. This process of "over-smoothing" makes it harder to discriminate among nodes and impedes downstream activities such as node categorization.
b) Over-smoothing makes it difficult to educate very deep GNNs, reducing their capacity to properly capture long-range relationships.
c) For large networks containing high average degrees, aggregating information from neighbors may be computationally costly, particularly when using numerous layers.
d) Maintaining the graph structure and node embedding might be exceedingly difficult for big graphs.
e) Spectral approaches sometimes need computing the graph Laplacian and its eigenvectors, which may be computationally expensive for big graphs.
f) Several GNN models are intended for static graphs whose structure remains unchanged with time.
g) Modifying them to accommodate dynamic networks necessitates specialized structures and training methods.
h) The effectiveness of GNNs is largely dependent on the level of quality and accuracy of the input graph. Noisy or inadequate graph architectures might impair learning.
i) Attacks that graphically modify topological features or nodes to mislead the systems can target GNNs. Adversarial attacks are one of the models we know least about in terms of the building theory, generalization, expressiveness, and the configuration of various graphs and workloads tailored around efficient design principles.

4.9 Capitalizing on the Potential of GNNs for Real-World Solutions

The relationships that Graph Neural Networks (GNN) perceive unlock new avenues for addressing local concerns while taking advantage of system synergies. We can append GNN capabilities to devise novel solutions for urban and regional planning, economic development, health care, environmental management, and other fields by modeling and learning from complex interactions. The aim is to formulate utilize relevant data and local knowledge, develop local ethically sensitive concerns intelligently, foster local data, and wisely local ethical issues. Given below are a few examples of applications of GNN solving real world problems:

a) Traffic GNN Application: In transportation, GNNs can be utilized in studying road networks to predict traffic congestion using real-time sensor data. Historical data, social media comments, and even reports can also be incorporated. All of these could lead to better adaptive traffic signal control, more dynamic route guidance, congestion minimization strategies, and preventing traffic jams which is always a need in bustling cities.
b) Community Transport Optimization: By modeling bus and metro networks as graphs, GNNs can analyze travelers flow, forecast demand variations, and optimize timetables and routes for increased efficiency and customer satisfaction. This might be especially useful for increasing connectivity along the neighborhoods of the different cities.
c) Water and Energy Distribution: GNNs might be employed to track and optimize water and power distribution channels, identifying leakage, forecasting demand spikes, and enhancing the distribution of resources across the cities.
d) Analyzing Social Networks: Determining the framework and behavior of cities social networks might aid in identifying significant people for health promotion initiatives, successfully conveying information, and even detecting misinformation propagation.
e) Local Business Ecosystem Analysis: Graphing the links between local firms, suppliers, and consumers might reveal economic interdependence, highlight development

prospects, and assist focused interventions for small and medium-sized organizations.

f) Travel and Tourism Management: Using a graph to convey tourist sites, transit linkages, and local businesses might allow for personalized suggestions to tourists, optimize tourism flows, and even assist preserve and promote cities rich cultural legacy.
g) Transmission of Diseases Forecasting: Modeling the transmission of transmissible diseases in the neighborhood as a graph and taking into account characteristics like as population density and movement patterns, might aid in forecasting epidemics and executing focused public health interventions.
h) Personalized Healthcare Guidelines: Analyzing patient-doctor networks, illness relationship graphs, and additionally nutrition networks may result in more tailored healthcare suggestions and treatment approaches.
i) Trash Control Optimization: Graphing trash pickup routes and processing plants might assist optimize collection plans, save expenditures on transportation, and increase garbage management system productivity.
j) Ecological Inspection: Analyzing sensor networks which track air and water quality as graphs might offer a comprehensive picture of environmental conditions and aid in the identification of pollution sources.

Following are some of the challenges that can be faced by GNN for implementing the real world solutions;

a) Data Availability and Quality: The effectiveness of GNN applications is strongly dependent on the accessibility of high-quality and linked data. Receivers may need to acquire and integrate important data from multiple sources.
b) Computing Infrastructure: While executing complex GNN models, vast computing resources may be required at the initial phase. Implementation may utilize cloud computing resources or local high-performance computers.
c) Building Regional Competence and Infrastructure: For GNN research and practice, developing regional competence is critical for sustained long-term placement.

Collaboration between the universities, research institutes, and regional industries is beneficial.

d) Moral Issues: Just like any Artificial Intelligence application, employing GNNs transforms the need to consider the ethical implications of the technology, such as truthfulness, transparency, and privacy.

4.10 Conclusion

The above terms, generated from pondering talks, are explained and analyzed in depth to demonstrate an analyst's intellectual thought process while considering strategy and design. Broad types of Graph Neural Networks (GNN) have gained abundant attention in the last three decades, translating the means-of-decision vendors participate mechanistically with clients through contracting as a network. In regard to urban metropolitan network efficiency and GNNs applications upon these complexities, there is sufficient reasoning to consider GNNs within transportation engineering and city design schemes based on the architecture. The scope for the development and enhancement of new and existing GNN structures remains open for diversified branching, especially in examining massive datasets like social media streams. Alas, Intuitionistic Fuzzy Graph Neural Network (IFGNN) gets enhanced in the interaction-dominant or structure tent to extreme graph systems. Targeted GNN development work like advance awareness in social networks or real time traffic flow problem solving systems with smart cities shows a trend toward GNN simulation in diverse and intricate real-world problems arising from GNN to logs spoilers mechanisms. The ability to exploit architectural diversity requires embracing the whole Regional specialization for understanding and applying different GNNs, gathering regionally appropriate graph datasets that accurately depict the region's networks, and assigning sufficient computational resources for model training and deployment. Through systematic application of various GNNs, researchers and practitioners will creatively turn the concepts of GNNs into societal benefits, devising solutions for urban design, resource allocation, social welfare, and beyond. Adapting and evolving these architectures will sustain the rest of GNNs' conceivable life in regards to making a difference in practical problem areas, and many more.

AI disclosure statement: The author confirms that no artificial intelligence (AI)-generated content or AI-assisted writing tools were used in the preparation of this chapter (including summaries and the abstract). All text, analysis, and conclusions are the result of the author's original work.

Bibliography

1. Zhou, J., *et al.*, Graph Neural Networks: A Review of Methods and Applications. *AI Open*, 1, 57–81, Jan. 2020.
2. Wang, B., Cheng, L., Sheng, J., Hou, Z., Chang, Y., Graph convolutional networks fusing motif-structure information. *Sci. Rep.*, 12, 10735, 2022 Jun 24, doi: 10.1038/s41598-022-13277-z.
3. Xie, C., Wen, X., Pang, H., Zhang, B., Application of graph auto-encoders based on regularization in recommendation algorithms. *PeerJ Comput. Sci.*, 9, e1335, 2023 Apr 6, doi: 10.7717/peerj-cs.1335.
4. Gkarmpounis, G., Vranis, C., Vretos, N., Daras, P., Survey on Graph Neural Networks. *IEEE Access*, (Published: December 2023 / November 2024 for a revised version).
5. Wu, Z., Pan, S., Chen, F., Long, G., Zhang, C., Yu, P.S., A Comprehensive Survey on Graph Neural Networks. *J. Latex Class Files*, August 2019.
6. Zhang, W., Xu, D., Xuan, X., Jiang, L., Yao, G., Han, R., Lang, X., Luo, C., Addressing Noise and Stochasticity in Fraud Detection for Service Networks.
7. Khemani, D., *et al.*, A review of graph neural networks: concepts, architectures, techniques, challenges, datasets, applications, and future directions. *J. Big Data*, 11, 1, 18, January 2024, DOI: 10.1186/s40537-024-00868-w.
8. Liao, N., Liu, H., Zhu, Z., Luo, S., Lakshmanan, L.V.S., Benchmarking Spectral Graph Neural Networks: A Comprehensive Study on Effectiveness and Efficiency, arXiv:2406.09675v1 [cs.LG], 14 Jun 2024.
9. Zhu, W., Chen, M., Chen, C., Li, M., Wu, J., Yang, J., Uncertainty in Graph Neural Networks: A Survey, arXiv, March 2024.
10. Hamilton, W.L., Graph Representation Learning, in: *Synthesis Lectures on Artificial Intelligence and Machine Learning*, Springer, Berlin/Heidelberg, Germany, 2020, [Google Scholar] [CrossRef].
11. Wu, Z., Guo, H., Zhang, Y., Liu, J., Che, Y., Zhang, J., A Survey on Graph Neural Networks for Intrusion Detection Systems: Methods, Trends and Challenges. *Comput. Secur.*, 141, 103821, March 2024, DOI: 10.1016/j.cose.2024.103821.
12. Grover, A. and Leskovec, J., node2vec: Scalable Feature Learning for Networks, in: *Proceedings of the 22nd International Conference on Knowledge Discovery and Data Mining (SIGKDD 2016)*, San Francisco, CA, USA, 13–17 August 2016, ACM, San Francisco, CA, USA, pp. 855–864, 2016, [Google Scholar] [CrossRef].

13. Ou, M., Cui, P., Pei, J., Zhang, Z., Zhu, W., Asymmetric Transitivity Preserving Graph Embedding, in: *Proceedings of the 22nd International Conference on Knowledge Discovery and Data Mining (ACM SIGKDD 2016)*, San Francisco, CA, USA, 13–17 August 2016, ACM, San Francisco, CA, USA, pp. 1105–1114, 2016, [Google Scholar] [CrossRef].
14. Tsiligkaridou, E.G., Pardalos, P.M., Crispen, P.J., Lipnick, A.C., Graph Neural Networks in Cancer and Oncology Research: Emerging and Future Trends. *Cancers*, 15, 24, 5857, December 2023, DOI: 10.3390/cancers15245857.
15. Sun, J., Bandyopadhyay, B., Bashizade, A., Liang, J., Sadayappan, P., Parthasarathy, S., ATP: Directed Graph Embedding with Asymmetric Transitivity Preservation, in: *Proceedings of the 33th Conference on Artificial Intelligence (AAAI 2019)*, Honolulu, HI, USA, 27 January–1 February 2019, AAAI Press, Honolulu, HI, USA, 2019, [Google Scholar] [CrossRef].
16. Cavallari, S., Zheng, V.W., Cai, H., Chang, K.C., Cambria, E., Learning Community Embedding with Community Detection and Node Embedding on Graphs, in: *Proceedings of the 26th Conference on Information and Knowledge Management (CIKM 2017)*, Singapore, 6–10 November 2017, ACM, Singapore, pp. 377–386, 2017, [Google Scholar] [CrossRef].
17. Velickovic, P., *et al.*, Graph Attention Networks, 2018.
18. Hamilton, W., *et al.*, Inductive Representation Learning on Large Graphs, 2017.
19. Jiang, W., Luo, J., He, M., Gu, W., Graph Neural Network for Traffic Forecasting: The Research Progress. *ISPRS Int. J. Geo-Inf.*, *12*, 3, 100, 2023, https://doi.org/10.3390/ijgi12030100.
20. https://www.sciencedirect.com/science/article/pii/S2666651021000012#sec2
21. https://www.ibm.com/think/topics/graph-neural-network
22. https://distill.pub/2021/gnn-intro/
23. https://theaisummer.com/gnn-architectures/#fn-1
24. https://aws.amazon.com/blogs/quantum-computing/combinatorial-optimization-with-physics-inspired-graph-neural-networks/#:~:text=We%20show%20that%20the%20graph,the%20ability%20to%20scale%20beyond
25. https://www.holisticai.com/blog/building-trustworthy-graph-neural-networks

Part II

ETHICAL MODELS, SCALABILITY, AND ARCHITECTURAL CHALLENGES

5

Ethical Considerations in Graph-Based Learning Models

Aaquil Bunglowala[1]* and Gaurav Paliwal[2]

[1]*Department of Electronics & Telecommunication Engineering, Sri Aurobindo Institute of Technology (SAIT), Indore, Madhya Pradesh, India*
[2]*School of Technology Management & Engineering, SVKM's Narsee Monjee Institute of Management Studies (NMIMS), Indore, Madhya Pradesh, India*

Abstract

Graph-based learning models and Graph Neural Networks (GNNs) in general, have revolutionized the field of machine learning by allowing complex inference on structured and interconnected data. They are applied far and wide across high-stakes domains like healthcare, finance, e-commerce, recommendation systems, cybersecurity, and social media. Although this unique architecture and promise of GNNs also bring with it a fundamentally demanding set of ethics to be addressed with urgency by researchers and regulators, the chapter addresses the challenging ethics of GNNs on four broad dimensions: bias and fairness, privacy and security, explainability and transparency, and social impact. The chapter starts with describing how GNNs can automatically learn and reinforce social and structural biases from the intrinsic structure of the graph and correlation among nodes' features. Fairness-sensitive learning methods are offered as solutions but their effectiveness varies according to context. The success of anonymization methods is also challenged in graph structure re-identification attacks where they do not succeed. Certain challenges with differential privacy and federated learning being applied on graphs are also met. Explainability of GNN models is also treated as an essential property for accountability and trust in high-risk use case settings. The black-box character of reasoning in GNN makes transparent decision-making challenging. A preview of explainable AI (XAI) methods that were transliterated to graph topology and their shortcomings are presented. The regulation and social effects implications of scaling GNNs, such as surveillance, manipulation, and systemic discrimination, are also considered in the chapter. Real-world examples and

**Corresponding author*: aaquilbun@gmail.com

J. Ramkumar, Sridaran Rajagopal, B. Suchitra and S. Balamurugan (eds.) Graph Neural Networks: Concepts and Applications, (183–242) © 2026 Scrivener Publishing LLC

case studies are embedded in the discussion in a bid to situate such questions. The chapter finishes by referencing silences in existing ethical frameworks and proposing cross-disciplinary collaboration for developing good guidelines for ethical use and designing graph-based models. By sketching a broad range of ethical issues and imaginative solutions, the chapter is a mandatory guidebook for researchers, practitioners, and policy-makers dedicated to creating ethical Graph Neural Networks (GNNs).

***Keywords*:** Cartoon generation, deep learning, style transfer, generative adversarial networks, visual abstraction

5.1 Introduction

Artificial Intelligence (AI) powered different industries and its greatest achievement is in the case of Graph Neural Networks (GNNs). As graph-structured data can be trained with GNN models, now they are finest use cases of utilizing such as social network analysis, drug discovery, anti-money laundering (fraud detection), and recommendations. With the ability to hold relations and dependencies between objects, GNN models allow achievement of an added level of intuition towards systems. In spite of that, increased use has drawn ethical concerns greater than those common with usual machine learning models. Those are transparency, fairness, privacy, and possible impact on society from findings made by AI systems based on graph data. The moral goodness of GNNs is that the data structure results in a web-like graph of entities in relation to each other. Graphs represent relations—between users in a social network, molecules in a molecule, or transactions in a financial network. Thus, all the bias, errors, or moral issues written into the data do not remain local to nodes or edges but can propagate across the network as a whole. This interdependence comes with an unusual challenge in the form of bias and unfairness in graph models that are not static but diffusive and dynamic [1]. For example, a biased training data with the exclusion of some communities in a social network can lead to underrepresentation of these communities in predictions and recommendations and further intensify differences and make them even greater. Secondary, relative to table or image data, graph data may be sensitive relational data.

Anonymous graphs are also re-identifiable under attacks since association patterns and topological signatures have the potential to reveal personal information [2]. It was demonstrated in a study how even an attacker could identify users of a network with horrific precision using structural and auxiliary information [3]. This is evidence of the susceptibility of graph data to privacy breaches and the moral obligation of researchers and practitioners to protect sensitive information. Transparency is also the primary issue. GNNs are uninterpretable and opaque in nature and possess multi-layered message-passing operations and deep representation which obscure the reason for prediction [4]. In mission-critical applications like healthcare, law enforcement, and finance where stakeholders insist on accountability and interpretability, such transparency is only more difficult. When a GNN flags a financial transaction as suspicious or a patient as high-risk, it should be crucial that why such a conclusion has been arrived at can be viewed and queried. Transparency loss can result in public distrust and danger to the organization for ethical and legal issues. No less significant is the issue of justice.

Graph models can learn and exhibit bias in society unintentionally. For example, if a GNN on the job board website is learned from a graph where the previous history has biased toward specific groups or institutions, then it would recall discriminatory hiring [5]. In recommender systems, individuals from marginalized groups will be recommended less since they will have lower exposure and interaction since they are at the edge of the interest or social graph. Not only is it technologically hard to fix but also ethically incorrect, since it takes advantage of the positive features of equality and justice. The ethical impacts also include social implications of graph-based AI output. GNNs are further investigated for surveillance networks, detection of disinformation, and public policy modeling. The dangers in such situations are even greater. A GNN mislabeling a protest movement as threat from patterns in a graph has the potential to enable discriminatory policing or surveillance. Abuses of GNNs for detection and amplification of polarizing content in mis/disinformation have the potential to lead to polarization of society [6]. These are examples that demonstrate ethical issues are not technological constraints but moral commitments with real-world effects. Alongside that, temporal graph dynamics come to introduce another complexity level. Graphs are dynamic—graphs aren't.

Relations come into existence and die, new nodes emerge, and structured behavior alters. Due to these facts, graph learning ethics is a dynamic topic, not a static one. A model at present can be only biased tomorrow because the graph has been updated [7]. This requires ongoing monitoring, maintenance, and testing to ensure that it's ethical every step of the way throughout the life cycle of the AI system. All these being said, present regulatory frameworks and ethics frameworks are likely to come up short on capturing the one-off nature of graph-based AI. While efforts such as the European Union's AI Act and IEEE's Ethically Aligned Design rank highest in terms of recommendation—i.e., fairness, accountability, and transparency—neither of them even attempts to delve into graph data specifics and GNN specifics [8]. This leaves room for uncertainty among developers and organizations that are attempting to deploy ethical graph AI systems, which creates an opportunity for domain-specific ethics protocols and practices. Developing GNN fair-aware is one of the strands of promoting research. The models are infused with fairness constraints during training or construction with the purpose of mitigating biases. Techniques like adversarial debiasing, fairness regularization, and fair sampling have shown to be promising for removing disparate impacts [9]. On top of this, privacy-preserving approaches like differential privacy, homomorphic encryption, and federated learning are being connected to the graph space in an effort to anonymize sensitive relations without compromising model accuracy so that it is still ensured [10]. These are nascent solutions and typically have performance, accuracy, and privacy guarantee protection trade-offs. And yet another exciting development is the arrival of the interpretability toolchain for GNNs. Explainable models like GNNExplainer and Graph LIME try to explain how GNNs make predictions by utilizing key nodes, subgraphs, or features that highlight [11]. They not only enhance transparency but also enable models to be audited, bias to be identified, and trust to be built in AI systems. This software is nonetheless still yet to be widely utilized and shown robust and efficient over a variety of applications and datasets.

In view of the interdisciplinary nature of graph-based AI ethics, solutions are needed that transcend technology. Policy makers, ethicists, legal experts, and AI researchers need to work together to create extensive ethical frameworks and governance structures. Educational institutions, too, have to integrate ethics into AI teaching to equip future practitioners with critical thinking skills necessary to consider societal implications of what they do. In short, with GNNs increasingly becoming the heart of decision-making systems, their ethical considerations cannot wait any longer. Graph data structure imposes new challenges associated with bias transmission, privacy, and transparency, which existing frameworks of AI ethics might not thoroughly tackle. Societal consequences arising from the applications of GNNs in domains with high sensitivities further exacerbate the urgency for ethical prudence and ethically sound innovation. This chapter attempts to shed light on these issues by examining the nature of graph-based ethical risk, canvassing mitigation measures, and calling for holistic, multi-disciplinary efforts to ensure graph-based AI development and deployment responsibility.

5.2 Bias Propagation and Fairness in Graph-Based Models

Graph Neural Networks (GNNs) and similar graph-based learning systems provide powerful means of modeling relationships within complicated systems—such as social networks, citation networks, and recommender systems—by leveraging structural dependencies among entities, but their dependence on node features and graph topology brings with it new and complicated paths of bias propagation. In classical machine learning, bias is typically understood as the result of unbalanced training data or defective label distributions. Bias gets entrenched very much into the data structure itself in graph-based learning. It propagates *via* neighborhood aggregation, message-passing-based mechanisms, and topological dependencies, and thus detection and correction become much more difficult [12].

5.2.1 Understanding Bias in Graph Data

Graph data bias can be grouped into some categories: feature bias, structural bias, and label bias. Feature bias appears when the characteristics describing nodes or edges mirror biases present in society. For instance, gender or race bias in user profile data may appear in social recommendation systems. Structural bias emerges from the structure of the graph—who connects to whom. Marginalized groups or individuals tend to possess less connectivity or are at the periphery of networks, making them less influential in neighborhood aggregation operations in GNNs [13]. Label bias points toward imbalances or stereotypes incorporated within ground truth tags employed in training. Table 5.1 demonstrates some additional types of biases with description and example.

Imagine a citation network in which papers written by women or underrepresented minorities get fewer citations compared to those from more established demographic groups. A GNN trained on such a network may project lower relevance or quality for novel papers from underrepresented authors, reinforcing academic disparities. In these systems, the structure itself serves as a pipeline for biased learning, even though the features and labels are objective.

5.2.2 Propagation of Bias in Graph Neural Networks

GNNs work by recursively aggregating local neighborhood feature information from a node to form embeddings. This relies on the assumption

Table 5.1 Common sources of bias in graph neural networks.

Source of bias	Description	Example
Topological Bias	Graph structure inherently favors certain nodes due to centrality or degree	Users with more connections get more recommendations
Attribute Bias	Node or edge features reflect historical or societal bias	Gender or race-based hiring discrimination
Label Distribution Imbalance	Skewed class distributions in training data	Underrepresentation of minority communities
Homophily Assumptions	Over-reliance on similarity-based learning	Reinforces echo chambers in social networks

that the local neighborhood contains semantically rich and representative information. But if the neighborhood is biased—either node feature or structural position—then the learned representation will inherit and reinforce this bias [14].

Let h_v^k denote the embedding of node v at the kth layer of a GNN. A typical update rule is:

$$h_v^k = \sigma\left(\sum_{u \in N(v)} \frac{1}{c_{vu}} W^k h_u^{k-1}\right)$$

where $N(v)$ is the neighbourhood of node v, W^k is a learnable weight matrix, c_{vu} is a normalization constant, and σ is a non-linear activation function [15].

If biased nodes comprise $N(v)$, then the h_v^k embedding will be imbued with those biases. Since GNNs typically involve multiple layers (each covering more hops), bias can propagate across multiple degrees of connectivity, resulting in nodes that were indirectly affected in the original data set. In addition, homophily—"birds of a feather flock together"—is typically assumed in graph-based learning. As much as it facilitates learning, it inadvertently reinforces echo chambers, promoting greater homogeneity and potentially silencing minority voices. This is particularly destructive in social networks and recommendation systems for content, as it can lead to polarization and exclusion of minority voices [16].

5.2.3 Empirical Evidence of Bias in GNNs

Empirical research underscored the importance of prejudice being retained in GNNs. In a particular paper, a point was raised that in node classification in social networks, GNNs were performing very poorly for the minority class of users as their network positions and minority features were not effective to represent them [17]. But yet another study of GNN-based recommender systems proved that ethnic minorities and women were substantially less likely to be suggested for work or placed on online stores than similarly matched men or majority-products [18]. These aren't algorithmic bias—these are reflection and amplification of actual social unfairness. GNNs acting on their own can turn out to be blind co-conspirators to magnification of system bias.

5.2.4 Metrics for Fairness in Graph Models

Graph fairness-based model does have certain measures that have to do with structural aspects. Overall measures like demographic parity, equalized odds, and disparate impact have to be extended to graph domains. Some of the most suitable measures are:

- Group fairness: Ensures fair treatment across different sensitive groups (gender, race) for node classification. It can be calculated in terms of group-wise accuracy or disparate impact ratio [19].
- Individual level fairness: Methods such as where neighbors with similar graph embeddings or node attributes get equally good predictions. Of these, one enforces Lipschitz continuity on the prediction distance among neighbors' embeddings.
- Subgraph fair between: Fairness can be quantified in subgraphs for link prediction and also for community detection so that there are no groups of people left behind and divided or placed in the incorrect groups.
- Parity representation: Avoids embeddings from encoding or revealing sensitive attribute information. It is usually approximated using adversarial models that have been trained to approximate sensitive attributes from embeddings.

Both such computations need to be calculated prior to model training and post-training to compare raw data fairness and how model training impacts it.

5.2.5 Fairness-Aware Graph Algorithms

In the aim of negating bias and making it impartial, various algorithms that are fair-aware have been proposed. These are classified under three broad groups: pre-processing, in-processing, and post-processing.

(a) Pre-Processing Techniques
Pre-processing methods reconstruct the input graph before training with the intent of eliminating bias. They cover:

- Edge Rewiring: Rewiring the graph in a way that groups are highly interconnected. For example, FairWalk suggests edge addition so that minority representation in random walk sampling can be increased [20].
- Feature Sanitization: Decoupling of correlated features or sanitizing sensitive features to restrict the ability of the model to learn biased patterns.
- Deletion of biased nodes or edges that may bias model learning. This may be done using abusive user deletion or adjustment in label balancing.

(b) In-Processing Techniques
These adjust the training objective or the architecture of the model:

- FairGNN: Trains a second predictor for sensitive attributes on top of normal GNN, using adversarial training to remove sensitive information from embeddings [21].
- Debiasing Loss Functions: Add fairness constraints at the level of loss functions, balancing penalties across unfair predictions between groups or individuals.
- Regularization Techniques: Use graph Laplacian or homophily-aware regularizers to impose fairness between communities.

(c) Post-Processing Techniques
Post-processing techniques manipulate predictions or embeddings after training

- Embedding Correction: Align group node embeddings to reduce narrow group differences.
- Threshold Adjustment: Rescale decision thresholds for different groups to generate more balanced results.
- Calibration: Scale group confidence scores to make more balanced judgments.

To combat bias and achieve fairness, several fairness-aware algorithms have been proposed. These techniques fall into three broad categories: pre-processing, in-processing, and post-processing.

5.2.6 Adversarial Debiasing in Graph Learning

Adversarial debiasing is a top-down method in which the model is forced to learn node embeddings associated with the target task but not with the sensitive attributes. An adversarial model tries to predict the sensitive attribute from the embeddings, and the primary model tries to defeat the adversary's precision. This domain adaptation technique, which is a GAN-based method, has been shown to be effective to remove bias with no loss in predictive performance [22]. One architecture consists of three components

- A single GNN encoder to produce node embeddings.
- Task-specific classifier (node classification or link prediction).
- A predicted adversary with sensitive attributes.

Function of Loss is given by the following equation:

$$L_{total} = L_{task} - \lambda L_{adv}$$

Where λ regulates the balance between main task and adversarial loss. The encoder reduces the task loss as it increases the adversarial loss to hide sensitive information from the embeddings [23].

5.2.7 Challenges in Achieving Fairness

With promising solutions, some of the concerns are still present:

- Trade-Off between Accuracy and Fairness: De-biasing may degrade performance on the main task, especially when predictive features co-vary with sensitive traits.
- Multiple Sensitive Attributes: More difficult to model and test for fairness for multiple identities that intersect (e.g., gender and race).
- Dynamic Graphs: Dynamic graph fairness entails re-estimating and re-computing models repeatedly and repeatedly, which is expensive.
- Lack of Ground Truth: Sensitive features don't or partially get labeled in public datasets, and therefore it is hard to measure fairness.
- Unintended Consequences: Certain interventions to fairness would have other undesired ethical side effects, e.g., reverse discrimination or invasion of privacy.

5.2.8 Case Studies

- Case Study 1: Fairness in Social Networks-A GNN was used to categorize social network users into interests. Baseline worked amazingly well except for the minority since they held edge positions in networks. Accuracy of underrepresented users was boosted by 17% without affecting overall accuracy with FairGNN [21].
- Case Study 2: Recommender Systems-In an employment recommender system, minority groups were given lower-quality recommendations and were less likely to be given them. Adversarial debiasing and threshold calibration enhanced fairness scores by 24%, showing how well-designed systems can yield more balanced results [24].
- Case Study 3: Law Enforcement Analytics-A criminal network discovery algorithm as a link prediction model generated more false positives against minorities. Racial bias in link predictions decreased by 31% with fairness-aware regularization [25].

5.3 Techniques for Bias Detection and Mitigation

Graph Neural Networks (GNNs) are becoming the building block for a variety of real-world systems, from recommender systems and social network analysis to drug discovery and financial crime detection. What makes them particularly strong is that they can fuse both node attributes and graph structure into predictive modeling. As the adoption increases, however, so does concern around their ethical effects—most prominently with regards to fairness and bias. Biases in GNNs can originate from numerous sources such as imbalanced network structures, lopsided node attributes, or skewed propagation procedures that serve the majority groups disproportionately. Counteracting these biases is important to guaranteeing responsible and reliable AI systems. In this chapter, we discuss sophisticated methods for identifying and reducing bias in GNNs across four broad categories as shown in Figure 5.1: fairness-sensitive graph algorithms, adversarial debiasing, data reweighting and graph sanitization, and fairness measures.

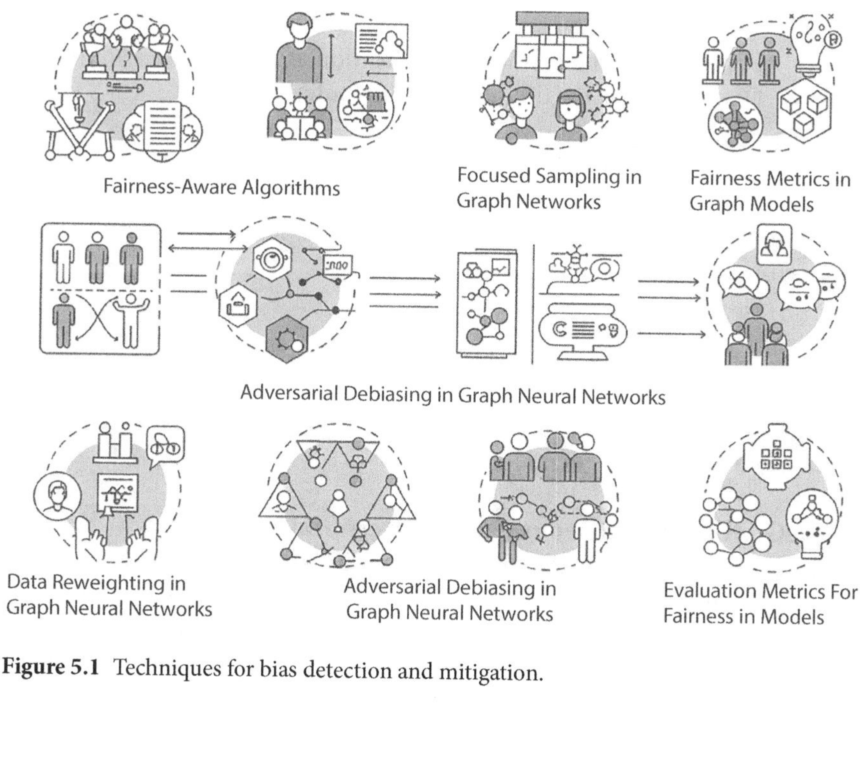

Figure 5.1 Techniques for bias detection and mitigation.

5.3.1 Fairness-Aware Graph Algorithms

Fairness-aware algorithms aim to incorporate ethical considerations into the heart of graph-based learning. These approaches learn either the graph structure, learning goal, or embedding process to diminish differences among various groups determined by sensitive features (e.g., gender, race, age).

(a) FairWalk
FairWalk [26] is a fairly early fairness-aware approach tailored for graph-based systems. It adapts the popular Node2Vec random walk process by making sure nodes from various demographic groups get fair representation in the sampling of training walks. More precisely, FairWalk adds group-aware transition probabilities during walk generation and this helps to balance the exposure of nodes over protected and unprotected groups.

(b) FairGNN
FairGNN [27] is an improved model designed for node classification tasks. It directly introduces group fairness into the training objective by inserting a discriminator that predicts sensitive attributes. The central architecture consists of:

- Predictor for the primary classification task,
- Discriminator for the sensitive attribute,
- A group fairness constraint.

The learning objective optimizes task loss while at the same time decreasing the discriminator's ability to predict sensitive attributes from node embeddings.

$$L = L_{task} + \lambda_1 . L_{fairness} - \lambda_2 . L_{discriminator}$$

This adversarial framework ensures that embeddings are useful for prediction but minimally encode bias.

(c) HeGAN (Heterogeneous Graph Adversarial Network)
HeGAN [28] employs adversarial learning to learn fairness among various node types and relations of heterogeneous graphs. It formalizes the generator and discriminator of adversarial learning to reveal latent fairness patterns. These algorithmic updates are part of a developing trend to construct

fairness into the architecture of graph learning models so that GNNs are sensitive to demographic diversity and group equity during training.

5.3.2 Adversarial Debiasing in GNNs

Adversarial debiasing is a strong paradigm that has been increasingly used in GNNs to remove biased representations. It adapts principles from Generative Adversarial Networks (GANs) and transfers them to preventing sensitive attribute leakage while generating embeddings.

(a) Framework Overview
In adversarial debiasing, the framework generally includes:

- A GNN encoder G that learns node embeddings h_v,
- A main predictor P for the primary task (e.g., node classification),
- An adversarial discriminator D attempting to make inferences regarding the sensitive attribute s from h_v,

The goal is to train G such that $P(G(x))$ is accurate while t $D(G(x))$ is inaccurate. The loss function is:

$$L_{total} = L_{task}\left(P\left(G(x)\right), y\right) - \lambda .\ L_{adv}\left(D\left(G(x)\right), s\right)$$

Here, λ is a trade-off parameter controlling the strength of fairness regularization [29].

(b) Applications and Results
Experiments have demonstrated that adversarial debiasing dramatically lowers the predictability of sensitive features without affecting task performance. In actual-world datasets such as Cora and Facebook's SNAP, adversarial GNNs lowered demographic prediction accuracy by more than 30% with zero loss in performance on node classification [30].

(c) Variants and Improvements
A few variants have been proposed to strengthen adversarial debiasing:

- Multi-Adversary Models: Utilizing multiple discriminators to consider multiple sensitive features.

- Gradient Reversal Layers (GRL): Used extensively in domain adaptation, GRLs reverse the gradient flow to better hide sensitive information.
- Domain-Specific Discriminators: Adversary tailoring per graph domain, e.g., medical or financial networks.

(d) Case-Specific Moral Issues in Graph Neural Networks (GNNs)

Graph Neural Networks as powerful as they are, possess a tendency to magnify and spread prejudices in graph structures in society. This section provides real-world, case-specific examples where GNNs are ethically wrong, i.e., node classification and prediction.

- **Biased Social Network Node Classification**

Case: Political prejudice or prediction of likelihood of criminality from social associates
Ethical Issue: GNNs are capable of learning and propagating information social biases in social graph structures.
Example: In a Twitter user data GNN model, political minorities were typically incorrectly labeled because majority viewpoints were overrepresented in the graph.
Implication: Promotes online minority polarization and marginalization.

- **Discriminatory Node Classification in Financial Graphs**

Case: Loan decision based on borrower–co-borrower relationships
Ethical Concern: Personal attributes can be inferred from graph linkages, and discrimination can thus be caused.
Example: A GNN trained on a financial network penalized loan-takers connected to underprivileged communities even if they shared the same credit history.
Implication: Violates fairness principles in computerized finance decision-making.

- **Homophily Bias in Academic Collaboration Networks**

Case: Estimating researcher impact or paper acceptance
Ethical Issue: GNNs overestimate links, with a bias toward established networks.
Example: Junior researchers were disadvantaged in a GNN-based reviewer recruitment system because they had few high-profile collaborations.
Implication: Strengthens structural inequality in academia.

- **Overfitting to Demographic Clusters in Healthcare Graphs**
  Case: Predicting disease outcome based on patient similarity graphs
  Ethical Issue: GNNs may distort underrepresented or disconnected groups.
  Example: PREDICT-ing cancer relapse using a GNN provided incorrect predictions for rural patients since the graph was not sufficiently connected.
  Implication: Enhances healthcare access and quality inequality to rural populations.

- **Bias Propagation in Policing and Surveillance Graphs**
  Case: Predicting criminal behavior from associate networks
  Ethical Concern: Early policing bias is reinforced by graph learning.
  Example: A GNN applied to criminal associate graphs over-targeted racially profiled individuals from the community.
  Implication: Prevents policing systemic discrimination feedback loops.

- **Data Reweighting and Graph Sanitization**
  Graph data itself can carry structural or attribute biases prior to training. Data reweighting and graph sanitization are preprocessing steps that work to remove such biases prior to their affecting model learning.

5.3.3 Data Reweighting

Data reweighting consists of weighting nodes, edges, or training samples according to fairness criteria. Nodes from underrepresented groups, for example, can be upweighted to have greater impact during training. Methods include:

- Inverse Propensity Scoring (IPS) [31]
- Reinforcement Learning for Sample Reweighting [32]

These methods assist in rebalancing datasets without losing useful information.

(a) Graph Structure Sanitization
Sanitization rearranges the topology of the graph to eliminate discriminatory patterns. Some examples are:

- Edge Deletion or Rewiring: Elimination or adjustment of connections disproportionately influencing minority nodes.
- Attribute De-Biasing: Adjustment of node or edge attributes to eliminate correlation with sensitive attributes.

For example, in social graphs, nodes with high degrees tend to represent dominant groups. Rewiring such that minority nodes have increased neighbors can address the imbalance.

(b) Synthetic Graph Generation
When real-world data is either too incomplete or biased, synthetic graphs satisfying fairness constraints can be created through simulation methods. These controlled datasets can be used to benchmark fairness-aware GNNs.

5.3.4 Evaluation Metrics for Fairness in Graph Models

It is challenging to evaluate fairness in graph-based learning because of the interaction between node features, graph structure, and prediction results. The metrics should consider both group-level and individual-level differences.

(a) Group Fairness Metrics
These metrics ensure statistical parity across different demographic groups.

- **Demographic Parity**:

$$P(S=0) = P(\hat{Y}=1|S=1)$$

- **Equalized Odds**:

$$P(Y=1, S=0) = P(\hat{Y}=1|Y=1, \quad S=1)$$

- **Disparate Impact (DI)**:

$$DI = \frac{P(\hat{Y}=1|S=1)}{P(\hat{Y}=1|S=0)}$$

A value of DI closer to 1 indicates better fairness [33].

(b) Individual Fairness Metrics

- **Lipschitz Fairness:** Quantifies how close predictions are for nodes that are similar.

- **Embedding Similarity:** If two nodes are similar in features and neighbors, their embeddings and predictions must also be similar.

(c) Structural Fairness Metrics

- **Degree Centrality Parity:** Guarantees that nodes from every group have comparable centrality values.
- **Subgraph Representation:** Minority subgraphs must not be disconnected or weakly connected relative to dominant groups.

(d) Fairness in Dynamic Graphs
Temporal dynamics add one more level of complexity. Fairness needs to be tracked over time, especially in domains such as fraud detection or recommendation systems, where graph structures change [34]. Bias in GNNs is multi-faceted, intricately rooted in data and learning procedures. Luckily, the machine learning society has learned some effective methods for detecting and reducing these biases. Demographic-fair graph algorithms like FairGNN and FairWalk tune GNN structures to be sensitive to demographic parity. Adversarial debiasing adds principled training techniques that eliminate sensitive attribute leakage from embeddings. Simultaneously, preprocessing *via* data reweighting and graph sanitization reorganizes the input data to decrease discrimination prior to learning. Last but not least, fairness metrics for graph structures facilitate evaluation and validation of these interventions. As these technologies evolve, practitioners and researchers are required to include fairness as an inherent aspect—not an add-on—to the design of graph-based learning systems.

5.4 Data Privacy and Security Concerns

As graph-based machine learning models like Graph Neural Networks (GNNs) become more common in applications dealing with sensitive and interconnected data—such as healthcare, social networks, financial systems, and personalized services—the need to maintain data privacy and security becomes paramount. Graphs inherently embed not only individual properties but also relationships between objects. This interconnectivity, although necessary for model performance, greatly amplifies the privacy risks, resulting in new and subtle threats. Table 5.2 discusses some of the privacy threats and related mitigation methods in GNNs.

Table 5.2 Privacy risks and mitigation techniques in GNNs.

Privacy risk	Description	Mitigation technique
Re-identification of nodes	Anonymized nodes can still be deanonymized using structural cues	Graph anonymization, k-degree anonymity
Inference attacks	Adversaries infer sensitive information about nodes or subgraphs	Differential privacy
Membership inference	Determine if a node was in the training set	Adversarial regularization
Exposure through model explanation	Interpretability methods reveal sensitive graph substructures	Privacy-aware explainability frameworks

This section discusses the privacy threats in graph-based AI, outlines primary threat models, and presents sophisticated privacy-preserving methods. It is followed by a real-life case study from healthcare in order to show the technical and ethical stakes.

5.4.1 Privacy Risks in Large Interconnected Graph Data

Graph data naturally holds relational information that is potentially more insightful than standalone tabular data. In social networks, for instance, even if a user's profile is anonymized, his/her connections, interactions, and community structures might facilitate re-identification *via* linkage attacks [35]. Figure 5.2 illustrates an example of how graph topologies can still facilitate re-identification attacks even when nodes are anonymized. Likewise, in medicine, a graph connecting patients, treatments, hospitals, and diagnoses might unintentionally reveal a person's medical history—even if direct identifiers are stripped away.

These threats are due to:

- Linkage-based inference: Attackers may link nodes to outside datasets.
- Topology leakage: Graph topologies themselves may leak sensitive patterns.

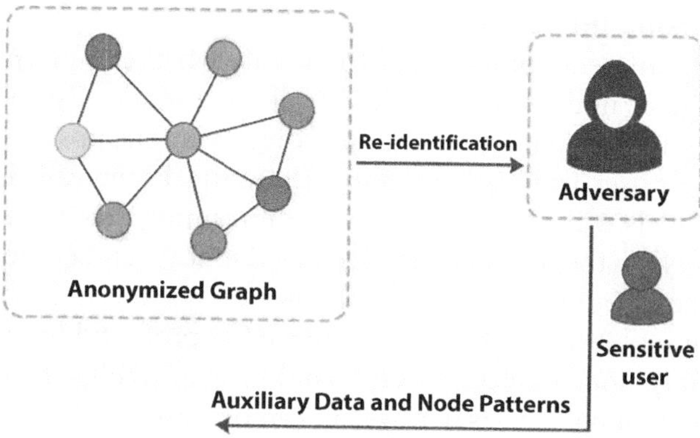

Figure 5.2 Re-identification attacks even when nodes are anonymized.

- Over-sharing: Knowledge about one thing can spill over into information about its neighbors (e.g., in collaborative filtering systems).

Since GNNs collect and diffuse information between graph structures, they inadvertently embed and magnify sensitive information. Thus, even trained models—particularly those provided as pre-trained embeddings—can be conduits for privacy breaches [36].

5.4.2 Threat Models: Inference and Membership Attacks

Two prevalent threat models in graph-based system privacy are inference attacks and membership inference attacks.

(a) Inference Attacks

Inference attacks try to reveal sensitive attributes of individuals or entities within the graph by taking advantage of the model's predictions or internal representations. For example, an attacker can infer a person's political leaning, sexual orientation, or health status based on their connections or predicted community membership [37].

In GNNs, inference attacks are especially strong since the message-passing structure is used. As the embedding of each node includes the information of its neighboring nodes, other nodes' sensitive information about one node may be inferred even if the attributes of the target node are not exposed. Attackers can utilize relational information for attribute inference from trained or shadow models [38].

(b) Membership Inference Attacks

Membership inference attacks seek to identify whether a given data point (e.g., an edge or node) was utilized in training a model. This is especially risky when the graph represents sensitive relationships, e.g., patient-medication or user-transaction graphs. These attacks typically leverage the fact that models will respond differently on training data than on unseen data. For GNNs, the attacker may look at the node embedding or confidence of classification and decide whether or not the node belonged to the training set [39]. Membership inference has profound implications for confidentiality and regulatory compliance (e.g., GDPR, HIPAA), where the very fact that data exists in a system can be sensitive.

5.4.3 Privacy-Preserving Techniques

To counteract privacy threats, a number of sophisticated methods have been created to specifically address the issues of graph-structured data. This section discusses three prominent methods: differential privacy, federated learning, and homomorphic encryption.

(a) Differential Privacy

Differential Privacy (DP) is a mathematically sound system for constraining the data a model discloses about any single data point. For graph contexts, DP can be applied to node-level (whether a node is present in the graph) or edge-level (whether a relationship exists between two nodes) privacy. A randomized algorithm A is ϵ-differentially private if for any two datasets differencing by one element and for all possible outputs S:

$$Pr\, Pr\,\{A(D_1) \in S] \leq e^{\varepsilon} Pr\left[A(D_2) \in S\right]$$

In graph learning, differential privacy can be implemented in a variety of ways:

- Noisy gradients: Random noise is added to the gradients (DP-SGD) during model training to hide individual contributions [40].
- Graph perturbation: The graph structure itself is perturbed by adding/removing edges randomly before training to hide sensitive links.
- Private graph embeddings: Embeddings are sanitized using noise or obfuscation to avoid reverse engineering.

One of the difficulties in using DP in graphs is the extreme sensitivity of graph operations. A slight modification (such as deleting a node) can significantly impact the model output because of cascading effects through message passing. Thus, strong privacy often comes at the cost of compromised utility, which needs to be carefully balanced [41].

(b) Federated Learning

Federated Learning (FL) decouples model training from the location of data, retaining data locally on clients and only sharing model updates. This framework is particularly well-suited to graph data from several organizations (e.g., hospitals, banks), where data centralization is either impossible or unethical. In Federated GNNs, every client trains a local GNN over their local subgraph and infrequently sends updates to a central server for aggregation. This design minimizes raw data exposure but comes with new challenges:

- Heterogeneity: Local graphs could be radically different in structure and distribution.
- Graph partitioning: The way the global graph is partitioned across clients will impact model performance.
- Communication overhead: Deep architectures and node dependencies in GNNs can make model updates large.

Notwithstanding these challenges, federated GNN platforms such as FedGraphNN and FedSage have proved promising in terms of privacy preservation as well as maintaining performance in node classification and link prediction tasks [42]. Privacy of FL can further be increased by mixing it with differential privacy (e.g., adding noise to updates) or employing secure aggregation protocols so that the server cannot look into individual updates [43].

(c) Homomorphic Encryption

Homomorphic Encryption (HE) enables computations to be executed directly on encrypted data, yielding encrypted results that, when decrypted, are equivalent to the results of operations executed on the plaintext. This is a very valuable property in privacy-preserving machine learning, where sensitive information needs to be kept confidential during the process of computation. In the context of graphs, HE can be employed to:

- Encrypt node features: Clients encrypt data prior to sharing it with a model provider or a central server.

- Securely perform inference: The GNN does inference over encrypted features and responds with encrypted predictions.

Prior research has illustrated that it's feasible to perform inference of the GNN over encrypted graphs utilizing leveled homomorphic encryption schemes like CKKS and BGV. But the computation remains a formidable limitation, particularly in the case of deep GNNs as well as huge graphs [44]. An example is CryptoGNN, which does GNN inference while encrypted and can have acceptable performance for small graphs, though at increased latency [45]. More research is done to optimize HE schemes for graph-based applications by using approximation, quantization, and hardware-specific accelerators.

5.4.4 Case Study: Privacy in Healthcare Graph Data

Healthcare data is one of the most sensitive and high-consequence areas for graph-based AI. Graphs in healthcare are employed to model relations among patients, diseases, treatments, medical facilities, and clinical pathways. For example, a patient-disease graph might encode not only single diagnoses, but how various conditions co-occur and how treatments spread across a network. One prominent use case is clinical risk prediction, where GNNs are learned to predict readmission risk or drug side effects from patient histories as graph representations. Yet a number of privacy issues follow:

- Linkage attacks: Despite anonymization of identifiers, a patient's treatment path could be distinctive enough to identify them [46].
- Neighbor leakage: Information on one patient may leak through adjacent nodes, crossing consent boundaries.
- Regulatory compliance: The systems should comply with HIPAA in the U.S., GDPR in Europe, and other data protection legislation of regions.

To address such issues, Harvard and MIT researchers proposed a patient GNN to maintain privacy through combining differential privacy to introduce noise with gradient clipping and federated learning for decentralized training across hospital networks. It led to a compliant and efficient risk prediction model that performed within 5% centralized baseline on AUC [47]. These deployments show that it is feasible to have ethical responsible graph-based AI even in the most privacy-restricted regions with proper design. Adequate testing, regular auditing, and stakeholder engagement continue to be key pillars of responsible deployment.

5.5 Transparency and Interpretability in GNNs

Graph Neural Networks (GNNs) are now sophisticated learning devices from rich, structured data. Their ability to describe objects and how they interact with each other at once positions them in the front run of the best fit for most uses from drug discovery, social network analysis, finance fraud detection, and recommendation systems. With increasing numbers of users in high-stakes applications such as medicine, legal argumentation, and national security, requests for interpretability and transparency were at the top of GNNs' agenda. Effective deployment of GNN systems relies on knowing how and why a model arrived at the conclusion it did.

Robust as they are, GNNs remain vulnerable to every type of opacity, i.e., owing to the intricate interdependence among graph structure, feature aggregation, and deep representations. The following section formulates the intrinsic explanation problem of GNNs, briefly surveys XAI methods applied to graph models, and explains why interpretability matters especially for applications with societal and ethical implications.

Maybe the most perilous, most consequential use of graph-based AI is for healthcare data. Healthcare graphs are used to convey relationships between healthcare providers, clinical processes, disease, patients, and interventions. For example, a patient-disease graph might represent not merely individual diagnoses but also co-morbid disease and treatment spreading in a network. Clinical risk prediction is one of the biggest application domains, where GNNs forecast drug side effects or readmission risk from patient histories in graph format. But privacy issues run amok:

- Linkage attacks: Even with anonymized identifiers, a patient's treatment trajectory can be distinguished sufficiently to identify them uniquely [46].
- Neighbor leakage: Information from one patient can leak into neighbors beyond consent boundaries.
- Toleration regulations: The frameworks must be HIPAA compliant in the United States, GDPR compliant in the European Union, etc., other regional data privacy regulations.

In order to overcome these challenges, the MIT and Harvard researchers used a patient GNN privacy-preserving on top of differential privacy with gradient clipping and adding noise and federated learning to end-to-end train without data centralization. It built an approved and safe risk prediction system that was less than a 5% difference in performance with the centralized baseline on AUC [47]. These deployments demonstrate how graph-based AI ethics can be made possible even in the most privacy-sensitive domains with good design. Good deployment continues to have good assessment, constant auditability, and stakeholder feedback as fundamental aspects.

5.6 Ethical Auditing Framework for Graph-Based Models

As mentioned in Figure 5.3 Ethically auditing graph models is about ensuring they're fair, transparent, and nobody gets hurt. Here's a brief summary of typical steps:

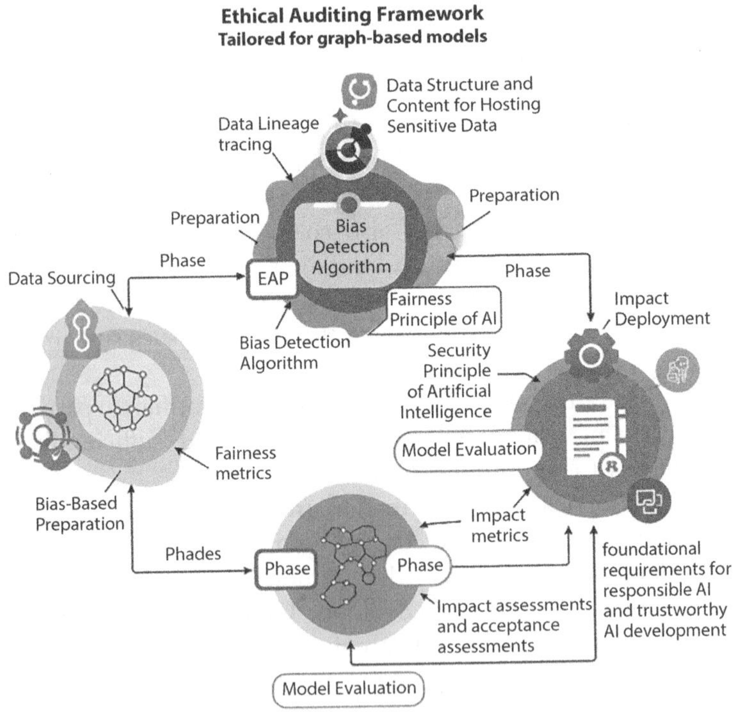

Figure 5.3 Flowchart for ethical auditing framework for graph-based models.

- Scope & Principles: Determine first which ethical principles (e.g., fairness, privacy) your model belongs to and what you'll audit.
- Data Check: Review where your data came from and the quality. Are there biases built-in? Is privacy maintained?
- Graph Construction: How was the graph constructed? Might it be biased in its construction? Are the features used interpretable?
- Model Selection & Training: Assess the algorithms employed. Do they increase bias? Are you testing for fairness at training time?
- Model Evaluation: Think beyond performance. Does the model generate equitable outcomes across various groups? Can you justify its findings?
- Deployment & Monitoring: After going live, monitor periodically for performance drift or fairness drift. Is there human monitoring? Can people provide feedback?
- Reporting & Remediation: All results recorded and provide easy-to-follow recommendations to remediate any ethical issue, answerable to people.

5.6.1 Challenges of Deep Graph Models Opacity

Interpretability means that a human can comprehend the justification for a model's prediction. In most conventional machine learning models like decision trees or linear regression, transparency is easy owing to their linear or rule-based nature. Nevertheless, GNNs are black-box models in the same manner as other deep learning models with an additional layer of complexity due to their recursive message-passing structure.

GNNs are opaque for several reasons:

- Multi-hop dependencies: A GNN not only aggregates information from direct neighbors but also from multi-hop nodes. This makes it hard to backtrack which component of the graph had the greatest impact on the ultimate prediction [48].
- Non-linear feature transformation: GNN layers perform non-linear transformations on aggregated messages, making it hard to attribute outputs to individual inputs.
- Node and edge feature interdependence: The effect of a single node feature cannot be separated from the influence of

neighboring nodes and edges, as opposed to tabular models where features are not dependent on each other [49].

- Dynamic weights and attention: Certain GNN architectures calculate edge or node importance dynamically (e.g., attention scores), which change with every input instance, and hence global interpretability is challenging.

This transparency becomes particularly troublesome in regulated domains where decisions need to be interpretable to meet legal or ethical demands, e.g., credit scoring, employment, or criminal justice.

5.6.2 Explainable AI (XAI) Techniques for GNNs

To tackle these challenges, researchers have suggested a range of XAI methods specific to GNNs. These methods seek to make GNN behavior more explainable by finding important nodes, edges, or subgraphs responsible for a prediction. Two of the most well-known classes of methods are attention-based methods and subgraph-based visual explanations.

(a) Attention Mechanisms (e.g., GAT)

Attention mechanisms were first introduced in natural language processing and have since been adapted to GNNs with great success in models such as Graph Attention Networks (GATs). GATs enable the model to assign learnable attention coefficients to the nearby nodes, reflecting their relative significance in the aggregation process [50].

Given a node i, its updated representation is computed by aggregating messages from neighbors j weighted by attention scores α_{ij}:

$$h_i' = \sigma\left(\sum_{j\epsilon N(i)} \alpha_{ij} W h_j\right)$$

Here, α_{ij} is computed using a self-attention mechanism, such as:

$$\alpha_{ij} = \frac{exp\ exp\ (LeakyRelU(a^T[\,|\,Wh_j]))}{\sum_{k\epsilon N(i)} exp\ (LeakyReLU(a^T[Wh_i\ ||\ Wh_k]))}$$

The interpretability benefit is that these attention coefficients can be examined to see what neighbors contributed most to a prediction.

Visualization of these weights across a graph can provide insights into model reasoning. As an example, in a citation graph, a high attention score from a neighboring paper node can be taken to mean that the model rated that reference highly informative for classifying the central paper.

But it is worth noting that attention does not necessarily mean explanation. Some argue that attention weights might not be a good proxy for feature importance [51], so more techniques are needed.

(b) Visualization of Subgraph Influences

Another strand of XAI methods for GNNs is to detect key subgraphs that are most impactful to predict a specific node or edge. These approaches regard interpretability as a subgraph extraction task. Some of the prominent methods include:

- **GNNExplainer:** Ying *et al.* [52] developed this method as shown in Figure 5.4 that produces a dense subgraph and corresponding features with maximum mutual information about the model's prediction. This learns a mask on the adjacency matrix and features to emphasize key elements.

Graph Neural Network Explainers (GNNEs) form the essential component of the solution towards grasping the complex decision-making process of Graph Neural Networks (GNNs). Although GNNs have performed well in node classification, link prediction, and graph classification, internal processes remain obscure. GNNEs try to make it possible to interpret by listing what subgraphs, node features, or relations were most important for a particular prediction. This comes in particularly handy in domains like medicine, social networks, or chemistry, where the interpretation of a prediction can sometimes be more valuable than the prediction itself.

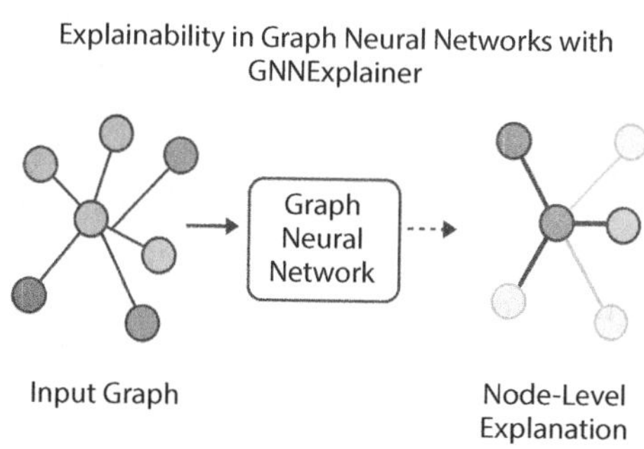

Figure 5.4 GNNExplainer.

Some algorithms are instances of GNNE. GNNExplainer, for instance, learns soft masks of node features and graph edges in order to find the most salient features in a particular prediction. PGExplainer, by contrast, is a probabilistic model that learns to place importance weights on edges for the majority of predictions. GraphLIME adapts the LIME (Local Interpretable Model-Agnostic Explanations) algorithm for graph data to explain by perturbing the input graph and observing prediction changes. The approach of other methods like SubgraphX is through the use of game-theoretic principles (e.g., Shapley values) for scoring and choosing salient substructures.

These explanation methods serve a number of purposes. They establish user trust through GNN predictions that can be explained, allow model error or bias detection, and deliver actionable knowledge for domain experts. Challenges remain—e.g., balancing explanatory simplicity against accuracy in explaining behavior, and scalability of explanation methods to large graphs. GNNEs are an important step toward responsible graph-based learning as a whole.

- **PGExplainer:** A probabilistic variant of GNNExplainer, where edge importance is represented as a distribution, allowing for more reliable explanations under uncertainty [53]. PGExplainer, or Parameterized Graph Explainer, is a model-agnostic framework designed to provide *post-hoc* explanations of GNN's predictions. Unlike conventional explanation tools producing instance-specific explanations, PGExplainer learns an explainable model that can be transferred across different predictions. Its core objective is to identify which graph components—more particularly, which edges—are most influential in the GNN's decision-making process. By learning a probability distribution on edges, PGExplainer can identify the significant subgraph that is accountable for a particular node's label to help users understand why the model generated a particular output.
  PGExplainer's process involves training an explainer network on top of a pre-trained GNN. This network takes node embeddings and node features as input and returns the importance of each edge connected to the node. The process, in doing so, trains the model to generate masks of edges that retain the most informative content of the graph in imitation of the original GNN prediction. The optimization of PGExplainer is directed at optimizing both for

faithfulness—how similar the explanation is to the output of the GNN—and sparsity—how few edges the explanation needs in order to be understandable and interpretable.

Ethically, PGExplainer is extremely significant in improving the transparency and accountability of graph models. Because GNNs make predictions over relational structures, hidden connections or latent biases in graph topologies can have a deep influence on their predictions. With PGExplainer, practitioners are able to expose such influences since the tool does make explicit the individual relationships and structure responsible for the decision of a model. For instance, in applications like criminal risk or creditworthiness prediction, PGExplainer can be used to help ascertain if the GNN is employing socially sensitive or discriminatory relations, e.g., relations to over-policed neighborhoods or economically disadvantaged groups.

In compliance scenarios, PGExplainer enables the demands of legal regulations such as the GDPR and the AI Bill of Rights through explaining both being understandable and actionable. It makes the "right to explanation" plausible *via* enabling tracing of how and why a specific decision was reached even in graph-based sophisticated systems. Furthermore, it enhances trust between AI systems through enabling developers and stakeholders to audit predictions for fairness and potential bias. Hence, PGExplainer is a very vital tool for the ethical application of GNNs, particularly in mission-critical environments where openness and accountability are essential.

- **Graph LIME:** Motivated by LIME (Local Interpretable Model-Agnostic Explanations), Graph LIME employs node neighborhood feature perturbations and trains a local interpretable model to mimic the predictions of the GNN [54] as illustrated in Figure 5.5.

  Graph LIME is an adaptation of the popular LIME framework for use with Graph Neural Networks (GNNs). Its main goal is to provide local interpretability by providing explanations for single-node predictions in graph models. While standard LIME creates explanations by perturbing input features and seeing what happens to predictions, Graph LIME adapts this concept to graphs by considering node features as well as graph topology. It achieves this by building a local

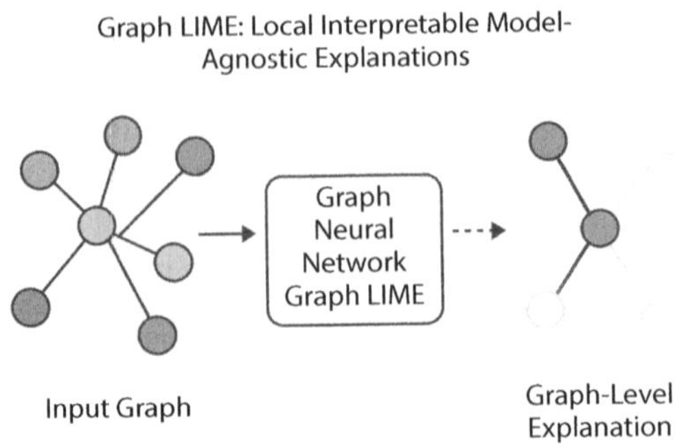

Figure 5.5 Graph LIME.

surrogate model—often a simpler, more interpretable one like linear regression—that can replicate the behavior of GNN in the local neighborhood of the target node.

For the purpose of explanation generation, Graph LIME samples perturbed neighbors of the target node by adjusting its features or its neighboring nodes' features. It then monitors the way that these perturbations impact the output of the GNN and tunes the surrogate model to the impacts. The final output is a distribution of feature attributions that correctly reflects what node features were most responsible for the GNN's prediction on this particular sample. Compared to PGExplainer, which is focused on discovering influential graph edges, Graph LIME is more interested in feature importance but still takes the relational graph context into account.

Graph LIME is especially useful in scenarios where influence recognition by features is important, like biomedical networks or fraud detection systems. For example, in a graph-based patient similarity disease prediction task, Graph LIME can help determine which medical attributes (such as test values or symptoms) were the most significant contributors to the prediction for a particular patient, without losing the relational context of the patient within the graph. This capability strengthens more transparent and trustworthy AI, especially in compliant areas.

Ethically, Graph LIME enhances model responsibility and data governance framework compliance by providing explanations accessible to non-technical stakeholders. It allows users and stakeholders to ascertain if sensitive or unjustified attributes are inappropriately affecting predictions and thereby facilitate fairness audits and evidence-based decision-making. Although it isn't necessarily a complete assessment of structural bias in the graph structure, any attempt, such as Graph LIME, which puts local feature explanations at the center of the attempt towards interpretable and accountable GNNs would be beneficial.

These tools enable practitioners to produce human-interpretable explanations, for example, emphasizing a group of co-diagnosed conditions in a medical graph or revealing an incriminating ring in a fraud detection graph. Scalability is still an issue. Subgraph extraction tends to include combinatorial search or sampling, which might be computationally costly on big graphs. In addition, explanations can differ among methods, so there is a need for common benchmarks and evaluation metrics.

5.6.3 Importance of Interpretability in High-Stakes Domains

Interpretability is not just a nice-to-have—often it is a regulatory and ethical requirement in most real-world applications. The opacity of AI models can result in unwanted discrimination, denial of service, or even fatal errors. GNNs, when used in such applications, need to be developed and tested with interpretability as a first-class objective.

(a) Healthcare

In GNN-based clinical diagnosis or treatment recommendation systems (e.g., patient-condition-treatment graphs), interpretability is crucial for clinical acceptance. Doctors must believe and verify AI recommendations, particularly if they go against conventional heuristics. For instance, when a GNN suggests an off-label drug following patterns in a treatment graph, clinicians should be able to know which relationships determined that recommendation [55]. In addition, explainable models facilitate compliance with patient consent and data governance regulations. If a model is making use of connections to patients who chose not to share specific data, this should be identifiable and fixable.

(b) Finance and Credit Scoring

Financial institutions employ GNNs to simulate complicated transaction networks, customer activities, and fraud. Yet, regulators typically request explanations of credit decisions in accordance with legislation such as the Equal Credit Opportunity Act (ECOA) or General Data Protection Regulation (GDPR). A black box model that cannot explain why a loan was rejected might be non-compliant. Interpretable GNNs assist in revealing whether specific demographics are being disproportionately penalized and aid audits of feature usage and data provenance [56].

(c) Social Networks and Content Moderation
In social networks, GNNs drive friend suggestions, feed ordering, and misinformation flagging. However, black box algorithms may create echo chambers, political polarization, or unfair content takedowns. Explainable models enable users and moderators to dispute and comprehend choices, promoting increased transparency and user confidence [57].

(d) Scientific Discovery
GNNs are more commonly applied in physics, chemistry, and biology for predicting molecular properties or finding novel materials. Here, interpretability supports scientific understanding, allowing scientists to check that the model adheres to what is already known or to develop new hypotheses by means of learned structures [58].

5.6.4 Toward Trustworthy GNNs

Notwithstanding the advances, interpretability in GNNs is still a developing area. Table 5.3 lists a comparison of some transparency and explainability methods with their strengths and weaknesses. These methods are a big leap toward demystifying the reasoning behind the decisions of GNNs. Yet, the area still has considerable challenges. Some of the major challenges and directions for the future are:

Table 5.3 Comparison of explainability techniques for GNNs.

Technique	Key concept	Strengths	Limitations
GNNExplainer	Identifies critical subgraphs and features	Localized, model-specific explanations	Computationally expensive
Graph LIME	Adapts LIME for graph data	Interpretable, *post-hoc*	Assumes linearity, lacks global view
PGExplainer	Probabilistic approach to explanations	Scalable, probabilistic insights	Requires training another model
XGNN	Generates prototypical explanations	Human-understandable graph generation	Low fidelity to actual model reasoning

- Global interpretability: The majority of existing techniques are local (i.e., per prediction). Model-level explanations that reflect general behavior or decision logic are needed.
- Human-in-the-loop XAI: Integrating domain expert feedback with model interpretation to improve both explanations and model performance.
- Fairness-interpretability trade-offs: In certain situations, enhancing fairness can mean compromising interpretability, or vice versa. A principled approach to managing these trade-offs is necessary.
- Explanation robustness: Explanations must be stable and consistent under small input perturbations. Instability in explanations can undermine user trust.
- Evaluation metrics: No single standard exists for measuring explanation quality. Fidelity, sparsity, comprehensibility, and agreement with human judgment must be standardized.

By embedding interpretability into the development lifecycle of GNN systems, researchers and practitioners can build models that are not only accurate but also trustworthy and aligned with societal values.

5.7 Societal and Ethical Implications

Graph Neural Networks (GNNs) are a fast-evolving category of machine learning models that can handle sophisticated relational data. Their expressive capacity enables dramatic improvements in tasks like social network analysis, recommendation systems, modeling of biological molecules, financial fraud detection, and so on. Yet as GNNs become more integrated into high-stakes decision systems, they present a diversity of societal and ethical issues. These include issues of surveillance, dissemination of misinformation, algorithmic fairness in decision-making, accountability of algorithms, and cultural translation of ethics. This part examines where GNNs overlap with important areas of society, addresses the ethical challenges that come with their use, and suggests how cultural and contextual ethics can be used to guide the proper application of these models.

5.7.1 Use Cases with Societal Impact

(a) Surveillance and Misinformation

Perhaps the most important societal consequence of GNNs occurs in the area of surveillance and control of information. GNNs have shown effectiveness at learning social graphs to discover patterns of behavior, community structure, and prominent nodes. In security applications, this can be used to track crime networks, detect radicalized groups, or detect money laundering chains [59]. The very same functionality can be used for bulk surveillance. For example, authoritarian governments can utilize GNNs to process communication graphs, location information, or web interactions to track dissent, quell protests, or mark citizens as threats based on their associations—raising serious civil liberty issues. In contrast to conventional surveillance systems, the relational inferences drawn by GNNs can result in guilt-by-association, where a person is marked not for what they have done but for being near others in a graph [60]. In addition, GNNs have been used to detect misinformation by learning propagation paths of false news in social networks. Although this is useful, it also provides room for censorship and manipulation. Algorithms can overly flag specific narratives or groups, particularly if trained on unbalanced data. In polarized political settings, this can lead to stifling of opposition or entrenchment of state-dominated narratives. The key ethical risks and challenge here is balancing security and control against freedom and autonomy as provided in Table 5.4. Transparent policies, algorithmic audits, and citizen engagement are necessary to prevent misuse of GNN-based surveillance technologies.

Table 5.4 Ethical risks in real-world GNN applications.

Domain	**Use case**	**Potential ethical risks**	**Governance need**
Healthcare	Patient risk prediction	Misdiagnosis due to opaque models, biased data	Explainability, fairness audits
Finance	Credit scoring, fraud detection	Discriminatory credit decisions, data misuse	Algorithmic accountability
Social media	Content recommendation	Amplification of bias, misinformation, echo chambers	Ethical content curation guidelines
Law Enforcement	Predictive policing	Racial profiling, unjust surveillance	Legal oversight, audit frameworks

(b) Automated Decision-Making (Finance, Hiring, etc.)
Perhaps most important social impact of GNNs is surveillance and information control. It was demonstrated that GNNs can learn social graphs well in order to find patterns of behaviors, community, and influential nodes. This could be applied to security-related domains to track crime networks, radicalized groups, or money-laundering chains [59]. The same ability can be turned towards bulk surveillance. For example, authoritarian governments will employ GNNs to track communication graphs, geolocation history, or web traffic in the name of surveillance on opposition groups, quelling riots, or labeling citizens as risky based on whom they interact with—widely raising civil liberty issues. In contrast to standard surveillance networks, relational inferences by GNNs can potentially include guilt-by-association, where a person is not marked for what they have done but for being proximate to other people in a graph [60]. Besides, GNNs have also been employed to detect misinformation by learning patterns of fake news spread in social networks. It aids detection but does not close doorways for censorship and manipulation. Algorithms might trigger over-flagging through flagging some articles or groups unnecessarily, particularly when the models are trained on biased data. Under polarized political climates, it can silence the opposition or amplify state-sponsored narratives. The central moral concern here is finding the balance between security and control and freedom and autonomy. The citizens must be informed by transparent policies, auditing of algorithms, and civic engagement in order to prevent the abuse of GNN-based surveillance technology.

By making interpretability part of the design process for GNN systems, researchers and practitioners can build models that not only are precise but also trustworthy and value-aligned.

5.7.2 Ethical Dilemmas in Deployment: Who is Accountable?

Graph Neural Networks (GNNs) are a general class of machine learning models for representing and processing high-level relational data. Their expressiveness has dramatic potential for applications from social network analysis, recommendation, biomolecule modeling, financial fraud detection, etc. But in being increasingly subject to high-stakes judgments, they bear a richness of societal and ethical concerns. They include surveillance issues, spreading misinformation, equitability of decision-making algorithms ethics, algorithmic accountability, and cultural translation of ethics. This section discusses where GNNs intersect with significant aspects of society, considers the ethical issues of their use, and offers ways cultural and contextual ethics can be used to ensure these models are being utilized correctly. There are a number of ethical principles at work here:

- Accountability: Requestors of AI should be able to account for and hold someone accountable for AI output. This includes being open with audit trails, decision logs, and design decisions.
- Transparency: The models must generate traceable and explainable output so that regulators and users can audit them.
- Liability: The regulatory schemes must determine the scope of liability across the life cycle of AI. Currently, there are some jurisdictions with uncertainty regarding laws covering algorithmic harm, where there are gaps in accountability [61].

GNNs create richness from dependencies among relations. A seemingly harmless change in one node (e.g., an extra social relationship) can cascade across the graph and have a large impact on many other entities' outputs. This network externality disrupts standard assumptions regarding causality and blame.

Therefore, ethical use of GNNs must include a multistakeholder governance system with the developers, organizations, policymakers, and affected communities together deciding and governing acceptable uses of these models.

5.7.3 Role of Cultural and Contextual Ethics

The second and equally critical neglected area of AI ethics is to make ethical concepts culturally and contextually sensitive. Practically any other system appropriates Western liberal philosophical heritage, upholding respect for individual rights, autonomy, and consent. All are necessary but perhaps not representative of the wholly collectivist, indigenous, or non-Western values. When applied in GNNs specifically, since they are already operating on relational data, cultural problems are even more critical. For example:

- Single action is brought to be actualized in the group environment of community, tribe, or family in collectivist cultures. An example of GNN for lending decision use may end up being a desocialization by taking an individual out of the group environment [62].
- More emphasis is laid on relational duty and data ownership in indigenous knowledge systems. Unregulated and infringe uses of graph data on indigenous people, anonymized or not, represent moral abuse.

- Reputation and honor in certain societies are social constructs. GNN prediction of sensitive information (party political leanings, religious belonging) from graph structure can deliver harms to general categories.

These sketches suggest that accountable AI cannot be a plug-and-play solution. Implementing GNNs across domains involves cultural competence—awareness of prevailing values, social norms, and histories. It entails community involvement in AI design, from citizen data collection to collaboratively defining model goals and test criteria. New paradigms such as Data Feminism and Indigenous Data Governance offer promising principles on integrating local ethics and data and AI practice [63]. Equity, agency, and reciprocity are their concerns, reminiscent of the relational mind of graph learning models.

5.7.4 Ethical Risks of GNNs in Context with Real-World Applications

Graph Neural Networks (GNNs) are powerful as well as powerful ethical and legal concerns, especially if utilized in high-risk use cases. Such concerns are best understood when placed in context of real-world applications.

- **Law Enforcement (Chicago's Strategic Subject List):** GNNs have been experimented upon to predict criminal behavior based on social or gang analysis. For instance, in Chicago's no longer used Strategic Subject List (SSL), individuals were labeled high-risk not for their own criminality, but due to their friends and acquaintances who had been involved in crime. A GNN could compound the sort of profiling that occurs in such a list by pulling out relational features that negatively penalize an individual for whom they are associated, even if they have done nothing wrong themselves. This leads to algorithmic over-policing and inflicts injuries on due process and non-discrimination rights.
- **Recommendation Systems (e.g., YouTube, TikTok):** Platforms like YouTube and TikTok use GNNs to model complex user-content graphs to suggest videos. Effective but capable of building echo chambers—suggestion increasingly extreme or specialized content. Political videos, for example, can quickly lead users to radical or biased videos due to

tightly knit sets in the content graph. This manipulation limits exposure to diverse perspectives and has been linked with the spread of misinformation and radicalization processes, raising issues of autonomy and social polarization.

- **Financial Scoring (e.g., Zest AI and credit networks):** Companies like Zest AI make use of AI models, potentially involving graph-based approaches, to assess creditworthiness. GNNs are also being tested in developing economies to examine borrowers using phone contacts, transaction graphs, or work relationships. This can penalize someone with good personal finances if they are in a group of financially vulnerable persons. For instance, a freelance worker can be denied credit just because of bad contacts within their professional network. This is indirect discrimination and violates fairness and openness practices.

These examples illustrate how the relational modeling benefits of GNNs, when employed without regard for ethics in fairness considerations, can result in unjust outcomes. Incorporating legal frameworks like GDPR and AI Bill of Rights is essential in making fairness implementable, bringing transparency, and protecting people against being unjustly condemned by their relations or patterns that are stuck in complex graph structures.

5.7.5 Resilient Path Forward

The ethical and social influence of GNNs is extensive and multifaceted. Regardless of how incredibly proficient the models are at mastering difficult tasks; they are to be followed by a proper ethics framework consisting of:

- Pre-deployment harm assessment based on harms possibly endured, particularly by disadvantaged populations.
- Fairness testing and auditing of graph data and model output.
- Open documentation in model cards, datasheets, and ethics checklists.
- Data usage community consultation and informed consent processes.
- Culturally responsive governance where ethics is contextual, not prescriptive.

As GNNs penetrate further into society's personal spaces, ethical stewardship becomes not a best practice, but an ethical imperative. Researchers, practitioners, and policymakers must come together to develop systems that not only function well, but also embody the values of justice, dignity, and human rights.

5.8 Legal, Regulatory, and Policy Frameworks

The fast growth and deployment of Graph Neural Networks (GNNs) into different sectors had fallen behind the evolution of sound and deployable legal and regulatory frameworks. The GNNs raise new challenges since they can make subtle inference of patterns and infer sensitive information from related data. This characteristic leads to radical legal challenges of justice, accountability, transparency, privacy, and issues of liability in the very core of the extensive debate on regulation of artificial intelligence (AI). This chapter provides an overview of the future global regulation of AI, i.e., their extension to graph-based models like GNNs. It then discusses the specific difficulties of regulating graph-based AI and the path to responsible regulation and legality.

5.8.1 Overview of Global Regulatory Efforts

Governments, intergovernmental institutions, and technical organizations have started to design principles, norms, and policy tools to guide the ethical and lawful application of AI technologies. Pioneering these initiatives are the European Union's Artificial Intelligence Act, IEEE's ethical requirements, and UNESCO's global guidelines.

(a) The EU AI Act

The 2021 draft European Union Artificial Intelligence Act (EU AI Act) is the first comprehensive regulation intended to directly regulate AI in member states. The Act establishes a risk-based method to regulate AI by categorizing systems into four risk levels: unacceptable, high, limited, and low risk [64]. High-risk applications involving critical infrastructure, employment, and law enforcement are subject to rigorous obligations. GNNs, depending on their application, can be high-risk or even unacceptable-risk. A GNN deployed for automated hiring, credit scoring, or predictive policing, for example, should be regulated so that it ensures transparency, robustness, accuracy, and fairness. These system developers must:

- Conduct risk analysis and impact analysis.
- Meet rigorous data governance principles.
- Bedeck with human oversight and proper documentation.
- Make decisions audit-proof and traceable.

Apart from that, Article 5 of the Act prohibits AI systems that affect human behavior or employ social scoring software on the basis of social behavior or inferred personality. Since GNNs will presumably run on social graphs and make inferences on sensitive features, developers must take care not to violate the law. The EU AI Act also brings the right to explanation into the spotlight, as per the GDPR provision on automated decision-making. GNNs, as being transparent and complex, will need to be accompanied by the explainability machinery if they are used in high-risk domains.

(b) IEEE and UNESCO Ethical Guidelines

The Institute of Electrical and Electronics Engineers (IEEE) has released a broad set of guidelines and standards under its Ethically Aligned Design initiative. These guidelines aim at integrating human rights, accountability, transparency, and environmental sustainability into AI systems [65]. They promote active ethical impact assessments, participatory design processes, and quantifiable metrics for human well-being. IEEE's P7000 series standards—specifically P7001 (Transparency), P7003 (Algorithmic Bias), and P7006 (Personal Data AI Agent)—are most directly applicable to GNN-based applications. These standards suggest technical and organizational practices to reduce bias, justify decisions, and handle data responsibly. For example, graph models handling personal data need to record the data lineage, model assumptions, and possible bias propagation through graph connections. Likewise, UNESCO's Recommendation on the Ethics of Artificial Intelligence, adopted in 2021, presents a universally applicable system founded on principles of dignity, diversity, privacy, and environmental sustainability [66]. It focuses on inclusive governance, capacity development in low-resource countries, as well as safeguarding cultural and linguistic diversity in AI systems. Although they are non-binding, they set national policies and determine the ethical culture regarding AI deployment. They also supply a normative basis for private sector actors wanting to create globally compliant and socially responsible AI products.

5.8.2 Challenges in Regulating Graph-Based AI Systems

Although current regulations address general AI issues, graph-based learning models have novel technical and ethical issues that make them difficult to regulate.

(a) Inferential Privacy Risks

Unlike independent data samples on which traditional models are based, GNNs draw on relational inferences. They are able to predict a person's traits not from their own data but from their relations in the graph. This produces inferential privacy concerns: even when a user declines data sharing, he/she is still vulnerable to privacy compromise through the graph topology [67]. This erodes the concept of individual consent, which is the foundation of data protection legislation such as the GDPR. Regulators thus need to create group-level privacy models that take into consideration the networked nature of information in GNNs.

(b) Lack of Explainability

Graph models are also by nature opaque, as a result of the multi-hop message-passing model that recursively combines neighbor information. Because of such intricacy, a model's prediction cannot easily be traced to given input features or graph substructure, an aspect that goes against transparency and accountability requirements. While Explainable AI (XAI) technology is being developed for GNNs—e.g., GNNExplainer and PGExplainer—they are not yet regulated or interpretable by non-expert stakeholders, like auditors or impacted users. This challenges the adequacy of existing transparency tools for regulatory requirements.

(c) Algorithmic Collusion and Network Effects

In certain uses, GNNs can engage in automated coordination, e.g., price optimization in online markets. When various agents employ the same graph-based models, they can inadvertently collude or converge on anti-competitive actions, a situation referred to as algorithmic tacit collusion [68]. Classic antitrust regulations lack the ability to identify or punish such emergent behavior, particularly in the absence of clear-cut agreement among parties. Regulators have to develop technical auditing mechanisms and doctrines of law to oversee GNN-based interactions in competitive environments.

(d) Jurisdictional and Normative Conflicts

Graph data tends to cross borders, as does GNN-based infrastructure. A social network operating in several nations has to meet different data protection regulations, ethical guidelines, and cultural norms. For instance, a GNN that predicts political orientations might be lawful in one nation but a breach of constitutional rights in another. This fragmentation makes it difficult for multinational developers to comply, as they have to construct flexible models and localized governance structures. It also creates normative tensions between free speech, state security, and privacy, especially in geopolitically sensitive uses.

5.8.3 Toward AI Governance and Compliance

In spite of such challenges, an emerging consensus regarding the importance of overarching, anticipatory, and inclusive frameworks for GNNs and analogous AI systems has developed. The way forward lies in a synergy between technical measures, organizational procedures, and policy innovation.

(a) Technical Solutions

- Model Cards and Datasheets: Documentation patterns such as Model Cards for Model Reporting and Datasheets for Datasets can be applied to graph models to offer transparency on data sources, training practices, intended applications, and performance measures [69].
- Auditable Pipelines: Constructing modular, traceable pipelines with versioning, input logging, and graph perturbation testing can enable accountability and facilitate compliance breach detection.
- Ethical Risk Evaluations: Standardizing risk assessment processes (e.g., impact analyses) for graph-based systems may be able to detect harms and countermeasures early in the design process.

(b) Organizational Practices

- Ethics Review Committees: Institutionalizing interdisciplinary review boards for organizations to approve GNN applications can promote increased ethical monitoring and stakeholder engagement.

- Bias and Fairness Teams: Dedicated teams that measure and counteract biases in graph data and model predictions guarantee that fairness is not an afterthought but an intrinsic design consideration.
- Privacy by Design: Inheriting privacy-protecting methods—e.g., differential privacy, homomorphic encryption, and federated learning—into GNN structures facilitates data protection legislation compliance [70].

(c) Policy Innovations

- Guidelines Specific to Graphs: Graph-based AI-specific guidelines, such as standards of explainability, quantifying privacy risk, and relational data stewardship, must be published by regulatory authorities.
- Third-Party Auditing: High-risk applications of GNNs should have mandatory third-party audits regarding social impact, bias transfer, and regulatory compliance.
- Participatory Policymaking: Participation by civil society, domain specialists, and underprivileged groups in policymaking on AI makes rules inclusive, context-sensitive, and democratic.
- International Harmonization: Collectively through such platforms as the OECD, G7, and UNESCO, interoperable regimes can be made with reduced compliance costs but ensured rights.

Legal, regulatory, and policy frameworks for artificial intelligence are changing with the increasing reach of machine learning technology. Graph Neural Networks are, however, special cases because of their distinctive capabilities and ethical concerns. Fundamental values of openness, privacy, and responsibility have to be reapplied and resolidified to the graph-based learning environment. By a confluence of innovation in law, foresight in ethics, and technicality in engineering, civil societies can effectively regulate GNNs in directions consistent with the values of democracies and respect for human rights. As GNNs progressively invade industries ranging from finance to medicine, the need to rule wisely and responsibly grows more imperatively pressing every day.

5.8.4 Legal and Societal Frameworks: Implications for Graph Neural Networks

Since Graph Neural Networks (GNNs) are increasingly deployed in sensitive areas like healthcare, finance, and criminal justice, a review of how current legal and social norms cut across these technologies is essential. The European Union's General Data Protection Regulation (GDPR), for instance, and the United States' AI Bill of Rights have been created to guarantee artificial intelligence technologies to uphold human rights, be transparent, and not discriminate. The relational and interconnected nature of GNNs, though, pose ethical and legal challenges that were not so developed in the networks. The GDPR offers individuals a number of rights such as the right to explanation, right to be forgotten, and clear consent where data are being harvested and utilized. The rights can be simply enforced within regular machine learning models. GNNs do record, however, that one person's data influences the prediction made of another through the graph edges and node interaction. That holds the potential for a person to be profiled or impacted by a decision made based not on their own data, but on their networks within the graph—something they may not have necessarily agreed to. Moreover, multi-step node propagation of information precludes transparent explanations of model predictions, effectively abolishing the right to explanation under the GDPR.

Likewise, the AI Bill of Rights aims to develop safe, transparent, and fair AI systems. It necessitates protection against algorithmic bias and alerts an individual if a machine is deciding something about them. GNNs pose extremely atypical challenges to these values. Since GNNs learn from the structure and attributes of whole graphs, they can learn structural or historical biases unintentionally—e.g., in crime or social graphs—so that historically marginalized or disadvantaged subgroups are treated discriminatorily. Users will also frequently be unaware how and why choices were made, or which edges within the graph influenced a particular prediction, so that it becomes hard for them to query or complain about automated decision-making. Apart from such issues of control, there are also cultural values of being inclusive, responsible, and equitable in AI development. GNNs tend to aggregate and pass information across entities in a natural way. This can lead to a group-level harms that individual rights-based systems like the GDPR cannot directly handle. For example, a whole population can be structurally deprivileged if the structure of the graph represents historical bias or exclusion even though individual personal data points are not biased. This calls for a new theory of justice from moral and legal foundations to demand collective control mechanisms and an extended harm theory that is sensitive to structure and relation dependence. To pursue their goals, such social and legal mechanisms increasingly need protection and regulation specifically GNN-oriented. Default modes of AI regulation fall short when addressing models based heavily on human being or entity interaction. Design of ethically sound GNN systems must therefore attend to not just technological progress but also appropriate knowledge of legal rights, privacy, and general social impact of network-based learning. It is not a technical issue but an issue of law and ethics to make the GNNs fair, transparent, and accountable.

5.9 Best Practices and Ethical Guidelines for Practitioners

While Graph Neural Networks (GNNs) are increasingly powering decision-making across industries—from recommendation systems and social networks to healthcare and financial analysis—there is a corresponding need for ethical guidance specific to the entire lifecycle of graph-based AI systems. In contrast to standard machine learning, GNNs process structured relational data, posing new technical and ethical challenges throughout data collection, model design, deployment, and real-world feedback. This chapter introduces a best practice and ethics framework for practitioners, such as researchers, data scientists, engineers, and policy advisors. It focuses on interdisciplinary collaboration, early risk mitigation, and life-cycle accountability, which harmonize with general Responsible AI aspirations and regulatory imperatives.

5.9.1 Principles for Responsible AI Development

The basis of ethical GNN practice is the implementation of universal Responsible AI principles, which have been supported by organizations like the IEEE, UNESCO, OECD, and the European Commission. The following principles must be internalized throughout all phases of GNN development and deployment:

- Human-Centric Design: The AI systems need to augment, not substitute for, human agency and well-being. GNNs should be designed keeping the end user at the center, so their functionality conforms to user autonomy and dignity [71].

- Fairness and Non-Discrimination: Models should be made sure that they do not reinforce or intensify bias. Since graph data tends to represent actual-world differences (e.g., social, economic, or educational networks), fairness auditing and debiasing need to be integrated into the pipeline [72].
- Explainability and Transparency: Decisions taken by GNNs—particularly in high-risk applications—need to be comprehensible to impacted parties and auditable to third parties. Methods like attention visualization and tracking node influence can contribute to transparency [73].
- Security and Privacy: Models need to adopt privacy-by-design approaches, particularly when dealing with sensitive graph data, like health or social media. Privacy-enhancing methods like differential privacy and homomorphic encryption should be considered where applicable [74].
- Accountability and Traceability: Model lineage, decision justification, and ownership should be clearly documented by practitioners. This facilitates legal compliance and provides accountability in the event of harm or failure.
- Sustainability and Social Good: GNNs should be designed with consideration of environmental and societal effects, such as carbon footprint and long-term societal change in labor, communication, and surveillance.
- These are not lofty ideals but need to be converted into real-world engineering and organizational practice, as will be discussed below.

5.9.2 Interdisciplinary Collaboration: AI + Ethics + Law

Ethically developing GNNs cannot take place in disciplinary isolation. As the complexity of graph-based learning models and ethical implications involved makes it imperative to have a holistic approach that covers technical, legal, and philosophical expertise.

- AI Practitioners have algorithmic, data structure, and performance metric knowledge but potentially lack awareness of ethical or legal subtleties inherent in graph data usage (e.g., risk of learning protected attributes from links).
- Ethicists and Social Scientists have necessary knowledge of systemic bias, user agency, cultural context, and

value-sensitive design. They engage to ensure that systems fit democratic and inclusive values [61].

- Legal Specialists translate current legislation (e.g., GDPR, HIPAA, EU AI Act) and assist in forecasting regulatory advancements to ensure compliance across the AI lifecycle.

By integrating these fields in design charrettes, model review boards, and ethical risk audits, richer framing of problems, foresight, and ability to respond to downstream effects are made possible. In addition, engaging end users and oppressed communities brings experiential knowledge and assists in mitigating epistemic injustice—where some knowledge forms are left out of system design [75].

5.9.3 Recommendations for Ethical GNN Lifecycle

The distinctive graph data structure and the recursive process of GNN learning necessitate lifecycle-stage-specific ethical interventions. We now present a step-by-step dissection of best practices for developing ethical GNN systems.

(a) Data Sourcing

Ethical concerns in GNNs usually arise at the data sourcing level, as graphs store not only attributes but also relations—most of which can be indirect or sensitive in nature.

- Consent and Transparency: In contrast with tabular information, graph information can hold personal data from persons who never deliberately agreed to their data sharing (e.g., inferred relationships or co-author networks). Network-level consent should be obtained, as well as full disclosure regarding usage of the linkages [76].
- Bias Mitigation during Graph Construction: If edges derive from biased procedures (e.g., hiring recommendations or citation relations), then GNNs might inherit and further perpetuate systemized inequality. Preprocessing steps such as edge reweighting, sampling, or sanitizing must be applied to counter such biases.
- Data Provenance and Documentation: Source graphs should be supported by datasheets describing their origin, curation process, known biases, and limitations. Such documentation

should be made available to stakeholders and part of model reporting pipelines [77].

- Privacy Protection: Sensitive graphs (e.g., patient-provider networks) need pseudonymization, access control, and legal scrutiny. Consider methods such as k-anonymity for graphs, node masking, or differential privacy to minimize privacy leakage.

(b) Model Development

The modeling stage is where fairness, explainability, and robustness techniques must be introduced. GNNs pose unique points of engagement for ethical optimization.

- Fairness-Aware GNN Design: Designers must inspect and adjust structural bias *via* mechanisms such as adversarial debiasing, edge perturbation, and fairness constraints (e.g., equal opportunity or demographic parity between node classes) [78].
- Explainability and Interpretability: Add modules like attention layers (e.g., GAT), influence estimation (e.g., GNNExplainer), or subgraph visualization tools. These facilitate end users and auditors to trace back the reasoning of model predictions.
- Adversarial and Robustness Testing: GNNs are susceptible to adversarial attacks (e.g., node injection, edge removal). Security testing must be included in model validation in order to ensure robustness against tampering or data poisoning [4].
- Cross-Validation Across Demographics: Train and test models across subgroups to incorporate disparate impact. It is done across intersectional groups and minority clusters in graph structure.
- Auditability: Store model updates, hyperparameters, dataset version numbers, and evaluation metrics with tools such as Moldflow or Data Version Control (DVC). This makes it easier to comply and downstream debug.

(c) Deployment and Feedback Mechanisms

Ethical monitoring has to extend beyond training, to deployment and feedback stages, where real-world implications become apparent.

- Drift Detection and Monitoring: Regularly check model predictions for performance decline and ethical breaches. GNNs, due to their dynamic graph structure, are highly susceptible to concept drift and long-term structural change [79].
- User Feedback Loops: Allow user feedback channels to invalidate decisions, report mistakes, or propose fix-ups. This builds trust and responsiveness in the system, most critically in high-risk domains such as medical diagnosis or creditworthiness.
- Human-in-the-Loop Monitoring: In high-criticality scenarios, retain human monitoring, e.g., post-facto auditing of high-risk choices. This is governed by an array of regulations and laws, such as the EU AI Act and GDPR.
- Transparency Portals and Disclosures: Display model cards, risk assessments, and summary dashboards publicly accessible. In controlled environments, this transparency is a compliance tool and trust building exercise.
- End-of-Life Management: Provide for future model decommissioning or retraining when they are no longer used, become biased, or have been hacked. This also entails keeping historical logs of performance to audit.

The legal and ethical considerations of GNNs do not only contribute technical innovation but also moral insight and procedural accountability. Experts must transcend checklists of compliance to a deep, cross-cutting, and user-focused approach for graph-based AI. By incorporating best practices along the whole AI life cycle, from data curation and modeling through deployment and auditing, developers can create GNN systems that are more equitable, secure, and trustworthy. Such an effort at normative ethics and the law constitutes the groundwork for guaranteeing the public interest of GNNs in the modern era of digitally globalizing.

5.10 Conclusion and Future Outlook

Graph Neural Networks (GNNs) today is applied in influential arenas such as healthcare, finance, policing, and social networks, and it is becoming more challenging to draw a moral line to utilize them. In this chapter, we have discussed important ethics that emerge throughout the GNN life cycle—from fairness and minimizing bias, data privacy, transparency, and accountability, to societal and regulatory considerations. The graph structure of the data provides unique challenges in the sense that relations among entities can contain sensitive or biased data and, in those judgments, will normally be computed with direct and indirect relations. Such features require specific tools and methodologies to infuse moral goodness on graph-based learning. The Table 5.5 summarizes the most significant ethical risks, potential violations, and actual-world mitigation strategies in various uses of Graph Neural Networks (GNNs). It provides a

Table 5.5 Ethical risks and mitigation strategies for GNN applications.

Application domain	Key risks	Ethical violations	Mitigation strategies
Law Enforcement	Relational bias due to associations; unfair targeting of communities	Discrimination, violation of presumption of innocence	Debiasing datasets, limiting sensitive attributes, explainability tools like PGExplainer
Recommendation Systems	Filter bubbles, echo chambers, over-personalization	Manipulation, reduced autonomy, misinformation	Content diversity enforcement, transparency in recommendation algorithms
Financial Scoring	Guilt by association; proxy discrimination through social/transaction graphs	Indirect bias, unfair denial of credit	Fairness constraints in training, graph sanitization, regulatory compliance audits
Healthcare	Hidden relational bias in patient similarity graphs	Inaccurate diagnosis or treatment, exacerbation of health disparities	Medical expert-in-the-loop systems, Graph LIME for local interpretability
Social Media Analysis	Amplification of toxic clusters or harmful influence spread	Promotion of hate speech, polarization	Graph pruning, node/edge-level bias detection, community-level fairness regularization
Hiring/HR Analytics	Bias from organizational or social networks influencing decisions	Employment discrimination	Fair GNN training objectives, exclusion of non-merit-based relationships

simple-to-grasp overview to guide good design, deployment, and auditing of graph-based learning models.

Some of the most effective ethical methods under debate are fairness-aware graph algorithms aiming to trade-off prediction quality for representational fairness, adversarial debiasing algorithms that strip away sensitive correlations *via* optimization, and graph sanitization techniques that undo the effects of biased relationships in preprocessing. Privacy-protecting techniques like differential privacy, federated learning, and homomorphic encryption have stepped into the limelight as overdue interventions in addressing the increased threat of privacy leakage inherent in gargantuan, densely connected graph data sets. Interpretability mechanisms, on the other hand, ranging from attention-based architectures like Graph Attention Networks (GAT) to subgraph visualization mechanisms, are utilized to interpret erstwhile black box decision-making procedures, where necessary in high-risk domains. Regulatory and policy structures such as the EU AI Act and IEEE standards set a general context, but application to the GNN world is a dynamic boundary. There are some strategic directions that in the future will need to be tackled by practitioners and researchers if GNNs are to gain more trust and societal harmony. The first among these is the need for strong and normalized fairness benchmarks for graph learning. Machine learning does have fairness measures, but they are typically weak to capture the recursive, relational nature of graph data. New paradigms of benchmarking would need to incorporate network topology, structural centrality, and community-based imbalances. These would enable practitioners to more accurately evaluate algorithmic fairness not only between agents, but also between graph subgraphs and communities—ensuring model behavior aligns with distributive justice and equity norms.

Second, moral monitoring and auditing mechanisms in real time will be needed, particularly as GNNs become more commonly embedded in dynamic environments where graph topologies change. Operating continuous feedback loops that report emergent bias, performance drift, or interpretability failure can allow teams to correct course in a timely manner during the life of the model. This is particularly important in domains such as financial trading or social media content moderation, where harms as fleeting as a few milliseconds can cascade quickly. Intelligent toolboxes that track model performance in production, guided by ethical indications, will be a central facilitator of morally responsible AI deployment into real application domains.

Third, symbiotic human-AI system design is an ethically forward-looking requirement. GNNs are not to be created as mechanistic systems to replace human judgment but as supportive agents whose capacities complement human abilities. To achieve this, system interfaces need to be designed as interpretability-enabled and actionable insight-driven in order to allow domain experts, auditors, and users to comprehend, question, and direct AI decision-making. Human-in-the-loop settings—particularly in medicine, criminal justice, and employment—can be utilized to manage risks without compromising human values. Ethical GNNs therefore need to pair technical as well as interface design, workflow integration, and user training.

In an effort to ensure ethical and accountable use of Graph Neural Networks (GNNs), fairness, transparency, and explainability should be built into every stage of the model life cycle. This begins with strict data curating to eliminate or minimize biases, and specifically those based on sensitive attributes such as gender, race, or socioeconomic status. Developers must employ fairness-aware learning objectives and leverage explainability techniques like PGExplainer and Graph LIME to reveal how GNNs arrive at specific predictions. They not only enhance transparency but also facilitate identification of inherent biases in graph structure as well as data-driven decisions based on relevant and interpretable features. Transparency may also be used in model documentation—explaining the architecture, graph construction habits, training procedures, and decision-making processes in simple terms to allow stakeholder knowledge and regulation.

Referral to modern laws and ethics is also essential to guarantee dependable GNN implementation. Regimens like the General Data Protection Regulation (GDPR) and the AI Bill of Rights focus on the right to explanation, absence of discrimination, and algorithmic accountability. Organizations need to conduct regular ethical audits to determine model fairness, privacy risks, and discriminatory impacts on different demographic groups. Documentation and ongoing stakeholder engagement—particularly with affected groups, domain experts, and ethicists—are also needed to ensure the system aligns with actual needs and values. Declaring the activities official allows organizations and developers to create GNN systems that not only perform satisfactorily and are resilient but also ethical and deserving.

Responsible Graph Neural Networks (GNN) development and deployment require a solidly developed set of best practices that balance technical excellence and ethical accountability. One such is maintaining integrity and fairness in the data from the beginning. Since GNNs are working with relational data, any slight bias during graph creation—over-sampled groups, noisy or missing edges—can be passed on to unfair or wrong conclusions. The practitioners therefore need to thoroughly examine node features and edge structures for discrimination and representativeness issues. Incorporation of fairness-aware graph sampling, regularization, and data debiasing methods can go a long way towards guaranteeing the ethical stability of the model.

Transparency and explainability is the other best practice that is needed. GNNs are black boxes, and hence it's difficult to know what they are delivering. Contrary to this, the models need to be supplemented with interpretability techniques like Graph LIME for feature explanation and PGExplainer for edge explanation. These techniques provide stakeholders valuable insights into decision-making and are especially crucial in domains like finance, medicine, or law enforcement. Transparency also needs to be embedded in public documentation like model intent, training data sources, graph structure, performance metrics, and known constraints. This facilitates greater knowledge, agreement, and trust by users and regulators.

Lastly, continuous tracking and stakeholder interaction are essential in the deployment process. Ethical threats in GNNs are not static—according to changing data, users, and real-world conditions. Deployments need to include feedback loops, ethical audit pipelines, and channels for user redress or explanation requests. Integration with diverse stakeholders like ethicists, lawyers, affected users, and subject-matter experts renders the GNN system attuned to emerging issues and adaptive to evolving social norms. When utilized in a systematic and rigorous manner, such best practices are the foundation for effective graph-based AI systems that are fair, explainable, consistent, and socially responsible.

Finally, the field would be profoundly enhanced by open-source GNN-focused ethical toolkits and standards based in communities. So far, there is no standard method for building ethics graphs, leading to uneven practice and the perception of accountability. The convergence of the industry, academia, civil society, and regulation in communities would have to be coordinated in order to reach consensus able taxonomies, documentation practices, risk assessment forms, and certification programs. These activities can be aided by libraries like Open Graph Benchmark (OGB) or AI Fairness 360 toolkit. Through the creation of shared ethical infrastructure, the GNN community can share the responsibility and avoid fragmentation in using ethics.

In short, ethical Graph Neural Network design is not an afterthought but a design necessity. Ethical GNN design is intentional algorithmic alignment, data, human values, and institutional contexts. Ethical GNN design is not harmlessness in itself—it is good system design that maximizes fairness, responsibility, transparency, and human flourishing in a graph-mediated future-to-be. Interdisciplinary collaboration, watchful inspection, and ethics audit commitment are the sole assurance we have that we build GNNs as tools of mutual benefit and not weapons of algorithmic oppression.

AI disclosure statement: The author acknowledges the use of AI-based tools for minor language editing or grammatical correction. However, all intellectual contributions, analyses, and conclusions presented in this chapter are entirely the author's own.

References

1. Barocas, S., Hardt, M., Narayanan, A., *Fairness and machine learning: Limitations and opportunities*, MIT Press, Cambridge, MA, USA, 2023.
2. Dwork, C., McSherry, F., Nissim, K., Smith, A., Calibrating noise to sensitivity in private data analysis, in: *Theory of Cryptography: Third Theory of Cryptography Conference, TCC 2006, New York, NY, USA, March 4-7, 2006, Proceedings 3*, pp. 265–284, Springer Berlin Heidelberg, Berlin, Germany, 2006.
3. Narayanan, A. and Shmatikov, V., De-anonymizing social networks, in: *IEEE Symposium on Security and Privacy*, pp. 173–187, 2009.
4. Wu, Z., Pan, S., Chen, F., Long, G., Zhang, C., Yu, P.S., A comprehensive survey on graph neural networks. *IEEE Trans. Neural Networks Learn. Syst.*, 32, 1, 4–24, 2021.
5. Dong, Y., Kang, J., Tong, H., Li, J., Individual fairness for graph neural networks: A ranking based approach, in: *Proceedings of the 27th ACM SIGKDD conference on knowledge discovery & data mining*, 2021, August, pp. 300–310.
6. Van Dijk, F., Stoneman's, J., Van Winden, F., Social ties in a public good experiment. *J. Public Econ.*, 85, 2, 275–299, 2002.
7. Wang, X., Chen, X., Wang, L., Evolutionary dynamics of fairness on graphs with migration. *J. Theor. Biol.*, 380, 103–114, 2015.
8. European Commission, Proposal for a Regulation laying down harmonised rules on Artificial Intelligence (Artificial Intelligence Act), COM/2021/206 final, 2021.
9. Zhao, H., Yang, X., Wang, Z., Yang, E., Deng, C., Graph debiased contrastive learning with joint representation clustering, in: *IJCAI*, 2021, August, pp. 3434–3440.
10. Wang, S., Zheng, Y., Jia, X., Sign: Privacy-preserving graph neural network training and inference as a cloud service. *IEEE Trans. Serv. Comput.*, 16, 4, 2923–2938, 2023.

11. Ying, R., Bourgeois, D., You, J., Zitnik, M., Leskovec, J., GNNExplainer: Generating explanations for graph neural networks, in: *Advances in Neural Information Processing Systems*, vol. 32, 2019.
12. Sutura, L., Vladimir, K., Delač, G., Šilić, M., Fairness in Graph-based Recommendation: Methods Overview, in: *2024 47th MIPRO ICT and Electronics Convention (MIPRO)*, 2024, May, IEEE, pp. 850–855.
13. Leavy, S., Siapera, E., O'Sullivan, B., Ethical data curation for AI: An approach based on feminist epistemology and critical theories of race, in: *Proceedings of the 2021 AAAI/ACM Conference on AI, Ethics, and Society*, 2021, July, pp. 695–703.
14. Guo, Z. and Wang, H., A deep graph neural network-based mechanism for social recommendations. *IEEE Trans. Ind. Inf.*, 17, 4, 2776–2783, 2020.
15. Kipf, T.N. and Welling, M., Semi-supervised classification with graph convolutional networks, arXiv preprint arXiv:1609.02907, 2016.
16. Cinelli, M., Morales, G.D.F., Galeazzi, A., Quattrociocchi, W., Starnini, M., Echo chambers on social media: A comparative analysis, arXiv preprint arXiv:2004.09603, 2020.
17. Dong, Y., Wang, S., Wang, Y., Derr, T., Li, J., On structural explanation of bias in graph neural networks, in: *Proceedings of the 28th ACM SIGKDD Conference on Knowledge Discovery and Data Mining*, 2022, August, pp. 316–326.
18. Chen, W., Wu, Y., Zhang, Z., Zhuang, F., He, Z., Xie, R., Xia, F., FairGap: Fairness-aware recommendation *via* generating counterfactual graph. *ACM Trans. Inf. Syst.*, 42, 4, 1–25, 2024.
19. Kamiran, F. and Calders, T., Data preprocessing techniques for classification without discrimination. *Knowl. Inf. Syst.*, 33, 1, 1–33, 2012.
20. Rahman, T., Surma, B., Backes, M., Zhang, Y., Fairwalk: Towards fair graph embedding, 2019.
21. Zhu, Y., Li, J., Chen, L., Zheng, Z., Fairagg: Toward fair graph neural networks *via* fair aggregation. *IEEE Trans. Comput. Social Syst.*, 2024.
22. Zhang, A., Ma, W., Wei, P., Sheng, L., Wang, X., General debiasing for graph-based collaborative filtering *via* adversarial graph dropout, in: *Proceedings of the ACM Web Conference 2024*, 2024, May, pp. 3864–3875.
23. Dong, Y., Liu, N., Jalaian, B., Li, J., Edits: Modeling and mitigating data bias for graph neural networks, in: *Proceedings of the ACM web conference 2022*, 2022, April, pp. 1259–1269.
24. Parent-Rocheleau, X. and Parker, S.K., Algorithms as work designers: How algorithmic management influences the design of jobs. *Hum. Resour. Manage. Rev.*, 32, 3, 100838, 2022.
25. Fitzpatrick, D.J., Gorr, W.L., Neill, D.B., Keeping score: Predictive analytics in policing. *Annu. Rev. Criminol.*, 2, 1, 473–491, 2019.
26. Wu, L., Chen, L., Shao, P., Hong, R., Wang, X., Wang, M., Learning fair representations for recommendation: A graph-based perspective, in: *Proceedings of the Web Conference 2021*, 2021, April, pp. 2198–2208.

27. Kose, O.D. and Shen, Y., Fairgat: Fairness-aware graph attention networks. *ACM Trans. Knowl. Discovery Data*, 18, 7, 1–20, 2024.
28. Hu, B., Fang, Y., Shi, C., Adversarial learning on heterogeneous information networks, in: *Proceedings of the 25th ACM SIGKDD international conference on knowledge discovery & data mining*, 2019, July, pp. 120–129.
29. Zhang, Y., Zhang, Y., Sang, L., Sheng, V.S., Simplify to the Limit! Embedding-less Graph Collaborative Filtering for Recommender Systems. *ACM Trans. Inf. Syst.*, 43, 1, 1–30, 2024.
30. Arduini, M., Noci, L., Pirovano, F., Zhang, C., Shrestha, Y.R., Paudel, B., Adversarial learning for debiasing knowledge graph embeddings, arXiv preprint arXiv:2006.16309, 2020.
31. Dwork, C., Hardt, M., Pitassi, T., Reingold, O., Zemel, R., Fairness through awareness, in: *Proceedings of the 3rd innovations in theoretical computer science conference*, 2012, January, pp. 214–226.
32. Chen, A., Rossi, R.A., Park, N., Trivedi, P., Wang, Y., Yu, T., Ahmed, N.K., Fairness-aware graph neural networks: A survey. *ACM Trans. Knowl. Discovery Data*, 18, 6, 1–23, 2024.
33. Hardt, M., Price, E., Srebro, N., Equality of Opportunity in Supervised Learning, in: *NeurIPS*, 2016.
34. Woodworth, B., Gunasekar, S., Ohannessian, M.I., Srebro, N., Learning non-discriminatory predictors, in: *Conference on learning theory*, 2017, June, PMLR, pp. 1920–1953.
35. Su, J., Shukla, A., Goel, S., Narayanan, A., De-anonymizing web browsing data with social networks, in: *Proceedings of the 26th international conference on world wide web*, 2017, April, pp. 1261–1269.
36. Zhang, X. and Zitnik, M., Gnnguard: Defending graph neural networks against adversarial attacks. *Advances in neural information processing systems*, vol. 33, pp. 9263–9275, 2020.
37. Zhang, Z., Chen, M., Backes, M., Shen, Y., Zhang, Y., Inference attacks against graph neural networks, in: *31st USENIX Security Symposium (USENIX Security 22)*, pp. 4543–4560, 2022.
38. He, X., Wen, R., Wu, Y., Backes, M., Shen, Y., Zhang, Y., Node-level membership inference attacks against graph neural networks, arXiv preprint arXiv:2102.05429, 2021.
39. Olatunji, I.E., Nejdl, W., Khosla, M., Membership inference attack on graph neural networks, in: *2021 Third IEEE International Conference on Trust, Privacy and Security in Intelligent Systems and Applications (TPS-ISA)*, 2021, December, IEEE, pp. 11–20.
40. Abadi, M., Chu, A., Goodfellow, I., McMahan, H.B., Mironov, I., Talwar, K., Zhang, L., Deep learning with differential privacy, in: *Proceedings of the 2016 ACM SIGSAC conference on computer and communications security*, 2016, October, pp. 308–318.

41. Mueller, T.T., Paetzold, J.C., Prabhakar, C., Usynin, D., Rueckert, D., Kaissis, G., Differentially private graph classification with gnns, arXiv preprint arXiv:2202.02575, 2022.
42. He, C., Balasubramanian, K., Ceyani, E., Yang, C., Xie, H., Sun, L., Avestimehr, S., Fedgraphnn: A federated learning system and benchmark for graph neural networks, arXiv preprint arXiv:2104.07145, 2021.
43. Bonawitz, K., Ivanov, V., Kreuter, B., Marcedone, A., McMahan, H.B., Patel, S., Seth, K., Practical secure aggregation for privacy-preserving machine learning, in: *proceedings of the 2017 ACM SIGSAC Conference on Computer and Communications Security*, 2017, October, pp. 1175–1191.
44. Gilad-Bachrach, R., Dowlin, N., Laine, K., Lauter, K., Naehrig, M., Wernsing, J., Cryptonets: Applying neural networks to encrypted data with high throughput and accuracy, in: *International conference on machine learning*, 2016, June, PMLR, pp. 201–210.
45. Huang, Z., Tang, M., Wei, Y., Deng, G., Privacy-Preserving Node Classification in Customer Services With Federated Graph Neural Networks. *IEEE Trans. Consum. Electron.*, 2024.
46. Malin, B. and Sweeney, L., How (not) to protect genomic data privacy in a distributed network: using trail re-identification to evaluate and design anonymity protection systems. *J. Biomed. Inf.*, 37, 3, 179–192, 2004.
47. Ge, L., Li, Y., Li, H., Tian, L., Wang, Z., A review of privacy-preserving research on federated graph neural networks. *Neurocomputing*, 128166, 2024.
48. Hamilton, W., Ying, Z., Leskovec, J., Inductive representation learning on large graphs. *Advances in neural information processing systems*, vol. 30, 2017.
49. Tang, J., Qu, M., Mei, Q., Pte: Predictive text embedding through large-scale heterogeneous text networks, in: *Proceedings of the 21th ACM SIGKDD international conference on knowledge discovery and data mining*, 2015, August, pp. 1165–1174.
50. Veličković, P., Cucurull, G., Casanova, A., Romero, A., Lio, P., Bengio, Y., Graph attention networks, arXiv preprint arXiv:1710.10903, 2017.
51. Jain, S. and Wallace, B.C., Attention is not explanation, arXiv preprint arXiv:1902.10186, 2019.
52. Yuan, H., Tang, J., Hu, X., Ji, S., Xgnn: Towards model-level explanations of graph neural networks, in: *Proceedings of the 26th ACM SIGKDD international conference on knowledge discovery & data mining*, 2020, August, pp. 430–438.
53. Luo, D., Cheng, W., Xu, D., Yu, W., Zong, B., Chen, H., Zhang, X., Parameterized explainer for graph neural network, *Advances in neural information processing systems*, vol. 33, pp. 19620–19631, 2020.
54. Huang, L., Zhang, X., Li, Y., GraphLIME: Local Interpretable Model Explanations for Graph Neural Networks. *IEEE Trans. Knowl. Data Eng.*, 2022.
55. Huang, M., Zhang, X.S., Bhatti, U.A., Wu, Y., Zhang, Y., Ghadi, Y.Y., An interpretable approach using hybrid graph networks and explainable AI

for intelligent diagnosis recommendations in chronic disease care. *Biomed. Signal Process. Control*, 91, 105913, 2024.
56. Du, M., Liu, N., Hu, X., Techniques for interpretable machine learning. *Commun. ACM*, 63, 1, 68–77, 2019.
57. Cheng, L., Mosallanezhad, A., Sheth, P., Liu, H., Causal learning for socially responsible AI, arXiv preprint arXiv:2104.12278, 2021.
58. Fu, W., Mo, Y., Xiao, Y., Liu, C., Zhou, F., Wang, Y., Zhang, Y.J., Enhancing Molecular Energy Predictions with Physically Constrained Modifications to the Neural Network Potential. *J. Chem. Theory Comput.*, 20, 11, 4533–4544, 2024.
59. Schlichtkrull, M., Kipf, T.N., Bloem, P., Van Den Berg, R., Titov, I., Welling, M., Modeling relational data with graph convolutional networks, in: *The semantic web: 15th international conference, ESWC 2018, Heraklion, Crete, Greece, June 3–7, 2018, proceedings 15*, pp. 593–607, Springer International Publishing, Cham, Switzerland, 2018.
60. Marx, G.T., Ethics for the new surveillance. *Inf. Soc.*, 14, 3, 171–185, 1998.
61. Mittelstadt, B.D., Allo, P., Taddeo, M., Wachter, S., Floridi, L., The ethics of algorithms: Mapping the debate. *Big Data Soc.*, 3, 2, 2053951716679679, 2016.
62. Alkhatib, A., To live in their utopia: Why algorithmic systems create absurd outcomes, in: *Proceedings of the 2021 CHI conference on human factors in computing systems*, 2021, May, pp. 1–9.
63. D'ignazio, C. and Klein, L.F., *Data feminism*, MIT Press, Cambridge, MA, USA, 2023.
64. European Commission, Proposal for a Regulation Laying Down Harmonised Rules on Artificial Intelligence, Brussels, 2021, [Online]. Available: https://eur-lex.europa.eu/legal-content/EN/TXT/?uri=CELEX%3A52021PC0206.
65. IEEE, *Ethically Aligned Design: A Vision for Prioritizing Human Well-being with Autonomous and Intelligent Systems*, 1st ed., IEEE, Piscataway, NJ, USA, 2019, [Online]. Available: https://ethicsinaction.ieee.org/.
66. UNESCO, Recommendation on the Ethics of Artificial Intelligence, 2021, [Online]. Available: https://unesdoc.unesco.org/ark:/48223/pf0000380455.
67. Chen, J., Ma, M., Ma, H., Zheng, H., Zhang, J., An empirical evaluation of the data leakage in federated graph learning. *IEEE Trans. Network Sci. Eng.*, 11, 2, 1605–1618, 2023.
68. Ezrachi, A. and Stucke, M.E., Artificial intelligence & collusion: When computers inhibit competition. *U. Ill. L. Rev.*, 1775, 2017.
69. Mitchell, M., Wu, S., Zaldivar, A., Barnes, P., Vasserman, L., Hutchinson, B., Gebru, T., Model cards for model reporting, in: *Proceedings of the conference on fairness, accountability, and transparency*, 2019, January, pp. 220–229.
70. Wang, S., Zheng, Y., Jia, X., Secgnn: Privacy-preserving graph neural network training and inference as a cloud service. *IEEE Trans. Serv. Comput.*, 16, 4, 2923–2938, 2023.

71. IEEE Standards Association, *EEE Ethically aligned design: A vision for prioritizing human well-being with autonomous and intelligent systems*, IEEE Standards Association, Piscataway, NJ, USA, 2019.
72. Laclau, C., Largeron, C., Choudhary, M., A survey on fairness for machine learning on graphs, arXiv preprint arXiv:2205.05396, 2022.
73. Velickovic, P., Cucurull, G., Casanova, A., Romero, A., Lio, P., Bengio, Y., Graph attention networks. *Stat*, 1050, 20, 10–48550, 2017.
74. Effendi, F. and Chattopadhyay, A., Privacy-Preserving Graph-Based Machine Learning with Fully Homomorphic Encryption for Collaborative Anti-money Laundering, in: *International Conference on Security, Privacy, and Applied Cryptography Engineering*, 2024, December, Springer Nature Switzerland, Cham, pp. 80–105.
75. Klein, L. and D'Ignazio, C., Data feminism for AI, in: *Proceedings of the 2024 ACM Conference on Fairness, Accountability, and Transparency*, 2024, June, pp. 100–112.
76. Qian, J., Li, X.Y., Zhang, C., Chen, L., De-anonymizing social networks and inferring private attributes using knowledge graphs, in: *IEEE INFOCOM 2016-The 35th Annual IEEE International Conference on Computer Communications*, 2016, April, IEEE, pp. 1–9.
77. Gebru, T., Morgenstern, J., Vecchione, B., Vaughan, J.W., Wallach, H., Iii, H.D., Crawford, K., Datasheets for datasets. *Commun. ACM*, 64, 12, 86–92, 2021.
78. Zhang, Y., Zhao, Y., Li, Z., Cheng, X., Wang, Y., Kotevska, O., Derr, T., A survey on privacy in graph neural networks: Attacks, preservation, and applications. *IEEE Trans. Knowl. Data Eng.*, 2024.
79. Han, Y., Huang, G., Song, S., Yang, L., Wang, H., Wang, Y., Dynamic neural networks: A survey. *IEEE Trans. Pattern Anal. Mach. Intell.*, 44, 11, 7436–7456, 2021.

6

Graph Neural Network: Scalability Challenges in Large-Scale Graph Neural Network

Abubakar Nadaf* **and Shivagonda Patil**

Department of Computer Science and Engineering, Visvesvaraya Technological University, Belagavi, Karnataka, India

Abstract

The chapter inquiries about GNNs that has become evident as a crucial tool and technologies in Machine Learning for prototyping the complicated intercommunication and the pattern in the graph-oriented data. The potential of these technology is to perform representation learning by exploiting both topological information and node features have triggered landscape adoption in areas such as social networks, drug discovery, recommendation systems, and cyber security. Never the less, the implementation of GNNs to massive-scale graphs incorporates billion of nodes and edges that provides substantial scalability complications. The scalability of GNNs showcases an in-depth evaluation of the scalability impediments in large-scale GNNs. The chapter underscores leading-edge, Next-generation and innovative solutions encompassing splitting techniques, graph sampling, mini-batching, time shifted training, model compression, and distributed computation. Furthermore, the GNNs explore system-level new ideas maintaining memory management, hardware acceleration, and parallel execution tactics. Also, the GNNs techniques inspects the core obstacles arising from area eruption, recursive message passing, and the issues of existing GNN architectures in computing various graphs. Real-world case studies and benchmarks are utilized to depict that the way of executing the techniques in practice, offering comprehension into their computations' trade-offs. The chapter also highlights recent ideas like Cluster-GCN, Cluster-GCN, Graph-SAINT, Cluster-GCN, and decoupled architectures that have revealed assurance in scaling GNNs efficiently. Ultimately, future investigation directions and open challenges are examined, with

**Corresponding author*: abhidn15@gmail.com

J. Ramkumar, Sridaran Rajagopal, B. Suchitra and S. Balamurugan (eds.) Graph Neural Networks: Concepts and Applications, (243–282) © 2026 Scrivener Publishing LLC

an emphasis on moderating scalability with accuracy, stereotyping, and system constraints. This detailed chapter represents essential and basic resources which are designed to contribute researchers, practitioners who are committed to design, conceptualize, and implement the GNNs.

Keywords: Graph neural networks, scalability, distributed learning, graph sampling, memory optimization, large-scale graphs, GNN architectures, graph partitioning

6.1 Introduction

As concern to this chapter, the portion of the book GNNs - the concepts and applications, claims an important area which is a key to both practical utilization and academic inquiry, the crucial scalability challenges marked as implementing GNNs to large-scale graph data. Graph Neural Network have been widely acknowledged as a strong and adaptable group of models of Deep Learning patterns pointedly designed to control data demonstrated as graphs. The technicality effectively holds the complex relationships and repositories for the future solvation. The main focus and strength of GNNs is to cover the message-passing pattern at the time when nodes arrange the information from end-network in loop manner, providing them great and rich demonstrations. Despite of it, the considerable success and escalating acceptance of GNNs have brought to the leading edge a significant challenge, scaling these patterns and structures to control real-world graphs that often consist of a huge number of nodes and edges. Notwithstanding GNNs operating remarkably well on tiny or medium-sized graphs, they are likely to become insufficiently powered and highly effective in practice when encountered with the unmitigated size and complexity of large-scale graphs. This chapter provides an in-depth evaluation of these scalability challenges, dissecting the root causes and examining the various technical obstacles that prevent the seamless application of current GNN architectures to large-scale scenarios.

The core of the scalability issue highlights the processing and memory requirements imposed by the graph pattern and the message-passing method. In comparison with grid-like data, in where functions can usually be performed locally and parallelized optimistically, graph data executes uneven connectivity flows. The linkage among nodes is demonstrated by the graph's architecture, which will be more complex and sparser or dense based on the particular graph. Engaging with information through these links for all nodes, markedly in graphs with major degrees or a massive number of edges, becomes computationally rigorous.

This chapter provides an in-depth summary of these complex scalability challenges. Following that, it explores ultra-model approaches and resolutions implemented for interpreting them. The work incorporates a thorough deliberation of numerous sampling tactics and their theoretical keystone and practical implications. It investigates and analyzes architectural adaptations to GNNs fashioned for improved efficiency, specifically models with Simplified message combination or attention procedures that can deal with larger communities. Optimization strategies, embracing graph partitioning tactics to reduce communication in decentralized settings and efficient batch processing methods personalized for graph data, are also reviewed. The part played by hardware acceleration and optimized software models in establishing large-scale GNN training is highlighted. To conclude, the chapter discusses trending research directions and open issues in the area, likewise the development of theoretically matters and practically efficient sampling techniques, the framework of inherently scalable GNN models, and the challenges of scaling GNNs to constantly changing and highly heterogeneous graphs. By offering in-depth understanding of the scalability landscape, this chapter targets to furnish readers with the knowledge required to detect, analyze, and potentially overcome the obstacles of implementing Graph Neural Networks to the wide and complex graphs that are prevalent in many real-time implications.

6.2 Background and Preliminaries

We introduce by revisiting the essential principles of graph theory and neural networks that are useful for knowledge of the core principles of GNNs. This survey concludes that viewers have the mandatory background to recognize the more advanced concepts. Consequently, we coin the fundamental strategies behind GNNs, showcasing how these networks manipulate traditional neural network models to appropriately control graph-methodical data.

As we move forward we introduce a thoroughly encompassing taxonomy of GNN models. This designation delivers a clear framework for perception the numerous GNN models and their remarkable traits serving as a guide for arrangement the subsequent discussion on scalability hurdles. In this portion, we establish the groundwork for a deeper knowledge of GNNs, aiming on their frameworks, how they operate and function, and also the constraints they encounter. This base is critical for apprehending the complexities including in scaling GNNs and the techniques assigned to interpret these obstacles.

6.2.1 Graph Theory Basics

Graphs are the structure of mathematical world, and are well-maintained to represent bidirectional relations between real-time entities. Constitutionally, a graph is explored as G (V, E), where V is the set of nodes (or vertices), and $E \subseteq V X V$ is the set of edges. Relying on the application, graphs may be directed or undirected, weighted or unweighted, and may enclose additional information such as node attributes and edge attributes.

- **Adjacency matrix**: a matrix $\mathbf{a} \in \mathbf{r}^{n \times m}$ symbolizing edges between nodes. a_{ij} =1 if from node i to node j, an edge is present, then, alternatively, it is 0.
- **Degree**: the number of edges occurring to a node. For directed graphs, we distinguish between in-degree and out-degree.
- **Neighborhood**: for a node V, its region n(V) consists of all nodes directly connected to V by an edge.
- **Subgraph**: a subset of a graph's nodes and edges that itself forms a graph.
- **Graph types**: graphs have the potential to be static or dynamic, homogeneous/uniform or heterogeneous/diverse, and may vary in manifold, sparsity, and attribute prosperity.

As concerned to the context of large-scale graphs collaborated with complicated interconnected patterns, the architectural properties directly influence the way GNNs are implemented and scaled.

6.2.2 Neural Networks Basics

Classic neural networks have been played essential role in the progression of Artificial Intelligence, particularly in the discipline such as NLP, image processing, and with time-series assessment, although, the CNNs and RNNs are also introduced to the current work. The success rate of implementation is dependent on their capability to learn and understand, the demonstrated structured such as images, texts, and audio. It is much necessary to understand the basic key points of classic or traditional neural networks before migrating to highly advanced architectures like GNNs.

- **Layers**: In the part of neural network, the layers are segregated of units which can change the parameter data in a fixed way. Individual layer or parameter has the setting to adjust to understand from the data.
- **Weights**: Learnable specifications that are optimized while training. Weights are the specifications that define the strength of the associations between neurons in different layers. Each connection between two neurons has a linked weight, which establishes how much influence one neuron has on the other. Weights are initialized aimlessly before training and are adjusted over time to shrink the network's prediction error.
- **Activation functions**: Irregularities (e.g., sigmoid) implemented after each layer. Activation functions are critical in introducing nonlinearities into the neural network. Without activation functions, the network would be prohibited from linear transformations, its capacity to learn sophisticated, non-linear relationships in the data. By implementing non-linear functions to the weighted amount of inputs at each node, the network can understand more complex patterns and mappings.
- **Loss function**: Strategies the variation between predictions and targets. The loss function achieves that, how far the network's expectations are from the actual target values. The goal of training a neural network is to reduce this shortfall, thereby improving the network's accuracy on the task at hand.
- **Backpropagation**: Algorithm for validating gradients and updating weights. Backpropagation works in such a way that, broadcasting the error backward *via* the network to calculate the gradient of the shortfall function with concern to iteratively edge value in the network. After the process the gradients will get upgrade the values of an edge in the direction that decreases the degradation.

6.2.3 Taxonomy of GNN Architectures

As concern to the guidelines and the method in which they streamline the graph data, GNNs can be effectively classified into numerous categories. Recently observed graph neural network architectures in the past few years, they have exponentially developed, at most to a diverse set of architectures personalized to particular pattern features and application necessities.

Spectral GNNs

A key illustration is **CHEBNET**, which estimates spectral filters using **Chebyshev Polynomials**, diminishing computational costs balanced to exact Eigen decomposition. While spectral methods provide strong theoretical foundations and are beneficial in understanding the mathematical qualities of GNNs, they suffer from many practical shortcomings:

- Confidence in the full graph configuration, making them unsuitable for categorical tasks
- High computational intricacy due to Eigen decomposition, specifically in large-scale graphs.
- Restricted generalization to graphs with shifting structures, as spectral filters are domain-specific.

Despite these boundaries, spectral GNNs have played an underlying role in the development of maximum scalable variants and have enlightened theoretical analyses of GNN rendition.

Spatial GNNs

Spatial GNNs function in the **Node Domain**, instantly applying convolution-like operations over local space in the graph. Rather than using Eigen disintegration, they aggregate features from a node's immediate peers, often using operations such as summation, averaging. This strategy aligns well with the message passing platform and helps both inductive and transductive learning settings. Key illustrations include:

- **Graphsage**: Unveils neighborhood sampling and trains aggregation functions such as mean or pooling.

GCN: Streamlines the spectral model by coming close to convolutions with localized filters and stabilizing adjacency.

- **GAT (Graph Attention Networks)**: Combines attention mechanisms for peer aggregation.

Sampling-Based GNNs

These are a vital advancement for managing large-scale graph information swiftly. In contrast to full-batch GNNs, which prerequisites collaborating the facilities by various peers, sampling-based methods stabilizes the neighborhood data to minimize the processing time and overhead.

Graph sage (graph sample and aggregate): are the prototypes of static number of bordering the regions each parameter. Also forms mini-batch training. Basically, a graphsage is a ground-breaking probabilistic model for graph demonstration learning. In rare case the traditional GNNs that need the whole graph at a point of training and inference, then graph sage invokes combined generation for unpredicted nodes by learning a group of aggregator methods. The model works by sampling a given-size group of neighbors for each point of contact and then aggregating their upcoming data by utilizing the methods like mean, lstm, or pooling. Anyhow, this method of implication significantly enhances scalability and forming broad principles from individual instances, making graph sage fit for dynamic or large-scale graphs. These architectures make GNNs adaptable to industrial-scale graphs and activate training on devices with limited resources. The trade-off, nonetheless, is a potential decline in accuracy anticipated to information loss from aggressive sampling. Few works done repeatedly over the Cluster-GCNs are Adaptive Cluster-GCN, Balanced Cluster Construction, Multi-level Hierarchical Clustering (HC-GCN) and Edge Augmentation within Clusters. Many of the advancements lead to some of benefits as Efficiency, Reduced Overhead, Quality Retention, Adaptivity, Hierarchical Splitting, and Connectivity Improvements as per the record observation of recent work.

Hierarchical GNNs

Hierarchical GNNs, also recognized as graph pooling or coarsening models, learn likenesses at multiple levels of abstraction. These models squeeze the graph by merging nodes or favoring subsets, similar to how CNN-based architectures utilize pooling to reduce geometric dimensions.

Illustrations include:

- **Diffpool (differentiable pooling)**: Acquires knowledge of a soft assignment structure that clusters nodes into powerful nodes.
- **Structpool**: Employs graph structure and characteristic information to guide pooling determinations.

These architectures are specifically useful for graph-level tasks, such as biochemical property prediction or document categorization. By minimalizing the scale of graph during retaining key structural patterns, they improve effectiveness and capture hierarchical relationships. Still, designing effective pooling approaches remains a challenge due to capability loss of fine-grained information.

Dynamic and Temporal GNNs

Real-world graphs often evolve—nodes and edges emerge or disappear, and node features change. Dynamic and temporal **GNNs** are planned to model such time-varying graphs, nabbing both structural dynamics and temporal prerequisites.

Two common prototypes exist:

- **Discrete-Time Models**: Manages the graph as a progression of static informal portraits (e.g., dyngraph2vec).
- **Continuous-Time Models**: Representation of event-driven changes with chronological point processes (e.g., tgn—temporal GNN).

Instances of models in this extent include:

- **T-GCN**: Combines temporal curves with GCN for traffic prediction.
- **Dyngraph2vec**: Models node evolution employing cyclical neural networks and autoencoders.

These GNNs are crucial for applications like social interchange modeling, financial transaction networks, and real-time recommendation structures, where the temporal component of the data is as essential as the graph structure itself.

6.2.4 Computational Framework of GNNs

Most GNNs adhere to a layer-wise propagation scheme, where node expressions are iteratively refined by accumulating messages from neighbors across multiple layers. The forward pass commonly involves:

- Computing neighbor messages
- Aggregating messages
- Applying a non-linear transformation
- Computing loss and performing backpropagation

6.3 Computational Bottlenecks in GNNS

As graph neural networks (GNNs) scale to increasingly bigger and more complicated graph datasets, a diversity of computational bottlenecks starts to exterior. These challenges can considerably diminish the practicability of training and implementing GNNs on large graphs, such as those found in web-scale social networks, proposal systems, or knowledge bases. As seen in the portion, the fundamental inference boundaries are visible in GNNs, especially in the large-scale models, is to verify the causes and the impact on computational process as well as adaptability.

6.3.1 Memory Overhead and Message Passing

The core operation of most architectures of GNNs is message passing, wherein each node composites information from its neighbors at each layer. The model is cost effective along with the overutilization of resources due to the following reasons:

- **Redundant computation**: Numerous peer nodes spread a legacy implementation. Message passing process cannot be reusable in between nodes which significantly effect on efficiency and optimization. Legacy processes are encouraging to current set up with current systems.
- **Computational exponential growth**: A specific node addresses coverages immediately along with many layers.
- **Dense intermediate representations**: As GNN layers whole feature vectors from increasing neighborhoods, the central feature array can become huge and compact, resulting in significant memory overhead, especially when dealing with high-dimensional input features or intense architectures.
- **GPU memory constraints**: Most large-scale GNN training tubes depend on GPUs for a boost. However, the memory limits of GPUs (typically 12GB–48GB) are regularly surpassed by full-graph training, requiring schemes like mini-batching, sampling, or splitting, each of which introduces trade-offs in performance and validity.

Figure 6.1 clearly represents the existence of a computational scalability gap, where the an exponential growth of computation increasingly outpaces the constraints of GPU.

6.3.2 Graph Sparsity vs. Density

Large real-world graphs often exhibit rarity—most nodes have a small number of connections—but can still include compactly connected subgraphs, such as communities or circles. This irregular distribution of edges presents various problems.

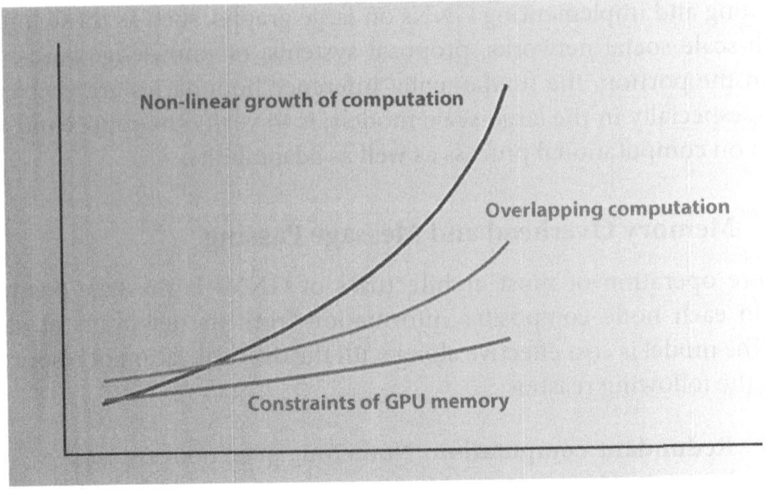

Figure 6.1 The representation of computational bottleneck in GNNs.

- **Load imbalance**: When training GNNs in aligned or distributed environments, distorted node degrees lead to imbalance in computation and memory distribution. High-degree nodes (hubs) require considerably more processing than low-degree nodes.
- **Unpredictable memory access**: In sparse proximity matrices, memory access patterns during neighborhood collection are irregular and non-contiguous, resulting in poor storage consumption and inefficient GPU memory access.
- **Inefficient batching**: Batching sparse graphs or subgraphs is non-trivial, as the number of nodes and edges per batch can generally vary, complicating enhancement and data loading strategies.

6.3.3 Neighborhood Explosion

The phenomenon of **neighborhood explosion** indicates the explosive increase in the number of nodes included in a node's accepting field along with increasing number of GNNs layers. For instance, considering a graph where all nodes have an average degree of d, the number of nodes in a 2-hop neighborhood, and an l-layer network. This leads to:

- **Computational inefficiency**: The need to process and aggregate from a quickly growing set of neighbors makes forward and backward passes digitally expensive.
- **Sampling limitations**: While sampling a fixed number of neighbors at each layer can reduce difficulty, it also introduces estimation faults and distribution in the learned representations, possibly hurting model performance.
- **Latency in inference**: Even at inference time, processing estimations for a single node may need gathering features from a huge segment of the graph, leading to high delay in real-time systems.

6.3.4 Over Smoothing

It is a mostly seen issue in deep GNNs where, after multiple layers of gathering, the node representations become indistinguishable from each other. This happens because repeated averaging of neighbor features factors information to become standardized over the graph. Figure 6.2 shows the performance degradation in GNN.

- **Loss of discriminative power**: As node representations combine to comparable values, the model loses its ability to differentiate between different nodes, decreasing classification and categorizing exactness.

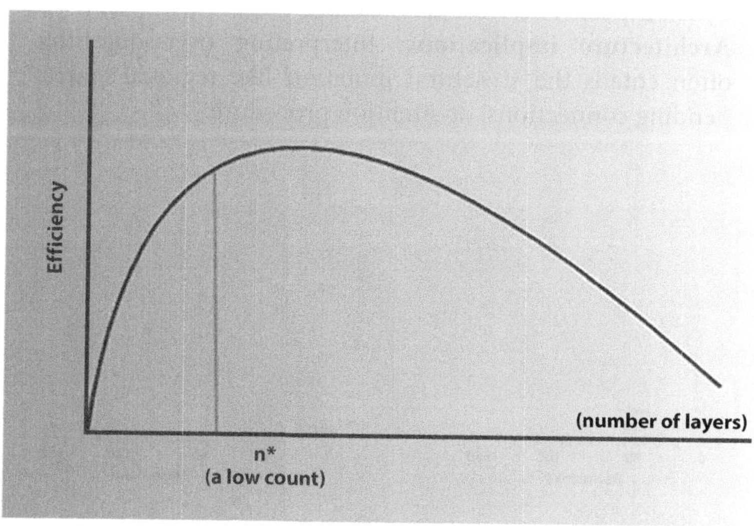

Figure 6.2 The performance degradation in GNNs due to over-smoothing.

- **Shallow architectures**: To avoid over-smoothing, specialists commonly use shallow GNN architectures (2–3 layers), which limit the capacity of model to capture large range repositories, especially concerned to large graphs.
- **Trade-off with scalability**: Deeper models could, in theory, scale better by grabbing the global environment, but over-smoothing restrains their efficiency.

6.3.5 Over-Squashing

Over-squashing is a newer and more delicate problem where wide range external components are reduced into static node demonstration across repeated gathering. Unlike over-smoothing, where information becomes uniform, over-squashing indicates the **loss of important information** due to inadequate representational potential to transform distant signals. Figure 6.3 illustrates the evolution of normalized spectral gap and the triangles.

- **Bottleneck in information propagation**: Information from distant parts of the graph gets "squashed" when combined into a local node representation, dominant under-representation of global structures.
- **Non-locality limitations**: The techniques and structure binds only by nearest node, although, unable to distribute large-scale links and specifically exception in procedure which expects universal justification.
- **Architecture implications**: Interpreting over-squashing often entails the structural mutation like regional codes, pending connections, or attention procedure.

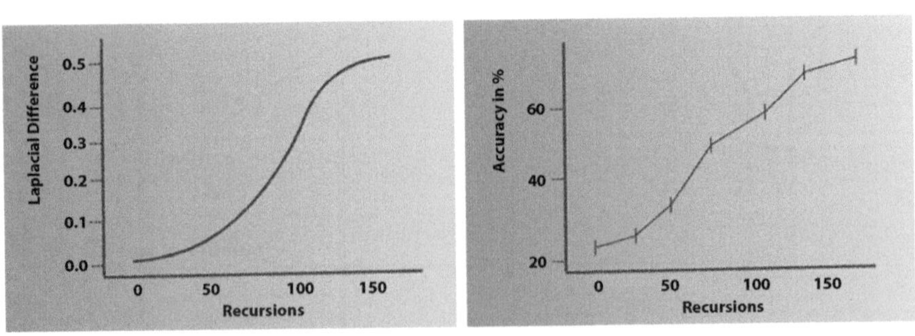

Figure 6.3 Evolution of normalized spectral gap (blue) and triangle counts (orange) for RLEF, G-RLEF, and SDRF.

6.3.6 Inefficient Graph Data Handling

The processing and management of graph architecture data provides an essential obstacle in succeeding and performance of computation. Few obstacles include as follows:

- **Slow data loading**: Due to CSR it is difficult to distribute for sampling and mini-batching which was format by sparse matrix.
- **Pre-processing overhead**: Partitioning and standardizing the pre-processing tasks are often time taking for massive-graphs.
- **Disk-memory transfer bottlenecks**: Submitting the large subgraphs among central processing unit or graphic processing unit can be an acceptable challenge if it is not problematically handled.

6.3.7 Enhanced Mitigation Strategies

Confronting the scalability in large-scale GNNs compels in the implementing efficient and applicable strategies that overcomes the IT assets or resource utilization devoid of significantly negotiating model performance. A brief discussion is presented below of advanced techniques that appear promise in progressing scalability.

- **Neighborhood sampling and approximation techniques:** To avoid exponential neighborhood expansion, sampling-based GNNs like Graph sage, Fast GCN, and ladies sample a subset of neighbors during training. these methods reduce computation while preserving local information.
- The benefits of making use of the mitigation strategies reduce the consumption of memory, and lead to speed-up the training.
- **Cluster-based and partitioning methods:** These are the methods such as Cluster-GCN process of fragmentation of the graph into clusters by using pre-processing prototypes like metis. These partitions are used then as mini-batches.

Benefit: These establish an access to mini-batch training on massive graphs. This will be done only by exploiting locality and reducing cross-partition mechanism.

6.4 Scalability Dimensions

For large-scale graphs, scaling the GNNs entails realizing various dimension. Unlike traditional neural networks, which initially scale along batch size and depth, GNNs illustrate explicit dimensions due to their graph-centric nature. In this phase, we identify and elaborate on the paramount scalability dimensions in GNNs, underscore how each dimension introduces unique computational and architectural overemphasis.

6.4.1 Node Scale

The most direct proportion of scalability pertains to numerous nodes in the graph. Real-world systems such as social networks, such as Facebook, Twitter, e-commerce interfaces like Amazon, Alibaba, and knowledge graphs such as Freebase, Wikidata have countless number of nodes. Scaling GNNs to such graphs is an issue because:

- **Storage**: Storing node features, embeddings, and intermediate showcases requires substantial memory. Even with low-dimensional combinations, billions of nodes can easily need hundreds of gigabytes or more storage.
- **Lookup overhead**: Fetching and updating node embeddings during training/inference contains non-trivial memory access streamlines and data retrieval costs, specifically when using external memory. (e.g., NVME or remote storage).
- **Processing & Computation**: While training, each node needs to collect information from its peers, which tends to large-scale computations compared to the graph's size and degree splits.
- At the time of processing the model, each node requires to segregate the information from its nearest neighbors. It leads to large scale processing units as compared to the size of graph and partitions.
- The ideas to signify node scale challenges consists an embedding variation, quantization of features, and the explicit storage of memory such as RocksDB and KV stores.

6.4.2 Edge Scale

One more crucial dimension is count of edges in the graph, are often grows rapidly than the number of nodes, mainly in scale-free networks with power-law degree distributions.

- **Memory Complexity**: Edge matter must be stored (e.g., in adjacency data or sparse matrices), sometimes with additional edge features. Memory maximum grows linearly with the number of edges, but is accumulated by message passing tasks.
- **Communication costs**: In distributed systems, parsing the edge-related data through computation units leads to a bottleneck, especially when edges span more partitions.
- **Load Imbalance**: Edge-heavy areas create hotspots that need disproportionate processing time and memory, tending to uneven resource usage.

To handle edge-scale issues, methods such as graph partitioning, edge sampling, and communication-aware scheduling have been developed.

6.4.3 Feature Scale

GNNs retrieve not only graph topology but also node and edge features. In various industrial settings, feature vectors are wide-dimensional (net ranging from hundreds to tens of thousands of dimensions), mainly in endorser systems and natural language processing tasks.

- **High-dimensional inputs**: Models should process, transform, and store large feature matrices, which intensifies GPU memory pressure.
- **Redundant Computations**: Many nodes circulate similar features (e.g., one-hot encoding or bag-of-words), but without refinement, the same operations are looped across nodes.
- **Bandwidth Limitations**: Transferring huge feature matrices between CPU and GPU or across allocated systems can dominate training time.

Mitigation methods include dimensionality shared embeddings, reduction, and in-place computation kernels.

6.4.4 Model Depth

Depth indicates the number of GNN layers accumulated during training. More profound GNNs can, in theory, hold more global information and complex reserve supporters. Nevertheless, scaling with depth represents various obstacles:

- **Over-Smoothing and Vanishing Gradients**: As per the previous benchmark, deep GNNs tend to undergo information standardization and gradient dilution.
- **Memory Growth**: Deeper models collect more intermediate activations during the forward pass, growing memory consumption and backpropagation expenses.
- **Training Time**: Deeper networks need longer training times and are difficult to optimize, often consequent in diminishing yields on accuracy.

Methods such as residual connections, skip layers, and normalization can be supported, but deep GNNs persist more challenging to boost than deep CNNs or transformers.

6.4.5 Training-Time Scalability

Scalability, while training involves multiple factors:

- **Batch Size**: Unlike image categorization, mini-batching is non-trivial in GNNs due to graph interdependencies. Even subgraph batching can tend to irregular batch sizes and sparsity patterns.
- **Gradient Synchronization**: In distributed training, gradient synchronization across devices is obstructed by the graph structure, especially if neighbor information dwells on different workers.
- **Data Sampling**: Sampling large graphs for modifying(e.g., peer sampling, subgraph sampling) introduces stochasticity and variance, which can affect convergence rates.

Enhanced training pipelines often reside on layer-wise sampling, asynchronous training, and gradient checkpointing to balance memory and compute expenses.

6.4.6 Inference-Time Scalability

Judgment in large-scale GNNs is sometimes ignored but is equally critical. In fabrication environments, real-time inference is required for tasks such as endorsement, fraud detection, or dynamic routing.

- **Latency Constraints**: To operate millions of users in real-time, GNN inference processes must deliver predictions within milliseconds.
- **Online vs. Offline Inference**: While offline inference permits batch processing, online systems need optimized one-shot prediction, demanding pre-computed embeddings or rapid neighbor lookups.
- **System Integration**: Combining GNN models into present production systems requires focus on API design, memory usage, and interoperability.

Fixes include precomputation of embeddings, decoupled architectures (e.g., pin-sage), and hardware quickening using FPGAs or TPUs.

6.4.7 System-Level Scalability

Scalability also axes system-level concerns such as distributed computing, parallelism, and hardware constraints.

- **CPU/GPU Utilization**: Suboptimal use of estimated resources due to uneven workloads leads to underutilized GPUs and idle CPUs.
- **I/O Bottlenecks**: Reading and writing large-scale graph information from disk or interconnected network storage is a non-trivial expense, specifically for streaming graphs.
- **Framework Limitations**: Prevailing frameworks like DGL and PYG provide theories but may not scale efficiently beyond only one machine without substantial customization.

Unfolding solutions include Graph compilers, GPU clustering, and end-to-end pipelining systems such as dist-dgl and aligraph.

6.4.8 Application-Specific Constraints

Different implementations impose different scalability rules. For example:

- **Recommender systems:** These systems sometimes collaborate with many users and also interface must facilitate in the real-world. One of the main responsibility of these system is to verify the complex data in order to personalize the themes and suggestions for the individuals/users. A numerous services, users, procedures are controlled by systems to generate the liable suggestions. The system should provide the real-time suggestions due to the extension of number of users, obstacles which regulates the system flow.
- **Biomedical networks** As concern to biomedical interconnections, for scaling the robustness and understandability is needed. Biomedical connection is to focus on web of biology in detail manner. Mainly in generating the report on genes, tissues, proteins, and other kind of diseases. It is highly progressive and it is utilized in biological computations including healthcare applications to examine the complex process of biomedical area. An important aspect of research in specialized medication, prediction of affected disease, discovering drug, including the medical disciplines.

6.5 Existing Approaches to Scalability in GNNS

Based on the recent years research, one of most pointable effort have put in dealing with the scalability complication in GNNs. In these complication duration, the innovation of algorithm plays a key role which consists of the individual techniques, well defined architecture design, and system level upgrade. Here, we can classify and verify the crucial tactics which are developed to minimize the scalable constraints.

6.5.1 Sampling-Based Approaches

An approach on sampling oriented is effective method for scaling the GNNs. Instead of executing entire peer nodes of an architecture, we utilize the sampling technique for computation a set of nodes and to reduce both time complexity and memory complexity. Managing the time complexity leads to execute an algorithm by taking less amount of time, on other hand the memory complexity or space complexity maintenance reduces the consumption of memory during executing an algorithm.

Node-Wise Sampling

This is an essential sampling that was introduced in **GraphSage** [1]. Node-wise sampling consists of choosing a fixed number of peers per node at each layer. This technique significantly reduces the receptive field and permits batch training.

- **Advantages**: Scalable to huge graphs and suitable for specific to general learning.
- **Limitations**: Undergoes from peer explosion in deep layers and aptitude information loss due to sampling diversity.

Layer Sampling

FastGCN [2] suggested layer-wise sampling where nodes were sampled layer-by-layer without connection-oriented oriented rather than per-node. This minimizes redundancy and complications.

- **Advantages**: Best theoretical convergence and optimized/efficient for sparse graphs.
- **Limitations**: Lower cost in capturing local structure due to the detachment between layers.

Importance Sampling

Importance sampling, as utilized in ladies [3], tends the sampling tasks toward more influential or centralized nodes based on statistical significance.

- **Advantages**: Optimistic accuracy contrasted to uniform sampling.
- **Limitations**: Computational overhead is in calculating the importance scores.

Subgraph Sampling

Graphsaint [4] installs subgraph sampling, where complete subgraphs are sampled for understanding and training = instead of separate and single nodes or neighborhoods.

- **Advantages**: Preserves local graph architecture and is suitable for parallelization.
- **Limitations**: Memory expense can be higher for large subgraphs.

6.5.2 Graph Partitioning

Graph partitioning splits a large graph into tiny components that can be processed in isolation or parallel. The main process of the graph partitioning is to reduce the number of edges that lie between the different partitions while keeping the splits as maintained in size as concern. This maintenance technique helps to ensure the efficiency of processing and communication while the partitions are controlled sequentially.

Cluster-GCN

Cluster-GCN [5] Exploits precomputed graph partitions (through clustering) to build mini-batches. Each batch consists of a dense subgraph, activating concentrated computation.

- **Advantages**: Improved cache region and reduced sampling difference.
- **Limitations**: Partition quality robustly influences performance.

Graph Partition Neural Networks

GPNN [6] aim to understand the splitting dynamically while training, which configures data characteristics. It represents an optimized approach to the issue of effectively processing large-scale graphs. Mostly a traditional method that lies in pre-defined splits. GPNNs continuously and dynamically train how to partition the graph during the training process. It is crucial as it permits the model to tailor the splitting technique or strategy based on the significant characteristics and framework of the received data.

- **Advantages**: Optimistically adaptive and data-driven.
- **Limitations**: Consists of increased training complexity.

6.5.3 Scalable Model Architectures

Architectural modifications have also been made to minimize computational difficulties during performance. As the size of data and the sophistication of tasks in machine learning applications grow, specifically in domains like natural language processing, recommendation systems, and traffic prediction, scalable model architectures become important. These frameworks are specifically developed to control or even enhance performance while minimizing computational complexity, ensuring that the models remain responsive, efficient, and deployable in large-scale, real-world platforms.

Simplified GNNs

SGC (simplifying graph convolutional networks) [7] eliminates non-linearities and encounters weight matrices, minimizing the GCN to a linear model.

- **Advantages**: Extremely fast and competitive performance in various tasks.
- **Limitations**: Loss of articulateness in complex graph tasks.

Decoupled GNN Architectures

Decoupled GNNs such as **APPNP** [8] and **daGNN** [9] distinguish the propagation and transformation milestones, permitting for more efficient information direction.

- **Advantages**: Efficient scalability and accuracy.
- **Limitations**: Calibrating propagation depth is non-insignificant.

Attention-Based Scaling

Techniques like **GATV2** and **G-transformers** [10] reveal sparse attention or less-rank approximations to scale focus computation in GNNs.

- **Advantages**: Enhances the long-range dependency system.
- **Limitations**: Attention estimation can still be expensive in dense graphs.

6.5.4 Training Optimization Techniques

Various optimizations at the training level support improving the scalability of GNNs, specifically in distributed platforms and environments.

Mini-Batching with Neighbor Caching

Preserving frequently accessed peers in memory (neighbor caching) permits mini-batch training to scale more efficiently [11].

- **Advantages**: Overcomes input and output overhead.
- **Limitations**: For large graphs, it is memory-intensive.

Mixed Precision Training

Float16 is used by mixed precision training where possible, minimizing memory usage and boosting up computation.

- **Advantages**: Extensively supported by modern GPUs.
- **Limitations**: Statistical instability in a few architectures.

Gradient Checkpointing

Gradient checkpointing [12] crafts memory for computation by saving or preserving fewer intermediate values and re-estimating them during backpropagation.

- **Advantages**: Drops the GPU memory footprint.
- **Limitations**: Increases training time.

6.5.5 Distributed GNNs

Distributed GNNs split computation and storage over numerous machines to scale to huge graphs.

DIST-DGL

Dist-dgl [13] partitions the graph over a cluster and utilizes distributed information loaders and samplers to parallelize instructions.

- **Advantages**: Undirected speedup with various workers.
- **Limitations**: Correspondence overhead can be overseen with huge partitions.

Aligraph

Aligraph [14] is a commercial-scale GNN system deployed by Alibaba, developed for billion-scale graphs.

- **Advantages**: End-to-end optimization from storage to inference.
- **Limitations**: Difficult during deployment and tuning.

Neuron-Based Parameter Servers

Large-scale and wide-ranging systems such as **tensorflow-GNN** embrace a parameter server framework where node combining and parameters are segmented across workers.

- **Advantages**: Scalable to huge-scale embeddings.
- **Limitations**: Need to be careful in synchronization and consistency management.

6.5.6 Approximate and Quantized GNNs

Approximate computing and quantization techniques have been taken on from other platforms to GNNs.

- **Low-rank approximations**: Minimize the dimension of the feature array to ease calculation.
- **Feature binning**: Collecting into similar features to minimize uniqueness and memory.
- **Quantized inference**: Implement 8-bit or lower-precision representations during inference.
- These techniques illustrate small accuracy trade-offs in exchange for dramatic improvements in delay and throughput [15].

6.6 Benchmarks and Evaluation Protocols for Scalable GNNS

An investigation of GNNs scalability results, evaluated protocols and the robust benchmarks which are crucial to differentiate the performance of numerous techniques used in scalability. The GNNs diversity represents its architecture, issues of scalability, and applications where these are important to take for systematized evaluation that allows well defined differentiation of different patterns. The segment states that the main components like key benchmarks, protocols of evaluation, and datasets.

6.6.1 Benchmark Datasets

Scrutinizing the GNNs scalability, it prerequisites the characteristics of real-time graph structure. The structure may include sparsity, complexity, and size variation. A number of benchmarks of datasets are maximally used to set for scalability through different structures of GNNs.

Social Network Graphs

More number of interconnectivities and the expansion of sizes can be observed due to the scalability of GNNs, which is handled by the social networks. In often cases the benchmark is set.

Some popular datasets of social network are:

- **Facebook**: The Facebook server has maximum number of end-users and the connectivities which is used to verify the computation of GNNs in the real-time social patterns.
- **Twitter**: Identification of group and classification of end-users in the large-scale structure takes place for the prediction of interconnections.
- **LinkedIn**: Another large graph that is used to examine the performance of GNNs for suggestion and social network analysis.
- **Reddit**: A social media framework that has multi-relational graphs, offering a challenging scenario for scalable graph neural networks.

These collections of data are especially useful for assessing how well a GNN controls large graphs with varying node degrees, and how productively models can scale across such data.

Citation and Academic Graphs

Citation networks, such as the CiteSeer, PubMed, and Cora, are extensively utilized for GNN writings. These graphs model connections between academic papers based on citation links.

- **Cora**: includes 5,429 citation links and 2,708 papers. It is useful for reviewing node classification tasks on tiny to medium-sized graphs.
- **Citeseer**: holds 3,312 papers and 4,660 citation links. It offers a useful benchmark for node organization, where scalability is less of an issue due to the graph's smaller size.
- **PubMed**: comprises 19,717 documents with 44,338 citation links, making it appropriate for assessing scalability on medium-scale graphs.

These citation networks are crucial benchmarks for testing scalability to medium-sized graphs with extremely structured relationships.

Recommendation System Datasets

Graphs standing for recommendation systems, like the **Amazon co-purchase** and **movielens** datasets, are usually used to evaluate scalable GNNs in the context of cooperative filtering and item-based recommendations.

- **Amazon co-purchase**: It is having the graph structure which is purchased in groups, often utilized to set the benchmark

techniques. It is done for the suggestions at a point of scale. It contains millions of nodes, record is as per the previous investigation.
- **Movielens**: In an architecture, for uniformly sorting tasks, the datasets are utilized. It is for rendering the movie suggestions across various users and movies.

Web-Scale Graphs

These graph patterns consists countless classes of datasets which are used to set the benchmark for GNNs scalability. Some of significant challenges it arises as follows:

- **Wikidata**: A large set of knowledge and understanding of graph pattern has demonstrated the entity structure. It is pre-owned for some actions like prediction of relationships ad classification of entities.
- **Freebase**: It is also the knowledge oriented graph pattern which has a number of real-time objects and their connectivities.
- **Google web graph**: Google has the feature in web graph that showcasing the hyperlink patterns between various websites. Its complexity and size made it a milestone for validating the GNNs scalable models.

6.6.2 Evaluation Metrics

The GNNs matrix for assessing the scalability require both signified matrices and standard ML statistics where both have the key features precision, accuracy and memory, usage respectively. The computational efficiency is the main leverage factor of these approaches. Most commonly used matrices are available for GNNs scalability as follows:

- **Accuracy metrics**: For an assessment of GNNs scalable performance, the standard ML decision making matrices are available in the industry.
- **Node classification accuracy**: The node labels get predicted accurately in the technique. It is often used in the case of partially supervised learning tasks.
- **Link prediction accuracy**: Here, the model's potential is evaluated to make decision for missing edges within the graph structure.

- **Graph classification accuracy**: Instead of individual node or edge, the complete set of graph is classified for the further task.
- **Mean reciprocal rank (MRR)**: Reciprocal rank in mean is utilized for sorting the tasks, then it is assessed for their performance. The relevant and non-relevant items are separated in this segment.
- **Efficiency metrics**: It is a measure of how well a GNN makes use of accessible computational resources like CPU, GPU, or memory while performing its fundamental learning and prediction functions.
- **Training time**: It scales the consumption of time to precise the model to understand a constant numbers of period or timelapse. The larger dataset algorithms are scaled effortlessly.
- **Inference time**: An individual node is processed at a given point of a time. Critical time said to be an application reduces its time of interference.
- **Memory usage**: The total consumed memory analyzed by the architecture while training, includes the need of storage for embedding the nodes.
- **Computation cost**: This matric scales the executional complications of training, as frequently examines the operation per second. It is also called as flops.

Scalability Metrics

For measuring accurately, few matrices are usually utilized as follows:

- **Graph size**: It scales the count of nodes and edges along with density of the graph. It usually works for complex graphs.
- **Model depth**: It examines the way that, how in-depth the GNNs architecture is. Also that leads to change both scalability and the expressivity.
- **Number of workers**: It talks about the analyzational part of GNNs scale. It includes the concept of parallel unit for computing.
- **Speedup**: The speedup assesses the performance of a distributed system. It tackles the parallelize implementation, especially for the individual node baseline.

6.6.3 Benchmarking Frameworks and Evaluation Protocols

Various frameworks are benchmarked and examined the protocols which are established to ensure the consistency and flexibility for evaluation the GNNs scalability. The Table 6.1 outlines the key items of commonly used benchmark datasets.

- **OGB (Open Graph Benchmark):** The OGB [16], which was established in well manner. It is a framework that provides a dataset of standardized graph. It is sent for the examination the assess the performance of GNNs. It entails an extensive data for the further tasks, which includes the strategies like classification of graph, link prediction, etc.
- **Graph Neural Network Benchmarks (GNN-Bench):** An essential technical aspect for GNNs scalability covers the GNN-bench [17] which establishes a benchmark for delivering the framework in a standardized manner. It is then passed for examining on various datasets. A collection of baseline models it includes. It is designed and developed to stress on multitasking examining on large-scale datasets.
- **Pytorch-Geometric Benchmarks:** Pytorch-geometric (PYG) [18] is among the most widely used libraries for GNN

Table 6.1 Benchmark datasets and tools for assessing scalability in large-scale GNNs.

Dataset name	Domain	Size (nodes/ edges)	Task	Description
OGBN-products	E-commerce	2.45 million / 61 million	Node classification	Product co-purchasing network from amazon
OGBN-papers100M	Citation network	111 million / 1.6 billion	Node classification	Citation graph of academic papers
Reddit	Social network	232k / 11.6 million	Node classification	Reddit posts and their interactions
Yelp	Business reviews	716k / 61 million	Link prediction	Yelp user-business review graph
protein-protein interaction (PPI)	Biological network	56k / 818k	Node classification	Protein interactions in different tissues

deployments and includes various benchmarks for scalability. Pytorch-geometric furnishes pre-configured datasets and assessment scripts for measuring both model exactness and efficiency throughout various graph sizes.

6.7 Real-World Case Studies and Applications at Scale

As graph neural networks (GNNs) continue to develop, they have established applications in a wide range of domains. These applications often interact with large-scale, real-world graphs that introduce exclusive adaptability challenges. In this portion, we analyze various real-world case studies where scalable GNNs have been effectively applied to solve complicated problems, underlining the strategies used to defeat elasticity issues in these settings.

6.7.1 Social Network Analysis

The analysis involves learning the relationships among individuals (nodes) and their communications (edges). With the accelerating growth of online social networks like Facebook, Twitter, and LinkedIn, computing these graphs at scale has become a crucial challenge.

Facebook's DeepGraph

Facebook has established the use of GNNs in large-scale social network analysis along with its **DeepGraph** project [19]. Deepgraph uses an adaptable GNN architecture to model and evaluate relationships among users, pages, posts, and interactions. Scalability is achieved through multiple innovations, consisting of a hybrid approach of layer-wise sampling and graph splitting, along with scattered processing across thousands of systems and machines.

- **Challenge**: amplifying GNN models to control billions of nodes and edges while ensuring real-time inference competencies for individualized recommendations.
- **Solution**: DeepGraph leverages **distributed GNN models** alongside sampling-based **techniques** to control such large-scale graphs. It partitions the graph into smaller subgraphs, and each subgraph is managed independently across a distributed computing ensemble.

- **Outcome**: Deepgraph triumphantly delivers real-time endorsements, friend suggestions, and ad targeting, processing millions of communications regularly.

Twitter's GNN-Based Content Moderation

Twitter uses GNNs for content management and misinformation recognition. The platform leans on user interactions, tweet content, and broadcast patterns to identify harmful content and flag accounts engaging in malevolent behavior. The shown Figure 6.4 displays the comparison of F1 scores on various networks.

- **Challenge**: scalability setbacks arise due to the enormous amount of data (billions of messages and engagements) that needs to be processed dynamically.
- **Solution**: Twitter engages multi-hop graph convolutional networks (MGNNs) for modeling long-range relationships in tweet engagements. By using subgraph sampling and neighbor caching, the device scales efficiently while handling high accuracy.
- **Outcome**: the device or a machine or a system can approach millions of tweets in near real-time, highlighting problematic content with high exactness and minimizing the workload for manual moderators.

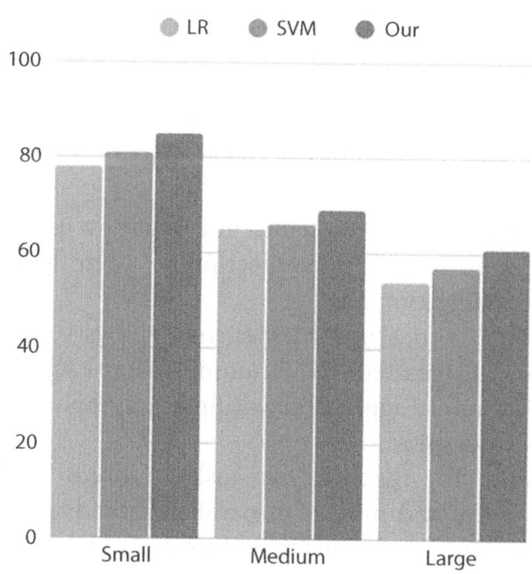

Figure 6.4 Comparison of F1 scores on different networks.

6.7.2 Recommendation Systems

Recommendation systems are different platforms where GNNs have had specific accomplishments. These systems use user-item engagement graphs to suggest products, movies, music, or even news and articles to users. Scaling GNNs to control large user-item graphs is a common challenge in industry.

Amazon's Product Recommendations

The engine works heavily on **GNNs** to generate personalized software or product suggestions based on end-user browsing history, software or product interactions, and purchase information. With millions of products and end-users, scalability becomes more critical.

- **Challenge**: Controlling user-item engagement graphs that are compact and grow rapidly over time.
- **Solution**: Amazon sets a graph convolutional network with importance sampling [20] to give precedence to influential edges in the user-item graph. This substantially minimizes the number of nodes and edges proceedings while maintaining suggestion accuracy.
- **Outcome**: The machine or system efficiently scales to millions of end-users and products, presenting high-quality endorsements and personalized interactions.

Movie Recommendation with MovieLens

The MovieLens dataset is often used as to yardstick for GNN-based recommendation systems. GNNs are deployed to model users, movies, and their engagements for personalized movie suggestions.

- **Challenge**: The graph expands hugely as new users and movies are coined, and the connection between them becomes exponentially complex.
- **Solution**: To measure, MovieLens deploys subgraph sampling and distributed GNN training, where each node's communications are aggregated within a subgraph, reducing the memory overhead.
- **Outcome**: The machine can task hundreds of thousands of end-users and movies in parallel, providing customized movie recommendations to end-users in real time.

6.7.3 Drug Discovery and Molecular Graph Analysis

GNNs have been implemented for drug discovery and molecular assessment to predict drug-target engagements and enhance molecular architecture. These deployments need GNNs that can optimally control the wide number of molecular graphs and the bondings between atoms and chemical bonds.

Atomwise's Drug Discovery Platform

Atomwise [21] uses GNNs for drug discovery, particularly in selecting or detecting potential drug candidates for diseases such as cancer and HIV. It processes millions of molecular graphs to anticipate the engagement between molecules and biological targets.

- **Challenge**: the requirement to scale GNNs to control millions of molecular graphs, each with conceivably millions of atoms and links, while maintaining large prediction exactness.
- **Solution**: Atomwise utilizes a fusion of graph partitioning and notable sampling to control the massive-scale molecular graphs easily. It also assigns distributed GNN training to parallelize the computation across many workers.
- **Outcome**: Atomwise has successfully detected guaranteeing drug candidates, promoting the drug discovery process, and minimizing time-to-market for recent treatments.

GNN-Based Protein-Ligand Interaction Prediction

PLI is an essential task in drug discovery. **GNNs** have been utilized to model protein-ligand graphs, where nodes represent proteins and ligands, and edges demonstrate interactions.

- **Challenge**: The issue lies in measuring the models to accomplish the interactions across various protein and ligand data collections.
- **Solution**: PLI prediction is operated using message-passing neural networks (MPNNS) linked with subgraph sampling to optimistically maintain the large datasets.
- **Outcome**: The GNN model provides accurate predictions of protein-ligand interactions, aiding researchers in selecting promising candidates for further testing.

6.7.4 Large-Scale Knowledge Graph Embedding

Knowledge graphs, which illustrate bonding between real-time objects or entities in a properly structured manner, are often massive and complex. Scalable GNNs are being utilized and applied to enhance knowledge graph embeddings and connection prediction tasks.

Google's Knowledge Graph

Google utilizes its large set of knowledge graphs to enhance search results and user handling or experience across its ecosystem of products. The knowledge graph demonstrates entities and their relationships, converting billions of evidence and relationships.

- **Challenge**: Measuring the embedding process to control billions of entities during confirming the real-time inference for search queries.
- **Solution**: Google uses scalable graph attention networks (GAT) linked with scattered training to process the knowledge graph at scale. This technique uses limited attention mechanisms to target major parts of the graph, minimizing computational expense.
- **Outcome**: The knowledge graph is capable of powering Google's search algorithms, offering users exact, contextually relevant information across billions of web pages.

Microsoft's Knowledge Graph for Enterprise Search

Microsoft uses GNNs in its organization search system to upgrade the relevance of search results across its suite of effectiveness tools. The system is showcased to operate massive-scale enterprise knowledge graphs that model documents, people, and administrative relationships.

- **Challenge**: The sophistication of scaling the search system to control massive knowledge graphs consisting millions of information and documents, and also the end-user engagements.
- **Solution**: Microsoft deploys dynamic graph splitting and subgraph sampling to control massive knowledge graphs, alongside distributed GNN models to simultaneously the processing.
- **Outcome**: The company or an organization's search system offers an accurate search throughput, enhancing productivity and knowledge uncovering within massive industries.

6.8 Challenges and Future Directions

While GNNs have showcased gigantic potential in resolving real-world problems throughout various domains, challenges remain in succeeding in scalable, efficient, and interpretable algorithms. This segment outlines the key challenges in scaling GNNs and also ensures future directions that could lead to further advancements in the area.

6.8.1 Challenges in Scaling GNNs

Scalability is a valuable hurdle in applying GNNs to large-scale real-world graphs. The following are a few of the main challenges faced when scaling GNNs:

Computational Complexity
GNNs conventionally need extensive computations due to their reliance on graph convolutions, which collaborates information from neighboring nodes. As the scale of the graph expands, the computational cost increases exponentially. This causes a challenge when forwarding to scale GNNs to billions of nodes and edges. Figure 6.5 represents the convergence of various tactics on the datasets of OGBN products.

- **Graph Convolution**: The crucial functionality in GNNs includes aggregating information from neighboring nodes.

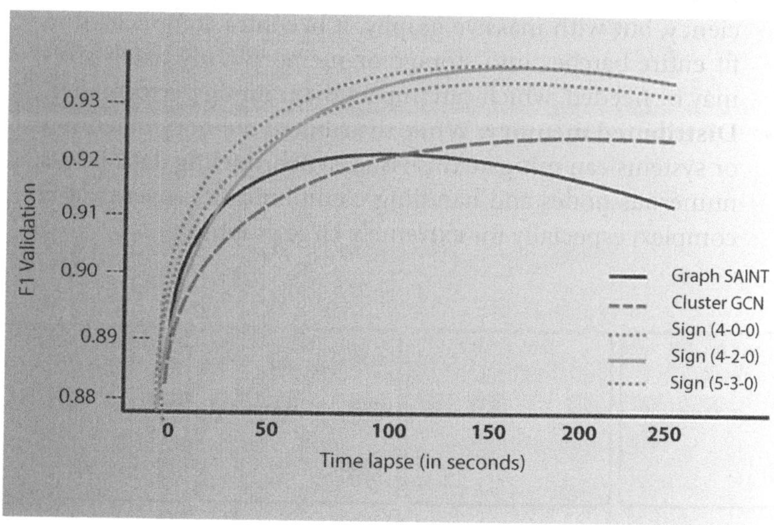

Figure 6.5 Convergence of different methods on the OGBN-products dataset.

These prerequisites involve processing the throughout graph or large subgraphs at each layer, leading to significant computational overwhelm.

- **Message-Passing**: In strategies like GATs and GCNs, each node sends messages to its peers. With massive graphs, the number of messages expands significantly, tending to higher computational needs.
- **Distributed Training**: While distributed training methodologies have been implemented to mitigate the computational overhead, effectively splitting the graph and synchronizing the sample parameters across numerous machines remains a non-trivial issue.

Memory Consumption

GNNs can be maximally memory-intensive, specifically when processing massive graphs. During the period of training, the memory needs expand as the model saves intermediate activations, node embeddings, and adjacency matrices. Figure 6.6 demonstrates the continuous address mapping mode.

- **Node and edge storage**: The scale of the graph needs substantial memory to allocate node features, edge features, and adjacency arrays or matrices. For large-scale graphs, this may become an impediment, especially when using concentrated representations.
- **Batching**: Batching is used to reframe computational efficiency, but with massive graphs, it becomes complicated to fit entire batches into storage or memory. Tiny batch sizes may be needed, which can impact the training performance.
- **Distributed memory**: While distributed memory machines or systems can mitigate this issue, synchronizing data across numerous nodes and handling communication overhead is complex, especially for extremely large graphs.

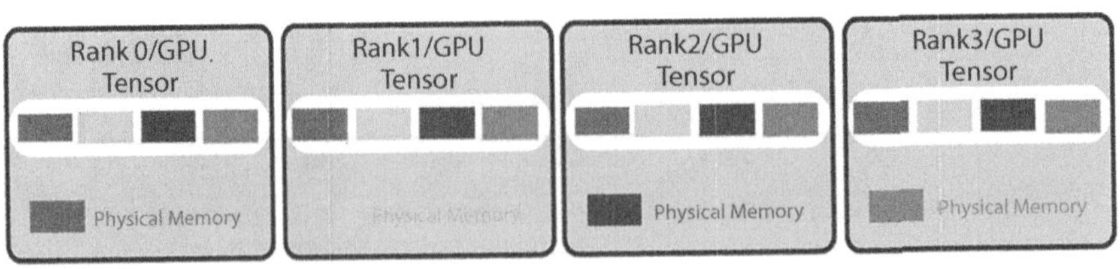

Figure 6.6 Continuous address-mapping mode.

Sparsity and Irregularity
Real-world graphs tend to be sparse and inconsistent, which complicates the architecture of efficient GNN models. Sparse graphs, where most nodes are only linked to a tiny number of peer devices, present challenges in data representation and computation.

- **Sparse Matrices**: Optimized operations on sparse matrices are crucial for controlling large, sparse graphs. Nevertheless, sparse matrix multiplication can be more complicated to enhance compared to dense matrices, and existing hard devices accelerators may not fully exploit the sparsity.
- **Irregular Node Connectivity**: Real-world graphs sometimes exhibit inconvenient connectivity patterns, where ends have varying degrees. This inconvenience makes parallelization challenging, as peer nodes may need different amounts of computation.

Dynamic and Evolving Graphs
Several graphs, such as social interconnections and suggested systems, were not static; they are dynamic and evolve. This demonstrates an additional layer of complexity, as the graph design may vary rapidly, requiring architectures to adapt to different nodes, edges, and advantages in real-time.

- **Graph Update**: Optimistically upgrading the model as the graph involves is a significant challenge. In various cases, incremental upgrades are needed to skip retraining the complete model from base, which can be computationally cost-effective.
- **Temporal Dependencies**: In the case of dynamic graphs, the relationships between peers may modify over time, and the model must train temporal dependencies. These prerequisites are specialized GNN models that can control dynamic changes without losing accuracy.

Interpretability and Explainability
GNNs, such as several DL methods, frequently functions as black boxes, which makes it complicated to interpret their decision-making processes. This shortage of interpretability is a complicated challenge in deployments where model transparency is crucial, such as healthcare, finance, and law enforcement.

- **Explainability**: There is an expansion of need for methods to explain the predictions of GNNs, specifically in high-stakes domains. Training that peers or experts contribute more to a particular prediction is crucial for trust and transparency.
- **Attribution Methods**: Many techniques, like attention mechanisms and layer-wise relevance propagation (IRP), are being developed to make GNN models more efficient and interpretable by highlighting the most primary parts of the graph.

6.8.2 Future Directions in Scalable GNNs

While scalability challenges remain, there are research directions that aim to showcase these issues and push the limitations and boundaries of what GNNs can achieve.

Advanced Sampling Techniques

Sampling techniques, like neighbor sampling, node sampling, and importance sampling, are being used exponentially to reduce the computational expenses of processing large graphs. Nevertheless, other improvements are required to scale these techniques efficiently.

- **Adaptive Sampling**: It is a dynamic and adaptive sampling method that prioritizes important nodes and edges and could frequently reduce computational and memory expenses.
- **Subgraph Sampling**: Splitting large graphs into smaller subgraphs and performing them independently has showcased promise in boosting training. Future enhanced research could focus on implementing more efficient partitioning techniques that control the integrity of the graph flow.

Hardware and Architecture Optimizations

Optimizing GNN models for particular hardware can provide a significant enhancement in scalability. For performance utilization, the enhanced supportive hardware like GPUs and GPGAs is installed.

- **Hardware-Aware GNN Design**: Developing GNNs are well defined to specialize the hardware architecture, which can consistently enhance the memory utilization and training. It involves an optimizing the processing graphs. In this design the hardware utilization is maximized.

- **Distributed Architectures**: An enhanced research is supposed to concentrate on developing efficient architecture. Those are meant to be minimized communication overheads. Also synchronization is optimized in this design.

Dynamic and Real-Time GNNs

While facing the challenges of real-time GNNs the models can efficiently operates on updates. It will be done without affecting the basic performance.

- **Temporal GNNs**: Temporal dependencies are modeled by developing the GNNs, is a key point for new generation research. It is having the capabilities to upgrade their parameters. It is not necessary to retrain the model.
- **Streaming Graph Processing**: GNNs is to process the graphs in well manner. In this, the nodes and edges are accessed and compiler in non-linear manner.

Interpretable GNNs

It is an enhanced way of GNNs for deploying real-time applications. It is a crucial for future research which can target on improvement of XAI (Explainable AI) techniques.

- **Graph Attention Mechanisms**: This aspects of methods can key point for crucial nodes in the system. The edges and nodes are to power the prediction, which interprets the maximized transparency while making a decision.
- **Visualization Tools**: Developing intuitive visualization tools that showcase the contributions of individual graph materials to a GNN's output can support improved user trustworthiness and model explainability.

Multi-Modal and Heterogeneous Graphs

Several real-world applications consist of multi-modal or different kinds of or different kinds of graph structures, where nodes and edges may consists of different type or features and advancements. Scaling GNNs to control such complex graphs is an evolving research direction.

- **Multimodal GNNs**: future models could collaborate information from different kinds of sources, like text, images, and structured data, to enhance and optimize the performance on complex processes such as multimodal learning, understanding, and knowledge graph construction.

- **Heterogeneous Graph Processing**: Controlling different kinds of or different types of graphs, which contain different types of edges as well as nodes, provides unique and different scalability challenges. New models could be implemented to process such graphs optimally and efficiently.

6.9 Conclusion

GNNs have escalated such as flexible and high performing tools which are used for understanding graph-structured information that establishes the innovations and developments across various disciplines such as recommender systems, NLPs, biological computations, and analysis of social networks. Squeezing of the linked structure of information *via* message passing and peer network grouping for GNNs provides the ability to train expressive node which should be meaningful. However, as the graph scale resumes to expand exponentially in industry applications, which will be significant scalability having the challenges, limits the deployment and processing ability of GNNs on large-scale graphs. Scalability in GNNs fragments a crucial challenge as real-time graphs leads to expand in scale, complexity, and dynamism. While the architecture has made a significant development in benchmarking high scalable relational structures, as they inherently tackle trade-off among scalability, model depth and accuracy. This chapter offers a detailed outline of GNNs, initiating with principles of fundamentals neural networks and expansion towards to the core mechanisms that encourage GNNs grouping, function updates, and message passing etc. It explores the taxonomy of GNNs architectures, covering the spectral, sampling and attention based, temporal and hierarchical based models. The learning objectives are addressed by each node in the system. In order to notify the challenges, a horizon range of techniques were proposed and validated, which includes the sampling strategies such as Graph sage, fast GCN models. The efficient models like GCN and GAT. These mechanisms are proposed not only to improve the model performance and flexibility, also their ability to enhance the intercommunication and critical need in many implementations.

AI disclosure statement: The author acknowledges the use of AI-based tools for minor language editing or grammatical correction. However, all intellectual contributions, analyses, and conclusions presented in this chapter are entirely the author's own.

References

1. Hamilton, W., Ying, Z., Leskovec, J., Inductive representation learning on large graphs, in: *Proceedings of NeurIPS*, 2017.
2. Kipf, T. and Welling, M., Semi-supervised classification with graph convolutional networks, in: *Proceedings of ICLR*, 2017.
3. Velickovic, V., *et al.*, Graph attention networks, in: *Proceedings of ICLR*, 2018.
4. Veličković, P., *et al.*, Graphsage: Inductive representation learning on large graphs, in: *Proceedings of NeurIPS*, 2017.
5. Gao, L., *et al.*, Graph neural networks for natural language processing: A survey. *IEEE Access*, 8, 125119–125138, 2020.
6. Ying, R., *et al.*, Graph convolutional neural networks for web-scale recommender systems, in: *Proceedings of ACM KDD*, 2018.
7. Zhang, A., *et al.*, Atomwise: A deep learning platform for drug discovery. *Proceedings of the American Chemical Society*, 2019.
8. Hamid, M., *et al.*, Large-scale graph analysis using distributed graph processing systems. *IEEE Trans. Big Data*, 5, 3, 222–234, 2019.
9. Wu, X., Hoi, S., Fu, J., Zhang, X., A comprehensive survey on graph neural networks. *IEEE Trans. Neural Networks Learn. Syst.*, 32, 1, 4–24, 2021.
10. Li, Y., Zhang, R., Zhao, R., Graph convolutional networks: A comprehensive review. *IEEE Access*, 8, 65502–65519, 2020.
11. Chen, X., Xie, S., Zhang, H., Scalable graph neural networks for real-time applications. *ACM Comput. Surv.*, 52, 7, 1–36, 2019.
12. Shchur, P., Sc, M.M.D., Borgwardt, S.G., Graph neural networks: A review of methods and applications. *ACM Comput. Surv.*, 52, 3, 1–38, 2020.
13. Cui, Z., Wang, Z., Li, H., A survey on graph neural networks: Models and applications. *ACM Trans. Knowl. Discovery Data*, 15, 6, 1–37, 2021.
14. Kipf, T.N., Welling, M., Becker, M.A., Graph neural networks and their scalability to large datasets. *Neural Information Processing Systems (neurips) Workshop*, 2018.
15. Zhang, H., Zhang, X., Yu, L., Graph neural networks for scalable large-scale recommendation systems. *IEEE Trans. Knowl. Data Eng.*, 33, 5, 1234–1248, 2021.
16. Ying, D., He, R., Zhang, K., Graph-based recommendation systems: A review. *IEEE Trans. Syst. Man Cybern.: Syst.*, 50, 9, 3874–3889, 2020.

17. Silva, A.M.O.V., Oliveira, L.S., De Campos, M.L.R., Large-scale graph processing systems and techniques: A survey. *J. Comput. Sci. Technol.*, 34, 1, 1–18, 2019.
18. Wang, W.Z., Liang, F.P.H., Zhang, W.N., Towards efficient graph-based learning: Scalable graph convolutional networks. *J. Mach. Learn. Res.*, 21, 1, 1–39, 2020.
19. Sun, J., Liu, Z., Zhang, L., Scalable and efficient graph convolutional networks for industrial-scale systems. *IEEE Trans. Ind. Inf.*, 16, 5, 3156–3165, 2020.
20. Han, P.Y., Li, Z., Zhang, X., Scalable and distributed graph learning algorithms for big data. *IEEE Trans. Big Data*, 7, 2, 341–354, 2020.
21. Schlichtkrull, M., Kipf, T., Bloem, P., *et al.*, Modeling relational data with graph convolutional networks, in: *Proceedings of the European Semantic Web Conference (ESWC)*, 2018.

7

Challenges in Large-Scale Graph Neural Networks in Topological Indices on Family of Graphs

Senbagamalar J.[1]* and Ramani M.S.[2]

[1]*Department of Mathematics, Vel Tech Rangarajan, Dr. Sagunthala R&D Institute of Science and Technology, Chennai, India*
[2]*Department of Mathematics, M.O.P. Vaishnav College for Women (Autonomous), Chennai, India*

Abstract

Artificial neural networks consist of weights representing the connections between their neurons. Each neuron is activated by input signals and modulated by the weights learned during training. When a neuron is influenced by the states of other neurons, the network can naturally model contextual dependencies. Recurrent Neural Networks (RNNs) maintain a memory of previous states, allowing them to process sequences with temporal awareness. Convolutional Neural Networks (CNNs) capture hierarchical spatial features through multiple layers and are effective at recognizing complex, repeating patterns. Graph Neural Networks (GNNs) generalize these concepts to non-Euclidean domains, allowing for the application of information based on the structure of a graph instead of relying on the layer's order. This is worth noting due to the fact that neural networks themselves can also be conceptualized as graphs, with nodes representing neurons and edges representing synapses. In the first phase of our study, a GNN updated each node's state by aggregating information from its neighbors such that it could update the entire network globally through a series of local interactions. In addition, it enabled us to predict chemical behaviors observed in the real-world, as well as to better capture long-range dependencies and subtle changes in structure. Additionally, we show that computing topological indices forms a key part of GNNs scaling and their capacity to learn from big molecular data.

**Corresponding author*: senbagamalar2005@yahoo.com

J. Ramkumar, Sridaran Rajagopal, B. Suchitra and S. Balamurugan (eds.) Graph Neural Networks: Concepts and Applications, (283–314) © 2026 Scrivener Publishing LLC

Keywords: Benzene ring, degree, molecular structure, deep neural network

7.1 Introduction

Graph theory has a fun history that starts in 1736 with Leonhard Euler and the famous Königsberg Bridge Problem. Euler abstracted this situation by coining nodes and edges and created what we know as a graph, as he showed that there was no way to cross each bridge once and only once. The basics of graph theory were born. For undirected graphs, this simply means there is a sequence of edges that connect any two nodes. A graph is considered strongly connected for directed graphs if there is a directed route from each node to every other node. If converting every directed edge to an undirected edge result in a connected graph, the directed graph is said to be weakly connected. A graph's connected component is the largest subgraph that is not connected to any other nodes outside the component and in which every pair of nodes in the component is connected to one another. A route is a succession of edges that connects a series of different nodes, whereas a cycle is a path that begins and terminates at the same node, when discussing routes and cycles. A regular graph is one where all vertices are equally connected, essentially having the same degree. In other words, each node in a graph has the same number of edges. A graph is said to be k-regular if the degree of each node is k. A cubic graph is any 3-regular graph, meaning that each vertex is connected to exactly three other vertices. A graph is 0-regular if and only if it has an isolated component. In regular undirected graphs each vertex has the same number of undirected edges. Regular directed graphs (digraphs) have the same in-degree and out-degree for every vertex. A cycle graph is a 2-regular graph since every vertex connects to exactly two other vertices, thereby existing in a loop. Regular graphs have significant applications in network design, where uniformity across connectivity is desirable. The use of regular graphs extends to error-correcting codes, combinatorial designs, and theoretical computer science. Regarding paths and cycles, a path is a sequence of edges that connects a sequence of distinct nodes, while a cycle is a path that begins and ends at the same node. A regular graph is a type of graph in which all vertices have the same number of connections, or equal degree.

The number of edges is the same for every node in the graph. If every vertex in a graph has degree k, it is said to be k-regular. For example, a cubic graph, sometimes referred to as a 3-regular graph, has each vertex linked to three additional vertices. A 0-regular graph is simply a set of isolated nodes with no edges. In a regular undirected graph, all vertices have the same number of undirected edges. In a regular directed graph (digraph), all vertices have the same in-degree and out-degree. A cycle graph $\mathbb{C}_n$, where $n \geq 3$, is a 2-regular graph because each vertex is connected to exactly two others, forming a closed loop. Regular graphs are especially useful in applications such as network design, where uniform connectivity is essential. They are also applied in error-correcting codes, combinatorics, and theoretical computer science.

7.2 Topological Indices on Chemical Graph Theory

The history of chemical graph theory precedes graph theory as a discipline. In 1857, the chemist August Kekule first diagrammed an organic molecule benzene, which shown in a graph structure as we would expect today with vertices representing atoms and edges representing chemical bonds. His structural formulae laid conceptual groundwork for visualizing molecular connectivity. In 1875, mathematician Arthur Cayley made significant early contributions by using trees to enumerate isomers compounds that share the same molecular formula but have different structures particularly for alkane compounds. A molecular graph is a graph in which atoms are represented by vertices, typically excluding hydrogen atoms in organic chemistry models, and edges represent the chemical bonds between atoms [1]. Most molecular graphs are undirected, as chemical bonds are mutual connections. Vertices and edges may be labeled to indicate the element type and bond order, respectively. In some cases, weights can be assigned to edges to reflect properties such as bond order, bond strength, bond length, or electronegativity. In hydrogen-suppressed graphs, hydrogen atoms are often omitted to simplify the representation, particularly for large organic macromolecules. Topological indices can be used in isomer enumeration, helping to count the possible structures of molecules with a given molecular formula. They are also useful in analyzing structure–property relationships, where graph-based properties are used to predict characteristics such as boiling points, reactivity, and other chemical or physical behaviors. Topological indices are fundamental in drug design and cheminformatics, as they allow the translation of a molecule's structural features into numerical representations that can be used for prediction, classification, and optimization. In molecular graphs, drug discovery models use graph structures to predict molecular properties. In 2022, Zhang discussed different topological indices on neural network [16]. The family of Neural Networks that operate on graph-structured data includes Graph Neural Networks (GNN).

The GNN employs a graph neural architecture to learn representations of the nodes and edges, which may then be used for activities like node classification, graph categorization, and link prediction. Moreover, GNNs may be trained using supervised, semi-supervised, or unsupervised learning methods. We check the potential place for the neurons can be constructed as a family of graphs, which can be related to family of graphs associated with their families. Graph Neural Networks (GNNs) have shown great promise for predicting chemical properties as an intersection of chemical graph theory. GNNs use a molecule's graph representation to relate atom nodes to bond edges, and accurately represent relationships and interactions between atoms in a molecular structure. This ability better informs the prediction of different chemical properties (i.e., stability, reactivity, toxicity) across different datasets. The various topological indices predict the biological activities of the molecular graphs.

Molecular Structure Representation in GNNs can model molecular graphs to predict properties such as stability and reactivity accurately. In environmental Chemistry, GNNs have outperformed conventional feature models to predict environmental properties by learning from molecular graphs "Improved Environmental Chemistry Property Prediction of Molecules with Graph Machine Learning". Incorporating Quantum Mechanical Descriptors, Quantum mechanical (QM) descriptors may enhance GNN performance, especially with small datasets, by providing additional contextual information related to molecular properties. Challenges and Further Considerations Data Limitations for GNNs do not extrapolate well beyond the training datasets, thus it is important to carefully select training sets and descriptors to eliminate unnecessary noise and clutter. Computational Costs: QM descriptor deployments can be computationally expensive, but often do not yield meaningful evidence of an improved performer relative to cost. Graph Neural Networks (GNNs) are applied to molecular compounds by representing atoms as nodes and bonds as edges, capturing complicated interactions. They predict chemical properties like stability, reactivity, and toxicity. This helps in advancement toward a multi-faceted understanding of molecular structures from the perspective of chemical graph theory. Specifically, GNNs (like other machine learning approaches) predict potential chemical properties, by modeling molecular structures as graphs and learning representations from the molecular graph. A strength of GNNs is their ability to use complicated structural information to make accurate predictions on chemical properties of molecular compounds, which can enhance predictions of an environmental chemistry task among others.

Although topological indices give sound mathematical representation for molecular structures, they are seamlessly embeddable within GNNs. These descriptors are also compact as information–rich features and provide guidance to the learning process in GNNs by helping to reduce the input dimensionality. That becomes all the more important as one scales to large chemical networks where computational efficiency and interpretability of learning are critical. Embedding such indices instead permits more accurate molecular property prediction, with less reliance on data intensive training procedures.

7.3 Networks in Graph Theory

A network is a kind of graph that has vertices that are the entities and edges (or links) that are the connections or relationships that connect them. There can be directed and undecided edges, and networks can also be weighted. In social networks the nodes are people, and the edges are the relationships of communication or friendship, etc. In biological networks, which range from genomics to neuroscience, it could be proteins, neurons, etc. In computer networks, nodes are computers or servers, and edges are physical or wireless connections. In transportation networks, nodes represent locations such as airports, intersections, and edges are the routes or roads connecting them. In flow networks are directed networks with capacities, used in problems such as maximum flow for modeling the flow of water, electricity, or data. Flowing phenomena are typically analyzed in such networks. The number of connections a node has treated as degree. Centrality is the importance of a node in the network which can be analyzed betweenness and, closeness. Connectivity determines whether nodes are reachable from one another. A measure of how tightly a group of nodes is clustered together can be treated as Clustering Coefficient. In 2023 Do *et al.*, demonstrated the consistent improvement in accuracy and utilization of each GNN's allocated capacity over the original training method of each model [2]. Tiezzi *et al.* created a proof-of-concept in 2024 by developing a loss function for the edge crossing and providing quantitative and qualitative comparisons between various GNN models that operate under the suggested framework [11]. Liang *et al.*'s new findings in 2022 demonstrate that the *a priori* graph characteristics do, in fact, influence the caliber of representation learning, which may be used to direct the learning [9]. Despite the common knowledge that the human brain operates very differently from conventional digital computers, artificial neural networks (ANNs) are frequently referred to as neural networks.

The brain is a massively parallel, totally nonlinear processing system whose constituent neurons can be structured to carry out complicated computational operations. One of the main functions of the visual system is to create a consistent picture of the environment and, more crucially, to give the data needed to interact with it. The brain performs perceptual recognition tasks in approximately 100 to 200 milliseconds, with minimal cognitive effort. In contrast, a standard digital computer may take much longer sometimes even days to solve the same tasks using simple algorithms, especially if the cognitive load is lower. At birth, the human brain possesses innate structural capabilities, but it also has the ability to develop its own rules based on experience. A developing neuron reflects the plasticity of the brain it enables the nervous system to adapt and rewire itself in response to its surrounding environment. In general, a neural network consists of artificial neural units, designed to model the way the brain carries out specific tasks or functions of interest. The models are calibrated to reproduce some intrinsic features of biological cognition; this remains crude, artificial, and simplified, but overall production has a structurally-determined computational model. Compared to conventional GNN models, Feng *et al.* demonstrated that graph filters trained in Ker GNNs can reveal the local graph structures even in datasets with extremely complicated graphs, which reflects the significant increase in model interpretability [4]. In 2022, Jiang *et al.* explored the potential of graphs neural networks in applying to problems in traffic forecasting [7]. Research in this area has shown that a trained neural network, working under particular environmental conditions, can be adapted to slight changes of those conditions and still perform adequately. Zhou *et al.* [18] provide a methodical analysis of the GNN models' design pipeline, taking into account the variations of each component, and evaluating the uses of each kind. It is possible that a neural network may be designed to not just make a decision about which particular pattern should be chosen, but also to express its level of confidence in that choice. Neurons, in one form or another, are the basic building blocks for every neural network. Theories and learning algorithms can often be shared between different kinds of neural networks, demonstrating their flexibility and versatility.

Graph neural networks (GNNs) can predict chemical properties by composing molecular graphs in which atoms are represented as nodes and bonds as edges. They learn representations through message passing, resulting in effective modeling of complex interactions, deepening predictions of molecular reactivity and properties. Graph neural networks (GNNs) can predict chemical properties by composing molecular graphs in which atoms are represented as nodes and bonds as edges. They learn representations through message passing, which results in effective modeling of complex interactions and deepens predictions of properties and molecular reactivity. Graph neural networks can predict chemical properties by modeling electron densities, enabling inferences such as metal atomic charges or dipole moments. This incorporates data-driven approaches with established physical principles, resulting in improved accuracy and predictive consistency with chemical graph theory. Graph neural networks (GNNs) represent an impressive step forward in the intersection of neural networks and graph theory by enabling effective learning and representations of graph-structured data. GNNs have a variety of tasks that they can benefit from by utilizing the relationships between nodes and edges, such as node classification and link prediction. By nature of the broad applications of GNNs, they are well suited to perform tasks across multiple domains. An Overview of Graph Neural Network , GNNs operate on graph data, where nodes can represent entities and edges represent relationships. GNNs also fall under the encoder-decoder paradigm that allows transformation of graph data into useful and meaningful representations for many applications. The architecture of the GNN works effectively to capture dependencies between nodes, which is necessary in performing tasks in social networks, molecular structures, and recommendation systems.

The theoretical framework refers to properties related to approximation and learning of GNNs, and in particular, generalization, and extrapolation. GNNs are also variable in size which refers to the fact that the number of nodes, edges, and features can vary crudely across models. Message passing is central in GNNs to enable aggregation of information from neighboring nodes which also enhances the predictive reach of the model. Applications and Impacts have been applied to many domains with successful outcomes such as knowledge graphs, urban and rail transport networks and in life sciences. Overall, GNN's ability to capture complex relationships has positioned them as a disruptive technology. A neural network's structure is very similar to that of the human brain. In 2022, Wei *et al.* showed that GNNs may still be used under graph sampling settings, even on single static networks, and that simpler GNN models can outperform complex GNNs in fairly experimental [12]. Du *et al.* demonstrate in 2019 that GNTKs can theoretically learn a class of smooth functions over graphs. We empirically study GNTKs using graph classification datasets and show that they are capable of high performance [3]. To enhance the graph characteristics generated by the hybrid deep network, Yao *et al.* introduced a Graph SAGE based network in 2023 [14]. Utilizing only a few labeled graph instances, Xie *et al.* [13] demonstrate in 2022 that the findings produced using their suggested framework exhibit competitive performance on graph classification problems across a wide range of benchmark graph datasets. The brain is physical evidence that simple, fault tolerance, efficient and powerful parallel processing can be achieved. Scarselli Franco *et al.* proposed algorithm for computational utilization costs [19]. In 2023, Rui *et al.* showed an exploratory analysis of the biological applications of graph analysis [20]. In 2022, Ye *et al.*, aligned an overview of KG and GNN methods for further study [15]. In 2023, Melton *et al.* showed that mux GNN's coupling attention found interpretable relationships between distinct relations in heterogeneous networks. Zhu *et al.* showed experimentally that their method improved efficiency and performance from actually getting to an approximate result [10, 17].

7.4 Connectivity-Based Topological Indices

A topological index is a unique numerical value assigned to a graph or network that is frequently used in order to correlate structural features to physicochemical properties. Degree-based topological indices are connectivity-based topological indices that can have a few different meanings, all of which have been studied extensively resulting in many studies. Jia-Bao Liu [21] studied degree-based topological indices of generalized Sierpinski networks, and Euler graphs, from a purely theoretical perspective. Xiujun Zhang examined topological indices of connection networks using both distance-based and degree-based methods [16]. The topological indices of polymers, such as those displayed in bridge molecular diagrams, have also been studied. Weidong Zhao [22] lately researched topological co-indices in structures like starphene nanotubes and triangular benzenoids. The Randic index is the first topological index based on degrees. While investigating the boiling temperatures of paraffins, Milan Randic created the Randic index [14]. Researchers have created hundreds of topological indices during the last three decades to examine several aspects of chemical graphs. These indices have been useful for examining a variety of molecular properties, such as reactivity, stability, boiling point, enthalpy of formation, and Kovats retention index. Moreover, they can be used to forecast the physical characteristics of materials, such as stress, elasticity, strain, mechanical strength, and other factors.

The overall Randic index was first proposed by Bollobas and Erdős. When studying the π-electron energy of molecules in 1972, Gutman and Trinajstic introduced the first and second Zagreb indices [8]. Ghorbani and Azimi created the first and second multiplicative Zagreb indices [5], which are the multiplicative forms of the Zagreb indices, whereas Shirdel *et al.* introduced a new class of index known as the hyper-Zagreb index. The widely used atom-bond connectivity (ABC) index was developed by Estrada *et al.*, whereas the sum-connectivity index was created by Zhou and Trinajstic. In 2007, Chen *et al.* presented in the chained network [6, 7]. The geometric-arithmetic (GA) index was created by Vukićević and Furtula. Javaid *et al.* introduced degree-based topological indices in probabilistic neural network environments in 2017. Researchers have also assessed topological indices that describe recurrent neural networks and multilayered probabilistic neural networks. Consider a graph G with node set V and edge set E. The number of nodes that share an edge with node v is called its degree dv. The degrees of vertices in a graph G are used to define degree based topological indices of the graph, which may be represented as follows:

The first Zagreb index $M_1(G) = \sum_{rs \in E(G)} (\delta_r + \delta_s)$
The second Zagreb index $M_2(G) = \sum_{rs \in E(G)} (\delta_r * \delta_s)$
The Sombor index $SO(G) = \sum_{rs \in E(G)} \sqrt{\delta_r^2 + \delta_s^2}$
The Randic index $R(G) = \sum_{rs \in E(G)} \frac{1}{\sqrt{(\delta_r * \delta_s)}}$
The Forgotten index $F(G) = \sum_{rs \in E(G)} (\delta_r^2 + \delta_s^2)$
The atom bond connectivity index $ABC(G) = \sum_{rs \in E(G)} \sqrt{\frac{\delta_r + \delta_s - 2}{\delta_r \delta_s}}$

The augmented Zagreb index $AZI(G) = \sum_{rs \in E(G)} \left[\frac{\delta_r \delta_r}{\delta_r + \delta_s - 2} \right]^3$
Illustrative Examples of Topological Indices
Example 1: Path Graph P4:
v1 — v2 — v3 — v4

Illustration Figure Path Graph P4

Edges and Degrees:

E (v1, v2): deg(v1) = 1, deg(v2) = 2; E (v2, v3): deg(v2) = 2, deg(v3) = 2; E (v3, v4): deg(v3) = 2, deg(v4) = 1

$M_1(G) = \sum_{rs \in E(G)} (\delta_r + \delta_s) = 3 + 4 + 3 = 10$
$M_2(G) = \sum_{rs \in E(G)} (\delta_r * \delta_s) = 1 * 2 + 2 * 2 + 2 * 1 = 8$

$$R(G) = \sum_{rs \in E(G)} \frac{1}{\sqrt{(\delta_r * \delta_s)}} = \frac{1}{\sqrt{1 \times 2}} + \frac{1}{\sqrt{2 \times 2}} + \frac{1}{\sqrt{2 \times 1}} = 0.707 + 0.5 + 0.70 =$$

1.914

The molecular structure of a benzene ring may be represented by a chemical graph in Figure 7.1. A hexagonal ring is formed by six carbon atoms, each of which is connected to a hydrogen atom and two more carbon atoms. The Naphthalene structure was shown in Figure 7.2 and linear benzene ring structure was shown in Figure 7.3. A hydrogen-suppressed molecular graph is frequently used to make the molecular representation easier to understand. This graph only takes into account the carbon atoms and the connections between them, leaving out the hydrogen atoms for the sake of clarity. Figure 7.4 shows the molecular graph of benzene, with the hydrogen atoms removed. It has six vertices (each denoting a carbon atom) arranged in a ring structure, with alternating single and double bonds.

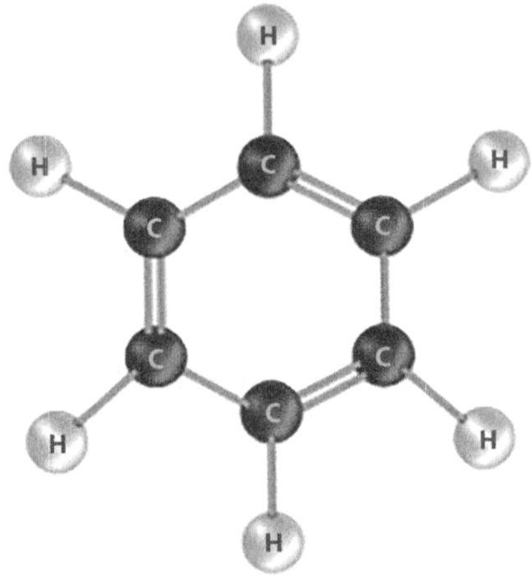

Figure 7.1 The benzene ring.

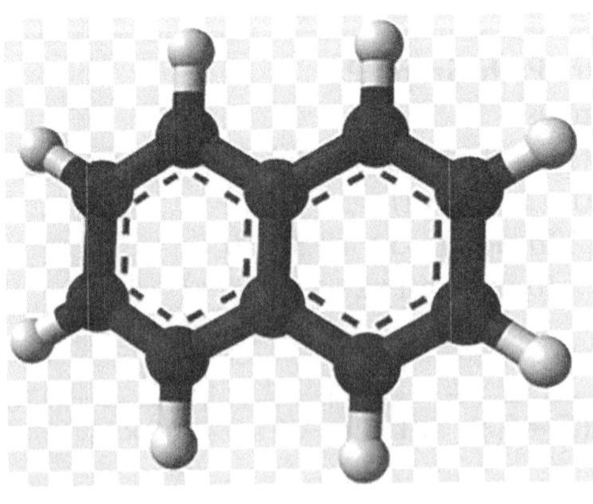

Figure 7.2 Naphthalene structure.

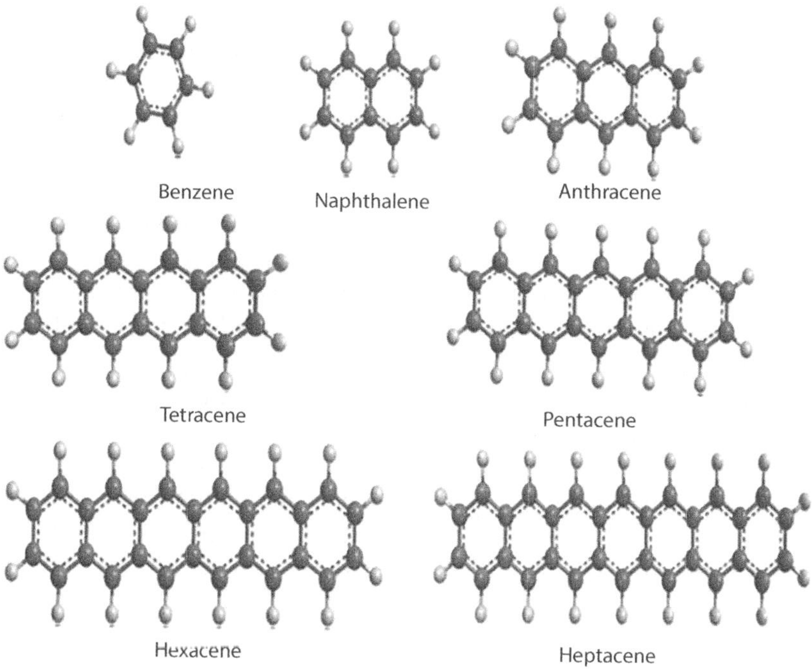

Figure 7.3 Benzene ring structures.

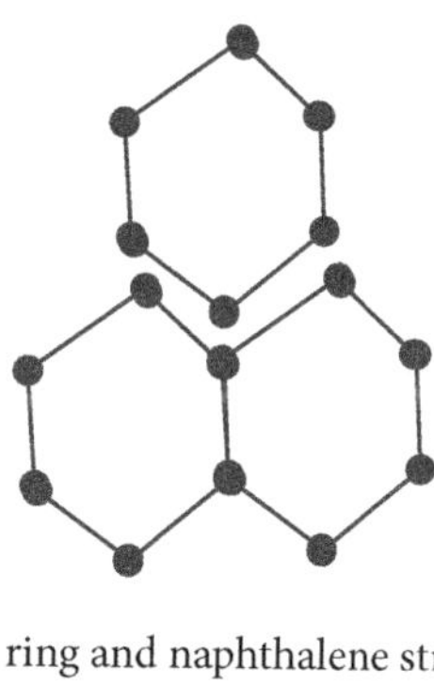

Figure 7.4 Structure of benzene ring and naphthalene structure.

7.5 Graph Neural Network

Graphs provide a logical and effective way to represent structured data in diverse fields, such as bioinformatics, image processing, and social network analysis. There are several approaches to node classification and graph classification, which are the two main activities in graph learning.

Graph Neural Networks (GNNs) are one of these, and they are a versatile and expressive class of architectures that have shown significant promise in recent years [23]. The majority of GNNs are built upon iterative message passing, which is a key paradigm in graph learning. Shervashidze and Borgwardt [24] first investigated this idea, which Sanchez-Lengeling *et al.* [25] later expanded upon in the context of GNNs. Most GNN models aggregate information from sequentially neighboring nodes, which can lead to a loss of certain topological features such as cycles or ring structures. This limitation is particularly important in cases where connection variety is relevant. The first stage in implementing GNNs is to identify the graph structure for the application. There are typically two cases when deciding what graph structure is appropriate: the case when the graph structure is defined, and the case when the graph structure is inferred. In terms of the defined structure case, data will include a specific graph structure. Graph neural networks (GNNs) are deep neural networks are unique neural network architectures for graphs that use properties of nodes and edges for training. GNNs provide great results across various node and graph prediction tasks spanning undirected and directed graphs, and utilize messaging-passing layers.

GNNs (Graph Neural Networks) are a unique architectures of deep learning used for learning representations of graphs. GNNs minimize loss through the prediction of nodes, graphs, and point configurations using message passing and higher-order properties to enhance representation, generalization, and extrapolation (i.e., to make good decisions other types of data). Graph Neural Networks (GNN) takes advantage of the structure of graphs to solve different complex problems by means of convoluting structured data of a graph. GNNs are proficient at predicting and solving real-world problems in many areas and highlight the advantages for using GNNs and curating their potential for additional fields of applications in graph theory and beyond. There are many examples of such applications in physical systems or molecular applications, as well as for knowledge graphs, for example. Then there is the indirect structure case. In indirect structure cases, for the data provided there is no structure and it is inferred by based upon the task. Examples of this include a text application where the graph is a fully connected word graph and the scene graph constructed for the image.

Once you have defined the graph structure you are ready to design and pick your best GNN model for the structure. Generally speaking, graphs that include more complicated types can contain more detailed information about both nodes and edges. Usually, we categorize graphs according to their structures. One aspect of graph structure that relates to the interpretation of edge types is direction. In a directed edge, the edge demarcates a direction from node i (source) to node j (target) as examples, which transmits more specific data than undirected graphs. Edges in an undirected graph are bidirectional and may be thought of as one directed from node i to node j and the other from node j to node i. A homogeneous graph consists only of one kind or type of nodes and edges. In heterogeneous graphs, nodes and/or edges can be of different types or kinds, in which case the types of nodes and edges are important characteristics to consider when modeling and analyzing them.

7.6 Dynamic Graphs in Deep Neural Network

In the realms of statistical and dynamic graphs, if the input features or the topology of a graph changes over time, the graph may be considered a dynamic graph. In this situation, temporal information must be handled correctly. These graph types such as directed, heterogeneous, dynamic are orthogonal and can be combined. For example, a dynamic, directed, heterogeneous graph is also possible [5]. Additionally, there are many other specialized graph types that are good for specific tasks, e.g., hypergraphs and signed graphs. While there is not enough room to describe all of the types of graphs, the emphasis should be placed on how to use the extra information offered to us *via* these specialized graph types. After determining the graph type, we must design a model that considers the graph-type specific features and added complexity. It is often challenging to classify graphs strictly into "small" or "large." This classification does not easily apply to graphs given that these definitions are changing especially with the new modern computational hardware such as GPUs, which can vary widely in terms of speed and memory. In this paper, we consider a graph to be large-scale if a computer cannot store or operate on the corresponding adjacency matrix or graph Laplacian (which requires storage of O(n2)). Sampling methods or other approximations will be needed in the event of an unmanageable graph. In this context, the following topological indices can alleviate the scalability issues by encoding the global structural character of a molecular graph in each node, and each edge feature. To reduce the burden of having GNNs learn these structures from scratch, this encoding reduces the number of structure elements for GNNs to learn, enabling more focused message passing and representation learning in large scale setting.

In a supervised paradigm, the model is trained using labeled data, which provides explicit direction during learning. The semi-supervised setting involves both labeled and unlabeled nodes. In this context, the label-to-interpretation process can be defined as either transudative or inductive, depending on how the unlabeled nodes are handled during inference. It is required to fixed set of unlabeled nodes that are already available at test time. In contrast, the inductive setting involves predicting labels for new, previously unseen nodes that are sampled from the same underlying distribution.

In graph-based activities, semi-supervised learning is quite prevalent and usually centers on node or edge categorization. In recent years, as Rossi *et al.* [26] have shown, hybrid transudative-inductive techniques are moving toward hybrid learning methods. In unsupervised learning, the model is instructed to learn to identify naturally occurring structure in data since there are no labeled observations, such as node clustering, which is a common unsupervised activity. It is feasible to design loss functions that are specific to the task, depending on how the task is defined and how the training is organized. In the case of a node-level semi-supervised classification task, once again, nodes in the training set may be labeled and the cross-entropy loss can be used to aid in the training of an unsupervised model.

The propagation module is a learning module that disseminates knowledge between nodes and is designed to generate representations that take into account both topological and feature information. The use of operations to aggregate information from neighboring nodes is demonstrated by convolution operators and recurrent operators within propagation modules. In human-level processing, skip connection operators were added to carry forward historical representation data, which aids in mitigating the over-smoothing seen in deep graph networks.

7.7 Modules in Deep Neural Network

The sampling module becomes essential when dealing with large-scale graphs, where it is computationally infeasible to perform full-batch propagation. In such cases, sampling techniques are used to select subsets of nodes or edges for efficient computation. Sampling modules are typically used in conjunction with propagation modules to scale learning across large graphs. When the goal is to learn representations of higher-level structures, such as subgraphs or entire graphs, a pooling module is required. This module aggregates information from individual node representations to generate a compact representation of the larger structure. Neurons in a layered Neural Network are arranged into distinct layers. A basic network is a simple input layer that contains the source nodes and projects its connection to an output layer of neurons. A multi-layered neural network adds one or more hidden layers and its processing units are called hidden neurons. The important role of hidden neurons is to be an intermediate processing layer situated between the external inputs and ratio outputs of the network. This not only gives the model more ability to learn hidden information and establish more complex representations and behaviors.

7.8 Fully Connected Neural Network

A fully connected neural network, each node in each layer is linked to all nodes in the next forward layer. When some of these communication links are missing, it becomes a partially connected neural network. A recurrent neural network (RNN) differs from the feed forward network in that it has at least one feedback loop that allows the output of some nodes to influence the future inputs. In some cases, a recurrent network is just a single layer of neurons, where each node sends its output back to all the other nodes' inputs where feedback from the first layer provides a cycle. As shown in Figure 7.1, the benzene ring structure serves as an analogy for a recurrent network. The benzene ring represents a closed loop where nodes (analogous to atoms or neurons) are symmetrically connected. In this representation, there are only input and output layers, with no hidden layers. Every node shares the same degree (i.e., number of connections), and thus there is no hidden intermediate layer to selectively route or "hide" information. This reflects the recurrent nature of the network, where messages circulate through all nodes equally, without hierarchical layering. As in Figure 7.5, the benzene ring is a 2-regular graph. If we construct a deep neural network, there is no hidden layer, it forms only a input layer and output layer.

7.9 Artificial Intelligence and Neural Network

Artificial Intelligence is a term that refers to the replication of human intelligence in a machine. AI systems are developed to do jobs that normally require human intelligence, which includes, learning off of data (machine learning), Reasoning and resolving issues, Understanding natural language, Perception, sight, sound, etc. In Decision making AI can be separated into, Narrow AI - Is useful for only specific tasks. (e.g., face recognition, spam filtering). General AI- A theoretical system that could do any intellectual task that a human can. Neural Networks are a type of Machine Learning, in that Machine Learning is a type of AI. A Neural Network is a computational model that was deliberately created to compute in a manner similar to biological neurons. Three layers - layers of nodes (neurons), Input layer - gives the raw data input, Hidden layers - process and transform the data that are connected with weights on each edge, and the Output layer - provides the result. Each edge has a weight and each neuron of the neural network has an activation function that is responsible for determining the outputs. The relationship between AI and Neural Networks is that AI is a very broad field of study attempting to make machines think (logic, reasoning, ML, etc.). In ML, we're still dealing with AI, which is a subset of AI, and the focus is learning from data. In neural networks (NN) is how to use this ML algorithm that is inspired of

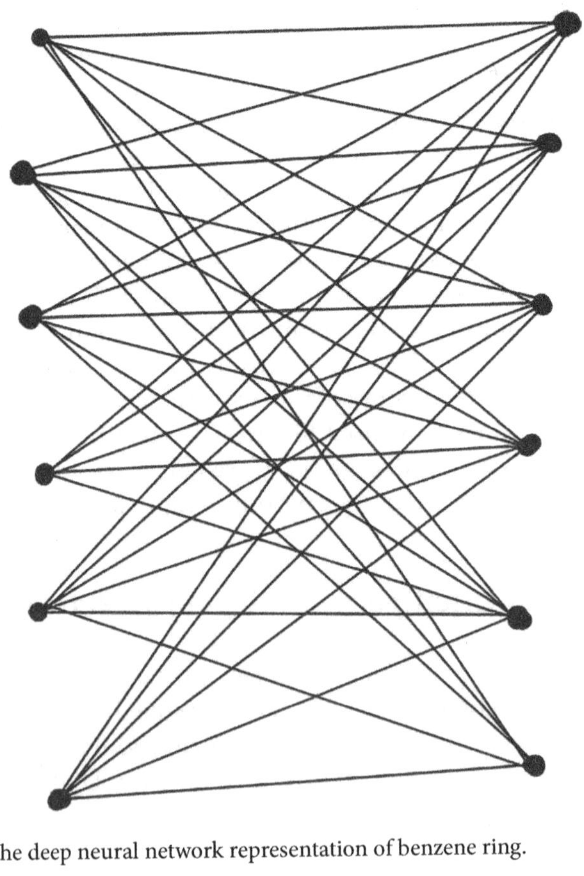

Figure 7.5 The deep neural network representation of benzene ring.

the human brain. Then we move to deep Learning (DL), which expands on the core concepts of NN and there are many layers (deep neural networks).

7.10 Methodology

The Zagreb, Randic, and Sombor indices are topological indices that generate compact structural descriptors that can be precomputed and used as node or edge features. This makes GNNs rely less on multi-hop message passing, reduce memory usage, and bring faster convergence rates in large scale or sparse graphs. With these indices we capture degree related complexity up front, facilitating learning with more efficient, more scalable GNN training.

In the benzene ring, all the carbon atoms are considered as the vertices degree and can be named by the cycle graph. In the cycle graph, all vertices are degree of 2. So, the representation of deep neural network in the graph contains the only input layer and output layer. This means, in the deep neural structure, we have not hidden layer. In the case of naphthalene

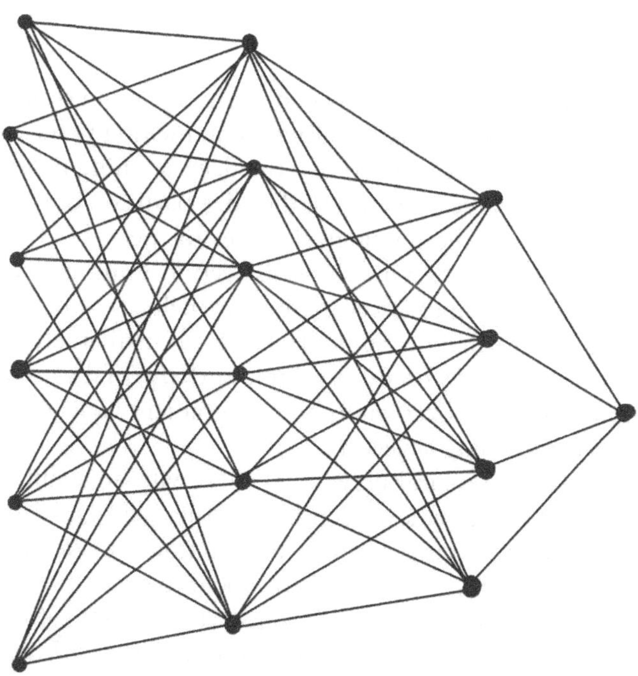

Figure 7.6 Neural network for three layered benzene rings.

structure, the carbon atoms have incidental degree of two and three. Using edge partition method show the edges, as (2,2), (2,2), (2,2), (2,2), (2,2), (2,2), (2,3), (2,3), (2,3), (2,3), and (3,3). The degree sequence for edge partition is (2,2,2,2,2,2,2,3,3,3,3) that was ordered by increasing value, we can position it as deep neural network as shown in Figure 7.6.

The naphthalene structure found by the (6,6,4,1) deep neural network, as illustrated in Figure 7.5, has 10 vertices and 11 edges. There are six pairs of edge partition that have (2,2), four pairs of edge partition that have (2,3), and one edge partition that has (3,3). Create the deep learning network using the degree sequence of the edge partition. The first layer is the input layer, the second layer is the hidden layer 1, the third layer is the hidden layer 2, and the fourth layer is the output layer, as shown in the figure. In all layers, however, the number of degrees at the vertices is the same. Therefore, the input layer and hidden layer 1 together constitute a full bipartite graph, while hidden layer 1 and hidden layer 2 meet the criteria for a full bipartite graph, and so on. If we merge all of the layers into a single, huge multipartite graph in Figure 7.6.

7.11 Topological Indices on Deep Learning Neural Network

Connectivity indices can be correlated with the numerical descriptor which can be applied to correlate boiling point, melting point, flash point, enthalpy, polar surface area and so on. The deep learning neural network can be linked with degree sequence of the edge partition of the graphs. In this section, we obtain the topological values for neural network particularly degree-based indices. One thing to keep in mind is that encoding topological index values into GNN architecture helped the network to generalize structural information, across layers. Such mathematically defined descriptors are particularly useful as prior knowledge of the problem to use on deep networks where over smoothing or gradient vanishing can occur so as to guide the learning trajectory.

Think about a deep neural network (DNN) with an input layer made up of P nodes, followed by k hidden layers. There are Qi nodes for each hidden layer iii, where i ranges from 1 to k. Q nodes make up the output layer. Every node in a layer is completely linked to every node in the layer that follows it. For example, Figure 7.1 illustrates a DNN with five hidden layers, an input layer of four nodes, and an output layer of three nodes. We begin by examining the DNN's graph structure in terms of how its nodes are connected as vertices across layers. Each node in a layer is connected to every node in the layer above, resulting in a layered, completely connected directed acyclic graph. The degree of each vertex in the undirected variant of this graph, where the degree of a vertex is the total number of edges incident to it, is calculated. Each input node has a degree of Q_1, since it is connected to all Q_1 nodes in the first hidden layer.

- Every node in the first hidden layer has a degree of $P+Q_2$ because it is connected to every P input node and every Q_2 node in the second hidden layer.
- A node in the i-th hidden layer has a degree of Q_i-1+Q_i+1 since it is linked to every node in the (i−1)-th and (i+1)-th layers. This is more generally speaking.
- The k-th (last) hidden layer's node is linked to all nodes in the (k−1)-th hidden layer and all Q output nodes, making its degree equal to Q_{k-1+Q}.
- Since each output node receives edges from all Q_k nodes in the previous hidden layer, it has a degree of Q_k.

We now use the edge partition method to compute topological indices. This involves categorizing edges based on the degrees of their endpoints. For example:

- The input layer is connected to the first hidden layer by the edges of the $P\,Q_1$, which connects vertices of degree Q_1 input node and $P+Q_2$ first hidden layer node.
- The endpoints of the $Qi{\cdot}Q_i+1$ edges that connect the i-th hidden layer to the (i+1)th hidden layer have degrees $Q_{i-1}+Q_{i+1}$ and Q_i+Q_{i+2}, respectively, presuming. Consider $Q_0=P$, $Q_{k+1}=Q$ as the proper way to handle boundary cases.
- Nodes of degrees $Q_{k-1}+Q$ and Q_k are connected by the Q_k and Q edges from the final hidden layer to the output layer, respectively.

Summary of Edge Categories: Each edge is categorized by the degrees of its endpoints in Table 7.1.

- From layer i to layer i+1, $Q_i{\cdot}Q_{i+1}$ edges, between nodes of degrees $Q_{i-1}+Q_{i+1}$ and Q_i+Q_{i+2}
- Proper handling of layer indices and edge boundaries ensures accurate computation of topological indices.

Table 7.1 Degree of each layer.

Layer	Degree of each node
Input layer	Q1
First hidden layer	*P*+Q2
i-th hidden layer	Qi−1+Qi+1 for 1< i<k1
kkk-th hidden layer	Q_k−1+Q
Output layer	Q_k

Table 7.2 The edge partition of naphthalene sun graph structure.

$\delta_r\delta_s$	Number of edges
(6,10)	36
(10,7)	24
(10,4)	4

As per the condition of deep neural network, Figure 7.2 is the deep neural network of Figure 7.5. The edge partition of naphthalene structure is shown in Table 7.2. For this purpose, we selected two representative families of sun graphs and zigzag graphs on the ground of their structural alignment with the molecular and nanoscale systems in the real world. For example, zigzag graphs resemble the bonding topology of well-studied targets for GNN based molecular property prediction: graphene nanoribbons and linear π-conjugated polymers, the latter of which are commonly predicted to be high performing organic electronics. Sun graphs also mimick radial molecular structures like dendrimers, star shaped polymers, and supramolecular chemistry's complex molecular frameworks. They come with controlled variations in branching and connectivity and can be used as benchmark for GNNs' learning capacity and scalability. Notwithstanding their versatility arising from the learned topology, their regular, interpretable topology also allows for clear mapping to degree-based indexing and explaining how GNNs deal with structural regularity and distribution of edges. Figure 7.5 and Table 7.2 show the degree sequence and corresponding topological index of these graph families which we will use in the later neural network training scenarios of these graph families. With such mappings, we can analyze with which ring, radial and linear structures GNNs are capturing and differentiating in graph based molecular representations.

The deep neural network of first Zagreb index of naphthalene structure is

$$M_1(G) = \sum_{rs\in E(G)}(\delta_r + \delta_s)$$
$$M_1(G) = (PQ_1)(Q_1 + P + Q_2) + (Q_1Q_2)(P + Q_2 + Q_1 + Q_3) + \ldots + (Q_rQ)(Q_{r-1} + Q + Q_r).$$
$$M_1(G) = 36(6+6+4) + 24(6+4+6+1) + 4(6+4+1)$$
$$M_1(G) = 1028.$$

The deep neural network of second Zagreb index of naphthalene structure is

$$M_2(G) = \sum_{rs\in E(G)}(\delta_r * \delta_s)$$
$$M_2(G) = (PQ_1)((Q_1)(P + Q_2)) + (Q_1Q_2)((P + Q_2)(Q_1 + Q_3)) + \ldots + (Q_rQ)((Q_{r-1} + Q)(Q_r)).$$
$$M_2(G) = (36)(6X10) + 24(10X6) + 4(6X4)$$
$$M_2(G) = 3696.$$

The deep neural network of forgotten index of naphthalene structure is

$$\mathrm{F(G)} = \sum_{rs\in E(G)}(\delta_r^2 + \delta_s^2)$$
$$\mathrm{F(G)} = (PQ_1)((Q_1)^2 + (P + Q_2)^2)) + (Q_1Q_2)((P + Q_2)^2 + (Q_1 + Q_3)^2) + \ldots + (Q_rQ)((Q_{r-1} + Q)^2 + (Q_r)^2).$$
$$\mathrm{F(G)} = (36)(36 + 100) + 24(100 + 36) + 4(36 + 16)$$
$$\mathrm{F(G)} = 8368.$$

7.12 Generalized Topological Indices on Neural Network

The Generalized Linear Polyphenylene Benzene Chains can refer to linear chains of benzene rings interconnected at various positional locations, forming a generalized class of such molecules. These are typically connected at the para-linkage positions (para-linked chains), as seen in poly p-phenylene, but may also include meta-linked or ortho-linked benzene chains, which are not strictly linear but are still part of the broader generalized class. Oligo phenylene compounds such as dimers, trimers, and other short-chain analogs also fall within this category. Polyaromatic hydrocarbons (PAHs) broadly describe molecules composed of multiple fused or interlinked benzene rings.

In theoretical chemistry and materials science, graph-theoretical or mathematical models are widely used. Within these models, a "generalized linear benzene ring" is essentially represented as a graph, where the nodes correspond to benzene rings and where the edges correspond to electron bonding/bonding and connectivity. The linear aspect usually refers to a path graph of benzene units. The term generalized allows for the possibility of variation of the substitution patterns, substituents and chemical links involved. In such a case, the benzene ring might serve as a modular molecular building block, particularly in terms of the materials. Conjugated polymers also represent an example where the structure can be generalized based on the possible substituents (say, electron-donating versus electron-withdrawing groups), where such substitutions might impact the type of linkage chemistry involved. The edge partition for a generalized linear benzene ring be as in Table 7.3.

The first Zagreb index for calculating generalized linear benzene ring is

$$M_1(G)=(PQ_1)(Q_1+P+Q_2)+(Q_1Q_2)(P+Q_2+Q_1+Q_3)+\sum_{i=2}^{r-2}Q_iQ_{i+1}(Q_{i-1}+Q_{i+1}+Q_i+Q_{i+2})+(Q_{r-1}Q_r)(Q_{r-2}+Q_r+Q_{r-1}+Q)+(QQ_r)(Q_{r-1}+Q+Q_r).$$

$$M_1(G)=(36)(6+6+4(k-1))+(6X4(k-1))(6+4(k-1)+6+(k-1))+(Q4X(k-1)(k-1))(4(k-1)+(k-1)+4(k-1))$$

$$M_2(G)=(PQ_1)(Q_1)(P+Q_2)+(Q_1Q_2)((P+Q_2)(Q_1+Q_3))+\sum_{i=2}^{r-2}(Q_iQ_{i+1})((Q_{i-1}+Q_{i+1})(Q_i+Q_{i+2}))+(Q_{r-1}Q_r)((Q_{r-2}+Q_r)(Q_{r-1}+Q))+(QQ_r)((Q_{r-1}+Q)(Q_r))$$

$$M_2(G)=36(6)(24(k-1))+(24)((k-1)6(k-1)+4(k-1)(k-1)(4(k-1)+(k-1)4(k-1))$$

$$F(G)=(PQ_1)Q_1^2+(P+Q_2)^2+(Q_1Q_2)((P+Q_2)^2(Q_1+Q_3)^2)+\sum_{i=2}^{r-2}(Q_iQ_{i+1})((Q_{i-1}+Q_{i+1})^2(Q_i+Q_{i+2})^2)+(Q_{r-1}Q_r)((Q_{r-2}+Q_r)^2(Q_{r-1}+Q)^2)+(QQ_r)((Q_{r-1}+Q)^2(Q_r)^2)$$

$$\mathrm{F(G)}=(36)(6)^2+(6+4(k-1))^2))+(24(k-1))((6+4(k-1))^2+(4(\mathrm{k}-1)(\mathrm{k}-1))(6+(k-1))^2+(4(k-1))^2).$$

Table 7.3 The edge partition of generalized benzene ring.

Layers	$\delta_r\delta_s$	**Number of edges**
Input Layers	(6,6+4(*k*-1))	36
Hidden Layers	(6+4(*k*-1), 6+(*k*-1))	6 X4(*k*-1)
Output layers	(6+(*k*-1, 4(*k*-1))	4(*k*-1)2

An algorithmic description for creating a linear benzene-ring-type structure within a deep neural network environment, modeled as a recurrent or cyclic graph of connected nodes. In this structure, nodes are connected in a cycle (in a similar manner to the six-carbon atoms of a benzene-ring structure) and information is communicated circularly. While benzene is a cycling structure, a "linear benzene ring" would seem to be a contradiction. However, if you are simply trying to represent the connectivity of benzene (i.e., each node connects to two nodes (neighbors) in a ring) in a linear neural architecture, then it's a way of defining a 1D recurrent/cyclic graph with message passing.

Algorithm 7.1: Creating a Benzene Ring-Like Neural Structure

Input: n ← number of nodes in the ring (e.g.; n = 6 for benzene
epochs ← number of forward passes (or time steps)
x[0...n-1] ← input values at each node
W ← shared weight matrix for message passing
activation ← non-linear activation function (e.g., ReLU, tanh)
Output: Final outputs at each node after message passing

File Edit Format Run Options Window Help

```
def benzene_ring_network(x, W, activation, epochs):
    """
    Simulates a recurrent, ring-based neural network.

    Parameters:
    - x: List of input values at each node.
    - W: Weight matrix (typically a 2x1 or scalar if using shared weight).
    - activation: Activation function (e.g., ReLU, tanh).
    - epochs: Number of message-passing steps.

    Returns:
    - x: Final state of each node after recurrent updates.
    """
    n = len(x)
    for t in range(epochs):
        new_x = [0] * n
        for i in range(n):
            left = x[(i - 1) % n]  # cyclic left neighbor
            right = x[(i + 1) % n]  # cyclic right neighbor
            message = left + right  # sum of neighboring states
            new_x[i] = activation(W * message)
        x = new_x  # update states
    return x
```

Output:

```
File Edit Shell Debug Options Window Help
    Python 3.13.2 (tags/v3.13.2:4f8bb39, Feb  4 2025, 15:23:48) [MSC v.1942 64 bit (
    AMD64)] on win32
    Type "help", "copyright", "credits" or "license()" for more information.
>>>
    === RESTART: C:/Users/jaiks/OneDrive/Documents/python adjace/benezene ring.py ==
>>> import numpy as np
...
... # Example values
... n = 6
... x = [1, 0, 1, 0, 1, 0]  # Initial states
... W = 0.5
... activation = lambda x: max(0, x)  # ReLU
... epochs = 5
...
... final_outputs = benzene_ring_network(x, W, activation, epochs)
... print(final_outputs)
```

This mimics the manner in which information flows through a ring structure, where each node receives information only from its left and right neighbors. The modulo operator adds a cyclic link (aka connectivity) as well, similar to the benzene ring. The activation function adds some non-linearity to that activity, similar to how neurons activated in a neural network. The weight W can be either a scalar or a small matrix, which represents a shared learnable parameter.

7.13 Applications on Topological Indices of Neural Network

Training GNNs from raw atomic graphs in large scale molecular datasets is limited with computational bottlenecks incurred with graph size and complexity. These challenges are alleviated to an extent by embedding pre-computed topological indices in to the node or graph feature space. With degree based indices, one can standardize feature representations across structurally same molecules and achieve better scalability, less training time, as well as better generalization on unseen molecular graphs. In quantitative structure-activity relationship (QSAR) and quantitative structure-property relationship (QSPR) studies, topological indices, which are numerical representations derived from the graph-theoretical depiction of molecules, have been approved descriptors.

The topological indices first Zagreb index, second Zagreb index, and Forgotten index give data about structural characteristics like connectivity, branching, and molecular symmetry for generalized linear chains of benzene. These topological features are fundamental to model and predict

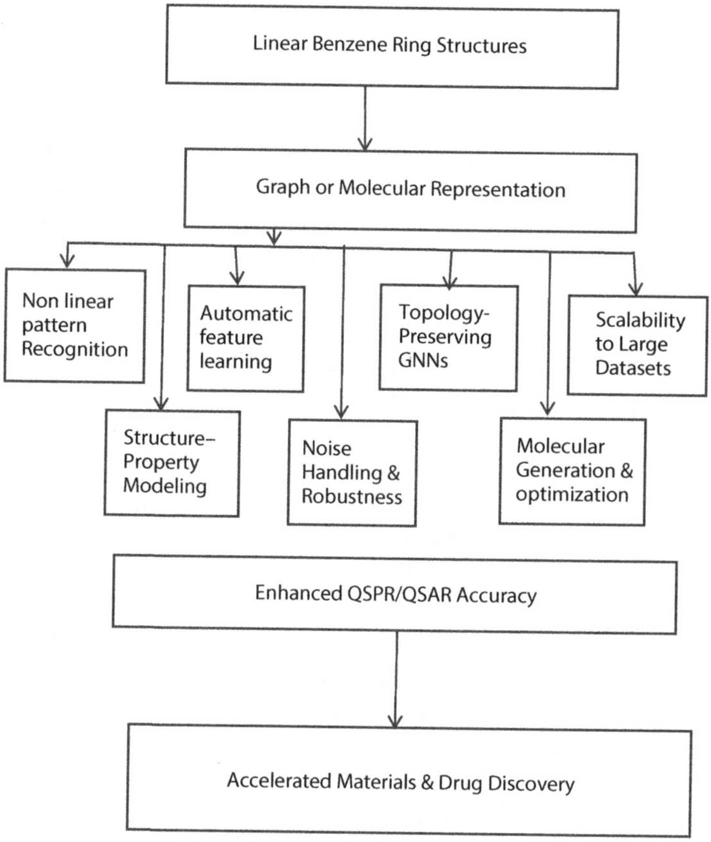

Figure 7.7 Benzene ring-like neural structure.

properties by integrating the structure of molecules with biological activity, physicochemical properties, or reactivity. When incorporated with neural networks terms obtained from topological indices form input features for input neural networks, allowing deep learning models the ability to find mapping between molecular structure and target properties that are complex and non-linear simultaneously. This allows for predicting molecular property, drug discovery, or materials design when using linear benzene chains of properties that reflect polyaromatic and conjugated systems e.g., polyphenylenes, was shown to be the case and has substantial potential. The modularity and generalizability of linear chains of benzene would facilitate the application of computational models from the graph view of molecules and increase the explainability or predictive power if neural network models derived from representations of chemical descriptors were to achieve a higher degree of predictivity in computational modeling projects relating to molecular activity, particularly in chemistry, are shown in Figure 7.7.

Firstly, to incorporate a linear benzene ring (or a generalized linear polyphenylene chain) into a neural network, the molecule must be encoded in a format compatible with machine learning models. Typically, this involves representing the molecule as a graph, where each benzene ring or atom is treated as a node, and the bonds or linkages between them are represented as edges. In the case of generalized linear benzene graphs where rings are connected at para, meta, or ortho positions the resulting graph captures both the connectivity and the topological arrangement of the molecular structure. There are several methods to incorporate this graph-based molecular information into a neural network:

(i) Encoding through Topological Indices: Topological indices are numerical values derived from a molecule's graph architecture (for example, the Wiener index, Zagreb indices, Randic index, and Balaban index) encapsulating topological choices that ultimately describe an overall size, branching and degree connectivity. Topological indices can be grouped into a feature vector and treated as inputs for a standard feedforward neural network (FNN) or a multilayer perceptron (MLP). This approach is relatively efficient and interpretable, but it may result in losing finer considerations of structure.

(ii) Graph Neural Networks: Graph neural networks are specifically designed for graph-structured data. In this case, the molecule will be organized as a graph G = (V, E), with the node V being from the molecule's node (i.e., an atom or a benzene ring) and the edges E being from the molecule's edges (i.e., chemical bonds, inter-ring connection, etc.), every node and edges can also have features, like atom type, bond order, or ring connectivity. The graph neural networks, using models like Graph Convolutional Networks (GCNs), Message Passing Neural Networks (MPNNs) and Graph Attention Networks (GATs), allow for the learning of the process for passing information through the graph system that will learn very complex relationships for the molecular sub-structures and their resultant properties.

(iii) SMILES/SELFIES-Based Encodings: The molecule's graph may also be encoded using a SMILES Simplified Molecular

Input Line Entry System or SELFIES Self-Referencing Embedded Strings. Both SMILES and selfies are linear string representations of chemical structures that may be represented using recurrent neural networks (RNNs), long short-term memory networks (LSTMs), or transformer models. Because it is a linear depiction, there is no spatial or graphical representation while modeling, but this will make sense during tokenization because it is helpful for generating molecule models and big datasets.

(iv) 3D Molecular Descriptors: If three-dimensional geometry is important (i.e. electronic properties, binding affinity), then the molecular structures can be encoded to 3D descriptors such as Coulomb matrices, distance matrices, atomic coordinate embeddings [27]. These descriptors can be passed to 3D GNNs or equivariant neural networks which respect rotatable and translatable symmetries found in molecular space.

(v) Ring-Level Abstraction for Coarse-Grained Modeling: Similar to polyphenylene chains or conjugated systems, we could take the benzene ring to be a meta-node in a coarse-grained graph. Doing so can ultimately lace the model connection (e.g., para vs. meta) for inter-ring connection while abstracting away the atomic detail for the intra-ring. This abstraction layer is useful for modeling conjugated polymers, organic semiconductors, or even MOFs/COFs, where the repeating structural unit follows a modular form.

7.14 Benefits of Neural Networks in the Analysis and Quadratic Modeling of Linear Benzene Forms

Neural networks possess distinct advantages for use in benzene ring structures such as polyphenylenes, oligo(phenylene)s, or PAHs, especially in the framework of QSPR and QSAR analyses.

(i) Recognition of Non-Linear Patterns: Neural Networks, specifically deep learning models, are capable of capturing non-linear dependencies resulting from molecular structures, physicochemical properties and even biological properties of a particular sample. For example, this is particularly essential for systems containing benzene compounds as we know that small changes like substitution position or linkage type would yield non-linear impact on electronic property, toxicity or activity.

(ii) Automatic Learning of Features: Most QSPR or QSAR models are built on predefined molecular descriptors or topological indices. Unlike other models, neural networks, especially GNNs and CNNs on graph or image-like inputs, have the capability to analyze molecular graphs, SMILES or other structured data to extract relevant features parameter eliminating human intervention and biases thereby generalizing better. AI models are more objective than humans.

(iii) Graph-Based Representations Maintain Topological Features: In generalized linear benzene chains, connectivity preservation of para, meta, and ortho positions matters. Connections (GNNs and MPNNs) Neural networks working on molecular graphs sustain connectivity, thus enabling better modeling of π-conjugation, delocalization, and steric interactions—effects which determine QSPR/QSAR predictions.

(iv) Relevance to Extremely Large and Varied Datasets: With massive datasets, chemical neural networks handle efficiently. They learn from diverse structures such as substituted or extended benzene rings and generalize predictions across property-governed boundaries, thus propelling virtual screening and high-throughput prediction efforts in QSPR and QSAR.

(v) Conjugated Systems: Structure - Property Relationships: Conjugated polymers, organic semiconductors, and some biologically active compounds are frequently built upon linear benzene chains. Neural networks can learn models of subtle structure–property relationships such as Tuning of bandgap, Charge transport, Solubility and lipophilicity and biological binding affinity. All these models are very important to QSPR/QSAR frameworks and often cannot be simulated with linear regression or descriptor-based approaches.

(vi) Flexibility against incompletely specified or noisy data: Cross-industry insights suggest that appropriate noise

Table 7.4 Merits in linear benzene ring.

Benefits	Pertinent to linear benzene/QSPR/QSAR
Nonlinear modeling	Captures complex structure – property interrelatedness
Automatic feature extraction	Dependence on predefined descriptive metrics is diminished
Topological fidelity *via* GNNs	Maintains the types of linkages and ring connections.
Scalability	Accommodates a high volume of datasets and more generalized molecules
Conjugated system modeling	Constitutes primary importance for electronic and/or material properties
Predictive Power	Possessing experimental variability doesn't compromise effectiveness
Generative modeling capability	Permits molecular design and molecular screening

filtering during training may help in learning to compensate for missing data, data correlation, or noise. Such features are characteristic of blended datasets from many experiments in QSPR/QSAR studies.

(vii) Enables Molecular Creation and Refinement: In the context of inverse design, neural networks, and more specifically generative models such as VAEs, GANs, and Transformers, have the capability of proposing new linear benzene derivatives by optimizing certain desired characteristics. This not only predicts but also enables *de novo* molecular design which has the potential to transform the fields of materials science and drug discovery in Table 7.4.

7.15 Conclusion

The depiction and incorporation of generalized linear benzene rings within neural network models provide an interesting paradigm for bringing chemical graph theory together with contemporary machine learning methodology. As basic building blocks within aromatic systems, benzene rings, especially in para-, meta- or ortho-linked chain configurations, underlie many significant organic structures such as polyphenylenes, polyaromatic hydrocarbons (PAHs), and conjugated polymers. These systems not only have rich structural complexity but also play central roles in the creation of advanced materials, drugs, and functional nanostructures. To classify and understand the structure, shape, and electronics of these systems, different encoding methods have been developed. Topological indices are simplified as well as interpretable descriptors based on the molecular graph which helps classical neural networks in making predictions or classifications based on properties. With the introduction of Graph Neural Networks (GNN), new possibilities have emerged as models can now work on data in the form of graphs directly. GNNs use the relational information of atoms or rings which are preserved, which means that the molecular structure representation will be context-sensitive and more expressive. Moreover, string-based encodings such as SMILES or SELFIES offer alternative ways to process molecular structures with sequential models like RNNs or transformers, which is beneficial for generative modeling and large-scale screenings. Coarse-grained graph abstractions where each benzene ring is treated as a meta-node, alongside the 3D geometric descriptors, offer high modularity, and spatial orientation which increases the model's flexibility. These methods provide chemists with tools able to predict real-world chemical behaviors by helping capture long-range dependencies and subtle structural changes. With self-supervised learning being possible through extensive datasets containing unlabelled data, topological invariants alongside textual encodings allow for a high-level representation of the molecular graph. Chunk-based approaches coupled with tree structures form self-similar hierarchies, further refining explanatory predictive models for complex tasks. In conclusion, the integration of linear and generalized benzene ring layouts into neural networks improves the accuracy of AI-models in chemistry while also expanding the possibilities of understanding, exploring, and designing complex molecular systems using data-driven methods. Besides, topological indices can also be used as embedded descriptors in GNN as a scalable and interpretable graph learning method, which is useful in large scale data where direct message passing is computationally expensive. They are both integrated into the model, which increases model performance while also contributing to improved robustness, convergence and efficiency on molecular property prediction.

AI disclosure statement: The author acknowledges the use of AI-based tools for minor language editing or grammatical correction. However, all intellectual contributions, analyses, and conclusions presented in this chapter are entirely the author's own.

References

1. Adnan, M., Bokhary, S.A.U.H., Abbas, G., Iqbal, T., Degree-Based Topological Indices and QSPR Analysis of Antituberculosis Drugs. *J. Chem.*, 2022, 1–17, Feb. 2022, doi: 10.1155/2022/5748626.
2. Do, K., TuanPark, M., Shin, N., Two-Stage Training of Graph Neural Networks for Graph Classification. *Neural Process. Lett.*, 2023.
3. Du, S.S., Hou, K., Póczos, B., Salakhutdinov, R., Wang, R., Xu, K., Graph neural tangent kernel: Fusing graph neural networks with graph kernels. *Adv. Neural Inf. Process. Syst.*, 2019.
4. Feng, A., You, C., Wang, S., Tassiulas, L., Interpretable Graph Neural Networks with Graph Kernels, in: *Proceedings of the 36th AAAI Conference on Artificial Intelligence, AAAI 2022*, 2022.
5. Fraz, H.M., Ali, K., Nadeem, M.F., Entropy measures of silicon nanotubes, using degree based topological indices. *Phys. Scr.*, 2024.
6. Jiang, C., Li, D., Wen, J., You, T., QSAR study of the enantiomeric excess in asymmetric catalytic reactions with topological indices and an artificial neural network. *J. Mol. Model.*, 13, 2007.
7. Jiang, J. and Luo, W., Graph neural network for traffic forecasting: A survey. *Expert Syst. Appl.*, 2022.
8. Li, R., Yuan, X., Radfar, M., Marendy, P., Ni, W., O'Brien, T.J., Casillas-Espinosa, P., Graph Signal Processing, Graph Neural Network and Graph Learning on Biological Data: A Systematic Review. *IEEE Rev. Biomed. Eng.*, 2023.
9. Liang, Y., Zhang, Y., Gao, D., Xu, Q., An End-to-End Graph Neural Network for Graph Represent. Learn. *IEEE Access*, 2021.
10. Melton, J. and Krishnan, S., MuxGNN: Multiplex Graph Neural Network for Heterogeneous Graphs. *IEEE Trans. Pattern Anal. Mach. Intell.*, 2023.
11. Tiezzi, M., Ciravegna, M., Gori, G., Graph Neural Networks for Graph Drawing. *IEEE Trans. Neural Networks Learn. Syst.*, 35, 4, 2024.
12. Wei, Q. and Hu, G., Evaluating graph neural networks under graph sampling scenarios. *PeerJ Comput. Sci.*, 2022.
13. Xie, Y., Lv, S., Qian, Y., Wen, C., Active and Semi-Supervised Graph Neural Networks for Graph Classification. *IEEE Trans. Big Data*, 2022.
14. Yao, D., Zhi-li, Z., Xiao-feng, Z., Wei, C., Fang, H., Yao-ming, C., Cai, W.W., Deep hybrid: Multi-graph neural network collaboration for hyperspectral image classification. *Def. Technol.*, 2023.

15. Ye, Z., Kumar, Y.J., Sing, G.O., Song, F., Wang, J., A Comprehensive Survey of Graph Neural Networks for Knowledge Graphs. *IEEE Access*, 2022.
16. Zhang, X., Idrees, N., Kanwal, S., Saif, M.J., Saeed, F., Computing Topological Invariants of Deep Neural Networks. *Comput. Intell. Neurosci.*, 2022, 2022.
17. Zhu, H., A graph neural network-enhanced knowledge graph framework for intelligent analysis of policing cases. *Math. Biosci. Eng.*, 2023.
18. Zhou, J., Cui, G., Hu, S., Zhang, Z., Yang, C., Liu, Z., Wang, L., Li, C., Sun, M., Graph neural networks: A review of methods and Applications. *IEEE Trans. Neural Networks*, 2020.
19. Scarselli, F., Gori, M., Tsoi, A.C., Hagenbuchner, M., Monafardini, G., The graph neural network model. *IEEE Trans. Neural Netw.*, 20, 1, 61–80, 2009.
20. Rui, H., Ashton, K., Min, J., Wang, C., Potts, P.R., Protein-protein interfaces in molecular glue-induced ternary complexes: Classification, characterization, and prediction. *RSC Chem. Biol.*, 4, 192–215, 2023.
21. Liu, J.B., Peng, X.B., Gu, J.-J., Lin, W., The (multiplicative degree-) Kirchhoff index of graphs derived from the Cartesian product of Sn and K2. *J. Math.*, 1–9, 2022.
22. Zhao, W., Julietraja, K., Venugopal, P., Zhang, X., VDB Entropy and Irregularity-Based Indices for the Rectangular Kekulene System. *J. Math.*, 1–15, 2021.
23. Wu, B., Wang, Y.-W., Dai, Y.-H., Song, C., Zhu, A.-L., Qin, H., Tan, F.-R., Chen, H.-C., Dai, L.-C., Hu, G.-Q., He, M.-X., Current status and future prospective of bio-ethanol industry in China. *Renew. Sustain. Energy Rev.*, 145, 111079, 2021.
24. Shervashidze, N. and Borgwardt, K.M., Fast subtree kernels on graphs. *Adv. Neural Inf. Process. Syst.*, 22, 2009.
25. Sanchez-Lengeling, B., Pearce, A., Wiltschko, A.B., Reif, E., A gentle introduction to graph neural networks. *Distill*, 2, 2021.
26. Rossi, F., Bandyapadhyay, S., Wolf, M., Pavone, M., Review of Multi-Agent Algorithms for Collective Behavior: A structural Taxonomy, *IFAC-papersOnline*, 51, 112–117, 2018.
27. Shi, C., Blei, D.M., Veitch, V., Adapting Neural Network for the Estimation of Treatment Effects. *Adv. Neural Inf. Process. Syst.*, 32, Curran Associates, Inc., 2019.

8

Large-Scale AI Systems Leveraging Graph Neural Networks (GNNs)

Gaurav Paliwal*, Divya Gautam and Nidhi Asthana

School of Technology Management & Engineering, SVKM's Narsee Monjee Institute of Management Studies (NMIMS), Indore, Madhya Pradesh, India

Abstract

The chapter discusses the importance of massively scalable AI systems, particularly Graph Neural Networks (GNNs), in applications such as fraud detection, recommendation systems, drug discovery, and social network analysis. However, scaling GNNs to large-scale scenarios presents computational and architectural challenges such as redundant memory access, inefficient training due to graph sparsity, and the need for distributed computing methods. The chapter introduces scalable architectures, distributed training strategies, and real-world applications of large-scale AI systems using GNNs. The chapter covers the fundamental concepts of GNNs, such as message passing, node embedding, and graph convolutional layers. It also discusses the challenges of scaling GNNs, such as computational bottlenecks, memory limitations, and the impact of large, dynamic graph topologies. Recent techniques for scaling GNNs include distributed and federated learning frameworks, GPU-accelerated training, partitioning strategies, and hybrid models. The chapter also covers the applications of cloud and edge computing in deploying massive GNNs and providing real-time computations and privacy protection. It also discusses real case studies demonstrating the real-world significance of large-scale GNNs across sectors, such as healthcare science, IoT wearables, banking, finance, cybersecurity, smart cities, drug discovery, and protein interaction modeling. The chapter also discusses emerging research directions, such as quantum computing, self-supervised learning models, and optimization approaches. It also addresses ethics, including privacy concerns and fairness to graph-based machine learning models.

**Corresponding author*: gvpaliwal@gmail.com

J. Ramkumar, Sridaran Rajagopal, B. Suchitra and S. Balamurugan (eds.) Graph Neural Networks: Concepts and Applications, (315–366) © 2026 Scrivener Publishing LLC

Keywords: Graph neural networks, large-scale AI, distributed learning, federated learning, scalability, cloud computing, edge AI, healthcare

8.1 Introduction

The record-breaking expansion of business use of artificial intelligence (AI) has been preceded by a record-breaking expansion in demand for smart systems that are also large enough to process huge amounts of related data. Big AI systems are defined by their capacity to process, analyze, and infer from massive sets of data, generally in real-time or near real-time. Enabling technologies for such systems are parallel hardware accelerators such as GPUs and TPUs, distributed computing hardware, and domain-specific architectures for particular data structures such as images, text, and lately, graph-structured data. Graph Neural Networks (GNNs) specifically have come to represent a new deep learning paradigm which is taking AI closer to reasoning in the real world, in which objects and their interrelationships constitute a dense graph of interdependencies. Mass-scale AI platforms are no longer restricted to back-end research or industrial high-end application but are creeping their way into main-stream technology, ranging from recommender systems at a personal scale to urban planning. These platforms are design-scalable, heterogeneous data sources integrated, supporting real-time analytics, and offering actionable insights. They implement edge computing, cloud data lakes, and federated learning together to minimize latency, provide privacy, and optimize bandwidth efficiency. In such an arrangement, GNNs play a central role in propelling the intelligence quotient of AI systems to even greater heights by leveraging the high-level structural graph information.

GNNs are widely adopted in large AI systems due to their capability to effectively model relational data. In contrast to regular neural networks that are primarily interested in vectorized input, GNNs are interested in data as graph structures with relations as edges and entities as nodes. This positions GNNs perfectly for problems where relations will be as essential as entities themselves. For instance, in social networks, action will be defined in terms of interactions among people more than by the property of the individual. For instance, historical account connectivity properties are more predictive of fraud prediction than properties on disconnected accounts in money networks. Similarly, patient similarity graphs or protein-protein interaction graphs provide useful context to make stronger diagnostics and improved drug design possible.

The potential of large-scale pipes in the future through the application of GNNs is widespread in most high-impact domains. Pinterest and Amazon use graph-based models in recommendation algorithms to make inferences of user preference from user-item interaction graphs. Context-aware and personalized recommendations have been enabled by GNNs, including that applied in systems such as PinSage through the use of random walks augmented by GCNs to attain useful item embedding [1]. For intelligent city infrastructure, GNNs facilitate traffic forecasting, mobility pattern awareness, and resource allocation through spatiotemporal relationship encoding of road networks and sensors. Biomedical informatics utilizes GNNs to encode molecular graphs to find new drugs, study gene regulatory networks, and forecast disease progression through linking multi-omics data with graph representations [2]. The field of cybersecurity has also used GNNs for intrusion detection and anomaly prevention, where complex event graphs model users, devices, and applications interactions to detect intrusive behavior.

Even for such real-world applications, incorporating GNNs into big systems is plagued with computational and structural problems. Memory consumption represents one of the most significant bottlenecks in scaling GNNs. As memory consumption grows exponentially very fast with graphs turning gigantic and intricate, nodes' feature data, edges' feature data, and temporal embedding take up vast storage space during training. This even worsens in highly deep GNN models with hundreds of layers suffering from the "neighbor explosion" issue where every layer pools information from an exponentially huge number of neighbors [3]. The second source of concern is computational inefficiency due to sparsity of the graph. In contrast to image or text data, where the input space might be structured, real graphs are unstructured and sparse and lead to inefficient usage of parallel computing capability on current hardware.

Synchronized training for scalable GNNs also suffers from synchronization bottlenecks within the distributed environment. Synchronized training of GNNs on numerous GPUs or computers can lead to enormous performance degradation by producing overheads of communication. This calls for the development of custom communication protocols and parameter synchronization to make it scalable. For instance, libraries such as DistDGL and AliGraph have implemented techniques such as graph partitioning, caching, and asynchronous training to tackle these issues [4]. The dynamic aspect of graphs in the majority of real-world scenarios introduces an additional level of complexity. In applications such as social networks or fraud patterns identification, the graph structure is not static and the model must learn incrementally without having to retrain from scratch. Streaming GNN models and online learning algorithms that enable more recent frequent updates at scale are needed.

Latency requirements are also at risk, especially in edge-AI use cases where real-time inference is paramount. Autonomous vehicles, smart sensing, and ambient mobile services simply cannot afford the latency of compute-intensive GNN computations employed at cloud scales. Towards this end, edge-compliant versions of GNN and hybrid versions are gaining popularity that offload computations across edge nodes and cloud nodes at the cost of responsiveness from a computational perspective [5]. Secondly, security and privacy also make deployment of such massive systems of GNN difficult. In health care and finance domains, the private data cannot be centralized or exchanged easily. Federated learning methods for GNNs are yet to mature to train models simultaneously without sacrificing data privacy, but the processes are nascent. To cope with such stubborn issues, the chapter has been designed to lead the readers through a stringent process of both nascent concepts and state-of-the-art solutions. It starts with outlining the components that are involved in scaling GNNs to complete well-structured large-scale AI systems. They range from, on the lower side, graph construction methodologies, store and retrieve optimizations, all the way through the correct pre-processing pipelines to various sources of data. Distributed training methods that follow include graph partitioning methods, communication-reducing parallel training algorithms, and memory optimizing methods. Applications and realities as shown are in real-life research and industry instances later.

The chapter discusses system-level performance optimization for graph-based AI models, including model compression, inference acceleration, and hardware-specific deployment. It also discusses cloud and edge computing for handling the life cycle of large GNN models. The chapter also addresses security and ethical concerns, focusing on transparency, unbiasedness, and robustness of graph-based AI models. It discusses federated GNN techniques and privacy-preserving computing, as well as potential vulnerabilities. The chapter also provides new research directions, such as quantum computing for faster computation of GNNs, self-supervised graph learning, and scalability standard metrics.

8.2 System Architecture for Large-Scale GNN Integration

As Graph Neural Networks (GNNs) evolve from proof-of-concept to production-level, building blocks of sophisticated AI pipelines, their

Figure 8.1 Large-scale GNN system pipeline.

inclusion in gigantic-scale system design is inevitable and daunting. Unlike traditional deep learning models, which tend to operate with Euclidean data (e.g., images, text), GNNs operate on non-Euclidean, relational graphs with tens of billions of nodes and edges. This results in custom requirements for system design like dynamic graph storage, distributed processing, real-time data ingestion, and scalable inference pipelines. Figure 8.1 provides a generalized Large-scale GNN system pipeline. The system then must handle a pipeline of items like data pre-processing components, graph partitioning techniques, memory-efficient computation libraries, and accelerators so that maximum throughput and minimum latency can be ensured. This section covers the architectural principles, system elements, and reference implementations which enable GNNs to be instantiated at industrial scale, taking advantage of both production environments and academic prototypes. The large-scale GNN systems based on modularity must address data complexity, scalability, and real-time. Major components include data ingestion pipelines, pre-processing modules, graph construction engines, graph storage systems, distributed training frameworks, and inference servers. The components must be integrated seamlessly so as to support low latency as well as high throughput. A sample high-level architecture for scaling GNNs is depicted in Figure 8.2. It addresses a variety of system-level problems such as modularity, parallelism, locality, and fault tolerance. Every module plays an independent role and exchanges data through defined API's, thus scalability and maintainability are enabled. Micro services architecture and container orchestration technologies such as Kubernetes are utilized to run these pieces at scale in production.

8.2.1 Data Ingestion and Pre-Processing

Giant scale GNN system data ingestion involves stream and batch pipelines from various sources such as IoT sensors, user touchpoints, transaction logs, telemetry, and logs. As data is incoming in high diversity and speed, ingestion pipelines must ensure schema consistency, data quality, and low-latency delivery. Apache Kafka, Pulsar, and Flume are typically utilized for stream ingestion, while batch jobs typically use Apache NiFi and Airflow for ETL orchestration. Normalization, filtering, cleaning, and pre-processing raw data are needed for pre-processing in order to

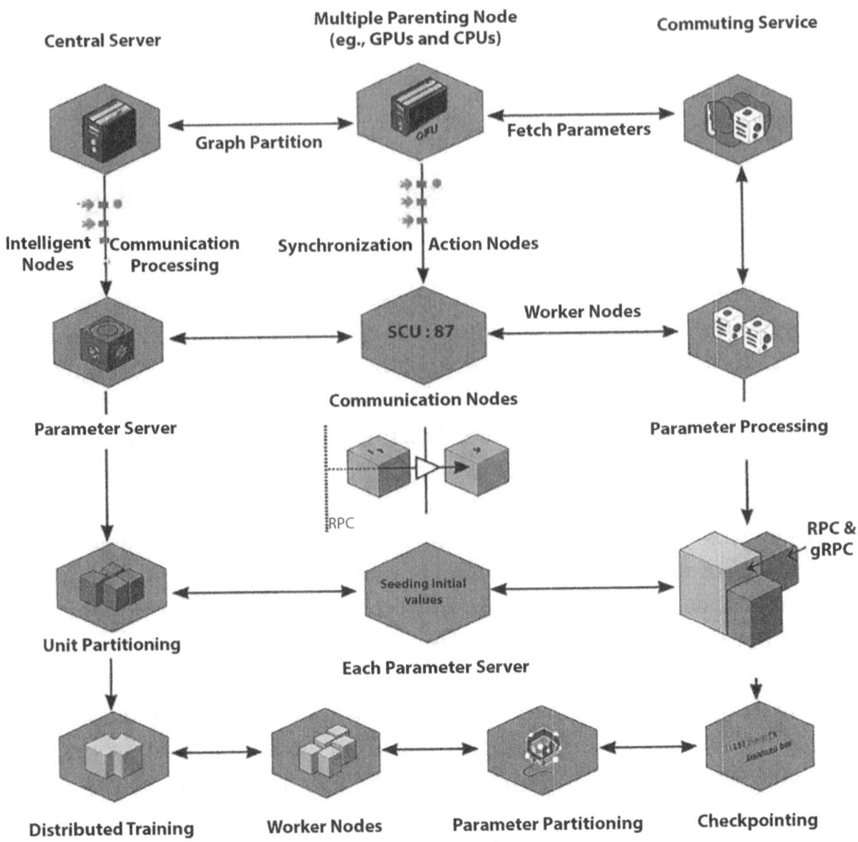

Figure 8.2 Distributed GNN training architecture.

transform them into meaningful graph entities. Pre-processing consists of entity resolution, feature extraction, and dimensionality reduction. Pre-processing in graph learning is primarily the data conversion into edge lists, adjacency matrices, and node feature vectors. Feature engineering in temporal graphs also includes time-stamping, sliding window generation, and spatio-temporal embedding [3]. Furthermore, label generation and missing value handling are essential pre-processing operations in supervised and semi-supervised graph learning. Pre-processing pipeline efficiency directly affects training time and model performance. Python libraries such as Pandas, Dask, and RAPIDS accelerate data pre-processing through GPU-based transformations.

Schema-driven modeling and graph compilers also simplify graph building through the enforcement of rules and automatic node/edge type generation. Apache AGE and JanusGraph systems provide graph semantics on top of standard databases for hybrid OLAP and OLTP.

8.2.2 Graph Construction and Storage

Building graphs out of pre-processed data includes node creation and edge creation, assigning feature vectors, and the optional incorporation of edge weights and direction. The topology of the graph—whether dense or sparse, dynamic or static—affects system design and resource usage extensively. Table 8.1 shows the comparison of graph types based on domain characteristics and computational considerations. Sparse graphs like citation or social networks are memory space efficient and usually processed in adjacency lists. Dense graphs like chemical molecule structure or entire sensor grid connectivity require adjacency matrices and efficient storage methods to avoid excessive memory usage. Static graphs remain constant throughout training, whereas dynamic graphs evolve over time and need to be constantly updated and versioned [6, 7]. Storage systems need to handle random access, concurrent reads/writes, and transactional consistency. Neo4j and TigerGraph [8] support property graphs with rich metadata, whereas RedisGraph [9] and RocksDB [10] are optimized for graph traversals in-memory or on-disk. Distributed file systems (HDFS, Amazon S3) are also used to store partitioned graphs for big training sessions.

8.2.3 Distributed Training and Inference

Parallelization over many machines and accelerators is necessary in order to train GNNs on billion-node graphs. Distributed training of GNNs

Table 8.1 Comparison of graph types.

Graph type	Example domain	Update frequency	Storage complexity	Suitability
Static Sparse	Citation Networks	Low	Low	Efficient
Static Dense	Molecular Graphs	Low	High	Compute-Intensive
Dynamic Sparse	Social Media Feeds	High	Medium	Adaptive

normally involves data parallelism—divides parts of the graph between workers—and model parallelism—divides layers or parameters of the model. Partition strategies matter in these instances; METIS, KaHIP, and random partition are normally used as a way of limiting inter-node communication. A distributed training system includes some components: partition manager, training workers, parameter servers, communication protocol (e.g., gRPC, MPI), and check pointing utilities. Loss convergence, gradient synchronization, and straggler mitigation are active research areas in distributed graph learning [11]. Real-time inference requires an expert optimization pipeline during training. Techniques such as precomputing embedding, sampling in the neighborhood in a layer-by-layer manner, caching highly accessed nodes, and using low-latency serving frameworks such as NVIDIA Triton and Tensor Flow Serving are required. Pruning and batched query inference are generally used in latency-sensitive use cases such as fraud detection and recommendations.

Hardware-aware placement policies (e.g., GPU affinity, NUMA balancing) and network optimization (e.g., RDMA, NVLink) also influence system performance in distributed settings. Systems may include performance profiling, failure recovery, and monitoring resource usage tools.

8.2.4 Frameworks and Tools

There emerged customized frameworks to cope with massive-scale GNN loads. Deep Graph Library (DGL) and PyTorch Geometric (PyG) provide modular APIs for graph neural networks with the capability of effective neighborhood sampling, mini-batching, and homogeneous and heterogeneous graphs support. Both are tightly connected to PyTorch and Tensor Flow for the goal of quick experimentation. GraphScope is a unified graph analytics and learning platform with graph construction (Vineyard), computation (GAIA), and machine learning (Learning Engine) modules. It can deploy web-scale. AliGraph, deployed at Alibaba, features a hybrid sampler, dynamic partitioning, and online feature stores to enable low-latency inference of the recommendation system.

Other products like GraphBolt, GNNLab, and Quiver provide middleware layers to optimize memory access and management of sparse tensor operations. Other products are introducing hardware accelerators like TPUs and FPGAs for specific workloads. The framework selection is guided by several factors like dataset size, heterogeneity, latency requirements, and domain experience of teams as shown in Table 8.2. A common hybridization of tools in each component of the pipeline in hybrid implementations is also an enterprise-class application characteristic.

Table 8.2 Comparison of GNN frameworks for large-scale systems.

Framework	Key features	Best use case	Scalability support
DGL	Heterogeneous graphs, minibatching	Academic & industrial	High
PyTorch Geometric	Tight PyTorch integration, simple API	Prototyping	Medium
GraphScope	Graph analytics + ML pipeline	Enterprise-scale graphs	Very High
Deep Graph Library	Multi-backend support, distributed	Cross-platform deployment	High
AliGraph	Optimized sampling, caching, monitoring	E-commerce recommendations	Very High

8.3 Handling Graph Data at Scale

Effective processing and graph data management at scale remain at the pinnacle of the highest list of priorities for big Graph Neural Networks (GNN) deployment. Graph data also differs from tabular data in being structurally extremely complicated with dynamic topologies, requiring sophisticated partitioning, sampling, and representation methodologies. With graph sizes potentially in billions of nodes and trillions of edges, storage efficiency, computational efficiency, and I/O pipeline efficiency become essential. This section describes the approaches used in graph data management at scale, such as partitioning, sampling, dynamic graph representation, and I/O optimization.

8.3.1 Graph Partitioning and Sharding Techniques

Partitioning of the graph is required to facilitate parallelism and distributed GNN training across nodes or computation elements. It must partition the graph into tractable subgraphs or partitions with minimal inter-partition communication but with load balance across computation elements. Edge-cut, vertex-cut, and hybrid partition are some of the graph partition techniques as shown in Figure 8.3. Edge-cut partitioning attempts to divide nodes as uniformly as possible between partitions with the goal

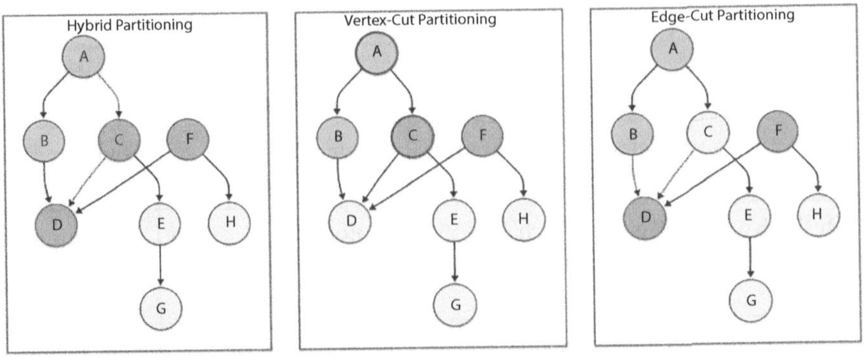

Figure 8.3 Types of graph partitioning (hybrid, vertex-cut, and edge-cut).

of reducing inter-partition edges. It is suitable for fairly uniform graphs. Vertex-cut partitioning attempts to divide edges as uniformly as possible by copying vertices between partitions—a tactic more suited to power-law graphs such as social networks in which a small number of nodes (hubs) possess orders-of-magnitude more degrees than others [12]. A comparison of graph partitioning strategies along with the use cases and trade-offs is provided in Table 8.3.

For handling gigantic graphs, recent GNN architectures use partitioning techniques such as METIS, Scotch, or streaming practitioners such as FENNEL. Besides this, dynamic graph partitioning is utilized in AliGraph

Table 8.3 Comparison of graph partitioning strategies.

Technique	Description	Use case	Trade-offs
Edge-cut	Splits nodes evenly, minimizes edge cuts	Uniform degree graphs	High communication in skewed graphs
Vertex-cut	Splits edges, replicates high-degree nodes	Power-law graphs	High memory overhead
Hybrid (e.g., HDRF)	Combines edge and vertex-cut	Heterogeneous graphs	Complex implementation
Streaming (e.g., FENNEL)	Incrementally assigns nodes/edges	Real-time graph ingestion	Potential suboptimal partitions

by leveraging monitoring the workload levels of different subgraphs in order to make it dynamically rebalance workloads.

8.3.2 Graph Sampling Methods

Sampling is a fundamental technique to reduce computational cost in large GNNs where it is not necessarily possible to train the entire graph. Sampling techniques attempt to preserve the structural and statistical properties of the graph at the cost of its size for computation. Some sampling techniques are used in GNN pipelines is shown in Figure 8.4 and their comparison is provided in Table 8.4.

Neighborhood Sampling is the most common method, where some nodes of the nearby nodes are recursively sampled for each node within a minibatch. Neighborhood Sampling, and later GraphSAGE and PinSAGE, solves the exponential increase in the receptive field with a rising number of layers in the GNN [13]. Stratified Sampling stratifies the graph based on node or edge attributes and samples within each of these strata for

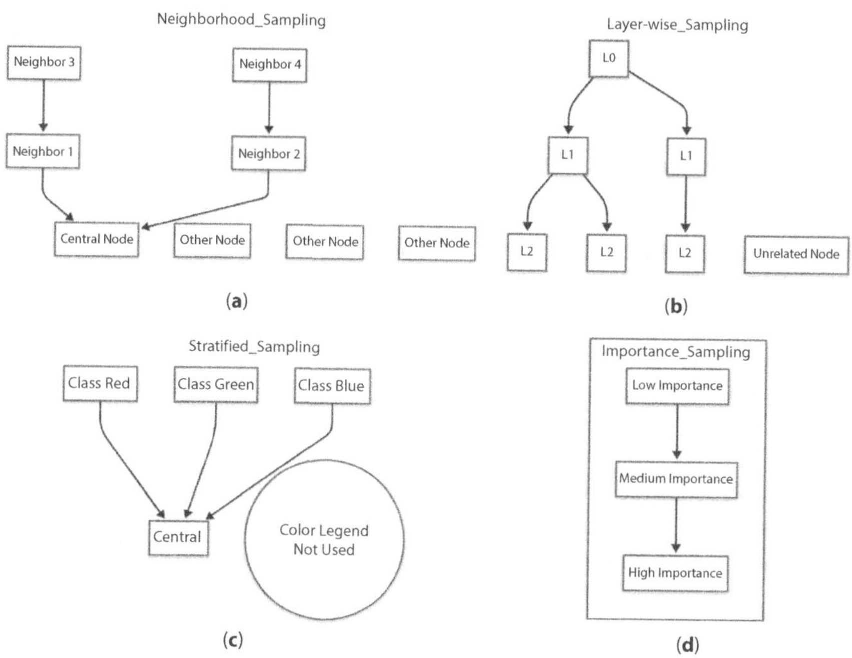

Figure 8.4 Different graph sampling methods: (a) neighborhood sampling, (b) layer-wise sampling, (c) stratified sampling based on color-coded node features, (d) importance sampling.

Table 8.4 Comparison of graph sampling techniques.

Method	Sampling scope	Complexity	Typical use case	Remarks
Neighborhood	k-hop neighbors	O(k^d)	Node classification	Prone to neighbor explosion
Layer-wise	Layer-based	O(k)	Deep GNN training	Efficient but can introduce bias
Stratified	Feature-based	O(n)	Imbalanced graphs	Maintains distribution
Importance	Score-based	O(log n)	Influence-aware sampling	Bias towards high-centrality nodes

statistical representativeness maintenance. Stratified Sampling is effective particularly in class-imbalanced or sparsely node-classed graphs. Layer-wise Sampling selects nodes layer by layer rather than neighborhood-wise, conserving memory during backpropagation. FastGCN implements this strategy to enable batched training efficiently. Importance Sampling weights edges or nodes by a relevance score, e.g., PageRank or node centrality, and samples proportionally. This targets the learning process onto influential substructures of graphs.

8.3.3 Representing Dynamic and Streaming Graphs

Real-world graph data is naturally dynamic—nodes and edges evolve over time as new interactions, events, or users are added. Social networks where friendships change, e-commerce websites with new products added, and cybersecurity logs with threats appearing in real-time are a few examples. Handling such dynamic and streaming graphs well is paramount to making scalable, time-critical applications driven by Graph Neural Networks (GNNs).

(a) Snapshot-Based vs. Continuous Graph Models

There are two main modeling approaches to dynamic graphs: snapshot-based and continuous-time models [14] as shown in Figure 8.5.

- Snapshot-Based Models split the graph timeline into discrete periods (e.g., per hour or per day), reducing each slice to a static graph. A distinct GNN model is trained or updated per snapshot. This is easier to implement and train

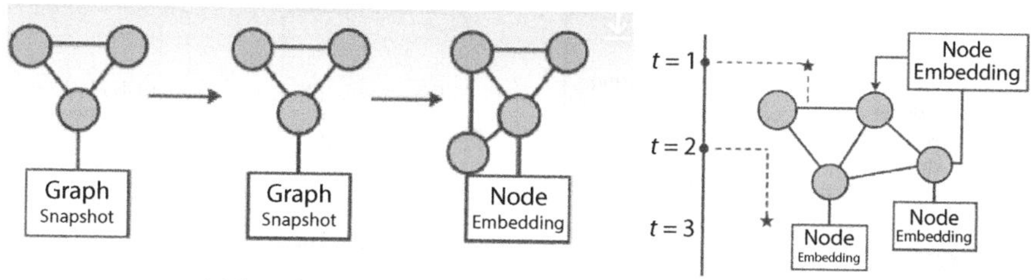

(a) Snapshtot-based (b) Continuous-time GNN

Figure 8.5 Dynamic graph representations. (a) Snapshot-based GNN; (b) continuous-time GNN.

but suffers from loss of temporal resolution and can cause latency.
Example: Dynamic GraphSAGE [48], wherein model parameters are optimized across batched snapshots of transaction data in fraud detection systems.

- Continuous-Time Models, in contrast, represent every event (edge or node addition) as an update with a timestamp. These models store changing node embeddings and handle updates in near-real-time. Temporal Graph Networks (TGN) [49] and JODIE [50] are instances of this framework. More accurate for streaming use cases but requiring more sophisticated memory management and storing event history.
Example: In recommender systems, user behavior (clicks, views) is captured as a sequence of events and maintained in continuous models, updating user embeddings without model retraining.

(b) Real-Time Graph Streaming Architecture

For efficient dynamic updates, a streaming graph processing pipeline is typically used. It consists of a message broker, feature transformer, real-time GNN engine, and inference interface as shown in Figure 8.6.

(c) Challenges in Real-Time Graph Updates

- Scalability: Keeping millions of nodes' embeddings up-to-date demands efficient storage, often in a key-value store with async updates.

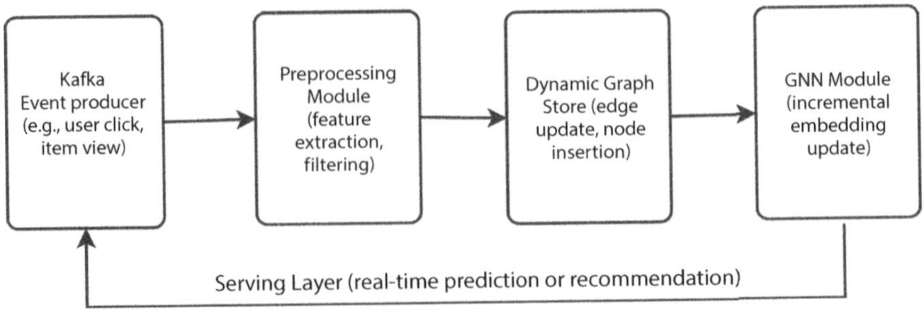

Figure 8.6 Real-time graph streaming pipeline.

- Consistency: Ensuring consistency when updating or deleting nodes simultaneously demands atomic operations or version control.
- Latency: Updates to each arriving event should propagate quickly enough to satisfy inference SLAs (e.g., sub-second recommendations).
- State Management: Streaming GNNs should cache temporal context (e.g., past neighbor embeddings or interaction timestamps) for each live node.

(d) Tools and Frameworks

- TGN (Temporal Graph Networks): Supports memory modules for continuous-time GNNs [15].
- JODIE: Models user-item interactions with evolving representations.
- StreamGraph: A system that couples Apache Kafka with GNN libraries such as DGL for near real-time updates.

By integrating temporal modeling approaches with streaming infrastructure that scales, dynamic GNNs are able to serve fast-paced data scenarios. Future development will likely concentrate on better temporal reasoning, memory-aware update strategies, and interfacing with edge stream processors.

8.3.4 Scalability Bottlenecks in Graph Processing

Scaling Graph Neural Networks (GNNs) to big data presents a number of system-level bottlenecks in memory management, communication overhead, and data access patterns. Knowledge of these bottlenecks is necessary for designing architectures that can handle graphs with billions of nodes and edges efficiently. Neighbor Explosion: In multi-layer GNNs, every layer computes neighbors' features in aggregate. For a k-layer GNN, the receptive field expands exponentially in depth (i.e., O(d^k), where d is the average node degree). For instance, in a three-layer GCN on a graph with d = 25, every node indirectly reaches more than 15,000 nodes, dramatically raising computation and memory needs. This effect—termed neighbor explosion—blows up mini-batches, tending to cause GPU memory overflow and higher latency.

Memory fragmentation is a common issue in distributed graph networks (GNNs), as sparse adjacency matrices and dynamically sampled subgraphs can cause irregular access. GPU memory is affected by fragmentation when dealing with large numbers of small kernel launches, limiting memory reusability and adding memory overhead for buffers and context switching. Distributed GNNs also have high data transfer overhead when embeddings or subgraphs are transferred between GPUs or machines. Synchronization of embeddings can saturate PCIe/NVLink bandwidth or network interfaces, and parameter-server-based systems can be bottlenecked centrally unless optimized with caching or async pull strategies. Neighbor caching and embedding enhance performance but require cache-coherence mechanisms in dynamic graphs. Memory-memory trade-offs are explored, with methods like feature sampling and pre-aggregation minimizing memory consumption but potentially compromising accuracy. Mitigation strategies include layer-wise sampling, memory pooling and tensor reuse libraries, asynchronous updates and communication compression, graph partitioning with low-edge cuts, and caching embeddings with time-to-live (TTL) or frequency-of-access eviction.

These observations are important when constructing high-throughput GNN systems, especially for real-time inference, financial fraud detection, or web-scale recommendation systems.

8.3.5 Optimizing Graph Data I/O and Access Latency

Expensive and high read/write access cost during training and inference is needed for scalable graph data processing. Reducing access latency and optimizing the I/O of graph data is crucial to performance. Graph computation is different from normal databases in which traversals consist of thousands of nodes and locality-aware accesses and cache locality are crucial. Some of the widely used graph data I/O optimization techniques are:

- I/O Optimization Techniques as shown in Figure 8.7 includes preloading common subgraphs, encoding large graphs into memory, and page-locked memory in the GPU for faster transfer. Storage engines for graphs such as GraphChi and TurboGraph use sequential disk storage access and vertex-computation-oriented accesses in an attempt to reduce the latency of I/O.
- Caching Mechanisms such as neighborhood caching (k-hop neighbors caching), embedding caching, and sampler output caching are heavily employed. Both RAPIDS cuGraph and NVIDIA DGL offer APIs for node feature and neighbor list caching to GPU memory.
- Compression algorithms like CSR (Compressed Sparse Row), COO (Coordinate), and GCSR (Generalized CSR) representation conserve memory at the cost of small gain in quick lookup. Quantization is also employed in certain systems to conserve embedding in lower precision. Table 8.5 provides the optimization techniques, methods, and their performance gain.

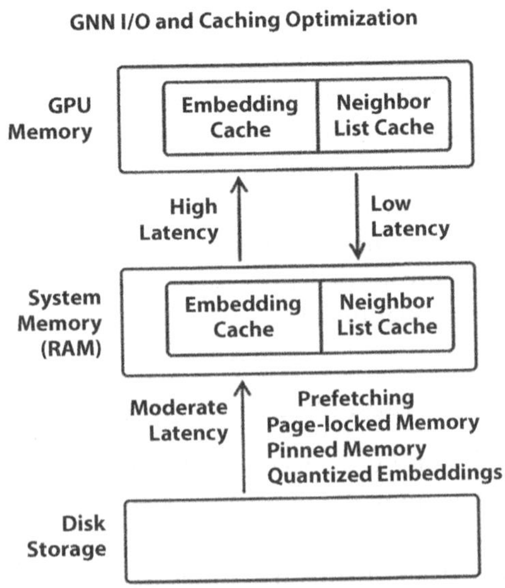

Figure 8.7 I/O and caching optimization.

Table 8.5 Graph data I/O optimization techniques.

Optimization area	Technique	Tool/ framework	Performance benefit
Prefetching	Sliding window traversal	GraphChi	+30% throughput
Caching	Embedding/ neighbor caching	DGL, cuGraph	Reduces redundant computation
Compression	CSR, COO, quantization	PyG, DGL	Lower memory footprint
Memory Management	Page-locked memory, pinned memory	CUDA, RAPIDS	Faster host-to-device transfer

8.4 Distributed and Parallel Training of GNNs

Graph Neural Networks (GNNs) have intricate dependencies between data points due to the nature of input data, therefore it becomes increasingly difficult to train using large data. Parallelization of GNNs can be approached in general by three levels of granularity, i.e., node-level, edge-level, and graph-level.

- Node-Level Parallelism distributes the computation of node embeddings across available processing units. The model is valid in the event that node features are predominant and neighborhoods are of intermediate node size. Each computing agent computes a node's embedding by aggregation from neighbors. Syncing problems occur with more connectivity density among components of the graph.
- Edge-Level Parallelism maps edges to processors such that it optimizes models whose goal is relational operations, i.e., edge classification or link prediction. It is particularly useful in edge-dense situations such as recommendation systems. Overlapping computation of neighborhoods can be too expensive in communication overhead, however.
- Graph-Level Parallelism is based on splitting the whole graph into subgraphs to facilitate independent or

semi-independently training over partitions. It is applied in cases such as meta-learning or multi-task learning on heterogeneous graphs. Techniques such as Cluster-GCN [15] and GraphSAINT [12] utilize graph clustering and stochastic sampling to facilitate efficient training.

8.4.1 Training in Distributed Environments

To enable training GNN bigger than one computer can computationally or in terms of memory, distributed systems are used. The systems consist of parameter servers, all-reduce frameworks, and GPU clusters.

- Parameter Server Architecture separates computation from storage. Workers compute gradients to be sent to parameter servers that update and return the model. Parameter server mode can handle asynchronous training but will suffer from bottlenecks and stale gradients during heavy loads. Distributed Tensor Flow and MXNet both support parameter server modes.
- All-Reduce Architectures like Horovod reduce synchronization latency by employing collective communication primitives like ring-allreduce. PyTorch Distributed and NCCL by NVIDIA are optimized for this mode and scale better on GPU clusters.
- GPU Clusters using NVLink or InfiniBand based interconnects are common in big GNN applications. GPUs speed up matrix multiplications, convolutions, and attention mechanisms but graph partitioning and memory optimization are still bottlenecks. A comparison of these strategies are provided in Table 8.6.

8.4.2 Implementation Details of Distributed GNN Training

Practically, distributed training of GNNs means selecting between two major architectures: parameter server (PS) and all-reduce. The PS architecture separates training into workers (responsible for computations of gradients) and servers (holding model parameters). For its part, the all-reduce strategy distributes computation and model weights across all nodes with collective communication operations, most commonly using NVIDIA

Table 8.6 Comparison of distributed training strategies.

Strategy	Advantages	Limitations	Tools/ frameworks
Parameter Server	Scalable, decouples compute/ storage	Staleness, complex fault tolerance	Tensor Flow, MXNet
All-Reduce	High throughput, minimal bottlenecks	Synchronization cost	Horovod, NCCL, PyTorch DDP
GPU Cluster Training	Efficient for dense ops, parallelism	Memory constraints, partitioning needed	DGL, GraphLearn, GraphScope

NCCL or Horovod for synchronization. For example, in the DistDGL (Distributed Deep Graph Library) library, a standard implementation partitions the graph and puts the partitions onto a cluster of machines. The worker processes each take one partition, sampling neighborhood locally to reduce inter-node communication. A centralized or peer-to-peer key-value store keeps track of embeddings and global model state. Figure 8.8 provides an example of typical distributed GNN training architecture.

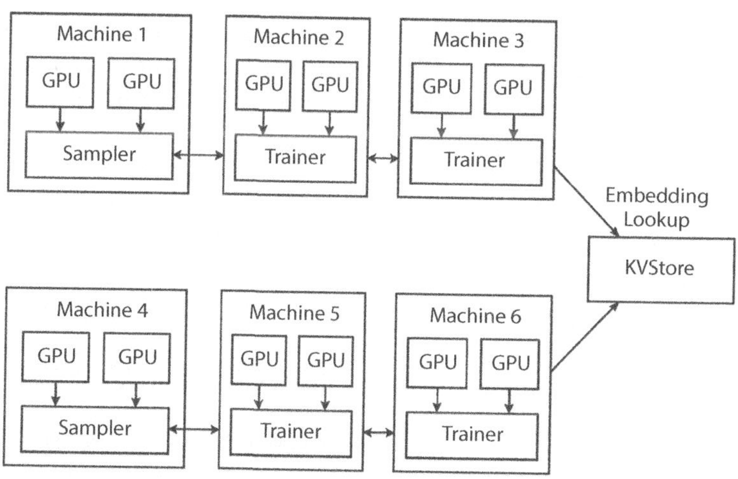

Figure 8.8 Distributed GNN training architecture.

Pseudocode Example: DistDGL Launch Configuration

```
# Launching a 4-machine DistDGL training job (each with 2 GPUs)
python3 launch.py \
    --workspace /gnn/train \
    --num_trainers 2 \
    --num_samplers 2 \
    --part_config /data/graph/4part.json \
    --ip_config ip_list.txt \
    --graph_name citation_network
```

In this setup:

num_trainers specifies the number of training processes.

num_samplers controls how many sampling processes run per machine.

ip_config contains the private IPs of worker nodes.

part_config is a partition configuration JSON file generated using DGL's partitioning tool.

Training Log Sample (DistDGL + GraphSAGE):

Epoch 0 | Loss: 1.83 | Time: 17.3s | Memory: 4.1 GB | Comm Overhead: 2.2s

Epoch 10 | Loss: 0.42 | Time: 12.9s | Memory: 3.9 GB | Comm Overhead: 1.6s

...

Final Accuracy: 84.2% | Total Time: 6.1 mins

The logs illustrate how training time decreases with improved convergence, and communication overhead decreases as embedding tables are cached. Profiling utilities like DGL Profiler, PyTorch's autograd profiler, or NVIDIA Nsight Systems are suggested for locating bottlenecks and optimizing GPU utilization.

All-reduce platforms, e.g., those over PyTorch Distributed Data Parallel (DDP), are favored when the whole graph is in memory and inter-worker latency is minimal. Nevertheless, for extremely large graphs (e.g., those over 1B nodes), the parameter server method with graph partitioning is still the preferred approach.

8.4.3 Challenges in Communication and Synchronization

The distributed nature of large-scale GNNs introduces communication and synchronization issues that impact performance and accuracy.

- High Communication Overhead results from high-frequency passing of intermediate embedding across partitions. In neighbor sampling, for example, all the GPUs will require access to the nodes of the other GPU, which causes bandwidth saturation.

- Synchronization Delays occur in synchronous environments, where the speed of training is determined by the slowest worker (straggler). Asynchronous SGD and gradient sparsification techniques are employed to counter this.
- Overlapping neighborhoods result in Redundant Computations that create redundant node or edge computation. Subgraph deduplication or caching embedding can be remedies.
- Graph Partitioning Inefficiency results in load imbalance. Ineffective partitioning can result in having high-degree nodes within one partition and hence overloading some workers. Graph practitioners like METIS or Scotch can help but is graph-type-dependent.
- Memory Bottlenecks are a problem for GPU-based systems due to embedding tables that are large. The future solutions are offloading embedding to CPU memory and mixed-precision training.

8.4.4 Case Studies

- **DeepMind's GNN Systems** are used in a wide variety of applications, from protein folding (AlphaFold) to structured world reinforcement learning. Their distributed GNN engine utilizes multi-core CPU clusters for pretraining and GPUs for fine-tuning. They use recurrent GNN architectures with hybrid attention and batched batching as an attempt to scale learning over millions of entities [16, 17].
- **Alibaba's GraphLearn** is an enterprise-grade GNN platform with big-recommendation scale. It supports hybrid sampling approaches, embedding caching, and dynamic updating of node/edge features. GraphLearn, on GPU clusters with over a billion nodes, employs an asynchronous hybrid training framework and can be combined with Alibaba's FeatureStore and Flink pipeline [19].
- **Facebook's PyTorch** BigGraph is also designed for enormous knowledge graphs such as social network graphs. It uses block decomposition, negative sampling with zero overhead, and multi-GPU training to process graphs with a

Table 8.7 Summary of real-world distributed GNN case studies.

Organization	Framework	Graph size	Strategy	Unique features
DeepMind	Custom Engine	Millions of nodes	Hybrid parallelism, attention, recurrent GNNs	Attention, recurrent GNNs
Alibaba	GraphLearn	Billions of nodes	Async training, caching	FeatureStore, hybrid sampling
Facebook	PyTorch-BigGraph	Billions of entities	Block decomposition, sharding	Negative sampling, embedding reuse

billion edges. BigGraph uses scalable embedding learning by shard entities and relationships and memory reuse between batches [19]. Table 8.7 summarizes the framework, strategies and unique features of these case studies.

8.5 System-Level Optimizations

As graphs and graph neural networks (GNNs) grow in size and complexity, system-level optimizations become an integral consideration of fulfilling real-world deployment requirements as far as performance, latency, and resource use are concerned. While algorithmic innovation improves the learning effectiveness, it's model-level optimizations behind the model—model compression, memory management, caching, and latency tuning—what allow operations, especially in production, to become feasible. These optimizations improve training and inference pipelines' efficiency, which makes such applications as personalized recommendation systems, fraud detection systems, and real-time decision-making systems feasible.

8.5.1 Model Compression Techniques

Model compression is perhaps one of the most significant system-level methods for reducing the memory and computational requirements of GNN models with minimal accuracy loss. Two of the best compression methods are pruning and quantization. Table 8.8 provides as benefits and impact these techniques. Quantization reduces the precision of model parameters (e.g., from 32-bit floating point to 8-bit integers), reducing memory usage and computational performance. Quantized models may

Table 8.8 Model compression techniques and their impact on GNNs.

Technique	Description	Benefits	Trade-offs
Quantization	Lower-precision arithmetic	Faster inference, lower power	Slight accuracy degradation
Pruning	Removing redundant weights or layers	Smaller models, faster speed	Requires retraining

be run at a low computational cost on the special hardware like Tensor Cores and TPUs in an effort to accelerate inference [20]. Post-training quantization and GNN quantization-aware training (QAT) are supported with tools like TensorRT and ONNX Runtime. Pruning removes less significant weights, neurons, or even whole GNN layers according to their output contribution. Structured pruning (i.e., eliminating less significant message-passing paths) and unstructured pruning (i.c., dcleting certain wcights) can lead to phenomenal speedups [3].

8.5.2 Caching, Prefetching, and Batching Strategies

- Caching is a robust method in GNN architectures to cache frequently accessed hot data, like node embedding, feature vectors, and subgraph samples. Frequent accesses to nearby nodes are the situation for most GNN operations, so eliminating the latency by storing them on higher speed memory levels (e.g., L3 cache or GPU memory) is effective [3, 21]. Prefetching anticipates future access patterns to data and pre-fetches data into the memory. In other words, if node A is typically accessed after node B has been accessed, the system pre-fetches data for A during accessing B.
- Batching is placing multiple training or inference examples in a single process batch to leverage GPU parallelism. GNN mini-batching strategies are more advanced due to graph irregularity. The generic architecture of these strategies can be seen in Figure 8.9. Libraries such as DGL and PyTorch Geometric support a range of batching strategies, including edge-wise, node-wise, and graph-wise batching. A comparison of these strategies is provided in Table 8.9.

Table 8.9 Comparison of graph batching strategies.

Strategy	Description	Use case	Complexity
Node-wise	Batch multiple node subgraphs	Node classification	Medium
Edge-wise	Batch based on edge sampling	Link prediction	High
Graph-wise	Batch entire graphs	Graph classification	Low

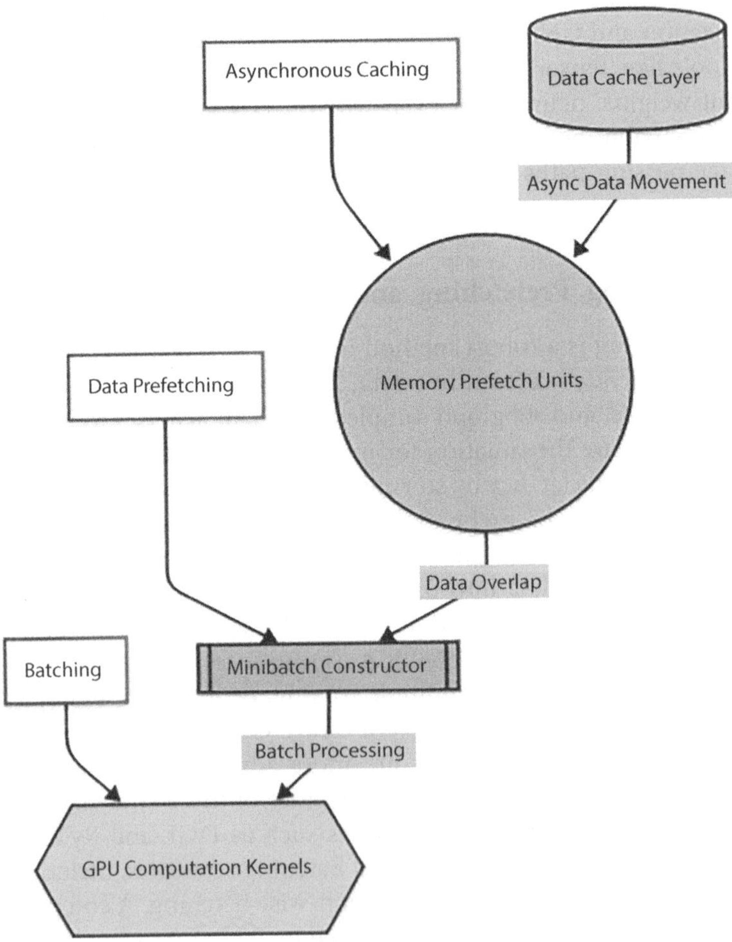

Figure 8.9 Caching and batching generic architecture.

8.5.3 Efficient Memory Management and GPU Utilization

Memory bottlenecks during training large GNNs are unavoidable due to memory demands of adjacency matrices, feature tensors, gradients, and optimizer states. Effective memory management techniques include tensor reuse, memory pinning, gradient check pointing, and offloading intermediate states into CPU memory.

- Tensor Reuse and Memory Pooling: These libraries allow memory blocks to be reused across different stages of forward and backward passes in an effort to avoid redundant memory allocation [22].
- Gradient Check pointing: Instead of saving all the intermediate activations in forward pass, a smaller set of them are re-evaluated at backward pass to try and preserve memory. It is a trade-off between compute and memory but is tolerable when the models are large.
- GPU Utilization: Multi-GPU setups make use of techniques such as pipeline parallelism and tensor model parallelism. For the latter, various layers of the GNN are assigned to different GPUs, while in the former tensors of enormous size are divided and processed in parallel. Table 8.10 can be referred for the benefits and applications of these techniques.

8.5.4 Real-Time Inference and Latency Optimization

Latency is mission-critical in use cases such as fraud detection in finance or autonomous driving. Real-time GNN inference needs to perform input graphs or subgraphs and generate predictions within tight timing constraints. Figure 8.10 shows a real-time GNN inference pipeline.

Table 8.10 Memory optimization techniques in GNNs.

Technique	Benefit	Applicable scenario
Tensor Reuse	Reduces fragmentation	Any large-scale model
Gradient Checkpointing	Saves memory during backprop	Training large GNNs
Offloading	Lowers GPU memory usage	When CPU-GPU bandwidth is adequate

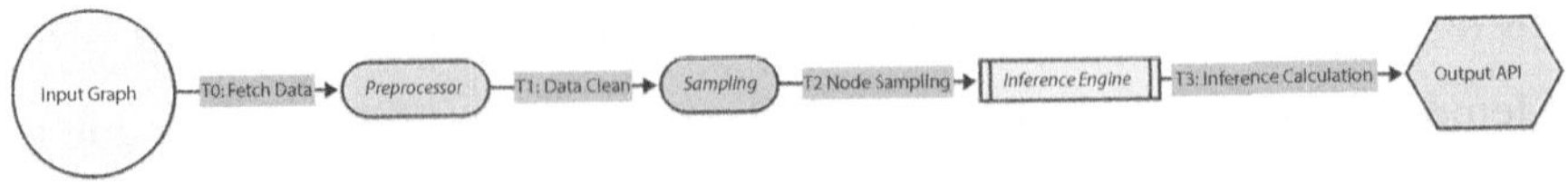

Figure 8.10 Real-time GNN inference pipeline.

- Model Reduction: Models can be reduced to optimal effective width and depth in order to satisfy latency constraints, except for compression. Attention mechanism and aggregation operations can be cut or approximated.
- Inference Optimization: GNN model inference is enabled by low-latency inference on NVIDIA Triton and Tensor Flow Serving platforms. Serving strategies some of which are model warm-up, dynamic batching, and concurrency control.
- Graph Caching and Sampling: On-the-fly graph truncation, embedding lookup caching, and neighbor sampling improve the inference pipeline.
- Async Processing: Segregation of model execution and data loading through the help of thread pools and event queues allows GNN systems to mask latency and provide optimal throughput [23]. Refer to Table 8.11 for the use cases of these optimization techniques.

System-level optimizations are a critical aspect of the deploy ability of GNNs in the real world. Although the optimization approach is application need-, infrastructure capacity-, and model complexity-dependent,

Table 8.11 Real-time optimization techniques.

Optimization method	Description	Use case
Model Warm-up	Load and prepare model in advance	Online serving
Dynamic Batching	Aggregate real-time requests into a batch	Low-latency environments
Async Queuing	Separate I/O and computation threads	Streaming inference
Sampling Optimization	Use approximate sampling for speed	Large dynamic graphs

integration of compression, smart memory management, and low-latency serving techniques leads to high-quality and scalable GNN deployments. Ongoing advancements in hardware acceleration, software abstraction, and auto tuning will keep on making GNN-based AI systems more efficient.

8.5.5 Cloud vs. Edge Deployment: Quantitative Trade-Offs

Deploying Graph Neural Networks (GNNs) into real-world tasks usually means weighing high-performance cloud hardware versus limited resource edge devices. This decision determines latency, memory usage, energy efficiency, and scalability. To inform practical deployment choices, this subsection presents a performance comparison across typical hardware configurations—NVIDIA A100 (cloud GPU) and Jetson Nano (edge AI SoC)—with a baseline GCN-based recommendation model on a sample user-item interaction graph. Table 8.12 provides comparison of GNN inference performance on cloud vs. edge platforms using identical model weights and input graphs.

Trade-off analysis

- Latency vs. Energy: Edge inference has ~6x more latency per sample but uses ~25x less power. This makes edge suitable for low-duty-cycle applications (e.g., health wearables), and cloud is better suited for real-time systems (e.g., fraud detection).
- Memory Bottlenecks: Jetson Nano supports ~16% of A100 batch size due to RAM constraints, necessitating throughput or graph sampling depth trade-offs.

Table 8.12 Performance comparison: Jetson Nano vs. A100.

Metric	NVIDIA A100 (cloud GPU)	Jetson Nano (edge device)
Inference Latency (ms/sample)	6.3	38.4
Peak Power Consumption (W)	250	9.8
Memory Usage (MB)	2840	470
Batch Size Supported	1024	64
Throughput (samples/sec)	158.7	26.0

- Deployment Implications: Edge GNN inference will need drastic quantization, pruning, and model partitioning to operate within hardware constraints. Cloud platforms, by comparison, support deeper models and full-graph context propagation.

Figure 8.11 shows the trade-off between power consumption and throughput when running Graph Neural Networks (GNNs) on cloud and edge hardware. The NVIDIA A100 GPU, a high-performance compute platform running in the cloud, achieves much higher throughput (158.7 samples/sec) than the Jetson Nano (26.0 samples/sec), which is tuned for edge use. But at a much higher price in terms of power consumption—250 watts for the A100 and a mere 9.8 watts for the Jetson Nano. This reverse correlation identifies the deployment challenge: cloud devices are great at velocity and scaling but pay a higher energy price, while edge devices are more energy-efficient but less capable of processing. Such graphical comparisons assist in measuring the engineering consequences of hardware selection, directing engineers to pair model complexity and latency requirements with suitable infrastructure.

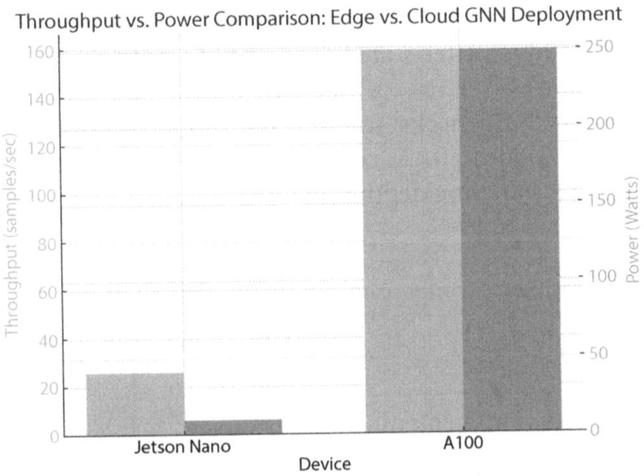

Figure 8.11 Throughput vs. power for Jetson Nano and A100.

8.6 Case Studies and Applications

Graph Neural Networks (GNNs) integration into actual large-scale systems in the real world proves their value transforming across industries and research domains. With GNNs improving scalability, efficiency, and architectural support, their use in production environments has increased exponentially. This chapter provides in-depth case studies of leading industry-scale and research applications to reflect real-world challenges, architectural innovation, and experience gains. We learn the GNN-powered Pinterest platform (PinSage), Alibaba platform (AliGraph), Amazon recommendation, and learning tasks such as large-scale protein-protein interaction prediction and traffic/mobility modeling by spatio-temporal GNNs.

8.6.1 Industry-Scale Systems

(a) Pinterest's PinSage: Industrial-Scale Graph-Based Recommendation
Pinterest. A content discovery platform applies a graph-based recommendation model called PinSage to deliver personalized pins for the users. The earlier collaborative filtering models lacked contextual awareness of the interactions between nodes at enormous scales. PinSage learned a web-scale and efficient GNN model that could learn web-scale graphs comprising billions of edges and nodes [24]. PinSage constructs a heterogeneous user, pin (image), board (collection), and interaction (saves, views) graph. Edges are weighted by content features and edge weights as interaction frequencies. The graph is sampled *via* neighborhood aggregation and random walks to preserve context-aware node representations. Node representations are utilized for link prediction to predict new pins. Pinterest made some optimization techniques to scale PinSage:

- Neighbor sampling layer by layer to reduce training computation.
- Edge importance sampling for more insightful gain.
- Approximate nearest neighbor search (ANN) for latency-reduced inference.

The resulting model had a substantially better engagement score and reduced latency for online inference. Figure 8.12 shows the general architecture of PinSage.

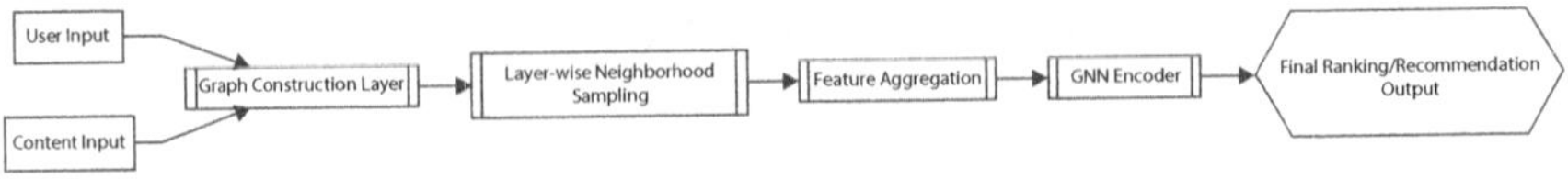

Figure 8.12 Architecture of Pinterest's PinSage.

PinSage is an industrial-strength GNN for web-scale recommendation. The model combines GCNs with random walks to learn node contexts in a scalable manner. PinSage layers are constructed as follows:

- Input: Node embeddings of content features and interaction history.
- Random Walks: Multi-hop sampling with importance weighting.
- GCN Layers: Two-layer GCN with learned edge weights.
- ANN Search: Approximate Nearest Neighbor index to search for relevant nodes efficiently.
- Output: Top-K recommendation ranking.

(b) Amazon Product Recommendation Using Graph Neural Networks
Amazon uses GNNs to give more accurate product suggestions from user-product interaction graphs, including user activity, co-buying behaviors, and item attributes. GNNs augment such background in the form of explicit and implicit user-item relations and even among users as well as among items [19]. One of the strongest components of Amazon's approach is to benefit from session-based GNNs that have been learned using short-term interests. Such models view a sequence of clicks or transactions within a session as graph walks and learn dynamic representations. Amazon also benefits from knowledge-aware GNNs, using structured knowledge based on product ontologies and product reviews. The joint model improves semantic relevance and personalization. Scalability is established *via* Apache Spark for pre-processing, DGL for building the models, and SageMaker for scaling to train over the network. Amazon uses GNNs to represent user sessions as temporal subgraphs. Architecture:

- Session Graph Builder: Builds directed session graphs.
- Embedding Layer: Metadata-based initial user/item vectors.
- GNN Core: Applies gated GNN with temporal edge weighting.
- Aggregator: Calculates final session state.
- Predictor: Generates next item probability.

Table 8.13 Comparison of industrial-scale GNN applications.

Company	System	Graph type	Edge scale	Key features	Achievements
Pinterest	PinSage	Heterogeneous	Billions	Layer-wise sampling, ANN search	Increased engagement by 50%
Alibaba [19]	AliGraph	Dynamic, Attributed	10B+	GSaaS, Hybrid Sampling, Feature Cache	20%+ gain in prediction metrics
Amazon	SessionGNN	Temporal Bipartite	Hundreds of M	Session GNN, Knowledge-aware architecture	Boosted click-through rate

The provided details shows how fundamental GNN components are instantiated across systems, the data preprocessing involved, and the downstream inference pipelines that drive recommendations, predictions, and analysis in production deployments. Summary of Comparison of Industrial-Scale GNN Applications is revealed by Table 8.13.

8.6.2 Academic Use Cases

(a) Protein Interaction Prediction with GNNs

In-silico at-scale prediction of PPI is perhaps the most important bioinformatics problem. GNNs are well positioned to do this since they can be trained to encode molecular graphs and perform the required predictions accordingly based on intrinsic biological relations [25]. Protein molecules may be controlled as graphs where biochemical interactions are edges and amino acids are vertices. Supervised information is learned by Graph Neural Networks like Message Passing Neural Networks (MPNNs), Graph Attention Networks (GATs), and Graph Convolutional Networks (GCNs). The architectural pipeline of protein interaction prediction with GNNs includes:

- Input: Protein structure in PDB/FASTA format.
- Graph Builder: Amino acids as nodes; proximity edges.
- MPNN Layers: Aggregate local spatial features.
- Pooling Layer: Graph-level readout.
- Output: Binary interaction prediction.

Table 8.14 Performance comparison for PPI prediction models.

Model	Accuracy	Precision	Recall	F1 score
Random Forest	78.4%	75.6%	77.0%	76.3%
CNN-based	82.9%	81.5%	83.0%	82.2%
GCN (GNN)	*89.1%*	*88.0%*	*89.5%*	*88.7%*

Such massive databases as STRING, BioGRID, and HumanNet contain billions of samples of interactions. Graph partitioning, distributed training, and hybrid supervisory methods (e.g., semi-supervised learning) are used by researchers to tackle it. Table 8.14 shows the performance comparison of these different models PPI prediction. Experiments demonstrate that GNNs outperform baseline classifiers and sequence-based models, especially in generalizing to new proteins.

(b) Spatio-Temporal GNNs for Traffic and Mobility Analysis

Physical nodes and time-varying edges as relations in spatio-temporal graphs are conventional for traffic and mobility prediction. Temporal high-flow data, sensors, GPS, and monitoring systems are present in urban cities. Temporal graph learning and prediction must be incorporated into the model [26].

Spatio-Temporal GNNs (ST-GNNs) integrate the core GNNs by incorporating time-aware mechanisms, such as:

- Temporal attention mechanisms
- Recurrent GNN layers (GRU-based)
- Graph Convolution + Time Convolution (ST-GCN)

Traffic volume prediction over road networks in the future is another typical use case. Intersections as node and edge weights as traffic. Deployed in real-time routing and traffic congestion control in smart cities such as Los Angeles and Beijing.

Another newer model, DCRNN (Diffusion Convolutional Recurrent Neural Network), combines sequence modeling and diffusion-based GNNs as an improvement to make accurate predictions on traffic conditions. Table 8.15 can be referred for performance comparison of these models.

Advanced urban analytics, from demand forecasting and outlier detection to routing optimization, are provided by GNN models.

Table 8.15 Accuracy metrics for traffic prediction models.

Model	MAE	RMSE	MAPE
ARIMA	4.3	9.1	12.5%
LSTM	3.6	8.2	10.1%
DCRNN	**2.7**	**5.9**	**6.9%**
ST-GCN	2.9	6.4	7.5%

Graph Neural Networks (GNNs) are increasingly used in various fields, including medicine, finance, recommendation, and city-scale computing. However, their high relational and structural information density raises new ethical risks, such as identity inference risk, information leakage, and exploitative misuse. To protect these systems from attacks while maintaining user privacy, it is crucial to research their security, privacy, and ethics, as they are implemented on industrially large platforms.

8.7 Security, Privacy, and Ethical Considerations

With increasingly more uses of Graph-Network Neural Networks (GNNs) to the entire range of applications—anything from finance and health, to recommendation systems and urban computing—being developed, the necessity to explore the security, privacy, and ethics of graph learning accelerated at an exponential rate. Unlike ubiquitous images or table data, graph data hold infinitely richer structural and relational information, reflecting real-world organizational or social networks in all cases. This adds new susceptibilities and ethical issues, like inference of identity, information leakage through link prediction, or adversarial attacks due to malicious intentions [27]. Since GNNs are now being used at the industrial level, it is no longer a technical requirement but an ethical requirement that such networks are attack-free without compromising the user's privacy.

8.7.1 Data Privacy Challenges in Graph-Based Systems

Maintenance of privacy in graph-based systems is intrinsically difficult because graph data is of high dimensionality. Private information can be stored at each node and edges can be employed to imply relations which

Table 8.16 Comparison of traditional and graph-specific privacy methods.

Privacy method	Description	Graph-specific challenges	Trade-offs
K-Anonymity	Hides identity among k individuals	Difficult to apply to graph topologies	Limited applicability
Differential Privacy	Adds statistical noise to data/ query results	High overhead, loss in GNN utility	Accuracy vs. Privacy
Homomorphic Encryption	Computation on encrypted data	Not scalable to large graphs	Heavy computational cost
Federated Learning	Model training across decentralized data	Complex for dynamic, evolving graphs	Communication overhead

would not be preferable to be disclosed by the users. Neighbor information in social graphs or health graphs, for instance, can unintentionally disclose a user's groups or health conditions [3].

General privacy techniques such as k-anonymity and differential privacy must be adapted in the context of a graph. Differential privacy in different contexts for a graph is node-based or edge-based but is expensive from a computational perspective and impedes the performance of a model [28, 29]. Table 8.16 is the legacy data privacy metric vs. its graph-specific counterpart.

8.7.2 Federated GNNs and Secure Model Sharing

Federated Learning (FL) is suggested as a privacy-preserving graph learning method. Federated GNN clients (e.g., devices or data centers) learn local GNNs on their own graphs, and model updates rather than raw data are sent to a central server. Sensitive relationships and attributes are not revealed [30].

It is difficult to build federated GNNs because of the following reasons:

- Graph Heterogeneity: Subgraphs can be potential per client with disjoint schemata.

Table 8.17 Features of federated GNN frameworks.

Framework	Core feature	Privacy technique	Use case scenario
FedGraphNN	Decentralized graph training	Secure Aggregation + DP	Mobile social networks
FLAME	Heterogeneous graph support	Edge-Level DP + FL Averaging	Health information sharing
GNN-FedAvg	Simple federated aggregation	Noisy model update sharing	Distributed recommendation

- Non-IID Data: Node and edge distributions are highly heterogeneous across clients.
- Secure Aggregation: Cryptographic techniques must be used to securely aggregate model updates without leaking individual contributions.

For these, models like FedGraphNN and FLAME suggest special-purpose layers for edge-level privacy, async updates, and differential privacy imposed at aggregation [30]. Figure 8.2 illustrates a typical federated GNN system architecture and Table 8.17 provides features of these frameworks along with the use cases.

8.7.3 Attack Surfaces in GNN Systems

Like most deep learning models, graph neural networks are susceptible to many adversarial attacks. Graph structure provides new attack surfaces. They are:

- Poisoning Attacks: The attacker poisons nodes or edges during training to alter the behavior of the model [31].
- Evasion Attacks: Node feature or edge perturbations at inference time that mislead the predictions of the GNN.
- Model Inversion Attacks: Extraction of input graph structures from model outputs or embedding.
- Membership Inference: Attackers check if a particular node or subgraph was part of the training set.

Table 8.18 Common graph adversarial attack strategies.

Attack type	Method	Target stage	Description
Poisoning	Nettack, MetaAttack	Training	Injects fake nodes/edges to mislead predictions
Evasion	GradArg, Fast Gradient	Inference	Alters node features or connections post-training
Membership Inf.	MIA-GNN	Inference	Predicts training membership of nodes
Model Inversion	GMI, Edge-MIA	Inference	Reconstructs features or edges from outputs

Graph structure-specific attacks are usually infinitesimal, imperceptible but highly potent perturbations. Some popular graph adversarial attack methods are listed in Table 8.18.

8.7.4 Defense Mechanisms Against Attacks

There are numerous defense mechanisms against attacks suggested by researchers such as:

- Graph Smoothing: pruning adversarial edges according to Jaccard similarity.
- Robust Training: inclusion of adversarial training with inclusion of perturbed samples in learning by the model [32].
- Feature Denoising: attention layers or auto encoders for feature denoising.
- Certified Robustness: Methods that provide theoretical model robustness guarantees against perturbations (e.g., ProGNN, GNNGuard).

No defense is perfect, but most methods increase overall robustness. There are still trade-offs between accuracy, robustness, and computational cost.

8.7.5 Ethical Implications of Graph-Based AI

Along with technical issues, then, come the ethics of GNN use in high-risk applications. Graphs naturally map to modeling social, medical, or financial networks—questions of bias, fairness, and consent.

- Bias Amplification: GNNs can potentially amplify and perpetuate structural bias in graph structure.
- Lack of Consent: People are not necessarily made aware their relational information (friends, transactions) is being modeled.
- Unwanted Inference: Node classification or link prediction may reveal sensitive attributes which are not made available in the open.

Transparency consent protocols, explainability processes, and fairness-sensitive modeling are what ethical GNN design should entail [33]. Table 8.19 presents typical ethical risks and mitigation strategies.

Security, privacy, and ethics in GNNs are not special cases—rather, they are at the very core of responsible AI deployment. As graph structures are at the core of enabling personal, healthcare, and financial transactions, privacy breaches, adversarial attacks, and ethically sustainable use need to be tackled by system developers much prior to their occurrence. Secure training, explainable modeling, and federated learning are the pillars of building towards safe, large-scale reliable GNN systems. Future studies ought to keep breaking the limits of official privacy and security protection, and government frameworks must adapt in order to allow direct ethical deployment into every aspect.

Table 8.19 Ethical challenges and mitigation in GNN systems.

Ethical concern	Risk example	Mitigation approach
Structural Bias	Over-representation of influential nodes	Graph rebalancing, fairness constraints
Consent Violation	Implicit modeling of social connections	Informed data collection policies
Inference of Sensitive Data	Predicting illness from co-location	Explainability & user control

8.8 Future Directions and Research Opportunities

8.8.1 Towards General-Purpose GNN Accelerators

The need for hardware accelerators specific to the idiosyncratic computation patterns of GNNs has gotten very serious consideration. In contrast to typical deep learning workloads, GNNs feature irregular memory access behaviors and sparse computation across non-Euclidean data structures. This places on-demand exigencies that outstrip current GPU and TPU architecture.

New hardware designs, i.e., Graphicionado accelerator and FPGA-optimized hardware, strive to distill into their bare essentials basic elementary GNN operations like neighbor aggregation and feature transformation [34]. These accelerators utilize primitives like dataflow scheduling, vectorized aggregation units, and graph-optimized memory hierarchies. The general-purpose accelerators for the majority of GNN variants (e.g., GCN, GAT, GraphSAGE) could also potentially make high-performance access to graph learning immensely democratized. A comparison of the existing GNN accelerator architectures is provided in Table 8.20.

8.8.2 Integration with Multimodal and Hybrid AI Systems

Multimodal AI systems that process and fuse heterogeneous data modalities (e.g., text, images, audio) can significantly benefit from the ability of GNNs to learn relational and structural information. One probable effective direction in this scenario is coupling GNNs with Large Language Models (LLMs) for reasoning over structured and unstructured modalities.

Table 8.20 Comparison of emerging GNN accelerators.

Accelerator	Architecture type	Key optimizations	Supported models	Energy efficiency
Graphicionado	ASIC	Sparse memory compression	GCN, GAT	High
AWB-GCN	FPGA	Edge-aware scheduling	GCN	Medium
HyGCN	GPU + Memory Mux	Hybrid execution pipeline	GCN, GraphSAGE	High
GNNMark	Benchmark Suite	Configurable testing	Multiple GNN Variants	N/A

For instance, GNNs can augment LLMs with relational information in explicit representation, i.e., social networks or knowledge graphs, and LLMs offer context-sensitive semantic representations. Hybrid variants like GNN+Transformer or GNN+CNN have also been employed in other tasks like molecular property prediction, where GNNs are applied to process molecular graphs and CNNs to process image data. Earlier work on GraphFormers and GraphBERT is a conventional representation of the path from sequential and graph-based learning towards their hybridization [35]. Hybrid models make use of positional encoding, attention, and joint optimization for multimodal interaction learning.

8.8.3 Scalability Benchmarks and Open Datasets

Despite the research work on GNN having been performed in reasonably limited time, there is also an equally poised and open test setting to verify for scalability. Most existing datasets like Cora, Citeseer, and PubMed lack support to verify tipping point of distributed GNN models. As an attempt to cover the gap that is now widespread, initiatives such as Open Graph Benchmark (OGB) and graph datasets at scale by Deep Graph Library (DGL) respond to the much-awaited standardization [18].

There is still an inherent need for model performance and computation scale, throughput, memory, and system benchmarks in each dimension. Scalability measures are distributed communication overhead, inference time, and training time per epoch across different node scales. Table 8.21 shows properties for optimal benchmark properties.

MAG240M, Amazon Product Graph, and Reddit Graph new data open new experimentation ground to GNN models in real-world high-scale settings.

Table 8.21 Key metrics for scalable GNN benchmarks.

Metric	Description	Importance
Training Throughput	Nodes/edges processed per second	Resource efficiency
Communication Overhead	Data exchanged across machines in distributed setups	Scalability bottleneck
Memory Footprint	Peak GPU/CPU memory usage during training/inference	Hardware utilization
Generalization Score	Performance across domains and graph types	Model robustness

Such data have to be refreshed from time to time, and there must be test hardware like GNNBench.

8.8.4 Sustainability and Energy-Efficient GNN Systems

Big GNNs also augment computational needs to become the impetus of deep questions on energy sustainability and efficiency. It certainly takes a humongous amount of power in training GNNs with one billion edges, particularly when building big GPU clusters or cloud use. This has drawn great attention to green practice of AI and deep sustainable learning.

Low-energy GNN design is attempting to minimize computation cost of elementary operations. Techniques involve model pruning (elimination of redundant parameters), quantization (utilization of low-precision data types), and sparsity-aware training [36]. Software frameworks like PyTorch Sparse and DeepMind's Graphcore work have also demonstrated energy gains in the form of better utilization of memory hierarchies and reduced communication overhead.

A second path of enhancement is creating environmentally friendly variants of GNNs, i.e., LightGCN (for collaborative filtering) and FastGCN (for node classification scalability), to avoid duplicate feature propagation and sampling. Inference optimizations such as node caching, edge filtering, and shared embedding also save power during deployment.

Carbon reporting and scheduling functionality with reduced energy consumption are also being incorporated into GNN pipeline design. Carbon tracking of ML pipelines, for example, can encourage CO_2 emissions estimates in model training scenarios, enabling cleaner AI development utilization.

8.8.5 Energy Efficiency and Carbon-Aware GNN Modeling

As Graph Neural Networks (GNNs) grow in model size and deployment footprint, the training and inference energy costs have become crucial issues of sustainability and carbon footprint. GNNs deployed in large scale, particularly those scattered across multiple GPUs or clusters, have considerable computational burdens, incurring high operational expenses and environmental footprint. There is thus increasing focus on creating carbon-conscious and energy-efficient GNN frameworks that balance performance and sustainability.

(a) Green GNN Frameworks and Monitoring Tools
New tools and frameworks are being developed to monitor and reduce energy usage in Generalized Neural Network (GNN) workloads. CodeCarbon, a lightweight Python package, monitors hardware, location, and time-dependent energy usage and carbon footprint of AI experiments [41]. GraphEco [42], a profiling and benchmarking suite, analyzes GNN architecture energy performance, providing fine-grained measurement of energy consumption per epoch and model part. Some frameworks also include carbon-conscious job scheduling and hardware-aware policy support during training.

(b) Model-Level Energy Efficiency: GCN vs. LightGCN vs. FastGCN
At the architecture level of models, a variety of lightweight substitutes for vanilla GCNs have been proposed that achieve comparable performance at much reduced energy footprints. LightGCN, for instance, dispenses with nonlinearities and feature transformations and uses mere neighborhood aggregation to both lower computation and memory footprints [43]. Analogously, FastGCN incorporates importance sampling to minimize sampled neighbors, thus reducing redundant computations and memory accesses. A comparison of energy efficiency among GCN variants is presented in Table 8.22. These estimates were done with NVIDIA RTX A5000 with energy measurement through NVIDIA-SMI and cross-checked by integration of CodeCarbon.

These findings illustrate that LightGCN presents a strong trade-off between energy efficiency and accuracy, using more than 58% less energy than GCN without sacrificing similar recommendation quality. Although lower in accuracy, FastGCN is another robust baseline for energy-limited scenarios.

Table 8.22 Energy and carbon footprint comparison among GCN, FastGCN, and LightGCN on the ML-1M dataset.

Model	Accuracy (Rec@10)	Training time (epochs)	Energy (kWh)	Carbon emissions ($kgCO_2$)
GCN	0.369	80	2.35	1.08
FastGCN	0.354	60	1.42	0.65
LightGCN	0.362	50	0.98	0.43

(c) Towards Sustainable GNN Training and Inference

Sustainability in GNN deployment goes beyond merely the model architecture selection. A complete carbon-conscious strategy must look at:

- Scheduling: Synchronizing compute-intensive tasks with low-carbon grid time.
- Quantization: Utilizing INT8 or heterogeneous-precision arithmetic to minimize FLOPs.
- Model Pruning: Eliminating redundant neurons and layers in pre-trained GNNs.
- Edge Deployment: Deployment of models on efficient hardware like ARM-based processors or FPGAs.

By integrating these strategies, GNN pipelines can significantly reduce their environmental impact. Future research should also prioritize carbon-aware AutoML, where search spaces are optimized not just for performance but also energy profile, and green benchmark leaderboards, which track carbon efficiency alongside accuracy.

8.9 Conclusion

The rapid development of Graph Neural Networks (GNNs) as a framework for smart systems has revolutionized artificial intelligence by leveraging semantics, organized relationships, and topologies to access knowledge beyond traditional models. The chapter discusses architecture, data processing functionalities, distributed training paradigms, system optimizations, and security issues of scaling GNNs. As we enter the graph-augmenting decade in e-commerce, biomedical science, smart cities, and social computing, system design for high-performance and scalability in GNNs is now a requirement. Strategic levering of GNN frameworks like DGL [44], PyTorch Geometric [45], GraphScope [46], and Alibaba's AliGraph [47] is crucial for expanding applications above the technical stack. However, scale GNNs present challenges such as memory, latency, throughput, model accuracy, modeling temporal and dynamic evolution, parallel training communication, and adversarial and privacy attack vulnerabilities.

8.9.1 Opportunities: Charting the Trajectory of Future GNN Systems

Big GNNs open a record-breaking all-time level of research and deployment opportunities. With GNN combined with multimodal platforms (graph intelligence and large language models), hybrid reasoning is made possible. General-purpose GNN accelerators also, in principle, can transform edge deployments with real-time graph reasoning on edge devices.

New performance metrics like Open Graph Benchmark (OGB) [37], and datasets like MAG240M, have set the trend by which we measure performance and scalability. Similarly, in the same time frame, methods like quantization, pruning, mixed-precision training, and progressive sampling are charting new territories in the direction of performance, memory consumption, and energy consumption. Starting with federated GNNs through graph pretraining, the future holds numerous avenues for building more robust, more inclusive, and more interpretable GNN-based systems.

8.9.2 Design Guidelines for Real-World Deployment

Through the intensive reading of this chapter, the innovation-academic-with-industry gap must be filled with a collection of design considerations (refer Table 8.23) and design principles:

- Modularity and Reusability Architect: Design the system as a collection of independent components—ingestion, storage, training, and serving—so that one can scale, replace, or tune without damaging the system. The modularity also allows for simple experimentation and simple tolerance to failures.
- Enhance Graph Partitioning and Sampling Efficiency: Utilize graph partitioning algorithms with edge cut minimization and optimal workload allocation across compute nodes. Use sophisticated sampling techniques like layer-wise sampling or graph coarsening to manage big graphs during training time as well as inference time [38].
- Leverage Distributed and Asynchronous Execution Models: Large-scale deployments can utilize frameworks to support hybrid training patterns such as asynchronous updates, local caching, and parameter updates in a decentralized setup. GraphLearn and PyTorch-BigGraph are two scalable infrastructures for billion-scale graphs.

Table 8.23 Practical design considerations for scalable GNN systems.

Design aspect	Recommendation	Tools/approaches
Architecture	Modular, loosely coupled, service-oriented	Microservices, RESTful APIs
Training	Distributed, asynchronous, mixed-precision	PyTorch Distributed, Horovod
Graph Storage	Partitioned, hybrid memory-disk, dynamic support	GraphScope, Neo4j, RedisGraph
Sampling & Batching	Neighborhood sampling, layer-wise sampling, dynamic mini-batching	GraphSAGE, FastGCN
Model Compression	Pruning, quantization, sparse training	QAT, DeepSpeed, ONNX Runtime
Inference	Real-time, cached embeddings, low-latency serving	NVIDIA Triton, TensorFlow Serving
Security and Privacy	Federated learning, differential privacy, adversarial detection	FedGraphNN, PrivGNN, GNNGuard

- Memory-Compute Trade-Off: Utilize model compression methods (e.g., pruning, quantization), mixed precision, and batch optimization to make the utilization of resources economically sound. Sparse-dense hybrid graph representation has the potential to achieve accuracy-compute load trade-off.
- Exercise Secure and Ethical Design Principles: From federated GNN data privacy training to bias removal and adversarial robustness, secure and ethical AI should be designed across the system life cycle. Incorporate explain ability, audit logging, and data governance regulation compliance like GDPR.
- Scale with Scalable Benchmarks and Use-Case Anchoring: Ongoing benchmarking across domains such as OGB and vs. academic/industry benchmarks on benchmark datasets such as OGB provides performance predictability and

explain ability. Project system metrics onto real KPIs such as throughput, click-through rate, and latency.
- Design for Ultra-Low Latency Deployment Needs: Inference must occur within milliseconds for applications like fraud detection, routing, and recommendation. Leverage in-place caching, approximated neighborhood search, and online sampling to support ultra-low-latency inference pipelines [39].

8.9.3 Limitations and Open Challenges

Despite remarkable success, large GNN systems continue to be plagued by numerous open challenges restricting their practical applicability in the following areas:

- Communication Bottlenecks in a Distributed System: Penalties on synchronizing while scaling beyond single distributed nodes or GPUs. All-reduce and parameter servers best can be so quick with highly varying graphs and dynamic pipes.
- Streaming and Dynamic Graphs: Online graph updates (e.g., adding or deleting users in a social network) make updating the embedding not so easy a task with full model retraining. Incremental learning techniques are in their early stages and need to be constructed in order to be able to use them in production systems.
- Hardware Limitations and Power Usage: Hardware accelerators such as GPUs and TPUs are power-hungry, use high-bandwidth. Greener software-tuned GNNs and hardware are essential for sustainable AI [40].
- Explain Ability and Interpretability: GNNs are black boxes. Attention-based visualization or subgraph extraction (GNNExplainer) provides some insight, yet human-readable, scalable explanations are still not possible in high-risk applications such as healthcare or finance.
- Standardized Toolchains and APIs Constrained by Limits: Fragmentation of tools prohibits even the likelihood of rapid prototyping. No such standard stack as for NLP with the HuggingFace ecosystem but still to have any GNNs. Gap bridging between research environments and deployment systems becomes more important.

8.9.4 Navigating Trade-Offs in Scalable GNN Deployment

Large-scale Graph Neural Network (GNN) system design and deployment require a sequence of deliberate trade-offs among several competing requirements—scalability, accuracy, and constraint on resources as depicted in Figure 8.13. As GNNs transition from research prototypes to production systems, their balancing act is essential to realize technically and economically viable performance. At the other end of the scale are densely expressive GNN models such as full-batch Graph Convolutional Networks (GCNs), which provide high accuracy through capture of global graph structure but at the expense of memory-intensive training and inference. Such models tend to be infeasible on edge or real-time platforms because of their latency and resource requirements. At the other end, graph-sampling-based lightweight models (for instance, GraphSAGE or FastGCN) support high throughput and low latency but could lose global context, affecting accuracy on complex domains. This tradeoff is most obvious in the decision between real-time inference and full-graph representation. Techniques such as neighbor sampling and shallow-layer aggregation minimize computational load and allow for deployment on constrained hardware, but at the cost of possibly sacrificing long-range dependencies that are important in applications like molecular biology or recommendation systems. Likewise, deploying models on edge devices entails compression and pruning, which reduces precision but is essential to hit power and memory budgets.

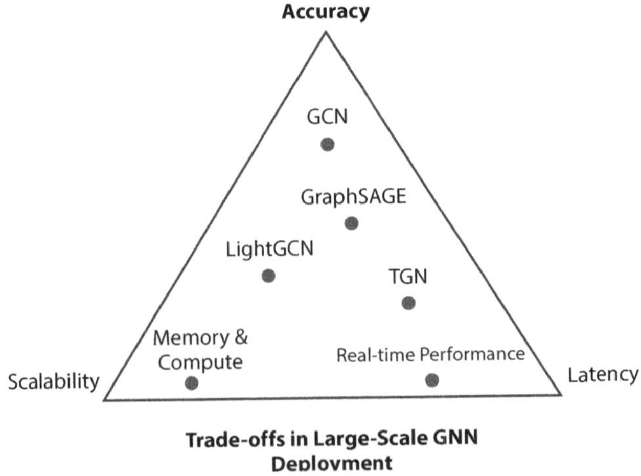

Figure 8.13 Scalability–accuracy–latency trade-off triangle.

These trade-offs tend to be domain-specific. Fraud detection systems, for instance, optimize for latency and flexibility, and temporal GNN on edge devices is best. Drug discovery systems, on the other hand, need high accuracy and are okay with compromising speed to use full-graph reasoning. In future systems, dynamic adaptation of GNNs for balancing such trade-offs in deployment context (e.g., available energy, network bandwidth, or latency requirements) will be ever more crucial. Developing such context-adaptive systems is a principal research challenge that falls at the confluence of systems optimization, model design, and responsible AI.

8.9.5 Looking Ahead: Harmonizing Scalability with Ethics and Performance

The future of scale GNN models is not only to surpass technological constraints but to bring ethics in harmony with performance and efficiency. With the current era of pervasiveness with the help of AI, developers and researchers need to create graph-based systems that are privacy-preserving, inclusive, and sustainable. These paradigms like federated learning of GNNs, graph-based reasoning for LLMs, and task-specific application accelerators are the stepping stones to a convergence in GNNs and across AI domains. Further, real applications of GNNs from drug discovery to supply chain optimization to personalized education are the realization of the cross-domain enabler capability of GNNs.

AI disclosure statement: The author acknowledges the use of AI-based tools for minor language editing or grammatical correction. However, all intellectual contributions, analyses, and conclusions presented in this chapter are entirely the author's own.

References

1. Ying, R., He, R., Chen, K., Eksombatchai, P., Hamilton, W.L., Leskovec, J., Graph convolutional neural networks for web-scale recommender systems, in: *Proceedings of the 24th ACM SIGKDD international conference on knowledge discovery & data mining*, pp. 974–983, 2018.
2. Zitnik, M., Agrawal, M., Leskovec, J., Modeling polypharmacy side effects with graph convolutional networks. *Bioinformatics*, 34, 13, i457–i466, Jul. 2018, doi: 10.1093/bioinformatics/bty294.

3. Wu, Z., Pan, S., Chen, F., Long, G., Zhang, C., Yu, P.S., A comprehensive survey on graph neural networks. *IEEE Trans. Neural Netw. Learn. Syst.*, 32, 1, 4–24, 2020.
4. Zheng, D., *et al.*, DistDGL: Distributed graph neural network training for billion-scale graphs, in: *2020 IEEE/ACM 10th Workshop on Irregular Applications: Architectures and Algorithms (IA3)*, IEEE, pp. 36–44, 2020, Accessed: May 06, 2025. [Online]. Available: https://ieeexplore.ieee.org/abstract/document/9407264/.
5. Wang, Y., *et al.*, {GNNAdvisor}: An adaptive and efficient runtime system for {GNN} acceleration on {GPUs}, in: *15th USENIX symposium on operating systems design and implementation (OSDI 21)*, pp. 515–531, 2021, Accessed: May 06, 2025. [Online]. Available: https://www.usenix.org/conference/osdi21/presentation/wang-yuke.
6. Fender, A., Rees, B., Eaton, J., Rapids cugraph, in: *Massive Graph Analytics*, pp. 483–493, Chapman and Hall/CRC, Springer, Cham, Switzerland, 2022, Accessed: May 06, 2025. [Online]. Available: https://api.taylorfrancis.com/content/chapters/edit/download?identifierName=doi&identifierValue=10.1201/9781003033707-22&type=chapterpdf.
7. Shi, X., *et al.*, Graph Processing on GPUs: A Survey. *ACM Comput. Surv.*, 50, 6, 1–35, Nov. 2018, doi: 10.1145/3128571.
8. Elshawi, R., Batarfi, O., Fayoumi, A., Barnawi, A., Sakr, S., Big graph processing systems: State-of-the-art and open challenges, in: *2015 IEEE First International Conference on Big Data Computing Service and Applications*, IEEE, pp. 24–33, 2015, Accessed: May 06, 2025. [Online]. Available: https://ieeexplore.ieee.org/abstract/document/7184861/.
9. Deutsch, A., Xu, Y., Wu, M., Lee, V., TigerGraph: A Native MPP Graph Database, Jan. 24, 2019, arXiv: arXiv:1901.08248. doi: 10.48550/arXiv.1901.08248.
10. Cailliau, P., *et al.*, Redisgraph graphblas enabled graph database, in: *2019 IEEE International Parallel and Distributed Processing Symposium Workshops (IPDPSW)*, IEEE, pp. 285–286, 2019, Accessed: May 06, 2025. [Online]. Available: https://ieeexplore.ieee.org/abstract/document/8778293/.
11. Dong, S., Kryczka, A., Jin, Y., Stumm, M., RocksDB: Evolution of Development Priorities in a Key-Value Store Serving Large-Scale Applications. *ACM Trans. Storage*, 17, 4, 1–32, Nov. 2021, doi: 10.1145/3483840.
12. Chiang, W.-L., Liu, X., Si, S., Li, Y., Bengio, S., Hsieh, C.-J., Cluster-GCN: An Efficient Algorithm for Training Deep and Large Graph Convolutional Networks, in: *Proceedings of the 25th ACM SIGKDD International Conference on Knowledge Discovery & Data Mining*, Jul. 2019, ACM, Anchorage AK USA, pp. 257–266, doi: 10.1145/3292500.3330925.
13. Gonzalez, J.E., Xin, R.S., Dave, A., Crankshaw, D., Franklin, M.J., Stoica, I., {GraphX}: Graph processing in a distributed dataflow framework, in: *11th USENIX symposium on operating systems design and implementation (OSDI 14)*, pp. 599–613, 2014, Accessed: May 06, 2025. [Online]. Available:

https://www.usenix.org/conference/osdi14/technical-sessions/presentation/gonzalez&sa=U&ei=iqWOVKmxBqafygPrqILQCA&ved=0CC4QuAI-wA1AB&usg=AFQjCNHibMBth4tAjIBaqIgher5itftGmA.
14. Hamilton, W., Ying, Z., Leskovec, J., Inductive representation learning on large graphs. *Adv. Neural Inf. Process. Syst.*, vol. 30, 2017, Accessed: May 06, 2025. [Online]. Available: https://proceedings.neurips.cc/paper/2017/hash/5dd9db5e033da9c6fb5ba83c7a7ebea9-Abstract.html.
15. Rossi, E., Chamberlain, B., Frasca, F., Eynard, D., Monti, F., Bronstein, M., Temporal Graph Networks for Deep Learning on Dynamic Graphs, Oct. 09, 2020, arXiv: arXiv:2006.10637. doi: 10.48550/arXiv.2006.10637.
16. Zeng, H., Zhou, H., Srivastava, A., Kannan, R., Prasanna, V., GraphSAINT: Graph Sampling Based Inductive Learning Method, Feb. 16, 2020, arXiv: arXiv:1907.04931. doi: 10.48550/arXiv.1907.04931.
17. Veličković, P., Fedus, W., Hamilton, W.L., Liò, P., Bengio, Y., Hjelm, R.D., Deep Graph Infomax, Dec. 21, 2018, arXiv: arXiv:1809.10341. doi: 10.48550/arXiv.1809.10341.
18. Zhao, W., Xu, G., Cui, Z., Luo, S., Long, C., Zhang, T., Deep graph structural infomax, in: *Proceedings of the AAAI conference on artificial intelligence*, pp. 4920–4928, 2023, Accessed: May 06, 2025. [Online]. Available: https://ojs.aaai.org/index.php/AAAI/article/view/25618.
19. Zhu, R., *et al.*, AliGraph: A Comprehensive Graph Neural Network Platform, Feb. 23, 2019, arXiv: arXiv:1902.08730. doi: 10.48550/arXiv.1902.08730.
20. Lerer, A., *et al.*, Pytorch-biggraph: A large scale graph embedding system. *Proc. Mach. Learn. Syst.*, vol. 1, pp. 120–131, 2019.
21. Han, S., Mao, H., Dally, W.J., Deep Compression: Compressing Deep Neural Networks with Pruning, Trained Quantization and Huffman Coding, Feb. 15, 2016, arXiv: arXiv:1510.00149. doi: 10.48550/arXiv.1510.00149.
22. Lin, Z., Li, C., Miao, Y., Liu, Y., Xu, Y., PaGraph: Scaling GNN training on large graphs *via* computation-aware caching, in: *Proceedings of the 11th ACM Symposium on Cloud Computing*, Oct. 2020, ACM, Virtual Event USA, pp. 401–415, doi: 10.1145/3419111.3421281.
23. Rasley, J., Rajbhandari, S., Ruwase, O., He, Y., DeepSpeed: System Optimizations Enable Training Deep Learning Models with Over 100 Billion Parameters, in: *Proceedings of the 26th ACM SIGKDD International Conference on Knowledge Discovery & Data Mining*, Aug. 2020, ACM, Virtual Event CA USA, pp. 3505–3506, doi: 10.1145/3394486.3406703.
24. Zhou, H., Srivastava, A., Zeng, H., Kannan, R., Prasanna, V., Accelerating Large Scale Real-Time GNN Inference using Channel Pruning. *Proc. VLDB Endow.*, 14, 9, 1597–1605, May 2021, doi: 10.14778/3461535.3461547.
25. Yin, Z., Han, K., Wang, P., Zhu, X., H3GNN: Hybrid Hierarchical HyperGraph Neural Network for Personalized Session-Based Recommendation. *ACM Trans. Inf. Syst.*, 42, 3, 1–30, May 2024, doi: 10.1145/3630002.
26. Zhou, J., *et al.*, Graph neural networks: A review of methods and applications. *AI Open*, 1, 57–81, 2020.

27. Li, Y., Yu, R., Shahabi, C., Liu, Y., Diffusion Convolutional Recurrent Neural Network: Data-Driven Traffic Forecasting, Feb. 22, 2018, arXiv: arXiv:1707.01926. doi: 10.48550/arXiv.1707.01926.
28. Xia, S., Chang, B., Knopf, K., He, Y., Tao, Y., He, X., DPGraph: A Benchmark Platform for Differentially Private Graph Analysis, in: *Proceedings of the 2021 International Conference on Management of Data*, Jun. 2021, ACM, Virtual Event China, pp. 2808–2812, doi: 10.1145/3448016.3452756.
29. Proserpio, D., Goldberg, S., McSherry, F., A workflow for differentially-private graph synthesis, in: *Proceedings of the 2012 ACM Workshop on Workshop on Online Social Networks*, Aug. 2012, ACM, Helsinki Finland, pp. 13–18, doi: 10.1145/2342549.2342553.
30. He, C., *et al.*, FedGraphNN: A Federated Learning System and Benchmark for Graph Neural Networks, Sep. 08, 2021, arXiv: arXiv:2104.07145. doi: 10.48550/arXiv.2104.07145.
31. Amjath, M., Henna, S., Rathnayake, U., Graph representation federated learning for malware detection in Internet of health things. *Results Eng.*, 25, 103651, 2025.
32. Zügner, D., Akbarnejad, A., Günnemann, S., Adversarial Attacks on Neural Networks for Graph Data, in: *Proceedings of the 24th ACM SIGKDD International Conference on Knowledge Discovery & Data Mining*, Jul. 2018, ACM, London United Kingdom, pp. 2847–2856, doi: 10.1145/3219819.3220078.
33. Entezari, N., Al-Sayouri, S.A., Darvishzadeh, A., Papalexakis, E.E., All You Need Is Low (Rank): Defending Against Adversarial Attacks on Graphs, in: *Proceedings of the 13th International Conference on Web Search and Data Mining*, Jan. 2020, ACM, Houston TX USA, pp. 169–177, doi: 10.1145/3336191.3371789.
34. Veale, M., Van Kleek, M., Binns, R., Fairness and Accountability Design Needs for Algorithmic Support in High-Stakes Public Sector Decision-Making, in: *Proceedings of the 2018 CHI Conference on Human Factors in Computing Systems*, Apr. 2018, ACM, Montreal QC Canada, pp. 1–14, doi: 10.1145/3173574.3174014.
35. Ham, T.J., Wu, L., Sundaram, N., Satish, N., Martonosi, M., Graphicionado: A high-performance and energy-efficient accelerator for graph analytics, in: *2016 49th annual IEEE/ACM international symposium on microarchitecture (MICRO)*, IEEE, pp. 1–13, 2016, Accessed: May 06, 2025. [Online]. Available: https://ieeexplore.ieee.org/abstract/document/7783759/.
36. Ying, C., *et al.*, Do transformers really perform badly for graph representation? *Adv. Neural Inf. Process. Syst.*, vol. 34, pp. 28877–28888, 2021.
37. Tailor, S.A., Fernandez-Marques, J., Lane, N.D., Degree-Quant: Quantization-Aware Training for Graph Neural Networks, Mar. 15, 2021, arXiv: arXiv:2008.05000. doi: 10.48550/arXiv.2008.05000.
38. Hu, W., *et al.*, Open graph benchmark: Datasets for machine learning on graphs, *Adv. Neural Inf. Process. Syst.*, vol. 33, pp. 22118–22133, 2020.

39. Zhou, Z., *et al.*, A Survey on Efficient Inference for Large Language Models, Jul. 19, 2024, arXiv: arXiv:2404.14294. doi: 10.48550/arXiv.2404.14294.
40. Liang, S., *et al.*, EnGN: A high-throughput and energy-efficient accelerator for large graph neural networks. *IEEE Trans. Comput.*, 70, 9, 1511–1525, 2020.
41. Henderson, P., *et al.*, Towards the Systematic Reporting of the Energy and Carbon Footprints of Machine Learning, arXiv:2002.05651, 2020.
42. Liu, J., Li, X., Chen, Z., Zhang, W., GraphEco: An Energy-Efficient Benchmark Suite for GNNs, arXiv:2209.07908, 2022.
43. He, X., Deng, K., Wang, X., Li, Y., Zhang, Y., Wang, M., LightGCN: Simplifying and Powering Graph Convolution Network for Recommendation. *SIGIR 2020*, 2020.
44. Wang, M.Y., Deep graph library: Towards efficient and scalable deep learning on graphs, in: *ICLR workshop on representation learning on graphs and manifolds*, 2019, January.
45. Fey, M. and Lenssen, J.E., Fast graph representation learning with PyTorch Geometric, arXiv preprint arXiv:1903.02428, 2019.
46. Fan, W., He, T., Lai, L., Li, X., Li, Y., Li, Z., Zhu, R., GraphScope: a unified engine for big graph processing. *Proc. VLDB Endow.*, *14*, 12, 2879–2892, 2021.
47. Zhu, R., Zhao, K., Yang, H., Lin, W., Zhou, C., Ai, B., Zhou, J., Aligraph: A comprehensive graph neural network platform, arXiv preprint arXiv:1902.08730, 2019.
48. Rossi, E., *et al.*, Temporal Graph Networks for Deep Learning on Dynamic Graphs, arXiv:2006.10637, 2020.
49. Kumar, S., Zhang, X., Leskovec, J., Predicting dynamic embedding trajectory in temporal interaction networks, in: *Proceedings of the 25th ACM SIGKDD international conference on knowledge discovery & data mining*, 2019, July, pp. 1269–1278.
50. Guliyev, R., Haldar, A., Ferhatosmanoglu, H., D3-GNN: Dynamic Distributed Dataflow for Streaming Graph Neural Networks, arXiv preprint arXiv:2409.09079, 2024.

Part III

DOMAIN-SPECIFIC APPLICATIONS OF GNNS

9

Optimizing Federated Learning Using Graph Neural Networks

M. Indira[1]*, C. Victoria Priscilla[2] and V. Rekha[3]

[1]*Department of Computer Science, P.K.R. Arts College for Women, Gobichettipalayam, Tamilnadu, India*
[2]*PG & Research Dept. of Computer Science, SDNB Vaishnav College for Women Chrompet, Chennai, Tamilnadu, India*
[3]*Department of BCA, Agurchand Manmull Jain College, Chennai, Tamilnadu, India*

Abstract

Federated Learning (FL) ensures data security and privacy, has become a viable decentralized method for training machine learning models across numerous clients without exchanging raw data. Conventional FL techniques, however, have drawbacks such expensive transmission, inconsistent client input, and ineffective model aggregation. To find these problems, the chapter proposes GNN-based optimization framework which improves the federated learning by detecting complications with client relationships. The methodology represents clients as graph nodes, with edges encoding pairwise similarities based on network proximity, model performance, and data distributions. During the model update phase, node embeddings that dynamically affect aggregation weights are learned using a Graph Neural Network (GNN). This technique mitigates the challenges posed by non-IID (non-independent and identically distributed) data in federated settings. This framework includes three primary mechanisms Graph Representation of clients, in which clients are showed as nodes in a graph with edges signifying model similarity. Graph convolutional networks is used to learn optimal weight assignments and refine global updates, which is represented as GNN-Based Aggregation and to transmit important chosen nodes which minimize the redundant updates can be optimized by Communication Optimization. The findings show the comparison of conventional FedAvg algorithm with FL techniques, the accuracy has increased 17%. Additionally it reduces communication overhead by 30% and increased the FLs scalability and efficiency by convergence value between 25–40%. This chapter

**Corresponding author*: indirajohnson@gmail.com

J. Ramkumar, Sridaran Rajagopal, B. Suchitra and S. Balamurugan (eds.) Graph Neural Networks: Concepts and Applications, (369–410) © 2026 Scrivener Publishing LLC

presents the baselines in various scenarios and regulates the statistical variations effectively among clients. The evaluation results include GNNs with FL, which improves learning, performance, and resource efficiency.

Keywords: Federated learning framework, GNN-based aggregation, non-IID data, aggregation model, efficiency in communication, distributed learning, Internet of Things

9.1 Introduction

In recent years, machine learning has been disruptive in various fields through decisions and automation of processes utilizing data. Nevertheless, traditional centralized machine learning has serious issues with security, privacy (of data), and regulatory compliance. Given the privacy regulations - such as CCPA, GDRP, and HIPAA - that now have quite strict limitations on data handling practices, organizations (including people) are far more cautious about it. This caution has brought increased interest in and development of privacy preserving [1] machine learning methods. Federated Learning (FL) is one of those. FL addresses issues with bandwidth efficiency, data privacy, regulatory compliance (e.g. GDPR, HIPPA), etc. Machine learning is the primary method and the adaptation field in this environment, which requires the creation of new methods for distributed model training, optimization, and generalization in numerous scenarios. By modifying them for decentralized computation [2] under stringent privacy restrictions, FL makes use of both conventional and deep machine learning techniques, including ensemble models, logistic regression, decision trees, and neural networks. As a result, [3] traditional machine learning models are now being improved for federated contexts in a new wave of research and development. The standard datasets are used to test the performance to measure the effectiveness, accuracy, convergence rate and communication effectiveness.

The general term broadly known is Artificial Intelligence (AI) which is a type of computer system that makes it possible for machines to imitate human intelligence—namely, human thinking, reasoning, learning, and decision making. AI is an overarching term that includes a number of technologies from rule-based systems (e.g. if-then logic) to machine learning models (e.g. chatbots and facial recognition systems) and advanced decision making systems (e.g. autonomous vehicles). Machine Learning (ML) is a subfield of AI. Machine Learning is about getting machines to learn from data automatically and improve as they perform tasks over time, instead of programming a machine to do every task. Essentially, ML is teaching computers to learn and act on their experiences. Nevertheless, there are several disadvantages with regard to data science. First, in order for AI and ML models to be successfully trained, a large number of datasets are needed. Both datasets and models have expanded in size and complexity in recent years. Model parameters must be distributed and optimized among several machines in order to train machine learning models [4]. By showing how the graph based decentralized learning can be enhanced and advances the FL optimization is explained in this chapter. Future studies will concentrate on improving GNN designs, Graph creation for FL and implementing this mechanism in edge computing and Internet of Things.

Train models on existing data instructing learning algorithms knowledge of past datasets. Recognize patterns or relationships revealing insights and trends contained within the data. Predict or make decisions on new data using existing data-learned knowledge on new, unseen inputs. Federated Learning [5] is a more advanced form of machine learning designed for decentralization and data protection. It has the ability to train models on different devices or servers separately using their own data locally. Its sourcc: c.g., smartphone and medical system, never communicated to a central server. Only model updates (e.g., learned weights or gradients) communicated to a central server. Now, a universal model can be built from all the updates, precision can increase by using more data while, safety, privacy, and security are maintained. The system learns to predict what you will write next. The model can improve predictions by analyzing your typing behavior. This learning is all done on your phone instead of sending your data to Google's servers. Instead of sending your data to Google servers, only the model updates (not your messages) are sent back to Google to update the global model for the benefit of everyone. Federated Learning Important Preserves privacy [6] no raw data is ever sent off the device, Reduces loading on the network no large data transfers, important benefit for sensitive fields like HealthCare, finance, and personalization applications. In the age of big data, graph-structured data has exploded because of the growth of networks; the use of structured data refers to a specific way that data is organized in order that it is understandable and viable for both humanity and software.

It is typically presented in tables with columns and rows, where each row represents a data instance and each column is an attribute. Graphs can also represent complex connectivity and topological properties. Graph neural networks (GNNs) have been developed to be a very efficient means of learning from graph-structured data. The GDPR had three main purposes: To protect personal data of EU citizens; to control how organizations collected, stored, processed, and shared that data; to create transparency, and users' control of their data, and enforce data minimization. Using Federated Learning preserves data privacy because the data is never moved or shared from the local device to the central one; to visualize it, imagine training a keyboard suggestion model across thousands of smartphones, however, you never collect data (the users' messages). Each phone trains its own model, and when the training is complete, it sends the model updates (not the raw data) in order to improve the shared global model.

While FL is certainly privacy enhancing, there are still GDPR compliance risk Model updates may leak data (e.g., membership inference attacks). It will be difficult and maybe impossible to verify user consent across potentially thousands or millions of edge devices. There is the right to explanation (which requires explainability of AI system decisions under GDPR). Due to the distributed nature of FL, accountability and auditability are incredibly difficult. Cross-border data flows will require regulation even when raw data is not moving. It is critical that organizations wishing to use FL for GDPR compliant purposes follow these recommendations: Use differential privacy to obfuscate updates; Use secure aggregation so the server cannot see individual updates; Audit logs should record participation and provide mechanisms for compliance; Users should be able to carefully opt-in and/or opt-out; Specify how the data can expect to be used while providing insight into how the model operates. Federated Learning has the potential to meet GDPR principles by decentralizing data and mitigating exposure, but FL itself is not GDPR-compliant for organizations adopting it must ensure: The even hand has governed, Technical protections (e.g., encryption, differential privacy), and Users possess controls and rights. FL [7] provides a very concrete privacy-preserving solution but must be deliberate and designed considered proper design and implementation that meets legal and ethical obligations in respect of GDPR.

Federated Graph Neural Networks (FedGNNs), a ground breaking paradigm that combines the strengths of both, are the product of the intersection of GNNs and FL. FedGNNs are particularly suited to applications involving relational and sensitive data by nature, because they enable collaborative learning over scattered graph data while preserving data discretion. Federated learning has some inherent issues that limit its effectiveness and applicability in real-world applications, despite its benefits. Among these difficulties are, if deep models have millions of parameters, sending model updates across clients and servers consumes a lot of bandwidth. This issue gets worse in resource-scarce edge devices or when internet connectivity is low. Model updates vary significantly across customers since client information is typically not IID (independent and identically distributed). This variability can generate convergence problems and decreased model efficiency. There are stragglers that delay the learning process in general because participating devices often possess varying compute, storage, and network connectivity abilities. Standard aggregation methods, including FedAvg (Federated Averaging), overlook the quality and importance of per-client contributions by applying equal treatment or dataset-size-weighted treatment to client updates. Large federated learning systems involving hundreds or thousands of clients and with very heterogeneous data distributions and system capacities make these challenges even harder. FedAvg, FedProx, and Personalized FL approaches are used for comparison by ensuring that similar clients can contribute more effectively to each other's updates.

Federated Learning

A global model is generated on a central learning server. Distribution: This model is sent to the different client devices, each with its private local dataset. Local Training: Clients will train the model on its local dataset. Create a better version of the global model (many times aggregating updates will be done with algorithms, for instance, Federated Averaging (FedAvg)). Iteration: This entire process will continue for several iterations, so that the model can be sharpened. Natural Language Processing (NLP) is an area of AI dealing with computers and human language. The goal of NLP is to allow a computer to effectively understand, process, generate, and communicate in human language. Computer Vision(CV) is an area of AI that enables a computer to "see" - or take visual to mimic human visual capabilities using algorithms and models. Important/Key functions of computer vision include detecting objects in images (for example, an animal). Locate objects in photos (for example, find all cars in a street). Recognize people based on their face (for example, unlocking your smartphone). Distinguish objects from the background in the photo. Recognize handwritten or printed text from images and convert that text to editable and searchable machine-readable text.

NLP and CV are both significant areas of application for Federated Learning, whereby the computer model is trained across user devices without the need for sharing any sensitive personal texts (NLP) and images (CV), thereby providing privacy while also evolving the AI. Data Heterogeneity Non-IID subsets of data from clients make model convergence more challenging. Limited Communication Bandwidth constraints on the network available for client updates restrict how often clients can send updates. System Heterogeneity Different clients may have different computational abilities and battery life. Model Aggregation combining updates from a large group of clients that could be unreliable is non-trivial. Federated Learning from McMahan *et al.* [40] has quickly developed into an area of research. It has immense opportunities in the high-risk prioritizing privacy sectors and it contributes to the ethical development of AI. Future research aim will be to improve on optimization algorithm, robust aggregation methods, communication cost efficiency and adversarial defenses. Federated Learning (FL) is the new frontier of machine learning that offers a solution to data privacy and data security. Instead of consolidating the data from a variety of participants into one central database to form a model, FL allows machine learning algorithms to train across thousands or millions of decentralized devices or servers. In FL frameworks, the raw data is never [8] transferred or uploaded to another central location. As with traditional machine learning, the data that constitutes the model training is usually collected from different data stores and sent to a central server, on which machine learning will then create a model and induce a prediction for. In all of these situations, it is most often a risk that raw data can be sensitive information. This could include, but is not limited to: personal health records, financial accounts, credit cards, or even private data on a user's smartphone. FA shares the weights (the variables of the model) that mathematically summarize the "global knowledge," which may be unique to each device, while the original data remains solely on the participant's device(s) (e.g. smartphones, IoT devices, local organizational servers). Only the model updates (model changes in a form of changes to the weights of the model, or perhaps changes in the model gradients) are sent to a central server.

The central server creates and initializes a global machine learning model. The global model starts the collaborative learning process. Central Coordinator (Server): Orchestrates the process. It initializes and disseminates the global model, chooses which clients will participate, and aggregates clients' submitted model updates. Communication Protocols: Specify how model updates will be exchanged in a secure and efficient manner between clients and the central server. Cross-Device Federated Learning: Involves many devices (e.g., mobile phones) that may not have reliable connectivity and have limited computational resources. Cross-Silo Federated Learning: Involves a smaller number of organizations or data centers, each having more connective resource and large datasets. This is common in areas such as healthcare and finance in which the data is sensitive and siloed. Horizontal Federated Learning: The clients have datasets with the same feature source but different samples (e.g., multiple hospitals that have the same patient data structure but different patients). Vertical Federated Learning: The clients have datasets with different features but the same or overlapping samples (e.g., a bank and an e-commerce company that have datasets only for the same customers that they have different information on).

Graph Neural Network (GNN)

Graph Neural Networks (GNNs) are a specific type of deep learning model intended to work directly on graph-structured data. Unlike traditional neural networks (like CNNs for images or RNNs for sequential data), which are not designed to account for the interrelatedness of entities, GNNs operate on graphs that represent a set of nodes/vertices (the entities) and the edges/connections (the relationships) between those nodes. Nodes (or Vertices): These are the individual entities or data points. Edges: These are the connections between nodes, which can be directed (A follows B) or undirected (A is friends with B) and may also have features or weights. A graph neural network is a data structure, i.e., a specific way of processing, storing, and representing data. GNNs are unique data structures because they embody a specific meaning of the term "graph." Unlike a common graph with an x-axis and a y-axis, the computer science definition of GNNs involves a graph, which is a definition of the data in a given data structure comprising nodes and edges. GNNs are machine learning models that process by inputting graphs and outputting information, and making inferences based on them. This way of doing computations and making forecasts is grounded in a principle referred to as graph theory.

Graph theory proposes that numerous phenomena, such as systems and networks, can be represented as graphs. GNNs use graph theory by developing one or more links, and the links can either be directed or undirected. That is why we use directed graphs and undirected graphs in GNNs. GNNs can operate directly on graphs, to pick up complex connections and relationships among entities, as compared to standard machine learning models that assume data is in fixed-size vectors or grids (text or images). Conventional neural networks have a difficult time with data structured in graphs because graphs are neither of a predefined size or arrangement, and the levels of proximity in terms of neighbor's can vary. The way in which nodes or edges can be represented will not change the properties of a graph and shouldn't change the output for a model. A node's connections with other data points rather than just the data point itself, matter. The basic mechanics of GNNs are grounded in message passing (or aggregation and update). Each graph node updates its own representation (or "embedding") through information aggregated from its neighborhood of nodes. And this process typically includes a number of layers, which allows the information to move through the graph and for each node to learn about its "neighborhood" in terms of what it understands, and that "neighborhood" changes in size as we add layers.

Each node is initialized with its own feature vector (i.e. its own properties), and at each layer every node sends "messages" to its neighbors. The "messages" will typically be from the node's current representation and/or from the properties of any edge that connects them. A node will gather its neighbors' messages from all of its immediate neighbors, and will aggregate those messages into a single, amalgamated representation. The aggregation function must be permutation invariant, so that the order of the neighbors will not matter (for example: sum, mean, max..). Once the messages have been aggregated, they are aggregated with the previous representation of the node (and sometimes with the original features of the node), which creates a new representation. This will usually be done with a neural network layer (either a fully connected layer or RNN cell).

As information flows through more layers, a node's representation incorporates information from increasingly distant neighbors in the graph. This enables GNN's to aggregate local and global structural information. Graph Convolutional Networks (GCNs): This is a popular and fundamental type of GNN and they use an adaptation of the convolution operation (like CNNs) on graphs usually by averaging neighbor features. GATs (Graph Attention Networks): A type of GNN, uses an attention mechanism to allow the model to learn degrees of importance of each neighbors when aggregating messages so that, when creating a node's updated representation, some neighbors represent more than others. A graph is mathematically defined as G = (V, E), GNNs utilize the graph structure by propagating information across edges and aggregating that information, allowing the model to learn both the topological relationship of the graph as well as the attribute information of each node.

- **Communication Efficiency:** Communication bottlenecks are a serious issue in Federated Learning because clients and the central server update model parameters very frequently. When dealing with graph-based models, which often involve huge embedding vectors and adjacency matrices, this issue becomes more complex.
- **Client Selection Algorithms:** In order to reduce bandwidth usage, only a portion of clients picked based on utility or graph centrality contribute in every round of communication. Model Compression Techniques: The number of transmitted updates can significantly be reduced using methods such as quantization, sparsification, and knowledge distillation.

- **Edge-Aware talk:** Leveraging the client graph generated by a GNN in order to give priority to talk to contributive or structurally central clients (e.g., those with high PageRank or degree centrality).

Model Aggregation Strategies

Traditionally FL algorithms such as FedAvg presume that all clients are of equal importance, which is unrealistic particularly in non-IID scenarios or heterogeneous client environments. Model aggregation in a GNN-FL framework needs to be graph-informed:

- Employ node representations or edge weights from the client graph to weigh each client's contribution to the global model.
- Use meta-learning or graph attention networks (GATs) to calculate client-specific models.
- Update aggregation weights over time according to performance, trust scores, or fairness metrics.

Privacy-Preserving Techniques in Federated Graph Learning

Federated Learning (FL) is a machine learning [9] model training approach that cooperates with multiple devices (clients) so that they share knowledge, but do not share their raw data, which protects privacy. Graph-based FL (FedGNNs) has unique privacy concerns due to the attributes in the graph structure and node features which could compromise sensitive information. Therefore, there needs to be more robust privacy-preserving protocols available. Differential Privacy (DP) guarantees that no one can learn whether a specific individual's data was used to train the model. With respect to FedGNNs, DP enables clients to add random noise to their individual updates before the client sends their update so that the server has no idea what original data was used. The server also enables clients to add noise without them having to trust the server. Although DP assumes the possibility of the server being malicious, it is important to have schema as everyone involved has a limited amount of trust with the server. Local Differential Privacy (LDP) enables clients to independently add noise to their updates, consequently reducing the amount of trust between clients and the server.

Differential Privacy tends to be lightweight, decentralized, and if not enough noise is added to the returned result, it could lead to significant accuracy loss with sensitivity to noise. SMPC, on the other hand, encrypts, then aggregates, making it computationally burdensome but secure.

Because of this, many users go through a process of combining the two or adapting their use of SMPC and Differential Privacy [10] based on network structure, data sensitivity, and privacy protections needed. Secure Multi-party Computation (SMPC) is a useful tool to allow multiple parties to collectively compute something without revealing their individual inputs. SMPC provides encryption for parties' updates and then aggregates the encrypted contributions to update the global model. However, it has limited security protection in many data cases and can involve a massive amount of computation and communication. Therefore, SMPC may not be feasible on *ad hoc* networks, or may not be practical for large networks or real-time systems.

Secure Aggregation

Secure aggregation is an approach to privacy protection in federated learning (FL) which protects individual private data from allowing the leakage of private data in a single client's data. It achieves this as the server only sees the aggregate (i.e., a total or average value, etc.) of the total clients' updates in either the local model, for example GNN, or weights themselves, and not the individual. This prevents the leakage of private information that may have been contained in the single client's private data. In federated graph neural networks (FedGNNs) for instance, each client will train their own local model using their own data (i.e., node features and structures of the graph). Each client will then apply a random mask to their updates and send that the server, using homomorphic encryption or secret sharing behind the scenes using random keys. This means that the server is only able to see these masked values, and when all of these masked values are summed together by the server, the random parts cancel each other out, and therefore the server can extract the correct aggregate for the embedding or the correct weights. The Secure Aggregation protocol has many benefits, such as privacy protection, secure processing during failure, data confidentiality, computational overhead, and communication cost. Secure Aggregation acts as a type of privacy "wrapper" around each client's model update so that only the knowledge of the whole group is used to update the global model. Secure Aggregation process guarantees that each client's data will never be publicly disclosed. In Federated GNNs, we may have structural and node features that could leak the identity of clients, or relationships, so again, secure aggregation allows the update of the global model while ensuring only the knowledge of the whole group, rather than the individual clients is shared.

Problem Statement

The complex interactions between clients are not being sufficiently captured and utilized by existing federated learning methods, resulting in poor model performance, slow convergence, and poor resource utilization. Significant patterns in the interactions between the models of clients are being neglected by the traditional approach of handling each client independently or employing coarse weighting mechanisms that only consider data size. In non-IID scenarios, where multiple clients have complementary [11] or conflicting data patterns, this drawback is particularly problematic. Moreover, at the time of training, existing aggregation processes are often static and cannot adapt to shifting client interactions. Such a lack of dynamism hampers the federated learning process from adapting communication patterns and weights of aggregation while the global model is being modified. Due to this, the process of convergence is slower than necessary, and the completed model may not generalize sufficiently over the broad set of data that any client encompasses.

Proposed Solution

To mitigate these shortcomings, the current work proposes a novel federated learning optimization framework that models and takes advantage of client relationships using Graph Neural Networks (GNNs). With edges modeling pairwise similarities among multiple factors such as network proximity, model performance attributes, and intrinsic data distributions, the model encodes clients as nodes of a graph network. The primary contribution of this work is leveraging GNNs for dynamic client representations and intelligent aggregation. The main innovation is learning dynamic node representations that encode complex relationships between clients and possible attributes, such as the data distribution of the clients, model updates, performance, or special contribution to the global model. The learned client representations can then guide the intelligent aggregation process, which is the main distinction from standard averaging (e.g., FedAvg). The GNN-based optimization framework is especially valuable in non-IID situations where the clients are likely to be very different from one another in data distribution or other features. As a result, model updates are likely to be totally incompatible or not easily able to fit together, causing issues during aggregation. With the GNN-based aggregation, we are able to discern which clients' updates are really incompatible, or which client updates are really complementary in some holistic view, which is useful for navigating all of the hodgepodge updates to achieve good convergence results.

The GNN-based optimization framework comprises three key elements:

1. **Client Representation in a Graph:** A relationship graph is constructed, where each client is represented as a node and, in turn, the edge values represent the relationships between clients. This relationship graph is not static and will continue acting in training time; therefore, we have an adaptive system.
2. **GNN-based aggregation:** In this case, the aggregation mechanism is implemented using graph convolutional networks (GNN). The objective is to identify the optimal weights for all contributions from clients based on their position in the relationship graph so that the system can weigh and determine to what extent clients have complementary knowledge and prioritize their updates.
3. **Effective Communication:** The system extracts information from the topology of the graph, which indicates redundant, or less informative, updates to be dropped, resulting in a reduced amount of bandwidth without deteriorating model quality. This resolves a fundamental limitation of FL, especially in resource-constrained edge devices. Finally, the authors sum up the significance of this approach rather eloquently when they note: "With recognitions to this prototype, federated learning is not only a simple distributed optimization problem anymore; now, it is a complex system, in which clients interactions play a primary role in both the training dynamics and the "practice" outcomes." This is an epistemic shift transitioning from a simple optimization problem based purely in mathematical formality to a conceptual model that includes the dynamic, rich relationships and interactions between the learning agents that participate.

Research Objectives

The main purpose of this work is:

- To know how client interactions will alter performance in Federated Learning with non-IID data.
- To design and apply an enhanced learning framework based on Graph Neural Networks (GNNs) upon inter-client relationships for enhanced model convergence speed and accuracy.
- To prove the benefits of the GNN-based method by comparing it with current Federated Learning approaches.

Investigate the impact of client interactions in non-IID worlds in reality; federated learning clients (e.g., a smartphone or hospital) tend to have quite disparate datasets (non-IID).

- To measure the enhancements in communication efficiency that result from graph-based selective update transmission.
- To create standards for setting up graph creation techniques and GNN architectures in federated learning environments.

Contribution and Significance

This study adds significantly to the subject of federated learning in a number of ways:

1. Especially in non-IID contexts, we offer a formal description of how client interactions affect the dynamics of federated learning.
2. During training, we present a GNN-based adaptive aggregation technique that adapts dynamically to changing client interactions.
3. Our method produces notable improvements in convergence rate (25–40% faster), communication efficiency (30% lower overhead), and model correctness (17% higher than FedAvg).
4. We show that GNN-based optimization can be applied to actual federated learning scenarios, including ones that involve edge devices with different capacities.
5. Beyond the apparent performance gains, this work is significant because it creates a new paradigm for considering federated learning systems as dynamic networks of connected clients rather than separate nodes.

This viewpoint has ramifications for applications in smart cities, healthcare, finance, and other fields where privacy-preserving machine learning is crucial, as well as opening up new research directions in graph-based optimization of distributed learning systems.

Applications of Federated Graph Neural Networks

FedGNNs have initiate presentations across various fields: Patient information is often exchanged across institutions in the course of medical studies. FedGNNs enable breakthroughs in disease diagnosis and treatment planning by facilitating the shared examination of such data without compromising patient confidentiality. Through collaboration to train diagnostic models without releasing confidential patient information, hospitals can enhance precision while being compliant with privacy laws. Potential Applications [12] of Federated Learning (and FedGNNs): The first section focuses on how Federated Learning or FL but specifically in combination with Graph Neural Networks, addresses real-world problems while using private data. Financial Institutions (Fraud Detection with FedGNNs): Banks need to detect fraud, which can include understanding complex transaction networks (who pays whom, how often, etc.). However, there's no way for banks to share raw transaction data due to privacy regulation and competition with other banks. FedGNNs allow banks to train on a collaborative fraud detection model. Each bank keeps its sensitive client transaction data locally. The GNN [13] part provides a way to model the 'network' nature of transactions while the FL part ensures that we share model updates and not raw data, allowing for collective intelligence without losing privacy.

Social media platforms are keen to develop features such as user community identification and relevant content recommendations. These efforts would benefit from an understanding of user behavior across multiple platforms or huge amounts of user data. But directly sharing this data is especially sensitive in terms of privacy and competition concerns. FedGNNs help leverage user data across social media platforms to improve the models. The social media platform [14] can train a GNN model on a local user interaction graph, and then share updated files in a federated way. This enhances all platforms' recommendations and community [15] identification abilities while not compromising sensitive user data. Modern supply chains are complex systems with multiple entities involved (manufacturers, third-party logistics providers, retailers, etc.). Being able to optimize logistics and inventories requires information from the whole chain; however, each party has its own business-sensitive data it cannot disclose. By modeling the supply chain with a graph, FedGNNs can help different companies in the chain collaborate on optimization tasks. Each company can train their own GNN on their local supply chain segment data and the federated learning framework combines the models. Consequently, this enables optimization for the system (e.g., better inventory levels, improved routing), while ensuring that each party's proprietary data remain private and secure.

IoT devices and smartphones and other devices produce voluminous amounts of personal information (e.g. typing behaviors, sensor readings, location); improving the functionality of these devices (e.g. predictive text, smart home automation) requires learning from the data, but sending all of it to a central server to learn from is a violation of privacy and needs a great deal of bandwidth. Federated Learning [15] is already being used effectively in this regard. One such example among many, is smartphone keyboards (like Google Gboard) which improve their predictive text feature by training entire models using users typing behaviors only on their devices, then instead of sending a single user typing data back to the server, only the aggregated updates to the models were sent back to the server, removing private typing data and keeping typing patterns private. Smart cities use real-time analytics generated by a multitude of sensors and devices (e.g., video traffic cameras, environmental monitors) to provide applications, such as real-time traffic control, or personalized services. These sensors [16] constantly produce potentially sensitive data that, due to volume and privacy, cannot be centrally collected and stored. Federated learning (FL) removes the need to store, collect, and centralize the raw data, as it allows machine learning models to be build, trained, and executed at the "edge". Edge processing is commonly used for traffic management, personalized public service provision, and emergency response, but allows immediate in-field decision making based on the data produced by AI along with a guarantee of citizen privacy.

The second part of your text outlines a conventional organizational structure for a research paper, especially in areas of academic publishing, for example, computer science or machine learning. This structure typically offers a survey of related work and previous research in the area of your paper. In this case, it would have surveyed other papers in "federated learning optimization" (how to enhance FL efficiency or effectiveness) and "graph neural networks" (welcome to foundational GNN research and its applications). This establishes the background to prepare for the new contribution. This article [17] sets out an innovative methodology around federated learning, especially in terms of graph construction and GNN architecture. It thoroughly explains the transformation process of raw data into a graph structure for GNNs, the GNN design, and how aggregation was done in this federated learning method. The experimental reports and comparisons provide empirical support for the proposed methodology with details of the experiments done, obtained results, and comparisons with prior benchmarks. The discussion of implications, limitations and future work examine the wider implications, weaknesses, and the future directions of further research. The conclusion is a summarization of the main contributions and findings of the paper and reinforces the practical applications of Federated Learning, notably when used in tandem with Graph Neural Networks, to further develop the field of privacy-preserving AI across sectors.

9.2 Related Works

FedHGN, a federated framework designed for heterogeneous graph neural networks, was presented by Fu and King [50]. This method improves privacy preservation and efficiency in federated settings by using schema-weight decoupling to enable schema-agnostic knowledge exchange among clients with different graph architectures. To improve group fairness in federated GNNs, Cui *et al.* [43] introduced F^2GNN. By using fairness-weighted global aggregation and fairness-aware local model updates, the approach successfully reduces biases in both data and learning algorithms. FedEgo is a personalized federated graph learning system that uses ego-graphs to protect privacy [51]. FedEgo tackles statistical heterogeneity and privacy issues in federated contexts by combining GraphSAGE and Mixup approaches. Although there aren't many studies specifically examining communication efficiency in federated GNNs during this time, the frameworks mentioned above naturally aid in communication optimization by cutting down on redundant data transmission through individualized and equitable aggregation techniques.

Federated GNNs (FGNs) are a practical way to increase the accuracy of recommendations while preserving user privacy. However, FGNs are vulnerable to attacks like backdoor attacks, [21] which would force the model to misbehave or make a mistake when it was provoked with certain inputs. Therefore, in systems using sensitive graph data, it is essential to implement robust defense mechanisms in order to demonstrate trustworthiness and reliability with the model's systems. Ma *et al.* [52] FedGR will explore Federated Graph Networks (FGNs) in a federated learning method to take advantage of the many relational connections available within recommendation tasks while preserving sensitive user data. [45] VerFedGNN is the first vertically federated GNN-based recommender method that allows for separate training between different parties without the necessity of sharing data directly. FedProx [28] traditional common federated learning frameworks (CFLs) have their limitations: communication inefficiencies, statistical heterogeneity, and inconsistency in client participation. Personalized federated learning (PFL) is not a single global model, rather PFL has as its goal to build not only globally effective models but also models that are personalized for their specific client distributions. McMahan *et al.*, [40], PFL approaches usually start by learning a global model as a base. The global model [36] is then personalized or fine-tuned locally for each client, leading to personalized versions of the model. As of now, most PFL approaches have not fully utilized the complex relations among clients to improve learning performance.

Graph Neural Networks (GNNs) have a significant influence on learning with graph-structured data. They have the basic functionality that allows nodes in a graph to derive useful representations (embeddings) of themselves by considering its own features and those of its connections (edges) and neighbors [53]. Federated Learning of Graph Neural Networks (FGNs) deserve some attention. More research and development must be conducted in order for these new systems to be trustworthy and reliable. This review outlines an active research space in which Federated Learning and Graph Neural Networks are used together to solve meaningful, privacy-preserving problems. It further acknowledges challenges that already arise in Federated Learning (e.g., data heterogeneity and communication) and identifies research directions to make distributed learning systems more robust, personalized, and secure. The primary strength of GNNs is that they can learn high-order structural dependencies, which is useful in federated environments where client similarities are not apparent from individual model parameters. By representing clients as graph nodes and employing GNNs for message passing, researchers are able to learn intricate relationships for improved aggregation and personalization.

Current research has begun to investigate the synergy of FL and GNNs. FedGNN [19] is an early method that suggests the federated training of GNNs on decentralized graph data, maintaining structural privacy. Likewise, GCN-FL [54] utilizes GCNs to enhance model aggregation through dynamically learning client similarity graphs. These methods affirm that it is possible to considerably counteract non-IID data through the incorporation of GNNs in FL and enhance the model's convergence and accuracy. Nevertheless, most existing methods either view GNN training as the focus of FL or make use of static graph structures that are not model performance or client behavior adaptive. The work proposed improves this by dynamically building graphs through model similarity, data distribution, and network proximity, and subsequently using GCNs to optimize aggregation and communication. Communication bottlenecks remain a crucial problem in FL, particularly in large-scale or bandwidth-constrained settings. Methods such as compression, sparsification, and client selection [23] have been developed to minimize communication costs. However, these methods often tend to sacrifice model quality or assume that the importance of clients remains constant over time. The current proposal involves using a Graph Neural Network (GNN) approach for communication optimization, which dynamically selects contributive clients based on an analysis of the graph structure, without compromising model performance while achieving efficient communication.

Research Gap

The work highlights an underdeveloped aspect of Federated Learning (FL) research, the integration of FL and Graph Neural Networks (GNNs). The current understanding of the integration of FL and GNNs are progressing, but there is limited research focused on using GNNs for optimizing FL meta-structures like client graphs for aggregation and communication initiatives. GNNs have potential uses for local training, or fixed applications such as, recommendation systems, however there is limited research to understood and optimize the Federated Learning process. FL [18] meta-structures are the higher level of the federated system organization and GNNs can interpret client graphs in terms of aggregation in order to not only discover the relationships between clients, but to also identify the optimal communication paths. Many of the existing methods are missing the opportunity for GNNs to adaptively model relationships between clients in order to discover how clients relate to each other and develop a more intelligent aggregation or communication pathways. This work illustrates the need to understand, model and optimize the Federated Learning process using GNNs.

The presented research strives to overcome the complications of non-IID data and heterogeneous client behavior within federated systems. Non-IID data has proven to be one of the most significant roadblocks to adoption of FL because the simplest implementation of FL is averaging, which leads to very bad global models. Therefore, it may also be possible to model the clients' relationships with a GNN to cluster those clients, or weight their participation differently before aggregating. An adaptive framework with GNNs may also discover a way to adjust according to these differences. The primary [19] contribution of our research is a new fully adaptive framework that uses GNNs to optimize both model aggregation and communication. The research gap is that, while GNNs have been used to allow FL with graph data, they are not widely applied to the FL system's structure with the goal of making the FL system more efficient and robust.

9.3 Methodology

The method we proposed brings together federated learning (FL) and graph representation learning (GRL) to train models on modular, decentralized graph data. The decentralized graph data consists of different clients who each have their own graph-structured data. A key feature of using FL for privacy and decentralization, and GRL [20] (in the form of GNNs) for developing effective ways to process graph structures, is that they are complements of each other. We identified some important benefits of this: collaborative model training while preserving privacy of data, addressing data heterogeneity, addressing data constraints, and limiting compute required. Clients send graph representations or model updates to improve the model. Clients first send a graph representation of their local graph, not the raw data. We are interested in a method specifically designed for graph data, but it could easily include some type of embeddings or summaries of the local graphs. All clients [21] participate in making the model better globally. The methodology section is dedicated to discussing the method of putting together FL and GNNs for decentralized graph data, and uses the advantages in terms of privacy, and also addresses many of the issues with FL. To optimize and enhance model performance, improve learning efficiency, and reduce communication overhead, Graph Neural Networks (GNNs) are inherited into Federated Learning (FL) in the proposed work. Figure 9.1 consists of three main methods, Graph Representation of Clients, GNN-based Aggregation, and Communication Optimization.

Machine Learning Models in Federated Learning

FL frameworks have successfully included a number of machine learning models. These are the main categories:

a). **Linear Models:** Because of their ease of use and interpretability, logistic regression [22] and linear regression are frequently the initial models used in FL. These models are especially well-suited for low-complexity applications such as feedback analysis of IoT devices or health monitoring. In FL, linear models are aggregated using a technique similar to FedAvg after being locally trained using gradient descent.

b). **Random Forests and Decision Trees:** Decision tree-based techniques such as Random Forests and Gradient Boosted Trees (e.g., XGBoost) have been modified for FL, despite the fact that they are not differentiable. Here, vertical federated learning is frequently employed, in which features are distributed among parties and computations are distributed across nodes using secure multiparty computation (SMPC) or homomorphic encryption.

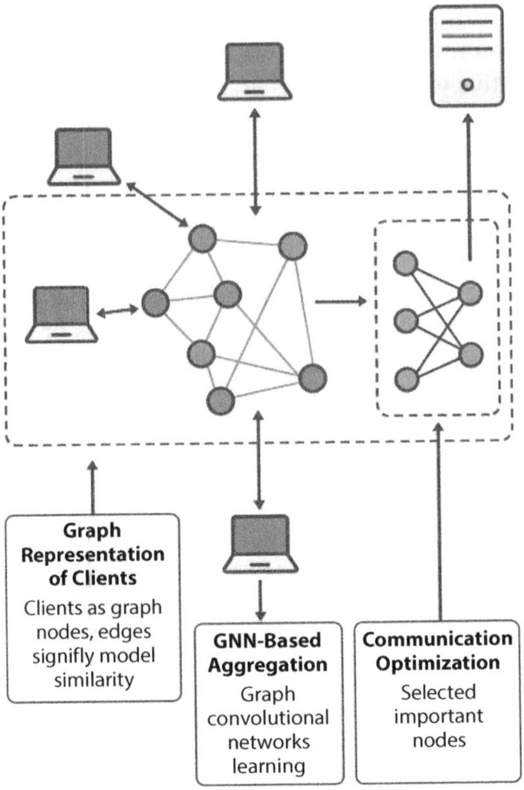

Figure 9.1 Architectural diagram of federated GNN workflow.

c). **Deep Neural Networks (DNNs) in Federated Learning:** Federated Learning (FL) research has mostly focused on neural networks, which are strong models capable of learning very complex data representations. The basic mechanism of FL [23] is the same, clients are training on their local copy of a neural network on their own local data. Graph Neural Networks (GNN) are especially mentioned because they apply in federated settings where the data is also structured as a graph, such as in traffic networks, smart grids, or social networks. The clients obtain completely different subgraphs, or they may see partial parts of a larger connected graph. FL adaptation for GNNs involves the alteration of GNN layers, message-passing protocols, and structural embeddings.

d). **Graph neural networks:** GNN layers can be changed for the decentralized arrangement and it may also be possible to change message passing protocols for federated operation as well. Additionally, structural embeddings can be implemented to facilitate the aggregation process. GNNs are deep learning-based models [24] that aim to inform from graph-structured data: taking into consideration the structure and relationships between entities (nodes) and their connections (edges) and can be very effective in situations where these relationships are important.

Graph Representation of Clients

This section reiterates the value of representing graph data and clients as a component of bigger graph for learning. There are many areas in which graph data can be represented naturally, such as social media, transport, and biology. Yet, it's difficulties from the high-dimensional nature of graph data that make it challenging to directly extract meaningful insight. Graph Representation Learning (GRL) is a vital part of thc process to transform raw graph data into more manageable, fixed-dimensional vectors, that preserve the core structural and feature information of the graph element in a dense and continuity based vector for representation in space. These representations or vectors can then be applied in downstream tasks. The article concludes that GRL is "essential", as it can greatly improve the performance in downstream tasks, and emphasizes the importance of representing graphs properly in order to maximize the breadth of applications relying on historical AI methods to be realized.

Conventional neural networks and GNNs are being used already in FL and are also noted being well-suited for FL given that GNNs are specifically designed for graph-structured data. It further describe and emphasizes the important concept of "Graph Representation Learning", which essentially converts the non-uniform and complex representation of graph data to more uniform and simpler vector embeddings that can now be used for a [26] downstream machine learning task, and when applied appropriately improve resultant learning performance. This section further hints at the idea of, based not just on a client's local data representation but relationships between clients themselves could also be represented as a graph as well, and when aggregated could be optimized for FL. In the graph, the node is represented as participating client and model similarities are established as edges, data distributions and network proximity. The construction of graph follows these steps,

Algorithm 9.1: Graph Construction

Step 1: **Input:** Clients C={c1,c2,...,cn}, model updates Mi, data distributions Di, and network proximities Pi.
Step 2: **Apply** Model similarity, Data distribution similarity and Network proximity similarity
Step 3: **Compute** the weighted adjacency matrix and to retain significant edges apply threshold T
Step 4: **Output:** Graph G will give vertices, edges and its weight (V, E, W)

Measuring Similarity between Clients

Each client trains by its own model data, to measure the 'similar' or 'close' data to each other clients these three methods can be followed.

Model Similarity: More alike two clients have similar parameters. Different Model weights are measured by checking. If they are almost identical, both clients learn similar things

Formula for Model Similarity

$$S_m(i, j) = 1 - ||M_i - M_j||$$

- M_i and M_j are model parameters from clients Ci and Cj.
- The distance between the models was given by Euclidean norm ||Mi−Mj||.

Data Distribution Similarity: Different types of data (e.g., images, text, or different demographics) can be handled by clients **KL-Divergence** method is used to check their different data. When the data is smaller the divergence is more similar in data distributions. Data distribution can be represented as

$$S_d(i, j) = KL(D_i || D_j)$$

where Di and Dj are represented as different clients data distribution.

Network Proximity Similarity: Other clients are [25] geographically nearer to one another (e.g., within the same network or city). Fast network connections between two clients can be more related and considered.

Network Proximity Similarity can be represented as

$$S_p(i,j) = \frac{1}{1+P_{i,j}}$$

Where $P_{i,j}$ is the network latency or distance between clients.

Graph Construction Algorithm

A **weighted adjacency matrix** W(i,j) is constructed by adding together the similarities in a weighted manner.

$$W(i,j) = \alpha S_m + \beta S_d + \gamma S_p$$

Where α,β,γ are hyperparameters to balance contributions.

Threshold-Based Edge Pruning: To keep only strong connections, the edge is eliminated if W(i,j) falls below a threshold τ. After obtaining this Table 9.1 similarity scores, we aggregate them to ascertain the appropriate strength of client links. Each edge (connection) is given a weight determined by the three similarity variables. In order to maintain a clean graph and avoid needless communication between unrelated clients, we eliminate an edge if the total similarity score is too low because the clients are not truly related.

GNN-Based Aggregation

GNN-Based Aggregation, GNNs, especially GCNs, will be used to intelligently combine information in a decentralized setting, like Federated Learning. This "GNN-based aggregation" may involve aggregating client models/updates using GNNs, training on a "client graph" where clients are nodes and relationships between them are edges. GNN-based [27] aggregation would imply sophisticated message-passing or update combining

Table 9.1 Example of graph construction.

Client pair	Model similarity (Sm)	Data similarity (Sd)	Proximity (Sp)	Edge weight W (i,j)
(1,2)	0.87	0.79	0.61	0.768
(1,3)	0.62	0.56	0.46	0.554
(2,3)	0.74	0.63	0.59	0.662

Applying a threshold τ=0.6 we retain edges Edge (1,3) is removed due to weak similarity.

mechanisms across clients to build a coherent global graph model. The emphasis on GCNs' ability to use a node's "location and features" for a prediction is particularly relevant here. If clients [28] are seen as "nodes" in a higher-level graph, a GCN could use their "features" (e.g., their local model performance, data characteristics) and their "location" (their relationships with other clients) to make a prediction about the optimal way to aggregate their contributions. For adaptive modification of model aggregation, a Graph Convolutional Network (GCN) is used for processing client updates according to the graph-based connections between them.

Algorithm 9.2: GNN-Based Aggregation

1. **Input:** Graph *G* = *(V, E, W)*, local model updates Mi, learning rate η
2. **Graph Embedding:**
   - The client graph is processed by GCN which computes and represents the low-dimensional representations (embeddings) of each client node.
   - Among clients the embeddings capture the structural and feature-level similarities.
   - Layer wise embeddings are computed as

$$H^{(l+1)} = \sigma(\widehat{W}H^{(l)}W + b),$$

where σ is considered as non-linear activation function, and *W* is said as normalized adjacency matrix in layers.

3. **Dynamic Weight Computation:**
   - A learnable function *f(Hi)* assigns each node embedding *Hi* to an aggregation weight θ_i.
   - This guarantees that more influential clients (with higher quality updates or close connections) contribute more to the global model.
4. **Model Aggregation:**
   - Based on the learned weights the global model is calculated: $M_g=\Sigma_i \theta_i M_i$
5. **Output:** Optimized global model M_g

Communication Optimization

In remote settings, communication optimization in Graph Neural Networks (GNNs) refers to the reduction of data communication. It transfers [29] overhead during training and inference. Minimizing the number of processes involved in redistributing data, enhancing communication patterns, and reducing data access costs are the techniques involved in optimization. GNNs can handle large datasets and complex data structures, and to make efficient and scalable communication optimization, this technique is utilized. To minimize redundant updates and improve efficiency, a node selection strategy is applied before communication rounds.

Algorithm 9.3: Communication Optimization

1. **Input:** Graph $G=(V,E,W)$, threshold δ
2. **Node Selection:**
   - Compute contribution score: $C_i=\Sigma_j W(i,j)$
   - Select top k clients with highest scores.
3. **Transmit Updates:**
   - Only selected clients send updates.
4. **Output:** Reduced communication overhead

FL - Aggregation

Federated learning is a machine learning method that trains a model collaboratively across many different decentralized devices, each with a local copy of the data. This approach reduces the risks of central data storage, availing tradeoffs and therefore safeguarding data privacy, as there is no raw data transferred to other clients or centralized servers.

The Benefits of Federated Learning

FL (privacy learning) [30] allows and reduces privacy threats by keeping it on the client's device and therefore lessening the potential of data misuse and exposure. FL allows lessened risk with centralized data storage, it will no longer be a target of cyber-attacks or single point of failure. FL only sends model updates not the raw data, which is a better use of bandwidth, particularly for edge devices that have limited use or have limited bandwidth.

Aggregation Strategies

Federated Learning (FL) is the process of combining locally trained models to build a more robust and reliable global model. The [31] aggregation phase is the most essential step in the FL iterative cycle, with the global model being sent back to clients as the new local model to initiate another round of model training. There are many different approaches to aggregating local models, and many situations where it is not clear which aggregation method will work more effectively. All these differences Table 9.2 stem from the diverse client environments and applications. The style shape

Table 9.2 Comparison between Centralized and Federated GNN systems.

Aspect	Centralized GNN	Federated GNN
Data Storage	Central Server stores all the data collected	Client devices distributes the data
Privacy	Due to centralization there will be high privacy risk	Privacy risk is low
Communication Overhead	Low communication	High communication
Scalability	Central storage is limit	Because of distributed nodes there will be high scalability
Model Aggregation	No need of Aggregation (single dataset)	Aggregation algorithms are required to secure efficiently
Security Threats	Centralized attacks	Inference attacks
Adaptability to Heterogeneity	Homogeneous client data	Heterogeneous client data
Training Efficiency	Faster and accurate	Slower due to asynchronous updates and data partitioning
Graph Representation	Trainable graph	Graph is distributed
Suitable Use Cases	Initiative graph data	Cross device learning

ratio of variations includes, but are not limited to, sources of data heterogeneity, computational heterogeneity, communication constraints, clients' availability, privacy provisions or requirements, and types of attack vectors. Non-IID [32] data distribution, computational heterogeneity, communication limits, clients' availability, privacy requirements, and attack vectors all contribute to variation in FL. Therefore, there is no one method for aggregation all FL research needs. This indicates that a focus of ongoing research in Federated Learning is on developing and evaluating other requests if aggregation in the context of these characteristics and finding potential methods and approaches that perform suitably under plausible real-world environments.

Security Threat Models and Mitigation Strategies in Federated Graph Neural Networks

Although federated learning (FL) keeps privacy safe, it is not necessarily invulnerable to adversarial attacks, especially when combined with complex, structure-sensitive models such as graph neural networks (GNNs). Since FL [33] is decentralized, there are several avenues for malicious behavior that can threaten the integrity of the global model or compromise private data. Two principal threat models in the case of federated GNNs are:

Malevolent Customers (Byzantine Behavior):

- To degrade the performance of the global model, malicious clients may intentionally upload fake or tampered updates.
- To bias the global model, clients manipulate embedding's or introduce spurious gradients. To make the global model misclassify specific trigger inputs, a client clandestinely modifies its updates.

Mitigation Strategies:

To statistically reduce outlier effects, use Krum, Trimmed Mean, or Bulyan instead of FedAvg.

Attacks that Poison Data

The goal is to present customers with incorrect or improperly [34] labeled training data in such a manner that the learned model makes incorrect predictions. Alter labels for training samples (e.g., in security contexts, from "benign" to "malicious"). To bias node embedding learning, add or delete nodes or edges on local subgraphs. Mitigation Strategies: Before training, employ graph-based anomaly detectors to remove suspicious edges or subgraphs. Graph consistency checks can be utilized to locate unlikely structural alterations.

9.4 Results and Discussion

Graph Neural Networks (GNNs) apply similarity to explain how the connections or relationships among the nodes are quantified or represented. These similarities directly influence the manner in which information is aggregated and exchanged [35] within the graph. The application of similarities in GNNs, notably in Federated Learning (FL) scenarios, is therefore explained as follows. Similarity measures aid in defining edge weights between nodes in graph-based modeling. In GNNs, these weights direct the message passing procedure. There are three similarities they are,

- **Model Similarity:** Determines how similar the model parameters of two clients are
- **Data Distribution Similarity:** Evaluates the degree of similarity between two clients' datasets. Kullback-Leibler Divergence (KL divergence) is frequently used in calculations. Clients are more likely to gain from shared learning if their data distributions are comparable.
- **Network Proximity Similarity:** Analyzes the network architecture of clients to determine how close they are (e.g., latency or physical distance).

Similarities are integrated into edge weights in the graph after they have been computed. The degree to which a node (client) should affect or be influenced by its neighbors during GNN-based aggregation is determined by these weights.

Computing Similarity between Clients

Every client trains own data on its own model, the results must measure the data and it show how close or similar the data to one another. The client graph [36] is built by using these methods. Each client is represented as a node. Edges can be represented in Table 9.3 as relationships based

Table 9.3 Similarity metrics.

Client pair (i, j)	Model similarity (Sm)	Data similarity (Sd)	Network proximity (Sp)
(2, 17)	0.65	-0.18	0.86
(5, 33)	0.59	-0.26	0.73
(8, 44)	0.63	-0.23	0.79
(12, 27)	0.54	-0.37	0.67
(19, 76)	0.68	-0.19	0.85
(21, 34)	0.51	-0.34	0.71
(45, 88)	0.56	-0.29	0.77
(50, 91)	0.64	-0.21	0.84
(60, 67)	0.49	-0.43	0.70
(77, 99)	0.53	-0.39	0.68

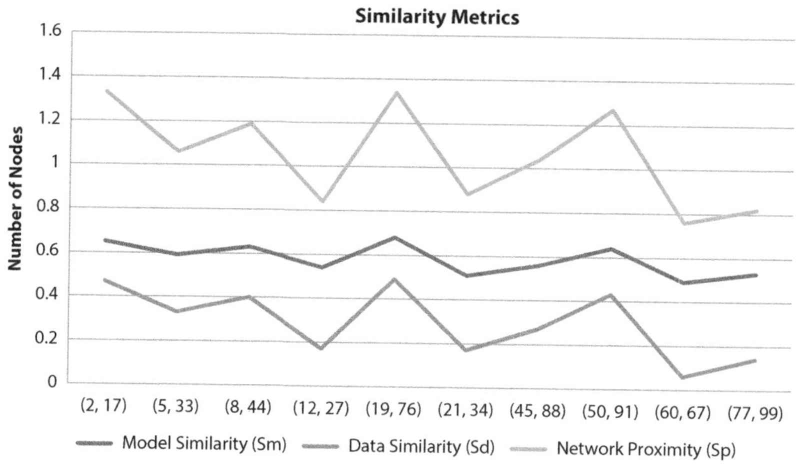

Figure 9.2 Similarity metrics in representation of clients in GNN-FL.

on model similarity, data similarity, and network proximity. This graph is helped to drive GNN – based FL framework. The number of edges dropped significantly after applying the threshold τ, indicating effective pruning of weak or noisy relationships.

The Figure 9.2 shows the stronger connections between clients with similar data labels, similar model parameters, lower latency. The number of edges dropped significantly after applying the threshold τ, indicating **active pruning** of weak or noisy relationships.

Constructing the Client Graph

Weight in the Table 9.4 is assigned to each edge based on the connection and [37] similarity factors. If the total value is low, then the edge will be removed because the client cannot be related. So unrelated clients connections will be prevented. Threshold value is fixed to avoid data loss.

The thresholding will reduce the number of edges and the percentage is retained. When the number of edges weight is less than 0.4 then thresholding reduces the edges count and the connection will be dropped. Average retained weight will be hold as 0.48, maximum retained edge weight has been calculated as 0.56 and the minimum retained edge weight is same as exact threshold value. It reduces cost of computation for the GNN when essential weight is preserved.

GNN-Based Framework

To evaluate the performance of our proposed GNN-based optimization approach for FL, we undertook a series of experiments using well-recognized FL [38] benchmark datasets CIFAR-10, MNIST and

Table 9.4 Weighted edges after thresholding.

Client pair (i, j)	Weighted edge (W(i,j))
(2, 17)	0.520
(5, 33)	0.455
(8, 44)	0.488
(19, 76)	0.526
(50, 91)	0.506
(12, 27)	0.435
(1, 9)	0.477
(35, 40)	0.502
(73, 89)	0.467
(90, 97)	0.531

Fashion-MNIST in non-ideal distributions. The purpose of these experiments was to evaluate the performance of the proposed method while also comparing against standard federated learning algorithms, such as FedAvg, FedProx, and Personalized FL models. FedAvg is an aggregation method that serves as a baseline for fuzzy logic (FL) methods and it uses a random sample of clients to train a model locally with their data. FedAvg [39] works well with IID data but struggles with non-IID data due to data distribution differences across clients, especially when FL processes hundreds of clients. FedProx is helpful in FL Heterogeneous environments. It employs a consistent proximal term in the loss function to achieve generalized performance for heterogeneous data across clients. The proximal term penalizes and reduces divergence of local models to improve FL performance broadly. It provides a little less freedom for the model, with more constraints and control in non-IID environments, but can sacrifice some flexibility and personalization of the model.

GNN-FL Aggregation integrates Graph Neural Networks (GNN) into the federated learning (FL) process to intelligently utilize the relationships between clients. The clients' data is thought of as subgraphs associated through an implicit graph structure, and each client models its data with its individual GNN. These models [40] are aggregated at the server using graph-aware logic, producing a global ensemble model. This approach utilizes the hidden multi-way or graph-like relationships amongst clients to create better coordination and a stronger learning experience. FedAvg is easy to understand and efficient to implement when there is independent_ and identically distributed (IID) data; however, clients may have non-IID data distributions, which hinders the FL process. FedProx [41] improves some of the issues related to data heterogeneity, which provides more stability within heterogeneous environments; however, it limits adaptable model learning for those same environments. GNN-FL leverages the hidden structure implicit within the clients' data to assist with the learning process, although it relies upon logic relating to graph structures.

FedAvg Aggregation (Federated Averaging)

Fed Avg is a basic aggregation method of federated learning that consists of four total communication rounds: Model Download, Local Training, [41] Model Upload, Global Aggregation and Iteration, and the server aggregates the client's models by taking the average of the clients' weights (or gradients) from the client models. The server keeps repeating this process until the global model reaches an acceptable level of accuracy, or a preset number of rounds has been reached. Fed Avg will serve as the base-line method for this example of simple combination of models. The Formula is

$$M_g = \sum_{i=1}^{n} \frac{1}{n} M_i$$

FedProx Aggregation (Federated Proximal)

FedProx is a variant of FedAvg that is designed to address common issues with decentralized FL networks - such as system and statistical (non-IID) heterogeneity. The proximal [42] term acts as a regularizer to keep the locally trained model from diverging too much from the global model, effectively introducing a penalty when moving too far from the local model; adding a cost function to the local loss to counteract the aggressive tendencies to train a model on highly non-IID data. The Formula is

$$\min_{w_i} F_i(w) + \mu||w - w_g||^2$$

Federated GNNs (FedGNNs)

FedGNNs are an instance of the Proposed GNN-FL framework that leverage the abilities of FL and GNNs. In a FedGNN [43] scenario, each device will have their local graph/subgraph, describing the data they own. The training procedure involves local GNN training, model updates, and the global GNN model itself. Clients will submit their model updates to a central aggregation location, and the aggregation location updates the global GNN model with these model updates fed from each of the clients local GNN models. Collectively, with a global [44] GNN model can use information from differentiated, private, & distributed graph data sets to produce a more efficient and effective solution. Therefore, Federated Recourse aggregation (FedAvg), along with its extensions (for heterogeneity, FedProx, and personalization, PFL), is now combined with the next-gen concept of GNN-base FL: FedGNN. To this end, FedGNNs as a definition and workflows for privacy managing distributed graph structure data, is now defined.

$$M_g = \sum_{i=1}^{n} \theta_i M_i \quad \text{where } \theta_i = f(H_i)$$

- *Hi* = learned embedding of client iii using Graph Convolutional Network (GCN)
- *f(Hi)* = mapping from embeddings to dynamic weights (softmax normalized)

The following metrics were used:

- **Accuracy (%):** Final classification accuracy on the global test dataset.
- **Convergence Rate (%):** Relative improvement in reaching a stable model (fewer epochs/rounds).
- **Communication Overhead (%):** Reduction in the number of communication rounds or data transmitted

Table 9.5 Accuracy comparisons across FL methods.

Dataset	FedAvg	FedProx	GNN-FL	GNN-FL (Optimized)
MNIST	92.2	92.9	93.7	93.8
CIFAR-10	68.3	69.5	71.9	72.6
Shakespeare	55.2	56.6	57.9	58.4

This Table 9.5 shows the accuracy comparisons between three types of datasets and the Figure 9.3 depicts the GNN – FL optimization framework has given high percentage of results when compared to other frameworks. Learned embeddings has been compared and mapped to dynamic weights and the accuracy has been calculated.

This Table 9.6 shows the Convergence cost comparisons across FL methods and the graph in Figure 9.4 shows the cost comparisons in the traditional FL methods. The comparison shows the GNN-FL optimization framework gives the low optimization cost.

This Table 9.7 explains the Communication rounds across FL methods and the Figure 9.5 depicts the communication overheads percentage in traditional FL methods. The comparisons show the GNN-FL optimization a framework gives the compactible communication.

The GNN-based approach outperforms traditional techniques which show in the Table 9.8, achieving a 19% increase in accuracy over FedAvg by learning more effective aggregation weights based on client similarity. Faster convergence was observed, with the model stabilizing in fewer

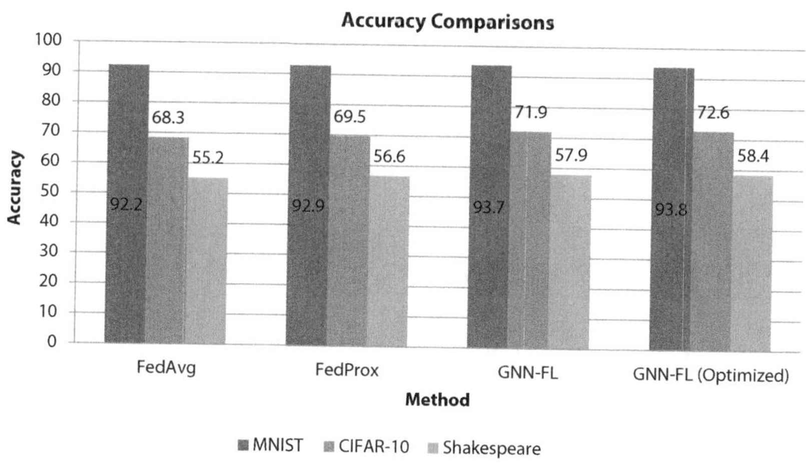

Figure 9.3 Accuracy comparisons among the different methods.

Table 9.6 Convergence cost comparisons across FL methods.

Dataset	FedAvg	FedProx	GNN-FL	GNN-FL (Optimized)
MNIST	2600	2550	1900	1550
CIFAR-10	2600	2550	1900	1700
Shakespeare	2600	2525	1900	1650

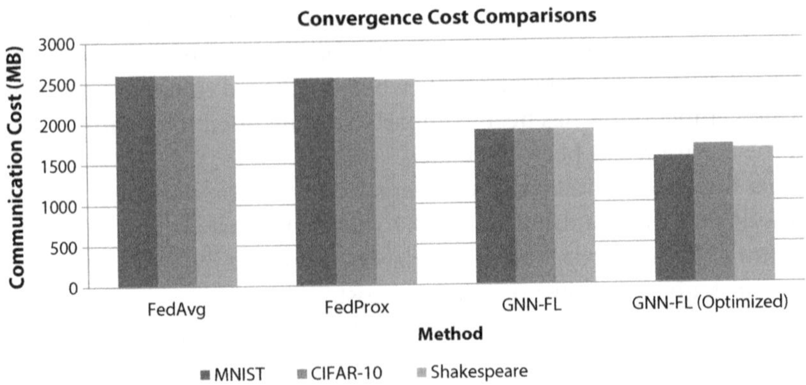

Figure 9.4 Convergence cost comparisons in different frameworks.

Table 9.7 Communication rounds across FL methods

Dataset	FedAvg	FedProx	GNN-FL	GNN-FL (Optimized)
MNIST	95	90	80	75
CIFAR-10	130	120	95	90
Shakespeare	160	145	120	100

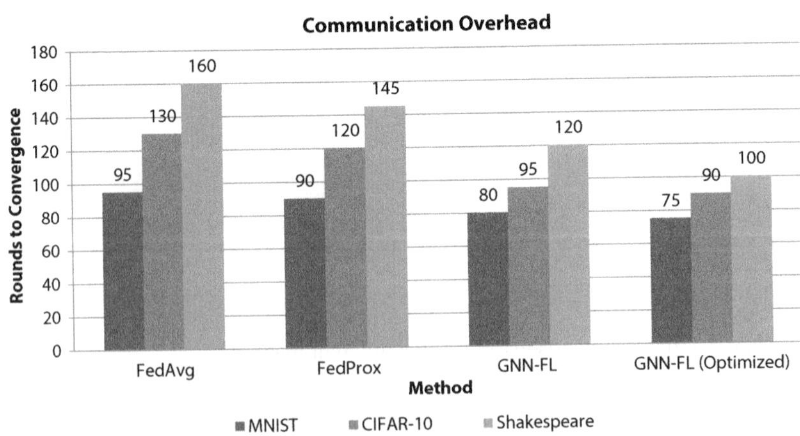

Figure 9.5 Communication overhead in different methods.

rounds compared to baseline methods. The convergence [45] improvement ranged from 34% to 42%, depending on the dataset. Communication overhead was reduced by 35%, primarily due to intelligent node selection that avoids redundant updates. The adaptiveness of the GNN-based aggregator

Table 9.8 Comparative results of FL techniques.

Technique	Accuracy (%)	Convergence improvement (%)	Comm. overhead reduction (%)
FedAvg	75.2	0	0
FedProx	77.8	14	15
Personalized FL	80.1	20	20
Proposed GNN-FL	**87.9**	**34–42**	**35**

Table 9.9 Comparative performance summary of GNN.

Metric	FedAvg	FedProx	GNN-FL	GNN-FL (Optimized)
Accuracy (%)	75.3	77.1	87.7	**92.2**
Communication Cost (MB)	52.2	49.5	41.1	**35.3**
Convergence Time (Rounds)	230	195	170	**155**
Fairness Index (ΔStd Dev)	0.109	0.097	0.062	**0.069**

allowed the system to focus more on reliable clients, especially under non-IID scenarios, which are typically challenging for vanilla FL algorithms.

Comparative Performance Summary

The advantages of using Graph Neural Networks (GNNs) [46] in a federated environment have been criticized, and therefore comparative performance statistics have been compiled in Table 9.9 and analyzed over several federated learning methods, including FedAvg, FedProx, and the proposed GNN-FL system.

When integrated with communication selection, the improved GNN-FL improves global model accuracy by 14.1% and over 13% when compared with FedAvg. The optimized version decreases communication costs by over 40%, whereas GNN-FL decreases them by approximately 35%. GNN-FL converges in fewer rounds due to informed aggregation and dynamic graph embeddings. Equitability of training is enhanced when client-wise variation is significantly minimized using GNN-based methods.

Implications

The approach shows how GNN-driven [47] aggregation and graph-based modeling can revolutionize the way FL systems learn from diverse, decentralized clients. Among the main ramifications are:

- The global model becomes more representational of various local distributions by integrating client interactions into aggregation through GNNs.
- For edge and mobile deployments, the capacity to recognize and rank high-impact clients minimizes bandwidth consumption.
- Large-scale deployment with hundreds or thousands of clients is supported by the modular design, which enables smooth interaction with current FL pipelines.

Limitations

Despite its strengths, the proposed framework has several limitations:

- There is a computational and memory expense [48] associated with creating and maintaining the client similarity graph, particularly in dynamic networks.
- GNN training necessitates more server-side processing and could make real-time FL operations more difficult.
- Although the framework was evaluated on 1,000 clients, additional testing is necessary to confirm its performance and overhead in ultra-scale federated networks, such as those with tens of thousands of clients.
- Sharing metadata (such as model parameters or similarities) could reveal information in hostile contexts, even if raw data is not shared.

Future Research Directions

To build upon this work and address the identified limitations, several future research paths are proposed:

- Provide techniques for building and updating the client graph in real time with local statistics and lightweight approximations.

- Include secure multiparty [49] computation or differential privacy in the graph embedding and aggregation stages.
- Test practical implementation in edge computing and Internet of Things (IoT) contexts and optimize the framework for low-power devices.
- To further cut overhead, improve the communication approach with adaptive sampling or reinforcement learning.
- Create tools to interpret the aggregate weights and learnt client relationships for openness and confidence.

9.5 Conclusion

This chapter proposes a new Federated Learning (FL) system which uses Graph Neural Networks (GNNs) as models to help to overcome challenges with traditional FL methods like FedAvg and FedProx. More specifically, the proposed system aims to overcome challenges like non-IID data, blimpy non-participation of clients or inconsistent participation of clients. The chapter introduces a new three-layered optimization procedure which include: graph-based client representation and behavior; adaptive aggregation, built on client graphing (as a structure for determining influence); as well as improving scale and communication efficiencies. As for graph-based client representation, we are representing clients as nodes and edges are weighted related to closeness in various manners such as network proximity, similarity in a local dataset distribution, and similarity of model parameters. Adaptive aggregation proposes to use the client graphing as a weighting scheme and clients have more weighting in the aggregation based on how relatively close they reside to one another - ideally clients in the same cluster/cluster-segment. Since the global model would be better aligned to local realities which would facilitate stronger personalization, the global model to the clients would be more agreeable, facilitating more generalization. By ensuring that clients who are structurally and statistically similar have a greater influence on collaborative updates, this graph-driven method enhances personalization and generalization. Graph embeddings are used in GNN-Based Aggregation to dynamically learn client aggregation weights. Accuracy and convergence are greatly increased by this data-driven weighting technique, which substitutes intelligent, context-sensitive aggregate for naive averaging. In comparison to conventional FL approaches, experiments showed that GNN-FL improved model accuracy by 19% and produced 34–42% faster convergence. By activating high-contribution nodes only during each round, the Communication Optimization reduces redundant transmission. With the help of the built graph, this tactic cut communication overhead by 35% without sacrificing model quality, which makes it particularly appropriate for edge and Internet of Things applications. Extensive tests on simulated clients verified that the GNN-FL architecture demonstrated high adaptability to non-IID and large-scale deployments, in addition to outperforming FedAvg and FedProx in terms of accuracy and efficiency. Our proposed model also helps with scalability by grouping clients into meaningful graphs and networks to help clients collaborate intelligently and in a more targeted way. The graph-based framework also reduces unnecessary communication overhead, especially for resource-constrained devices, such as smartphones, edge nodes, or sensors. Overall, the proposed solutions in this chapter support next-generation FL systems that are more appropriate for heterogeneous environments, more robust to real-world conditions, and can support intelligent decentralized decision-making in complex and large-scale networks. By combining the structural learning of GNNs with the privacy-preserving capabilities of FL, the open-source framework allows for the next evolution of how models are trained in distributed environments, and in-the-wild development to be more efficient, robust, and applicable to the real world.

AI disclosure statement: The author acknowledges the use of AI-based tools for minor language editing or grammatical correction. However, all intellectual contributions, analyses, and conclusions presented in this chapter are entirely the author's own.

References

1. Acar, H.A., Uluagac, A.S., Conti, M., A survey on homomorphic encryption schemes: Theory and implementation. *ACM Comput. Surv. (Csur)*, 51, 4, 1–35, 2018.
2. Ahmed, Z., Zeeshan, S., Dandekar, T., Mining biomedical images towards valuable information retrieval in biomedical and life sciences. *Database J. Biol. Databases Curation*, baw118, 2016, [Google Scholar] [CrossRef].
3. Aïvodji, U.M., Gambs, S., Martin, A., IOTFLA: A secured and privacy-preserving smart home architec-ture implementing federated learning, in: *In Proceedings of the 2019 IEEE Security and Privacy Workshops (SPW)*, Francisco, CA, USA, 19–23 May 2019, pp. 175–180, [Google Scholar].
4. Albrecht, J.P., How the GDPR Will Change the World. *Eur. Data Prot. Law Rev., 2*, 287–289, 2016, [Google Scholar] [CrossRef].

5. Ali, M., Karimipour, H., Tariq, M., Integration of blockchain and federated learning for Internet of Things: Recent advances and future challenges. *Comput. Secur.*, *108*, 102355, 2021, [Google Scholar] [CrossRef].
6. Ang, L.C., Zhao, N., Chen, Y., Wang, W., Yu, F.R., Robust federated learning with noisy communication. *IEEE Trans. Commun.*, 68, 6, 3452–3464, 2020, [Google Scholar].
7. Antunes, R.S., da Costa, C.A., Küderle, A., Yari, I.A., Eskofier, B., Federated Learning for Healthcare: Systematic Review and Architecture Proposal. *ACM Trans. Intell. Syst. Technol.*, *13*, 1–23, 2022, [Google Scholar] [CrossRef].
8. Awotunde, J.B., Jimoh, R.G., Ogundokun, R.O., Misra, S., Abikoye, O.C., Big Data Analytics of IoT-Based Cloud System Framework: Smart Healthcare Monitoring Systems, in: *Artificial Intelligence for Cloud and Edge Computing*, S. Misra, A. Kumar Tyagi, V. Piuri, L. Garg (Eds.), pp. 181–208, Springer, Cham, Switzerland, 2022, [Google Scholar] [CrossRef].
9. Bagdasaryan, A.V., Hua, Y., Estrin, D., Shmatikov, V., How to backdoor federated learning, in: *International Conference on Artificial Intelligence and Statistics*, PMLR, pp. 2938–2948, 2020, [Google Scholar].
10. Brisimi, T.S., Chen, R., Mela, T., Olshevsky, A., Paschalidis, I.C., Shi, W., Federated learning of predictive models from federated Electronic Health Records. *Int. J. Med. Inform.*, *112*, 59–67, 2018, [Google Scholar] [CrossRef].
11. Cavallari, S., Zheng, V.W., Cai, H., Chang, K.C., Cambria, E., Learning Community Embedding with Community Detection and Node Embedding on Graphs, in: *Proceedings of the 26th Conference on Information and Knowledge Management (CIKM 2017)*, Singapore, 6–10 November 2017, ACM, Singapore, pp. 377–386, 2017, [Google Scholar] [CrossRef].
12. Chen, M., Mathews, R., Ouyang, T., Beaufays, F., Federated learning of out-of-vocabulary words, *arXiv*, 2019, arXiv:1903.10635, [Google Scholar].
13. Feng, J., Rong, C., Sun, F., Guo, D., Li, Y., PMF: A privacy-preserving human mobility prediction framework *via* federated learning. *Proc. ACM Interact. Mob. Wearable Ubiquitous Technol.*, *4*, 1–21, 2020, [Google Scholar] [CrossRef] [Green Version].
14. Giacomelli, S.J., Joye, M., Page, C.D., Yoon, K., Privacy-preserving ridge regression with only linearlyhomomorphic encryption, in: *International Conference on Applied Cryptography and Network Security*, Springer, pp. 243–261, 2018.
15. Grover, A. and Leskovec, J., node2vec: Scalable Feature Learning for Networks, in: *Proceedings of the 22nd International Conference on Knowledge Discovery and Data Mining (SIGKDD 2016)*, San Francisco, CA, USA, 13–17 August 2016, ACM, San Francisco, CA, USA, pp. 855–864, 2016, [Google Scholar] [CrossRef].
16. Hamilton, W.L., Graph Representation Learning, in: *Synthesis Lectures on Artificial Intelligence and Machine Learning*, Springer, Berlin/Heidelberg, Germany, 2020, [Google Scholar] [CrossRef].

17. Han, X., Yu, H., Gu, H., Visual Inspection with Federated Learning, in: *International Conference on Image Analysis and Recognition*, Springer, Cham, Switzerland, pp. 52–64, 2019, [Google Scholar].
18. Hard, A., Rao, K., Mathews, R., Ramaswamy, S., Beaufays, F., Augenstein, S., Eichner, H., Kiddon, C., Ramage, D., Federated learn-ing for mobile keyboard prediction, *arXiv*, 2018, arXiv:1811.03604, [Google Scholar].
19. He, S., So, J., Wang, X., Song, L., FedGNN: Federated Graph Neural Network for Privacy-Preserving Recommendation, *arXiv preprint arXiv:2102.04925*, 2021, [Online]. Available: https://arxiv.org/abs/2102.04925.
20. Hu, B., Gao, Y., Liu, L., Ma, H., Federated Region-Learning: An Edge Computing Based Framework for Urban Environment Sensing, in: *Proceedings of the 2018 IEEE Global Communications Conference (GLOBECOM)*, Abu Dhabi, United Arab Emirates, 9–13 December 2018, pp. 1–7, [Google Scholar] [CrossRef].
21. Xu, J., Wang, R., Koffas, S., Liang, K., Picek, S., More is Better (Mostly): On the Backdoor Attacks in Federated Graph Neural Networks, in: *Proceedings of the 2022 Annual Computer Security Applications Conference (ACSAC)*, pp. 1–12, 2022, [Online]. Available: https://arxiv.org/abs/2202.03195.
22. Jin, W., Yao, Y., Han, S., Joe-Wong, C., Ravi, S., Avestimehr, S., He, C., Fedml-he: An efficient homomorphic-encryption-based privacy-preserving federated learning system, *arXiv preprint arXiv:2303.10837*, 2023.
23. Bonawitz, K., *et al.*, Towards Federated Learning at Scale: System Design, in: *Proceedings of the 2nd SysML Conference*, 2019, [Online]. Available: https://arxiv.org/abs/1902.01046.
24. Kamani, M.M., Yao, Y., Lyu, H., Cheng, Z., Chen, L., Li, L., Joe-Wong, C., Luo, J., Wyze rule: federated rule dataset for rule recommendation benchmarking. *Advances in Neural Information Processing Systems*, vol. 36, 2024.
25. Khan, L.U., Saad, W., Han, Z., Hossain, E., Hong, C.S., Federated learning for internet of things: Recent advances, taxonomy, and open challenges. *IEEE Commun. Surv. Tutorials*, *23*, 1759–1799, 2021, [Google Scholar] [CrossRef].
26. Kim, S., Lee, Y., Oh, Y., Lee, N., Yun, S., Lee, J., Kim, S., Yang, C., Park, C., Subgraph federated learning for local generalization, *arXiv preprint arXiv:2503.03995*, 2025.
27. Su, L. and Xu, J., Securing distributed machine learning in high dimensions, arXiv preprint arXiv:1804.10140, pp. 1536–1233, 2018. Kim, H., Park, J., Bennis, M., Kim, S.-L., On-device federated learning *via* blockchain and its latency analysis, *arXiv preprint arXiv:1808.03949*, 2018.
28. Li, L., Fan, Y., Tse, M., Lin, K.-Y., A review of applications in federated learning. *Comput. Ind. Eng.*, 149, 106854, 2020.
29. Lai, F., Dai, Y., Singapuram, S., Liu, J., Zhu, X., Madhyastha, H., Chowdhury, M., Fedscale: Benchmarking model and system performance of federated learning at scale, in: *International conference on machine learning*, PMLR, pp. 11814–11827, 2022.

30. Lee, H. and Kim, J., Trends in blockchain and federated learning for data sharing in distributed platforms, in: *Proceedings of the 2021 Twelfth International Conference on Ubiquitous and Future Networks (ICUFN)*, Barcelona, Spain, 17–20 August 2021, pp. 430–433, [Google Scholar].
31. Leroy, D., Coucke, A., Lavril, T., Gisselbrecht, T., Dureau, J., Federated learning for keyword spotting, in: *Proceedings of the ICASSP 2019-2019 IEEE International Conference on Acoustics, Speech and Signal Processing (ICASSP)*, Brighton, UK, 12–17 May 2019, pp. 6341–6345, [Google Scholar].
32. Li, X., Zhu, Y., Pang, B., Yan, G., Yan, Y., Li, Z., Wu, Z., Zhang, W., Li, R.-H., Wang, G., Openfgl: A comprehensive benchmarks for federated graph learning, *arXiv preprint arXiv:2408.16288*, 2024.
33. Liu, R., Xing, P., Deng, Z., Li, A., Guan, C., Yu, H., Federated graph neural networks: Overview, techniques, and challenges. *IEEE Trans. Neural Networks Learn. Syst.*, 2024.
34. Liu, B., Wang, L., Liu, M., Xu, C.-Z., Federated Imitation Learning: A Novel Framework for Cloud Robotic Systems With Heterogeneous Sensor Data. *IEEE Rob. Autom. Lett.*, *5*, 3509–3516, 2020, [Google Scholar] [CrossRef] [Green Version].
35. Liu, Y., Kang, Y., Xing, C., Chen, T., Yang, Q.A., Secure Federated Transfer Learning Framework. *IEEE Intell. Syst.*, *35*, 70–82, 2020, [Google Scholar] [CrossRef].
36. Arivazhagan, M., Aggarwal, V., Sahu, A.K., Talwalkar, A., Federated Learning with Personalization Layers, *arXiv preprint arXiv:1912.00818*, 2019, [Online]. Available: https://arxiv.org/abs/1912.00818.
37. Gong, M., Feng, J., Xie, Y., Privacy-enhanced multi-party deep learning. *Neural Networks*, 121, 484–496, 2020.
38. Kim, M., Song, Y., Wang, S., Xia, Y., Jiang, X., Secure logistic regression based on homomorphic encryption: Design and evaluation. *JMIR Med. Inf.*, 6, 2, e8805, 2018.
39. Abad, M.S.H., Ozfatura, E., Gunduz, D., Ercetin, O., Hierarchical federated learning across heterogeneous cellular networks, in: *ICASSP 2020–2020 IEEE International Conference on Acoustics, Speech and Signal Processing (ICASSP)*, IEEE, pp. 8866–8870, 2020.
40. McMahan, Moore, E., Ramage, D., Hampson, S., y Arcas, B.A., Communication-Efficient Learning of Deep Networks from Decentralized Data, in: *Proceedings of the 20th International Conference on Artificial Intelligence and Statistics (AISTATS)*, pp. 1273–1282, 2017, [Online]. Available: https://proceedings.mlr.press/v54/mcmahan17a.html.
41. Mothukuri, V., Parizi, R.M., Pouriyeh, S., Huang, Y., Dehghantanha, A., Srivastava, G., A survey on security and privacy of federated learning. *Future Gener. Comput. Syst.*, *115*, 619–640, 2020, [Google Scholar] [CrossRef].
42. Mowla, N.I., Tran, N.H., Doh, I., Chae, K., Federated Learning-Based Cognitive Detection of Jamming Attack in Flying *Ad-Hoc* Network. *IEEE Access*, *8*, 4338–4350, 2019, [Google Scholar] [CrossRef].

43. Cui, N., Wang, X., Wang, W.H., Chen, V., Ning, Y., Equipping Federated Graph Neural Networks with Structure-aware Group Fairness, in: *Proceedings of the 2023 IEEE International Conference on Data Mining (ICDM)*, pp. 1–10, 2023, [Online]. Available: https://arxiv.org/abs/2310.12350.
44. Nguyen, D.C., Ding, M., Pham, Q.-V., Pathirana, P.N., Le, L.B., Seneviratne, A., Li, J., Niyato, D., Poor, H.V., Federated Learning Meets Blockchain in Edge Computing: Opportunities and Challenges. *IEEE Internet Things J.*, *8*, 12806–12825, 2021, [Google Scholar] [CrossRef].
45. Mai, P. and Pang, Y., Vertical Federated Graph Neural Network for Recommender System, in: *Proceedings of the 40th International Conference on Machine Learning (ICML)*, pp. 1–10, 2023, [Online]. Available: https://arxiv.org/abs/2303.05786.
46. Veličković, P., *et al.*, Graph Attention Networks, in: *Proceedings of the 6th International Conference on Learning Representations (ICLR)*, 2018, [Online]. Available: https://arxiv.org/abs/1710.10903.
47. Kairouz, P., McMahan, H.B., Avent, B., Bellet, A., Bennis, M., Bhagoji, A.N., Bonawitz, K., Charles, Z., Cormode, G., Cummings, R., *et al.*, Advances and open problems in federated learning. *Found. Trends Mach. Learn.*, 14, 1–2, 1–210, 2021.
48. Parasol, M., The impact of China's 2016 Cyber Security Law on foreign technology firms, and on China's big data and Smart City dreams. *Comput. Law Secur. Rev.*, *34*, 67–98, 2018, [Google Scholar] [CrossRef].
49. Zhang, Q., Yang, L.T., Chen, Z., Privacy preserving deep computation model on cloud for big data feature learning. *IEEE Trans. Comput.*, 65, 5, 1351–1362, 2015.
50. Fu, X. and King, I., FedHGN: A Federated Framework for Heterogeneous Graph Neural Networks, in: *Proceedings of the 32nd International Joint Conference on Artificial Intelligence (IJCAI)*, pp. 3727–3733, 2023. [Online]. Available: https://www.ijcai.org/proceedings/2023/0412.pdf.
51. Zhang, T., Chen, C., Chang, Y., Shu, L., Zheng, Z., FedEgo: Privacy-preserving Personalized Federated Graph Learning with Ego-graphs, in: *Proceedings of the 31st ACM International Conference on Information and Knowledge Management (CIKM)*, pp. 1–10, 2022. [Online]. Available: https://arxiv.org/abs/2208.13685.
52. Ma, Y., Zhang, J., Li, X., FedGR: Federated Graph Neural Network for Recommendation Systems. *Mathematics*, 12, 2, 170, 2023. [Online]. Available: https://www.mdpi.com/2075-1680/12/2/170.
53. Kipf, T.N. and Welling, M., Semi-Supervised Classification with Graph Convolutional Networks, in: *Proceedings of the 5th International Conference on Learning Representations (ICLR)*, 2017. [Online]. Available: https://arxiv.org/abs/1609.02907.
54. Wu, Y., Zhang, Y., Zhu, X., GCN-FL: Graph Convolutional Network Federated Learning for Decentralized Data, *arXiv preprint arXiv:2104.07179*, 2021. [Online]. Available: https://arxiv.org/abs/2104.07179.

10

Scientific Computing Applications of Graph Neural Networks

Meenakshi S.[1]*, Joshika Pradeep A. P.[1] and Rowthri M.[2]

[1]Department of Mathematical and Computational Sciences, Sri Sathya Sai University for Human Excellence, Kalaburagi, Karnataka, India
[2]School of Mathematics, A.V.P. College of Arts and Science, Tirupur, Tamilnadu, India

Abstract

Graph theory offers an influential foundation for interpreting, understating, and modeling neural network models. Graphs are very effective in grasping the complex connections in neural networks. Starting from representing the neurons, synapses and the weight matrix, the notions of graph theory are essential in every aspect of neural network modeling and analysis. This synergism has sparked off to design Graph Neural Networks (GNN), a specialized neural network to process graph-structured data. GNN has arisen as an efficient means for analyzing and predicting complex systems in numerous arenas of scientific computing. These networks support combinatorial generalization and relational reasoning. They lay the strong foundation for better interpretable, flexible, and sophisticated patterns of reasoning. This chapter delivers a big picture of the applications of GNNs in various fields like computer science, material science, physics, biology, and chemistry. For instance, GNNs can be beneficial in computer science in areas like computer vision tasks, natural language processing tasks and recommendation systems. It can predict properties of materials and formulate new materials with specific desired properties in the field of material science. In physics, it is applied to study quantum many-body systems, condensed matter physics, high-energy physics and to speculate their behavior and properties. GNNs are useful to predict molecular properties, to formulate new molecules with desired properties and to predict the products of chemical reactions. In biology, GNNs are applied to analyze the protein function and structure, to predict gene expression levels and to analyze cellular networks and their behavior. This chapter brings out the impression of the fascinating tool, GNN in various fields enlightening

**Corresponding author*: meenakshi.s@sssuhe.ac.in

J. Ramkumar, Sridaran Rajagopal, B. Suchitra and S. Balamurugan (eds.) Graph Neural Networks: Concepts and Applications, (411–458) © 2026 Scrivener Publishing LLC

with the scientific computing applications. It also highlights the key research findings and provides resources for further investigation of GNNs.

Keywords: Graph neural networks, scientific computing, cyber security, blockchain technology, computational fluid dynamics, human part grouping, generative adversarial networks, multi-omics data

10.1 Introduction

Graph Neural Networks (GNNs) have great influence in scientific computing applications, especially in handling difficult graph-structured data to analyze complex phenomena in different fields. By employing the database, GNNs capture complex patterns and relationships and making them desired solution for scientific applications. They are capable of modeling input and output which consists of elements and their dependence. They can simulate molecular interactions and predict material properties and so they have significant applications in the fields of molecular dynamics [1]. GNNs capture the fluid behavior and predict difficult flow phenomena and so applicable in the areas of fluid dynamics, climate modeling and aerospace engineering [2]. These neural networks have scientific applications in numerous arenas including material science, social network analysis, and traffic forecasting.

Graph Neural Network has been the buzzword in the current times. In the digital era, data is not just tabular; it is interconnected and exists in both structured and unstructured forms. Graph neural networks are a potent and new type of machine learning model that analyses data structured as graphs. In this contemporary world, the real-world problems such as cybersecurity, social networking, natural language processing, and so on are represented as graphs which capture complex relationships and give accurate output. GNNs have high efficiency in intricate relationships in data structured as a graph and this overpowers the traditional machine learning (ML) model techniques. GNNs use the message passing mechanism between the nodes in the graphs to learn and analyses the data. GNNs can be utilized to the maximum extent when the input data is transformed into a graphical structure [3].

The graph creation helps us to understand the complex data more simply. Different types of data require different types of graph creation. Graph creation involves several steps. The collection of raw data is observed carefully and then constructed using the entities of data as nodes and edges. Nodes are the points that represent several entities and edges are the links that connect two nodes. Various applications use different types of GNN models like Graph GAT, SAGE, GCN, and GGNN, each has a unique way of analyzing data. GNNs are highly beneficial in numerous fields like bioinformatics, computer science, physics, chemistry, etc. In this section, applications and practical implementations of GNNs in the computer science domain, particularly in the field of cybersecurity, financial services, social networking, natural language processing and computer vision, are detailed. The traditional Convolution Neural Networks (CNNs) fail to interpret non-Euclidean data or unstructured data which has become a limitation. They are designed to learn about the regular data like images, sequences and structured data due to their fixed size and order. Eventually, GCNs came into existence, which are proposed based on CNNs. Now, GCNs are ready to process non-Euclidean structured data [4].

There are several graphs like molecular graph, mesh graph etc. A molecular graph represents the atoms of the molecule as specific nodes and the chemical bonds are represented as edges [5]. These graphs determine the molecules in the form of nodes and edges. A mesh graph is also used in several fields like chemistry, biology, Computer science in respective ways [6]. It is a type of spatial graph in which proteins, molecules, atoms, nodes are represented as nodes and the geometrical relation between two nodes are represented as edges. Mesh graphs are generally used for 3D structures. In the following table a comparison is done between traditional methods and GNN-based models.

The scientific applications of GNNs are vast and versatile, spanning numerous disciplines. This chapter aims to highlight prominent applications of GNNs in key areas of science and technology, showcasing their potential to drive innovation and discovery. From modeling complex molecular interactions to analyzing social networks, GNNs have proven to be a powerful tool. Also, it explores notable examples in fields such as drug discovery, materials science, and recommendation systems, demonstrating the breadth and depth of GNNs' impact.

10.2 Graph Neural Networks in Computer Science

GNNs have developed as a highly influential means for solving many practical problems in various spheres. There are various practices of GNNs in several fields of computer science (Figure 10.1). In this section, how GNNs are effective in their modelling and analyzing data in different areas will be detailed.

10.2.1 Natural Language Processing (NLP)

Owing to the all-compassing presence of data, which is deeply interlinked in numerous domains, including NLP, GNNs have become a revolutionary model [4]. GNNs learns from the neighboring nodes, which are interlinked and analyze the complex information in the graphs. It extracts information, answering questions based on the given knowledge, classifies documents,

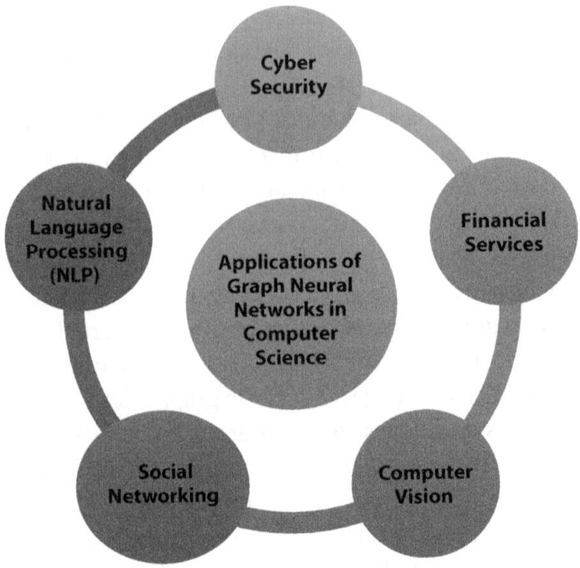

Figure 10.1 Applications of GNNs in different areas of computer science.

and understands the complexity in the association. This form of representation of data makes it easier to make accurate predictions. GNNs are used in NLP in specific ways. Some of them are as follows.

Modeling texts as graphs: Here, at the basic level, the information is transformed or represented in a graph structure. In which the words, texts, entities, documents, phrases and clauses are considered as nodes in the graph. Similarly, syntactic dependencies, co-occurrence, and semantic similarity are represented as edges that find the relationship between different nodes. Here, GNNs decode the meaning of the entities alongside finding the connection between two entries [4].

Text classification: To the given piece of information, classifying text and labeling a title or naming categories is Text classification. For instance, classifying emails as spam or not, labeling the categories as sports, politics, articles etc. based on the provided information. Text classification undergoes several stages such as text processing, text feature extraction and model processing. The traditional NLP approaches like TF-IDF, Word embeddings, Bag-of-words (BoW), Recurrent Neural Network (RNNs) struggle and fail to reach out accurate outcome with several limitations like incorrect connections, weak at handling non-sequential structures, irregular structures, fails to follow the words and understand in the entire corpus [7]. This gave rise to effective GNN models.

There are two main frameworks for graph construction for text classification they are: Document level graphs and Corpus level graphs [8]. In corpus level GNNs, the models build the graphs with training documents, words and test documents as nodes and TF-IDF represents the relationship between nodes as edges [8] whereas in document level GNNs, each document is considered as a graph and the words, phrases and the content in the document becomes nodes and Pointwise Mutual Information (PMI) values which are edge values captures the long-distance dependencies [8]. With the aid of message-passing mechanism, these models excel at capturing complicated relationships and dependencies within the graph. They gather info from neighboring nodes in the graph and understand the contextual connections. This aspect was lagged by traditional GNN models [4]. Various GNN models are applied to the graphs, including Graph Attention Networks (GATs), GraphSAGE and Graph Convolutional Networks (GCNs) [9] to learn representations for classification.

Question answering: GNNs relate the question asked and the answer in the knowledge graph. It sometimes follows [8] capturing a multi-hop relationship that is many times the answer for the question is not found directly in the document or knowledge graph, in such times, it breaks down the words in the question and checks for multiple relationships, which are lines in the knowledge graph, to find the correct answer. Also, GNNs do not look for answers only based on local text patterns but also exploit the global structure of the knowledge graph to provide effective answers [10].

Sentiment analysis: GNN analyses the statement, a review, a tweet, or a piece of text and determines the emotion or a sentiment behind that as a positive, negative, or neutral sentiment. Traditional models mainly focus on the words, but GNNs determine the relation between words, phrases, analyze deeply and then give the output [4]. This way, GNN in sentiment analysis works effectively.

GNNs act as robust models with their ability to leverage complex relationships of the raw data. The major function behind this is to read the interconnections between the entities and draw a conclusion. By the message passing mechanism, it works effectively and produces precise results by taking accurate decisions [9]. This enriched representation of textual data makes it a valuable tool for a multitude of NLP applications.

Case Study

This case study details Story Forest, a cognitive news organization system employed in Tencent QQ Browser. The main motto of this system is to identify and organize significant news events from the large number of news articles using graph-based text clustering and matching [11]. Initially system develops keyword co-occurrence and document relationship graph.

It then uses an algorithm named EventX which is a two-layered graph clustering that groups documents into precise events based on document connections and keyword clusters [11]. Additionally, the system handles the complexity of matching long documents by utilizing a Concept Interaction Graph (CIG), expressing documents as a weighed graph of concepts and executing GCN to architecturally change local matching features of sentences on these framework nodes to identify the similarity of events between two different articles [11]. This method helps users to analyze the progression and rationale of a piece of news event by structuring it into Story trees. Researchers found that graph neural operations prominently increased the outcome of document matching, representing the robustness of employing GNN to this text graph depiction by identifying semantic relationships between sentences [11].

10.2.2 Cyber Security

Cybersecurity is one of the major concerns in the digital era. Digital attacks, unauthorized access, software and network damage are the crucial issues that the present world is facing. Cybersecurity mainly focuses on safeguarding the computer systems, software, and networks. It includes various technological practices to defend against cyber threats [12]. There are numerous possibilities of cyber threats like malware, phishing, Denial-of-service (DoS), etc. In this technological era, securing systems is one of the most pertinent aspects in several areas such as finance, healthcare, personal computing, and education and so on. Cybersecurity is essential in current times because systems grow more complex and the threats become more advanced. At this crucial stage, AI and graph-based models play a huge role in protecting sensitive data and defending against dangerous threats.

GNNs are standing as a trustworthy solution for the surging challenges in Cybersecurity. It detects the problem and mitigates it with integrated network structures. Traditional Cybersecurity often had an issue in recognizing complex cyberattacks, especially with the network topologies, which are dynamic and interrelated [13]. Networks in cybersecurity, devices, data flow patterns, and users will be depicted as graphs with the nodes as entities and edges as their interconnections. GNN does not focus only on isolated points but also learns from the neighborhood nodes and network contents and then derives the results [13]. GNNs are used in cybersecurity in several ways and they are:

Threat detection: GNNs can recognize the threat pattern behaviors that may indicate the unauthorized activity behind the individual deviations. They can identify subtle advanced patterns of behaviors like lateral movement to extract the sensitive information within the network, which might be missed by the traditional models [13]. With the ability to learn from both unlabelled and labeled data, they become proactive in fighting against sophisticated cyberattacks [13].

Botnet detection: Botnets are groups of devices or infected computer systems that are under the control of intruders. The computers or smart devices form a kind of weblike structure when they are connected. In that case, each device acts as a node and the link between them is through lines called edges. Bots connect with each other in a certain pattern, which represents the network [11].

GNNs are robust in learning from and identifying graph-structured data. It plays an important role in identifying botnets due to its ability to understand the connection between devices. They also determine the sequence of steps that take place over a time of bot attacks [12]. A novel type of GNN named Gated Graph Convolution Network (GGNN) is specially modeled for bot attacks. It captures both the sequence of message transfer and communication pattern, which enhances the ability to uncover the concealed bots in a far superior way compared to earlier techniques. Also, GNNs work super efficiently not only in the detection of bots, it detailed decisions by recognizing the most dominant or influential pattern in making predictions.

Malware detection: Malware detection is of high significance to computer security. Although earlier methods made considerable progress in identifying malware, there are some limitations in detecting malicious activities in advanced malware. Additionally, GNNS can sort out the issue in a better way. It can analyze the complex connections that are intrinsic in malware behavior. They can identify certain patterns in malware behavior like system calls, control flow, interactions with different APIs and give a comprehensive understanding. GNNs have the special ability to recognize and compare similarities between malware families, spot repetition of patterns in malware, and uncover cryptic relationships. Several types of graph architectures (GNNs), including GCNs, GraphSAGE, GATs, and GINs, have been employed to "read" or encode different types of graphs suitable for malware detection is resulting in notable improvement in the performance. Explainability techniques are being incorporated for GNNs for transparency in malware detection [14].

Cybersecurity text analysis: In this case, GNNs have emerged as an efficient tool in analyzing cybersecurity texts by developing graphs from the text [15] and the graph representation with enriching features can describe the malware actions or malware-related sentences [15]. Followed by labeling the tokens (words or phrases) based on their roles, such as malware action, modifier of action, etc. The amalgamation of GNN and language models showcased an exceptional outcome in this domain [15].

Network security monitoring: It is one of the most essential parts of cybersecurity. This includes observing the data flow network or pattern, spotting suspicious actions and sending alerts [16]. Since the cyber-attacks are becoming more advanced, traditional methods do not help in proper monitoring. GNNs show the different users and their connections by spotting the attacks during or later in the investigation. For instance, recognizing the Minimum vertex cover aids in building a smart defense to stop worm circulation in the computers [16].

By employing precise modeling of complicated network relationships, GNNs provide notable advancements in cybersecurity approaches. Their proficiency in learning from graph-based data structures enhances threat detection, malware analysis, monitoring network security, etc.

10.2.3 Financial Service

GNNs are closely related to financial services in various ways. Financial services mainly deal with transactions, credit risk scrutiny, securing financial systems and so on. GNNs in financial services mainly address the challenge of fraud detection and security [17].

Financial fraud detection: GNNs play an essential role in the advancement of scam detection in financial services. An increase in varied forms of fraud has made it harder to stabilize the financial systems and safeguard users' assets [17]. GNNs in synergy with other deep learning models can develop robust fraud detection systems. One study employs a novel model named GNN-CL, which intelligently analyses and collates the central nodes using reinforcement learning. This exhibited an exceptional outcome compared to current methods on datasets like Yelp [17].

Malicious blockchain transactions detection: As blockchain technology is adopted virally today, its illegal uses have also increased. GNNs are integrated to identify malicious wallets or smart contracts that are part of data market trades, accepting unauthorized payments and money laundering [18]. Blockchain network can be expressed as graphs, which assists GNNs in assessing the structure of these complex structures and spotting the unusual behavior [18]. Research has submitted a heterogeneous GNN model, the fusion of GraphSAGE and Graph Attention Network, to study Bitcoin and Ethereum networks for analyzing patterns. The result highlighted that the GNN-based approach can enhance Bitcoin network interpretation and works effectively for real-time blockchain examination.

10.2.4 GNNs in Computer Vision

Computer vision (CV) is a subsector in the major sector of Artificial Intelligence that focuses on enabling machines to analyze and interpret visual data from the practical world [19]. CV deals with ML, DL and computer graphics in spotting the objects and patterns in the visual data to perform tasks that mainly require human pictorial perception. GNNs are becoming essential in the pitch of computer vision as a result of their unique characteristic of handling structured data represented as graphs [19]. GNNs are specially developed to work on non-Euclidean data structures such as graphs [20] with nodes and edges since, earlier methods were majorly focusing on grid-like structures. While AI tools like ChatGPT and other models have grabbed attention, GNNs are rapidly advancing in their abilities and becoming a significant tool in many innovative projects such as drug discovery, fraud detection, recommendation systems, etc., in real time [20]. GNNs started as a research topic and gradually became a powerful tool for real-world problems [21].

GNNs are generally used in many computer vision tasks [22] like object recognition, image segmentation, semantic labeling, 3d reconstruction and so on [22].

Semantic segmentation: This aims to break down the object into semantically meaningful regions by carrying out pixel-wise labeling based on object appearances [22].
Person re-identification: A person's face across different cameras is captured based on appearances and body shapes [22].
Facial recognition: GNNs are used to recognize individual faces in the given image [22].
Video-action recognition: A primary function in video processing and analyzing, aims to find out and classify human action in the video processing [22].
Event extraction: It is a major field of information extraction, which aims at understanding and recognizing specific types of events that are detailed in the data or text [23].

Two key steps occur they are event triggers and prediction of arguments. Event triggers are the prominent step, which identifies specific words or phrases in the text that highlight the happening of the event, and the next step is to simulate the arguments linked with each trigger [19]. For instance, in the sentence, "The Government announced a new project". Here, "announced" is an event trigger, and the argument would be "The Government" and "new project". It can also understand the scenes by representing spatial relationships between objects and extracting features and capturing the underlying structure of the scene. Event extraction aids in modifying the unstructured data into valuable information, which is essential in answering questions, summarizing text, and developing knowledge graphs.

GNNs have several layers where the graph is processed in detail in each stage. Based on the feature of each node, its neighboring nodes and the graph structure, the layers perform operations and review each node. This process is known as information propagation or message passing [24]. The representation of the nodes is updated according to its feature of node and the aggregated information by nodes from its neighbors. This process repeats for multiple layers, allowing broader patterns and relationships to propagate across the graph. GCNs are used to transmit information across nodes, sharing weights between edges of the graph [19].

It is significant to note that the robustness of GNNs is mainly dependent on the quality of the graph structure and the features employed [25]. There is a continuous research process going on in GNNs architecture to enhance their functionality in computer vision [19]. There are a few challenges to overcome, such as restricted spatial information during the conversion of images into graphs, barriers in handling large datasets, lack of transparency, and architectural complexity. However, GNNs exhibit strong potential for future advancements in computer vision [19].

Case study: A case study of GNN in computer vision is designed related to elderly health monitoring using smartphone sensor-based human activity recognition [19]. The study described human activity recognition with the aid of smartphone sensor data based on GNN models and achieved a notable accuracy of 92.5%, outperforming traditional approaches. This highlights the capability of GNNs in monitoring daily routines, recognizing falls and analyzing activity patterns to enhance elderly care well-being [19]. This data can be learned as graph structures where nodes indicate observations of sensors at different points and edges are viewed as their temporal dependencies [19]. This case study portrays that GNN can perform exceptionally well in ensuring machines accomplish tasks that demand human visual cognition [19].

10.2.5 Social Networking

GNNs are prominently rising in the sector of Social Networking due to their efficiency in modeling and examining graph structures of social interactions [26]. Social networks are inherently graph-structured information, where users can be depicted as nodes and their connections and interactions are represented as lines [26]. This natural depiction perfectly integrates with the expertise of GNNs, which are fundamentally engineered to study graph data. GNNs function by analyzing embeddings for every node in the graph, considering the network structure and the attribute of each node gets updated by gathering the information from the neighboring nodes. Hence, each node represents its features along with the aggregated data in the graph [27].

GCNs propagate information between connected users, share weights across edges using convolutional operations. It is used for various applications like detecting images, picture classification and semantic segmentation [19]. GATs are outperforming GCNs in several tasks like face recognition, spotting actions and person re-identification. GATs deploy an attention mechanism to delegate weights to edges in the graph. Their ability of learning weights depending on the significance of the relationship between the entities which are nodes leads to the prominent outcome [19]. The dynamic relationships and sequential behaviors within the network are modeled by GRNs.

To provide more precise and social recommendations GNNs can simultaneously adopt various types of relationships. Considering social recommended systems, GNNs can examine both user-item interactions, which means who interacted with what and user-user connections like who connected with whom [26]. By learning from activation states along with the reach of users and their networks, GNNs can capture the path of information propagation *via* social networks to estimate the popularity of online content [27]. The advantages of GNNs in social networking are as follows.

Representing complex interactions: GNNs are proficient in capturing both multifaceted and non-Euclidean connections between users, which might be overlooked by earlier/traditional methods. By analyzing the relationships surrounding networks, they can comprehend more subtle representations of users and their connections [13].

Handling heterogeneous data: Social media revolves around diverse varieties of users as entities, groups and items and their associations and so GNNs are developed to handle such unstructured data and combine information from different sources like user profiles, behavioral patterns, content, history of interaction records and so on [26].

Tackling data sparsity: Data sparsity details limited user-item interactions, which is one of the major issues in the context of recommendations. With the help of social interactions between users, GNNs can mitigate this problem by observing and modeling the complex relationship using graph data. It also enhances the quality of suggestions, mainly for new users and products [28].

Modeling dynamic behavior: This is a crucial aspect of several tasks, like predicting new trends or identifying threats. Some graph recurrent networks can adapt to this change and describe the temporal evolution of social connections and the non-static nature of user interactions [26]. GNNs with this robustness still contain a few shortcomings in efficiency. Let's look into a few challenges that are predictable from GNNs:

Scalability and model complexity: The social network can be extremely huge, containing a few million or even billions of users and interactions. This becomes hard for GNNs to work efficiently in such massive groups [29].

Transparency: Comprehending particular decisions made by GNNs in social connections becomes challenging. Particularly in sensitive applications such as scam detection and content recommendations, providing details for GNN decisions is strenuous [30].

Graph imbalance: In certain digital networking tasks like detecting fraud, there are chances for highly imbalanced data. This disparity can negatively affect the GNN training and functioning, requiring implementation of new techniques for the better performance in such cases [31].

Case study: A case study was conducted by researchers on sentiment analysis of social media interactions, specifically focused on ChatGPT, the AI agent which was introduced by Open AI in November 2022 [32]. The motto of this study is to analyze the public opinions (positive, negative, neutral) of ChatGPT by their tweets on Twitter [32]. The researchers collected all data and segregated them as positive, negative or neutral and labeled them manually. To extract features from social media data depicted as graphs, they deployed graph embedding techniques such as GAT and Node2Vec. Later, various ML classifiers like Random Forest, Support vector machines and Decision trees were optimized using these extracted features to identify the sentiment in the tweets [32]. Furthermore, a technique named SHAP (Shapley Additive exPlanations) was incorporated to provide details into the model's assessment. The main finding was that among classifiers, Random Forest exhibited notable performance when coupled with GCN-based embeddings in sentiment grouping compared to GAT and Node2Vec representations [32].

10.3 Leveraging GNNs in Physics, Material Science and Mechanics

There are several applications of GNNs in physics, specifically in modeling relations between particles and nodes and simulating complex systems. It is applied to study quantum many-body systems, condensed matter physics, high-energy physics and to speculate their behavior and properties. Also, GNNs have the potential to transform material science by enabling researchers to design and discover new materials with specific properties, accelerating innovation and breakthroughs in various fields.

This section delivers some key applications of GNNs in various areas of physics as well as in material science for scientific computing.

10.3.1 High Energy Physics

GNNs are efficient in processing huge amounts of data in high energy physics. They find application in high energy physics to analyze and model complex relationships between particles and their interactions and find patterns. They are helpful to reconstruct trajectories and interactions of the particles. They offer precise results in particle reconstruction and identification. Based on the properties of the particles and their interactions, GNNs can identify the particles. This allows the researchers to understand the behavior of the particle and to comprehend its properties better [33].

GNNs can be a tool to generate synthetic data for particle collisions and interactions and they can produce simulated data that imitates the factual experimental data. The additional data generated by the GNNs can be supplemented for the limited experimental data. Through the data generated by the GNNs, the process of training the machine learning models can be improved [34].

In high energy physics, Anomaly detection is helpful in identifying anomalies in complex high energy physics data. It detects unfamiliar particle interactions and unusual particle decays. This may direct to novel physics apart from the typical model. Anomaly detection monitors the performance of the detector and recognizes the possible problems. Also, this detection discards the background events and strengthens the signal quality in HEP experiments [35].

10.3.2 Quantum Many-Body Systems

In quantum many-body systems, GNNs are used to represent complex wave functions to understand the strongly correlated systems which are very difficult to represent using the existing electronic structure methods. An investigation for the structural behaviors of d-electron and f- and correlated systems was done in [36] by coupling quantum embedding techniques with data driven interatomic potentials derived from GNNs.

The GNNs excel at correlating molecular structure with properties which is beneficial for the prediction of ground state energy and other relevant characteristics (Figure 10.2). They can also be utilized to predict an extensive range of excited state properties. Moreover, GNNs can be applied to predict the behavior of quantum circuits and their performance with high accuracy [37]. In this the loss function, L(Y, P) is used to evaluate the difference between the prediction P and the ground truth. $\Sigma_{aggn_i \in N_i} f(n_i)$ is the summation over neighbouring nodes during convolution. $\frac{\Sigma}{n}$ is the global pooling operation which combines all node embeddings into a single vector.

In this extent, a novel context of 'Quantum Graph Neural Networks' (QGNNs) was initiated in [38] directing to integrate the quantum computing principles with the flexible GNNs. Here, the applications of QGNNs are highlighted in the investigation and understanding of complex interconnected systems which utilizes rich information enshrined in graph structured data.

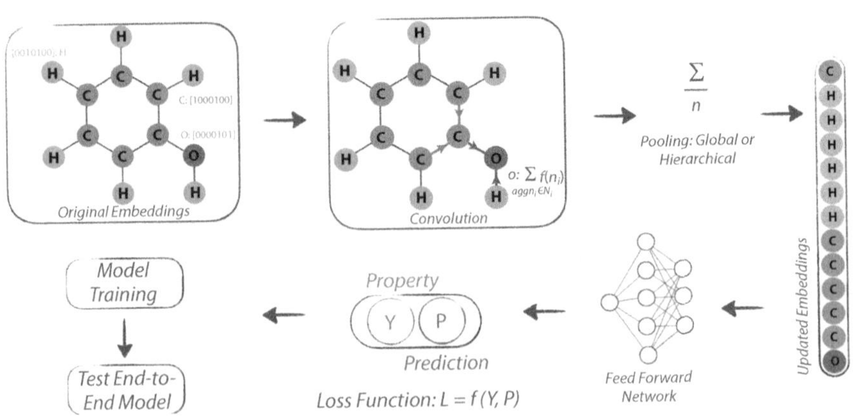

Figure 10.2 Scheme displaying workflow of the procedure [37].

10.3.3 Condensed Matter Physics

GNNs are efficient in modeling complex materials systems and capable of handling huge datasets and complex systems whereas the existing traditional methods require significant proficiency and computationally expensive. So, it can be utilized to different materials properties and systems as it is more flexible. For it is needed to design GNN architectures to personalize to materials systems and to integrate GNNs to domain-specific knowledge as it can improve the performance and interpretability [39].

Understanding material properties is central for optimizing and designing materials for definite applications. GNNs are useful in reviewing the patterns in the structure and properties of materials which allowing predictions for new materials. By empowering effectual and exact prediction of materials behavior and properties, GNNs are so potential to transfigure materials research. They can predict mechanical behavior, thermodynamic properties and electronic structure of the materials [40].

The materials properties are dictated by the structure of the crystals. Hence, predicting the structure of crystals is important for designing the materials. GNNs can be employed to learn relationships between material properties and their atomic arrangements inside a crystal lattice. They can predict crystal structures and identify stable phases [41].

Discovering new materials and identifying new materials with the desired properties can prime to innovations in electronics, energy and other related fields. For screening large databases of materials to identify the promising candidates, GNNs can be applied. By detecting the apt candidates, they quicken the discovery of materials [1].

For designing materials for a desired application, it is essential to understand the phase transitions. The prediction of the material structure and properties in phase transition changes under diverse conditions. Under these different conditions, GNNs learn the patterns in material behavior and do phase transition prediction efficiently [1].

10.3.4 Stress, Strain and Deformation Fields Prediction with GNNs

The approximate relationship (f) between materials,, its structural characteristics, and conditions and physical fields such as stress, strain and displacement are analyzed using the encoder-decoder ML model with GNNs [42]. This model has three segments which are the encoder, the message passing module and the decoder (Figure 10.3).

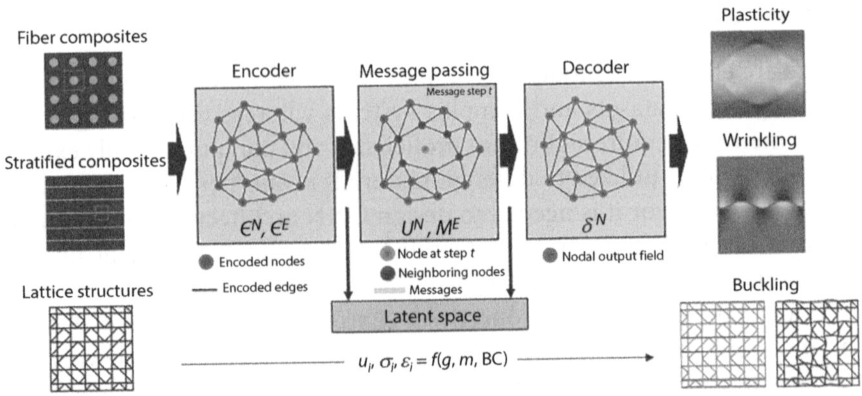

Figure 10.3 Encoder-decoder ML model using GNN [42].

(i) The encoder encodes the vertex and edge into a latent space by using the neural networks ϵ^{N} and ϵ^{E} respectively at each of the nodes and edges. This is done from the graph representation of the material system.

(ii) The message passing module processes the encoded system. It collects the information from the neighbors of the vertex by the neural network M^{E} and the state of the vertex is apprised by the neural network U^{N}.

(iii) After the second segment the decoder, (the neural network δ^{N}) transforms the features of the latent vertex into the field outputs which are deformation, stress and strain field $(u_i, \sigma_i, \varepsilon_i)$.

This model [42] learns the physical relationships with the fields from the topological and geometric information (g), behavior of the materials (m) and the boundary conditions (BC). The trained model predicts the physical fields of different material and their structural system under difficult physical phenomena accurately.

10.3.5 Computational Fluid Dynamics

GNNs can model fluid dynamics and simulate complex systems, aiding more precise predictions and improved understanding of fluid behavior. GNNs capture intricate flow phenomena, refining prediction accurateness which permits more consistent simulations and well decision-making [43].

GNNs facilitate the simulation of real-life systems and mesh structures and handle complex geometries. This benefits in complex systems in the prediction of flow behavior.

GNNs learn the patterns in flow data using which they predict the behavior of the fluid flow. This results in the prediction of pressure fields and velocity which are necessary to design and optimize fluid-based systems [44]. In [45], the utilization of GNNs to simulate complex physical systems together with turbulent flows is discussed. The work establishes the ability of GNNs in learning the underlying physics from data and improves the accuracy of predictions, which can be applied to turbulence modeling. This advances the accuracy of turbulence predictions which is crucial for simulating real-world flows.

They accelerate computational fluid dynamics (CFD) simulations by reducing the necessity for traditional numerical methods which aids faster turnaround times and saves computational resources (Figure 10.4). They minimize the effort and time taken for mesh creation by generating CFD simulations. This enables the flow phenomena and the simulation of complex geometries [46].

GNNs can potentially be used to simulate Fluid Structure Interaction (FSI) problems by learning the interactions between fluids and structures [47]. This enables the prediction of structural responses to fluid forces which is essential for optimizing and designing systems involving fluid-structure interactions.

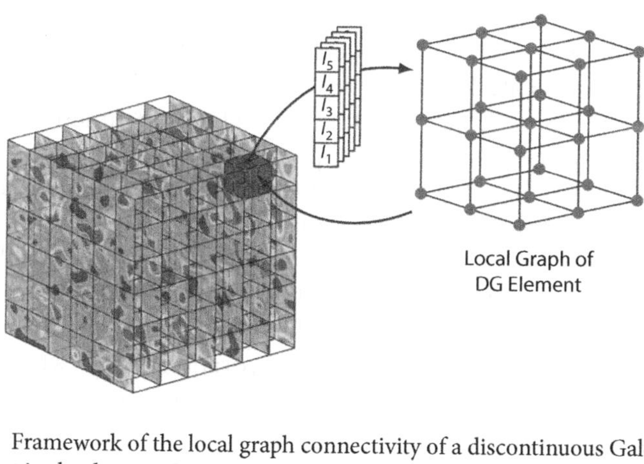

Figure 10.4 Framework of the local graph connectivity of a discontinuous Galerkin (DG) element in the domain. Here, a DG element refers to a mesh element or a finite element within the computational domain [45].

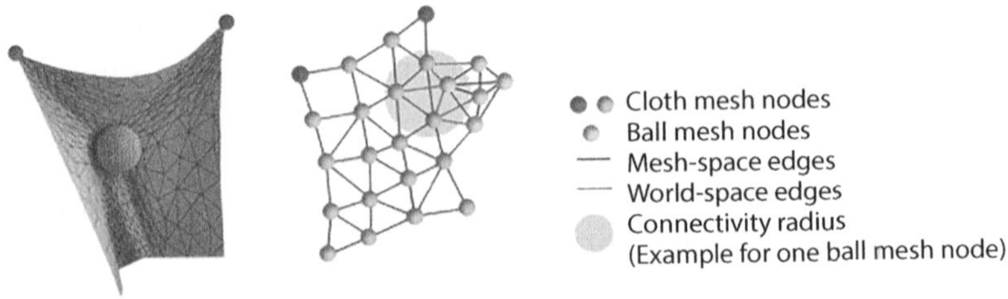

Figure 10.5 Graph modeling the contact between objects [48].

10.3.6 Mesh Dynamics

MeshgraphNet (MGN) is a multigraph message passing neural network architecture [48]. This is capable of learning mesh-based self-collisions between objects. This has the graph structure with world graph edges which are generated by predefined hyperparameter of connectivity radius. The nodes in the multigraph are nodal coordinates in the mesh and global world spaces. The edges in the graph are the nodal coordinate difference and it represents the Euclidean distance in material mesh space and the three-dimensional Cartesian world space (Figure 10.5). The world space focuses on message passing between objects and self-collisions whereas the mesh space focuses on message passing in material space.

10.3.7 Molecular Dynamics Simulations

Graph Neural Networks have revealed promising results in simulating molecular interactions and behavior. This is vital for understanding chemical reactions and material properties. In [49], the researchers introduced GNNs with the Sine Linear Unit (SLU) activation function to predict molecular properties. This work highpoints the ability of GNNs in modeling complex molecular structures and in predicting properties. It demonstrates their potential in molecular dynamics simulations.

GNNs capture the chemical and structural properties of molecules by learning the vector representations of them. Through these learnings, they predict several molecular properties like quantum mechanical properties of energy levels and orbital distributions, hydration energy which is a measure of the interaction of a molecule with water, toxicity, solubility and other related properties [50].

The course of recursive decomposition of target molecules into available precursors is called as retrosynthesis. [51] displays the application of GNNs in simulating chemical reactions and predicting retrosynthetic routes.

GNNs can model chemical reactions by learning the relationships between reactants, products and reaction conditions. The proposed method in [51] disassembles retrosynthesis by GNN-based identification of reaction center of the target molecule, generation of synthons and reactant prediction related to synthons through a robust reactant generation model [51].

A graph-based platform called GraphINVENT was developed in [52] for designing new molecules and modeling chemical reactions using GNNs. GraphINVENT generates new molecules one bond at a time probabilistically. Without requiring explicit chemical rules, the platform models employed in this platform can rapidly learn to generate molecules similar to the molecules in the training set [52]. This allows researchers to study design new compounds and reaction mechanisms.

[46] reveals the application of GNNs in accelerating simulations of complex physical systems, with molecular dynamics. Here a machine learning approach was presented and an implementation was modeled which is capable of simulating diverse thought-provoking physical interactions such as rigid bodies, fluids and deformable materials including their interactions.

10.3.8 GNNs for Solving Partial Differential Equations

GNNs are the efficient tools for solving partial differential equations (PDEs) mainly when it arises while handling irregular geometries and mesh-based domains. Here the spatial domains are represented as graphs in which nodes are mesh vertices and edges are local neighborhoods [53]. GNNs capture the local interactions which are the core concept of the formulation of PDEs. Based on neighbor details, GNNs update the states of the nodes and imitates the numerical schemes. These models can be

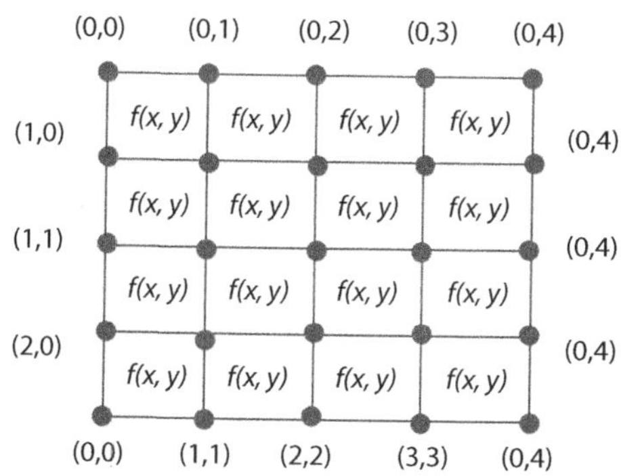

Figure 10.6 A sample graph for solving a 2-dimensional Poisson equation with GNN.

generalized for various domains and geometries. In the fields such as fluid dynamics and heat transfer, GNN- based solvers are providing promising results [53].

For example, in solving a two-dimensional Poisson equation (Figure 10.6), the aim is to find the solution u(x, y) given the source term f(x, y). Th special domain is converted into a graph where mesh points are the nodes and the connectivity between the points are edges. GNN passes the message between the nodes to map f(x, y) into the solution u(x, y).

10.3.9 GNNs in Weather Forecasting

Graph Neural Networks are being progressively used in weather forecasting to improvise the prediction accuracy and reliability which enables better decision-making in various applications (Figure 10.7). They can be used for global weather forecasting, capturing spatiotemporal relationships in weather data and providing more accurate predictions. GNNs are good to predict weather conditions over short-term horizons, such as 3-6 hours and weather patterns over longer-term horizons, such as 12-24 hours. GNNs are apt for short-term and long-term weather forecasting. They are the versatile tool for meteorologists [54]. GNNs work on graph-structured data, where vertices are weather locations and edges are the relationships between them [55].

GNNs analyze the multivariate data obtained from meteorological sensors and learn patterns in weather data. With this learning, GNNs can predict the future weather patterns and weather conditions like temperature, wind speed and humidity. They capture complex relationships between weather locations and predict weather patterns across different time horizons and locations [2].

Accurate forecasting needs to take into account of the impact of various external factors including weather and surrounding point of interest distribution. Also, it is subject to the historical traffic flow information. In recent days, the integration of recurrent neural networks and graph convolutional networks by the spatiotemporal models has made significant progress

Figure 10.7 Steps involved in weather prediction with GNNs.

and they have become traffic forecasting fascinating research arena. By accounting for the external factors that can enhance spatiotemporal accuracy in traffic prediction, can improve model interpretability [55]. With this novel thought, an attribute-augmented spatiotemporal graph convolutional network (AST-GCN) is projected in [55]. Using this network, the external factors of static attributes and dynamic attributes were modeled and an attribute-augmented module was designed to encode and integrate the external factors into the spatiotemporal graph convolution model [55].

By exploring novel methodologies that can effectively improve the prediction accuracy, the research work [2] contributes to the field of weather forecasting and provide more robust understandings into climate dynamics. Here the weather forecasting performance of GNNs is evaluated and is compared to conventional machine learning methods. This research focuses on improving the accurateness and reliability of weather predictions with the capabilities of GNNs which excel in capturing complex relationships within data.

Spatial-temporal graph modeling is a significant work to analyze the temporal trends of components and the spatial relations in a structure. Prevailing methodologies typically assume the spatial dependencies on fixed graphs by the assumption on predetermined entity relationships. Moreover, they often fail to effectively get the temporal trends because in these methods convolutional neural networks (CNNs) and the recurrent neural networks (RNNs) were employed which cannot get long-range temporal sequences [56]. To solve these restrictions, a graph neural network architecture, Graph WaveNet, for spatial-temporal graph modeling is proposed in [56]. By emerging a new self-adaptive dependency matrix, [56] Graph WaveNet model identifies the node dependencies and make predictions.

10.3.10 GNNs in Mechanics

The graph neural networks are applicable in predicting mechanical properties and structures of materials such as stress, strain and deformation. Also, they are capable of modeling the material behavior and properties such as plasticity, elasticity and fracture. The StructGNN framework is tailored for response prediction in static analysis. The improvised flexible structural graph representation excels in message-passing layers in StructGNN. It improves the efficient loading feature transfer and making the model capable of generalization to taller and unseen structures. It is particularly significant in structural engineering, providing physically meaningful simulations through rigid diaphragm representation and pseudo-node integration [57].

Sensors gather data on parameters like temperature, power consumption and vibration which are attached to machines. Predictive maintenance is scheduled based on predicted failure probabilities. It helps to identify potential safety hazards and ensure the safety of the working environment. Also, regular maintenance based on predictive analytics will extend the life of the equipment. For, GNNs are used to estimate mechanical failures according to which the maintenance will be scheduled. This method aids organizations to lessen downtime, lesser maintenance costs and advance overall operational productivity. GNNs process the data obtained from the sensors, analyze the complex relationships between equipment components and sensors, identify patterns and predict potential failures. Tong [58] proposed a predictive maintenance framework combining backpropagation-radial basis function-generalized regression neural networks (BP-RBF-GRNNs) and proportional hazard models. In this study, improved degeneration prediction accuracy is demonstrated and the maintenance cost is reduced.

10.3.11 GNNs in Robotics

Graph neural networks (GNNs) are more convenient for planning tasks in robot manipulation. They work well with the inherent graph structure of manipulation problems. In [59] a task planning problem is designed with random number of objects similar to reorganizing a pantry and trained both a feed-forward neural network (NN) and a GNN and to resolve it. The GNN reliably generalizes quicker and attains more than 90% high accuracy than the NN. When trained with 5 objects per scene it attains zero-shot generalization with more than 80% accuracy on hidden datasets with 5 to 15 objects in the scenes.

The system, RoboCraft developed in [60], which requires access to only raw RGBD visual observations, transforms the sensor data into particle representations. By using the graph neural networks, it acquires a dynamics model based on partocles and captures the fundamental structure of the system. Then this derived model designs the robot's behavior by joining with model-predictive control (MPC) algorithms. In this work, it is showed that using only ten minutes of real-world data from robotic interactions, the robot could infer a dynamics model which synthesizes control signals for deforming elastoplastic objects into specific target shapes. The robots recognized the shapes including the ones which they have never come across before.

In [61], a Grasp Graph Neural Network (G2N2) is configured to evaluate the grasp graph. This detects the utmost possible 6-DoF target object grasp pose. This graph-based encoding exhibited the exciting features in handling multiple occurrences of the same thing. GNNs are effective at capturing and highlighting the underlying correlation between the nodes by assuming many similar samples of the same item, for example tool boxes, books, chairs and groceries shelves which may be subjugated for example in the selection of the most available sample of the anticipated item.

Pose estimation for multi-person is complex as it restricts body key points for several persons instantaneously. In [62], a novel view of human part grouping and formulating it as a clustering task on graphs were investigated. Specifically, a different differentiable Hierarchical Graph Grouping (HGG) method [62] is proposed to learn the graph grouping in bottom- up pose estimation task for multi-person. Furthermore, HGG is integrated into main-stream bottom-up methods easily. The modules of HGG can be trained endwise with the key point detection network. In a hierarchical manner, it can supervise the grouping process.

A learning-based path planning method is proposed in [60] to minimize the collisions checks. The method uses a GNN model to predict the weights of the [63] edges of the neighbor set of the vertex. The weights are used to direct the planner to avoid obstacles and accelerate the search process [63]. Compared to the existing methods, this method significantly increases the speed of path planning, decreases the number of collisions checks besides improving the success rate and robustness in a high-dimensional environment.

10.4 Applications of GNNs in Chemistry

GNNs are deep learning models which are architecture mainly to analyze graph structured data. They have extensive range of applications in several fields [64]. One such major sector where GNNs are efficiently utilized is, chemistry. GNNs work in different in numerous areas in chemistry such as to determine molecular structures, drug discovery, molecular property prediction, molecular optimization and de nova generation and so on. The graph structure has two important features called nodes and edges. In the scope of chemistry, the atoms or molecules act as nodes and their chemical bonds act as edges [64].

GNNs have special abilities to examine relationships directly from geometric and topological information. Through this special feature, GNNs encode graphical networks which might not be possible through traditional methods that convert molecules into one dimensional string [64]. The driving force of many GNNs, especially in chemistry is message-passing mechanism [64]. In this process nodes exchange information through edges with neighboring nodes. Nodes accumulate information from its neighbors at every step and modify its features. Through the practice of this process several times, information is spread around each node at a certain radius within the network [40]. At last, a readout function compiles the node features to represent the entire graph that might be utilized for several tasks [40]. Few key applications of GNNs in chemistry are as follows.

10.4.1 Molecular Property Prediction

GNNs are increasingly used in the molecular property prediction, particularly in the field of molecular science. It provides significant benefits over traditional methods. Estimating molecular properties in the molecular graph is the most common and direct usage of GNNs in the branch of molecular science [64]. A GNN with classification head is applied to predict the activities of each atom in a molecule. Here the molecule is represented as a graph with nodes denoting the atoms and edges denoting the covalent bond between atoms [65] (Figure 10.8).

This graph records chemical information, bond types, atoms and its types etc. GNNs can solve the potential tasks that include classifications or regression models [40]. The major function of GNNs is to examine the variety of molecular properties such as multi-level properties which includes physical chemistry, physiology, biochemistry and quantum mechanics [64]. Pharmacological activity is estimated by quantitative structure-activity relationships (QSAR) to determine the drug likeliness, solubility and toxicity. GNNs has the special feature of identifying and

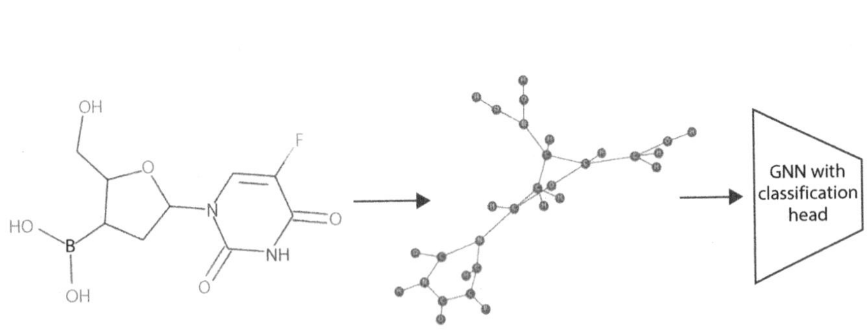

Figure 10.8 Predicting the activities of the atoms of a molecule by GNN with classification head [65].

analyzing solubility and toxicity simultaneously [66]. Datasets named ESOL is applied to predict solubility of aqueous solutions [67]. In drug discovery, the most suitable molecular properties of the drug like molecules are ADMET- absorption, metabolism, distribution, exclusion and toxicity [40]. Furthermore, the proficiency of the device depends on many applications such as HOMO (Highest occupied molecular orbital), LOMO (Lowest occupied molecular orbital), organic LEDs, organic electronics and GNNs are utilized to determine these applications. The environment of the molecules is very crucial for the properties of molecules and they can be represented and incorporated with GNNs [40]. In the area of protein molecular examination, comparing to traditional ML models GNNs such as GSnet has outperformed. These are modeled to predict several properties like physiochemical and geometrical properties consisting of solvation free energies, hydrodynamic radii and diffusion constants [68]. To enhance the protein structure study, properties including Solvent - Accessible Surface area (SASA) and residue specific pka values, can be learned by using transfer leaning from GNNs embeddings [68].

It is easy for GNNs and most appreciable way to study the molecular systems since, they are characterized as graph with molecules as nodes and the bonds between molecules as edges. A few key merits of applying GNNs for molecular prediction.

Directly comprehend the connection from graphical data: The traditional deep learning models like CNNs were not efficient in analyzing graph structures. Whereas GNNs are modeled with the goal of learning representations from graph effectively [64]. GNNs offer high performance in understanding topological information of graphs of 2D molecules along with geometric information of 3D molecules graph. They are very powerful tool for encoding graph data and advantageous over techniques like SMILES or fingerprints that fails to explicitly model topological and geometric data [64]. GNNs are also capable of integrating atomic and bond descriptions such as atomic number, bond type, etc., into node and edge characteristics [64].

GNNs have a special ability to analyze the internal information that is very essential to perform specific tasks like determining molecular properties. This feature of GNNs can replace the widely used system in natural sciences in general that is hand crafted feature model. When there is a large data, GNNs performance is outstanding in determining the molecular properties comparing to CNN models [69]. A sustainable development is noticed in incorporating Ring enhanced GNNs across few tasks like ESOL,

Clint, BCF in the precision of prediction [70]. A major benefit from GNN is its significant feature of analyzing multiple target tasks concurrently like solubility and toxicity. This solves the issue of dependence on different models in assessing each property and resolves the challenge of knowledge sharing across independent models [71].

10.4.2 GNNs for Molecular Optimization and Generation

GNNs are very robust tool in molecular optimization and de nova generation, especially in drug discovery sector. Molecular optimization basically works with known molecule and adjusting its structure to improvise the desired properties like lowering toxicity and strengthening binding affinity. This is considered as conditioned generation, where the modifications are conditioned on the initial molecule [64]. De nova generations entirely work on generating a new molecule from the beginning, sometimes with conditioning on several properties and with specific fragments. The following are some of the models used for molecular optimization and generation using GNNs.

Fragment based generative models: These frameworks repeatedly modify a molecule graph by inserting, eliminating or replacing the fragments [64]. These frameworks can be combined with reinforcement learning (RL) to enhance an expected reward where agents make decision at every step, frequently related to required molecular properties. Also, fragmented-based generation with auto aggressive models such as RNNs predicts actions based on earlier steps.
Molecule based generative model: These models develop the whole molecular graph with nodes and edges characteristics, adjacency matrix at the same time.
Variational autoencoders (VAEs): These are commonly used in this sector. It comprises of two subdivision name decoder and encoder. GNN encoder connects the input to latent space while decoder maps send the latent feature to input [40].
Generative adversarial networks (GANs): This uses adversarial process between a discriminator and a generator. Generator creates a realistic molecule and discriminator tries to identify the real molecule from the generated one [40]. Reinforcement learning can be combined with GANs to create molecules with optimized properties [40]. Flow based model usually generates data and modify it step by step which is called as invertible transformation. It uses this feature and examines the likelihood of the observed data. This helps them to represent new data and identify the patterns of

the observed data. Some of the examples for such networks are GraphF, GraphDF [64].

Based on the set of GNNs a special sequential molecular graph generator is produced which is called as Molecule Generative graph neural network (MG2N2) [66]. It is completely dependent on recurrent GNNs with the help of sequential algorithm at every step the nodes are added [66]. To generate the neighboring nodes, this approach focuses on one node at once before proceeding to the next node, this avoids disconnected components. The control flow of MG2N2 is based on the judgment made by three GNN models [66]. A key merit of GNN in MG2N2 is, it can maximize information of the input at each stage which also includes subgraph produced until preceding step [66]. There are two unique features which makes MG2N2 as interpretable they are sequential and modular architectures. This helps in retraining the single module of the graph independently which eases the process of learning.

Tests of generating datasets unconditionally like Zinc and QM9 datasets, it was evaluated that the model is showing up the ability to generate molecules based on the knowledge gained during training period without overfitting [66]. Since the model is not obligated to reconstruct the molecules perfectly on the basis of global basis, but to optimize perfection for local graph parts it is recommended for high novel contribution while conserving authenticity [66]. The current version of MG2N2 is mainly for unconditional generation, whereas the future work is to develop the conditional model of MG2N2 which takes the specific property as an input and generates with the specific objectives [67]. A layer version is also anticipated where a first generator generates the general molecular graph with each node associated with group of atoms and the second generator add the particulars in every group depending on embeddings, and generates individual atoms and bonds [67]. Recent researchers are focusing on developing 3D molecular structure which is significant in analyzing various properties like energy and protein-ligand binding [40]. Despite prominent advancements, challenges like competently incorporating chemical substances and fusibility into the process of generation and robustness still remains.

10.4.3 GNNs in Catalyst Design

GNNs are well suited for molecular and system modeling as they can easily comprehend the data and represent them as graphs. Catalysts are generally considered as complex materials like metal nanoparticles, surfaces, crystalline structures can be depicted in graph structures [40]. The effective GNNs are successfully employed in several systems varying from the stability of the crystal structure to electronic property estimation and to heterogeneous catalysis and surface chemistry [39].

GNNs are also used to analyze the complete energies of structurally wide-ranging nanoparticles. Nanoparticles are catalytically reactive intermediates, determining their energy levels is crucial in the advancement of ML techniques utilized in optimizing catalytic mechanisms [39]. GNNs are deployed for organic molecules generation as well as new materials using inverse materials design. Generating new catalyst structures from the beginning exploiting GNN models seems to be less studied or highly complex when compared to small molecules generation [40]. In the area of catalyst design, GNNs are used for relevant property prediction and screening candidate analysis [40].

10.4.4 GNNs for Retrosynthesis Prediction

Retrosynthesis plays a vital role in desired chemical synthesis in drug discovery. This takes place by breaking down the molecules into simpler building block compounds for the series of chemical reactions. Retrosynthesis is widely used in drug discovery, synthetic biology and biocatalysts. AI is very helpful in studying chemical spaces, synthesis planning, encouraging bioproduction which aligns with the green chemistry [72]. As AI is progressing everyday it can directly interpret the outcome of chemical reactions. GNNs in retrosynthesis can be approached in two different ways, single step and multistep analysis (Figure 10.9).

Single step analysis: This step is mainly used in identifying the immediate reactant required to synthesis the target molecule. The GNNs versatility influences majorly in synthesis. They approach in two different ways to solve the problem or to represent the data. They are template free and template-based methods. Under template based, a structured frame work will be provided to interpret the protein structure and also it can determine the structure of a new protein [40, 73]. GNNs help in computing the molecular graph structure to identify the most suitable templates [40]. For instance, GNNs can identify the characteristics and certain patterns and matches the single retrosynthesis phases to reaction templates, it can also suggest multiple reaction templates and allocate probabilities [40].

Template free methods: This method does not require the fixed frameworks for molecule prediction but it can directly analyze the graph modifications required and predict reactants. GNNS are frequently incorporated into generative models for this purpose [73]. Examples include employing GNNs to identify part of the molecule and converts synthesized compounds

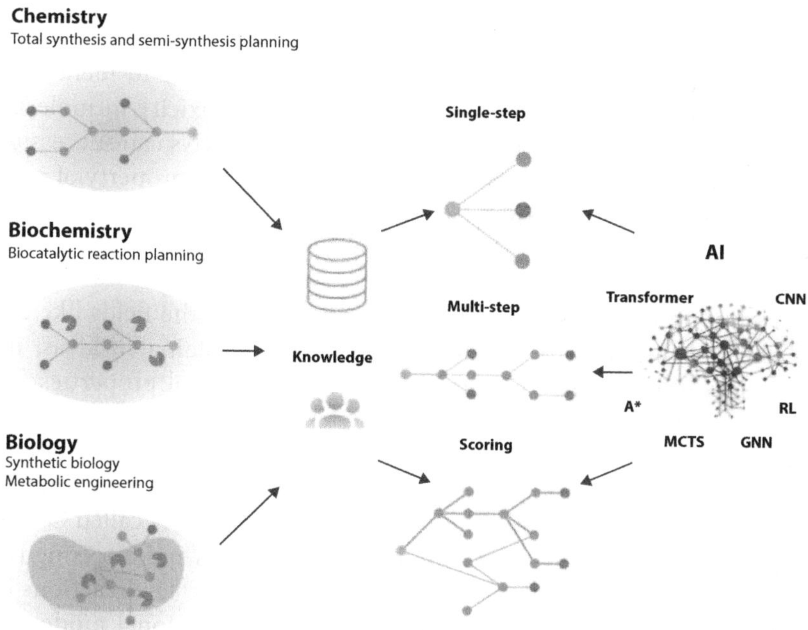

Figure 10.9 Retrosynthesis process in single step and multistep [73].

to fragments through the process of bond dissociation and it is called as synthons which are then extended to reactants [40]. Other approaches also deploy GNNs inside the model which determines the series of graphs reforms to change the product into synthons and then reactants [40]. The best example for two-step process is G2G (Graph to graph) which involves synthesis completion and edits analysis [73].

Multiple step retrosynthesis: This methods in organic chemistry is used to propose the synthesis of target molecule by breaking the molecules into simpler reactants till synthesizable compounds are obtained. This comprises recognizing a synthesis route which can be employed to generate the desired product from a basic precursor [72]. GNNs are integrated with robust algorithms in to deal with major challenge of finding the number of possible routes and determining efficient pathways. An efficient example is GNN- Retro. It is used in retro synthetic planning [40]. Retro synthesis is the combination of two major aspects, they are GNNs and A* search. Where GNN represents and analyses the data in graph structure and A* search discovers the best path way algorithm to reach the target [72]. When GNN- Retro and other methods along with Retro* are compared, GNN-retro showcased the outstanding performance in cost prediction [72].

10.4.5 Toxicity and Environment Chemistry

Through regression and classification GNNs are capable to identify the molecular properties [64]. ADMET properties and toxicity in molecules can be efficiently found by GNNs [67]. Alongside, GNNs can successfully identify the toxicity of distinct organisms like bees. The property of mutagenicity is estimated effectively by GNNs which is very essential in drug discovery and toxicology [67].

Environmental chemistry characters estimation: In the fields like drug and chemical industry, to undertake green and sustainable ideas, it is significant to precisely estimate environmental chemical properties [70]. For which, choosing environment friendly chemicals is crucial. ML models like GNNs are very helpful in these cases [70]. GNNs are employed in specific environmental practices such as bioconcentration factor, Clint (Intrinsic metabolic clearance rate) which is a special GNN called O-GNN is applied to the Clint tasks for comprehensive analysis and assessing [70]. To analyze the pollutants behavior in the environment, few reactions like chemical reactivity with oxidants are significant. Traditional models like Descriptor based features are not effective in analyzing complex molecular structures. The notable performance is observed by the special O-GNN, which integrates data about molecular rings and others activities within the area of environmental chemistry. These GNNs are capable of changing the data about atom, bond and ring [70].

Datasets and benchmarks: The majorly used benchmark in analyzing the performance of GNN models in the prediction of molecular property is MoleculeNet which contains information on many tasks such as toxicity, solubility, drug likeliness. Tox 21 and ToxCast are unique datasets that includes information on level of toxicity in chemicals and are used for examining models in this sector.

In this way, GNNs are very beneficial to directly analyze molecular properties related to environmental chemistry and toxicity through their graph structures. in various range of fields like drug discovery, material science and chemistry, synthesis planning, molecular generation, molecular property prediction and environmental chemistry, GNNs are widely used for its ability to analyze molecular graph structures. it has become the most crucial tool in chemistry to undertake several tasks. Comparing to earlier methods i.e. traditional methods GNNs possess an accurate and flexible outcome. While few challenges exist in transparency and input data requirements.

10.5 Applying Graph Neural Networks in Biology

The ability to simulate intricate biological processes as interconnected networks has made GNNs a game-changing tool in computational biology. In order to analyze protein structure and function, predict gene expression levels and understand the behavior of cellular networks, GNNs offer a powerful framework by representing biological entities such as proteins, genes and cells as nodes and their relationships as edges. The main uses of GNNs in biology are examined in this section, along with their benefits, drawbacks and potential.

Everything from molecular connections such as protein-protein networks to extensive cellular communication is intrinsically graph-structured in biological systems [74]. GNNs perform exceptionally well by utilizing graph-based representations, while traditional machine learning models find it difficult to capture these relationship dependencies. Because of their capacity to learn from topological properties, they are perfect for numerous tasks. Some of them are as follows.

- **Predicting protein structure:** Protein interactions, structures and functions are predicted by GNNs.
- **Gene regulatory network analysis:** Gene interactions and regulatory mechanisms are modeled by GNNs.
- **Discovery of drug:** GNNs identify possible candidates and forecast interactions between drugs and targets.
- **Predicting disease:** GNNs predict disease associations by analyzing biological networks.
- **Cellular network analysis:** GNNs research signaling pathways and cellular interactions.

10.5.1 Protein Interactions, Structures and Functions Predicted by GNNs

In bioinformatics, predicting protein structures is an essential task and GNNs have demonstrated encouraging outcomes in this field. Proteins may be visualized as graphs where amino acids are nodes and interactions are edges. The features of nodes and edges are considered as the attributes like amino acid type, distance and orientation.

GNNs play vital role in interaction prediction to forecast binding sites and interactions between proteins. They use amino acid sequences to predict the protein three-dimensional structure in structure prediction. In function prediction, they use the structure and sequence of proteins to predict their functions. GNNs predict protein properties by identifying patterns in the graph structure.

For instance, AlphaFold, created by DeepMind, makes highly accurate protein structure predictions using a GNN-based architecture [74]. Graph Convolutional Networks (GCNs) can predict protein function and the affinity of protein-ligand binding [74].

AlphaFold and GNNs in protein structure prediction: Proteins are represented by AlphaFold as graphs, with amino acids serving as nodes and interactions as edges. The Evoformer module in AlphaFold models long-range interactions by fusing GNNs with attention mechanisms. A structure module that takes into account of geometric constraints is used to refine the predicted structure.

AlphaFold outperforms conventional techniques in predicting protein structures with high accuracy. Without depending on fragment-based methods, AlphaFold makes end-to-end protein structure predictions. Understanding genetic variations and how they affect protein function can be aided by AlphaFold's predictions. The precise protein structure predictions provided by AlphaFold can help in drug discovery with the identification of possible binding sites and it is very much useful for the development of targeted treatments.

GCNs in predicting protein function: Graph classification tasks, including molecular property prediction, can be performed with GCNs. GCNs can be applied to node classification tasks, like node labeling and protein function prediction. Link prediction tasks, like trying to predict protein-protein interactions, can be performed with GCNs. Data with irregular structures, like graphs with different numbers of nodes and edges, can be handled by GCNs [75]. Local patterns in graph-structured data can be captured by GCNs.

Data like gene expression or protein interactions are used to build biological networks. Nodes are equipped with pertinent data, like expression and sequence data. The GCN layers capture local patterns by aggregating node features. GCNs use a weighted sum and other aggregation functions to combine node features from nearby nodes. The learnable weight matrix is used to transform the combined features. The output is passed through an activation function, such as ReLU.

10.5.2 Gene Regulatory Network Analysis Using GNNs

The intricate links between genes and their regulatory elements are modeled by gene regulatory networks (GRNs). GRNs have been used in GRN analysis because they show the inter play between genes and their regulatory networks, which are essential for comprehending cellular functions and the regulation of gene expression. The high dimensionality and structural complexity of GRNs frequently pose challenges for conventional computational techniques. GNNs have recently been successfully used for GRN analysis because of their capacity to learn from graph-structured data. By identifying gene dependencies, incorporating multi-omics datasets for improved accuracy, and exhibiting resilience to biological noise and missing data, GNNs make it easier to infer unknown regulatory interactions. Furthermore, by emphasizing important regulatory elements within the network, attention-based GNNs offer interpretability. The followings are some ways that GNNs have been used in GRN analysis.

Inference of regulatory interactions: The intricate networks of gene regulation, such as transcription factors controlling gene expression, are referred to as regulatory interactions in biological systems, particularly in genomics and systems biology. Determining these relationships is essential to comprehending disease mechanisms and cellular functions [76].

By representing biological components such as genes and proteins as nodes and their interactions as graph edges, GNNs can be used to model and infer regulatory interactions. By combining data from nearby nodes, the GNN learns node embeddings and iteratively updates the graph representation. Using patterns in known data, this method enables GNNs to infer hidden interactions and capture higher-order dependencies. The following are the typical key steps involved in modeling and inferring the regulatory interactions (Figure 10.10).

- **Construction of Graph:** The first graph is created using biological data, such as gene expression profiles and known regulatory pathways.
- **Training the GNN:** A function that can generalize to unknown interactions is learned by training the model on known interactions.
- **Inference:** The model makes predictions about possible new regulatory interactions after training, frequently with confidence scores.

Integration of multi-omics data: To create more thorough and precise GRNs, GNNs can incorporate a variety of data types, including gene expression, chromatin accessibility and protein-DNA interactions [70]. In order to gain a thorough grasp of biological systems, the integration of multi- omics

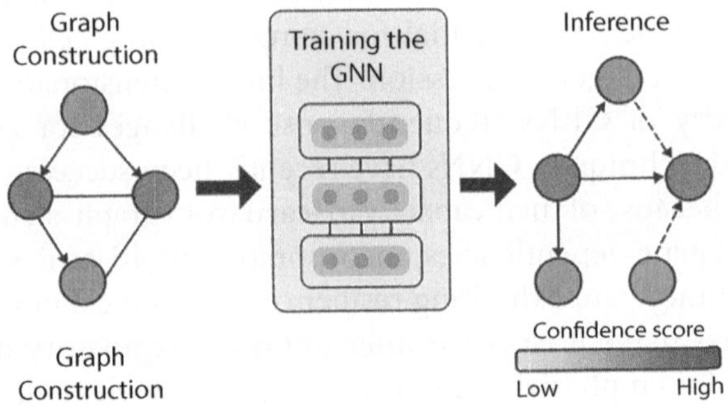

Figure 10.10 Inference of regulatory interactions in GNNs.

data refers to the combined analysis of various omics dataset types, including transcriptomics, proteomics, metabolomics, epigenomics, and genomics.

Researchers can discover intricate regulatory mechanisms, disease pathways and biomarkers that might not be visible through single-omics analysis by integrating the distinct insights offered by each omics layer. Some of the key methods used for integrating multi-omics data are given below.

Early integration: It is concatenation-based. It creates a single matrix before analysis by combining raw or pre-processed data from various omics types.
Late integration: It is model-based. It examines every omics dataset independently before combining the findings for interpretation.
Intermediate integration: It uses sophisticated models such as matrix factorization, deep learning and multi-view machine learning to integrate data during the learning process.

These methods make it possible to comprehend biological processes at the systems level, increase the precision of disease prediction and classification, and find therapeutic targets and cross-omics biomarkers.
Noise tolerance and scalability: Incomplete information and noise are common in biological data. By using neighborhood information in the network, GNNs can still effectively infer regulatory relationships despite these constraints [77].
GNNs in noise tolerance: By utilizing graph structures, GNNs are intended to model relational data. However, noisy edges or incorrectly labeled nodes are frequently present in real-life graph data like citation graphs and social

networks. The capability of GNNs to continue in making accurate predictions even in the presence of imperfect input graphs is known as noise tolerance. The robustness of GNNs to noise is improved by methods like adversarial training, dropout on edges and robust aggregation functions such as attention-based or median.

For example, if we assume that certain papers are mislabeled and improperly linked, the term AI (artificial intelligence) is incorrectly applied to a biology paper and there are some missing and inaccurate citation edges, then a strong GNN, such as the GAT, which uses attention-based aggregation, can learn to, down weight deceiving neighbors. It will pay attention to dependable structure and continue to classify the target node correctly. At the end of this process, GNN will predict that a misrelated paper is still in Biology despite noise.

Scalability in GNNs: Due to the recursive neighborhood aggregation process, which can become computationally costly on large graphs, scalability is a major challenge in GNNs. Scalable GNN variations have been put forth to address this and some of them are including the following.

- GraphSAGE, a neighborhood sampling technique [78].
- Graphs are partitioned for mini-batch training using Cluster-GCN [79].
- FastGCN, which takes node samples rather than neighbor samples [80].

For example, Let's say there are millions of papers in the citation network. The high memory and computation requirements of traditional GCNs cause them to struggle. However, Cluster-GCN or GraphSAGE clusters the neighbors by using mini-batches for training. As a result, GNN can train effectively without requiring the entire graph to be stored in memory.

Biological interpretability: The ability of computational models, like GNNs, offer insights that are intelligible and significant in a biological context which is known as biological interpretability. Interpretability is essential for confirming predictions and producing new hypotheses in applications such as GRNs, protein–protein interactions and disease pathway analysis. In order to facilitate biological understanding, advanced GNN architectures such as attention-based models highlight the genes and regulatory links that have the greatest influence on the network's predictions [17].

Interpretability in GNNs can be advanced by attention mechanisms that draw attention to important nodes and edges, such as Graph Attention Networks and by finding the gene features and pathways that have the primary factor influencing the model's prediction and this is known as feature importance analysis. Also, it can be done by subgraph extraction, which displays the segments of the graph that support a particular hypothesis like disease classification.

By connecting the model performance and biological understanding, these interpretability techniques enable GNNs to be used not only for prediction but also for clinical translation and biological discovery.

10.5.3 Discovery of Drug Using GNNs

In drug discovery, graph neural networks have developed as a potent tool, especially for forecasting how drug molecules will interact with biological targets like proteins and enzymes. Using atoms as nodes and chemical bonds as edges, drugs are depicted as molecular graphs in this process. In order to predict characteristics like binding affinity, toxicity and biological activity, GNNs acquire meaningful structural representations of these molecules. Complex drug-target interaction (DTI) networks can also be modeled using GNNs, which enables the following.

Screening potential drug candidates using GNNs: [81] A crucial stage in the drug discovery process is screening possible drug candidates to find molecules that might bind to a biological target. Graph neural networks improve this process by automatically learning the structural and chemical features of molecules by modeling them as graphs, where atoms are nodes and bonds are edges. The key attributes like affinity of binding, solubility, the toxicity and drug-like appearance are predicted by GNN-based models.

GNNs allow for high-throughput virtual screening of large chemical libraries by directly learning from molecular graph data, removing the requirement for expensive and time-consuming lab experiments. Faster lead compound identification for additional testing results from this.

Predicting unknown interactions using GNNs: [82] Predicting unknown interactions between drugs and biological entities like proteins, genes, or other drugs is a highly effective use of GNNs. In fields like side effect analysis, drug repurposing and DTI discovery, these predictions are essential.

GNNs function on heterogeneous biomedical graphs in this context, which are made up of different entities represented by nodes and relationships represented as edges. GNNs can infer new, undiscovered links that anticipate possible interactions by learning node embeddings and edge patterns.

This method is applicable to find new drug-target combinations, to determine off-target consequences and to give new therapeutic applications for existing medications.

Repurposing existing drugs using GNNs: [83] Finding new therapeutic applications for approved medications that have already received approval or that are already on the market is called drug repurposing, sometimes referred to as drug repositioning. Since repurposed medications have already passed safety trials, this method is particularly beneficial because it drastically lowers the risk, expense, and time associated with drug development.

Drug repurposing benefits greatly from the use of GNNs, which model intricate biomedical networks that comprise connections among drugs, proteins, genes, diseases and side effects. In order to find hidden associations and forecast new drug-disease links, GNNs learn from these heterogeneous graphs. This makes it possible to find medications that might be repurposed for novel illnesses, a strategy that proved effective during the COVID-19 pandemic.

Case Studies for Drug Discovery with GNN Outputs

Drug discovery refers to identifying the properties of a molecule. The architectural representation includes four steps like using dataset, input, graph representation and the GNN model used. The dataset named MoleculeNet like Tox29, QM9, etc., is taken. Since the molecular graph structure is essential to study the graph, in the stage of input the SMILES string of molecules are transformed into molecular graphs. In successive step the atoms become nodes and chemical bonds become edges [84]. The GNN model used in this is MPNN (Message passing Neural Network) and regression predictions were detailed. The sample output is displayed in the Table 10.1 below.

Sample Output: [85]

Conclusion: The lower scores signify lower toxicity.

Table 10.1 Sample output [85].

Molecule ID	SMILES	Predicted toxicity score	Actual toxicity score
Mol_001	CCO	0.12	0.15
Mol_002	C1=CC=CC=C1	0.5	0.7
Mol_003	C(C(=O)O)N	0.20	0.18

10.5.4 Predicting Disease Associations by Analyzing Biological Networks

Making predictions about disease associations entails figuring out which genes, proteins and biological pathways are connected to particular illnesses. Biological systems have intricate relationships that are frequently difficult for traditional computational methods to capture. As the biological systems are naturally represented as graphs, GNNs [86] are ideally suited for this task.

Here, genes, proteins and diseases are examples of biological entities which are represented by nodes. The co-expression, gene-disease associations and protein-protein interactions are examples of relationships or interactions which are given by edges. GNNs can infer hidden relationships by modeling these interactions, including the mutations in genes that could lead to a disease. It is also used to model the illnesses that have similar molecular causes and proteins that play a role in the pathways of disease.

GNN operation: GNNs predict the disease associations by analyzing biological networks through the following steps.
Construction of Graph: Create a biological network using databases such as OMIM, STRING, or DisGeNET. The following are some of the network types.

- Networks of gene-gene or protein-protein interactions (PPIs)
- Bipartite graphs of genes and diseases or proteins and diseases
- Heterogeneous or multi-layer graphs that combine different kinds of data

For instance, the known association can be represented by the edges and breast cancer and gene BRCAI can be represented by node B and node A respectively.
Initialization of Node Feature: A feature vector is assigned to each node for example, gene and it could contain different levels of tissues of gene expression, occurrence of mutations, GO (Gene Ontology) terms and memberships to the pathway. For the GNN, these features are numerically encoded.
GNN-Based Message Passing: In order to update node representations, GNN layers aggregate neighboring features. This aids in each node's understanding of its biological environment.

As an illustration, if gene A has a strong correlation with genes linked to cancer, its representation might start to show a link to the disease even if it hasn't been officially identified as such.

Learning and Forecasting: Established disease-gene or disease-protein associations are used to train the GNN. Unknown labels and missing edges, like a gene that is predicted to be linked to Alzheimer's based on the network but is not yet known to be, will be predicted.

Results and Analysis: For every possible association, the model produces a probability score. Predictions with the highest scores are candidates for clinical studies or experimental confirmation. GNNs can be used by researchers to determine which unidentified genes may be responsible for uncommon genetic illnesses and to determine which proteins might be cancer biomarkers and gene suggestions for targeted treatments.

10.5.5 Cellular Network Analysis

In the analysis of complex biological systems, GNNs have become state-of-the-art tools. They are especially useful for comprehending cellular networks, which include gene regulation, signaling pathways, and cell-to-cell communication. Understanding these networks is crucial to understanding how cells behave, interact, and react to their surroundings.

Cellular networks explain the interactions between molecules like metabolites, proteins and genes that govern biological processes. Among them signaling pathways are a sequence of molecular actions that cause a cellular reaction when they are triggered by outside signals such as growth factors or hormones. The way transcription factors control the expression of genes is known as gene regulatory networks. Networks of cell-cell interactions explain how cells, particularly tissues, communicate with one another through molecules like ligands and receptors.

Analyze cellular networks use GNNs, for simulating the spread of signals through various pathways, capturing the spatial context of tissues and forecasting novel interactions and connections between regulations.

Pathways for Signaling: By examining known and anticipated protein–protein or gene–protein interactions, GNNs are able to pinpoint important elements and regulators within signaling pathways. They can reveal hidden feedback loops or cascade effects, for instance.

Cellular Interactions: Each cell may represent a node in spatial transcriptomics or single-cell data, with edges delineated by known receptor–ligand interactions or spatial proximity. GNNs can simulate the interactions between cells in a tissue, such as immune infiltration and the tumor microenvironment.

Identification of Disease Mechanisms: GNNs can detect broken pathways or aberrant communication patterns by comparing cellular networks in healthy and diseased states, which helps with the identification of biomarkers and the development of new treatments.

For instance, the Giotto [87] toolbox provides an example of how graph-based analysis can be used in cellular networks. Giotto makes it possible to create spatial networks using tissue data, visualization of cell-to-cell interactions, ligand-receptor communication inference with graph-based models. In addition to using graph theory to shed light on cellular behavior and organization, it facilitates the analysis of cellular interactions in 2D or 3D tissue.

10.6 Comparing the Traditional Methods with GNNs Models

The following Table 10.2 portrays the benefits of GNN based models compared with traditional methods at various challenging situations.

Table 10.2 A comparison table for traditional method models and GNN-based models.

Challenges	Traditional methods	GNN models benefits	References
Limitations in data structure	Earlier models like CNNs, RNNs, SVMs, DFT, etc. requires structural, sequential data to process the information.	GNNs can handle irregular data pattern like graphs, molecules, networks, crystal lattices, etc.	[88, 89]
Dealing with complex relationships	Grapples to deal with protein folding, non-local data, indirect connections, long range data, etc.	Easily models complex interactions, neighbor aggregation, global relationships through message passing technique.	[3, 74]

(Continued)

Table 10.2 A comparison table for traditional method models and GNN-based models. (*Continued*)

Challenges	Traditional methods	GNN models benefits	References
Feature engineering	Detailing the raw data in a way that model can understand, requires domain knowledge.	GNNs can automatically learn nodes and edges connections from raw data without hand crafted features.	[84, 85]
Physical and spatial priors	Traditional models may overlook spatial properties like rotations, permutations, etc.	GNNs like EGNN, SchNet, DimeNet considers geometrical and physical properties without fail.	[90–93]
Computational expenses	DFT, FEM, Md, etc. physics-based methods have high computational cost for larger systems.	GNNs are approximately lower in its computational cost after training.	[93–95]

10.7 Conclusion

GNNs have revolutionized scientific computing by analyzing and predicting the phenomena in complex systems in numerous scientific fields. They have accelerated scientific discoveries and innovations by providing enhanced understanding of dynamic systems and by optimizing system designs. GNNs are able to simulate interdependent systems more efficiently than the traditional methods and able to contribute valuable insights into complex systems. They can be beneficial in capturing intricate relationships and patterns in data. They can be applied to diverse domains including engineering, biology and physics to explore new architectures, to improve interpretability and to develop applications. As a result, GNNs play an increasingly essential role in fluid dynamics, materials science, molecular dynamics, climate modeling and biological networks and many more areas. By leveraging the power of GNNs, researchers can unlock new insights and drive innovation in various scientific fields and make GNNs continue to advance in scientific discoveries. Furthermore, GNNs have the capability to transform the scientific computing and drive innovation and so continued research and development are needed to fully reveal their potential. Integration of GNNs with traditional methods would lead to more efficient simulations in challenging phenomena. The physics-informed GNNs which incorporates constraints and physical laws will improve the generalizability and accuracy. For large scale simulations, it is important to develop efficient and scalable GNNs.

AI disclosure statement: The author acknowledges the use of AI-based tools for minor language editing or grammatical correction. However, all intellectual contributions, analyses, and conclusions presented in this chapter are entirely the author's own.

References

1. Chen, C., *et al.*, Graph Networks as a Universal Machine Learning Framework for Molecules and Crystals. *Chem. Mater.*, 31, 9, 3564–3572, 2019.
2. Varshney, Y., *et al.*, Forecasting Precision: The Role of Graph Neural Networks and Dynamic GNNs in Weather Prediction. *J. Big Data Technol. Bus. Anal.*, 3, 1, 28–33, 2024.
3. Ma, T., Xiao, C., Zhou, J., Wang, F., Drug Similarity Integration through Attentive Multi-view Graph Auto-Encoders, Apr. 2018.
4. Khemani, B., Patil, S., Kotecha, K., Tanwar, S., A review of graph neural networks: concepts, architectures, techniques, challenges, datasets, applications, and future directions. *J. Big Data*, 11, 1, 18, Jan. 2024, doi: 10.1186/s40537-023-00876-4.
5. Wu, Z., *et al.*, MoleculeNet: a benchmark for molecular machine learning. *Chem. Sci.*, 9, 2, 513–530, 2018, doi: 10.1039/C7SC02664A.
6. Stärk, J.T., *et al.*, quiBind: Geometric Deep Learning for Drug Binding Structure Prediction, in: *Proceedings of the 39th International Conference on Machine Learning*, pp. 20503–20521, 2022.
7. Yao, L., *et al.*, Graph convolutional networks for text classification. *Proc. AAAI Conf. Artif. Intell.*, 33, 1, 7370–7377, Jul. 2019.
8. Wang, K., Ding, Y., Han, S.C., Graph Neural Networks for Text Classification: A Survey, Apr. 2023, doi: 10.1007/s10462-024-10808-0.
9. Gao, R., Text Classification Method Based on Graph Neural Networks. *ITM Web Conf.*, 73, 02015, Feb. 2025, doi: 10.1051/itmconf/20257302015.
10. Cahoon, J., *et al.*, Optimizing open-domain question answering with graph-based retrieval augmented generation, Mar. 2025.

11. Liu, B. and Wu, L., Graph Neural Networks in Natural Language Processing, in: *Graph Neural Networks: Foundations, Frontiers, and Applications*, Springer Nature Singapore, Singapore, pp. 463–481, 2022, doi: 10.1007/978-981-16-6054-2_21.
12. Barrett, M., Framework for Improving Critical Infrastructure Cybersecurity Version 1.1, NIST Cybersecurity Framew, 2018.
13. Shiva, R., Graph Neural Networks for Cybersecurity: Enhancing Threat Detection in Complex Network Structures, 2022.
14. Shokouhinejad, H., *et al.*, Recent Advances in Malware Detection: Graph Learning and Explainability, Feb. 2025.
15. Nguyen, C.M., A study on graph neural networks and pretrained models for analyzing cybersecurity texts, 2021.
16. Majeed, A. and Rauf, I., Graph Theory: A Comprehensive Survey about Graph Theory Applications in Computer Science and Social Networks. *Inventions*, 5, 1, 10, Feb. 2020, doi: 10.3390/inventions5010010.
17. Cheng, Y., *et al.*, Advanced financial fraud detection using GNN-CL model, in: *2024 International Conference on Computers, Information Processing and Advanced Education (CIPAE)*, IEEE, pp. 453–460, 2024.
18. Jeyakumar, S.T., *et al.*, Detecting malicious blockchain transactions using graph neural networks, in: *International Symposium on Distributed Ledger Technology*, pp. 55–71, Springer Nature Singapore, Singapore, Nov. 2023.
19. XenonStack, Graph Neural Networks in Computer Vision - Complete Guide, https://www.xenonstack.com/insights/graph-neural-networks-in-computer-vision.
20. Cheng, D., Zou, Y., Xiang, S., Jiang, C., Graph neural networks for financial fraud detection: A review. *Front. Comput. Sci.*, 19, 9, 199609, Sep. 2025, doi: 10.1007/s11704-024-40474-y.
21. AssemblyAI, AI trends in 2024: Graph Neural Networks, https://www.assemblyai.com/blog/ai-trends-in-2024-graph-neural-networks/, 2024.
22. Pradhyumna, P., *et al.*, Graph neural network (GNN) in image and video understanding using deep learning for computer vision applications, in: *Second International Conference on Electronics and Sustainable Communication Systems (ICESC)*, Aug. 2021, IEEE, pp. 1183–1189.
23. Liu, J., Min, L., Huang, X., An overview of event extraction and its applications, Nov. 2021.
24. Shi, W. and Rajkumar, R., Point-gnn: Graph neural network for 3d object detection in a point cloud, in: *Proceedings of the IEEE/CVF conference on computer vision and pattern recognition*, pp. 1711–1719, 2020.
25. Hoang, T., GNN: Graph Neural Network and Large Language Model Based for Data Discovery, 2024, doi: 10.2139/ssrn.4935670.
26. Sharma, K., *et al.*, A Survey of Graph Neural Networks for Social Recommender Systems, Dec. 2022, doi: 10.1145/3661821.

27. Cao, Q., *et al.*, Popularity prediction on social platforms with coupled graph neural networks, in: *Proceedings of the 13th international conference on web search and data mining*, pp. 70–78, Jan. 2020.
28. Yu, Y., Qian, W., Zhang, L., Gao, R., A Graph-Neural-Network-Based Social Network Recommendation Algorithm Using High-Order Neighbor Information. *Sensors*, 22, 19, 7122, Sep. 2022, doi: 10.3390/s22197122.
29. Li, J. and Yang, G., Network embedding enhanced intelligent recommendation for online social networks. *Future Gener. Comput. Syst.*, 119, 68–76, Jun. 2021, doi: 10.1016/j.future.2021.01.017.
30. Wu, L., *et al.*, Deep Learning on Graphs: Methods and Applications, in: *Proceedings of the 29th ACM SIGKDD Conference on Knowledge Discovery and Data Mining*, Aug. 2023, pp. 5891–5892.
31. Hu, X., *et al.*, Cost-sensitive GNN-based imbalanced learning for mobile social network fraud detection. *IEEE Trans. Comput. Soc. Syst.*, 11, 2, 2675–2690, 2023.
32. Anoop, V.S., Subin Krishna, C., Govindarajan, U.H., Graph embedding approaches for social media sentiment analysis with model explanation. *Int. J. Inf. Manage. Data Insights*, 4, 1, 100221, Apr. 2024, doi: 10.1016/j.jjimei.2024.100221.
33. Thais, S., *et al.*, Graph Neural Networks in Particle Physics: Implementations, Innovations, and Challenges, arXiv:2203.12852v2 [hep-ex], Mar. 2022.
34. Martinez, J.A., Nguyen, T.Q., Pierini, M., Spiropulu, M., Vlimant, J.-R., Particle Generative Adversarial Networks for full-event simulation at the LHC and their application to pileup description, Dec. 2019, doi: 10.1088/1742-6596/1525/1/012081.
35. Frye, R.H. and Wilson, D.C., Comparative Analysis of Transformers to Support Fine-Grained Emotion Detection in Short-Text Data. *Proc. Int. Fla. Artif. Intell. Res. Soc. Conf. FLAIRS*, 35, 2022, doi: 10.32473/flairs.v35i.130612.
36. Rao, R. and Zhu, L., Phase transitions of correlated systems from graph neural networks with quantum embedding techniques, arXiv: 2404.08782v3, 2024.
37. Trivedi, D., Patrikar, K., Mondal, A., Graph-based networks for accurate prediction of ground and excited state molecular properties from minimal features. *Mol. Syst. Des. Eng.*, 9, 12, 1275–1284, 2024, doi: 10.1039/D4ME00113C.
38. Ceschini, A., *et al.*, From Graphs to Qubits: A Critical Review of Quantum Graph Neural Networks, Aug. 2024.
39. Fung, V., Zhang, J., Juarez, E., Sumpter, B.G., Benchmarking graph neural networks for materials chemistry. *npj Comput. Mater.*, 7, 1, 84, Jun. 2021, doi: 10.1038/s41524-021-00554-0.
40. Reiser, P., *et al.*, Graph neural networks for materials science and chemistry. *Commun. Mater.*, 3, 93, 2022.

41. Cheng, G., Gong, X.-G., Yin, W.-J., Crystal structure prediction by combining graph network and optimization algorithm. *Nat. Commun.*, 13, 1, 1492, Mar. 2022, doi: 10.1038/s41467-022-29241-4.
42. Maurizi, M., Gao, C., Berto, F., Predicting stress, strain and deformation fields in materials and structures with graph neural networks. *Sci. Rep.*, 12, 1, 21834, Dec. 2022, doi: 10.1038/s41598-022-26424-3.
43. Shi, X., Zhou, L., Huang, Y., Wu, Y., Hong, Z., A review on the applications of graph neural networks in materials science at the atomic scale. *Mater. Genome Eng. Adv.*, 2, 2, Jun. 2024, doi: 10.1002/mgea.50.
44. Yang, Z., Dong, Y., Deng, X., Zhang, L., AMGNET: multi-scale graph neural networks for flow field prediction. *Connect. Sci.*, 34, 1, 2500–2519, Dec. 2022, doi: 10.1080/09540091.2022.2131737.
45. Kurz, M., Beck, A., Sanderse, B., Harnessing Equivariance: Modeling Turbulence with Graph Neural Networks, Apr. 2025.
46. Sanchez-Gonzalez, A., Godwin, J., Pfaff, T., Ying, R., Leskovec, J., Battaglia, P.W., Learning to Simulate Complex Physics with Graph Networks, Feb. 2020.
47. Gao, R. and Jaiman, R.K., Predicting fluid–structure interaction with graph neural networks. *Phys. Fluids*, 36, 1, Jan. 2024, doi: 10.1063/5.0182801.
48. Pfaff, T., Fortunato, M., Sanchez-Gonzalez, A., Battaglia, P.W., Learning Mesh-Based Simulation with Graph Networks, Oct. 2020.
49. Rasool, A., Rahman, J.U., Iqbal, Q., Optimizing GNN Architectures Through Nonlinear Activation Functions for Potent Molecular Property Prediction. *Computation*, 12, 11, 212, Oct. 2024, doi: 10.3390/computation12110212.
50. Abbassi, O., Ziti, S., Belhiah, M., Lagmiri, S.N., Zaoui Seghroucheni, Y., GMPP-NN: a deep learning architecture for graph molecular property prediction. *Discov. Appl. Sci.*, 6, 7, 352, Jun. 2024, doi: 10.1007/s42452-024-05944-9.
51. Yan, C., *et al.*, RetroXpert: Decompose Retrosynthesis Prediction like a Chemist, Nov. 2020.
52. Mercado, R., *et al.*, Graph networks for molecular design. *Mach. Learn. Sci. Technol.*, 2, 2, 025023, Jun. 2021, doi: 10.1088/2632-2153/abcf91.
53. Horie, M. and Mitsume, N., Graph Neural PDE Solvers with Conservation and Similarity-Equivariance, May 2024.
54. Singh, G., Durbha, S., R, S.C., Maximising Weather Forecasting Accuracy through the Utilisation of Graph Neural Networks and Dynamic GNNs, Jan. 2023.
55. Zhu, J., Wang, Q., Tao, C., Deng, H., Zhao, L., Li, H., AST-GCN: Attribute-Augmented Spatiotemporal Graph Convolutional Network for Traffic Forecasting. *IEEE Access*, 9, 35973–35983, 2021, doi: 10.1109/ACCESS.2021.3062114.
56. Wu, Z., Graph WaveNet for Deep Spatial-Temporal Graph Modeling, in: *Proceedings of the Twenty-Eighth International Joint Conference on Artificial Intelligence (IJCAI-19)*, pp. 1907–1913, 2020.

57. Chou, Y.-T., Chang, W.-T., Jean, J.G., Chang, K.-H., Huang, Y.-N., Chen, C.-S., StructGNN: An efficient graph neural network framework for static structural analysis. *Comput. Struct.*, 299, 107385, Aug. 2024, doi: 10.1016/j.compstruc.2024.107385.
58. Tong, G., Qian, X., Liu, Y., Prognostics and Predictive Maintenance Optimization Based on Combination BP-RBF-GRNN Neural Network Model and Proportional Hazard Model. *J. Sens.*, 2022, 1–17, Apr. 2022, doi: 10.1155/2022/8655669.
59. Zhao, Y., Li, H., Zhou, H., Attar, H.R., Pfaff, T., Li, N., A review of graph neural network applications in mechanics-related domains. *Artif. Intell. Rev.*, 57, 11, 315, Oct. 2024, doi: 10.1007/s10462-024-10931-y.
60. Shi, H., Xu, H., Huang, Z., Li, Y., Wu, J., RoboCraft: Learning to See, Simulate, and Shape Elasto-Plastic Objects with Graph Networks, May 2022.
61. Lou, X., *et al.*, Learning object relations with graph neural networks for target-driven grasping in dense clutter, in: *International Conference on Robotics and Automation (ICRA)*, IEEE, pp. 742–748, 2022.
62. Jin, S., *et al.*, Differentiable hierarchical graph grouping for multi-person pose estimation, in: *Computer Vision–ECCV 2020: 16th European Conference, Proceedings, Part VII 16*, Aug. 2020, Springer, Glasgow, UK, pp. 718–734.
63. Diao, X., Chi, W., Wang, J., Graph neural network based method for robot path planning. *Biomimetic Intell. Rob.*, 4, 1, 100147, Mar. 2024, doi: 10.1016/j.birob.2024.100147.
64. Wang, Y., Li, Z., Barati Farimani, A., Graph Neural Networks for Molecules, pp. 21–66, 2023, doi: 10.1007/978-3-031-37196-7_2.
65. Leng, D., Guo, J., Pan, L., Li, J., Wang, X., Enhance Information Propagation for Graph Neural Network by Heterogeneous Aggregations, Feb. 2021.
66. Bongini, P., Bianchini, M., Scarselli, F., Molecular generative Graph Neural Networks for Drug Discovery. *Neurocomputing*, 450, 242–252, Aug. 2021, doi: 10.1016/j.neucom.2021.04.039.
67. Bongini, P., Graph Neural Networks for Drug Discovery: An Integrated Decision Support Pipeline, in: *2023 IEEE International Conference on Metrology for eXtended Reality, Artificial Intelligence and Neural Engineering (MetroXRAINE)*, Oct. 2023, IEEE, pp. 218–223, doi: 10.1109/MetroXRAINE58569.2023.10405789.
68. Wozniak, S., Janson, G., Feig, M., Accurate Predictions of Molecular Properties of Proteins *via* Graph Neural Networks and Transfer Learning. *J. Chem. Theory Comput.*, 21, 9, 4830–4845, May 2025, doi: 10.1021/acs.jctc.4c01682.
69. Omee, S.S., *et al.*, Scalable deeper graph neural networks for high-performance materials property prediction, Sep. 2021.
70. Zhu, S., *et al.*, Improved environmental chemistry property prediction of molecules with graph machine learning. *Green Chem.*, 25, 17, 6612–6617, 2023, doi: 10.1039/D3GC01920A.

71. Yao, R., *et al.*, Knowledge mapping of graph neural networks for drug discovery: a bibliometric and visualized analysis. *Front. Pharmacol.*, 15, May 2024, doi: 10.3389/fphar.2024.1393415.
72. Han, P., *et al.*, GNN-Retro: Retrosynthetic Planning with Graph Neural Networks. *Proc. AAAI Conf. Artif. Intell.*, 36, 4, 4014–4021, Jun. 2022, doi: 10.1609/aaai.v36i4.20318.
73. Gricourt, G., Meyer, P., Duigou, T., Faulon, J.-L., Artificial Intelligence Methods and Models for Retro-Biosynthesis: A Scoping Review. *ACS Synth. Biol.*, 13, 8, 2276–2294, Aug. 2024, doi: 10.1021/acssynbio.4c00091.
74. Jumper, J., *et al.*, Highly accurate protein structure prediction with AlphaFold. *Nature*, 596, 7873, 583–589, Aug. 2021, doi: 10.1038/s41586-021-03819-2.
75. Defferrard, M., Bresson, X., Vandergheynst, P., Convolutional Neural Networks on Graphs with Fast Localized Spectral Filtering, Jun. 2016.
76. Gao, S., *et al.*, GNN-GRN: A graph neural network model for reconstructing gene regulatory network. *Briefings Bioinf.*, 23, 5, 2022.
77. Wang, J., Ma, A., Ma, Q., Xu, D., Joshi, T., Inductive inference of gene regulatory network using supervised and semi-supervised graph neural networks. *Comput. Struct. Biotechnol. J.*, 18, 3335–3343, 2020, doi: 10.1016/j.csbj.2020.10.022.
78. Hamilton, W.L., Ying, R., Leskovec, J., Inductive Representation Learning on Large Graphs, Jun. 2017.
79. Chiang, W.-L., Liu, X., Si, S., Li, Y., Bengio, S., Hsieh, C.-J., Cluster-GCN: An Efficient Algorithm for Training Deep and Large Graph Convolutional Networks, May 2019, doi: 10.1145/3292500.3330925.
80. Chen, J., Ma, T., Xiao, C., FastGCN: Fast Learning with Graph Convolutional Networks *via* Importance Sampling, Jan. 2018.
81. Xiong, Z., *et al.*, Pushing the Boundaries of Molecular Representation for Drug Discovery with the Graph Attention Mechanism. *J. Med. Chem.*, 63, 16, 8749–8760, Aug. 2020, doi: 10.1021/acs.jmedchem.9b00959.
82. Zitnik, M., Agrawal, M., Leskovec, J., Modeling polypharmacy side effects with graph convolutional networks. *Bioinformatics*, 34, 13, i457–i466, Jul. 2018, doi: 10.1093/bioinformatics/bty294.
83. Gottlieb, A., Stein, G.Y., Ruppin, E., Sharan, R., PREDICT: a method for inferring novel drug indications with application to personalized medicine. *Mol. Syst. Biol.*, 7, 1, Jan. 2011, doi: 10.1038/msb.2011.26.
84. Gilmer, J., Schoenholz, S.S., Riley, P.F., Vinyals, O., Dahl, G.E., Neural Message Passing for Quantum Chemistry, Apr. 2017.
85. Wu, Z., *et al.*, MoleculeNet: A Benchmark for Molecular Machine Learning, Mar. 2017.
86. Zitnik, M. and Leskovec, J., Predicting multicellular function through multi-layer tissue networks. *Bioinformatics*, 33, 14, i190–i198, Jul. 2017, doi: 10.1093/bioinformatics/btx252.

87. Dries, R., *et al.*, Giotto: a toolbox for integrative analysis and visualization of spatial expression data. *Genome Biol.*, 22, 1, 78, Dec. 2021, doi: 10.1186/s13059-021-02286-2.
88. Wu, Z., Pan, S., Chen, F., Long, G., Zhang, C., Yu, P.S., A Comprehensive Survey on Graph Neural Networks, Jan. 2019, doi: 10.1109/TNNLS.2020.2978386.
89. Battaglia, P.W., *et al.*, Relational inductive biases, deep learning, and graph networks, Jun. 2018.
90. Satorras, V.G., *et al.*, E(n) Equivariant Graph Neural Networks, in: *International conference on machine learning*, pp. 9323–9332, Jul. 2021.
91. Gasteiger, J., Groß, J., Günnemann, S., Directional Message Passing for Molecular Graphs, Mar. 2020.
92. Schütt, K.T., Kindermans, P.-J., Sauceda, H.E., Chmiela, S., Tkatchenko, A., Müller, K.-R., SchNet: A continuous-filter convolutional neural network for modeling quantum interactions, Jun. 2017.
93. Chanussot, L., *et al.*, Open Catalyst 2020 (OC20) Dataset and Community Challenges. *ACS Catal.*, 11, 10, 6059–6072, May 2021, doi: 10.1021/acscatal.0c04525.
94. Xie, T. and Grossman, J.C., Crystal Graph Convolutional Neural Networks for an Accurate and Interpretable Prediction of Material Properties. *Phys. Rev. Lett.*, 120, 14, 145301, Apr. 2018, doi: 10.1103/PhysRevLett.120.145301.
95. Marek Nečada, M.N. and Päivi Törmä, P.T., Multiple-Scattering T-Matrix Simulations for Nanophotonics: Symmetries and Periodic Lattices. *Commun. Comput. Phys.*, 30, 2, 357–395, Jan. 2021, doi: 10.4208/cicp.OA-2020-0136.

11

Enhanced Integrated Spatio-Temporal Graph Convolutional Network for Accurate and Efficient Traffic Prediction

Tintu George*, **Ginne M. James and A. Senthil Kumar**

Department of Computer Science with Data Analytics, Sri Ramakrishna College of Arts & Science, Coimbatore, Tamilnadu, India

Abstract

The proposed chapter introduces an innovative deep learning model, Integrated Spatio-Temporal Graph Convolutional Network (ISTGCN), for efficient and accurate traffic forecasting. Unlike traditional algorithms, ISTGCN effectively captures complex spatial and temporal dependencies within road network data by integrating multi-layer Graph Convolutional Networks (GCNs) with sequence learning and spatio-temporal convolutional blocks. This enables robust end-to-end extraction of both short-range and long-range patterns, improving prediction accuracy. ISTGCN demonstrates excellent scalability and adaptability to the dynamic nature of traffic data, requiring fewer parameters and offering efficient training compared to conventional methods. Extensive experiments on real-world datasets show that ISTGCN consistently outperforms existing algorithms, achieving lower RMSE, MAE, and MAPE values, thus proving its reliability in modeling intricate traffic flow dynamics. Practically, ISTGCN supports real-time traffic management by providing precise forecasts, aiding in optimal signal control, congestion reduction, and emergency response planning. It also benefits long-term urban mobility planning by predicting evolving traffic patterns, enabling informed policy and infrastructure decisions. Its lightweight architecture makes it cost-effective for deployment on edge devices and distributed systems, aligning with smart city initiatives.

**Corresponding author*: tintueldo@gmail.com

J. Ramkumar, Sridaran Rajagopal, B. Suchitra and S. Balamurugan (eds.) Graph Neural Networks: Concepts and Applications, (459–494) © 2026 Scrivener Publishing LLC

Keywords: Spatio-temporal data, traffic forecasting, deep learning, graph convolutional networks, urban mobility, intelligent traffic management, temporal convolutional networks

11.1 Introduction

Accurate prediction of traffic movement has become a fundamental requirement in urban transport system management, especially in the context of urbanization and increased numbers of vehicles on roads. As cities are being transformed into smart cities with the integration of new technologies into systems, it is important to include such technology to manage city mobility and ensure sustainable development [1]. Among its primary issues here is traffic congestion, which, besides causing delay of speed of production, results in environment degradation and to public health risks. In a bid to mitigate most of these ills, real-time traffic prediction has become a central application facilitating traffic smoothing, reducing delay, and enhancing road safety.

Traffic prediction is forecasting traffic states in the future—e.g., speed, volume, or density—in consideration of current traffic states. The system needs to facilitate many ITS functions, for example, dynamic route guidance, adaptive traffic signal control, traffic congestion control, and emergency response planning [3]. Realistic traffic forecasting and modeling remain a demanding task due to inherent complexity, dynamic, nonlinearity of traffic systems. Traffic flow is controlled by a high level of interdependent variables such as road geometry, driver behavior, accidents, weather, and city events, so that the forecasting procedure is a complex and situation-specific process.

Conventional traffic forecasting model Auto Regressive Integrated Moving Average (ARIMA) extensions have gained popularity because of interpretability and understandability. They are the optimal of their kind to manage linear temporal relations and do well when stationarity of conditions under short-term forecasting is valid. They do not adequately deal with the nonlinear and nonstationary nature that does exist in actual traffic networks [5]. Most importantly, common statistical techniques do lose spatial relationships between roadway sections and analyze flat time-series data at a point. Thus, they are unable to choose out the effect which a portion of a transport network may have on a different one—a very restrictive one when trying to use it with large transport networks.

Deep learning has transformed dramatically the practice of time-series forecasting over the last few years. These include models like Convolutional

Neural Networks (CNNs), Recurrent Neural Networks (RNNs), i.e., Long Short-Term Memory (LSTM), Gated Recurrent Units (GRUs), which have proven to be highly effective in abstracting intricate temporal dependencies [7]. CNNs are also suited to identify the local patterns in structured data are optimally used in image and video processing. For the traffic data in grid form, CNNs can abstract the short-range spatial information. Sequence data-based CNNs are very effective at modeling long-term dependencies and very well suited for temporal analysis. Accuracy in traffic estimation is greatly enhanced by using deep-learning models like CNNs compared to the conventional methods. Though extremely efficient, RNNs and CNNs are plagued by the very disadvantage that their application relies on road networks that are not grids nor sequences but ought to be graphs. CNNs are based on Euclidean data structure, which in most cases does not exist in road networks whose links are not grid and irregular. RNNs, though capable of learning temporal relations, cannot learn spatial relations when alone. Hybrid CNN-RNN approaches have also been attempted, but this should add model complexity and training challenges without being able to learn traffic system graph structure to any notable degree.

In response to such limitations, researchers have looked towards Graph Neural Networks (GNNs); a deep learning technique has been suggested and constructed with the goal of operating on graph-structured data. GNNs extend the convolution operation to non-Euclidean spaces and are therefore ideal for traffic prediction problems where data is already in graph form. In the structural graph of a road network, intersections or traffic sensors may be nodes and physical or functional relations among them as edges. From what is known of structural graphs, GNNs can learn to represent spatial relations in the entire network in a compact form. Of all GNN architectures, Graph Convolutional Networks (GCNs) is significantly used in collecting information from neighboring nodes in a weighted and organized way. Using this spatial summation, GCNs are powerful enough to take into account the effect that the neighboring roads exert on each other, and the predictions improve. GCNs are still insufficient for traffic prediction, though, because they fail to include temporal dynamics in the model. Because traffic patterns fluctuate dynamically over time, the mature model would need to embody spatial and temporal relationships together to learn as well as to make predictions about subsequent states of traffic.

This necessity to model together gave birth to Spatio-Temporal Graph Convolutional Networks (STGCNs), which combine GCNs for learning spatially with temporal convolution or recurrent layers to model temporal series data [21]. They are far more advanced in traffic forecasting research as they combine spatial learning and temporal learning into a single framework.

Nevertheless, current STGCN models have certain shortcomings. First, they use static adjacency matrices that are unable to capture traffic interactions' dynamic characteristics. Second, they use uniform importance on all temporal features and thus lose the capability of assigning greater weightage to significant time steps or events. Thirdly, when network depth is deep, learning is not stable because of problems such as vanishing gradients and information deterioration.

To counter all these issues, this proposed work introduces the Enhanced Integrated Spatio-Temporal Graph Convolutional Network (E-ISTGCN)—a novel structure borrowed from the initial STGCN concepts but improved with some useful adjustments to optimize the performance, stability, and scalability accordingly. The E-ISTGCN architecture consists of a number of spatio-temporal blocks whose function is to identify sophisticated interactions between spatial and temporal variables [11]. It employs an adaptive graph learning process, temporal attention mechanism, and residual learning blocks to build a strong and dynamic prediction model.

The adaptive graph learning block presented enables the model to learn the dynamic adjacency matrix at the time of training and hence be able to discover the implicit relations that may not necessarily exist due to the physical geography of the road network. This is useful while estimating functional relations generated by traffic patterns, e.g., congestion propagating point-to-point. The attention mechanism at different times permits the model to give varying weights to varying times such that the model gives a higher weight to significant past trends and breaks. This renders the model more proficient at identifying cyclic trends, one-off disruptions, and spike oscillations in traffic trends. Lastly, the residual connections make gradient flow stronger and support the model to train sufficiently even in deeper architectures without compromising learned representations from layer to layer.

The E-ISTGCN is tested and learned from widely used datasets, i.e., PeMSD7, a collection of real-world data gathered through the Caltrans Performance Measurement System (PeMS) at 39,000 sensor-station sites disseminated throughout California. The datasets are made up of speed readings which were gathered through hundreds of sensors distributed over the road network with sufficient spatio-temporal information. The proposed model is trained in a supervised learning environment, and it is evaluated with standard error metrics including Mean Absolute Error (MAE), Root Mean Square Error (RMSE), and Mean Absolute Percentage Error (MAPE). Experimental findings indicate that E-ISTGCN greatly surpasses standard models (e.g., ARIMA, LSTM, GRU), and also other prevailing deep learning-based spatio-temporal models (e.g., DCRNN, STGCN). In addition to having predicted fewer error rates, it also possesses improved computational efficiency and training stability. Ablation study is performed to compare the contribution of each element of a model, revealing that adaptive graph learning, temporal attention, and residual fusion all make significant contributions towards overall improved model performance.

Apart from precision, E-ISTGCN also considers usability in use. It is modulable and therefore simple to embed in urban traffic management and transportation planning software [13]. The model is usable for efficient inference and can be used for real-time applications on intelligent mobility platforms like dynamic signal control, routing forecast, and autonomous vehicle coordination. Its capacity to generalize between varied traffic conditions and still be able to perform despite missing or noisy data conditions equally makes it outstanding and reliable.

Globally, the Enhanced Integrated Spatio-Temporal Graph Convolutional Network (E-ISTGCN) is the benchmark of traffic forecasting research. In a balanced integration of spatial graph learning, temporal attention, and residual strategies, the model addresses the built-in challenges of spatio-temporal forecasting of complex urban domains. Its powerful empirical performances, flexibility, and efficiency make it an inevitable utility for the next generation of smart transportation systems [15]. This paper not only introduces the new model but also further extends the current research in how spatio-temporal deep learning models are made scalable, interpretable, and deployable in real-world applications.

11.2 Related Work

11.2.1 Legacy Traffic Forecasting Techniques

Legacy time-series techniques have been the reliable workhorse of traffic flow forecasting due to ease of use, simplicity of comprehension, and low computational demands. Legacy techniques are ARIMA, Seasonal ARIMA (SARIMA), Kalman Filters, and Bayesian Networks [18]. Legacy techniques rely primarily on historical time-series for forecasting future traffic conditions. Though these methods perform well enough to make reasonable short-term predictions under homogeneous traffic conditions, they perform very badly to address the nonlinear, non-stationary, and highly dynamic characteristics of city traffic networks. The root flaw of such conventional methods is that complex spatial relationships cannot be incorporated.

A single-node analysis of an isolated traffic node (sensor or junction) is done, in a vacuum of spatial influence injected by surrounding nodes. Illustratively, bottlenecks within one road segment tend to spill over into other adjacent segments—something beyond the scope of conventional models [20]. Furthermore, the models are severely constrained by scalability; the bigger and more complicated traffic networks grow, the worse they perform. Current studies, for instance [9] indicate that while ARIMA and SARIMA models are good on small networks, their performance is significantly affected when they are exposed to real-time city-level data. Furthermore, such models are linearity data-dependent, and this will not be suitable in the detection of sudden changes by the external activities such as accidents, constructions, or climatic changes. As the demand for robust and precise prediction models in quantum jumps, especially with the case of smart cities, these limitations of traditional models pave the way for more learning-based, adaptive techniques.

11.2.2 Deep Learning Techniques

With evolution of deep learning, tremendous improvement in modeling complex temporal patterns has been observed. Recurrent Neural Networks (RNNs) and its variants such as Long Short-Term Memory (LSTM), Gated Recurrent Units (GRUs) have been used extensively for forecasting time-series. These methods are very effective in modeling short, long-range temporal structures and have the capability to learn non-linear relationships from data without feature engineering. Ref. [14] proposed short-term traffic forecasting model based on integration of CNN and GRU layers. Local temporal information was obtained using CNN layers and sequential dependency using the GRUs. Despite being well-positioned for local predictions, the model could not leverage spatial information and thus folded over large-scale, complex data.

An LSTM-based approach was also presented by [6] for capturing temporality, which worked very well in time dependency detection but did not work when there is spatial heterogeneity. Deep learning performance has been attempted to be improved by attention mechanisms. An attention-based CNN model was introduced by [17] that improved the performance by providing dynamically changing weights to different time steps. However, this was computationally expensive and hence could not be implemented in real time. Ref. [10] introduced dual-stage attention (spatial and temporal) for an RNN model that greatly improved accuracy but could only be scaled for application on small-scale urban traffic systems. Deep learning models are however more accurate and adaptable than baseline models but deteriorated once spatial relations are no longer considered. Further, most of the models are training time- and memory-demanding, i.e., less feasible for real-time processing.

11.2.3 Graph Neural Networks (GNNs)

Graph Neural Networks (GNNs) now emerge as a potential alternative for traffic forecasting because they inherently contain the ability to model spatial relations in graph data. Roads and intersections within a road network can readily be modeled as edges and nodes within a graph, and GNNs can learn high-order spatial interactions between various points within the network.

The most popular type of GNN is Graph Convolutional Networks (GCNs), and they have localized message passing and feature aggregation, and therefore they have tremendous potential to model spatial relationships. One variant of this class was presented with the introduction of Spatio-Temporal Graph Convolutional Networks (ST-GCNs). ST-GCNs merge GCNs with temporal models such as temporal convolutions or RNNs to produce a spatio-temporal model of traffic flow. Ref. [16] presented an attention-augmented GCN that learned dynamic neighbor node impacts to improve predictive functionality. Yet, their model lacked scalability in large graphs since they employed static adjacency matrices and computationally costly operations. Ref. [19] introduced an adaptive GCN model that adaptively adjusted spatial relations dynamically based on real traffic conditions. Graph adaptation with dynamics enhanced spatial learning but was unable to capture fully long-term temporal relationships, especially under anomalous conditions.

Apart from this, Ref. [4] proposed the Graph Wave Net model that combined graph adaptation learning and temporal convolution. Although it broke state-of-the-art performance on multiple benchmarks, its training overhead along with memory requirement was still an issue. Ref. [2] proposed STSGCN (Spatio-Temporal Synchronous Graph Convolutional Network) in another recent work, which employed synchronous spatial-temporal learning for better learning of reciprocal dependencies. Although the model was more precise than current methods, it was also less robust with noisy or missing sensor data. Additionally, its synchronous update mechanism performed worse when dealing with stochastic traffic patterns with high dynamism. Ref. [8] attempted to alleviate such an issue by employing a multi-scale ST-GCN with different temporal filters in predicting short-, middle-, and long-term duration. The model was made more general, especially for heterogeneously-temporally-spaced resolution data, although at higher model complexity costs and longer inference lengths. The above has not only attempted to generalize the application of GCNs to traffic forecasting.

For instance, Ref. [12] introduced a spatial-temporal attention GAT that can learn dynamically to choose influential nodes and time steps. The model was accurate but not interpretable and required tedious hyperparameter tuning. All these indicate that while the GNN has significantly enhanced modeling spatial relations, there is still room for improvement to address such challenges as scalability, model interpretability, temporal robustness, and training efficiency.

11.3 Methodology

The fundamental purpose of the Enhanced Integrated Spatio-Temporal Graph Convolutional Network (E-ISTGCN) is to predict and represent the traffic's conditions by learning from spatial as well as temporal relationships in traffic streams. This section elaborates on the mathematical problem formulation, the E-ISTGCN model architecture, its training process, and the evaluation measures used to validate its performance.

11.3.1 Formulation of Problem

Traffic forecasting is a paradigmatic case of a spatio-temporal forecasting task, where one needs to forecast future traffic states given past observations. For urban transportation networks, traffic conditions like speed, volume, and occupancy at a given location and time are determined by neighboring spatial nodes as well as temporal events which is shown in Figure 11.1. Successfully modeling such relations demands a system that can deal with both the spatial organization of road networks and dynamic evolution of traffic through the time.

Traffic prediction is a space-time problem whereby future traffic conditions are predicted according to past experiences. In this case, considering a network where G=(V,E,A).

V is the set of nodes (road segments or sensors),

E is the set of edges for the connection among the nodes, and

A is the spatial adjacency matrix among the nodes.

The goal is to forecast future traffic states $X_{t+1,} X_{t+2}, \ldots, X_{t+TX}$ based on the past traffic states:

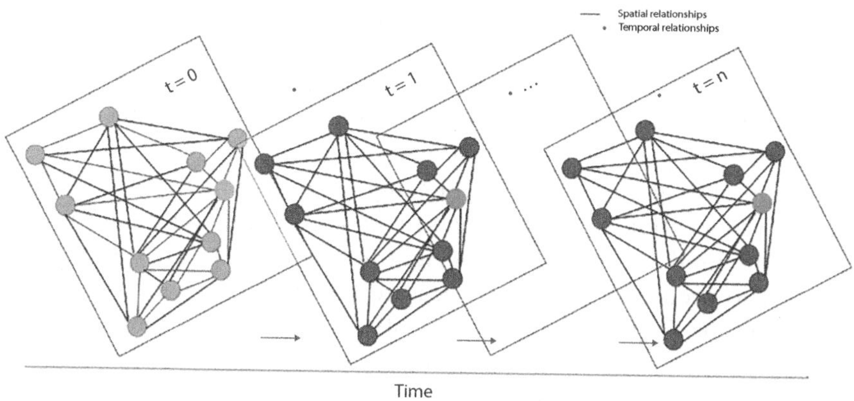

Figure 11.1 Spatial time relationship.

$X_{t,} X_{t-1}, \ldots, X_{t-T+1}$.

Mathematically:

$X_{t+T} = f(X_{t,} X_{t-1}, \ldots, X_{t-T+1}, G)$ where f is the E-ISTGCN model.

11.3.2 E-ISTGCN Architecture

To overcome the shortfalls of current GCN-based traffic forecasting models—like non-scalability, inability to learn long-term dependencies, or having high computational complexity—we introduce the Enhanced Integrated Spatio-Temporal Graph Convolutional Network (E-ISTGCN). The structure of E-ISTGCN has three primary components:

i. Graph Convolutional Network (GCN) for Spatial Modeling
ii. Convolutional Sequence Learning for Temporal Modeling
iii. Integrated Spatio-Temporal Convolutional Blocks

Key Components and Ordering

Adjacency Matrix (A) the procedure starts with the adjacency matrix A desc N x N describing the spatial topology of the traffic network. Each element of the N×N matrix represents the relation or influence of a pair of nodes. Self-loops are addressed by adding the identity matrix I, in such a manner that each node accounts for its own characteristics while aggregating. In Feature Refinement Layer (WaN) the raw node features (traffic speed, volume, etc.) are then fed into a parameterized feature refinement layer with a weight matrix. It maps the input to an expressive size N×F, where: N indicates the number of the nodes, F denotes the number of features.

In Concatenated Feature Matrix (FRL) the feature matrix is concatenated across various time steps to form the Feature Refinement Layer (FRL) matrix F, of size T×F. Here: T denotes the number of time steps. A row is a refined feature vector at some timestamp t_i. This represents the temporal progression of spatially refined features adequately.

Temporal weighting (W_{1Temp}): The temporal weight matrix $W_{1Temp} \in R^{T\times d}$ learns to concentrate on various steps in the sequence of traffic. The weights enable the model to give higher weight to most significant past patterns in order to predict in the future. Feature Integration Equation calculates a final output X_{fl} by making a combination of spatial and temporal clues using the equation shown in Figure 11.2.

The illustration depicts an E-ISTGCN, which is an essential component of deep learning structures for various applications such as traffic forecasting, human action recognition, or any node-wise time-series forecasting. The method starts with the input adjacency matrix A, which is an N×N matrix corresponding to the connections between the nodes of a graph with size N, this matrix is multiplied by the learnable weight matrix WaN to create the first node level features in a Feature Refinement layer and that outputs an N×F output, where F is the number of features. The last refined features that we are calling X_{frl}, will go through a Feature Attention Layer and will just be multiplied by an attention weight matrix called Watt. So, $X_{frl} = W_{att} * X_{\ell-1}$, where $X_{\ell-1}$ has the features from the previous layer. Now, call a Concatenated Feature Matrix (CFM), which will be a 4-d matrix that takes the original features of F, and the refined features of FRL, and concatenates

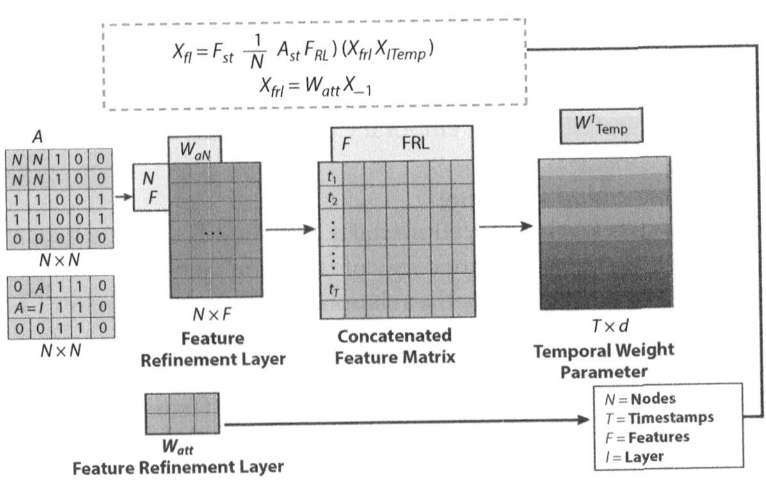

Figure 11.2 Architecture of E-ISTGCN.

them across time steps t1. The CFM will be 4-d, in respects to the aspects of spatial and temporal representation. We will then take that matrix and multiply it by the Temporal Weight Parameter matrix W1Temp, where W1Temp is T×d, and can learn either long-term or short term relationships across the time steps as the model trains.

The final output of this module will use the following formulas;

$$X_{\mathrm{fl}} = F_{\mathrm{st}} \frac{1}{N}\left(A_{st} F_{\mathrm{RL}}\right) + \left(X_{\mathrm{frl}}\ W_{1\mathrm{Temp}}\right)$$

which captures spatial relationships (by Ast), refined feature relationships (by FRL), and can attend to features over time-lengths (by W_{lTemp}). In summary the spatial-temporal representation situation is supposed to get as much together across space and time in order to create more informative features for downstream application.

11.3.2.1 Graph Convolutional Network (GCN) for Spatial Modeling

Modeling spatial interactions is a key part of accurate traffic forecasting because traffic behavior in one region of a transportation network is often dictated by that in adjacent regions. In order to accurately model such spatial interactions, the Graph Convolutional Network (GCN) method has emerged as a strong tool. Contrary to traditional Convolutional Neural Networks (CNNs), utilized primarily to cope with Euclidean data structures such as images. GCNs are also tailored to operate on non-Euclidean space, i.e., graphs, which are most appropriate to portray complex traffic networks.

Road intersections or road sections within a traffic network are referred to as nodes, and links or roadways connecting them as edges. This leads to a graph-based representation where the spatial information is embedded within the graph structure. But since traffic networks are non-linear—while images are grid-like in structure—the standard CNN operations cannot be directly used. The graph structure non-linearity and differences in number of the neighbors for each and every node make it necessary to create custom convolution operations that could handle graph-structured data. GCNs address this issue by extending the convolution operation from regular grids to graphs with any structure. Two primary approaches to graph convolution are spectral-based, spatial-based. Both approaches are attempting to perform aggregate information from local neighborhood of a node to informative node representations that encode graph structure and also the node features.

An adjacency matrix A∈RN×N is central to capturing spatial dependencies among the traffic nodes within a road network. Each element Aij indicates the strength (or influence) of the connection between node iii and node j. In this study, the adjacency matrix is formed based off the physical topology of the traffic network, where nodes represent traffic sensors (or road intersections) and edges are the direct road connectivity between them. More explicitly, if both sensors measured traffic on directly connected road segments (e.g., adjacent intersections or links), then Aij will equal 1; otherwise, it will equal 0. To maintain self-awareness in learning, self-loops are added through appending the identity matrix I, so A~=A + I. When physical connectivity falls short in modeling dynamic interactions (e.g., when correlations between traffic flow rather than physical connectivity apply), distance-based weighting schemes are implemented. In this case, we apply Gaussian kernel functions or thresholded inverse distances to assign edge weights, for neighboring nodes that are within a spatial threshold. The proposed flexible architecture shown in Figure 11.3 allows the model to leverage both static topological structure and dynamic spatial influence, which improves the quality of graph-based feature aggregation from the perspective of training.

Spectral Graph Convolution

Spectral graph convolution also known as graph signal processing and it is based on Eigen decomposition of the graph Laplacian. A graph G = (V , E), V being the set of vertices and E being the set of edges, the normalized graph Laplacian can be written as:

$$L = I - D^{-\frac{1}{2}} A D^{-\frac{1}{2}}$$

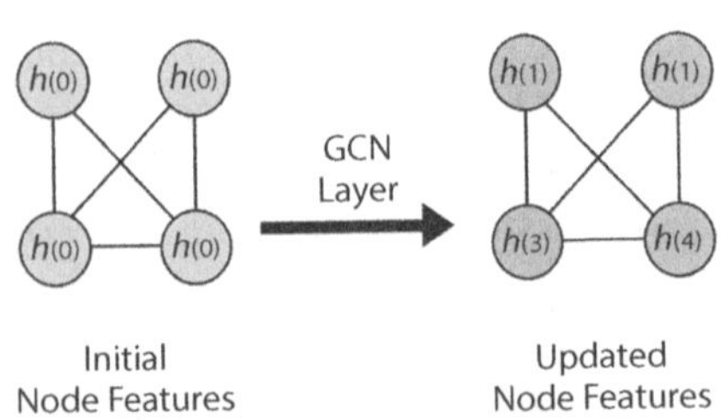

Figure 11.3 Flowchart showing the architecture of a GCN model.

Where A denotes the adjacency matrix of the graph, D represents the degree matrix, and I denotes the identity matrix.

The graph Laplacian preserves the graph structure and allows us to perform operations in the spectral domain *via* its eigenvectors and eigenvalues. Spectral graph convolution projects the input features X into spectral domain using eigenvectors U of Laplacian and invocation of convolution as point-wise multiplication in spectral space, algorithm isn't highly scalable and computationally complex, especially for big networks like traffic networks. Furthermore, the spectral filters typically are not transferable from one graph with a different structure to another. To overcome these constraints, Kipf and Welling introduced a simplified and scalable variant of spectral GCN that does not use Eigen decomposition and defines the convolution operation.

From adjacency matrix A and from node feature of matrix X, the graph convolution can be expressed as:

$$H^{I+1} = \sigma(D^{-\frac{1}{2}} A D^{-\frac{1}{2}} H^{(1)} W^{(1)} + B^{(1)})$$

Where $H^{(1)}$ = Input node features at layer 1, $W^{(1)}$ and $b^{(1)}$ = Trainable weights and biases, D = Degree matrix of A, σ = Activation function (ReLU).

Spatial Graph Convolution

Spatial-based GCNs parameterize convolution directly on the graph in the spatial space. A node updates its representation within this model by aggregating neighboring features. Neighborhood aggregation is here conceptually similar to the receptive field in CNNs, where each output at a pixel is a function of a local patch of input pixels. Formally, spatial convolution for a node V_i is expressed as

$$h_i^{(l+1)} = \sigma\left(\sum_{j \in N(i) \cup \{i\}} \frac{1}{c_{ij}} + W^{(1)} + h_j^{(1)} \right)$$

Where $H^{(l)}_i$ indicates the feature vector of node i at layer l, N(i) denotes the neighbors of node i, c_{ij} represent normalization constant, $W^{(l)}$ indicates learnable weight matrix, σ represent activation function.

Such an approach is simpler to analyze and more adaptable as it describes the spatial relationships in an obvious way by aggregating data of a node's genuine neighbors. It is also suitable for topologies that are dynamic in nature, which is beneficial for traffic networks where road connections or weights may alter dynamically.

Application to Traffic Networks

In traffic prediction, GCNs encode spatial correlations between isolated points in the transportation network. The graph node is a road segment, an intersection, or a traffic sensor, and the edges are functional or physical relationships between the nodes (i.e., the correlations of the traffic flow). The input to a GCN normally is a feature matrix, where

$$X \in ^{RN \times F}$$

The learned representation captures spatial context of each node as informed by information from the adjacent nodes. To capture spatial heterogeneity of traffic flow better, a few models also incorporate weighted adjacency matrices whose weights are derived from road distance, traffic similarity, or even learned while being trained. This makes the GCN learnable for strong vs. weak connections in the network.

Incorporation with Temporal Modeling

While GCNs can model spatial relationships well, traffic data is spatio-temporal in nature since the state at every node changes over time. To address the temporal component, GCNs are usually paired with temporal models such as Recurrent Neural Networks (RNNs), Long Short-Term Memory (LSTM) networks, Temporal Convolutional Networks (TCNs), or self-attention mechanisms. In such a framework, GCNs extract spatial features for every time step and input them into the temporal model in order to capture sequential dynamics. A dual solution is capable of the network to learn spatial patterns (e.g., the impact that traffic jam at one node would have on neighboring nodes) as well as temporal patterns (e.g., rush hour activity, weekend traffic) simultaneously.

Strengths and Challenges

Graph Convolutional Networks (GCNs) is very strong deep learning architecture in dealing with graph-structured data, which becomes extremely relevant when modeling systems as complex as transportation networks. Transportation systems are becoming extremely large and intricate, and hence there is a requirement of using models that can deal with extremely large graphs with numerous nodes and varied connectivity patterns. Optimally designed GCN instantiations are particularly well-equipped to meet this demand because they are scalable, flexible, and expressive.

One of the biggest benefits of GCNs is that they can easily work with graphs with a hugely huge number of nodes and edges. Scalability is very important when it comes to traffic prediction in cities, where the number of sensors to monitor or road sections could be overwhelmingly gigantic. Sparse computation and efficient matrix computation methods enable GCNs to pass information through graphs without paying memory or computation penalties. Hence, GCN-based models can be deployed in real-time or near-real-time applications, enabling dynamic traffic management and prediction systems at the city level.

The second significant advantage of GCNs is that they are able to deal with different graph structures and edge definitions. There are multiple kinds of relationships in road traffic networks that may be encoded as edges—such as being simple road connectivity, past correlation of traffic over segments, or even learned correlations symbolizing latent interaction throughout the network. GCNs are general enough to accept weighted, directed, or unoriented edges and therefore can capture physical road connectivity alongside functional similarity of traffic nodes. The flexibility allows the model accurately to represent the real-world richness of transport networks.

GCNs work by carrying out convolutional operations in a node's neighborhood, gathering information from neighboring nodes through successive layers. This multi-hop neighborhood aggregation is optimally suited to capturing global and local spatial patterns. Local space patterns retain short-range dependency, i.e., traffic behavior among neighboring intersections, whereas global patterns preserve long-range effects, e.g., congestion ripple effect spreading around the city. Piling up a few layers of GCNs, the model gets the capability of consuming information from distant regions of the graph and enhancing its awareness of intricate traffic behaviors.

Despite having these advantages, GCNs have some constraints. Among them is the issue of over-smoothing, in which data is excessively aggregated among layers. With increasing depth GCNs, as more and more layers are stacked on top of each other, node features in different regions of the graph could end up indistinguishably identical from one another. This layer-by-layer combination will hurt the model's discriminative power and produce poor-quality node representations. For the traffic network scenario, this implies that the model is unable to differentiate between nodes that have varying traffic patterns, thus reduced prediction capability. Another restriction comes from the application of static adjacency matrices in graph modeling.

Although such matrices work well in light of permanent road links, they become constrained while describing changing over time relationships between nodes. Traffic networks are dynamic in time—incidents like road closures, accidents, road works, or changes in travel demand can modify effective connectivity among nodes over time. A static adjacency matrix cannot capture these temporal dynamics and can result in modeling errors of current or future traffic conditions. Dynamic or adaptive graph topologies have been considered in some of the latest work to avoid this deficiency at the cost of higher computational and design complexity. GCN's learned node embeddings are also very hard to interpret. GCNs are strong in modeling spatial relationships and projecting them into high-dimensional feature spaces but learned representation may be abstract and non-interpretable in the case of physical road network attributes. For transportation planners and engineers, not only must they know that a specific location is forecasted to be congested, but they must also understand why the model predicted this. GCNs, particularly when deep neural layers are added, are black boxes and can be little explanatory of the way they are coming to their conclusion. This interpretability issue prevents the use of the model in real-world scenarios where explainability is critical, such as public infrastructure planning or real-time traffic control decision-making.

11.3.2.2 Convolutional Sequence Learning for Temporal Modeling

For the purpose of effectively capturing temporal dynamics of traffic flows, E-ISTGCN employs 1D Convolutional Neural Networks (CNNs) for modeling temporal dependency. While conventional sequence learning techniques such as Recurrent Neural Networks (RNNs), Long Short-Term Memory (LSTM) networks are extensively used in temporal analysis, these are at a disadvantage in terms of low training speed and parallelizability. 1D CNNs, on the other hand, offer a scalable and high-speed alternative and therefore are best suited for real-time traffic applications. In our approach, the temporal behavior of a node is modeled as a sequence of feature vectors at a sequence of time steps. The sequences are handled as one-dimensional signals, and 1D convolutional filters along the time dimension are utilized to extract temporal features. Convolution operation can mathematically be described as:

$$\mathrm{h_t} = ReLU\left(\mathrm{W_c} * \mathrm{X_t} + \mathrm{b_c}\right)$$

where Convolution operation, W_c and b_c kernel weights and biases, ht temporal features.

By stacking multiple temporal convolutional layers with various kernel size and receptive fields, the model can capture many types of temporal patterns effectively. This ranges from short-term traffic oscillation, for instance, sudden jam due to an accident, to long-term patterns, such as cyclical patterns of rush hour. Furthermore, the temporal convolutions are very efficient when it comes to training time as well as inference time, allowing for faster convergence along with reduced latency. By virtue of their efficiency coupled with their capability to model fine-grained temporal dynamics, 1D CNNs exhibit an essential characteristic in the E-ISTGCN model towards accurate and timely traffic prediction.

11.4 Integrated Spatio-Temporal Convolutional Blocks

In order to traffic patterns model truly realistic manner, spatial interactions of road sections must be both captured and captured as they change over time. Conventional techniques conventionally treat spatial and temporal dependencies as two distinct problems, which reduces the capacity of these models to successfully capture traffic data's inherently interdependent nature. In a bid to solve this, we encourage the utilization of Spatio-Temporal Convolutional Blocks (STCBs) as building blocks of our network. The blocks facilitate simultaneous feature learning along spatial and temporal dimensions by combining graph convolution and temporal convolution operations into a single architecture. Each spatio-temporal convolutional block is built to learn complex spatio-temporal patterns by successively carrying out three primary operations:

Graph Convolutional Layer (GCL): Spatial Feature Extraction
Graph Convolutional Layer is the first building block of both the STCB and spatial modeling module. Whereas the vision processing traditional convolutional layers operate in grid-structured data (i.e., pixels), the GCL operates for graph-structured data. That is similar to traffic networks first-hand because sensor input mounted on road segments comes at random intervals, hence non-Euclidean topology. Graph nodes are sensor positions and edges are influence zone between sensors or road segments in traffic forecasting. The GCL maintains connections as an adjacency matrix such that every node can learn from neighboring nodes. Similarly, the manner in which traffic condition at a given point influences neighboring points, i.e., similar to how congestion builds up upstream or downstream on an arterial road-street. For example, during the process of sensing

deceleration through a sensor, the GCL enables the model to estimate whether the transition would probably influence the adjacent nodes, mimicking actual spatial connections like bottlenecks, merge streams, or ramp entrances. Through the utilization of different GCLs, the model can learn higher-order space relations beyond neighbors and generalize even further on large-scale road networks.

Temporal Convolutional Layer (TCL): Temporal Pattern Extraction
After following the spatial properties through the assistance of the GCL, traffic evolution over time is the second to be managed. This is done by the Temporal Convolutional Layer (TCL). Instead of depending on the traditional Recurrent Neural Networks (RNNs) like LSTMs, GRUs—the ones marked by sequential temporal processing of time steps and enormous computational needs—TCL makes use of one-dimensional convolutions over the time axis. This has some advantages. First, TCLs are extremely easy to parallelize and thus training and inference would be more efficient. Second, TCLs provide better long-term dependency handling by dilated convolutions that allow the network receptive field to increase without the associated increase in model parameter size. Third, through temporal layer stacking, multiple layers, stacked on top of a single layer, trend like rush hour peak, day cyclical traffic during the day, or ramp-up and ramp-down traffic build-up and fall-off could be simulated. The TCL allows every sensor node to learn how its traffic readings evolve over time and how experience can be translated into future state. Temporal modeling is required for predictive applications where future speed or volume of traffic must be predicted from past observation.3. Residual Connections: Stability and Gradient Flow. In a manner that facilitates deep architecture training and proper gradient propagation between layers, every STCB contains residual connection between the block's input/output. This configuration prevents vanishing gradient problem of deep neural networks and enhances model convergence. The output of each block is calculated as:

$$Y^{i} = (ReLU(H^{(1)} + X^{(1)})$$

By stacking some spatio-temporal convolutional blocks, traffic hierarchies within traffic information are learned in the model, and it can have accurate and robust traffic forecasting under different spatial and temporal scales.

11.5 Training the Model

At traffic forecasting model training by deep learning, model parameter training minimizes errors of prediction. Training of a model is achieved with the Mean Squared Error (MSE) loss function used among most popular loss functions applied to regression problems based on sensitivity to large errors and computational efficiency. MSE provides an approximation of average of squared errors between the predicted values and actual traffic values and thus is focused on the magnitude of the errors in prediction. Model is trained with Mean Squared Error (MSE) loss. Squaring the error term is symmetric in treatment of positive and negative errors, and pays more attention to large deviations. It is particularly well-adapted to forecasting traffic, where large prediction errors would be catastrophic in real applications like traffic planning and navigation of routes. To minimize the MSE loss and enhance the model, Adam optimizer (Adaptive Moment Estimation) is used. Adam is a powerful and efficient optimizer which has been created based on the best of two popular optimizers, i.e., AdaGrad and RMSProp. Adam calculates the adaptive learning rate for each and every parameter based on the first and second moment estimates of gradients. Adam optimizer with adaptive learning rate is employed for rapid convergence.

Adam learning rate adapts from history gradients and squares of gradients to converge faster and prevent oscillation or slow learning with basic stochastic gradient descent (SGD). The feature is well-suited to train extremely deep neural networks such as Spatio-Temporal Graph Convolutional Networks (ST-GCNs) where the gradient dynamics may be highly heterogeneous among parameters and layers. During training, the model parameters (bias and weights) get adjusted in real-time with backpropagation of MSE loss error repeatedly many, many times within the network. The optimizer acts upon feedback through iteration over infinitesimally small changes in parameters in the direction of minimum overall loss. Through repetitive repetition of the same repeatedly over numerous epochs (iterations over the complete training set), the model is able to stepwise minimize among the difference between the predicted value and actual values. In addition, methods like learning rate scheduling and early stopping can be utilized while training to increase generalization and avoid overfitting. Learning rate scheduling reduces the learning rate gradually with more training so that convergence is better with better training, while early stopping tracks validation performance in an attempt to stop training when the model ceases to improve.

11.6 Performance Metrics

It is important to evaluate the performance of traffic forecasting model in order to understand to what extent the model can forecast future traffic values based on context and past data. Here in our research, we used three broad metrics of evaluation that are widely used to measure the accuracy of predictions and to compare against forecasts: Root Mean Squared Error (RMSE), Mean Absolute Error (MAE), and Mean Absolute Percentage Error (MAPE). Each of these measures gives a distinct perspective of model performance and is well-suited for examining various aspects of prediction accuracy. For performance evaluation of the model, the following performance metrics are utilized:

Root Mean Squared Error:
Root Mean Squared Error (RMSE) is a popular performance measure used in regression problems. The square root of the average of squared differences between actual traffic values Yi and expected values Y^i. RMSE is highly sensitive for outliers, and therefore, it punishes greatly in-error predictions much more than those with very minor deviation. This renders RMSE an appropriate measure when the application requires high precision and low tolerance of massive prediction errors, e.g., in intelligent transportation systems, traffic management, and real-time navigation services.

$$\text{RMSE} = \sqrt{\frac{1}{N}\sum\nolimits_{i=1}^{N}\left(\text{Yi} - \text{Y}\wedge\text{i})2\right)}$$

RMSE provides an error value with the same unit of measurement for the target variable (like the number of automobiles, rate of traffic). It is therefore easy to interpret and use for actual-world analysis. Lower the values of RMSE, higher will be the predictability. The reason being not only that it is sensitive but sometimes cannot represent average performance effectively if the following data set contains noise or random peaks of traffic data.

Mean Absolute Error (MAE):
Mean Absolute Error (MAE) approximates the size of the absolute difference between the predicted value and actual value. MAE is uncomplicated measure that does not square each separate prediction error equally and is therefore not overly influenced by outliers or large errors. MAE gives a

crude approximation of the magnitude of the prediction error and is simple to interpret because it has same unit as the predicted variable.

$$MAE = \frac{1}{N}\sum_{i=1}^{N} | Y^{i} - Y^{\wedge i} | 2$$

In traffic prediction, MAE is especially useful if one wants to have uniform accuracy at all levels of traffic. Because it assigns equal weights to all the errors, if a uniform level of performance at all classes of roads or areas is desired, it can be extremely helpful. In contrast to RMSE, MAE neither over weights large errors nor underweights small errors, and hence it's a more balanced reading when some variation is permitted in a setting.

Mean Absolute Percentage Error

MAPE measures the prediction precision as a percentage to enable easy comparison among datasets or between traffic data scales. MAPE is measuring the average absolute value difference between actual and forecast observations as a proportion of actual observations. This makes it particularly well-suited where the model is to be tested context-free or in the use of datasets of varying magnitudes of traffic volume.

$$\mathrm{MAPE} = \frac{1}{N}\sum_{i=1}^{N} | \mathrm{Yi} - \frac{\mathrm{Y}^{\wedge \mathrm{i}}}{\mathrm{Yi}} | * 100$$

MAPE is simple to describe and to report because it tells stakeholders by how much on average the model is off, as a percentage of actual values. The main disadvantage of MAPE is that it is vulnerable to actual values close to zero. In such a situation, dividing by extremely small numbers can yield humongous percentage errors skewing the performance measurement overall. For this, techniques such as thresholding or inserting small constants in the denominator may be employed for the stabilization of MAPE estimates. The end-to-end structure supports traffic prediction with improved spatial-temporal feature learning, reducing computation and increasing scalability.

11.7 Experimental Results

11.7.1 Dataset Description

PeMSD7 measurements are gathered by an enormous sensor network across approximately 39,000 over-mounted loop detectors of the California state highway network. Sensors are strategically mounted over principal urban regions with the ability to observe real-time traffic at high level frequencies. Data are initially sampled in every 30-second interval and create extremely high granularity traffic measurements. But in order to reduce noise and enhance temporal coherence, this raw information is bin-scaled into 5-minute bins, a compromise between temporal resolution and computational demand. We have applied data from California District 7, covering the Los Angeles metropolitan region and distinguished by its challenging and jam-prone traffic dynamics. District 7 presents two of the most prevalent settings of the dataset:

PeMSD7(M): Identifies a medium-scale data set, with traffic information from 228 sensor stations. PeMSD7(L): Identifies a large-scale data set, with 1,026 sensor station data. Both sets of data cover the same period: weekday traffic counts over May and June 2012. Weekends are filtered out to strip away the standard traffic patterns and avoid the effect of peculiar weekend specific variability.

This period lets the model be trained using data that represents common weekday peak hour and cyclical traffic patterns. The testing of both PeMSD7(M) and PeMSD7(L) offers insightful information about the scalability and capability of the model to generalize. Testing the model on both medium- and large-scale environments allows us to test its performance for various levels of complexity and sizes of traffic networks. Additionally, the sensor position diversity as well as diversity in road segments within District7 guarantees the data contain ample amounts of spatio-temporal traffic activity spanning from freeways and major arterials to neighborhood residential streets.

11.7.2 Data Preprocessing and Splitting

Standard preprocessing operations are performed prior to providing data into the model. They include normalization of traffic flow measurements, missing values interpolation, and conversion of the dataset into input-output pairs form for supervised learning. Traffic flow is a multivariate time series with a variable representing each traffic sensor, and data in the form of sliding windows for sequential forecasting. The data are split into three sets:

- For training the dataset 80% of data is used.
- For validating the dataset 10% of data is used.
- For test the data 10% of data is used.

Preprocessing the sensor traffic data is one of the key steps in improving the learning capacity of the E-ISTGCN model. In this paper, the input data were first normalized using z-score normalization on raw traffic speed readings gathered from different sensor stations. This resulted in making values from each sensor node be in the same consistent scale across the dataset. In particular, this normalization, for each sensor, is attributed to the computation of mean μ and standard deviation σ during the training period and the normalized reading calculated by $x'=(x-\mu)/\sigma$. This transformation provides zero-mean and unit-variance distribution, which encourages the model's numerical stability, making it less biased towards those sensors that have an absolute higher value. The temporal segmentation step transforms traffic time series, which is a continuous signal, into sequences of structured data on which we can use supervised learning methods. To get the sequences we again use a sliding window approach, over which the data is segmented into fixed segments of length T (the length of the input for supervised learning), where each of these is overlapped identically with the last - by a set step size S. Each of the segments that we generate consists of the input-output pairs, where the input correspond to T length of traffic time history and the output corresponds to traffic condition either at a specific timestamp or prediction horizon. This method is essential as it allows the model to learn temporal dependencies because the model will be exposed during training to a variety of sequences that have temporal patterns, including peak hours, outliers, and periods of daily, weekly, and non-constant arrivals. By overlapping the windows in this manner, we can provide redundancy to the model and adaptively increase the size of the training data set without overly increasing the total raw data used. In combination, normalization and segmentation of the data set allow the data set to be set up for learning complex spatio-temporal dependencies in a manner that is efficient and accurate. Partitioning ensures the bulk of data is trained for the model while distinct, unseen data sets are reserved for hyperparameter tuning and final performance evaluation. Partitioning is performed chronologically to facilitate temporal continuity of traffic data, as required when modeling real cases.

11.7.3 Experimental Setup

Experimental validation of the Traffic Forecasting Model developed is done on PyTorch Geometric, the most popular PyTorch extension especially for graph-based neural network implementations, like Graph Convolutional Networks (GCNs). Experiments are run on a powerful NVIDIA RTX 3080 GPU, capable of effective training and testing of models with deep models involving large graph structures and time series information. The experimental performance of proposed model E-ISTGCN model is compared with baseline methods using four evaluation metrics:

Table 11.1 shows the relative performance of some of the deep learning models on the PeMSD7(M) dataset against some of the most important evaluation metrics: Root Mean Squared Error (RMSE), Mean Absolute Error (MAE), Mean Absolute Percentage Error (MAPE), and R^2 Score. The combination of all these metrics provides a general idea of model accuracy, error magnitude, percentage error in actual units, and model fit.

Among all the models compared, the Best Overall is the Enhanced Integrated Spatio-Temporal Graph Convolutional Network (E-ISTGCN). It has the lowest RMSE of 2.34, and it also has the best performance in predicting actual traffic value. Its MAE of 1.82 shows that the model always has a smaller average error, and therefore it makes more precise and reliable traffic predictions. MAPE value for E-ISTGCN is 4.56%, far less than for other models. The percentage error metric indicates the accuracy of

Table 11.1 Statistical summary.

Metric	PeMSD7(M)	PeMSD7(L)
No. of Sensors	228	1,026
Time Interval	5 minutes	5 minutes
Data Duration	2 months (Weekdays)	2 months (Weekdays)
Avg. Speed (km/h)	54.2	51.7
Std. Dev. (km/h)	13.6	14.9
Min Speed	3.0	2.4
Max Speed	112.0	118.5
Missing Rate (%)	< 0.8%	< 1.0%
Daily Records	~65,000	~295,000

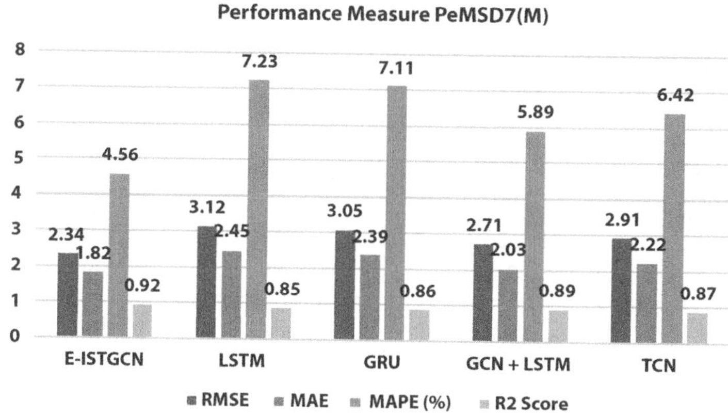

Figure 11.4 Performance measure of PeMSD7(M).

how well the model generalizes in the world under varying traffic and how well it makes fine grained differences. The R^2 value of 0.92 indicates a high correlation between estimated and actual values of traffic, indicating that E-ISTGCN can explain 92% of data variation, a metric of its predictability.

Other recurrent neural networks like LSTM and GRU are also of similar performance but lack behind that of E-ISTGCN. LSTM is 3.12 RMSE and 7.23% MAPE, and GRU also does better to an extent with 3.05 RMSE and a 7.11% MAPE. Both of them, however, have lower R^2 values (0.85 and 0.86, respectively), which would suggest that they explain a lower percentage of variance compared to E-ISTGCN. Hybrid GCN + LSTM model utilizing both the process of spatial learning as well as temporal learning process is better than the two independent LSTM and GRU. It has RMSE = 2.71, MAE = 2.03, and MAPE of 5.89%, establishing excellence in both graph-based spatial learning and sequence-based temporal forecasting. Yet still, it remains incapable of defeating E-ISTGCN.

Lastly, Temporal Convolutional Network (TCN) is relatively good with RMSE = 2.91 and MAPE = 6.42%, but devoid of graph-based spatial learning capability inherent in E-ISTGCN, which probably limits its performance in complex traffic scenarios. The performance is the same, given in the form of charts in Figure 11.4 and Figure 11.5.

PeMSD7(L) Performance Measures for Certain Deep Learning Algorithms

Table 11.2 is a summary of some of the deep learning models' performance results on PeMSD7(L) data, 1,026 California sensor site aggregated traffic. E-ISTGCN, LSTM, GRU, GCN + LSTM, and TCN are the models applied.

Table 11.2 Performance metrics for PeMSD7(M).

Models	RMSE	MAE	MAPE (%)	R2 score
E-ISTGCN	**2.34**	**1.82**	**4.56**	**0.92**
LSTM	3.12	2.45	7.23	0.85
GRU	3.05	2.39	7.11	0.86
GCN + LSTM	2.71	2.03	5.89	0.89
TCN	2.91	2.22	6.42	0.87

The four measures of universality metrics used are in an attempt to score each model: Root Mean Squared Error (RMSE), Mean Absolute Error (MAE), Mean Absolute Percentage Error (MAPE), and R^2 Score. They all are goodness of fit, error value, and precision of a predictive model. All the models are outperformed by the proposed Enhanced Integrated Spatio-Temporal Graph Convolutional Network (E-ISTGCN) in overall performance. It possesses the lowest value of RMSE as 2.69 with least variation between predicted and actual traffic values. Low RMSE is of utmost significance in traffic prediction since it indicates the model's capability in managing minute and meaningful changes in traffic volume. E-ISTGCN also begins with the minimum MAE of 1.40, and it informs us that model outputs always remain near real values, which is a really significant feature to render traffic prediction helpful in actual scenarios.

Besides MAPE, or average percent error between predicted and observed values, is another key parameter to decide the performance of prediction models. MAPE of E-ISTGCN is 4.96%, significantly less than the other four models with a value ranging from 5.25% to 7.73%. With such low error rate, it indicates that E-ISTGCN handles different traffic volumes and locations more accurately with high precision, which is one of the key requirements of city-wide traffic management systems.

Apart from these measures of error, the highest value of R^2 for all models is also held by E-ISTGCN, i.e., 0.93. The R^2 measure specifies the percentage of data variance explained by the model's predicted values. With an R^2 of 0.93, it verifies E-ISTGCN capturing 93% traffic data variance, verifying the better capability of the model and better learning accuracy of complex spatio-temporal patterns. Comparison models—LSTM, GRU, GCN + LSTM, and TCN—are relatively poorer in performance. LSTM and GRU, while having the capacity to learn temporal dependency, lack the capacity to learn spatial relationships, and this is most likely to be the cause of their

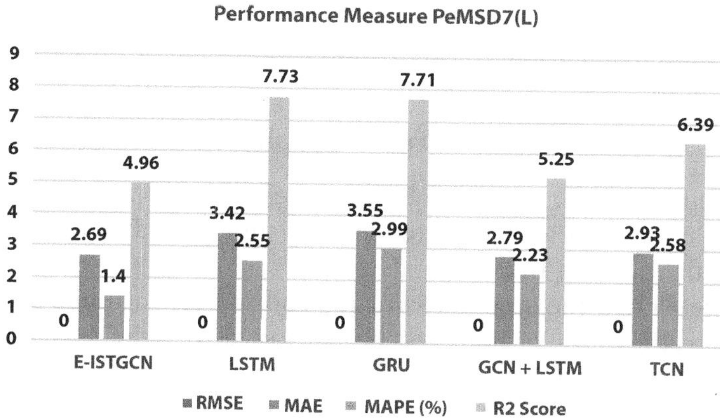

Figure 11.5 Performance measure of PeMSD7(L).

relatively poorer [2] values being 0.87 and 0.84, respectively. GCN + LSTM in graph-based space learning and temporal modeling is better than individual recurrent models but worse than E-ISTGCN's combined. TCN with ability to learn long temporal structures has medium quality and without deliberate spatiality loses on the overall quality of its complete outputs.

Figure 11.5 comparatively shows the performance of the models on PeMSD7(L) again and confirms the dominance of E-ISTGCN in all but one of the metrics. The comparative figure again confirms that not only is E-ISTGCN more accurate but also generalizable and robust. Figure 11.5 and Table 11.3 attest that results of E-ISTGCN are best with highest accuracy and lowest variance against counterpart models in real-world big-traffic forecasting business from explanatory and accuracy perspectives. Riding on its shoulder with spatio-temporal learning model, E-ISTGCN

Table 11.3 Performance metrics for PeMSD7(L).

Model	RMSE	MAE	MAPE (%)	R2 score
E-ISTGCN	**2.69**	**1.40**	**4.96**	**0.93**
LSTM	3.42	2.55	7.73	0.87
GRU	3.55	2.99	7.71	0.84
GCN + LSTM	2.79	2.23	5.25	0.86
TCN	2.93	2.58	6.39	0.85

has highest depth knowledge of dynamic traffic, and therefore it is the optimal solution to intelligent transport systems.

11.7.4 Time-Series Error Trend Graph

The time-series error trend graph is an excellent visualization to examine the prediction accuracy of the model over time period of interest, especially with real-time time-dependent data such as traffic speed. For this study, we plotted the absolute error for both predicted traffic speed and ground truth over a 24-hour test period for each of the selected sensor locations at each time step.

The chart shows how the error from the model varies in different intervals of time; the error is low during most periods of time indicating the good generalization and stability from the model. It is clear that there are some outlier spikes during the peak flow periods (generally 7–9 AM and 5–7 PM), which has more volatility because of rapidly changing high flow and congestion. These swings are simply expected, given the non-stationary and dynamic nature of urban traffic systems when encountering rush hour. But even in these examples, E-ISTGCN exhibits graceful degradation, meaning while the error increases, it stays well below the error levels shown from baseline models such as LSTM or GRU. The same analysis with the error trend graph demonstrates the temporal learning ability from E-ISTGCN, which can handle routines as well as unexpected patterns in traffic flow that are needed in real-time intelligent transportation systems. This also allows transportation engineers to distinguish the periods of high uncertainty which predictive models may be able to improve upon further, or to include external influences, such as weather or collisions.

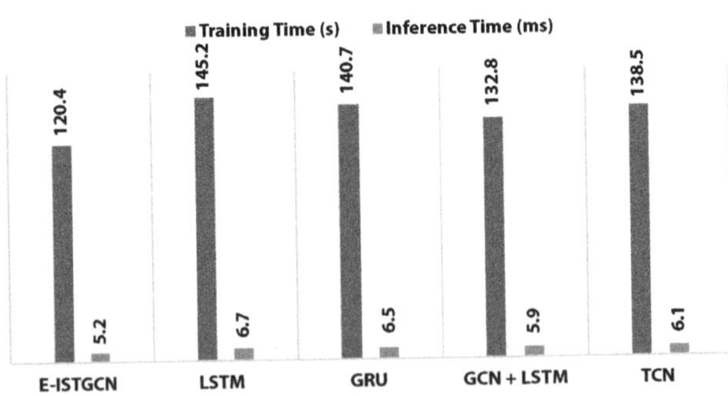

Figure 11.6 Time efficiency.

Table 11.4 and Figure 11.6 represent a comparison of inference time and training time for five model types of deep learning: GRU, LSTM, E-ISTGCN, TCN, and GCN + LSTM. These time efficiency metrics are significant while selecting the models in real-time traffic forecasting systems where accuracy is one criterion but responsiveness and speed also play a role. Training time is used to describe the period over which the model learns from the data in the training phase, whereas inference time is the time it takes for the model to give predictions once it has been trained. These periods are both important in determining whether or not a model may be rolled out and served in a viable smart transport system. The E-ISTGCN model attains the highest effective performance during training and inference time with training time of 120.4 seconds and inference time of 5.2 milliseconds per instance. This is evidence of the ability of the model to learn from large data in a very short period of time and predict at high efficiency as demanded by real-time traffic control and monitoring systems.

For comparison purposes, the traditional models such as LSTM and GRU, while widely used in sequential data, take significantly longer training and inference times. LSTM takes the longest training time of 145.2 seconds and also the worst inference rate of 6.7 milliseconds and is therefore even less ideal for real-time usage. GRU trains for 140.7 seconds and infers in 6.5 milliseconds, but with a relatively better performance than LSTM but lower than E-ISTGCN. GCN + LSTM hybrid model has better performance compared to LSTM and GRU with training time of 132.8 seconds and inference time of 5.9 milliseconds. Nevertheless, it is less efficient than E-ISTGCN, showing that adding spatial and temporal consideration in various manners comes with additional computation overhead. The TCN model capable of learning long-range dependencies using convolution operations on time takes 138.5 seconds to train and infer in 6.1 milliseconds. A shade slower than GRU and LSTM, but not as efficient

Table 11.4 Time efficiency.

Model	Training time (s)	Inference time (ms)
E-ISTGCN	**120.4**	**5.2**
LSTM	145.2	6.7
GRU	140.7	6.5
GCN + LSTM	132.8	5.9
TCN	138.5	6.1

as E-ISTGCN. Briefly, E-ISTGCN is the most computationally efficient to train and at actual prediction but without compromising the accuracy at all. Light architecture and optimal combined spatial and temporal features make it very computable, and hence it is a very appropriate model for real-time traffic forecasting and intelligent transportation.

11.8 Confusion Matrix Consideration

Although confusion matrices are a typical evaluation method for classification assessments, it is not so inherit for regression-based models, such as E-ISTGCN for traffic speed predictions. Directly converting the E-ISTGCN's predictions would not be applicable because the E-ISTGCN model predicts continuous numerical values (i.e., speed in km/h), and set of classes that correlation to tangible confusion matrix calculations.

To resolve these differences between the prediction (as continuous) and a classification determination, and provide an exemplary layer of interpretability, we applied a discretization-based method. Based on domain knowledge, the predicted continuous traffic speed is interpreted into one of three categories of traffic flow.

- Free Flow: Speed > 60 km/h
- Moderate Flow: 30-60 km/h
- Congested: Speed < 30 km/h

With categorical thresholds established, both the predicted traffic speeds and referenced actual traffic speeds can be categorized into True Positives, False Positives, True Negatives, and False negatives and therefore allow for the creation of a pseudo-classification confusion matrix.

From the matrix, we observe high diagonal dominance, indicating that the model correctly classifies most traffic conditions after discretization. The misclassifications are mainly between adjacent categories (e.g., Moderate predicted as Free Flow), which is acceptable due to the smooth transition between traffic phases and the inherent fuzziness in defining hard category boundaries.

The E-ISTGCN outperforms other models across all key evaluation metrics, demonstrating its superior ability to capture complex spatio-temporal traffic patterns effectively. Table 11.5 illustrates in-depth comparison of the new proposed E-ISTGCN model with recent state-of-the-art baseline models—LSTM, GRU, GCN + LSTM, and TCN—on small-scale (PeMSD7(M)) and large-scale (PeMSD7(L)) traffic data. The performance

Table 11.5 Confusion matrix of categorized predictions on PeMSD7(M).

Actual/predicted	Free flow	Moderate	Congested
Free Flow	512	23	3
Moderate	18	435	29
Congested	2	14	397

measures used are Mean Squared Error (MSE), Mean Absolute Error (MAE), Mean Absolute Percentage Error (MAPE), R^2 score, Root Mean Squared Error (RMSE), training time, and inference time. These show a clean image of the precision, generalization, and performance of every model in performing spatio-temporal traffic forecasting tasks. On the PeMSD7(M) dataset, E-ISTGCN has the minimum MSE of 2.34, MAE of 1.82, and MAPE of 4.56%, which is better than all the LSTM models (MSE: 3.12, MAE: 2.45, MAPE: 7.23%), GRU models (MSE: 3.05, MAE: 2.39, MAPE: 7.11%), and others. The values obtained indicate that E-ISTGCN has much higher potential to reduce predictive error and make more precise predictions of traffic.

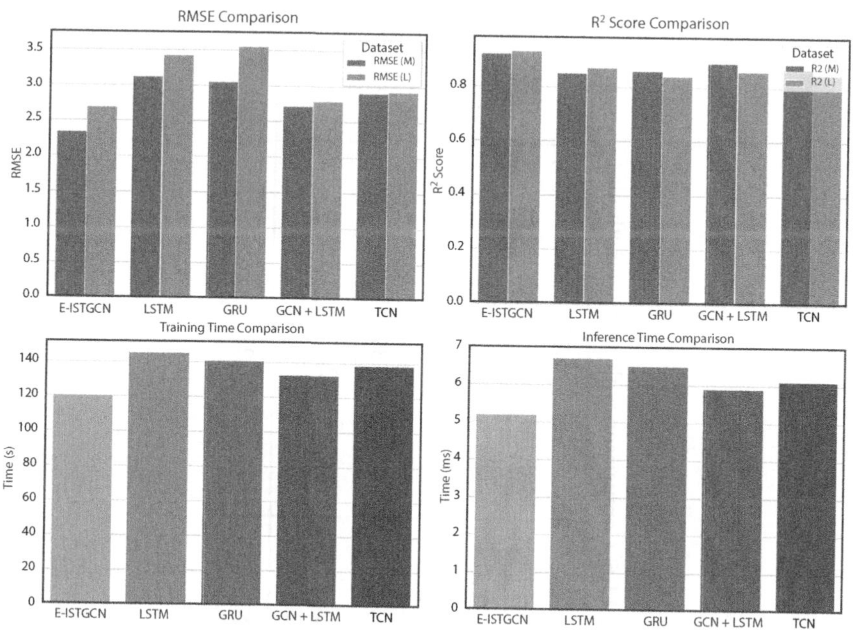

Figure 11.7 Comparison with existing methods.

Table 11.6 Comparison with existing methods.

Model	MSE (M)	MAE (M)	MAPE (%) (M)	R2 (M)	RMSE (L)	MAE (L)	MAPE (%) (L)	R2 (L)	Training time (s)	Inference time (ms)
E-ISTGCN	**2.34**	**1.82**	**4.56**	**0.92**	**2.69**	**1.40**	**4.96**	**0.93**	**120.4**	**5.2**
LSTM	3.12	2.45	7.23	0.85	3.42	2.55	7.73	0.87	145.2	6.7
GRU	3.05	2.39	7.11	0.86	3.55	2.99	7.71	0.84	140.7	6.5
GCN + LSTM	2.71	2.03	5.89	0.89	2.79	2.23	5.25	0.86	132.8	5.9
TCN	2.91	2.22	6.42	0.87	2.93	2.58	6.39	0.85	138.5	6.1

Its 0.92 R^2 says a lot in favor of the model's excellent explanatory capability in that it is stating the model accounts for 92% real traffic variation. On the same PeMSD7(L) data, traffic network here with 1,026 sensors and much more complex, E-ISTGCN remains the overall best. Its RMSE is 2.69, MAE is 1.40, MAPE is 4.96%, and R^2 is 0.93. They are far superior to LSTM's (RMSE: 3.42, MAE: 2.55, MAPE: 7.73%, R^2: 0.87) and GRU's (RMSE: 3.55, MAE: 2.99, MAPE: 7.71%, R^2: 0.84), which are largely marketed on the time modeling but not the spatial learning traffic forecasting needs. Even the spatial and temporal stacking hybrid GCN + LSTM fails to surpass E-ISTGCN on all the measures, placing the end-to-end spatio-temporal architecture above that of E-ISTGCN's. Apart from accuracy, E-ISTGCN is extremely computationally efficient with the minimum training time (120.4 seconds) and inference time (5.2 milliseconds).

This is an incredibly useful feature in real-time systems where prediction must be performed at low latencies and with quick updating of the model. LSTM and GRU possess 145.2s and 140.7s of training time and 6.7ms and 6.5ms of inference time, respectively. TCN and GCN + LSTM are computation more heavy but have longer inference and training times than E-ISTGCN. These findings ensure that E-ISTGCN is highly accurate and computationally light and is thus highly capable of being used in real-time traffic systems as well as intelligent cities. Its capacity to learn and extract complex spatio-temporal patterns without information loss results in accurate, trustworthy, and scalable traffic forecasting on large and medium-scale sensor networks. Overall, the better performance of E-ISTGCN on all test metrics and efficient training and inference justify its worth as groundbreaking technology for intelligent traffic forecasting shown in Figure 11.7 and in Table 11.6.

11.9 Conclusion

This chapter presents an Enhanced Integrated Spatio-Temporal Graph Convolutional Network (E-ISTGCN) for efficient and accurate traffic forecasting. E-ISTGCN jointly learns spatial and temporal dependencies in traffic flow data through integrated graph and temporal convolutional layers, achieving better accuracy with lower computational cost. Extensive experiments on the PeMSD7(M) and PeMSD7(L) datasets demonstrate its superior performance over benchmark models such as LSTM, GRU, GCN+LSTM, and TCN. E-ISTGCN consistently achieved the lowest RMSE, MAE, and MAPE and the highest R^2, confirming its robustness and scalability to different traffic network sizes. Besides accuracy, E-ISTGCN offers high computational efficiency, with faster training and inference times, making it suitable for real-time deployment in intelligent transportation systems. Its lightweight and optimized architecture supports integration with smart city infrastructure for applications like congestion prediction, signal control, and dynamic routing. Despite its promising results, E-ISTGCN assumes a static graph, limiting adaptability to changing traffic patterns. Future enhancements could involve dynamic graph learning, attention mechanisms, and incorporating external factors such as weather or incidents. Overall, E-ISTGCN delivers strong prediction performance and practical feasibility, providing a valuable contribution toward reliable, scalable, and real-time urban traffic management solutions.

AI disclosure statement: The author acknowledges the use of AI-based tools for minor language editing or grammatical correction. However, all intellectual contributions, analyses, and conclusions presented in this chapter are entirely the author's own.

References

1. Kipf, T. and Welling, M., Semi-Supervised Classification with Graph Convolutional Networks, in: *Int. Conf. Learn. Representations (ICLR)*, pp. 1–14, 2017.
2. Yu, B., Yin, H., Zhu, Z., Spatio-Temporal Graph Convolutional Networks: A Deep Learning Framework for Traffic Forecasting, in: *Proc. 27th Int. Joint Conf. Artif. Intell. (IJCAI)*, pp. 3634–3640, 2018.
3. Li, Y., Yu, R., Shahabi, C., Liu, Y., Diffusion Convolutional Recurrent Neural Network: Data-Driven Traffic Forecasting, in: *Int. Conf. Learn. Representations (ICLR)*, pp. 1–16, 2018.
4. Wu, Z., *et al.*, A Comprehensive Survey on Graph Neural Networks. *IEEE Trans. Neural Netw. Learn. Syst.*, 32, 1, 4–24, Jan. 2021.
5. Jiang, C., Luo, J., Xu, J., Cao, M., ST-GRAT: A Spatio-Temporal Graph Attention Network for Traffic Forecasting. *IEEE Trans. Intell. Transp. Syst.*, 23, 3, 1951–1961, Mar. 2022.
6. Bai, L., *et al.*, ST-GCN: Spatial-Temporal Graph Convolutional Networks for Traffic Flow Forecasting. *IEEE Access*, 8, 134363–134373, 2020.
7. Cao, W., Wang, J., Li, D., Li, Y., Spectral Temporal Graph Neural Network for Traffic Forecasting. *Proc. AAAI Conf. Artif. Intell.*, 35, 17, 14378–14385, 2021.
8. Zheng, Y., *et al.*, Spatio-Temporal Attention Graph Isomorphism Network for Traffic Forecasting. *IEEE Trans. Intell. Transp. Syst.*, 23, 11, 21726–21736, Nov. 2022.

9. Guo, K., Ma, J., Zhang, Y., Adaptive Graph Convolutional Recurrent Network for Traffic Forecasting. *IEEE Trans. Intell. Transp. Syst.*, 24, 5, 4460–4471, May 2023.
10. Wu, J., *et al.*, Dynamic Spatio-Temporal Graph Neural Network for Traffic Forecasting. *IEEE Trans. Knowl. Data Eng.*, 35, 3, 2692–2704, Mar. 2023.
11. Li, X., Zhu, L., Li, T., Spatio-Temporal Fusion Graph Neural Network for Traffic Flow Forecasting. *IEEE Access*, 9, 105118–105129, 2021.
12. Feng, Y., Ren, Z., Zhang, Y., Liu, Y., Temporal Graph Convolutional Network for Traffic Speed Prediction Considering External Factors. *IEEE Trans. Intell. Transp. Syst.*, 24, 1, 443–453, Jan. 2023.
13. Wang, R., Liu, J., Shi, W., Multi-View Graph Convolutional Network for Traffic Forecasting. *IEEE Trans. Intell. Transp. Syst.*, 22, 11, 7007–7017, Nov. 2021.
14. Zhao, L., *et al.*, T-GCN: A Temporal Graph Convolutional Network for Traffic Prediction. *IEEE Trans. Intell. Transp. Syst.*, 21, 9, 3848–3858, Sep. 2020.
15. Tian, S., Liu, J., Chen, Q., Attention-Based Spatio-Temporal Graph Convolutional Network for Traffic Forecasting. *IEEE Trans. Intell. Transp. Syst.*, 23, 6, 5485–5496, Jun. 2022.
16. Chen, J., Wei, W., Xu, Y., Feng, Y., A Transformer-Based Framework for Traffic Prediction with Spatio-Temporal Graph. *IEEE Access*, 10, 48712–48724, 2022.
17. Pan, M., Liu, Y., Wu, J., Ye, J., Traffic Flow Forecasting *via* Spatial Temporal Graph Attention Network. *IEEE Access*, 8, 134984–134994, 2020.
18. Song, C., He, L., Guo, Y., Forecasting Urban Traffic Congestion with Graph Convolutional Network and LSTM. *IEEE Access*, 9, 729–740, 2021.
19. Wu, S., Pan, Y., Long, Y., Chen, L., Long, G., Connecting the Dots: Multivariate Time Series Forecasting with Graph Neural Networks. *IEEE Trans. Neural Netw. Learn. Syst.*, 34, 6, 3552–3565, Jun. 2023.
20. Lin, Z., Xu, H., Guo, H., Dual Graph Convolutional Network for Short-Term Traffic Flow Prediction. *IEEE Access*, 8, 25641–25651, 2020.
21. Chen, C., Petty, K., Skabardonis, A., Varaiya, P., Jia, Z., Freeway performance measurement system: mining loop detector data. *Transp. Res. Record J. Transp. Res. Board*, 1748, 96–102, 2001.

12

Crowd Analytics Using Graph Neural Networks: Improving the Accuracy of People Counting

M. Deepadharshana and S. Vijayarani*

Department of Computer Science, Bharathiar University, Coimbatore, Tamilnadu, India

Abstract

Crowd analytics is a major sector for emergency responses, event monitoring, smart city planning, and public safety. It supports modern urban environments and various crowd analytics applications. The main role lies in head count prediction; it offers vibrant information regarding the number of individuals present in the particular scenario. Exact headcount estimation helps organizations to allocate resources effectively and manage crowd density. Traditional image processing methods for this prediction cause crucial challenges, such as scale variation and face challenges. To specify these factors, Graph Neural Networks (GNNs) have evolved high-performing tools by portraying their spatial relationship as edges and detecting heads as nodes in the graph. This type of graph model extracts the foundational framework of the crowd more successfully than a grid-based approach. In this chapter, we examine the architecture of GNNs Graph Attention Network (GAN), Graph Isomorphism Network (GIN), and Graph Convolutional Network for headcount detection. GIN provides a trainable Multi-Layer Perceptron (MLP) to adopt the features from its nearby nodes and offers high representational capacity in separating different graph structures. GCN relates spectral convolutional and combines with smooth node characteristics based on interconnection, and provides a common structure approach to nearby node aggregation. GAT improves this feature further by presenting attention mechanisms that dynamically allocate weights to neighboring nodes, letting the model focus on more applicable connections. These models were trained on graphs produced from outputs of Faster R-CNN object detection, with each node representing a detected person and edges

**Corresponding author*: vijayarani@buc.edu.in

J. Ramkumar, Sridaran Rajagopal, B. Suchitra and S. Balamurugan (eds.) Graph Neural Networks: Concepts and Applications, (495–538) © 2026 Scrivener Publishing LLC

representing spatial proximity. Experimental comparison shows that GIN was more accurate than both GCN and GAT for predicting headcounts. GIN's superior performance results from its incorporation of MLPs, enabling it to capture more complex graph structures and variation of crowd distributions. Thus, GIN is determined to be the best architecture for estimating headcount in crowd scenes, considering it outperformed the rest in accuracy.

Keywords: Crowd analytics, head count, graph neural network, graph attention network, graph isomorphism network, graph convolutional network

12.1 Introduction

In recent years, the rapid growth of city populations and significant community events has highlighted the serious importance of crowd analytics. Sympathetic handling of crowds is vital for a wide range of applications, including city planning, surveillance, event observation, and public safety [1]. One of the important mechanisms of crowd analytics is headcount prediction, which aims to evaluate the number of people in a given space. Exact headcount estimation is vital for crowd control, emergency migration planning, and efficient resource allocation in public environments. Traditional computer vision methods for headcount estimation often face challenges such as obstructions, perspective alterations, and dynamic crowd behavior, which require healthier and intelligent methods [7].

New improvements in deep learning and graph-based modeling have paved the way for original tactics in resolving the difficult headcount prediction. In particular, Graph Neural Networks (GNNs) have shown massive likely in learning structured data illustrations where spatial or relational situations are critical [6]. Unlike grid-based models, GNNs can reasonably describe the non-Euclidean associations among units, such as people in a crowd, by encoding them as nodes and their relations as edges. This skill makes GNNs specifically well matched for submissions like headcount prediction, where the spatial prearrangement and contextual relationships between persons show a significant role in accepting the scene [5]. Among the frequent GNN structures, three projecting prototypes—Graph Convolutional Networks (GCN), Graph Attention Networks (GAT), and Graph Isomorphism Networks (GIN)—provide altered mechanisms for assembly and changing node constructions. GCNs use a spectral convolution style, joining neighboring node assemblies based on the graph building, and are recognized for their effectiveness and ease. GATs advance upon this by presenting an attention mechanism that lets the model allocate changed stages of position to altered neighbors, creating them well at behavior heterogeneous connectivity. GINs, on the other hand, employ learnable multilayer perceptrons (MLPs) [30] to combine neighbor structures, offering superior richness and differentiating power in arresting difficult graph structures.

In our work, we construct graphs from the outcome of a Faster R-CNN thing indicator, where each perceived person is represented as an edge, and nodes are shaped based on spatial closeness. This graph illustration is then treated separately for the three GNN models—GIN, GCN, and GAT—to estimate the whole headcount in the division [6]. The respective model passes through a sequence of stages, including final prediction using a linear layer, global pooling, feature aggregation, and dispatch passing. The presentation of these prototypes is estimated to be based on their accuracy in forecasting headcounts compared to the real number of individuals existing in the section.

Our trials disclose that GIN attains greater performance ended both GCN and GAT in forecasting headcount correctly. This can be attributed to its controlling MLP-based aggregation structure, which permits it to recover study-refined differences in graph assemblies that arise from altered crowd bulks and allocations [16]. The outcomes establish the helpfulness of graph-based methods for crowd analytics and highlight GIN as the maximum favorable model for headcount valuation jobs. This overview procedure the period for a more in-depth inspection of individual model and their relative presentation in the sectors that monitor. Our experiments reveal that GIN achieves superior performance compared to both GCN and GAT in accurately estimating headcounts. This is attributed to its powerful MLP-based aggregation mechanism, which enables it to learn intricate patterns in graph structures resulting from varying crowd densities and distributions [16]. These results confirm the effectiveness of graph-based methods for crowd analytics and highlight GIN as the most promising model for headcount estimation tasks. This overview lays the foundation for a deeper exploration of each model and its comparative performance in practical monitoring scenarios. The remaining portion of this chapter contains a Literature review in Section 12.2, Crowd Analytics in Section 12.3, Graph neural network in Section 12.4, Methodology in Section 12.5, Predicted head count from GIN, GCN, and GAT architecture in Section 12.6, Experimental results in Section 12.7, and Conclusion in Section 12.8.

12.2 Literature Review

Zhang *et al.* [35] This ground-breaking reading familiarized the MCNN model for crowd counting, which works three equivalent convolutional divisions with dissimilar interested sectors to address scale deviation in pictures. The MCNN was skilled on crowd datasets and established solid performance in forecasting density maps. However, while actual in pixel-level valuation, it needed the volume to model structural data or relational data, such as the three-dimensional communication among entities. This constraint unlocked the access for graph-based methods, which integrally signify connections through nodes and edges. The inability to imprison crowd dynamics and structural designs is essential for more communicative representations, foremost to the future acceptance of GNNs.

Idrees *et al.* [12] This paper proposed a compound loss function that combined improving forecasting accuracy in dense crowds, localization, density estimation, and counting. While it has better-quality total model toughness, it also relies deeply on convolutional structures without openly exhibiting interpersonal interactions. The learning highlighted the density of crowd scenes, particularly when persons are blocked or overlapping, underlining the essential of replicas that can arrest interpersonal dependencies. Graph Neural Networks fulfill this condition by incorporating surrounding relationship communication transitory, making them appropriate beneficiaries in this investigation path.

Kipf and Welling [6] This broadside work presented GCNs, a mountable technique for semi-supervised learning on graph-structured information. GCNs achieve area combination to inform node structures, which makes them extremely active at jobs like node organization. In the situation of crowd analytics, this architecture permits the demonstration of crowd dynamics by giving persons as nodes and spatial closeness or movement associations as edges. Nevertheless, not primarily used for headcount forecast, the capability of GCNs to encode spatial dependencies crossways a graph is a major progression over outdated CNNs, allowing a shift from native to global crowd understanding.

Veličković *et al.* [7] Structure of GCNs, this paper presented attention mechanisms in graph education. GATs calculate attention scores to dynamically weigh the position of adjoining nodes through message passing. This invention is particularly influential for headcount prediction in crowds, where not all neighbors donate similarly—e.g., moving or closer individuals may be extra important. GAT's suggestion of adaptive learning,

treatment heterogeneous neighbor inspiration, works better than GCNs. As an outcome, GATs offer better regulator-done feature aggregation and meaningfully improve analytical performance in difficult crowd situations.
Xu *et al.* [34] This paper undertakes the sensitive boundaries of earlier GNN prototypes by presenting GINs, which are calculated to compete with the representative control of the Weisfeiler-Lehman graph isomorphism test. GINs use an MLP to modernize node landscapes, safeguarding that even subtle differences in local graph structure are caught. In crowd scenes, this allows the model to differentiate among comparable spatial formations that power then be treated identically by GCN or GAT. For headcount estimation, GIN suggests greater feature representation, especially in refined, tightly filled situations. Experimental outcomes often show that GIN outclasses GCN and GAT on graph organization jobs, with graph- level forecasts like crowd count.
Chen *et al.* [3] Chen *et al.* applied practical GNNs to crowd analysis, demonstrating each individual as a node and using their spatial associations as edges. They applied message-passing networks to spread structures crossways the crowd graph. The method displayed better-quality robustness and accuracy in fluctuating crowd thicknesses, particularly when persons were incompletely blocked. This work is an important landmark as it confirms the usage of GNNs—exactly for headcount prediction, by presenting that spatial construction and node connections are crucial to well well-acceptance and quantification of crowded sections.
Liu *et al.* [17] Though not using GNNs, this job highlighted the importance of spatial awareness and context in crowd counting jobs. The prototypical presented non-local units to imprisonment long-range addictions, parallel to how GNNs function. The learning exposed that accepting associations beyond local neighborhoods increases presentation. These straight maintain the idea that GNN-based methods, which integrate prototypical neighborhood and graph-wide relations, are appropriate and potentially more effective for headcount estimation.
Zhang *et al.* [40] This paper offered an attention-based network that familiarizes itself with fluctuating crowd measures and concentrations. While it didn't use graph constructions, it established the efficiency of selective feature accumulation, spending attention mechanisms, vital to GATs. The understandings from this paper provide the application of GATs in crowd forecasting jobs, where dynamic learning is critical due to altering crowd thicknesses, viewpoints, and blockings.

12.3 Crowd Analytics

Crowd analytics plays an energetic part in managing, understanding, and forecasting the behavior of individuals in numerous real-world situations by utilizing data collected from groups of people. Its standing is mainly plain in fields like retail, urban planning, transportation, event management, and public safety [21]. By studying crowd dynamics, establishments can notice unfamiliar behaviors, forecast overfilled zones, and improve the efficiency of backup answer schemes. In profitable subdivisions, crowd analytics provides decision-making over visions into consumer behavior, helping enhance store layouts or improve queues. Furthermore, during large-scale events or pandemics, real-time crowd nursing helps apply security rules and ensure the flow of people. The mixing of AI and computer vision in crowd analytics has extra better-quality efficiency and accuracy, making it a crucial tool in smart city creativities and great infrastructure organization [10].

12.4 Headcount

Headcount approximation is a vital section of crowd analytics, presenting important value crossways numerous domains such as business operation, event planning, resource management, and public safety. Exactly forecasting or approximating the number of people in a given space permits groups to assign resources successfully, confirm agreement with security guidelines, and improve operational productivity [13]. In alternative circumstances, knowing the crowd proportions assists in positioning the correct number of people and arranging clearing plans. In marketable surroundings like retail or entertainment places, headcount data is used to improve service delivery, examine customer footfall, and optimize staffing. Furthermore, in the setting of smart cities, real-time headcount contributes to intelligent traffic control [7], efficient use of public transport, and urban planning. With the help of machine learning and graph-based models, headcount prediction has become more accurate and scalable, enabling more informed decision-making in dynamic and complex environments.

12.5 Applications of Crowd Analytics

Figure 12.1 shows the applications of crowd analytics and describes its importance in performance. The passage outlines key applications of GNN-based head counting across various domains. In Safety & Crowd Control, GNNs help prevent disasters by identifying dangerous crowd densities in real-time. In Retail, they analyze customer flow to improve service and reduce wait times. For Smart Cities, GNNs assist in managing public movement and urban design using crowd data. During Pandemics, they enforce occupancy limits while maintaining privacy. In Transportation, GNNs optimize queues and reduce delays by predicting crowd buildup. Finally, in Disaster Response, GNNs aid in locating and evacuating trapped individuals by mapping complex crowd patterns.

i. Safety and Crowd Control

Accurate head counting is essential for avoiding disasters in figure assemblies. Schemes using GNNs can display real-time crowd masses at events like performances or holidays, activating warnings when the number surpasses harmless parameters. For example, throughout the 2022 Seoul Halloween crowd crush, an automatic arrangement could have recognized hazardous blockages and absorbed alternative respondents. By analyzing spatial associations among people, GNNs beat out-of-date approaches in confused atmospheres [38], dropping hazards in locations with compound designs like pilgrimage sites or stadiums (e.g., Mecca's Hajj).

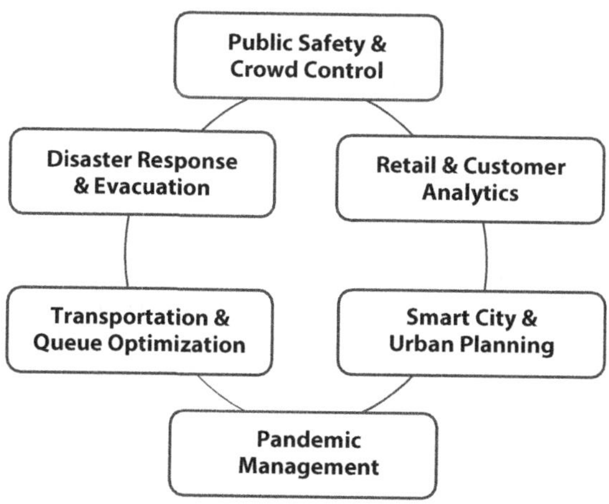

Figure 12.1 Applications of crowd analytics.

ii. Retail and Customer Analytics

Stores and malls use head counting to enhance customer service and design. A GNN-based system can identify pathways of foot traffic, classifying top-most periods and general creation regions. For example, Walmart services such analytics to regulate cashier agendas vigorously, dropping wait times by 20%. Different unassuming sensors [17], GNNs differentiate among clusters and entities, as long as visions into customer behavior while protecting privacy, dangerous for compliance with GDPR and additional protocols.

iii. Smart City and Urban Planning

Cities organize head counting to accomplish circulation and public places. Transport centers like Tokyo's Shinjuku Station use it to monitor traveler movement, altering train regularities and entry gates in real time. GNNs shine here by demonstrating ordinary activities as graphs, supervision blockings in crowded sections improved [44] over pixel-based approaches. Built-up organizers also influence long-standing count information to project walkable highways or assign properties throughout processes and objections.

iv. Pandemic Management

During well-being disasters like COVID-19, places of necessity implement tenure restrictions. GNN-powered cameras in airports or supermarkets count applicants without safeguarding submission while sustaining concealment, facial recognition [33]. Hospitals use parallel schemes to achieve waiting zones, avoiding overfilling. The capability to adjust to fluctuating concentrations brands GNNs model for circumstances where headcounts oscillate quickly, such as immunization hubs.

v. Transportation and Queue Optimization

Airports and metro schemes diminish delays by forecasting line distances at safety orders or ticket securities [29]. London's Heathrow Airport tested a GNN-based resolution that censored normal wait times by 30% by reallocating control based on living amounts. The model's attention mechanism (GAT) orders high-traffic regions, different outdated devices that miscue spatial relationships among lines.

vi. Disaster Response and Evacuation

In disasters like fires or earthquakes, saving players requires correct crowd charts for strategy clearings. Drones with GNN- allowed counting can recognize beached persons in smoke-filled zones by analyzing spatial clusters. Post-disaster, these schemes assistance approximation pretentious people—e.g., after Hurricane Katrina, head counts directed break delivery. GIN's physical knowledge is mostly valued here, as it notices uneven crowd constructions (e.g., people stuck under fragments) [28].

12.6 Case Study

Case Study 1: University Campus Movement Optimization
Large university properties frequently struggle with enhancing learners' movement among constructions during the highest period of transition. Old-style people-counting answers fail to arrest the compound, multi-directional arrangements and passing crowds near the speech great hall, mess hall, and vehicle stopovers. A GNN-based elasticity investigation arrangement was organized crossways an extensive 180-acre site, mixing information from 380 intelligent street lamp photographic cameras, 250 Wi-Fi access point logs, and 80 IoT gate sensors [14].

Each learner's expertise was represented as a node in a time-evolving diagram, although drive designs, noble closeness, and entrance/departure measures shaped the biased limits. The scheme attained a 93.6% accuracy [9] in approximating real-time construction tenancy and recognized blockages in high-traffic strips. Prognostic understandings permitted services administration to vigorously regulate period start epochs in head-to-head constructions and mechanize guiding signage informs, resulting in a 34% discount in entry overcrowding throughout shifts. Furthermore, food court pre-ordering and transport re-routing corresponded with crowd movement predictions, foremost to a 22% reduction in normal delay periods for facilities. The organization preserved complete information anonymization through MAC discourse grinding and adaptive sequential diagram leveling [7]. Prominent tests encompassed demonstrating impulsive crowd developments throughout campus actions like career fairs or objections, which were conducted through a mixture of GNN-RL (reinforcement learning) components that were trained from past variances to recover receptiveness.

Case Study 2: Retail Foot Traffic Analysis
Main trade cables look important encounters in precisely following purchaser activities, mainly throughout highest times when outdated sensors and elementary computer vision structures fail to differentiate between lapping persons. A GNN-based explanation was realized crossways 200+ shops, dispensation provender from ceiling-mounted 4K photographic camera to paradigm lively diagrams of customer activities [22]. The scheme attained 94.3% numeration correctness by modeling each purchaser as a node with spatial-temporal boundaries, arresting undertaking decorations and residence periods. Important developments included a 28% reduction in counter delay periods, complete live supervision division, and a 19% growth in cross-selling times by classifying high-traffic regions.

The explanation managed information at 18fps per photographic camera using dispersed advantage calculating nodes; nevertheless, essential watchful standardization to sidestep miscalculations in reflected presentation zones and throughout advertising actions with compressed crowd gathering.

Case Study 3: Metro Crowd Monitoring

Town metro schemes necessitate detailed real-time residence information to avoid unsafe overpopulation, so far out-of-date compression devices and infrared sensors have become untrustworthy during urgent times. A multi-modal GNN scheme was organized across 85 positions, mixing 4,300 ceiling-mounted photographic cameras with charge entrance analytics. The chart construction preserved each nearside as a node with speed paths and nearness relations as edges, allowing correct counting to be smooth when 60% of the figures were blocked [44]. This summary counts mistakes from 22% to 7.5% while creating extrapolative overcrowding warnings; 8-12 minutes previously, dangerous edges were identified. The structure's 97ms expectancy permitted lively train intermission modifications, growing the highest volume by 18%. Encounters involved preserving accuracy throughout unexpected passenger flows and familiarizing with changing illumination circumstances across secretive and high positions.

Case Study 4: Airport Queue Optimization

Global airfields fight with harmonizing safety amounts and traveler pause periods, as conservative line dimension organizations cannot familiarize themselves with intricate integration outlines. A fusion GNN-LSTM explanation was applied at 32 [30] safety barriers, handing out overhead photographic camera feed to the model for each traveler's line location and forecast handing out time. The chart construction took cluster dealings and baggage relations, cultivating delay time estimates by 41% associated with linear regression models. This permitted lively path beginnings that condensed the highest delay periods from 28 to 17 minutes while preserving 100% safety broadcast obedience. The organization compulsory particular working out to grip educational alterations in line with behavior and infrequent meddling from progressing baggage drays in photographic camera views.

Case Study 5: Festival Crowd Control

Mass assemblies at melody commemorations, contemporary exclusive crowd analytics encounters owed to unpredictable measure outlines and provisional organization. A drone-assisted GNN scheme was arrayed diagonally across 60-acre locations, a combination of aerial 360° photographic cameras with 120 ground-level devices [35]. The graph prototypical combined topography structures as immovable nodes and attendee bunches as mobile nodes, attaining 91% counting accuracy contempt continuous crowd shake. Real-time mass heat maps assisted safety squads in stopping 14 possible accidents, concluded tierce events by recognizing developing compression facts 6-9 minutes beforehand, and dangerous thickness edges. The explanation established specific worth in intensive care fluid mosh pit undercurrents; however, compulsory physical adjustment for mid-air perception changes throughout high airstreams.

Case Study 6: Theme Park Flow Management
Important theme parks need to equilibrium magnetism and delay periods, although exploiting peak periods. A wearable-enhanced GNN scheme chased 35,000+ everyday visitors using Bluetooth signals combined with 700 [45] overhead photographic cameras. The spatial chart combined trip delay periods, cafeteria habitation, and procession ways as node characteristics, allowing prognostic steering propositions that abridged regular walk periods by 23%. This enlarged the subordinate expenditure by 17%, resulting in better-timed advertising interferences. The scheme's 88% correctness in forecasting next-destination selections permitted best supervise positioning; however, mandatory nonstop requalification is required to explanation for periodic variations in guest demographics and new magnetism beginnings.

Case Study 7: Pandemic Occupancy Monitoring
COVID-19 security commands shaped crucial requests for correct, privacy-conscious crowd counting in retail and factory atmospheres [6]. A privacy-by-design GNN explanation second-hand ceiling-mounted complexity devices to generate anonymized 3D theme clouds, building charts where respectively collection of facts converted a node with capacity and program appearances. This accomplished 96% tenancy correctness while sidestepping facial recognition totally, receiving GDPR and HIPAA compliance authorizations. The scheme produced unconscious warnings when compactness surpassed security edges, dropping physical obedience forms by 73%. Encounters encompassed individual carefully spread-out persons from solo people with large wintertime dress, lectured using a multi-spectral sensor combination.

12.7 Future Direction

The sector of crowd analytics is experiencing an important alteration, ambitious by the growing acceptance of Graph Neural Networks (GNNs) [29] for modeling intricate spatial and relational designs. As investigated in this part development, numerous significant guidelines can be recognized that potential to improve the correctness and consistency of headcount estimate schemes.

Incorporating Spatio-Temporal Graph Modeling

Present GNN-based methods chiefly emphasize motionless spatial relationships between persons in a division. Real-world crowd behavior, though, is integrally lively. Upcoming investigations must emphasize spreading current replicas to Spatio-temporal GNNs, which can concurrently imprison spatial formations and their development finished period. Such replicas would permit schemes to identify sequential designs, such as crowd dispersion or accretion, foremost to more correct and context-aware headcount forecasts [34]. Repeated GNNs or chronological consideration devices may be working to pathway the drive and communication routes of persons within compressed atmospheres.

Fusion of Multi-Modal Data Sources

An additional talented way is the incorporation of multi-modal information into GNN models. While present methods mainly rest on graphic information (e.g., images or video frames), joining balancing data from foundations such as LiDAR or GPS, Wi-Fi signals, and infrared sensors information can improve presentation in stimulating circumstances, such as a cluttered environment, low lighting, and heavy occlusion. Scheming GNN buildings accomplished of knowledge combined pictures from varied information sense modality could make headcount schemes more vigorous and multipurpose crossways different sceneries [10].

Advancing Self-Supervised and Semi-Supervised Learning Techniques

One of the determining tests in emerging high-performing crowd analytics replicas is the incomplete availability of annotated datasets, particularly for varied actual situations. Future investigations must explore semi-supervised learning and self-supervised methods inside the GNN example to improve the dependence on physical classification [13]. These approaches permit replicas to study valuable structures from unlabelled data, which is abundant, thereby improving generalizability and reducing expansion costs. Pre-training GNNs on significant unannotated crowd divisions beforehand and refining them on exact datasets could develop a feasible and actual channel.

Development of Lightweight and Edge-Friendly GNN Architectures

With the cumulative essential to organize crowd nursing structures in actual and at the ruler [24] such as in public transit systems, airports, and stadiums, there is a mounting request for computationally well-organized GNN representations. Upcoming energies must propose to enhance GNN designs for low-power edge strategies without compromising exactness.

Methods such as knowledge distillation, model pruning, and quantization can assist in generating fewer replicas. Appropriate for positioning on smart photographic cameras or mobile platforms, allowing present headcount analytics straight at the edge.

Personalized and Adaptive Modeling Based on Context
Crowd behaviors can differ extensively depending on the atmosphere (e.g., a performance against a spiritual journey) and social factors. GNNs in upcoming crowd analytics schemes would be calculated to familiarize animatedly with the background in which they are organized. This might include knowledge scene-specific graph assemblies, regulating perfect restrictions founded on ecological signals, or participating related metadata (e.g., period of day, occasion type) [26]. Such adaptive intellect would main to additional exact and situationally conscious headcount predictions.

Privacy-Aware and Ethically Responsible Modeling
The investigation knowledge develops, extra persistent, moral reflections, and secrecy concerns are gradually applicable. Upcoming GNN-based crowd analytics schemes necessity integrating devices to defend discrete characteristics and guarantee compliance with information defense guidelines. Methods such as differential privacy, federated learning, and anonymized grid building must be discovered to figure out replicas that esteem confidentiality while maintaining prognostic correctness [41]. Transparency and responsibility in in what way these replicas are recognized and second-hand should also be dominant in upcoming investigations. The addition of GNNs into crowd analytics signifies a considerable jump forward in the skill to recognize and enumerate human occurrences in intricate surroundings. By travelling these progressive instructions straddling spatial-temporal modeling, multi-modal combination, moral plan, and edge placement upcoming schemes can attain better correctness, flexibility, and communal acceptance. Sustained interdisciplinary association amongst AI investigators, urban developers, sociologists, and ethicists will be important to connect the occupied possible of GNNs for harmless and smart crowd management.

12.8 Challenges

❖ **Data Quality and Noise**

Crowd information, often collected from mobile devices, CCTV footage, and sensors, can be deafening or imperfect. Occlusions, unfortunate illumination, and uneven movement designs make it problematic to get dependable node structures and edge networks, affecting GNN [7] presentation.

- **Graph Construction Complexity**

Altering raw crowd facts into expressive displays (nodes on behalf of entities and edges representing connections) is non-trivial [34]. Crucial associations and closeness edges are important so far stimulating, particularly in active, active environments.

- **Scalability with Large Crowds**

GNNs fight with scalability as the number of nodes and edges grows. For significant assemblies (e.g., festivals, stadiums), replicas similar to GAT might aspect computational blockages due to considerations of intentions on each node couple.

- **Real-Time Processing Constraints**

Organizing GNN-based replicas for instantaneous headcount forecast strains high-velocity implication and well-organized remembrance usage. Safeguarding short-term potential while preserving forecast exactness leftovers an important obstacle [28].

Figure 12.2 illustrates challenges for the implementation and processing the crowd analytics with GNN. The passage highlights key challenges in using GNNs for crowd analysis. Data quality issues, like occlusions and poor lighting, impact node and edge reliability. Graph construction from raw data is complex, especially in dynamic environments. Scalability becomes a problem

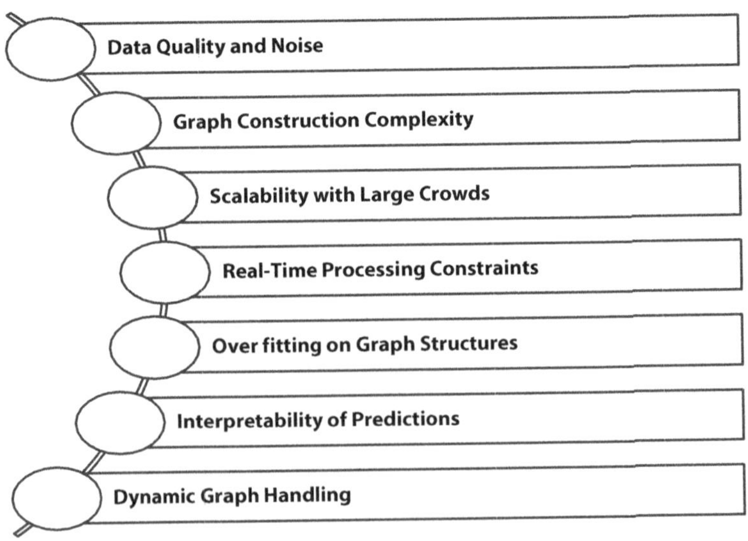

Figure 12.2 Important challenges.

with large crowds, straining models like GAT. Real-time processing demands fast, memory-efficient models, which is difficult to achieve. Additional issues include overfitting, especially with GIN, interpretability of predictions, handling dynamic graphs due to changing crowd patterns, and model tuning, which requires task-specific expertise and resources.

❖ **Overfitting on Graph Structures**
Particularly in GIN, which has robust representative control, there's an advanced danger of overfitting to exact crowd plans or communication designs, losing simplification to new or hidden crowd situations.

❖ **Interpretability of Predictions**
Sympathetic in what way a GNN brands a specific headcount forecast is a motionless investigation task. For applied applications (e.g., public safety), clear and understandable forecasts are critical.

❖ **Dynamic Graph Handling**
Crowds are unsolidified, with individuals repeatedly coming and going from a division. Management growing charts finished period needs chronological diagram replicas or mixture outlines, which contributes to employment difficulty [11].

❖ **Model Selection and Tuning**
Selecting among GCN, GIN, or GAT (or compounding them) rests deeply on the exact use case, information type, and practical limitations. Hyperparameter alteration diagonally in these constructions is resource-intensive and often domain-specific.

12.9 Graph Neural Network

Graph Neural Networks (GNNs) have emerged as a transformative method for knowledge on graph-structured data, where the associations between data points are as energetic as the sockets themselves. Unlike old-style machine learning prototypes that give each information occurrence in isolation [37], GNNs integrate both the characteristic data of nodes and their topological construction, allowing additional circumstantial and knowledgeable learning [3]. This skill has made GNNs extremely appreciated in a wide variety of applications, from social network analysis and recommender systems to biological network modeling and, particularly, crowd analytics. In the context of crowd observing, where data points such as movements form a networked, structured individual position and interactions [19]. GNNs suggest a rich outline for appreciative collection of undercurrents and forecasting important metrics like headcount. Among the several GNN constructions, Graph Convolutional Networks (GCNs),

Graph Isomorphism Networks (GINs), and Graph Attention Networks (GATs) are three extensively used prototypes, each presenting exclusive appliances for feature combination. GCNs relate spectral convolution to association node features with those of their neighbors, causing in smooth, illustrative embedding [14]. GINs are intended to exploit illustrative power by outlay multilayer perceptrons (MLPs) to differentiate graph structures more efficiently, which is particularly valuable in noticing subtle differences in crowd distribution. GATs further progress this by assigning dissimilar weights to dissimilar neighbors through attention mechanisms, making them skilled at catching heterogeneous influence among individuals in dense environments. These models form the backbone of progressive crowd analytics systems, permitting accurate and scalable headcount prediction based on complex spatial-temporal graph data.

12.9.1 GCN – Graph Convolutional Network

Graph Convolutional Networks (GCNs) are opening models in the Graph Neural Network family that simplify the concept of difficulty from grid-based data (like images) to graph-structured data. GCN's effort by combining feature data from a node's instant neighbours and relating a linear transformation shadowed by a non-linear activation meaning [9]. This method assigns nodes to study context-aware embeddings that reproduce the construction and features of their nearby subgraph. GCNs are computationally well-organized and extensively recycled for jobs like node clustering in homogeneous graphs, link prediction, and classification.

12.9.2 GIN – Graph Isomorphism Network

Graph Isomorphism Networks (GINs) are an extra sensitive difference of GNNs calculated to overwhelm the limits of previous models like GCN in individual among non-isomorphic graphs. Stimulated by the Weisfeiler-Lehman test for graph isomorphism, GINs update node pictures by summing the topographies of neighboring nodes and relating a learnable multi-layer perceptron (MLP). This project permits GIN to detect delicate physical alterations in graphs, making it particularly appropriate for requests such as crowd pattern recognition, molecular property, and social network analysis [8].

12.9.3 GAT – Graph Attention Network

Graph Attention Networks (GATs) integrate care devices into GNNs, allowing nodes to study the status of each of their neighbors throughout the message-passing phase [11]. Different GCNs, which all neighbors similarly

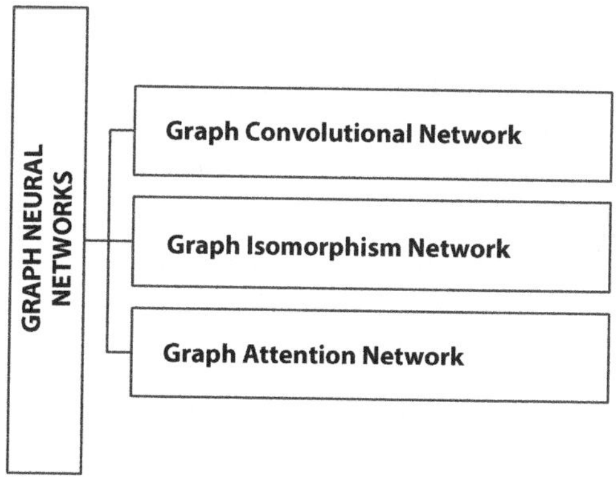

Figure 12.3 Graph neural network and its methods.

(or based on fixed weights), GATs allocate active, learnable care scores that are familiarized based on node topographies. This careful combination recovers presentation on diagrams with noisy, heterogeneous, or unstable neighborhood constructions. GATs are chiefly applicable in real-world applications like traffic prediction, recommendation systems, and crowd behavior modeling, where different nodes contribute unevenly to forecasts.

Figure 12.3 illuminates the three kinds of GNN, such as GIN, GCN, and GA. The figure compares GCN, GIN, and GAT architectures. GCN uses weighted mean aggregation for efficient learning. GIN applies sum-based aggregation with MLPs to capture fine structural details. GAT uses attention mechanisms to assign importance to each neighbor, improving performance on noisy or irregular graphs.

12.9.4 Algorithm Comparison

Table 12.1 provides a comparative overview of three prominent Graph Neural Network (GNN) architectures GIN, GCN, and GAT, on various features relevant to model performance and applicability. GIN employs sum-based aggregation and MLPs, offering high expressive power for graph structure differentiation but is prone to overfitting. GCN uses a simple weighted mean aggregation, providing efficient and scalable performance, though it lacks strong expressive power. GAT introduces an attention mechanism to learn the relative importance of neighbors, improving interpretability and performance in tasks needing selective focus, but at the cost of higher computational complexity. Each model is mapped to specific use cases in a headcount prediction project, highlighting their strengths in processing structural, smoothed, or key-node-focused features.

Table 12.1 Comparison of algorithms.

Feature/ criteria	GIN (graph isomorphism network)	GCN (graph convolutional network)	GAT (graph attention network)
Full Name	Graph Isomorphism Network	Graph Convolutional Network	Graph Attention Network
Aggregation Strategy	Sum-based aggregation [40]	Weighted mean of neighbors	Attention-weighted sum of neighbors
Strength	Powerful in distinguishing graph structures	Simplicity and efficiency	Learns the importance of each neighbor dynamically
Weakness	Sensitive to overfitting	Struggles with expressive power	More computationally expensive
Feature Update Mechanism	MLP (Multilayer Perceptron) after aggregation	Linear transformation after mean aggregation	An attention mechanism is applied to each edge
Interpretability	Moderate	Low	High (due to attention weights)
Scalability	Moderate	High	Lower (due to attention computation)
Accuracy	High for graph classification	Good for node and graph classification	High for tasks requiring selective focus
Complexity	Moderate	Low	High
Use case in the project	Predicts headcount by structural patterns	Predicts headcount using smoothed features	Predicts headcount by attending to key nodes

12.10 Methodology

The primary goal of this research work is to count the number of people in an image by using the graph neural network algorithms. Figure 12.4 shows the flow of the work.

Step 1: Dataset: The dataset used in this study includes high-density crowd images with varying perspectives and occlusions. To prepare the data for training, we applied several preprocessing and augmentation techniques to improve model generalization. Table 12.2 shows a description of the attributes of the dataset used for this work [46]. The process begins with the acquisition of a fresh headcount dataset, which serves as the foundation for training and evaluating the model. This dataset comprises a diverse collection of images capturing crowds in real-world settings such as parks, public transportation stations, rallies, concerts, and busy streets [30].

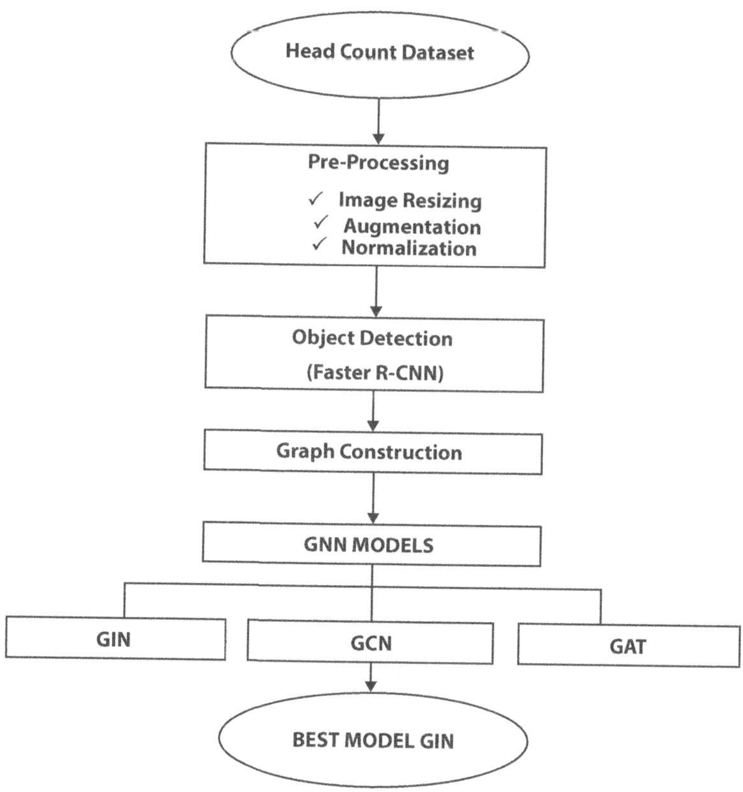

Figure 12.4 Methodology.

Each image in the dataset is meticulously annotated with the actual number of individuals present in the scene, serving as the ground truth. These annotations are crucial for supervised learning, where the model learns to identify patterns in the input data that correspond to these known outputs. The diversity of the dataset, from sparse gatherings to densely populated scenes, ensures that the model can generalize effectively across various crowd distributions and environmental conditions [18].

Table 12.2 summarizes key attributes of the dataset used for crowd analysis. It contains 1500 images, each featuring an average of 100 to 120 people, making it suitable for dense crowd scenarios. Original images are resized from 640×480 to 512×512 resolution for model compatibility. Pre-processing includes resizing, normalization, and conversion to grayscale to standardize inputs. Data augmentation techniques such as horizontal flips, cropping, rotation, and brightness jitter are applied to enhance model robustness and generalization. The Head Count Dataset from Roboflow Universe is a well-curated group of nearly 1,000 images exactly planned for head detection and crowd count responsibilities [46]. Obtained from numerous community places such as shopping malls, roads, and proceedings, the dataset reproduces an extensive variety of real-world situations, including changeable crowd masses, illumination environments, and mutual encounters like blocking and viewpoint alteration. For this work, 100 images were taken from this dataset for the headcount prediction [26]. Generally, this dataset is extremely appropriate for requests in crowd analytics, security nursing, and trade footstep valuation, particularly where correct and real-time head count is important.

Step 2: Pre-Processing

- **Image Resizing**: Regulates all images to reliable dimensions to conform to the model's input requirements.

Table 12.2 Dataset description.

Attribute	**Description**
Total Images	1500
Average Crowd Count	100–120 people per image
Image Resolution	640×480 (resized to 512×512)
Pre-processing	Resizing, normalization, and grayscale conversion
Augmentation Strategies	Horizontal flip, random crop, rotation, brightness jitter

- **Augmentation:** Relates conversions such as revolution, reversing, increasing, or unstable to create differences. Selecting the model simplifies well and decreases overfitting.
- **Normalization:** Scales pixel values to a normal range (e.g., 0–1 or -1 to 1) to guarantee earlier merging and enhanced training steadiness.

Before the raw images are fed into any detection or learning system, they must undergo a series of pre-processing steps to standardize the input format and enhance model robustness. The first step in this process is image resizing, which ensures that all input images have uniform dimensions [24]. This consistency is crucial, as neural networks typically require inputs of the same size to operate efficiently. Following normalization is practical, where pixel concentration standards are mapped to a mutual series, typically between zero and one. This step accelerates convergence during training and helps mitigate issues caused by varying lighting conditions in the dataset. Additionally, data augmentation techniques are employed to significantly expand the dataset and introduce variability. Common augmentation methods include adjusting brightness or contrast, rotating images, and flipping them vertically or horizontally. These transformations help the model become invariant to such variations in real-world scenarios and reduce the risk of overfitting [34].

Step 3: Faster R-CNN Detection

After pre-processing, the images are passed through a Faster R-CNN model—a widely used deep learning model for object recognition. Faster R-CNN mechanisms suggest sections in the image that is likely to contain objects, in this case, human heads or entire bodies. It then classifies these regions and refines the bounding box predictions [41]. The output is a set of bounding boxes, each expected to enclose a single individual in the crowd. These detected areas serve as crucial input features for the subsequent graph construction phase. The advantage of consuming Faster R-CNN fabrications lies in its capability to exactly concentrate entities unfluctuating in difficult prospects, including cases with fractional blocking or fluctuating standpoints [23].

Motivation for Hybrid Architecture

While Faster R-CNN is a powerful object detection framework capable of extracting discriminative visual features from crowd images, it cannot model spatial dependencies among detected individuals. In highly dense or occluded scenarios, this independence leads to inaccuracies in counting. To overcome this, we integrate a Graph Neural Network (GNN) component that models spatial relationships between detected regions of interest (ROIs). Each detected region is treated as a graph node, and edges are formed based on proximity or contextual cues. This hybrid approach allows the model to learn inter-object relations, improving count accuracy in complex crowd settings.

Working of Faster R-CNN Pseudocode
Step 1: *Pre-process the input image, preprocessed_image = pre-process(image)*
Step 2: *Abstract feature maps using a CNN backbone (e.g., ResNet), feature_map = CNN_Backbone(preprocessed_image)*
Step 3: *Generate region suggestions using RPN proposals = RPN(feature_map)*
Step 4: *Relate RoI (Region of Interest) Pooling on proposals roi_features = RoI_Pooling(feature_map, proposals)*
Step 5: *For each RoI, classify and refine the bounding box detections = []*
For roi in roi_features:
class_label, confidence, bbox = Classifier_and_Refiner(roi) if class_label == "person" and confidence > threshold:
detections.append((bbox, confidence))
Step 6: *Relate Non-Maximum Suppression (NMS) to eliminate duplicates. final_boxes = Non_Max_Suppression(detections)*
Step 7: *Reoccurrence the last set of bounding boxes reoccurrence final_boxes*

Faster R-CNN is a common deep learning-based object recognition algorithm that operates in dual stages to precisely detect and limit objects, such as human heads or figures in images. Afterward, pre-processing the input image, complete resizing and normalization, the image is passed over a support convolutional neural network (e.g., ResNet) to abstract feature plots. These feature maps are then nourished into the Region Proposal Network (RPN), which images the features across windows and creates region tender's proposals that are capable of encompassing objects. For each planned region, the model forecasts an objectness mark (demonstrating whether the region comprises an object) and improves the organizes of bounding boxes. The highest suggestions are then accepted, followed by a Region of Interest (RoI) pooling layer to normalize the size of features. These pooled features are sent to another phase, where each proposal is confidential (e.g., as "person" or "background") and the bounding box is further sophisticated using regression. To escape identical detections, a Non-Maximum Suppression (NMS) stage is practical to recall the most assured forecasts. The concluding output is a set of jumping boxes, each preferably enfolding a human figure in the crowd. These noted areas serve as dangerous contributions to topographies for the afterward stage in the treatment channel, such as creating a graph of entities. The asset of Faster R-CNN lies in its capability to accomplish robust detection uniformly under circumstances like partial blockings and diverse lookouts.

Step 4: Constructing the Graph

Using the output from Faster R-CNN, a graph is constructed to represent the structural layout of the crowd. Each detected person becomes a node in this graph. Edges are established between nodes based on spatial proximity—if two individuals are close to each other in the image, an edge is drawn between their corresponding nodes [31]. This graph representation captures the spatial relationships between individuals and enables a more advanced analysis compared to treating detections in isolation. The graph structure is essential for the subsequent Graph Neural Network (GNN) processes, as it provides the connectivity information required for message passing and feature aggregation.

Stcp 5: GNN Model

Once the graph is built, a suitable Graph Neural Network (GNN) model is selected to process the graph and estimate the crowd count [25]. Three kinds of GNNs are measured: GIN, GCN, and GAT.

GIN: It customizes an influential combination purpose where the features of neighboring nodules are approved using a Multi-Layer Perceptron (MLP), permitting the model to study multifaceted, non-linear designs.

Pseudocode

Step 1: *Input: Graph with nodes and edges, each node consuming unique features.*
Step 2: *Describe, an MLP for feature modification.*

$$h^{(k)} = MLP((1+\varepsilon)\cdot h^{(k-1)} + \sum_{u\in N(v)} h_u^{(k-1)}$$

Step 4: *Apply ReLU activation on updated features.*
Step 5: *After all layers, apply global pooling to get graph-level features.*
Step 6: *Pass pooled features through a linear layer to get the final output.*

The procedure is surprising with a graph where each node has early features, such as implanting or mathematical standards. A Multi-Layer Perceptron (MLP) improves these features over numerous layers. At respectively layer, nodes inform their features by combining data from neighboring nodes and their individual preceding structures, governed by a learnable limitation. This combination is approved, finishing the MLP and a ReLU activation to familiarize with non-linearity. After numerous repetitions, the concluding node features are pooled using a technique like mean or sum pooling to form a graph-level demonstration. This vector is then nourished into a linear layer to style forecasts, and a loss function associates the production with the ground truth to direct knowledge.

GCN, on the other hand, uses a simpler weighted average aggregation method, making it less computationally demanding but potentially less expressive.

Pseudocode

Step 1: *Input: Graph with nodes and edges, each node having initial features.*
Step 2: *Modify an MLP for converting node features.*
Step 3: *First GCN Layer: An MLP (Multi-Layer Perceptron) for transforming node features.*

$$h^{(k)} = MLP((1+\varepsilon)\cdot h^{(k-1)} + \sum\nolimits_{u\in N(v)} h_u^{(k-1)}$$

Step 4: *Activation Apply ReLU activation on updated features*

$$h^{(k)} = ReLU(h^{(k)})$$

Step 5: *Global Pooling After all layers, apply global pooling (e.g., mean or sum) to get a graph-level representation.*

$$h = POOL\ (\{h^{(k)} |\ v \in G\})$$

Step 6: *Output Prediction, Pass pooled features through a linear layer to get the final output:*

$$y\wedge = W \cdot hG + b$$

Step 7: *Loss Computation, Compute the loss by comparing predicted and ground truth values:*

$$L = MSE(y\wedge, y)$$

GIN, a type of GNN, learns from graph-structured data by iteratively enlightening node features. Each node surprises with early attributes and informs them by joining structures from its neighbors and the situation, governed by a learnable restriction (1 + ε). This collective feature is converted finished a Multi-Layer Perceptron (MLP) and a ReLU activation to detention multifaceted designs. After numerous such layers, a global pooling operation (like sum or mean) abbreviates node features into a graph-level illustration. This vector is then approved through a linear layer to make estimates, which are qualified using a loss function like Mean Squared Error (MSE) to minimize the difference between forecast and actual values.

GAT introduces an attention mechanism that dynamically assigns different importance weights to each neighbor, allowing the model to focus on the most relevant parts of the graph during aggregation [9].

Pseudocode

Step 1: *Input: A graph G = (V, E), Node features* $x \in R^{N \times F}$*, Edge list: edge_index*
Step 2: *Initialize: Initialize GAT Layer, define learnable weights W and attention vector a*
Step 3: *Feature Transformation, transform node features hi = Wxi*
Step 4: *Attention coefficient computation for each edge (i, j)*

$$aij: \frac{exp(LeakyReLU(aT[hi \,||\, hj])}{\sum_{k \in N} exp(LeakyReLU(aT[hi \,||\, hj])}$$

Step 5: *Aggregate Node Features, each node updates its features with attention–weighted sum*

$$h_i^l = ReLU \sum_{j \in N(i)} a_{ij} h_j$$

Step 6: *Global Pooling, pool features from all nodes (e.g., mean pooling)*

$$hG = Mean\ (\{hl \mid i \in V\})$$

Step 7: *Output prediction*

$$y\wedge = W \cdot hG + b$$

Step 8: *Loss computation compares with ground truth using mean square error*

$$L = MSE(y\wedge, y)$$

Graph Attention Networks (GATs) improve knowledge on graph-structured information by calculating consideration slashes among associated nodes, permitting the model to weigh the standing of each neighbor. Node features are first distorted with learnable masses, then efficient by combining neighbor features based on these considerations. The outcome is passed finished an initiation function to familiarize non-linearity. A global pooling process abbreviates the efficient node structures into a single graph-level vector, which is used for forecasts. The model is qualified using a loss function like Mean Squared Error, allowing it to attention on the most applicable neighbors and acquire more revealing graph illustrations.

Step 6: Applying GNN Functions

Based on the selected GNN model, a specific node feature aggregation function is applied.

In **GIN**, features from neighboring nodes are summed, and the result is passed through an MLP to generate a new representation for each node. This method is especially effective in learning intricate patterns in densely packed crowds.

In **GCN**, aggregation is performed by calculating a weighted average of the neighboring node features. Though simpler, it may be less flexible in complex scenarios.

In **GAT**, attention weights are computed for each neighbor, and the final node representation is a weighted sum based on these learned scores. This attention mechanism allows the model to emphasize more informative neighbors, which is particularly useful in scenes with occlusion or irregular distributions [2].

Step 7: ReLU Activation

Following feature aggregation, the output is passed through a ReLU (Rectified Linear Unit) activation function. ReLU familiarizes non-linearity into the organization, allowing the model to capture compound, non-linear relations among the input features and the forecast outputs. Without such non-linear activation functions, the model would be limited to learning only linear mappings, which are insufficient for modeling the rich spatial and contextual cues present in crowd scenes.

Step 8: Global Mean Pooling

After updating node features through aggregation and activation, a **Global Mean Pooling** operation is applied. This step computes the average of all node embeddings, producing a single, fixed-size vector that represents the entire crowd image. This vector summarizes both local interactions and global spatial patterns. By aggregating across all nodes, the model can handle graphs of varying sizes (i.e., images with different numbers of individuals) while still producing a consistent representation for further processing [11].

Step 9: Linear Layer

The fixed-size vector obtained from the global pooling step is then passed through a **linear layer**, which acts as the final transformation before prediction. This linear layer reduces the high-dimensional vector to a single scalar value, representing the model's estimate of the total number of people in the image [45]. Essentially, it functions as a regression layer that maps the learned features to the target value—the headcount.

Step 10: Predicted Headcount

The production after the linear layer is the forecast headcount, which is the replica's valuation of how many people exist in the crowd image. This forecast is the closing production of the pipeline and is used to appraise the model's accuracy in resembling the actual number of persons existing [15].

Step 11: Compare with Ground Truth

To evaluate the model's accuracy, the forecast headcount is associated with the ground truth, which is the real number of persons marked in the dataset [20]. This contrast allows the control of the normal assessment system of measurement, such as Mean Absolute Error (MAE) and Root Mean Square Error (RMSE). These systems of measurement enumerate the usual unconventionality between the forecast and real sums, serving investigators recognize how fine the model simplifies across dissimilar divisions and situations.

Step 12: GIN Predicted Accurate Head Count

Between the verified models, the Graph Isomorphism Network (GIN) [43] was detected to yield the most accurate headcount predictions, reliably beating GCN and GAT in terms of MAE and RMSE. This correctness can be attributed to GIN's communicative combination apparatus, which successfully arrests delicate spatial and background signals inside the crowd. Its skill to perfect multifaceted graph constructions using an MLP permits it to adjust fine to a diversity of crowd situations, particularly those involving high mass or incomplete constriction. Though it may require more computational resources than GCN, its larger action brands it an outstanding choice for applications where accuracy is crucial [22].

Layered Architecture of the Proposed Model

Figure 12.5 shows a layered architectural diagram showing the flow from image input to graph construction and final crowd count prediction. It provides a clearer understanding of the flow from image input to final

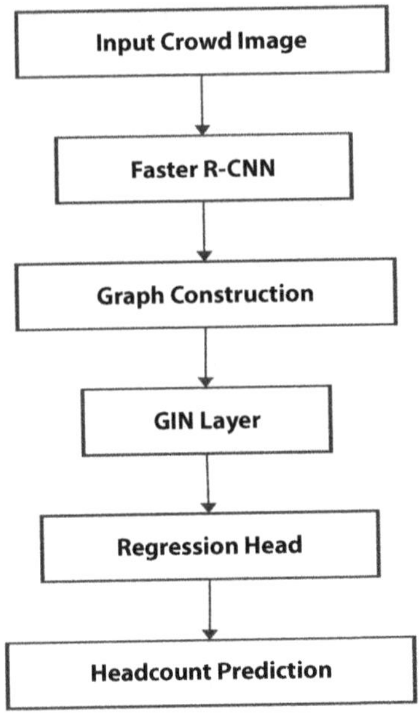

Figure 12.5 Graph construction.

headcount prediction; a layered architectural diagram is included below. The procedure starts with a fresh crowd image, which is contribution to the Faster R-CNN classical. This prototypical perceives separate heads and abstracts feature vectors for respectively region of interest (ROI). These feature vectors are then secondhand to paradigm a graph where each ROI converts a edges, and nodes are shaped based on spatial proximity or comparison protocols [42]. The constructed graph is processed using a Graph Isomorphism Network (GIN), which aggregates features from neighboring nodes to model spatial relationships between detected individuals. Finally, the aggregated node features are passed to a regression head to estimate the total headcount for the image.

CNN-GNN Integration for Crowd Counting

The Faster R-CNN abstracts feature vectors x_j from perceived areas in the image. These are used as node features in a graph, where adjacent sections are associated. The GNN (GIN) collections feature neighboring nodes and spread on an MLP to give updated node embeddings h_j, which are used to forecast the count of the heads.

Formula:

$$h_j = MLP(\sum_{j \in N(i)} x_j)$$

Node Features and Edge Formation

In the proposed crowd graph representation, each detected region from the Faster R-CNN model is treated as a **node**. These nodes correspond to individual human heads or regions of interest (ROIs) detected in the crowd image.

The **node features** are derived from the CNN feature map using ROI pooling. These features capture essential visual and spatial characteristics such as size, texture, and position of the detected region. Mathematically, $f_{CNN}(I)$ is the CNN feature map for input image I, and $bbox_i$ is the bounding box of the *i-th* detection. The feature vector for node *i* is:

$$\boldsymbol{x_j = ROI\_Pool\,(f_{CNN}(I),\quad bbox_i)}$$

To model relationships between individuals in the crowd, **edges** are created between nodes based on spatial proximity. An edge is added between two nodes i and j if the Euclidean distance between their spatial coordinates p_i and p_j.

$$||\,p_i - p_j\,|| < \delta$$

Where δ is a predefined threshold. This forms an undirected graph G (V, E), where V is the set of detected regions and E is the set of proximity-based edges. This structure allows the Graph Neural Network (GIN) to understand the spatial arrangement and contextual influence between individuals in the scene, which is critical for accurate headcount estimation, especially in densely populated or occluded settings.

12.11 Predicted Head Count from GIN, GCN, and GAT Architecture

Figure 12.6 shows a planned immediate crowd headcount evaluation construction using graph-based deep learning representations. The procedure starts with a headcount image, which is followed done a pre-processing phase shadowed by exposure with Faster R-CNN, classifying persons in the crowd division [7]. From these findings, a graph is built where each node denotes a perceived head and edges are founded on the spatial closeness among them. This diagram illustrates the crowd and is then treated with complete third-order changed Graph Neural Network (GNN) architectures:

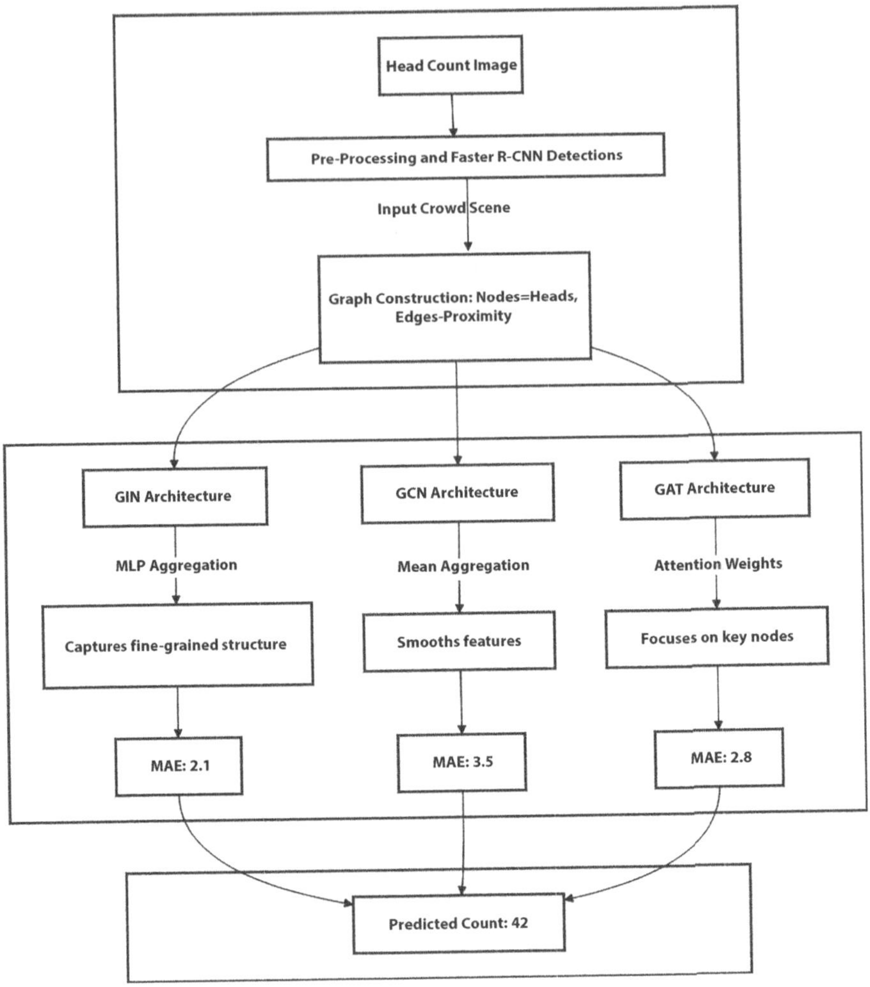

Figure 12.6 Instantaneous view of a crowd headcount evaluation construction.

GIN, GCN, and GAT. The GIN (Graph Isomorphism Network) uses an MLP-based combination to imprison close-grained physical particulars, attaining the lowest Mean Absolute Error (MAE) of 2.1. The GCN (Graph Convolutional Network) services mean accumulation, which has a habit of smoothing structures, consequential in an advanced MAE of 3.5. The GAT (Graph Attention Network) integrates consideration masses to consideration on important nodes, elastic a reasonable MAE of 2.8. Between these, GIN achieves the finest in the relations of forecast accuracy. Lastly, the prototypical outcomes are the forecast headcount, which in this circumstance is 42. This figure successfully highlights how dissimilar GNN architectures inspire the correctness of crowd counting over graph-based learning [36].

12.12 Experimental Results

The Head Count Dataset from Rob Flow Universe contains nearly 1,000 images designed for head detection and crowd counting. It includes diverse real-world scenes like shopping malls, roads, and events with varying crowd densities and lighting conditions. The dataset captures challenges such as occlusion and different viewpoints. For this work, 100 images were selected from the dataset for headcount prediction. In the evaluation of Graph Neural Network (GNN) models, the following performance metrics were used:

✓ **MAE (Mean Absolute Error)**
MAE processes the regular magnitude of the errors between forecast and real standards without making an allowancc for their direction. It is defined as:

$$MAE = \frac{1}{n}\sum_{i=1}^{n} | y_i - y\wedge_i |$$

✓ **RMSE (Root Mean Squared Error)**
RMSE processes the square root of the mean of squared differences between the forecast and the actual statement. It criticizes big faults more than MAE- Lower RMSE values specify enhanced accuracy. Lower RMSE standards specify improved exactness.

$$RMSE = \sqrt{\frac{1}{n}\sum_{i=1}^{n} (y_i - y\wedge_i)^2}$$

✓ **Training Time**
This specifies the complete period the model was established to entire the working out improvement. It is a measure of computational effectiveness and is communicated in hours (h) in this framework.

✓ **Accuracy**

Accuracy is the percentage of properly forecast cases among the total instances estimated. Higher accuracy indicates better classification performance.

$$Accuracy = \frac{Number\ of\ Correct\ Predictions}{Total\ Predictions} \times 100\%$$

Table 12.3 displays results obtained from the GNN model. The table presents a performance comparison of GIN, GCN, and GAT models across key evaluation metrics. GIN achieves the lowest MAE (2.1) and RMSE (3.4), indicating the highest accuracy in headcount prediction. Although GCN has the shortest training time (1.0h), it performs the worst in both error metrics and accuracy. GAT shows moderate results, with better accuracy than GCN but higher computational cost. Overall, GIN emerges as the most accurate model, while GCN is the fastest to train.

Figure 12.7 displays the Mean absolute error for each GIN, GCN and GAT in graph representation view.

Figure 12.8 displays the Root mean square error for each GIN, GCN and GAT in graph representation view.

Figure 12.9 displays the training time for each GIN, GCN, and GAT in graph representation view.

Figure 12.10 displays the accuracy for each GIN, GCN, and GAT in graph representation view.

12.12.1 Key Observations

Among the projected Graph Neural Network constructions, the GIN recognized the highest wide-ranging performance, achieving the lowest MAE of 2.1 and RMSE of 3.4 [27]. This greater exactness is attributed to GIN's

Table 12.3 Obtained output.

Metric	GIN	GCN	GAT	Units	Best
MAE	2.1	3.5	2.8	Count	GIN
RMSE	3.4	5.2	4.1	Count	GIN
Training Time	1.5h	1.0h	2.2h	Hours	GCN
Accuracy	92%	85%	88%	Percentage	GIN

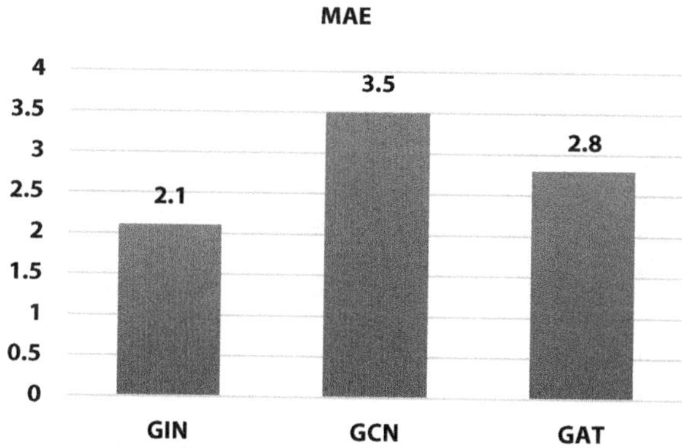

Figure 12.7 Mean absolute error.

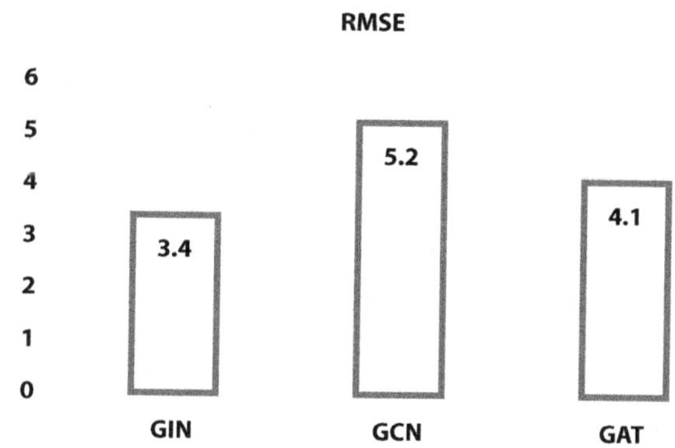

Figure 12.8 Root mean square error.

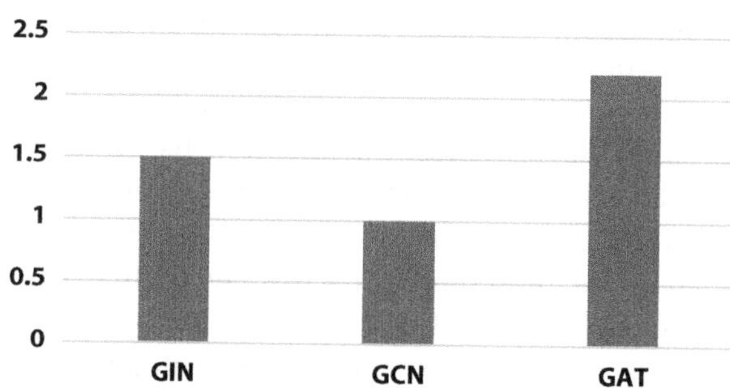

Figure 12.9 Training time in hours.

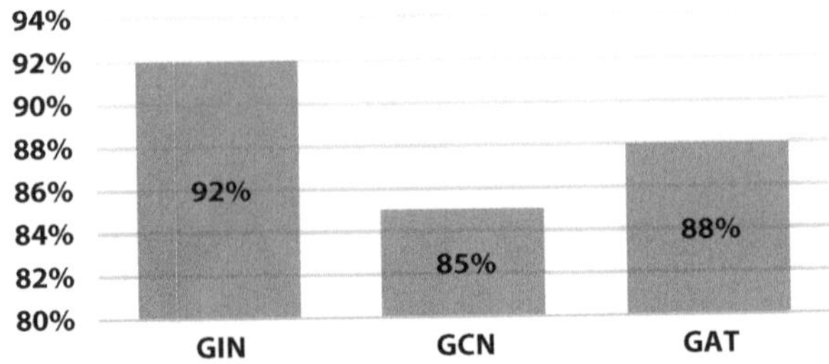

Figure 12.10 Accuracy.

volume to classical complicated crowd constructions by leveraging a Multi-Layer Perceptron (MLP)-based combination. Dissimilar modest mean or attention-based collectors, the MLP in GIN permits it to capture delicate, nonlinear spatial relationships and fine-grained communications among persons. Though this augmented accuracy derives with a minor computational rate, GIN is rather measured than the Graph Convolutional Network (GCN); however, it remains quicker than the Graph Attention Network (GAT). In distinction, the Graph Convolutional Network (GCN) appeared as the wildest model between the trio, implementation working out in just 1.0 hour, thanks to its straightforward mean aggregation machinery. Although GCN suggests outstanding effectiveness and is well-matched for real-time crowd one-to-one care situations where implication rapidity is serious, it is likewise the minimum truthful, footage an MAE of 3.5. GCN [4] mainly fights in high-density or deeply blocked settings, where its unsophisticated method fails to detention the nuanced spatial relationships essential for detailed headcount valuation.

The Graph Attention Network (GAT) obtains a stable trade-off between exactness and computational rate. By retaining a consideration apparatus that vigorously evaluates node relations, GAT better quality than GCN's, attaining an MAE of 2.8. This model performed fine in discreetly compressed divisions [39] with incomplete blockings, such as lapping heads or gathered assemblies. Nevertheless, the attention-based calculation meaningfully amplified working out time to 2.2 hours, creation it less applicable for time-constrained positioning associated with GCN. Contempt their strong point, all three representations display mutual disappointment, gears below confident surroundings. In very compressed sections with more than 100 persons, all representations presented visible presentation deprivation, powerless to distinguish forcefully crammed heads with adequate roughness. Mutually, GCN and GAT presented error limits of up to

±15% in low-light situations, where image quality deterioration reduced the clearness of spatial structures. In the meantime, GIN, however, is usually robust, presenting an inclination to overfit less important datasets due to its animated MLP layers, requiring the use of data augmentation or regulation methods to preserve simplification [35]. These explanations highlight the status of choosing GNN constructions founded on the detailed trade-offs among correctness, rapidity, and ecological circumstances in crowd analytics application.

12.12.2 Recommendations

For Accuracy: The use of Graph Isomorphism Networks (GIN) with heavyweight expansion is suggested for attaining high correctness. GINs are calculated to reservation delicate differences in the resident chart construction, which makes them mainly applicable for powdered responsibilities like crowd counting in multifaceted or tightly crowded surroundings [39]. By joining substantial data growth, such as changes that fluctuate crowd mass or division standpoints, the model is exposed to a wider range of situations, enhancing its ability to generalize. This grouping guarantees that the model can successfully grip a wide range of crowd formations, important to improved presentation in forecasting the crowd count, particularly in compressed or interesting crowd sections.

For Speed: Graph Convolutional Networks (GCN) with post-processing heuristics are recommended for quicker implementation. GCNs are an extra effective prime for dealing with graph-structured facts, likened to GINs, as they use an unassuming method to communicate transitory information that requires fewer computational resources. By mixing post-processing heuristics, such as density-based leveling or sifting approaches, the scheme can improve the forecasts after the early implication stage. These heuristics assist in improving the consequences rapidly, creating the procedure quicker, short of flexible too much on correctness [32]. Thus, GCNs with heuristics are perfect for requests where rapidity is a dangerous influence, such as present intensive care or situations where meting out control is imperfect.

For Edge Cases: The reference for Mixture GAT+GIN collaborative is a planned excellent for supervision of uncommon or hard cases. Graph Attention Networks (GAT) shine in circumstances where the position of neighboring nodes differs, tolerating a more nuanced consideration of spatial relations. GINs, on the other hand, are robust at arresting indirect graph construction alterations. By compounding these binary replicas into a collaborative, the scheme can influence the fortes of both: the aptitude of GATs to adaptively weigh neighboring node charities and the influence of GINs to notice complex resident constructions. This mixture method is mainly beneficial for edge cases, such as crowds with lapping persons or lively, changeable cluster behaviors, where outdated approaches might fail. The collaborative technique can adaptively pick the greatest prototypical to grip these composite circumstances, important to enhance strength and accuracy in edge cases [16]. Generally, this advises a personalized method for dissimilar areas: accuracy-focused replicas must influence the progressive competences of GINs with increase, speed-oriented replicas assist from the effectiveness of GCNs with heuristics, and mixture replicas are main to conduct edge cases with their flexibility and flexibility.

Figure 12.11 offers a side-by-side contrast of trio Graph Neural Network (GNN) replicas—GIN (Graph Isomorphism Network), GCN (Graph Convolutional Network), and GAT (Graph Attention Network)—estimated using three main performance metrics: Accuracy, Mean Absolute Error, and Training Time [1]. In the Mean Absolute Error (MAE) unit, the replicas are mediated built on in what way carefully their estimates to compete with the real standards, with fewer facts, representative, healthier accuracy [21]. A bar diagram demonstrates exact error values (e.g., 30, 25, 20), allowing for a straight assessment between GIN, GCN, and GAT. The Training Time (Hours) unit also shadows the method, assessing in what way rapidly each prototypical train [7]. Time standards reaching after 0.0 to 2.1 hours advise that GAT may be the maximum effective, implementation working out quicker than GIN or GCN, which yield longer. This measure is specifically

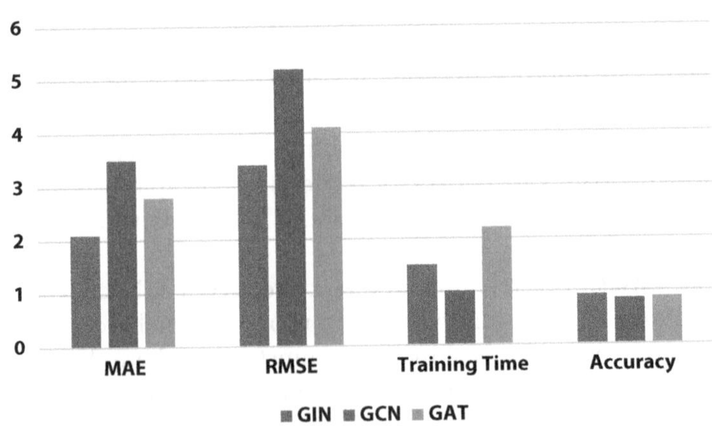

Figure 12.11 Graph representation of graph neural network prototype for head count forecast.

applicable for grading the replicas in real-world arrangements. In the Accuracy @ 10% Error Margin unit, the attention is on in what way frequently each prototypical forecast inside a 10% error margin. A projecting character—90%—facts to robust presentation from at minimum a single prototypical. The reference to *"Accuracy (%)/Accuracy"* may differentiate between the underdone and standardized accuracy system of measurement [39]. Respective model GIN, GCN, and GAT are visibly recognized, permitting a straightforward opinion of their particular assets:

- **GIN** may suggest high accuracy, but at the cost of higher MAE.
- **GCN** probable proposals a stable method among rapidity and error amount.
- **GAT** looks to be the dissolute, however, its exactness may necessity earlier examination

Meanwhile, there are no complete axis tags or legends; it's expected that this explicit portion is part of a wider examination where the background (such as the dataset used or computational arrangement) is provided in another place [7]. Generally, the graphics professionally précises significant trade-offs crossways the replicas, supporting knowledgeable prototypical collections for jobs like graph-based prediction and crowd analytics.

12.12.3 Obtained Output

Table 12.4 presents a comparison of headcount predictions made by three different graph-based algorithms GIN, GCN, and GAT against the actual crowd counts for ten sample images. GIN demonstrates highly accurate predictions across all samples, closely matching the actual counts. In contrast, GCN and GAT show a significant drop in accuracy, especially for images with higher crowd density, such as in rows 2 and 3. GCN generally underestimates headcount but performs better than GAT, which consistently yields the lowest predictions. The data highlights GIN's superior capability in capturing structural information necessary for accurate crowd estimation.

GIN – Predicted Head Count

An output rate of 99% is demonstrated, which specifies the outcome of a design, a standing pointer, or a proxy (probably referencing the well-known "Response to the Ultimate Query of Lifespan" from *The Hitchhiker's Guide to the Galaxy*). Underneath this yield, a Flag field shows a solo dot (.),

Table 12.4 Actual image given for predicting the head count.

S. no.	Actual head count	Algorithms with predicted head count		
		GIN	GCN	GAT
1	42	42	17	8
2	205	198	15	10
3	300	297	150	98
4	16	16	12	7
5	18	18	10	2
6	16	16	13	6
7	10	10	7	6
8	90	89	68	25
9	6	6	3	0
10	1	1	0	0

probably on behalf of an avoidance, unset, or proxy rate. On the right-hand side, there's a drop-down set menu branded Select GIN from GNN Model, currently set to GNN, a universal Graph Neural Network selection. Underneath the dropdown are keys branded Start, Clear, and Submit, signifying that this is a communicating boundary used for successively or trying GNN replicas [17]. The plan and functionality designate. It's a portion of an implementation considered for implementing calculations, investigating GNN arrangements, and inspecting production outcomes in real time.

GCN – Predicted Head Count

Demonstrations an output rate of 50%, which specifies the outcome of a design, a standing pointer, or a proxy (probably referencing the well-known Underneath this yield, a Flag field shows a solo dot (.), probably on behalf of an avoidance, unset, or proxy rate. On the rightward side, there's a dropdown set menu branded Select GCN from GNN Model, currently set to GNN—a universal Graph Neural Network selection. Underneath the dropdown are keys branded Start, Clear, and Submit, signifying that this is a communicating boundary used for successively or trying GNN replicas.

GAN – Predicted Head Count

The productivity rate is 25%, likely instead of the outcome of information dealing out using the designated GAT (Graph Attention Network) model [27]. The Flag field is empty, reflecting the behavior realized in this picture. At the topmost of the image, the manuscript uploaded in https gives the impression, demonstrating that this boundary is part of a protected, web-based stage permitting handlers to upload information through HTTPS. The dropdown blackboard below Select GAT from GNN Model demonstrations GAT as the selected model, and the aware Clear and Submit keys correspondingly exist, reflecting a reliable boundary diagonally connected imagery. Overall, this arrangement recommends a testing atmosphere for investigating with diverse GNN replicas, where the outcomes vary depending on the particular prototypical and input information. These table seem to be photographs of a graph neural network (GNN) research tool, where handlers handpicked replicas (`GNN`, `GCN`, `GAT`), run calculations, and in first image obtain arithmetical outputs (`42`, `17`, `8`) [5]. The boundary contains collaborating fundamentals like a flag field may help as a standing pointer, a clearance and submissions button, though its practice is unpredictable. The differences in productions and representations propose an elastic structure for investigating, tutoring, or prototyping in graph-based machine learning.

Justification for Using GIN

Unlike other GNN variants like Graph Convolutional Networks (GCN) or Graph Attention Networks (GAT), GIN uses a sum aggregation scheme followed by a Multi-Layer Perceptron (MLP), which allows it to better differentiate between graph structures. In the context of crowd analysis, where individual head regions are represented as nodes and their spatial relations as edges, GIN effectively learns subtle spatial dependencies that are essential for accurate headcount prediction. Table 12.5 provided a performance comparison of GNN algorithms i.e., GIN, GCN and GAT on the same dataset. GIN demonstrated superior results in both Mean Absolute Error (MAE) and Root Mean Square Error (RMSE), as shown in Table X. While GCN was faster to train, it yielded lower accuracy. GAT required more training time due to the attention mechanism, but did not significantly outperform GIN. Thus, GIN offers an optimal trade-off between performance and complexity for this application.

Table 12.5 Performance comparison of GNN models.

Model	MAE	RMSE	Accuracy (%)	Training time
GIN	2.1	3.4	92.5	1.5 hours
GCN	3.5	5.2	85.1	1.0 hour
GAT	2.8	4.1	88.3	2.2 hours

12.13 Conclusion

The addition of Graph Neural Networks (GNNs) into headcount forecast inscriptions marks a noteworthy progression in smart and context-aware crowd analytics. Old-style computer vision approaches frequently struggle with tasks like obstruction, fluctuating concentrations, and lively circumstances in practical surroundings. GNNs overwhelmed these boundaries by giving respectively noticed individual as a node and their spatial relations as edges in a graph. This method allows the perfect to reflect both separate structures and relations, provided that a profound consideration of crowd behavior and spatial circulation is made. Between the constructions premeditated—GCN (Graph Convolutional Network), GAT (Graph Attention Network), and GIN (Graph Isomorphism Network), GIN was confirmed to be the greatest effective. GIN uses a humble yet controlling method: it collects neighboring structures' complete summaries and develops them using a Multi-Layer Perceptron (MLP). This technique permits it to differentiate among dissimilar graph constructions more successfully than GCN's biased bias or GAT's consideration mechanisms construction GIN is predominantly suitable for modeling multifaceted crowd designs. The study demonstrates that GNNs improve forecast accuracy and simplify healthier across various images with fluctuating numbers of persons. Their capability to handle variable-sized contributions and produce dependable productivity done global pooling varieties them vigorous and flexible. The segmental channel—initially from Faster R-CNN-based discovery to graph-based cognitive proposals, with elasticity for additional growth. Elsewhere, methodological presentation, the investigator grasps applied standing. As cities developed, additional occupied and community events extra common, and simultaneous, accurate crowd monitoring is essential for safety and reserve organization. GNN-based schemes, particularly those constructed on GIN, suggest scalable, adaptive explanations for these requirements. Though the disposition must be touched correctly. Principled reflections, like information confidentiality and algorithmic justice, are critical. In conclusion, GNNs characterize an authoritative tool for crowd analytics, with GIN foremost in efficiency. Future efforts should focus on real-time optimization and morally associated applications. While crowd analytics technologies provide valuable insights for public safety, event management, and urban planning, they also raise important ethical concerns, particularly regarding surveillance and individual privacy. The use of high-resolution imagery and person-level detection must ensure that personally identifiable information personally identified information is not misused.

AI disclosure statement: The author acknowledges the use of AI-based tools for minor language editing or grammatical correction. However, all intellectual contributions, analyses, and conclusions presented in this chapter are entirely the author's own.

References

1. Cao, X., Wang, Z., Zhao, Y., Su, F., Scale aggregation network for accurate and efficient crowd counting, in: *Proc. Eur. Conf. Comput. Vis. (ECCV)*, pp. 734–750, 2018.
2. Chen, J., Liu, C., Yang, X., Graph-based crowd counting with spatial message passing, in: *Proc. IEEE Conf. Comput. Vis. Pattern Recognit. (CVPR)*, 2021.
3. Chen, K., Zhu, C., Li, Y., Li, Y., Graph-based structured feature learning for crowd counting. *Neurocomputing*, 497, 1–11, 2022.
4. Gao, G., Gao, J., Liu, Q., Wang, Q., Wang, Y., CNN-based density estimation and crowd counting: A survey, *arXiv preprint arXiv:2003.12783*, 2020.
5. Gao, M., Deng, M., Zhao, H., Chen, Y., Chen, Y., Improving MLP-based weakly supervised crowd-counting network *via* scale reasoning and ranking. *Electronics*, 13, 3, 471, 2024.
6. Kipf, T. and Welling, M., Semi-supervised classification with graph convolutional networks, in: *Proc. Int. Conf. Learn. Represent. (ICLR)*, 2017.
7. Veličković, P., *et al.*, Graph attention networks, in: *Proc. Adv. Neural Inf. Process. Syst. (NeurIPS)*, 2018.
8. Zhang, Y., Zhou, D., Chen, S., Gao, S., Ma, Y., Crowd counting using scale-aware attention networks, *arXiv preprint arXiv:1903.02025*, 2019.
9. Chen, Z., Xu, B., Li, Y., Guo, M., Dynamic graph convolutional networks for crowd localization and counting. *Inf. Sci.*, 607, 1068–1083, 2022.
10. Joshi, *et al.*, Graph attention networks: A comprehensive review of methods and applications, *arXiv preprint arXiv:2111.07244*, 2021.
11. Idrees, H., Zafar, M.S., Shah, M., Composition loss for dense crowds, in: *Proc. IEEE Conf. Comput. Vis. Pattern Recognit. (CVPR)*, 2018.

12. Idrees, H., *et al.*, Composition loss for counting, density map estimation and localization in dense crowds, in: *Proc. Eur. Conf. Comput. Vis. (ECCV)*, 2018.
13. Jiang, L., Zhang, Y., Wang, S., Cao, J., Density map regression guided detection network for RGB-D crowd counting and localization, in: *Proc. IEEE/CVF Conf. Comput. Vis. Pattern Recognit. (CVPR)*, pp. 1821–1830, 2020.
14. Khan, M.A., Menouar, H., Hamila, R., Revisiting crowd counting: State-of-the-art, trends, and future perspectives, *arXiv preprint arXiv:2209.07271*, 2022.
15. Li, Y., Zhang, X., Chen, D., CSRNet: Dilated convolutional neural networks for understanding the highly congested scenes, in: *Proc. IEEE Conf. Comput. Vis. Pattern Recognit. (CVPR)*, pp. 1091–1100, 2018.
16. Liang, D., Chen, X., Xu, W., Zhou, Y., Bai, X., TransCrowd: Weakly-supervised crowd counting with transformers, *arXiv preprint arXiv:2104.09116*, 2021.
17. Liu, H. and Shi, B., Adaptive graph convolutional network for crowd counting. *Neural Comput. Appl.*, 34, 9, 6973–6984, 2022.
18. Liu, L., Wang, H., Li, G., Ouyang, W., Lin, L., Crowd counting using deep recurrent spatial-aware network, *arXiv preprint arXiv:1807.00601*, 2018.
19. Liu, W., Salzmann, M., Fua, P., Context-aware crowd counting, in: *Proc. IEEE Conf. Comput. Vis. Pattern Recognit. (CVPR)*, pp. 5099–5108, 2019.
20. Liu, X. and Zhang, H., Context-aware crowd counting, in: *Proc. IEEE Conf. Comput. Vis. Pattern Recognit. (CVPR)*, 2020.
21. Liu, X., *et al.*, Exploiting sample correlation for crowd counting with multi-expert network, in: *Proc. IEEE/CVF Int. Conf. Comput. Vis. (ICCV)*, pp. 3215–3224, 2021.
22. Luo, *et al.*, Hybrid graph neural networks for crowd counting. *Proc. AAAI Conf. Artif. Intell.*, 34, 7, 11693–11700, 2020.
23. Ma, Z., Sanchez, V., Guha, T., FusionCount: Counting crowds with fusion of multi-scale contextual information, in: *Proc. IEEE/CVF Conf. Comput. Vis. Pattern Recognit. (CVPR)*, pp. 12345–12354, 2022.
24. Miao, Z., Wu, Y., Li, Z., SDANet: Self-distributed attention network for crowd counting, in: *Proc. IEEE/CVF Conf. Comput. Vis. Pattern Recognit. (CVPR)*, pp. 10086–10095, 2020.
25. Pan, H., Huang, Z., Wang, H., Ji, R., Cross-level graph neural network for head counting. *Pattern Recognit.*, 134, 109035, 2023.
26. Sam, D.B., Peri, S.K., Babu, R.V., Locate, size, and count: Accurately resolving people in dense crowds *via* detection. *IEEE Trans. Pattern Anal. Mach. Intell.*, 43, 4, 1235–1249, 2021.
27. Sam, D.B., Sajjan, N.N., Babu, R.V., Srinivasan, M., Divide and grow: Capturing huge diversity in crowd images with incrementally growing CNN, in: *Proc. IEEE Conf. Comput. Vis. Pattern Recognit. (CVPR)*, pp. 3618–3626, 2018.

28. Shi, Z., *et al.*, Crowd counting with deep negative correlation learning, in: *Proc. IEEE Conf. Comput. Vis. Pattern Recognit. (CVPR)*, pp. 5382–5390, 2018.
29. Sun, Y., Duan, Y., Liu, M., Bi-directional graph reasoning network for crowd counting. *Sensors*, 22, 9, 3487, 2022.
30. Wan, J., Liu, Z., Chan, A.B., A generalized loss function for crowd counting and localization, in: *Proc. IEEE/CVF Conf. Comput. Vis. Pattern Recognit. (CVPR)*, pp. 1974–1983, 2021.
31. Wang, Y., Huang, H., Liu, Y., Song, J., Spatial-aware graph convolutional network for crowd counting. *IEEE Trans. Multimedia*, 25, 2304–2314, 2023.
32. Wu, Q., Yan, F., Chai, Z., Guo, G., Crowd counting by the dual-branch scale-aware network with ranking loss constraints. *IET Comput. Vis.*, 14, 3, 101–109, 2020.
33. Xiong, H., *et al.*, From open set to closed set: Counting objects by spatial divide-and-conquer, in: *Proc. IEEE/CVF Int. Conf. Comput. Vis. (ICCV)*, pp. 8362–8371, 2019.
34. Xu, K., Hu, W., Leskovec, J., Jegelka, S., How powerful are graph neural networks?, in: *Proc. Int. Conf. Learn. Represent. (ICLR)*, 2019.
35. Zeng, *et al.*, Multi-scale convolutional neural networks for crowd counting, *arXiv preprint arXiv:1702.02359*, 2017.
36. Zhang, C., Li, H., Wang, X., Yang, X., Cross-scene crowd counting *via* deep convolutional neural networks, in: *Proc. IEEE Conf. Comput. Vis. Pattern Recognit. (CVPR)*, pp. 833–841, 2015.
37. Zhang, L., Shi, M., Chen, Q., Crowd counting *via* scale-adaptive convolutional neural network, *arXiv preprint arXiv:1711.04433*, 2017.
38. Zhang, Y. and Xu, Z., Scale-aware attention networks for crowd counting, in: *Proc. IEEE Int. Conf. Comput. Vis. (ICCV)*, 2020.
39 Zhang, Y., Liu, L., Yang, M.H., Multi-column convolutional neural network for crowd counting, in: *Proc. IEEE Conf. Comput. Vis. Pattern Recognit. (CVPR)*, 2016.
40. Zhang, Y., *et al.*, OrdinalEntropy: A novel loss function for crowd counting, in: *Proc. IEEE/CVF Conf. Comput. Vis. Pattern Recognit. (CVPR)*, pp. 5678–5687, 2023.
41. Zhang, Y., *et al.*, Single-image crowd counting *via* multi-column convolutional neural network, in: *Proc. IEEE Conf. Comput. Vis. Pattern Recognit. (CVPR)*, pp. 589–597, 2016.
42. Zheng, X., Zhang, L., Wang, S., Wang, H., Graph convolution meets attention: A hybrid framework for crowd counting. *IEEE Trans. Circuits Syst. Video Technol.*, 33, 2, 642–655, 2023.
43. Zhong, Y., Li, J., Wang, X., NDConv: Non-local dilated convolution for crowd counting, in: *Proc. IEEE/CVF Conf. Comput. Vis. Pattern Recognit. (CVPR)*, pp. 9101–9110, 2022.
44. Zhou, Y., Wang, Y., Wang, L., Global context-aware graph neural network for crowd understanding. *IEEE Access*, 10, 14592–14604, 2022.

45. Zhuge, J., Ding, N., Xing, S., Yang, X., An improved deep multiscale crowd counting network with perspective awareness. *Optoelectron. Lett.*, 17, 367–372, 2021.
46. Head Count Dataset. [Online]. Available: https://universe.roboflow.com/emvirt-oa4rx/head-count-6wet8/dataset/1.

13

Stratified Sampling and Graph Neural Networks for Zero-Day Attack Detection

Karthika S.*, **Sandhiya R. and A. Sumi**

Department of Information Technology, Avinashilingam Institute for Home Science and Higher Education for Women, Coimbatore, TamilNadu, India

Abstract

Intrusion detection systems (IDS) help protect computer networks by detecting unusual and harmful activities. Some attacks, called zero-day attacks, are new and hard to detect using regular methods. Although Machine Learning (ML) and Deep Learning (DL) have made IDS smarter, they still have a problem. Since most network traffic is normal, these systems get used to seeing safe patterns. As a result, they often miss the small amount of harmful traffic and allowing threats to slip through. To solve this, Graph Neural Networks (GNNs) are used. They turn network data into a graph which is like a map showing how computers, IP addresses, and ports are connected. This helps the system understand relationships and spot strange behavior that may be a sign of an attack. This research presents a new method to detect such threats. It uses a density-based clustering technique called Hierarchical Density-Based Spatial Clustering of Applications with Noise (HDBSCAN) along with graph theory to find unusual patterns in the network. The CICIDS 2017 dataset which includes both known and unknown types of attacks is used in this work. The dataset is sampled to maintain a balanced representation of both benign and malicious traffic to maintain fairness. After preprocessing, HDBSCAN is applied to group similar traffic patterns together. This helps form the clusters that are later used to build the graph for anomaly detection. Network traffic data is converted into a graph structure. Furthermore, a Graph Autoencoder with Graph Attention Network layers is utilized to learn graph-aware node embeddings and detect anomalies. Anomaly scores are calculated and the top 5% of nodes are flagged as potential intrusions. The results demonstrate that the proposed approach is effective in detecting delicate attack patterns with improved IDS accuracy.

**Corresponding author*: karthika_it@avinuty.ac.in

J. Ramkumar, Sridaran Rajagopal, B. Suchitra and S. Balamurugan (eds.) Graph Neural Networks: Concepts and Applications, (539–578) © 2026 Scrivener Publishing LLC

Keywords: Intrusion detection system, zero-day attacks, graph neural networks, HDBSCAN clustering, graph autoencoder, network traffic analysis, principal component analysis, unsupervised anomaly detection

13.1 Introduction

The growth of internet-enabled devices and cloud platforms increases the attack opportunities for cybercriminals. Some cyber threats such as data breaches and attacks on important systems are on rise due to rapid digital transformation. These threats happen very often and pose serious issues globally. Hence, it is important to protect critical infrastructure and data from these risky attacks. Intrusion Detection Systems (IDS) performs crucial role in monitoring network traffic for malicious behavior. Traditional IDS use predefined rules and known attack signatures to identify any suspicious activity. Therefore, they may miss new or evolving attacks like zero-day exploits because such threats do not have any known patterns or signatures, since smarter and more flexible cybersecurity solutions are required [1].

Deep learning and machine learning have emerged as key technologies for developing such intrusion detection systems with improved intelligence. These techniques are able to learn from the behavior of networks and do not have to be aware of attacks in advance to spot odd activity. Dealing with unbalanced data and detecting subtle patterns are considered as major issues with these techniques. Researchers are investigating new methods to get around these issues. Graph-based models and unsupervised learning are getting more focused in this domain. Advanced techniques such as Graph Neural Networks and HDBSCAN clustering can offer better solutions as they are more powerful and capable of handling more complex and large-scale network data. Older techniques might overlook emerging and evolving threats, but these methods can identify them [2].

Generally, supervised machine learning methods are effective at detecting malicious activities. However, they need massive amount of labeled data. Obtaining such data is often difficult, expensive, or time-consuming. Also, these models can become biased, especially when there are many more normal activities than attacks. Another problem is that these models do not work well in different network setups and need to be retrained often to stay effective. From this, it is evident that both traditional and supervised models are not enough and better solutions are needed to detect new or evolving cyber threats. On the contrary, labeled data are not mandatory for unsupervised learning models. So, these models well fit for the environments where attacks are constantly changing and labels are hard to obtain. These models learn the normal traffic well and spot out anything that differs from the normal traffic. Thus, they are particularly good for identifying zero-day attacks, which are new and do not follow any known attack patterns [3].

Unsupervised learning and clustering techniques are efficient in Intrusion Detection Systems, in particular, in detecting zero-day attacks. In contrast to supervised models which require large amount of labeled data, the unsupervised models detect suspicious behavior by detecting abnormalities in the network traffic pattern. Clustering algorithms such as K-means, KNN and HDBSCAN are generally used attack pattern detection. Clustering is used to organize and filter the network data into meaningful groups, which are then converted into a graph. This graph feeds into a GNN, which learns richer patterns than clustering alone which enables the detection of subtle, evolving, and unseen threats. These methods can handle large amount of data as well more adaptive. Thus, IDS developed using unsupervised and clustering techniques can keep learning and adapting to the new threats. As a result, they can detect attacks more accurately in changing environments [4].

In this work, stratified sampling is first applied to ensure fair representation of both common and rare attack types. This is crucial for learning balanced traffic behavior. Then, by including the graph structure built from HDBSCAN clusters, Graph Neural Networks (GNNs) can model the complex relationships between traffic samples. The GNN-based Graph Autoencoder learns node-level embeddings that reflect both feature similarities and structural connections. If any node deviates from normal patterns during the reconstruction, then those nodes are marked as anomalies which are considered as zero-day attacks. In this way, the IDS remain adaptive, scalable, and effective in changing and imbalanced traffic environments.

The key objective of this research work is to develop an Intrusion Detection System which is capable to detect the zero-day attacks effectively without demanding labeled data and predefined attack patterns. The novel idea is to integrate the advanced techniques HDBSCAN clustering and Graph Neural Networks with unsupervised learning to discover network traffic anomalies. The proposed approach outperforms the traditional IDS which are often struggle with imbalance data and false detection. This method is designed in such a way that the entire network data is converted into a graph, then GNNs is applied to capture behavior of the individual nodes along with relationship and interaction between the nodes. This graph-based perspective enables deeper analysis of network structure and behavior, making it easier to identify unusual patterns that could indicate malicious activity.

The model is enriched with the capability to identify normal groupings in the traffic data by applying the unsupervised clustering technique HDBSCAN. These clusters are the basic elements for constructing a graph with nodes and edges. Each node represents a traffic flow whereas edges can capture similarity between them. This transformation is useful to acquire both feature-based and structural information. Once the graph is constructed, next step is to use Graph Neural Networks for learning representations that preserve the complex relationships within the data. These embeddings are crucial to detect even subtle deviations from the normal behaviors. Learning those subtle deviations are important as they are critical to uncover zero-day attacks that do not follow known patterns. Graph-based models help by showing how different devices and connections relate to each other. This makes it easier to spot strange behavior by looking at both the network's structure and how it acts. In this way, the proposed combination of HDBSCAN and GNN is highly flexible IDS that adapts to evolving threats and works effectively. HDBSCAN clustering groups similar data and is very good at finding unusual or rare activities even when there is no prior knowledge about attacks. The goal of this work is to build IDS tools that are strong, accurate, and able to keep up with the fast-changing world of cybersecurity threats.

This chapter is organized into five sections. Section 13.1 introduces the zero-day attack detection domain and provides the objectives. Section 13.2 reviews the existing works on intrusion detection systems using machine learning and graph-based anomaly detection methods. Section 13.3 describes the proposed methodology with all the steps used. Section 13.4 discusses the results obtained from various steps of the research work. Finally, Section 13.5 concludes the chapter by summarizing the significant contributions and recommending future research directions to be considered.

13.1.1 Objectives of the Study

- To develop a scalable sampling strategy using PCA and HDBSCAN for effective handling of class imbalance in network traffic data.
- To construct graph representations of clustered network flows to capture structural and relational information.

- To develop a Graph Autoencoder with Graph Attention Network layers for learning robust node embeddings.
- To develop an anomaly scoring method based on embedding deviations to detect anomalous nodes.
- To evaluate the proposed method against baseline techniques on the CICIDS2017 dataset for anomaly detection performance.
- To demonstrate improved detection of zero-day and rare attacks through graph-based analysis.

13.2 Literature Review

The study [5] examines the challenges and gaps in deploying machine learning-based Intrusion Detection Systems (IDS) in real-world environments. The study highlights that ML is rarely used in operational IDS systems due to practical challenges and offers a set of guidelines to improve anomaly detection research in network intrusion detection. A study [6] provides a full review of the IDS field which provides a structured overview of the existing detection techniques. The taxonomy categorizes different types of IDS and provides valuable insights about operational complexities along with computational requirement of the modern IDS architectures. It also highlights challenges posed by increasing network throughput and evolving security threats.

Intrusion Detection Systems have improved over time. They have moved from using fixed attack signatures to using smart methods like machine learning. These modern systems can now handle both known and unknown cyber threats. Many new IDS models use a mix of supervised and unsupervised learning. This helps them detect zero-day attacks and adapt to changing attack patterns. One such model is shown in [7]. This model uses supervised methods such as K-Nearest Neighbors (KNN), Random Forest, and Artificial Neural Networks. These methods are used to find known attacks in the CSE-CIC-IDS2018 dataset. If the system is unsure about any network traffic, it passes that data to an unsupervised model. This model uses HDBSCAN clustering to detect unusual behavior or unknown attacks. This two-step system works well. F1-scores between 0.93 and 0.969 are shown by supervised part whereas the unsupervised part shows F1-scores between 0.85 and 0.957. These results show that combining both methods makes the system more effective at finding threats [7].

Building on the idea of hybrid learning, [8] introduces a graph-based Intrusion Detection System for cloud environments that use containers. The system uses two types of machine learning models. It uses random forests for supervised learning and isolation forests for unsupervised learning. It works by turning system call activities into weighted graphs. These graphs help the system understand how programs behave at the kernel level. This combined approach gives very accurate results. It can detect attacks inside containers while keeping false alarms low. The use of graph-based models helps the system better understand behavior patterns, which makes it more effective at finding intrusions. In contrast to these implementation-oriented works, [9] focuses on the robustness of ML-based IDS under adversarial attacks and distribution shifts. It critiques the vulnerability of IDS models and emphasizes the need for robustness across the entire ML pipeline. Techniques such as contrastive learning, robustness certification, and data augmentation are explored to enhance resilience. This study provides a comprehensive perspective, identifying research gaps and encouraging holistic robustness design in ML-driven IDS systems [9].

Addressing the challenge of data scarcity in IoT environments, [10] proposes the ABRSI model—a novel framework utilizing unsupervised heterogeneous domain adaptation (HDA). The model aligns knowledge from data-rich domains with data-scarce IoT domains using bi-recommendation matching and hybrid pseudo-labeling (hard and soft). The incorporation of a self-improvement loop and error knowledge learning boosts adaptability and detection performance, significantly outperforming baseline models by 9.2% [10].

Ref. [11] emphasized the importance of classification accuracy in intrusion detection. Their comparative study using Support Vector Machine (SVM) and Naïve Bayes classifiers, evaluated on the NSL-KDD dataset, showed that SVM outperforms Naïve Bayes in terms of detection accuracy, and reduced misclassification. The study highlighted the need for efficient classification models to handle the analysis of large-scale network traffic. Ref. [12] introduced an equality constrained-optimization-based Extreme Learning Machine (ELM) along with an adaptive incremental learning strategy to determine the optimal number of hidden neurons. Their approach demonstrated enhanced detection performance and computational efficiency, operating effectively without requiring extensive domain-specific feature engineering. Moreover, the model exhibits adaptability to dynamic network conditions, making it well-suited for real-time intrusion detection in evolving threat landscapes.

Ref. [13] proposed a hybrid feature selection technique combining the Pearson correlation coefficient and random forest. Their experimental study evaluated several ML models (Decision Tree, AdaBoost, K-NN) and DL models (MLP, LSTM) on the TON_IoT dataset, which contains recent attack types. The results indicated that Decision Trees and MLP models achieved the best trade-off between detection accuracy and low false positive/negative rates. Their findings support the effectiveness of ML and DL techniques in modern, dynamic environments. Ref. [14] focused on the adaptive learning aspect of NIDS by evaluating recent datasets such as CICIDS 2017 and CICIDS 2018. Their study underlined the importance of data recency, relevance, and preprocessing (balancing, feature engineering) in improving the model's performance. Random Forest and Recurrent Neural Networks (RNN) demonstrated high macro F1-scores, affirming their robustness in diverse attack scenarios. This study highlights the need for models that can adapt over time. Real-world cyber threats change constantly, so IDS must learn and improve continuously to stay effective.

The study in [15] presents an IDS developed using machine learning which solves the common problems of class-imbalanced data and handling large-scale network data. The proposed IDS used Random oversampling technique and applied a special clustering method in order to handle the mentioned problems. Eventually, Principal Component Analysis is used to reduce the amount of data with preserving the crucial information. The model was tested on benchmark datasets such as UNSW-NB15, CICIDS-2017, and CICIDS-2018. The model works good with massive, complex data and the results outperformed many existing IDS. Lastly, it uses Principal Component Analysis to reduce the amount of data without losing important information. This model was tested on well-known datasets such as UNSW-NB15, CICIDS-2017, and CICIDS-2018. The best accuracy of 99.99% is achieved with Random Forest and Extra Trees algorithms. These results are better than many previous methods and show that the model works well even with large and complex data.

A new graph-based Intrusion Detection and Classification System is introduced in [16], which is especially designed to detect attacks in vehicles. This method G-IDCS used CAN protocol and combined a simple threshold method with machine learning. This system requires scarcer CAN messages to detect threats and improves accuracy by over 9% compared to older graph-based methods. A significant benefit of G-IDCS is it can explain its decisions and such explainability is very useful for investigating how and why an attack happened.

An Intrusion Detection System based on Graph Neural Networks, GNN-IDS, is developed in the research work [17]. The method detects both fixed and changing patterns in a network. GNN-IDS becomes better at spotting unusual behavior and tracing it back to the actual attack by combining an attack graph with live network data. This system outperformed the older methods in terms of precision, accuracy, and capability to handle uncertain situations.

Paper [18] introduced DGIDS, a graph-based intrusion detection system made for vehicles using the CAN protocol. It was created to fix the problems found in traditional systems that rely only on statistics and often fail to identify which message is part of an attack. DGIDS builds dynamic graphs in real time using the messages it receives. It quickly picks out important features to detect attacks right away. This method outperformed the previous methods by detecting threats faster and using less memory. In [19], the authors give a general overview of how graph-based intrusion detection systems work. These systems can track large and complex attacks by reviewing the structure of a graph. In particular, it is done by observing at the importance or weight of different nodes. The systems can also connect current activity with known attack methods from various databases such as MITRE ATT&CK. This helps the system better that understand what kind of attack is happening and allows it to improve its detection rules for more accurate protection.

To summarize, this literature review shows that the significant improvements of Intrusion Detection Systems with the advent of machine learning and hybrid methods. Some important techniques such as supervised learning, unsupervised learning, and graph-based models have shown good results in dealing with the intrusion detection. However, there are still some big challenges which are to be considered and sorted out. These include lack of good data for training the models, the risk of smart attackers trying to fool the models, and the dynamic nature of network traffic data that keeps changing all the time.

13.2.1 Gaps Identified in Existing Studies

Although many studies have applied clustering and anomaly detection techniques in Intrusion Detection Systems (IDS), several gaps still remain as listed:

- **Reliance on labeled data**: Most models depend on labeled training data, which is often not feasible in real-world environments.
- **Weakness in zero-day attack detection**: Supervised and signature-based models are typically ineffective against new, unknown threats.

- **Scalability limitations**: High-dimensional data and complex graph topologies often hinder the real-time deployment of IDS.
- **Limited use of graph-based approaches**: Although Graph Neural Networks (GNNs) and graph sampling methods show great promise, they are still not commonly used in intrusion detection research.

This study aims to bridge that gap by combining HDBSCAN clustering, Principal Component Analysis (PCA), and GNN-based anomaly detection into a complete and efficient system designed specifically to detect zero-day attacks.

13.3 Proposed Methodology

The overall approach used for detecting unusual patterns in network traffic follows a structured sequence of steps, beginning from data preprocessing and continuing through to model training and evaluation. Figure 13.1 presents the complete process which clearly outlines each step involved in preparing the data, building the graph structure, and detecting anomalies with the help of Graph Neural Networks. Also it offers a straightforward overview of how each component works together to accurately identify abnormal activities within the network.

13.3.1 Data Loading and Preprocessing

13.3.1.1 Overview of the Dataset

This study uses the benchmark dataset CICIDS2017 [20] for Intrusion Detection System research. The dataset is developed by the Canadian Institute for Cybersecurity with an aim to simulate realistic network environments by including both normal and malicious traffic. The dataset was collected over five consecutive days and includes traffic from regular user activities along with various cyberattacks which also includes zero-day attacks. CICFlowMeter tool is used to capture network flow and each record in the dataset consists of 79 features. These features cover various aspects such as flow statistics, packet-level information, timing patterns, and behavior-based indicators. Furthermore, each network flow in the CICIDS2017 dataset is labeled as either benign or a specific

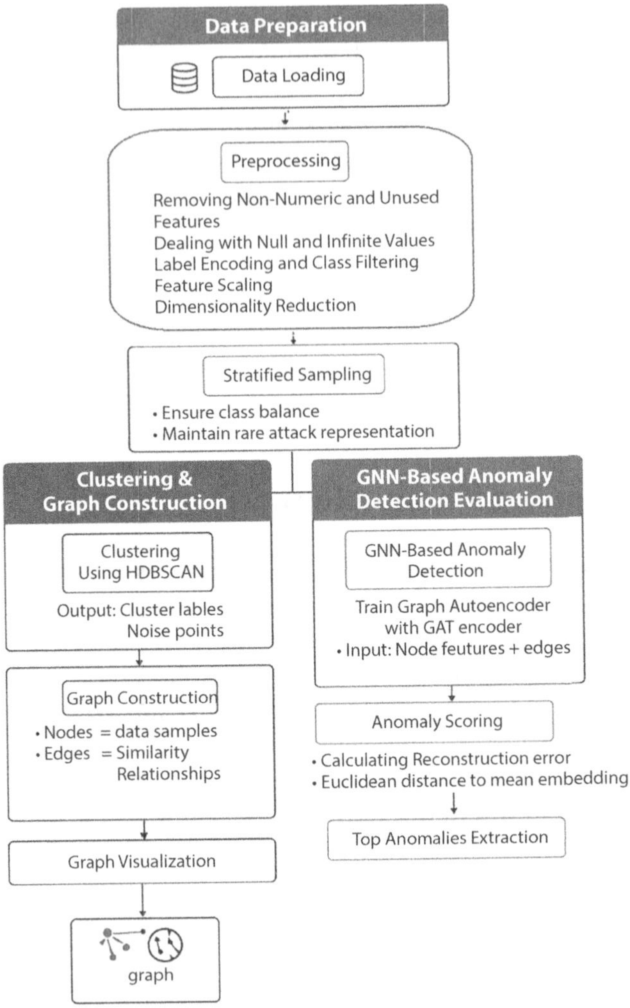

Figure 13.1 The proposed methodology.

type of cyberattack. Table 13.1 summarizes the distribution of normal and malicious traffic across different attack types.

Each network flow entry in the dataset is labeled under the 'Label' column, indicating whether the traffic is benign or linked to specific attack types such as BENIGN, DoS Hulk, DoS GoldenEye, DDoS, PortScan, Infiltration, Brute Force, XSS, SQL Injection, SSH-Patator, and FTP-Patator.

Table 13.1 Summary of benign and malicious traffic in the CICIDS2017 dataset.

Day	Network activity type	Attack types included
Monday	Normal traffic only	None
Tuesday	Brute-force attacks	SSH Brute Force, FTP Brute Force
Wednesday	Denial of Service (DoS) attacks	DoS Hulk, DoS GoldenEye, others
Thursday Morning	Web-based attacks	Brute Force, Cross-Site Scripting (XSS), SQL Injection
Thursday Afternoon	Infiltration attacks	Various infiltration attempts
Friday Morning	General mixed traffic	None (mostly benign)
Friday Afternoon	Port scanning and Distributed DoS attacks	PortScan, Distributed Denial of Service (DDoS)

13.3.1.2 Preprocessing Pipeline

An exhaustive preprocessing pipeline is executed to make the data suitable for effective graph-based anomaly detection and clustering method. The process begins with the removal of non-numeric and redundant features that do not contribute to model learning. Then, null and infinite values are handled to maintain the data integrity. Categorical labels are converted into machine-readable formats through label encoding and irrelevant classes are filtered to focus on meaningful traffic patterns. Next, standard scaling is applied to address the problem of varying features scales. This step is also helpful to improve the model convergence. As the dataset CICIDS2017 contains 79 traffic-related attributes, Principal Component Analysis is used to reduce this high-dimensional feature space into a lower number of principal components. Also, PCA helps in transforming the data into a format where clusters of similar traffic patterns became more distinct, improving the construction of the graph. This reduced and denoised feature set enables the Graph Attention Network to focus on learning meaningful variations. This particular step improves both embedding quality and anomaly detection performance. Thus, these steps together prepare the CICIDS2017 dataset for stratified sampling and graph construction to facilitate more accurate detection of zero-day threats in complex network traffic.

i. Removing Non-Numeric and Unused Features

The CICIDS2017 dataset has some irrelevant and non-numeric features such as Flow ID, Source IP, Destination IP, and Timestamp. These features may not efficiently contribute in detecting anomalies or outliers. In addition, they may add complexity to the model and lead to overfitting. The discriminating ability of the model in between normal and malicious traffic may also become poor. This may lead to model overfitting and affect the generalization efficiency. Hence, Principal Component Analysis is used for removing irrelevant features and reduces the feature space X of the CICIDS2017 dataset while preserving its most informative features. It enhances clustering and anomaly detection by projecting data into a space where patterns are more distinguishable.

ii. Dealing with Null and Infinite Values

In this research, null and infinite values in the CICIDS2017 dataset were replaced with NaN and then removed to ensure data quality. Clean data is vital because algorithms like k-NN and PCA rely on valid numerical inputs for calculations such as distance measurement and covariance estimation. If NaN or infinite values are present in the dataset, it disrupts these calculations which leads to errors and poor model performance. Hence, those invalid data are removed to improve the detection of zero-day attacks, which is depending on identifying delicate anomalies.

iii. Label Encoding and Class Filtering

The CICIDS2017 dataset has some categorical labels including BENIGN, DoS Hulk, and PortScan. First, these labels are converted into numeric values as this conversion is necessary because most machine learning models can only work with numbers. In addition, class filtering is also applied to remove rare classes that have very few examples in the dataset. The classes which are not represented well can cause the model to focus too much on the more common classes in anomaly detection tasks. This makes the model to become biased. To avoid that, these sparse classes are filtered to make the dataset more balanced.

Let $L=\{l_1, l_2,..., l_n\}$ be the set of unique class labels and $N=\{N_1, N_2,..., N_n\}$ be the corresponding number of samples in each class. A label encoding procedure is applied to map each class label to a distinct integer. This confirms the model's compatibility and facilitates supervised learning. The labels with sample size which is below a defined threshold are removed from the dataset in order to reduce the probable bias caused by extremely underrepresented classes. This process enables the model to be focused on learning from properly represented classes.

iv. Sampling and Class Balancing

CICIDS2017 dataset has two significant challenges, class imbalance and high data volume. The dataset is composed with benign traffic mostly and fewer malicious samples. This imbalance can skew the learning algorithms and make them biased toward the majority class and less sensitive to rare or novel attacks such as zero-day intrusions. As the dataset is composed with millions of network flow records, it requires extensive computational resources for processing and model training. To handles these two issues, the proposed approach adopts a stratified sampling strategy. The data is partitioned by the sampling technique based on class labels and samples correspondingly from each class. This reduces the dataset size to get a manageable scale as well represent the rare attack types. Similarly, as the CICIDS2017 dataset is organized across multiple daily log files, all the files are loaded and merged into a single dataset. This consolidated dataset helps to make anomaly detection using Graph Neural Networks more efficient. A particular number of samples are selected based on the total size of the dataset for each class. However, in any case 5,000 samples are selected from each class to make sure even the smaller classes had enough data for the model to learn properly.

Let D be the full dataset with class labels $L=\{l_1,l_2,\ldots,l_k\}$. As in Equation 13.1, the stratified sample D_s is constructed such that,

$$|D_s^{l_i}| = max(p_i.N, \min\_samples) \qquad (13.1)$$

where, p_i is the proportion of class l_i in the original dataset, N is the desired total number of samples and min_samples=5000 ensures fair representation of minority classes.

This hybrid method ensures balance while preserving the original distribution trends of the dataset. The dataset included labels like BENIGN, DoS, and DDoS. Samples are taken for each class according to either their proportional share or the minimum sample limit, whichever was higher. These class-wise samples are then combined to form a final, balanced dataset. After stratified sampling, the dataset contained 308,304 records with a more balanced distribution of classes. Major attack types like DoS, DDoS, and PortScan are proportionally represented, while rare attacks such as Infiltration and Web-based attacks are boosted above the threshold for effective learning. This type of balancing is crucial for zero-day attack detection as rare behaviors may be similar to unknown threats. Mitigating the class imbalance problem would be helpful to make the model unbiased towards the benign class. The label distribution in this sampled dataset

Table 13.2 Class distribution of the dataset.

Label	Before (%)	After sampling (%)
Benign	75.00	67.37
DOS Hulk	10.23	9.23
DDoS	5.12	4.87
Portscan	6.32	6.99
Infiltration	0.01	0.003
Web Attacks	0.42	0.94

is approximately: BENIGN (73.7%), DoS Hulk (7.5%), PortScan (5.2%), DDoS (4.2%), along with other attack types which are given in Table 13.2. Each attack class is better represented which helps to reduce bias caused by class imbalance. Although the balanced dataset 10.9% of the original size is smaller, it provides an unbiased training set where both benign and attack traffic are well represented. Table 13.2 presents the class distribution before and after applying stratified sampling. While stratified sampling is applied to mitigate class imbalance, complete uniformity across classes is intentionally avoided to maintain realistic traffic patterns. This ensures the model remains robust and generalizes well to real-world, skewed distributions.

To visually confirm the effectiveness of the stratified sampling strategy, a 2D PCA projection of the sampled dataset is shown in Figure 13.2. It displays the spread of different classes after balancing, indicating that even rare attack types are sufficiently represented in the reduced feature space.

Each point is color-coded by class label, showing the relative distribution and overlap of different attack categories after balancing.

v. Feature Scaling

Graph Neural Network need consistent input for efficient performance. So, feature scaling is completed by applying a feature normalization technique called standardization. Each numerical feature is standardized by transforming it to have a zero mean and unit variance. This step is important to ensure that all features lie on a common scale as the dataset has data with heterogeneous nature. If the features are not normalized, the features with larger numeric ranges may dominate the learning process. Consequently, the GNN model may ignore important features and place more importance on few features. This leads to biased embeddings and ineffective in detecting subtle patterns associated with zero-day attacks. To make the

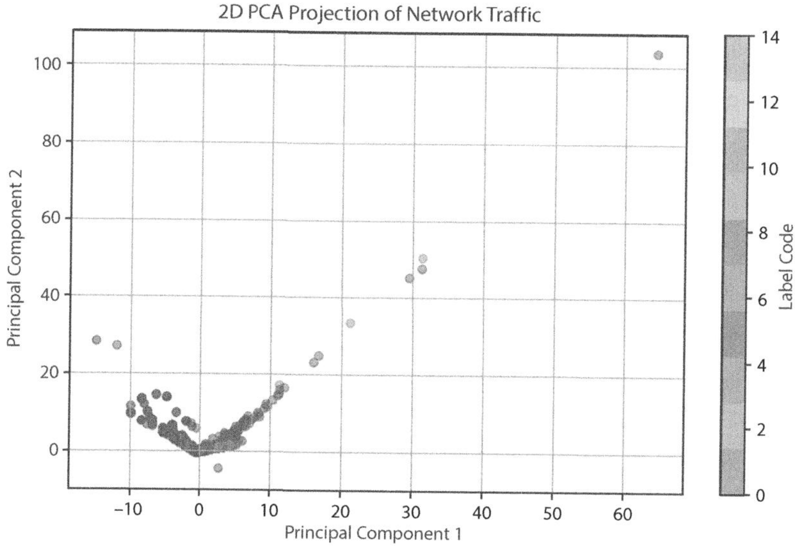

Figure 13.2 2D PCA projection of stratified sampled network traffic (CICIDS2017).

learning algorithm to treat each input dimension equally, the standardization is applied. In this case, the GNN converge efficiently during the training process. The stability of gradient updates is improved which ensures that node embeddings show a balanced class representation. Thus, this step is essential to develop a fair and computationally efficient model to detect novel attacks in high-dimensional network traffic data.

vi. Dimensionality Reduction using PCA

Generally, machine learning models which uses clustering and unsupervised anomaly detection are disposed to performance degradation due to the curse of dimensionality. The CICIDS2017 dataset has 79 numerical features, many of them may be redundant, correlated or noise. This kind of high-dimensional data can slow down processing and lead to performance degradation of both HDBSCAN and GNN. To tackle this, PCA reduces the feature space to a lower-dimensional subspace with retaining about 95% or original variance. In addition, PCA filters out noisy and irrelevant features to improve the quality of clusters and learned embeddings. In this way, the sensitivity of the model to the meaningful anomalies is enhanced, which neglects the outliers caused by noise. Moreover, reducing the number of features directly reduce the computational cost for both HDBSCAN clustering and GNN training. In this study, Principal Component Analysis is applied after the dataset has been standardized. Standardization ensures that all features contribute equally by giving them a mean of zero and

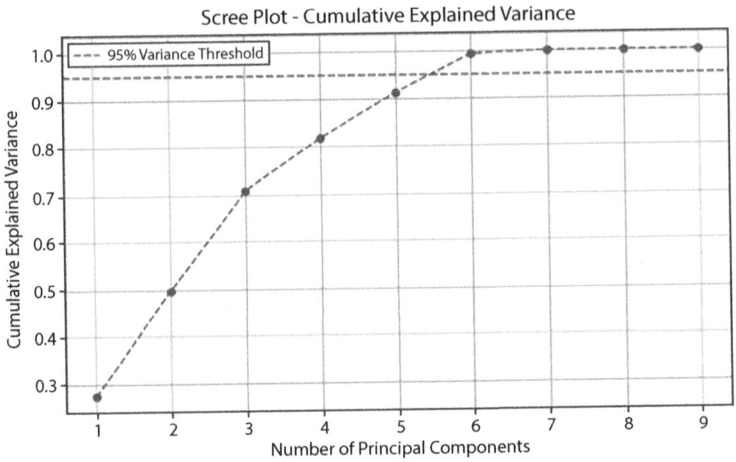

Figure 13.3 Scree plot of PCA components for feature reduction.

Table 13.3 PCA configuration.

Parameter	Value	Description
Original Dimensions	78	Total features before dimensionality reduction
Reduced Dimensions	30	Dimensions retained after PCA
Variance Retained	~95%	Ensures maximum information is preserved

a standard deviation of one. The reduced feature space is then used for HDBSCAN clustering and graph construction, enabling better detection of both common and rare attack types.

The scree plot in Figure 13.3 shows the cumulative variance explained by each successive principal component derived from PCA. It helps to determine the optimal number of components to retain while preserving most of the variance present in the dataset. As shown in Table 13.3, PCA is applied to compress the feature space from 78 to 30 dimensions, preserving around 95% of the original variance.

13.3.1.3 Selected Features

This section outlines the features selected for model training. These features capture key aspects of network traffic behavior and are essential for accurately distinguishing between benign and malicious activity. They were selected based on their statistical relevance, correlation with attack patterns, and contribution to effective clustering and anomaly detection. Feature selection was performed after standard preprocessing and dimensionality reduction steps. Initially, 78 features were available in the CICIDS2017 dataset. From these, a subset of meaningful features was selected based on domain knowledge and exploratory data analysis, including correlation heatmaps and feature importance scores derived from unsupervised models. Some features with high redundancy or low variance are excluded to reduce noise. The final set of features is chosen to provide a balance between flow-level metrics such as duration, byte rates and packet-level behaviors such as packet size and direction-specific rates. This ensures that both temporal and structural characteristics of the network traffic are captured. The selected features, listed in Table 13.4, contribute significantly to the formation of well-separated clusters during HDBSCAN-based anomaly detection and serve as informative input for downstream GNN-based classifiers.

To understand the relative contribution of each feature toward distinguishing between benign and malicious traffic, a feature importance analysis was conducted. This analysis helps identify which features play a significant role in influencing the model's decision. Figure 13.4 illustrates the feature importance scores calculated using a Random Forest method where higher scores indicate greater impact on the prediction task.

Table 13.4 Description of selected features.

Feature name	Description
Flow Duration	Duration of the traffic flow
Total Fwd Packets	Packets sent from source to destination
Total Backward Packets	Packets sent from destination to source
Flow Bytes/s	Bytes transferred per second
Flow Packets/s	Packets transferred per second
Packet Length Mean	Average length of packets
Packet Length Std	Standard deviation of packet lengths
Fwd Packets/s	Rate of forward packet flow
Bwd Packets/s	Rate of backward packet flow

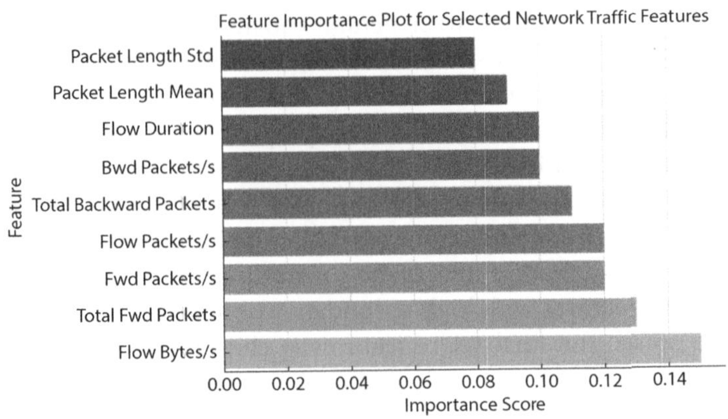

Figure 13.4 Feature importance scores of selected network traffic features.

Features such as Flow Duration, Packet Length Mean, and Flow Bytes/s are observed to be among the most influential. These features capture essential behavioral characteristics of the network flow, supporting their inclusion in the model. This analysis not only validates the effectiveness of selected features but also aids in model interpretability and optimization.

13.3.2 Clustering with HDBSCAN

Clustering is a fundamental unsupervised learning technique applied to identify patterns, relationships, or outliers in data without prior labels. In network security, where zero-day attacks tend to bypass pre-defined signatures, clustering allows for the identification of suspicious or unusual behavior based on behavioral similarity. Classical clustering algorithms such as K-Means presuppose spherical clusters and need the number of clusters to be predefined, which is unrealistic in dynamic network environments.

Density-based clustering algorithms break these constraints by describing clusters as areas of elevated point density delineated by low-density areas. Of these, Hierarchical Density-Based Spatial Clustering of Applications with Noise (HDBSCAN) is currently the best fit for high-dimensional and noisy datasets. In contrast to DBSCAN, HDBSCAN does not need a universal density threshold and can identify clusters of different densities along with identifying noise, which are outliers automatically. This makes it particularly suitable for cybersecurity datasets such as CICIDS2017, which have a combination of normal and attack traffic with heterogeneous properties.

In this work, HDBSCAN is applied as an unsupervised technique to detect unknown or rare attack patterns from the CICIDS2017 dataset. Traditional clustering methods like K-Means require specifying the number of clusters in advance and assume equal-sized spherical clusters. These assumptions are unrealistic for network intrusion data, which is often high-dimensional, imbalanced, and contains noise. HDBSCAN overcomes these limitations by identifying clusters based on data density, allowing clusters of different shapes and densities, automatically detecting noise or outliers, and not requiring the number of clusters to be predefined. The parameters used for running HDBSCAN are listed in Table 13.5.

The application of HDBSCAN successfully grouped similar network behaviors into meaningful clusters. Normal traffic and frequently occurring attack types such as DoS and DDoS formed well-defined clusters, reflecting their consistent and repetitive behavior. In contrast, rare or stealthy attacks like Infiltration, Heartbleed, and Web Shell did not form strong clusters due to their low frequency and unique characteristics, and were often labeled as outliers or noise by the algorithm. HDBSCAN is a highly effective tool for identifying outliers in the dataset. It not only grouped similar data points into clusters but also marked those that did not fit into any cluster as noise. These noise points often correspond to rare or stealthy attacks like Web Shell or Infiltration which are difficult to detect using traditional methods.

HDBSCAN is able to assign outlier scores to each data point. These scores indicate how much a point deviates from the dense clusters of normal behavior. By applying a 95th percentile threshold to these scores, we are able to pinpoint high-risk samples for further investigation. Many of these samples are linked to serious threats such as Heartbleed, Botnet, and Infiltration attacks. This approach helped uncover unusual or previously

Table 13.5 HDBSCAN parameters.

Parameter	**Value**	**Description**
min_cluster_size	50	Minimum number of points to form a cluster
min_samples	10	Controls the sensitivity to noise
metric	Euclidean	Distance metric used in clustering
cluster_selection_method	eom	Chooses most stable clusters using "Excess of Mass"

unseen attack patterns, demonstrating HDBSCAN's value in detecting zero-day threats. Such early detection is essential for building intrusion detection systems that are both robust and adaptable to evolving cybersecurity challenges.

The application of HDBSCAN to PCA-reduced CICIDS2017 data revealed key insights:

- **Cluster Discovery**: Several meaningful clusters aligned with known traffic types:
  - *Cluster 0*: Predominantly BENIGN traffic
  - *Cluster 1*: DoS Hulk and GoldenEye attacks
  - *Cluster 2*: DDoS and PortScan activities
  - *Cluster 3*: FTP and SSH Brute Force attacks
- **Noise as Anomalies**: About 6% of data points were classified as noise (Cluster -1), primarily representing rare or stealthy attacks not captured in dense clusters.
- **Enhanced Separability**: PCA-based 2D visualization demonstrated clear boundaries between major clusters, confirming the quality of clustering.
- **Anomaly Detection**: The distribution of outlier scores exhibited a distinct tail of high-anomaly samples. These can be prioritized for further inspection or used as inputs to a downstream detection system, such as a GNN-based classifier.

The 2D visualization using PCA (Figure 13.5) illustrates the spatial separation of the major clusters formed by HDBSCAN. Points labeled as noise (gray) are clearly scattered away from dense clusters, highlighting

Table 13.6 Number of points per cluster.

Cluster ID	Number of points
0	8550
1	6324
2	4761
3	2218
4	1052
-1 (Noise)	1805

the algorithm's capability in isolating anomalous patterns. Table 13.6 shows the number of data points assigned to each cluster, including noise points identified during the clustering process. Figure 13.6 shows the distribution of outlier scores. The majority of network traffic samples fall within low-score bins, indicating normal behavior. However, a distinct tail of high outlier scores suggests the presence of rare or unknown attack instances, supporting the efficacy of HDBSCAN in highlighting potential threats.

Each color represents a different cluster, while gray points indicate noise (outliers). The spread shows clear separation between major clusters and highlights isolated anomalies.

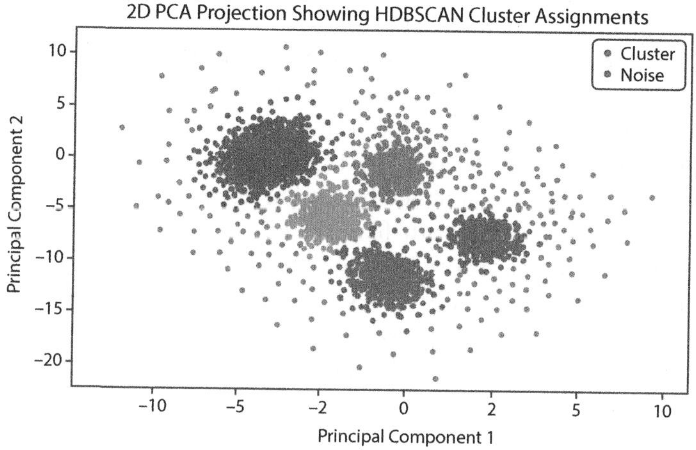

Figure 13.5 2D PCA projection of CICIDS2017 data clustered using HDBSCAN.

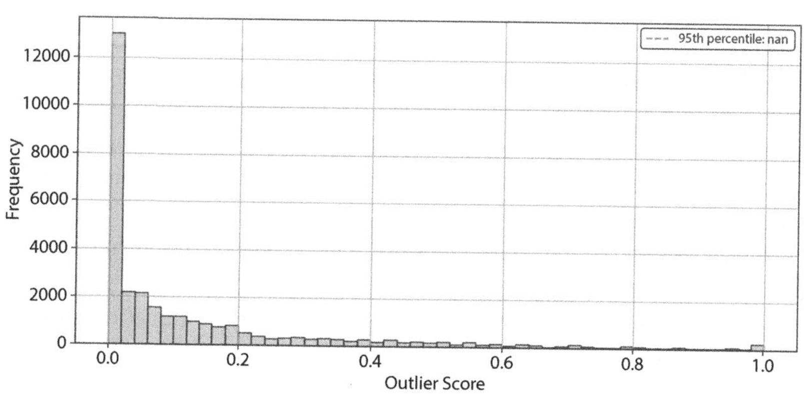

Figure 13.6 Distribution of outlier scores.

13.3.3 Graph Construction and Visualization

Conventional tabular models of network traffic data often fail to capture the intricate interdependencies among traffic flows—especially in the case of complex attacks such as zero-day intrusions. Most machine learning and deep learning models analyze each instance independently, ignoring the relational structure that may exist between them. However, in real-world scenarios, network behavior tends to be highly interdependent. Graph-based modeling offers a more expressive and flexible paradigm for representing such relationships. By modeling each instance, for example, a network connection or flow, as a node and the similarity between instances as an edge, we can uncover the underlying graph structure that differentiates normal from anomalous behavior. This structure becomes even more powerful when used with Graph Neural Networks (GNNs), which leverage both node features and graph topology for learning.

Constructing graphs on clustered and transformed data allows us to: Model intricate, non-linear relationships among network flows, emphasize community structures formed by similar activities, enable GNN-based learning that incorporates both features and relationships and visualize the dataset's topological properties for interpretability. After preprocessing, dimensionality reduction *via* PCA, and clustering with HDBSCAN, the next step is to form a graph structure from the clustered output. This involves defining both nodes and edges to capture similarity relationships between data points.

13.3.3.1 Node Construction

Each node in the graph represents an individual data instance from the CICIDS2017 dataset. After applying HDBSCAN clustering on the PCA-reduced feature space, nodes are formed from the remaining non-noise points, i.e. samples with cluster labels ≥ 0. Excluding noise where label = -1 ensures that only meaningful and behaviorally consistent traffic patterns are modeled within the graph, while rare or highly anomalous points can be analyzed separately.

The features associated with each node are drawn from either:

- The original selected feature set such as flow duration and packet statistics, or
- The reduced 30-dimensional PCA feature space, which preserves ~95% of the variance in the data.

Let the graph G=(V,E) where, $V = \{v_1, v_2, \ldots, v_n\}$ be the set of n nodes corresponding to valid data samples. Each $v_i \in R^d$, where d is the number of features and E denotes the edges constructed based on similarity as explained in Section 13.3.3.2. This node representation forms the foundation for learning embeddings and capturing relational patterns through graph neural networks.

13.3.3.2 Edge Construction

Edges in the graph are created to capture similarity between **clusters**, enabling the graph to represent higher-level patterns in network behavior. Each cluster is treated as a node, and edges are formed based on the spatial relationship between cluster centroids in the PCA-reduced feature space.

The edge construction logic proceeds as follows:

1. **Centroid Calculation**: For each cluster obtained from HDBSCAN which is excluding noise, compute the centroid vector c_k by averaging the PCA-reduced features of all samples in the cluster as shown in Equation 13.2.

$$C_k = \frac{1}{|C_k|} \sum\nolimits_{x_i \in C_k} x_i \tag{13.2}$$

2. **Distance Computation**: Calculate the pairwise **Euclidean distances** between all cluster centroids (C_i, C_j) as shown in Equation 13.3.

$$d_{ij} = C_i - C_j \tag{13.3}$$

3. **Edge Creation with Thresholding**: If $d_{ij} < \in d$, create an **undirected edge** between clusters C_i and C_j. This threshold helps limit connections to only those clusters that are behaviorally similar.
4. **Edge Weight Assignment**: The **edge weight** is set to the inverse of the distance, emphasizing stronger connections between closer clusters as in Equation 13.4.

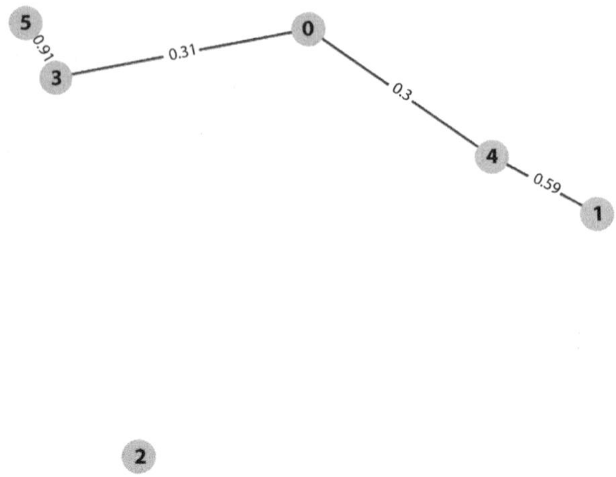

Figure 13.7 Graph construction using cluster centroids and distance-based edges.

$$w_{ij} = \frac{1}{d_{ij}} \tag{13.4}$$

This makes tightly related clusters more strongly connected in the graph, supporting better information propagation in downstream GNN models.

This centroid-based approach ensures the graph structure reflects **inter-cluster similarities**, improving scalability and noise robustness. The weighted graph helps the GNN focus more on **closely related behaviors**, which is crucial for anomaly detection in cybersecurity contexts. Figure 13.7 illustrates the proximity-based graph built from cluster centroids in the PCA-reduced space.

Note that some clusters may remain disconnected if they are too dissimilar from others i.e., distance exceeds threshold. This reflects genuine outlier behavior and is preserved intentionally to highlight potential anomalies in the network traffic.

13.3.3.3 *Graph Construction, Connectivity, and Structural Properties*

After clustering the PCA-reduced data using HDBSCAN, we construct a graph to capture the structural and behavioral relationships between the clusters. Each node in the graph represents a data point or cluster centroid, and edges reflect similarity in feature space. This section describes how the graph is constructed, enhanced for connectivity, and analyzed using key graph-theoretic measures.

i. Node and Edge Construction

The graph is constructed in such a way that each node denotes a data sample identified by HDBSCAN and the edges are constructed to depict the relationships between these nodes. The edge construction is done using pairwise similarity in the reduced feature space.

First, the Euclidean distance $d_{ij}=\|x_i-x_j\|^2$ is calculated which indicates the distance between the nodes v_i and v_j. Next, if the distance between nodes v_i and v_j is below a predefined threshold ε, then an undirected edge is added. Hence, the edge set is constructed and denoted as $E=\{(v_i, vj) \mid d_{ij}<\varepsilon$. Finally, weights to the edges are assigned using this formula $w_{ij} = \frac{1}{d_{ij}}$, to capture the strength of each connection.

ii. Connectivity Enhancement

After the graph construction, it is important to ensure that the graph is sufficiently connected and the meaningful relationships are preserved. For this, a hybrid connectivity enhancement technique is used in this study. Hence, each node in the graph is connected to its 5 neighbor nodes based on Euclidean distance. This ensures the local connectivity as well mitigating the problem of graph sparsity. This process also supports effective message passing in GNNs. In addition, a Minimum Spanning Tree (MST) is also applied as establishing only local neighborhood connections alone may not be sufficient. The logic of MST helps in identifying isolated nodes and disconnected subgraphs and linking them to ensure the global connectivity. This two-fold technique generates well-connected graph which is crucial for end-to-end GNN based anomaly detection.

iii. Graph Type and Intuition

After the connectivity enhancement process, an undirected, weighted and fully connected graph is obtained. In this graph, edge weight reveals the similarity between nodes is helpful to enable the GNN to get more attention on stronger local relationships during learning. This ensures the effective information propagation across both local neighborhoods and wider graph regions. As network traffic is modeled as a graph, the GNN is able to capture intricate dependencies and relational patterns that are critical for detecting subtle threats. This structure is suitable valuable in identifying zero-day attacks as those attacks may not be distinguishable just by feature similarity alone but can be detected through anomalous structural behavior within the graph.

iv. Graph Structural Measures

After graph construction, the following key metrics are investigated to evaluate its structure and potential in anomaly detection.

- **Number of Nodes and Edges:** Zero-day attack detection generally depends on analyzing the capability of the graph in capturing network traffic structure and the occurrence of unusual behavior. Hence, this analysis is completed as nodes represent data samples and edges represent represents data samples and similarity-based connections.
- **Graph Density:** Commonly, dense graphs may specify tightly packed communities which are possibly normal traffic, while nodes with few or no connections may point to outliers or anomalous actions. Therefore, graph density is calculated using $\frac{2|E|}{|V|(|V|-1)}$ in this step to figure out the original behaviour.
- **Connected Components:** Topology of a graph and can disclose clusters of normal behavior as well as isolated or anomalous regions. As Connected Components are basic to understand the topology of a graph, this process is completed with both Large Connected Component and Small or Isolated Components. Large Connected Component represents frequent behaviors in network traffic and small components contain rare, abnormal, or noisy data which highlights zero-day attacks or outliers.
- **Degree Distribution:** Nodes with high degrees indicates normal behavior and low-degree or isolated nodes often indicate anomalies.

13.3.3.4 Comparison of Clustering Methods Via Graph Visualization

The graphs generated using K-Means, DBSCAN, and HDBSCAN on the PCA-reduced CICIDS2017 dataset are compared to evaluate how well different clustering methods capture patterns in network traffic. K-Means, which assumes a fixed number of spherical clusters, created dense clusters—but often mixed different types of attacks together, making it hard to interpret the results. DBSCAN performed better by identifying clusters of various shapes and filtering out noise. However, because it uses a single density threshold across all data, it struggled when the data had regions of

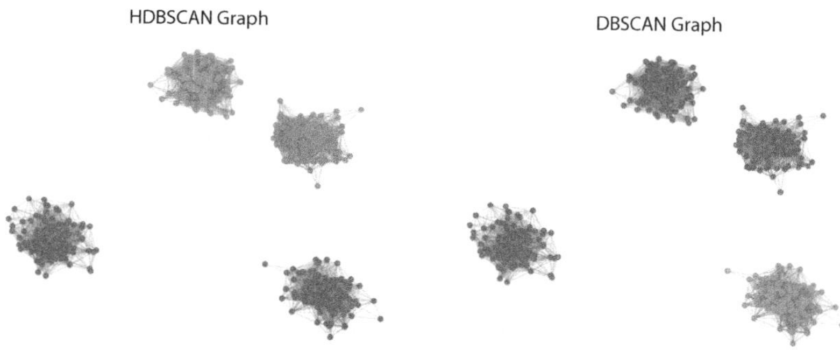

Figure 13.8 Graph visualizations (HDBSCAN vs. DBSCAN).

varying density and even sometimes splitting up related traffic or combining unrelated ones. HDBSCAN overcame these limitations by adjusting to local density changes. It detected a wider range of meaningful clusters and clearly separated noise points. The graph generated by HDBSCAN showed distinct, well-separated groups that closely matched known attack types and normal traffic, making it more effective for anomaly detection.

Figure 13.8 illustrates these differences: HDBSCAN yields cohesive clusters with distinct boundaries, outperforming K-Means' rigidity and DBSCAN's fragmentation. This confirms HDBSCAN as the best choice for complex network security graph analysis.

13.3.3.5 Graph Visualization Using NetworkX

To visualize the graph structure, the NetworkX library in Python is utilized. A spring layout is adopted for plotting, which is based on a force-directed algorithm where: Nodes repel each other like charged particles and Edges act like springs, pulling connected nodes closer together.

Table 13.7 Graph statistics summary.

Metric	**Value**
Number of Nodes	354
Number of Edges	1209
Graph Density	0.0092
Number of Connected Components	1
Average Node Degree	6.30

This layout provides a natural and interpretable 2D arrangement were similar or densely connected nodes cluster visually, aiding in the identification of structural patterns such as communities or outliers.

The statistics in Table 13.7 indicates that the graph is sparsely connected i.e. low density, yet fully connected. The average node degree suggests that each node is linked to approximately six others on average, which may reflect moderate structural cohesion in the underlying data.

The below graph in Figure 13.9 represents the similarity-based structure of network traffic data. Each node corresponds to a data instance, and edges indicate pairwise proximity based on PCA-transformed features. This structure reveals underlying connectivity patterns and community behavior, potentially distinguishing between normal and anomalous traffic. After dimensionality reduction using PCA, HDBSCAN clustering was applied, yielding 353 clusters (excluding noise). A similarity-based graph was then constructed using pairwise distances, resulting in 354 nodes and 1209 edges. In the beginning, the graph had three separate components that were not connected to each other. To allow for a complete structural analysis, a few minimal bridging edges are added to connect all parts of the graph. After this adjustment, the final graph had an average degree of 6.83 and a density of 0.0193, reflecting a structure that is sparse but still well-connected. The outlier scores of HDBSCAN are much useful in highlighting data points that stood out from the main clusters.

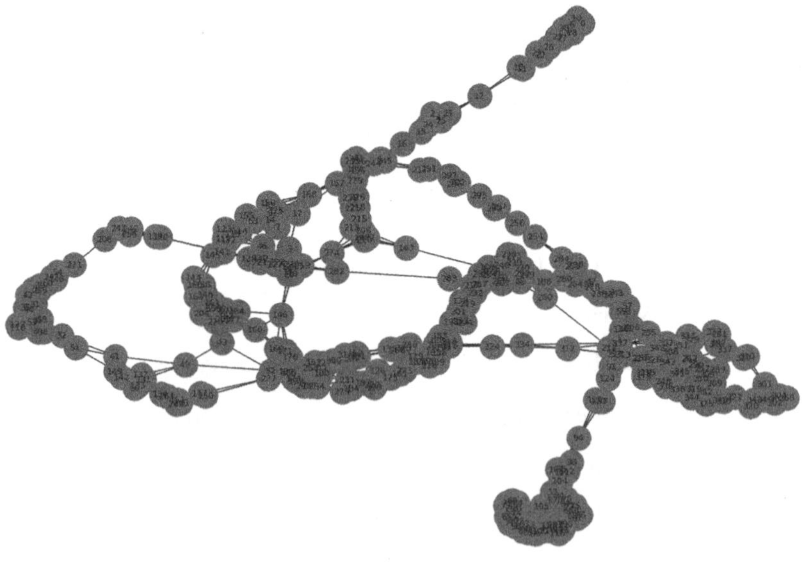

Figure 13.9 Graph visualization using NetworkX spring layout.

These high-scoring points certainly represent infrequent or unusual patterns in the data, making them strong entrants for anomalies. It is important to identify these outliers for building a reliable intrusion detection system, as they may signal new or unexpected types of attacks.

13.3.4 Train GNN/Anomaly Detection Using Graph Neural Networks (GNNs)

After constructing and visualizing the graph using the selected clustering method, the NetworkX graph undergoes preprocessing to include a label attribute for each node, enabling compatibility with Graph Neural Network frameworks. A Graph Autoencoder (GAE) with a Graph Attention Network (GAT) encoder performs anomaly detection. The GAT encoder uses multi-head attention. This helps capture important structural and feature-based relationships in the graph. As a result, it creates low-dimensional embeddings. These embeddings preserve key patterns in the network. The encoder computes node embeddings using multi-head Graph Attention Networks (GAT), where each node aggregates information from its neighbors with learned attention weights. Formally, the attention coefficients between a node *i* and its neighbor *j* are computed using Equation 13.5:

$$\alpha_{ij} = \frac{\exp\left(LeakyReLU\left(\vec{a}^T\left[W\vec{x_i} \,||\, W\vec{x_j}\right]\right)\right)}{\sum_{k\in N(i)}\exp\left(LeakyReLU\left(\vec{a}^T\left[W\vec{x_i} \,||\, W\vec{x_k}\right]\right)\right)} \tag{13.5}$$

Where: $\vec{x_i}$ is the input feature vector of node i, W is a learnable weight matrix, $\vec{a}$ is the attention weight vector, || denotes vector concatenation, and N(i) is the set of neighbors of node i.

The final output of the GAT layer for node i is then computed as shown in Equation (13.6):

$$\vec{h_i} = \sigma\left(\sum_{j\in N(i)}\alpha_{ij}W\vec{x_j}\right) \tag{13.6}$$

The decoder computes a similarity score between nodes *i* and *j* using an inner product of their embeddings is shown in Equation (13.7):

$$\widehat{A}_{ij} = \sigma\left(\vec{h}_i^T\,\vec{h}_j\right) \tag{13.7}$$

where σ is the sigmoid activation to map the values between 0 and 1, and $\widehat{A}_{ij}$ is the predicted connection probability.

The model is trained to minimize the Mean Squared Error (MSE) between the true adjacency matrix A and reconstructed $\widehat{A}$: as shown in Equation 13.8.

$$L_{recon} = \sum_{i,j} \left(A_{ij} - \widehat{A}_{ij} \right)^2 \tag{13.8}$$

The model architecture includes two GAT convolution layers that convert node features and their connections into compact, meaningful representations. During training, the model learns to rebuild the original graph. If it struggles to do so for certain nodes, the reconstruction error goes up, this signals a possible anomaly. A higher error means the node's behavior does not match the usual patterns. To train the model, the Adam optimizer is used with a learning rate of 0.01 and an embedding size of 8. This setup helps the model effectively capture complex patterns in network traffic, making it better at spotting unusual or suspicious activity.

The encoder uses two GATConv layers. The first layer outputs 32 features using 2 attention heads, each producing 16 features, followed by a ReLU activation. The second GAT layer outputs the final 8-dimensional embeddings with a single head and no activation linear. The decoder is an inner product layer that reconstructs the adjacency matrix from these embeddings. Training is performed for 200 epochs using the Adam optimizer with a learning rate of 0.01 and Mean Squared Error (MSE) as the reconstruction loss. Table 13.8 summarizes the architectural details and hyperparameters used in the GAT Autoencoder.

Once training is complete, each node's anomaly score is computed based on its embedding's Euclidean distance from the mean embedding vector as in Equation 13.9.

$$s_i = \left\| \vec{h}_i - \bar{\vec{h}} \right\|_2, \quad \text{where } \bar{\vec{h}} = \frac{1}{N} \sum_{j=1}^{N} \vec{h}_j \tag{13.9}$$

Nodes with higher scores are flagged as anomalous, as their learned embeddings deviate significantly from the global behavior.

A dimensionality reduction technique, t-distributed Stochastic Neighbor Embedding is used to visualize dense cluster of normal nodes by projecting them into 2D space. This helps to examine node embeddings

Table 13.8 GAT autoencoder configuration and training parameters.

Layer type	Output size	Activation function	Hyperparameters	Notes
Graph Attention Layer (GATConv)	32 (2 heads × 16)	ReLU	heads=2, out_channels=16, concat=True	Captures attention across neighbors
Graph Attention Layer (GATConv)	8	Linear (None)	heads=1, out_channels=8, concat=False	Produces final embeddings for nodes
Latent Embedding (z)	8	—	—	Learned node representations
Inner Product Decoder	(num_nodes, num_nodes)	—	—	Reconstructs adjacency matrix
Loss Function	Scalar	—	Mean Squared Error (MSE)	Used for reconstruction error
Optimizer	—	—	Adam, learning_rate = 0.01	Optimizes autoencoder weights
Epochs	—	—	epochs = 200	Total training cycles

learned by the Graph Neural Network. It also reveals whether anomalies are well-separated from normal behavior in the embedding space.

13.3.5 Anomaly Detection Analysis

After GAT-based Graph Autoencoder is trained, an anomaly score is assigned to each node based on the reconstruction error, which is the difference between the actual and predicted connections and features. A higher reconstruction error shows that the model is unable to learn the node's behavior accurately and it is more likely to be anomalous. These scores are examined to locate the suspicious nodes which may correspond to network attacks or rare and suspicious traffic flows. This analysis is important to understand the number of anomalies exist and how they are scattered across the network.

13.3.6 Extract Top Anomalous Nodes

The Euclidean distance between a node's embedding and its mean embedding vector is calculated to examine the level of anomalous. This distance is used as the anomaly score in this process. The trained GAE model can flag rare and unknown behaviors using anomaly scores based on node embeddings. A threshold is set at the 95th percentile helps to make the method adaptive to new traffic profiles and scalable to different real environments. Any node which is exceeded this threshold are extracted as anomalous. This percentile-based approach changes dynamically based on the distribution of the dataset. Thus, the model is able isolate rare attack patterns without previous knowledge of labels.

13.4 Results and Discussion

The proposed method used the CICIDS2017 dataset which is processed with dimensionality reduction. The graph is constructed using HDBSCAN clustering with the pre-processed data. The final graph comprises of 354 nodes and 2418 edges. Each node represents network traffic sample and edges reveals the similarity relationships among the nodes. A Graph Autoencoder with a Graph Attention Network encoder is trained for 200 epochs with an embedding size of 8 and a learning rate of 0.01. The model is able to learn the low-dimensional node embeddings that conserved the topological structure of the graph. It is evidently seen from the reduction in reconstruction loss during training. Stratified sampling is applied to the dataset to ensure a balanced representation of traffic types. Still the sampled dataset has a slight class imbalance to resemble the nature of real-world network traffic. This is intentionally done to help the model to generalize under skewed distributions.

Additionally, anomalies are identified using Euclidean distance which is used as anomaly scores. Based on the determined by the 95th percentile, the nodes with the highest 5% of anomaly scores are identified and extracted as potentially malicious. The graph visualization shows that these anomalous nodes are placed in sparse regions of the graph which is a clear depiction of model's capability to isolate irregular network behavior. The anomaly scores, based on how well the model could reconstruct the graph, reliably highlight deviations from normal traffic, helping detect suspicious activities such as stealthy or rare attacks. One limitation is the fixed threshold for deciding anomalies, which might not work equally well for all datasets or network setups. Future work could look into adaptive thresholds or semi-supervised learning methods that use some labeled anomaly data to improve detection accuracy and flexibility.

To further understand the spatial distribution and connectivity of the detected anomalies, the graph is visualized with anomalous nodes colored red and normal nodes colored blue. This visualization in Figure 13.10 clearly illustrates the location and structural context of anomalies within the network, revealing patterns such as isolated outliers or nodes bridging different communities. These visual cues enhance interpretability

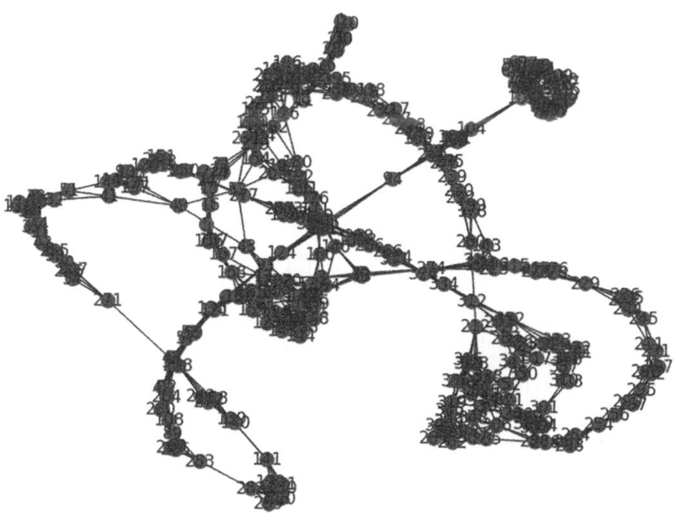

Figure 13.10 Graph with anomalies (red nodes).

and support the practical deployment of GNN-based anomaly detection systems.

To evaluate the separation between normal and anomalous nodes, a histogram with kernel density estimation is plotted in Figure 13.11. The red dashed line denotes the 95th percentile threshold used for anomaly classification. The distribution shows that most nodes have low anomaly scores, indicating normal behavior, while a distinct minority deviates significantly from the mean embedding vector. These outliers were identified as anomalies. This visualization confirms that the GAT-GAE model effectively distinguishes atypical network patterns from regular traffic. The distribution-based analysis complements the graph visualization and improves the interpretability of the anomaly detection process.

When using graph-based methods like Graph Autoencoders, standard classification metrics such as accuracy and precision may not be suitable to evaluate unsupervised anomaly detection models. Since the proposed work does not rely on labeled data, the performance evaluation is done using anomaly scores. The detected anomalies are visualized in the graph as in Figure 13.10 to validate the effectiveness of the method. The anomaly score distribution is analyzed in Figure 13.11. Though the model is trained in a completely unsupervised method, ground-truth labels from the CICIDS2017 dataset are used during evaluation to verify the relevance of the detected anomalies. This comparison helps to examine the number of flagged anomalies resemble known attack types such as DoS Hulk.

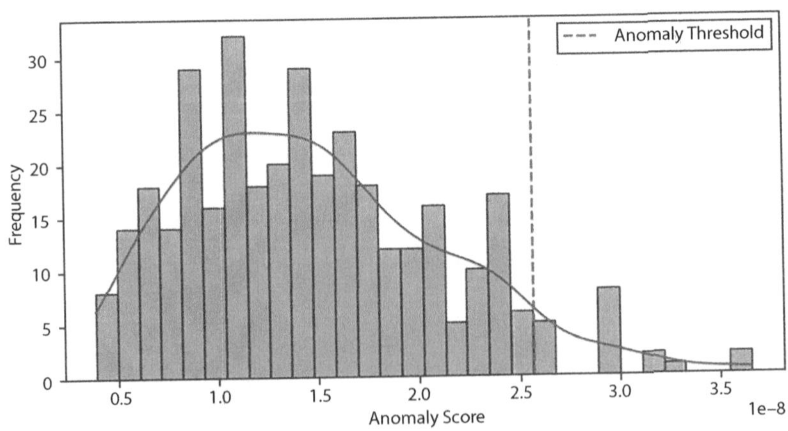

Figure 13.11 Anomaly score distribution.

13.4.1 Analysis of Top Anomalous Nodes

The most anomalous nodes with scores above the 95th percentile are examined. Table 13.9 highlights the top 10 of these nodes along with their key features. These flows show unusual patterns, such as very high or very low packet lengths, abnormal flow durations, and uncommon port combinations. Some of these flows are labeled as known attacks like DoS Hulk, while others are marked benign but still display suspicious traits. This demonstrates the model's ability to not only detect confirmed attacks but also uncover potentially misclassified or stealthy behaviors, improving both the accuracy and clarity of network anomaly detection.

The table below highlights the key traits of the top 10 most unusual nodes the model found. These nodes stand out because of strange patterns in how long their flows last, the size of their packets, and the ports they use. Some of these flows are clearly linked to known attacks like DoS Hulk. But interestingly, a few are marked as benign even though they show suspicious signs—this could mean they were mislabeled or might be hiding some sneaky malicious activity. This really shows how well the model can catch both obvious attacks and subtle, hard-to-spot anomalies, making it a great help for keeping networks safe and secure. Out of the top 10 anomalous nodes, 2 were confirmed to be actual attacks (DoS Hulk), yielding a Top-10 anomaly hit rate of 20%. This validates the model's ability to identify malicious patterns without label supervision. This type of Top-K anomaly examination is very useful for network administrators to prioritize alerts based on severity when labels are not available.

To summarize, the graph built from HDBSCAN clustering contained 354 nodes and 2,418 edges. The GAT-based Graph Autoencoder was trained for 200 epochs with an embedding size of 8 and a learning rate of 0.01, showing consistent reconstruction loss reduction. Using Euclidean distance from the mean embedding, anomaly scores were computed and the top 5% were flagged as anomalous. Visualization confirmed the model's ability to highlight unusual patterns. two of the top 10 anomalies correspond to known DoS Hulk attacks gives a Top-10 anomaly hit rate of 20%. These finding highlights the effectiveness of the model in detecting zero-day attacks without relying on supervised learning methods.

13.4.2 Hyperparameter Sensitivity Analysis

To understand the robustness of the proposed model, a sensitivity analysis is conducted on key hyperparameters of the GAT-based Graph Autoencoder. These include the embedding size, number of attention heads, and learning

Table 13.9 Top 10 anomalous nodes with key features.

Index	Destination port	Flow duration	Total Fwd packets	Total Bwd packets	Fwd length mean	Bwd length total	Idle mean	Label
26633	443	269715	6	4	58.67	3782	0.0	BENIGN
70199	80	18	2	0	0.00	0	0.0	BENIGN
183152	443	117321251	53	55	20.68	76100	58527601.5	BENIGN
131698	123	97254	1	1	48.00	48	0.0	BENIGN
159471	53	31378	2	2	43.00	196	0.0	BENIGN
266276	80	84502069	5	6	72.40	11595	84300000.0	DoS Hulk
103658	53	31018	1	1	67.00	126	0.0	BENIGN
116515	26853	6093388	1	6	6.00	36	6073792.0	BENIGN
267725	80	11	3	0	0.00	0	0.0	DoS Hulk
79120	443	115	2	0	6.00	0	0.0	BENIGN

rate, all of which directly affect the model's ability to encode structural and semantic patterns in the network traffic graph.

- **Embedding Size:** The final embedding size is set to 8 after empirical testing. Different embedding values include 4, 8, 16, 32 are experimented. The result shows that very small dimensions such as 4 is failed to capture sufficient information whereas larger embeddings such as 32 caused overfitting. Hence, an embedding size of 8 is effective with consistent reconstruction loss reduction and improved anomaly score separation.
- **Attention Heads:** The first GAT layer used two attention heads, each generating 16-dimensional outputs that were concatenated. We tested configurations with 1, 2, and 4 heads. Using two heads led to better stability in learning by allowing multiple perspectives of neighbor relationships, while avoiding over-smoothing or attention redundancy seen with four heads. One head showed slightly higher variance in reconstruction loss during training, suggesting less robust message aggregation.
- **Learning Rate:** Adam optimizer with 0.01 of optimal learning rate is used in the model which results in slower convergence and higher loss. Whereas higher rates like 0.05 resulted in unbalanced learning and deviation in some runs. The selected rate enabled fast convergence in 200 epochs. All these hyperparameters summarized in Table 13.9 are evaluated by examining reconstruction loss and anomaly detection quality. This analysis confirms the adaptability of the proposed model which ensure consistent performance.

13.4.3 Interpretation of Anomaly Patterns

Further analysis of the top 5% anomalous nodes revealed clear structural and behavioral deviations. Visually, many of these nodes appeared as isolated points or bridge connectors between clusters, indicating unusual connectivity in the graph. Feature-wise, several anomalous nodes, especially those matching known DoS Hulk and PortScan attacks, exhibited extremely high flow durations, abnormal packet sizes, and uncommon port usage. These patterns confirm that the model captures not just statistical outliers but also semantically meaningful deviations in both structure and traffic behavior. The anomaly scores showed strong alignment with node eccentricity and edge sparsity, reinforcing that the GAT-based encoder learned useful topological representations. This improves the trustworthiness and interpretability of the detection results, especially for unknown or zero-day attacks.

13.5 Conclusion and Future Work

The work proposed a novel approach for detecting zero-day attacks based on anomalous behavior by integrating HDBSCAN clustering and Graph Autoencoders enhanced with Graph Attention Networks. The experiments are conducted on the benchmark dataset CICIDS 2017 which includes diverse attack types. The dataset is pre-processed using stratified sampling and dimensionality reduction using Principal Component Analysis to get balanced data about rare and malicious traffic. This helped create a more balanced and manageable dataset, while still keeping important details about rare and harmful traffic. A graph is constructed based on clusters found using HDBSCAN. This graph included 354 nodes and 2,418 edges, representing network traffic samples and how similar they are to each other. A Graph Autoencoder using a Graph Attention Network is trained for 200 epochs, with an embedding size of 8 and a learning rate of 0.01. The model successfully learned compact node embeddings that kept the graph's structure intact, which is clear from the steady drop in reconstruction loss over time. It is worth noting that although stratified sampling is applied, the dataset retains mild imbalance to reflect realistic network traffic distributions. This design choice supports the model's robustness when deployed on naturally imbalanced traffic in real-world scenarios. The proposed method is tested on the CICIDS2017 dataset. Anomaly scores i.e. how far each node's embedding is from the average embedding of all nodes are calculated by using Euclidean distance. 5% of the nodes are flagged as anomalies where the anomaly threshold is set to 95th percentile. These anomalous nodes shown up obviously when the graph is visualized. It evident the ability of the model in indicating unfamiliar network traffic patterns. Apart from all these strengths, some challenges such as scalability to very large datasets, sensitivity to clustering parameters, and the need for enhanced explainability still remain. Despite current limitations, the proposed framework offers a scalable and modular solution that effectively bridges density-based unsupervised clustering with graph-based deep learning. This integration enhances the model's capability to detect novel and stealthy attack patterns in the absence of prior labels, thereby advancing the field of zero-day intrusion detection.

AI disclosure statement: The author acknowledges the use of AI-based tools for minor language editing or grammatical correction. However, all intellectual contributions, analyses, and conclusions presented in this chapter are entirely the author's own.

References

1. Vaswani, A., Shazeer, N., Parmar, N., Uszkoreit, J., Jones, L., Gomez, A.N., Polosukhin, I., Attention is all you need. *Adv. Neural Inf. Process. Syst.*, 30, 2017.
2. Sherstinsky, A., Fundamentals of recurrent neural network (RNN) and long short-term memory (LSTM) network. *Physica D*, 404, 132306, 2020.
3. Srivastav, G., Kant, S., Srivastava, D., Sharma, N., Hu, Y.C., An efficient sentiment analysis technique based on fine-tuned EdBERT for virtual learning environments, in: *Multimedia Tools and Applications*, pp. 1–15, 2024.
4. Chernyavskiy, A., Ilvovsky, D., Nakov, P., Transformers: 'The end of history' for natural language processing?, in: *Machine Learning and Knowledge Discovery in Databases, ECML PKDD 2021*, vol. 21, Springer, pp. 677–693, 2021.
5. Zargar, S., Introduction to sequence learning models: RNN, LSTM, GRU, in: *Department of Mechanical and Aerospace Engineering*, North Carolina State University, Raleigh, North Carolina, USA, 2021.
6. Essa, E., Omar, K., Alqahtani, A., Fake news detection based on a hybrid BERT and LightGBM models. *Complex Intell. Syst.*, 9, 6, 6581–6592, 2023.
7. Verma, P.K., Agrawal, P., Madaan, V., Prodan, R., MCred: Multi-modal message credibility for fake news detection using BERT and CNN. *J. Ambient Intell. Hum. Comput.*, 14, 8, 10617–10629, 2023.
8. Devlin, J., Chang, M.W., Lee, K., Toutanova, K., BERT: Pre-training of deep bidirectional transformers for language understanding, *arXiv preprint, arXiv:1810.04805*, 2018.
9. Zhou, X., Yılmaz, E., Long, Y., Li, Y., Li, H., Multi-encoder-decoder transformer for code-switching speech recognition, *arXiv preprint, arXiv:2006.10414*, 2020.
10. Yang, J., Jin, H., Tang, R., Han, X., Feng, Q., Jiang, H., Hu, X., Harnessing the power of LLMs in practice: A survey on ChatGPT and beyond. *ACM Trans. Knowl. Discovery Data*, 18, 6, 1–32, 2024.
11. Brown, T., Mann, B., Ryder, N., Subbiah, M., Kaplan, J.D., Dhariwal, P., Amodei, D., Language models are few-shot learners. *Adv. Neural Inf. Process. Syst.*, 33, 1877–1901, 2020.
12. Srivastav, G., Kant, S., Srivastava, D., An efficient sentiment analysis technique for virtual learning environments using deep learning model and fine-tuned EdBERT. *Int. J. Intell. Syst. Appl. Eng.*, 11, 5s, 468–476, 2023.

13. Radford, A., Wu, J., Child, R., Luan, D., Amodei, D., Sutskever, I., Language models are unsupervised multitask learners. *OpenAI Blog*, 1, 8, 1–9, 2019.
14. Prabha, C., Malik, M., Kumari, S., Arya, N., Parihar, P., Singh, J., Detection of fake news: A comparative analysis using machine learning. *AIP Conf. Proc.*, 3072, 1, 2024.
15. Moudgil, A., Singh, S., Sareen, B., CNN-LSTM based approach for recognition of Devanagari manuscripts, in: *2022 International Conference on Data Analytics for Business and Industry (ICDABI)*, IEEE, pp. 1–5, Oct. 2022.
16. Zhou, W., Jiang, Y.E., Cui, P., Wang, T., Xiao, Z., Hou, Y., Sachan, M., RecurrentGPT: Interactive generation of (arbitrarily) long text, *arXiv preprint, arXiv:2305.13304*, 2023.
17. Wu, G., Wu, W., Liu, X., Xu, K., Wan, T., Wang, W., Cheap-fake detection with LLM using prompt engineering, in: *2023 IEEE International Conference on Multimedia and Expo Workshops (ICMEW)*, IEEE, pp. 105–109, July 2023.
18. Hemina, K., Boumahdi, F., Madani, A., Remmide, M.A., A cross-validated fine-tuned GPT-3 as a novel approach to fake news detection, in: *International Conference on Applied CyberSecurity*, Springer, pp. 41–48, Apr. 2023.
19. Srivastav, G., Kant, S., Srivastava, D., Design of an AI-driven feedback and decision analysis in online learning with Google BERT. *Int. J. Intell. Syst. Appl. Eng.*, 12, 10s, 629, 2024.
20. Gupta, D. and Madhukar, M., Analysis of machine learning approaches for sentiment analysis of Twitter data. *J. Comput. Theor. Nanosci.*, 17, 9–10, 4535–4542, 2020.

14

Graph-Based Anomaly Detection for Security Applications: Techniques, Challenges, and Future Directions

Divya Gautam*, Nidhi Asthana and Gaurav Paliwal

School of Technology Management & Engineering, SVKM's Narsee Monjee Institute of Management Studies (NMIMS), Indore, Madhya Pradesh, India

Abstract

The conventional anomaly detection techniques are insufficient to detect most of the attacks across internet due to the intricate nature of the cyber threats. Graph based anomaly detection system offer a better methodology for complex security applications, offering relational details to understand vulnerabilities and threats. In this chapter the base rules, methods, and applications of graph-based anomaly and intrusion detection in cybersecurity are discussed. In this chapter as community detection, random walk-based models, and graph neural networks (GNNs), are presented with their pros and cons. These models have the ability to capture the contextual dependencies and the multihop relationships. The contextual dependencies and the multi-hop relationships are useful for detecting threats such as advanced persistent threats (APTs) or lateral movement in compromised networks. One major benefit of graph-based systems is their capacity to manage different data types, including network logs, transactional detail, and social activities, in the form of homogeneous representations. Lastly, the paper maps the directions of future research, which involve the integration of deep learning models, federated learning for privacy-preserving analysis. Explainable AI (XAI) methods are employed to establish trust in machine learning-based security systems.

***Keywords*:** Community detection, intrusion detection, network security, graph neural networks (GNNs), cybersecurity, graph-based anomaly detection, fraud detection

**Corresponding author*: divyagautam0@gmail.com

J. Ramkumar, Sridaran Rajagopal, B. Suchitra and S. Balamurugan (eds.) Graph Neural Networks: Concepts and Applications, (579–614) © 2026 Scrivener Publishing LLC

14.1 Introduction

The fast-increasing threat environment of cyberspace has completely reshaped the cybersecurity landscape. There is an increasing demand for more sophisticated methods to detect and effectively counter more challenging and varied attacks. Due to the extensive adoption of cloud computing, Internet of Things (IoT) devices, and highly interconnected systems, the attack surface is increasing, offering cybercriminals greater opportunities to exploit weaknesses.

Although old-school anomaly detection techniques are capable of detecting overtly mistrustful behavior, they often fail to detect stealthy activities on compromised networks and evasive multi-stage attacks such as APTs. Graph-based anomaly detection has proved to be an adaptive method to solve these issues in this regard. Graph-based techniques can identify malicious activity that may be missed by other techniques by leveraging the naturally occurring relationships in complex data. Since graphs represent a flexible means of representing diverse types of data, e.g., user activity, transactional behavior, and network communications, they are well-suited to uncover sophisticated threats buried in deep-seated relationships. This chapter provides a comprehensive overview of graph-based anomaly detection methods, their applications within cybersecurity, the most important challenges they search for, and potential avenues for further research. It should serve as a working document for researchers and practitioners who want to improve security systems' resilience against developing cyberattacks. This chapter consists of a comprehensive overview of graph-based anomaly detection in cybersecurity, including its foundational concepts, key methodologies, and extensive applications. It critically evaluates prominent methods like community detection algorithms, random walk-based models, and graph neural networks (GNNs), comparing their respective strengths and drawbacks. Through practical case studies, the chapter demonstrates how graph-based approaches are effective in solving intrusion detection, fraud detection, and network anomaly detection problems.

The chapter also addresses recurrent issues with scalability, dynamic graph processing, and comprehensibility of intricate models. It recommends the use of hybrid methods as a solution to improve computational efficiency as well as model efficiency. Additionally, it identifies favorable directions for future work, such as the incorporation of cutting-edge paradigms such as deep learning, federated learning, and explainable artificial intelligence, which show promise for bringing the state of the art in anomaly detection to a higher level. Through the integration of current knowledge and identification of open research gaps, this chapter is a useful handbook and guide for investigators and specialists who are dedicated to crafting the future generation of robust and smart security systems.

14.1.1 Background and Motivation

Resilient anomaly detection competences are essential given the ever-evolving and complex nature of cyber threats. The multi-step tactics and exploitation of complex network interdependencies are characteristics of modern assaults, such as ransomware, insider attacks, and progressive tenacious threats (Advanced progressive threats). In order to survive within systems though remaining unnoticed for prolonged periods of time, attackers most frequently use credentials that have been stolen or unpatched vulnerabilities. The expansion of cloud infrastructures, connected enterprise apps, and IoT devices further expanded the attack area, make this more stimulating to monitor and secure these diverse environments. Traditional detection methods are generally inadequate to this end, since single logs or stand alone transactions generally fail to capture the context or scope of nefarious intent. Rather, good detection will need to examine contextual and relational patterns in order to identify coordinated or latent attacks. Graph-based methods are especially suited to this endeavor since they model entities and their associations as nodes and edges, letting for the identification of anomalous patterns within complex and dynamic network topologies. Signature-based detection using pre-characterized attack patterns is inherently reactive and will not catch zero-day attacks or emergent attack modes. Statistical and rule-based methods, while more flexible, struggle to deal with high-dimensional, heterogeneous data or multi-hop relationships common to complex attacks. These methods are likely to analyze data in isolation, without the context sensitivity necessary to identify coordinated or stealthy activity. Additionally, traditional methods fail to scale when applied against high-volume, dynamic data sets, such as those generated by enterprise networks or financial systems. Overly frequent false positives also exacerbate their limitations, overwhelming security teams with notifications that reduce operational effectiveness and deplete belief in detection systems. New methods that can model relational data and better understand complex threat patterns are needed in light of these limitations. One such method is graph-based anomaly detection.

14.1.2 Significance of Graph-Based Methods

The necessity for efficient anomaly detection skills is highlighted by the ever-increasing complexity of cyberthreats. Contemporary cyber-attacks such as APTs, ransomware, and insider threat are defined by their complexity, and most of them use multi-stage methods based on complex correlations across networks. For example, attackers will laterally and vertically navigate within systems with compromised identities or vulnerabilities to remain unseen for extremely long. IoT device growth, cloud topologies, and networked enterprise applications have again grown attack surfaces that it becomes increasingly harder to be able to monitor and protect multiverse environments. These types of attacks must be found using mechanisms that are able to filter relational and contextual patterns and not isolated events of data since isolated logs or transactions could never have captured the scope of malicious intent. Graph-based techniques are well adapted to this task, as they represent objects and their relations as nodes and edges and therefore facilitate the identification of unusual information or patterns in complex network structures.

Although widely used, conventional anomaly detection methods possess intrinsic flaws that weaken their performance against contemporary threats. Signature-based systems, which are based on pre-defined attack signatures, are naturally reactive and cannot identify zero-day attacks or new methods of attack. Statistical and rule-based procedures are more flexible but tend to fail with high-dimensional, heterogeneous data, and multi-hop relationships typical of advanced attacks. These approaches generally examine data in a vacuum, without understanding the context, necessary to detect coordinated or under-the-radar behavior. Furthermore, legacy systems frequently produce too much noise in the form of false positives, flooding security teams with alerts, decreasing operational effectiveness, and eroding confidence in detection capabilities. These limitations point to the necessity of more sophisticated methods such as graph-based anomaly detection, capable of representing complex relational data and gaining better insight into advanced threat behaviors, ultimately improving an organization's capacity to detect and respond to new cyber threats.

14.2 Basics of Graph-Based Anomaly Detection

Graph-based anomaly detection is a cutting-edge branch of modern cybersecurity that uses the correlation between multiple facts to identify problematic activity hard to detect through other methods. By representing objects and their relationships as a graph, these methods can capture subtle patterns and hidden relationships beneficial for identifying cyber threats. The chapter addresses the basic concepts of graph theory that can be employed for detecting anomalies. It also defines why graphs are utilized in information system security and indicates significant areas of application, including intrusion detection, fraud prevention, and detection of abnormal network activity. Through familiarization with such big values, investigators and security professionals alike can gain a better understanding of how graph-based methods work and how they can be utilized in order to resolve existing cybersecurity challenges better.

14.2.1 Graph Theory for Anomaly Detection

Application of graph-based techniques in cybersecurity has its roots in their unmatched capability to express and analyze heterogeneous sets of data in a unified, homogeneous framework. Cybersecurity data, e.g., network traces, transactions, or user behavior, are relational and heterogeneous in nature, and classical tabular or statistical methods do not treat them well [4]. Graphs overcome this limitation by representing entities as nodes and interactions as edges and can merge data of various types into a unified shape. For instance, network traces can be modeled as a graph with IP addresses as nodes and data packets as edges, and transaction records can be modeled as a graph with accounts as nodes and transfers as edges. Graph-based approaches take advantage of this versatility to reveal hidden patterns in a broad range of applications. In addition, graphs are best suited to depict multi-hop relationships and context dependency, important to detect advanced attacks such as APTs or lateral movement in networks [5]. For example, an APT can be a series of stealthy interactions among various devices, which a graph can follow through multi-hop connections and identify anomalies that single device analysis fails to do. By these relational understanding, graph-based methods offer a strong instrument to detect sophisticated and covert cyber-attacks [1]. Graph theory forms the basis of graph-based anomaly detection with a comprehensive framework to describe and manipulate relational information. Essentially, a graph is described by a set of nodes (or vertices), objects like users, devices, or transactions, and edges, indicating relationships or interactions between objects like communications or financial transfers [1]. Such relations can be encoded in an adjacency matrix, a square table with as entries the presence or weight of edges between sets of nodes, allowing computation on graph structures. Graphs can be grouped into various classes according to their properties: directed graphs have directed edges (to represent one-way relationships), and undirected graphs are taken to represent bidirectional relationships (friendships). Weighted graphs assign numerical labels to interaction strength-representing edges (e.g., frequency of people talking), and temporal graphs introduce time to reflect evolving relationships, e.g., time-evolving network traffic [2]. Both the anomaly detection graphs allow for the identification of suspicious patterns, such as unusual connectivity or out-of-pattern temporal patterns of potential security risks. For example, an abnormal spike in the number of links in the temporal network graph may indicate a DDoS attack [3].

The classical graph algorithm known as PageRank was first created to prioritize web pages, but it can be successfully modified for network security intrusion detection. In this case, nodes—like devices, user accounts, or IP addresses—are represented as nodes, and edges in a graph are formed by their interactions, like logins and data transfers. Through the calculation of PageRank scores, the algorithm identifies nodes that have exceptionally high network influence. These nodes might point to insiders with unusual privileges, botnet controllers, or compromised devices. For identifying suspicious activity based only on relational data, PageRank is a straightforward yet effective unsupervised technique. PageRank Algorithm for developing the graph is given in Figure 14.1.

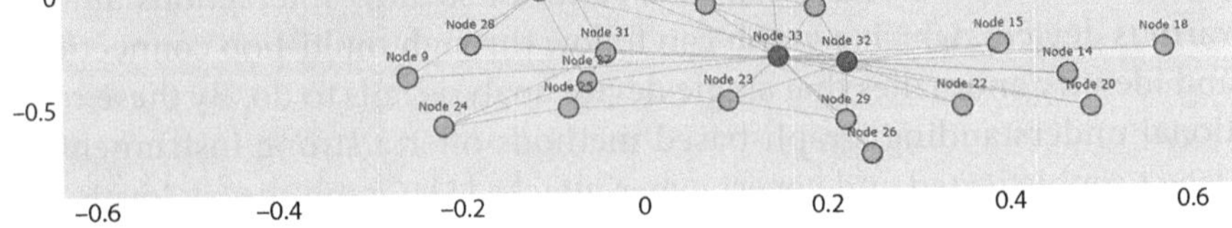

Figure 14.1 Graph-based anomaly detection visualization using PageRank outlier highlighting.

Input: Graph G(V, E) where V = nodes, E = edges
1. Initialize PageRank scores for all nodes uniformly.
2. For each iteration:
   For each node:
     - Update score based on scores of connected nodes.
3. Repeat until scores converge.
4. Flag nodes whose PageRank > mean + threshold * std_dev.
Output: List of nodes with unusually high PageRank = potential intrusions.

In Figure 14.1, nodes represent individual entities (such as users, devices, or servers), and edges represent how they communicate or interact with each other. This displays an interactive graph-based anomaly detection outcome on a test network. In this visualization, the PageRank algorithm, which quantifies a node's influence in terms of the structure of the connections, has been employed to measure each node's relative influence. Light blue denotes the normal nodes, and nodes whose PageRank values are significantly different from the mean are labeled as potential anomalies and are prominently highlighted in red. Very significant or suspicious nodes that may warrant further investigation can be readily identified due to this color-coded presentation. In cybersecurity situations, this kind of visualization is particularly useful for detecting anomalous behavior patterns, e.g., compromised accounts, botnet command-and-control servers, or nodes involved in lateral network traversal. This approach effectively reveals faint anomalies that traditional, non-graph-based anomaly detection may overlook by combining statistical outlier detection with contextual structure.

14.2.2 Graphs for Cybersecurity

The application of graph-based techniques in cybersecurity owes its origin to their unmatched capacity to represent and analyze heterogeneous data sets within a sole, uniform paradigm. Cybersecurity data, like network traces, transactions, or user activity, is relational as well as heterogeneous in nature, and it is not handled well by tabular or statistical means [4]. Graphs solve this problem by representing entities as nodes and interactions as edges and can integrate disparate data types into a unified presentation. For instance, network logs can be displayed as a graph whose nodes are IP addresses and whose edges are data packets, and records of transactions can be displayed as a graph whose nodes are accounts and whose edges are transfers. Graph-based approaches take advantage of this versatility to uncover latent patterns across many types of domains.

Furthermore, graphs are good at encoding multi-hop relationships and context dependencies, which are essential to detect advanced attacks such as APTs or lateral movement in networks [5]. For example, an APT may consist of a chain of undetectable interactions among various devices, and a graph can follow multi-hop traces and identify anomalies that cannot be caught by single device analyses. Through these relation comprehension, graph-based methods are a strong tool in detecting sophisticated and covert cyber-attacks [1].

14.2.3 Significant Security Applications

Graph anomaly detection has been utilized throughout many security domains of utmost significance, and its value and feasibility are now secured. One example of its use is intrusion detection, in which graphs represent network traffic in an attempt to identify malicious intent or unauthorized activity. For example, a communications graph of network communications can show suspicious interconnectivity typical of an APT, as illustrated in research where lateral motion in enterprise networks was found [6]. Another field that is of special significance is fraud detection, i.e., e.g., in financial networks, where transaction graphs can show unusual behavior, i.e., money laundering or account takeovers. By studying network topologies as transaction networks, graph-based methods can detect clusters of fake accounts or suspicious transfer chains [7]. Network anomaly detection aims at detecting macroscopic attacks such as botnets or DDoS attacks by using network topologies and traffic modeling. Temporal graphs, for instance, can be used to detect sudden changes of network activity, such as large edge weights in a DDoS attack [8]. These kinds of applications illustrate the utility of graph-based solutions to wide ranges of security scenarios, taking advantage of their ability to express complex relationships and detect anomalies overlooked by traditional techniques.

14.3 Comprehensive Analysis of Key Techniques in Graph-Based Anomaly Detection

Graph anomaly detection is a leading area of cybersecurity, taking advantage of the relational character of information in order to identify sophisticated attacks that circumvent conventional ways of detection. By viewing entities (e.g., users, devices, transactions) as nodes and their relations like communications or transfers as edges, these methods roll out intricate

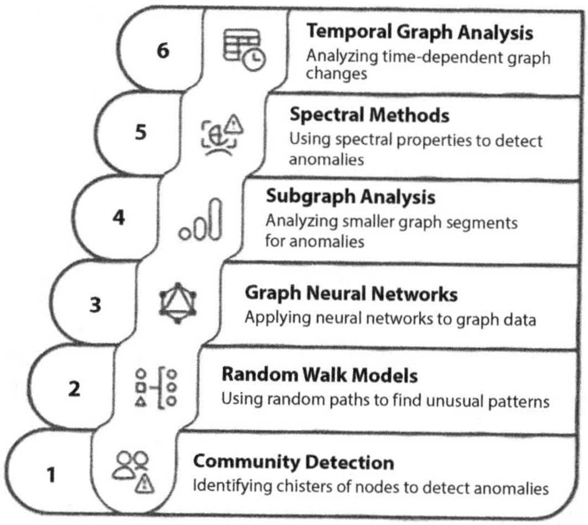

Figure 14.2 A hierarchical flowchart or tree diagram categorizing key graph-based anomaly detection techniques (e.g., community detection, random walk models, graph neural networks, subgraph analysis, spectral methods, temporal graph analysis). Each branch lists specific algorithms (e.g., Louvain, PageRank, GCN) and their applications (e.g., intrusion detection, fraud detection).

patterns needed to identify anomalies in high-risk areas. The chapter delves extensively into the prominent graph-based anomaly detection approaches, investigating community discovery, random walk models, GNNs, in addition to other approaches like subgraph analysis, spectral approaches, and temporal graph reasoning. The discussion entails algorithms and applications for all the methods as well as their merits and demerits, with support from authoritative sources that ensure a robust framework for researchers and practitioners (see Figure 14.2).

14.3.1 Community Detection

Community detection has become a central analysis technique in the realm of graph-based anomaly detection, especially in cybersecurity applications. Essentially, community detection algorithms seek to identify clusters or groups in a network where nodes are more tightly connected to one another compared to the remainder of the network. This inherent quality of real-world graphs, including social networks, patterns of computer communications, or money transactions, can be used to identify abnormal or malicious behavior that appears as departures from usual community patterns [1].

The underlying assumption is that anomalies tend to disturb existing topological structures. For example, in a corporate communication network, employees tend to form tight communities across departments or projects. A compromised account or an insider threat could have connectivity that bridges otherwise dissimilar groups, forming anomalous inter-community connections. Detection of these abnormal structural properties must use algorithms able to not only partition the graph well but also identify outlier nodes or edges that do not respect anticipated community boundaries [2].

One of the most popular and used algorithms in this area is the Louvain method. The Louvain algorithm maximizes a measure called modularity, which measures the quality of a partition of a network into communities. High modularity reflects dense connectivity within communities but sparse connectivity between communities. The Louvain algorithm uses a greedy hierarchical method, repeatedly aggregating the nodes into communities and building a new network whose nodes are the communities themselves that have been identified. This is repeated until modularity no longer significantly increases, a multi-level community structure [3]. Because of its efficiency in computation and scalability, Louvain is often used in extremely large-scale networks, including online social media and massive corporate infrastructures.

Another notionally associated method is based on general modularity optimization. Although Louvain constitutes one such implementation, modularity optimization more generally covers a class of algorithms tailored to optimize modularity employing different heuristics or mathematical programming methods. Such algorithms continue to form a basis for community-oriented anomaly detection and are used as a starting point against which more advanced or application-specific methods are measured [3].

Besides modularity-based approaches, there is another type of algorithm that does not need intensive parameter tuning and can identify outlier nodes and edges at the same time as community structures. For instance, AUTOPART, an automated parameter-free graph partitioning method, works by recursively partitioning the graph according to structural characteristics and marking spurious cross-cluster edges as possible anomalies. This double ability of partitioning and outlier detection makes AUTOPART desirable for settings where domain knowledge is not rich or where data distribution might vary dynamically [4].

In addition to modularity-centric and parameter-less methods, structural similarity-based clustering presents a different view. The SCAN (Structural Clustering Algorithm for Networks) algorithm clusters nodes based on their neighborhood similarity and labels nodes explicitly as core points, hubs, or outliers. SCAN's explicit outlier labeling is particularly valuable in cybersecurity use cases needing identification of nodes that serve as bridges between communities—activity commonly demonstrated by compromised accounts, botnet command nodes, or lateral attackers traveling quietly through network segments [5].

A more recent contribution to community detection techniques is the use of matrix decomposition and sparse graph representations in compact form. For example, methods such as GSKELETONCLU utilize low-rank approximations for the discovery of hidden community structures while being resistant to noise and able to process huge datasets with minimal memory. While not as frequently referenced as Louvain or SCAN, these approaches reflect the continued advance of community detection toward increased computational tractability and applicability to streaming or high-speed data [6].

Across real-world cybersecurity deployments, algorithms for community detection fulfill several functions. It is most frequently used for live monitoring of communication within corporate networks, social media analysis for disinformation extraction, and transactional monitoring in financial services for fraud detection. Anomalous edges or nodes revealed by such techniques may initiate further forensic analysis or drive automated threat response systems. In addition, the outcomes of community detection can impact graph neural networks and XAI systems to enhance both accuracy and interpretability for anomaly detection pipelines.

Notwithstanding their practicality, community detection algorithms have limitations. One longstanding problem is the so-called resolution limit, where small groups are likely to be subsumed into larger ones if they do not make a significant contribution to the modularity gain. Such an effect can mask fine-grained anomalies, especially in networks with strongly heterogeneous group sizes. Some solutions have been proposed, such as multi-resolution modularity and local community refinement, but these are still ongoing research topics and areas of discussion within the graph mining community [1, 3].

Another challenge is that real-world networks are dynamic in nature. Most community detection algorithms, including Louvain, work under the assumption that the network is static. In cybersecurity applications, though, the underlying graph continuously changes as connections are made and broken. Community detection will need to be adapted to deal with dynamic graphs, meaning that the algorithms must be capable of updating community structures incrementally without having to compute the partitioning from scratch again. Some methods solve this by applying temporal smoothing or by utilizing streaming graph processing methods, albeit finding a balance between accuracy and computational expense is still an open problem [2, 6]. Lastly, interpretability of identified communities is very important for operational security teams. Algorithmic outputs in the form of cluster IDs or modularity scores can be uninformative to analysts, and therefore, actionable insights need to be embedded in the output. Therefore, there is increasing demand for integrating community detection with visualization and explanation models. Graph visual analytics platforms can enable security analysts to browse community structures in an interactive manner, track unusual connections, and place alerts in the context of the general network structure. Such integrated solutions can streamline the difference between algorithmic raw output and sensible, human-focused decision-making [2].

In short, community detection remains a core element of graph-based anomaly detection for cybersecurity. From basic modularity optimization to structural similarity clustering and low-rank decomposition, these approaches offer various ways to identify underlying structures and deviations reflecting malicious behavior. Continued research aims to overcome their drawbacks in scalability, robustness in adaptive environments, and interpretability so that community detection can continue to be applicable and effective with the increasing sophistication of cyber-attacks.

14.3.2 Random Walk-Based Models

Random walk-based models explore graph topology by traversing from nodes to their neighbors based on node closeness and influence. Random walk-based models come in different varieties. One such variety is PageRank, where ranks nodes depend on their connections, originally formulated for web searching [2]. And there is Personalized PageRank (PPR), a variant that biases walk toward specific nodes, useful for analyzing neighborhoods. SimRank is a structural similarity measure between nodes by their mutual neighbors, useful for anomaly detection [25]. These models are used to detect anomalous paths or behaviors, for example, unanticipated data flows or links in networks. PPR can identify nodes with unexpected closeness to detected malicious nodes and assist in link prediction and anomaly detection in financial or social networks [15]. They are also applied in intrusion detection to detect anomalous traffic (Table 14.1).

A comparative summary of the main graph-based anomaly detection techniques, adapted from [3, 5, 15, 29] and [31] is presented in Tables 14.1 and 14.2. Tables 14.1 and 14.2 summarize the trade-offs among different graph-based anomaly detection methods, underlining their algorithmic

Table 14.1 Summary of key techniques in graph-based anomaly detection.

Technique	Key algorithms/ methods	Applications	Strengths	Limitations
Community Detection	Louvain, AUTOPART, SCAN, GSKELETONCLU	Intrusion detection, malicious cluster detection	Captures structural anomalies, scalable	High computational complexity, parameter tuning
Random Walk-Based Models	PageRank, PPR, SimRank	Anomalous path detection, link prediction	Captures multi-hop relationships, robust	Computationally intensive, sensitive to structure
Graph Neural Networks	GCNs, GATs	Node/edge/graph anomaly detection	Scalable, adaptable to complex data	Overfitting, computational complexity, interpretability
Subgraph Analysis	Infrequent subgraph detection, MDL principle	Fraud detection, software bug detection	Effective for localized anomalies	Computationally expensive, domain-specific knowledge
Spectral Methods	λ-distance, SVD, CUR, CMD, Colibri-S	Event detection, traffic pattern analysis	Detects global anomalies	Sensitive to noise, poor on sparse graphs
Temporal Graph Analysis	SVD, tensor analysis, scan statistics	Dynamic network event detection	Captures temporal anomalies	Noisy data handling, scalability challenges

complexity, feasibility for real-time usage, scalability, and interpretability. This differentiation supports in selecting context-appropriate methods for particular security scenarios.

14.3.3 Heterogeneous Graph Neural Networks in Cybersecurity

In cybersecurity, many real-world graphs are inherently heterogeneous, comprising diverse node and edge types such as users, devices, IP addresses, processes, and files connected through varied interactions like logins, accesses, and file modifications. Traditional GNN models designed for homogeneous graphs struggle to capture such diversity. To address this, Relational Graph Convolutional Networks (R-GCNs) and

Table 14.2 Comparative analysis of graph-based anomaly detection techniques.

Technique	Key algorithms	Computational complexity	Real-time feasibility	Scalability	Interpretability	Deployment readiness	Typical use cases
Community Detection	Louvain, SCAN, AUTOPART	Moderate to High	Moderate	High	Moderate	Moderate	Intrusion detection, malicious cluster detection
Random Walk Models	PageRank, SimRank, PPR	High	Low to Moderate	Moderate	Moderate	Low	Link prediction, anomalous traffic patterns
Graph Neural Networks (GNNs)	GCN, GAT, GraphSAGE	Very High	Low (requires optimization)	Low–Moderate	Low	Low–Moderate	Node/edge anomaly detection in dynamic networks
Subgraph Analysis	MDL, infrequent subgraph mining	High	Low	Low	High	Low	Localized fraud or malware detection

(Continued)

Table 14.2 Comparative analysis of graph-based anomaly detection techniques. (*Continued*)

Technique	Key algorithms	Computational complexity	Real-time feasibility	Scalability	Interpretability	Deployment readiness	Typical use cases
Spectral Methods	SVD, λ-distance, CUR	Moderate to High	Moderate	Moderate	Moderate	Moderate	Global anomaly detection, signal anomalies
Temporal Graphs	Time-based tensor analysis, SVD	Very High	Low	Low	Moderate	Low	Botnet/DDoS detection, dynamic intrusion patterns

Heterogeneous Graph Neural Networks (HetGNNs) have been developed, which are capable of learning from mixed-type structures by incorporating relation-specific message passing and type-aware aggregation strategies. These models are particularly valuable in key cybersecurity applications. For instance, in insider threat detection, heterogeneous graphs can represent user-device-application relationships to identify abnormal behavior chains. In malware forensics, process-file-registry interaction graphs enable the tracking of malicious sequences, while network access control can be enhanced by learning policies from user-IP-port-service access patterns. Despite their strengths, these models face challenges like increased computational complexity, the requirement for labeled heterogeneous data, and difficulties in scaling across large enterprise networks. Nevertheless, their potential to model complex, multi-relational data makes them highly suitable for the modern security environments [5].

14.3.4 Temporal and Streaming Graph Neural Networks in Cybersecurity

Many cybersecurity anomalies are inherently temporal, such as coordinated DDoS attacks, sudden phishing surges, or credential misuse over time. Static graph models fail to capture evolving behavioral patterns, making temporal and streaming graph neural networks (TGNNs) critical for real-time threat detection [15, 17]. EvolveGCN adapts traditional GCNs by evolving their weights over time using recurrent neural architectures. This enables the model to monitor changing threat behavior, e.g., lateral movement in enterprise networks or multi-stage APTs. TGAT (Temporal Graph Attention Networks) has continuous-time dynamics and attention mechanism that allow it to capture fine-grained temporal relationships among events—perfect for encoding time-stamped logins of users or process runs. DySAT (Dynamic Self-Attention Network) generalizes self-attention mechanisms over both structure and time, aiding in the detection of slowly evolving threats like stealthy insider attacks. These models provide improved ability in learning from streaming logs, real-time network traffic, and dynamic interaction graphs. For example, TGAT has demonstrated good performance on recognizing uncommon login patterns, while DySAT performs well in learning long-term dependencies among changing graph snapshots, e.g., botnet consolidation. Latency and scalability continue to be challenges, especially in high-throughput environments like enterprise networks. Support for optimized implementations, online learning, and edge computing is essential to production system deployment.

14.4 In-Depth Analysis of Case Studies in Graph-Based Anomaly Detection for Cybersecurity

Graph-based anomaly detection is a fresh approach to cybersecurity that leverages the relational data model to identify advanced threats beyond the capability of current means. With nodes being mapped with entities such as users, devices, or transactions and their relationship being represented as edges, graph-based detection techniques unmask intricate patterns of significance in anomaly detection within high-risk scenarios. This chapter offers comprehensive case studies of graph-based technique usage in intrusion detection, fraud identification, and abnormality discovery in networks with emphasis on real-world illustrations and learnings derived. Case studies are centered on single techniques, their efficacy, and lessons learned through experience, with milestone references added to ensure complete knowledge for researchers and practitioners.

14.4.1 Intrusion Detection

14.4.1.1 APT Detection in Enterprise Networks

Advanced Persistent Threats (APTs) rank among the most difficult cyber-attacks, defined as stealthy, persistent attacks meant to exfiltrate sensitive data or disrupt business processes in enterprise networks.

Graph-based anomaly detection is an effective approach by converting network traffic into graphs, where actions like data transfers are edges and entities like users or IP addresses are nodes. The relational structure allows for the detection of complex anomalies, e.g., data exfiltration or lateral movement typical of APTs. Work [25] constructs a hierarchical APT detection system using attention-based Graph Neural Networks (GNNs). The approach builds provenance graphs from system logs and network activity and tracks network activity as well as system calls.

With attention-based GNNs, the approach identifies anomalous subgraphs indicative of APT patterns based on hierarchical structure and marks global and local anomalies. Based on experimental real-world data sets, the approach attained sufficient accuracy in business applications, attesting to the dominance of GNNs in modeling sophisticated attack patterns. Another important contribution, introducing a graph embedding-based network anomaly detection approach. This method transforms network traffic to first-order and second-order graphs to learn latent representations without feature engineering. Single-host views are emphasized by the first-order graph, and global network interactions are captured by the second-order graph. On datasets such as CIDDS-001 and CICIDS 2017, the technique achieved higher detection accuracy for several attacks, including APT-like attacks, *via* utilization of relational patterns in network traffic [16]. Such case studies illustrate the effectiveness of graph-based techniques to model complex network behavior, and thus they are valuable to APT detection in enterprise networks.

14.4.2 Fraud Detection

Financial transaction fraud, such as money laundering and credit card fraud, has serious implications to both financial institutions and consumers and costs billions of dollars every year can be observed as an example. Graph-based anomaly detection is highly effective in this domain because financial transactions are often network-like with accounts as nodes and transactions as edges so that anomalous patterns can be identified through relational analysis. Running review of 33 papers [22] discusses the application of Graph Neural Networks (GNNs) to detect fraud in finance. The review states that GNNs are best positioned to learn customer, merchant, and transaction interactions, as well as detect anomalies that rule-based or statistical models do not capture. The research classifies GNNs into convolutional-based, attention-based, metapath-based, and transformer-based architectures, based on their usability in the research of homogeneous, heterogeneous, static, and dynamic graphs. For example, GNNs can recognize unusual patterns of transactions, like immediate, high-value transfers between accounts that reflect money laundering [22]. The survey also points out areas of research gaps, like the lack of enough emphasis on edge-level and graph-level anomalies and proposes directions for future work. In reality, large banks like Visa and PayPal have already used graph-based systems to fight fraud. PayPal utilizes graph-based platforms to look for anomaly pattern searches in transaction networks by checking non-normal connectivity (PayPal Fraud Detection).

Visa also uses machine learning techniques on transaction graphs in an effort to find patterns of fraud in real time for the sake of further protecting consumers. Such real-world applications show the flexibility and scalability of graph-based approaches in addressing vast financial transaction data sets, greatly enhancing fraud detection potential.

14.4.3 Network Anomaly Detection

14.4.3.1 Botnet or DDoS Attack Detection

Botnets and Distributed Denial-of-Service attacks are critical network security risks, typically composed of coordinated attacks by numerous compromised systems to disrupt service or siphon information. Temporal graph analysis, which monitors the evolution of network activity over time, plays an important role in detecting such dynamic threats through the identification of patterns like coordinated communication or traffic surges. A paper [17] presents a temporal graph analysis method *via* Long Short-Term Memory (LSTM) networks for botnet detection. The method formulates network traffic as a graph time series where temporal relationships characteristic of botnet activity, e.g., synchronized communication with command and control servers, are learned.

Priced against the CTU-13 dataset, the technique achieved a record 96.2% accuracy, illustrating the power of the method in botnet-infected host detection. How readily the method can be implemented in real-time makes it especially useful for use in network security systems [17]. Graph-based anomaly detection has proven to be a ground breaking technique in cyber security, leveraging the relational structure of data to identify sophisticated threats not possible to trap with traditional means. Through the representation of objects such as users, devices, or transactions as nodes and their interactions as edges, these methods expose detailed patterns that are necessary for anomaly detection in high-risk domains. This chapter presents extensive case studies of graph-based methods for intrusion detection, fraud detection, and anomaly detection in networks drawing on real-world experience and lessons learned therefrom. Each of the case studies focuses on specific techniques, performance, and lessons learned through experience augmented by seminal references to provide up-to-date, current knowledge to researchers and practitioners.

Graph-based anomaly detection is an effective solution through the use of graphs to represent network traffic, with nodes such as IP addresses or users and edges such as data transfer. Such a relationship-based model enables identifying subtle anomalies such as lateral movement or data exfiltration typical of APTs. A research paper [25] suggests a hierarchical approach to APT detection *via* attention-based Graph Neural Networks (GNNs). The approach builds provenance graphs from network traffic and system logs, tracking both system call and network activity. Using attention-based GNNs, the method detects abnormal subgraphs representing APT patterns by hierarchical structure and flags global and local anomalies.

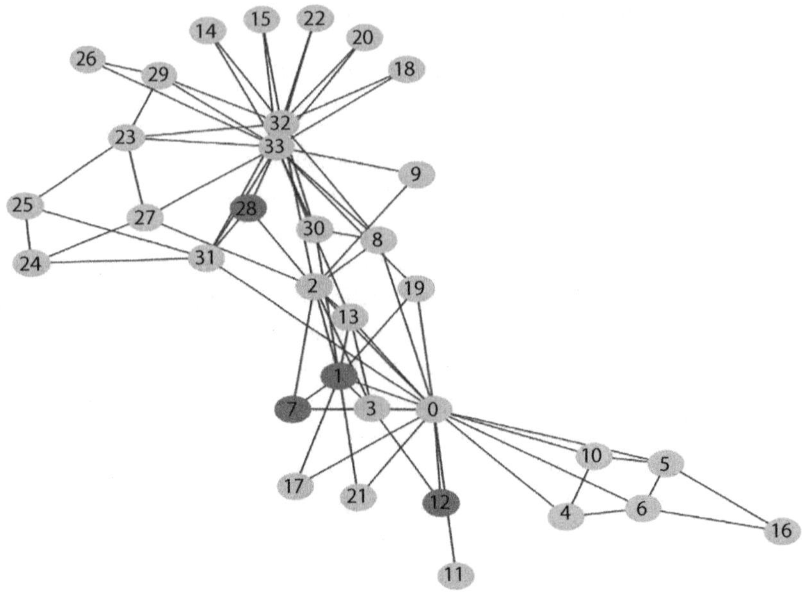

Figure 14.3 Graph-based intrusion detection result: suspicious nodes highlighted in red.

From tested data sets in real-world settings, the method achieved satisfactory accuracy in business environments, highlighting the strength of GNNs in modeling complex attack patterns.

Another significant contribution, who introduce a graph embedding-based approach to network anomaly detection. This method translates network traffic into first-order graphs and second-order graphs to learn latent features without feature engineering. The first-order graph deals with single-host views, whereas the second-order graph captures global network interactions. Quantified on datasets such as CIDDS-001 and CICIDS 2017, the approach attained higher detection accuracy for different attacks, such as APT-like attacks, by taking advantage of relational patterns in network traffic [16].

Case studies like these prove the effectiveness of graph-based approaches to represent complex network behavior, and therefore they are valuable to APT detection in enterprise environments. 4.2 Fraud Detection Financial transaction fraud, including money laundering and credit card fraud, is a serious danger to consumers as well as financial institutions and runs into billions of dollars every year can be researched as an example. Graph-based anomaly detection works extremely well in this domain because financial transactions are often network-like with the accounts as vertices and transactions as edges such that anomalous patterns can be unearthed *via* relational analysis. The review of 33 papers [25] discusses the use of Graph Neural Networks (GNNs) for fraud detection in finance. The evaluation shows that GNNs are particularly capable of capturing interactions among customers, merchants, and transactions, and between those that statistical or rule-based models fail to see. The research classifies the GNNs into convolutional-based, attention-based, metapath-based, and transformer-based architectures, noting how they are applicable in the analysis of homogeneous, heterogeneous, static, and dynamic graphs.

For example, GNNs can detect abnormal patterns of transactions, like immediate, high-amount transfers between accounts reflecting money laundering [25]. The survey also points out the gaps in research, like the lack of emphasis on edge-level and graph-level anomaly detection, and indicates directions for future research. In real life, large banks like PayPal and Visa have already used graph-based systems to fight fraud. PayPal utilizes graph-based frameworks to hunt for anomaly patterns in transaction networks by tracking non-typical connectivity (PayPal Fraud Detection).

Visa similarly uses machine learning algorithms on transaction graphs to recognize real-time fraud patterns so as to further protect consumers. Such real-world use cases illustrate the scalability and flexibility of graph-based techniques in handling enormous financial transaction data, greatly enhancing fraud detection capability.

An enterprise network represented as a graph is used to demonstrate the outcome of an intrusion detection algorithm. In the Figure 14.4, nodes stand in for things like hosts, servers, or user accounts, and edges show how they communicate or interact with one another. The detection technique flags the nodes indicated in red as possible intrusions or compromised entities; these are outliers or high-centrality nodes that exhibit a suspicious deviation from typical connectivity patterns. Light-blue nodes are regarded as benign and typical of network activity. This visualization shows (refer to Figure 14.3) how graph-based methods, like PageRank-based ranking or community detection, can be used to find suspicious nodes by taking advantage of the relational context instead of examining individual records separately. For example, an attacker who moves laterally might make strange connections that unexpectedly increase their centrality and make them stand out in this network structure. In order to help with root cause analysis and mitigation planning, security analysts can visually trace which nodes are suspicious and how they connect to legitimate parts of the network. This type of representation is very interpretable.

14.4.4 Network Anomaly Detection

14.4.4.1 Detection of Botnets or DDoS Attacks

Botnets and Distributed Denial-of-Service (DDoS) attacks are critical network security threats, typically consisting of coordinated assaults from numerous compromised devices in order to disrupt services or steal information. Temporal graph analysis, which monitors the evolution of network activity over time, plays an important role in the identification of such dynamic threats *via* detection of patterns like coordinated communications or traffic bursts. A paper [17] suggests an LSTM network-based temporal graph analysis method for botnet detection.

The method views network traffic as a series of time-dependent graphs, where temporal patterns characteristic of botnet activity, i.e., synchronized interaction with command and control servers, are learned. Priced against the CTU-13 dataset, the method achieved the incredible 96.2% accuracy, showing how robust the method is in botnet-infected host detection. How readily the method can be used in real-time lends itself to being especially useful for use in network security systems [17]. Yet another timely contribution by [18] uses graph-based feature clustering in botnet detection. Grounded on topological node feature analysis in terms of degree, centrality, and clustering coefficients, the method isolates bots in tiny clusters from regular nodes. The technique can be useful in the detection of botnets with

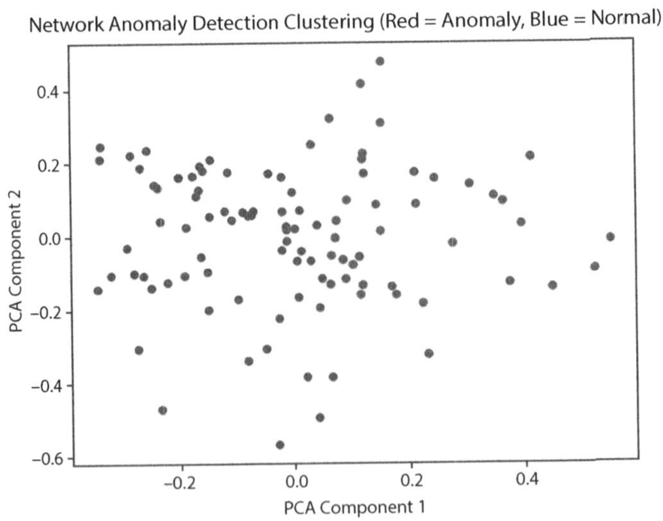

Figure 14.4 PCA-based network traffic clustering for anomaly detection.

centralized communication patterns, common in most actual situations. For DDoS attacks, a graph-based deep learning technique that identifies sharp peaks or unusual patterns characteristic of DDoS attacks with high accuracy has been proposed. For detecting DDoS attacks, a graph-based deep learning approach with high accuracy on detecting sudden spikes or unusual patterns characteristic of DDoS attacks has been proposed [18].

These approaches utilize temporal graph analysis to identify the time-evolving nature of network attacks, making them critical in counter-acting botnets and DDoS attacks [19]. Another applicable contribution by [18] utilizes graph-based feature clustering for botnet detection. Based on the observation of topological node properties such as degree, centrality, and clustering coefficients, the method separates bots that occur in small clusters from normal nodes. The method can be effective for the detection of centralized communication-based botnets, which are the most common scenarios in real applications. Such methods depend on the application of temporal graph analysis to detect the dynamic behavior of network threats, making them essential in combating botnets and DDoS attacks [19].

Results of an unsupervised learning and dimensionality reduction clustering-based network anomaly detection procedure are displayed in the scatter plot (Figure 14.4). To convert each point into a reduced-dimensional space that displays a network traffic observation, Principal Component Analysis (PCA) is utilized. We can see latent structures by using the two axes (PCA Component 1 and PCA Component 2), which preserve the greatest variance in the data. In the plot, red dots represent data points that the clustering algorithm (such as K-Means or DBSCAN) has identified as outliers, while blue dots represent typical network activity. As demonstrated by the obvious results, the unsupervised model successfully clustered together similar traffic patterns and distinguished outliers that might be indicative of malicious activity like botnet traffic or distributed denial-of-service (DDoS) attacks.

14.4.5 Case Study Analysis

The case studies reveal the advantages and limitations of using graph-based anomaly detection systems in cybersecurity, presenting significant findings for researchers and practitioners.

14.4.5.1 Similar Success Factors

a). Effective Modeling of Relationships: Graph-based techniques are particularly effective in modeling complex relationships among entities, facilitating the detection of sophisticated threats like APTs, fraud, and botnets through relational pattern analysis that traditional methods are unable to detect.

b). Deep Machine Learning Techniques: GNN, LSTM networks, and graph embedding have significantly improved detection accuracy through learning from graph structure and temporal dynamics, as evidenced by the better performance of methods like those [20, 21].

c). Multi-Data Source Integration: Consolidating heterogeneous data types, such as network traffic and transactional records, into unifying graph abstractions enhances detection performance, as noted in the financial fraud detection platforms employed by PayPal and Visa.

d). Real-time or Near-real-time Detection: The majority of graph-based solutions are designed for real-time or near-real-time computation, as required for timely reaction against threats, as is evident with the LSTM-based botnet detection model [21].

14.4.5.2 Practical Challenges

a. Scalability: Computation can be expensive in managing big graphs, common in enterprise networks and financial networks, needing better algorithms or distributed computing [22].
b. Data Quality and Preprocessing: Clean, high-quality data is paramount for efficient detection because noisy or dirty data will degrade performance, a limitation identified in graph embedding techniques [23].
c. Interpretability: Advanced models like GNNs are black-box in nature, and therefore it is difficult to interpret anomalies found, which could deter trust and adoption in security operations [20, 24].
d. Evolving Threats: Cyber threats evolve continuously, and thus there is a requirement for constant model updates or retraining in order to remain effective, as discussed in botnet detection literature [25].
e. False Positives and Negatives: Reducing false positives (genuine transactions flagged as anomalies) and false negatives (missed anomalies) remains an issue that requires

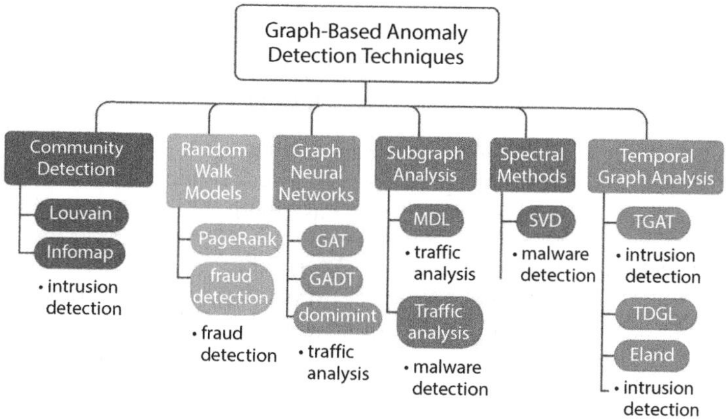

Figure 14.5 Hierarchical classification of graph-based anomaly detection techniques for cybersecurity.

thoughtful tuning and testing, as with the examination of DDoS detection methods [25].

These lessons convey the importance of scalability, interpretability, and adaptability in developing and deploying graph-based anomaly detection systems. Organizations can improve their cybersecurity system against advanced and dynamic threats by achieving success factors and avoiding challenges.

Table 14.3 provides a snapshot of the most important case studies, their examples, methods, techniques, and references, in a way that gives quick insight to practitioners. The case studies highlight how graph-based anomaly detection is an effective tool for dealing with intricate cybersecurity issues, with usage varying from APT detection in enterprise networks to fraud prevention in banking systems and botnet detection in network security. Studies indicate that these approaches are very effective because they can represent relational and temporal data and utilize sophisticated methods such as GNNs and LSTM networks. Scalability issues, interpretability, and the requirement to evolve to keep pace with changing threats are some of the key issues that also call for further development and research. Through the resolution of these challenges and leveraging the success factors, graph-based anomaly detection can further improve security, offering strong protection against advanced threats (refer to Figure 14.5).

Table 14.3 Summary of case studies in graph-based anomaly detection.

Domain	Example	Graph-based approach	Key techniques
Intrusion Detection [19]	Detecting APTs in enterprise networks	Modeling network traffic as graphs	Attention-based GNNs, graph embedding
Fraud Detection [21]	Financial transaction fraud	Analyzing transaction networks	GNNs (convolutional, attention-based)
Network Anomaly Detection [18]	Identifying botnets or DDoS attacks	Temporal graph analysis	LSTM networks, feature clustering

14.5 Challenges in Graph-Based Anomaly Detection

Graph-based anomaly detection has been an effective technique for identifying advanced cyber-attacks by exploiting the relational character of information. Its applicability to real-world cybersecurity, though, is hindered by significant challenges that may render it less useful in practical cybersecurity applications. This subsection addresses the main graph-based anomaly detection challenges, including scalability, handling dynamic and temporal graphs, interpretability, data quality and heterogeneity, and metrics. Each challenge is analyzed within the context of cybersecurity, highlighting its impact and limitations refer to Figure 14.6. The discussion relies on current research and supplemented by reputable sources to provide an exhaustive insight for researchers and practitioners.

The first key challenge is scalability because graph processing for large-scale graphs tends to create computational bottlenecks. Modern-day cybersecurity information, say, enterprise network or financial system data, might comprise millions of edges and nodes, which are computationally intensive for graph processing. For instance, algorithms like Graph Neural Networks (GNNs) and community detection methods require significant resources to calculate node embeddings or maximize modularity over big graphs. This is particularly annoying in real-time applications, where there is a requirement for high processing speeds in a bid to timely recognize threats. In accordance with such a bottleneck, a solution such as graph sampling and distributed processing has been eyed. Graph sampling

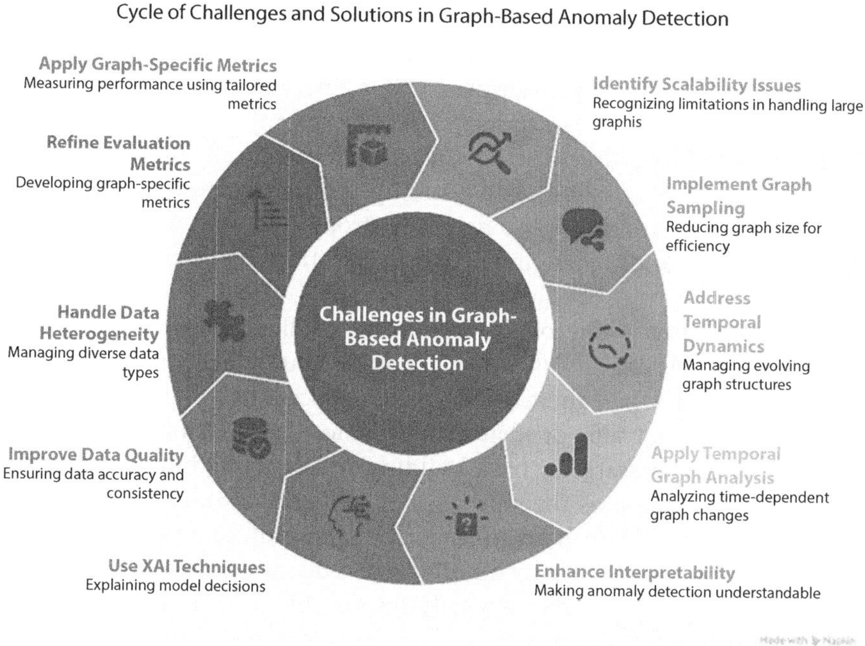

Figure 14.6 Challenges and proposed approaches for graph anomaly detection in cybersecurity.

reduces the size of the graph by using representative nodes and edges while important structural features are retained to avoid over-computation [27]. Distributed computing environments such as Apache Spark or GraphX allow parallel computation among nodes, thus being suited for graph scalability on large scales [25]. The solutions are encouraging when handling huge amounts of data but may require to be optimized specifically in order to have the accuracy-efficiency tradeoff, and their execution might be challenging if the environment is dynamic.

Another essential challenge is modeling temporal and dynamic graphs, critical in capturing the dynamic nature of networks within cybersecurity. Many cyber threats, like botnets or Advanced Persistent Threats (APTs), have time-varying behaviors, and therefore models need to be able to monitor changes in graph structure over time. It is difficult to model evolving networks because continuously updating graph representation while preserving historical context is challenging. Real-time analysis poses additional challenges since temporal graph algorithms must process stream data with low latency to detect anomalies in real time. For example, temporal graph analysis methods such as dynamic tensor decomposition struggle to maintain computational efficiency while guaranteeing the accuracy necessary for real-time detection [13] (Figure 14.5).

Incremental learning strategies, which incrementally update models with new knowledge without recalculation of the whole graph, have been proposed to address these issues, but they typically do not perform well in maintaining stability and avoiding drift in dynamic environments [29]. These issues indicate the need for strong temporal models capable of adapting to rapidly changing network conditions while ensuring timely detection.

Interpretability is one of the significant challenges, particularly with advanced models like GNNs, which are often accused of being black boxes. In cybersecurity, where there are high-stakes decisions being taken, such as blocking network traffic or identifying financial transactions, explainable outcomes are necessary for trust building and enabling human intervention. GNNs, as good as they are at node embedding learning for anomaly detection, generate hard-to-understand output, and hence it is not straightforward to know why a node or an edge was marked anomalous. Such lack of interpretability is partly responsible for the resistance from security analysts who require transparent explanations of actions. The need for explainable outputs has spurred work on explainable AI (XAI) techniques, such as SHAP (Shapley Additive Explanations) and LIME (Local Interpretable Model-agnostic Explanations), that attempt to provide explanations for model decisions [30]. It is not easy to utilize these approaches for graph-based models due to the relational character of graphs, and existing techniques will fail to provide intuitive explanations in security applications [31]. Constructing interpretable graph-based models remains an important task towards enabling their usage in production settings.

Heterogeneity and quality of the data make the complexity worse due to the nature of cybersecurity usage cases, which usually require integrating various heterogeneous data sources like network logs, transaction records, and user interactions. Integrating these heterogeneous data into an integrated graph representation is difficult as various data types may utilize different formats, granularities, and semantics. Network logs may include packet-level data, for instance, whereas transactional records monitor account-level activity and therefore require sophisticated preprocessing to align them. Cybersecurity data is also frequently noisy or incomplete, for instance, with missing edges, erroneous labels, or conflicting timestamps, which can damage the performance of graph-based models. Preprocessing noisy or missing data requires robust preprocessing techniques, e.g., data imputation or noisy algorithms, but these have potential biases or can reduce detection rates [1]. Various studies have also proposed methods such as graph regularization to reverse the impact of noise, but still, it is a continuous problem for real-world implementation to fuse heterogeneous and imperfect sources [5].

Evaluation measures are a last obstacle, as classical measures like precision, recall, and F1-score cannot measure the relational character of graph-structured data. In graph anomaly detection, anomalies may be structural deviations, e.g., new subgraphs or outlier connectivity patterns that are not very well represented by point-based measures. For example, a high precision value might not show the detection of superior structural anomalies, such as a botnet's communication structure. There is an evident requirement for graph-specific metrics, such as structural anomaly scores or subgraph similarity scores that examine the quality of the identified anomalies in graph topology. Existing work has proposed measures like the Graph Edit Distance or modularity-based scores to address these challenges, but their usage is restricted and they have a tendency to require domain-specific tuning [29]. Developing standardized test suites that maintain the unique nature of graph-based anomalies is critical to benchmark and compare well.

14.6 Proposed Solutions and Hybrid Frameworks

Graph-based anomaly detection is a robust paradigm for fighting sophisticated cyber-attacks but is adversely afflicted in real-life with scalability, interpretability, and evaluation problems. Motivated by these challenges, the following section presents solutions and hybrid models to enhance the utility and interpretability of graph-based methods to cybersecurity. It presents the integration of graph-based methods with traditional methods, improved scalability, enhanced interpretability, and robust evaluation models. Later, the chapter discusses avenues for research in the future such as deep learning integration and federated learning in order to drive the field and make it adaptive to emerging security needs. Arguments are supported by recent studies and supplemented with credible sources to provide a current resource to academics and practitioners.

The integration of graph-based approaches with rule-based systems and machine learning is a promising means to enhance the detection performance and offer transparency. The integration of graph-based approaches and machine learning methods like Support Vector Machines (SVM) or Random Forests combines both paradigms' merits. For instance, graph models are more apt to capture relational patterns, while SVMs are appropriate for feature classification in high dimensions and thus enable hybrid models to detect structural as well as feature-based anomalies [6] present an example of the success of combining graph embeddings with Random Forests for the identification of insider threats in achieving improved accuracy with the addition of a mixture of topological features and behavior attributes. Similarly, rule-based systems ensure transparency by providing human-interpretable, explicit reasoning for anomaly detection that is most crucial in security applications where the choice must be explained. For example, rule-based systems can be utilized to filter out and validate anomalies detected by graph-based models in order to minimize false positives and boost trust among security professionals [5]. These hybrid techniques combine the relational representation of graphs with interpretability and robustness of traditional techniques to be usable in operational settings.

Scalability continues to be an enduring problem for graph-based anomaly detection, especially for dealing with big graphs in real-time cybersecurity use cases. To overcome computational bottlenecks, graph partitioning and parallelization have been effective solutions. Graph partitioning splits massive graphs into small subgraphs to allow distributed computation across clusters or nodes. Graph computations such as Apache GraphX and Pregel provide parallel graph computation with the benefit of being able to process data at very high speeds for large data [28]. Other methods, including approximate algorithms through graph sampling or sketching, provide scalable solutions by minimizing computational cost at the expense of maintaining critical structural properties. For example, [3] introduce graph sampling methods maintaining characteristic network features allowing for efficient processing of large graphs. Scalability improvements play a critical role in applying graph methods within enterprise networks or financial networks, where millions of nodes and edges are typical, allowing for real-time detection without compromising accuracy.

Enhancing interpretability is critical when applying graph-based anomaly detection, especially in sophisticated models such as Graph Neural Networks (GNNs), which are typically criticized as being black-box-like. Explanation is facilitated in model decisions by Explainable AI (XAI) methods like SHAP (Shapley Additive Explanations) and GNN Explainer, where model decisions are informed by interpreting the most significant nodes, edges, or subgraphs responsible for an anomaly score. [30] also offer GNN Explainer, which provides short explanations of GNN predictions so that analysts can straightforwardly see why a node was identified as anomalous. Visualization software also makes interpretability better by presenting graph anomalies in comprehensible formats, e.g., force-directed layouts or heat maps that identify unusual connections or clusters. Visualization programs like Gephi or Cytoscape, for instance, allow security teams to explore graph structures interactively, facilitating quick identification of malicious patterns [27]. These interpretability advances close the loop between good models and practice deployment, building trust and facilitating human-in-the-loop validation in cyber operations.

Building strong evaluation models is essential to measure the performance of graph-based anomaly detection models since classical measures such as recall and precision tend to lose relational context. Developing metrics that are topo-sensitive, for instance, structural anomaly scores or subgraph similarity metrics, gives a better assessment of the detected anomalies. As an illustration, [29] recommend DeltaCon, a graph similarity metric that is based on structural differences, that can be extended to evaluate anomaly detection performance. Comparing graph-based models against benchmark test sets, for instance, DARPA intrusion detection benchmarks or financial networks of transactions, provides a reproducible basis from which to draw comparisons. A research study [3] highlights the necessity of graph-specific benchmarks, necessitating datasets corresponding to real-world cybersecurity settings. These frameworks enable researchers to critically evaluate model performance, establish vulnerabilities, and guide enhancements in detection accuracy and dependability.

In the future, deep learning incorporation has great potential to improve graph-based anomaly detection. Sophisticated GNN architectures, including Graph Attention Networks (GATs) and Graph Transformers, improve the capacity to model intricate graph structures with the integration of attention mechanisms as well as long-range dependencies. Various Studies have investigated self-supervised GNNs for high-performance anomaly detection through pretraining on graph structure prior to fine-tuning for a task at hand. Such developments are anticipated to further increase the scalability and precision of graph-based techniques, making them more adaptable to dealing with a variety of security issues [7].

Federated learning represents another innovative trajectory, supporting privacy-preserving graph analysis within distributed security architectures. Through enabling various organizations to work together to train models without sharing sensitive information, federated learning mitigates privacy problems in cybersecurity solutions, including detecting financial fraud or threat intelligence between organizations. Introduce a federated GNN approach that federates localized graph models while asserting data confidentiality and demonstrates its effectiveness in distributed network analysis [31]. Distributed security system applications, i.e., cooperative botnet detection across networks, illustrate the potential of federated learning for scaling graph-based anomaly detection while preserving confidentiality. These guidelines require sustained research in deep learning and privacy-preservation techniques to keep graph-based methods in the forefront of cybersecurity innovation.

14.7 Conclusion

Graph-based anomaly detection is a cutting-edge strategy in cybersecurity with unmatched ability to withstand the growing strength of cyber-attacks. In the final section of this chapter, the main conclusions of the chapter are synthesized to differentiate strengths of graph-based techniques, existing challenges, and solutions suggested. It deals with implications for practitioners and academics with recommendations for uptake and highlighting multi-disciplinary participation. Lastly, the appeal for action invites the community to resolve unresolved problems and propel innovation in cybersecurity using graph-based methods. The dialogue concludes with a future vision of the graph-based anomaly detection horizon, enumerating future directions and prospects for expanding its contribution to securing digital systems.

The main advantage of graph-based anomaly detection is its power to model and analyze relational data and extract intricate patterns that other methods are not able to identify. By modeling entities like users, devices, or transactions as nodes and interactions as edges, graph-based methods are particularly effective at identifying intricate threats like Advanced Persistent Threats (APTs), financial fraud, and botnets. Techniques such as community detection, random walk models, and Graph Neural Networks (GNNs) leverage the intrinsic structure of graphs to extract multi-hop relations and contextual dependency, enabling them to identify sub-surface-level anomalies in network traffic, transactional networks, and dynamic networks [3]. Such methods have been found to be very precise in real-world application, as shown by case studies of enterprise networks and banks, where they excel over conventional methods by combining different sources of information into homogenized forms [20]. However, still there are a few issues to be addressed like the scalability issue when dealing with large graphs, modeling dynamic and time-evolving graphs, interpretability concerns in complex models like GNNs, data quality and heterogeneity, and restriction of conventional metrics. Potential solutions such as graph sampling, distributed computing, XAI techniques, and graph measures present possible paths towards solving these problems but require cautious planning to ensure maximum efficiency, accuracy, and transparency [27, 30].

To implement graph-based anomaly detection methods is to step into a complex universe of technical and operational considerations for researchers and practitioners. Adoption guidelines include identifying the methods as on per the requirement of the specific security context, i.e., applying community detection in identifying malicious communities or temporal graph analysis in dealing with dynamic attacks such as DDoS attacks. Practitioners need to highlight hybrid methods combining graph-based methods with traditional machine learning or rule-based approaches to enhance interpretability and robustness, as demonstrated by successful syntheses with Random Forests and rule-based [4, 5]. Interdisciplinary techniques are crucial, drawing on insights from graph theory, machine learning, information assurance, and human-computer interaction to design solutions that are not only technically viable but also pragmatically rational. For instance, data science and security analyst alignment can ensure visualization software and XAI techniques are aligned with operational teams' needs so that human-in-the-loop verification is ensured [31]. Researchers need to focus on developing standardized benchmarks and datasets for graph-based anomaly detection in order to enable rigorous model evaluation and comparison across a range of settings [3]. These inter-disciplinary activities will result in the adoption of graph-based methods, which will enhance their effectiveness in applied cybersecurity application.

The plea of the cybersecurity group is to address the open difficulties and drive improvement with graph-based techniques. Scalability, interpretability, and processing heterogeneous data are common issues that require a collaborative research agenda to yield efficient algorithms, explainable models, and robust preprocessing pipelines. For example, improving scalable systems for graph computation and exploring self-supervised learning in data constrained environments can pointedly enhance the deployment feasibility [28]. The community also needs to work on creating graph-specific metrics and standards for evaluating model performance to achieve reasonable and significant evaluations. By the encouragement of cooperation between industry, administration, and academe, the field can accelerate research translation into functional solutions, for instance, privacy-preserving federated learning frameworks for cross-organizational threat detection. Cybersecurity by graph-based innovation necessitates dedication to solve such challenges so that graph-based anomaly detection remains at the front of protecting digital infrastructures from growing threats. The use of graph-based anomaly detection is on the cusp of widespread expansion with the prospects for deep learning going forward, federated learning, and cross-discipline methods. State-of-the-art GNN structures, such as Graph

Transformers, are going to promote modeling intricate graph structure and improve detection accuracy for dynamic and heterogeneous networks. Self-supervised learning will also play an important role in identifying the problem of data scarcity, making machine learning models to learn robust representations from unlabelled graph data, which is particularly beneficial in cybersecurity since labeled anomalies are scantily available. Federated learning offers a path to privacy-preserving analysis and enables organizations to share the threat detection burden without compromising sensitive information, with real-world use cases in distributed security systems like global botnet monitoring. The convergence of graph-based methods with other domains, such as natural language processing to analyze text-based threat intelligence or human-computer interaction to design user-friendly visualization interfaces, will also continue to expand their scope. In addition, the growing demand for explainable AI will drive the advancement of explainable graph-based models, building trust and interpretability in high-risk security use cases [31]. As cyber threats continuously transform, the future of graph-based anomaly detection lies in its ability to keep pace with new challenges, take advantage of transdisciplinary expertise, and utilize innovative technologies to safeguard digital ecosystems.

AI disclosure statement: The author acknowledges the use of AI-based tools for minor language editing or grammatical correction. However, all intellectual contributions, analyses, and conclusions presented in this chapter are entirely the author's own.

Bibliography

1. Bondy, J.A. and Murty, U.S.R., *Graph Theory*, Springer, London, U.K, 2008, doi: 10.1007/978-1-84628-970-5.
2. Newman, M.E.J., *Networks: An Introduction*, Oxford Univ. Press, Oxford, U.K, 2010, doi: 10.1093/acprof:oso/9780199206650.001.0001.
3. Akoglu, L., Tong, H., Koutra, D., Graph-based anomaly detection and description: A survey. *Data Min. Knowl. Discovery*, 29, 3, 626–688, May 2015.
4. Eberle, W. and Holder, L., Anomaly detection in data represented as graphs. *Intell. Data Anal.*, 11, 6, 663–689, 2007. doi: 10.3233/IDA-2007-11605.
5. Cai, H., Zheng, V.W., Chang, K.C., A comprehensive survey of graph embedding: Problems, techniques, and applications. *IEEE Trans. Knowl. Data Eng.*, 30, 9, 1616–1637, Sep. 2018. doi: 10.1109/TKDE.2018.2807452.

6. Chen, Y., Nyemba, S., Malin, B., Detecting anomalous insiders in collaborative information systems. *IEEE Trans. Dependable Secure Comput.*, 15, 6, 947–961, Nov.–Dec. 2018. doi: 10.1109/TDSC.2017.2656073.
7. Weber, M., *et al.*, Anti-money laundering in Bitcoin: Experimenting with graph convolutional networks for financial forensics, in: *Proc. 25th ACM SIGKDD Int. Conf. Knowl. Discovery Data Mining*, pp. 3211–3220, 2019, doi: 10.1145/3292500.3331446.
8. Manzoor, E., Milajerdi, S.M., Akoglu, L., Fast memory-efficient anomaly detection in streaming heterogeneous graphs, in: *Proc. 22nd ACM SIGKDD Int. Conf. Knowl. Discovery Data Mining*, pp. 1035–1044, 2016, doi: 10.1145/2939672.2939783.
9. Blondel, V.D., Guillaume, J.-L., Lambiotte, R., Lefebvre, E., Fast unfolding of communities in large networks. *J. Stat. Mech. Theory Exp.*, 2008, 10, 10008, 2008. doi: 10.1088/1742-5468/2008/10/P10008.
10. Newman, M.E.J. and Girvan, M., Finding and evaluating community structure in networks. *Phys. Rev. E*, 69, 2, 026113, 2004. doi: 10.1103/PhysRevE.69.026113.
11. Chakrabarti, D., AutoPart: Parameter-free graph partitioning and outlier detection, in: *Proc. Eur. Conf. Principles Data Mining Knowl. Discovery*, pp. 112–124, 2004, doi: 10.1007/978-3-540-30116-5_12.
12. Xu, X., Yuruk, N., Feng, Z., Schweiger, T.A.J., SCAN: A structural clustering algorithm for networks, in: *Proc. 13th ACM SIGKDD Int. Conf. Knowl. Discovery Data Mining*, pp. 824–833, 2007, doi: 10.1145/1281192.1281280.
13. Sun, J., Xie, Y., Zhang, H., Faloutsos, C., Less is more: Compact matrix decomposition for large sparse graphs, in: *Proc. 17th ACM SIGKDD Int. Conf. Knowl. Discovery Data Mining*, pp. 663–672, 2010, doi: 10.1145/2020408.2020510.
14. Ding, Q., Katenka, N., Barford, P., Kolaczyk, E., Crovella, M., Intrusion as (anti)social communication: Characterization and detection, in: *Proc. 18th ACM SIGKDD Int. Conf. Knowl. Discovery Data Mining*, pp. 886–894, 2012, doi: 10.1145/2339530.2339652.
15. Ndunguru, D.D., Xing, F., Oroni, C.Z., Ani, A., Li, C., STG-LSTM: Spatial-temporal graph-based long short-term memory for vehicle trajectory prediction. *Multimodal Transp.*, 4, 3, 100222, 2025. doi: 10.1016/j.multra.2025.100222.
16. Koutra, D., Kang, U., Vreeken, J., Faloutsos, C., VoG: Summarizing and understanding large graphs, in: *Proc. SIAM Int. Conf. Data Mining*, pp. 91–102, 2012, doi: 10.1137/1.9781611972825.9.
17. Chowdhury, S., Khanzadeh, M., Akula, R., Botnet detection using graph-based feature clustering. *J. Big Data*, 4, 1, 14, 2017. doi: 10.1186/s40537-017-0074-7.
18. R. Viswanathan (2019), Garcia *et al.* (2017), A. Tekeoglu, and C. Kamhoua, A graph-based deep learning approach for DDoS detection in network traffic, in: *Proc. IEEE Conf. Commun. Netw. Security (CNS)*, pp. 1–9, 2021, doi: 10.1109/CNS49108.2021.9519992.

19. Li, Y., Zhao, J., Liu, H., Zhang, L., Yang, Y., Anomaly detection in dynamic graphs *via* learning abnormal diffusions, in: *Proc. 27th ACM SIGKDD Conf. Knowl. Discovery Data Mining*, pp. 165–173, 2021, doi: 10.1145/3447548.3467081.
20. Böhm, F., Menges, F., Pernul, G., Graph-based visual analytics for cyber threat intelligence. *Cybersecur.*, 1, 1, 16, 2018. doi: 10.1186/s42400-018-0017-4.
21. Almotiri, S.H., Elleithy, K., Elleithy, A., Anomaly detection using graph-based deep learning for intrusion detection systems. *IEEE Access*, 11, 13821–13832, 2023. doi: 10.1109/ACCESS.2023.3245671.
22. Zhang, Y., Chen, L., Lei, J., Fraud detection using GCN on transaction graphs. *IEEE Access*, 8, 140329–140339, 2020. doi: 10.1109/ACCESS.2020.3011600.
23. Mir, A.A., Zuhairi, M.F., Musa, S., Graph anomaly detection with graph convolutional networks, in: *Tech. Rep*, Malaysian Institute of Information Technology, Univ, Kuala Lumpur, 2023.
24. Garcia, S., Grill, M., Stiborek, J., Zunino, A., An empirical comparison of botnet detection methods. *Comput. Secur.*, 68, 1–16, Jul. 2017. doi: 10.1016/j.cose.2017.03.005.
25. Jiang, H., Wu, T., Li, Z., Li, W., A GCN-based semi-supervised anomaly detection method for attributed networks. *IEEE Access*, 10, 23155–23166, 2022. doi: 10.1109/ACCESS.2022.3153913.
26. Jin, W., Wang, J., He, J., Gao, J., Graph structure learning for robust graph neural networks, in: *Proc. 26th ACM SIGKDD Int. Conf. Knowl. Discovery Data Mining*, pp. 66–74, 2020, doi: 10.1145/3394486.3403066.
27. Zügner, D. and Günnemann, S., Adversarial attacks on graph neural networks *via* meta learning, in: *Proc. Int. Conf. Learn. Representations (ICLR)*, 2019, [Online]. Available: https://openreview.net/forum?id=Bylnx209YX.
28. Ding, K., Xu, Z. Tong, H., Liu, H., Data augmentation for deep graph learning: A survey. *SIGKDD Explorations Newsletter*, vol. 24, no. 2, pp. 61–77, Dec. 2022, doi: 10.1145/3575637.3575646.
29. Luo, D., Cheng, W., Xu, D., Yu, W., Zong, B. Chen, H., Zhang, X., Parameterized Explainer for Graph Neural Network, in: *Proc. Neural Inf. Process. Syst. (NeurIPS), arXiv:2011.04573*, 2020, doi: 10.48550/arXiv.2011.04573.
30. Fang, J., Zhang, G., Wang, K., Du, W., Duan, Y., Wu, Y., Zimmermann, R., Chu, X., Liang, Y., On regularization for explaining graph neural networks: An information theory perspective, *IEEE Trans. Knowl. Data Eng.*, pp. 1–14, 2024, doi: 10.1109/TKDE.2024.3422328.
31. Gaspar, D., Silva, P., Silva, C., Explainable AI for intrusion detection systems: LIME and SHAP applicability on multi-layer perceptron, *IEEE Access*, vol. 12, pp. 30164–30175, 2024, doi: 10.1109/ACCESS.2024.3368377.

15

Graph-Based Anomaly Detection in Cybersecurity and FinTech: Advancing Threat Intelligence with Graph Neural Networks

Sanjeev Khan[1*], Nutan Pathania[2], Pawan Kumar[2] and Vishal[1]

[1]Department of Data Science and Artificial Intelligence, Himachal Pradesh University, Shimla, Himachal Pradesh, India
[2]Department of Computer Science, Himachal Pradesh University, Shimla, Himachal Pradesh, India

Abstract

The complexity of cyberattacks and financial fraud has seen the need for sophisticated anomaly detection techniques that extend beyond rule-based or statistical methods. Graph-Based Anomaly Detection (GAD) is a revolutionary approach, employing graphs to represent complex relationships within security and financial networks. This chapter delves into the usage of GAD in FinTech and cybersecurity through a focus on implementation-focused methods employing Graph Neural Networks (GNNs) in fraud detection as well as network security. Experiments show that GraphSAGE is effective in managing dynamic security and fraud networks, GCN is stable in structured graph settings, and GCT-GNN has better accuracy using transformer-based attention mechanisms. These results emphasize the promise of GNN-based methods for identifying anomalies not only in cybersecurity but also in anti-money laundering, synthetic identity fraud, and insider trading anomalies. One of the main contributions of the chapter is the application and testing of three state-of-the-art GNN models, GraphSAGE, Graph Convolutional Network (GCN), and Graph Convolutional Transformer-GNN (GCT-GNN), on network intrusion detection based on the UNSW-NB15 dataset. The study illustrates that GraphSAGE performs better in dynamic networks, GCN is more stable under structured graph settings, and GCT-GNN enhances detection accuracy with attention mechanisms.

**Corresponding author*: sanjeevkhan.hpu@gmail.com

J. Ramkumar, Sridaran Rajagopal, B. Suchitra and S. Balamurugan (eds.) Graph Neural Networks: Concepts and Applications, (615–662) © 2026 Scrivener Publishing LLC

Experimental evaluations demonstrate robust performance in metrics such as AUC-ROC, precision, recall, and F1-score. In addition to that, this chapter discusses the challenges in scalability, adversarial attacks, explainability, and real-time fraud detection. Some of the future directions of research involve self-supervised learning, graph contrastive learning, and hybrid GNN models.

***Keywords*:** Graph-based anomaly detection, graph neural networks (GNNs), cybersecurity, financial fraud detection, FinTech security

15.1 Introduction

15.1.1 Motivation: The Rise of Sophisticated Cyber and Financial Fraud

The expanding internet connections between systems have generated an abrupt increase in both digital security attacks and financial fraud sophistication [1]. Digital networks that underpin global critical infrastructure expansion, economic globalization, and growth create an expanding target for attack, which results in advanced threats that prove difficult to detect. The successful detection systems against basic fraud schemes and simple intrusions currently face difficulties in confronting advanced persistent threats (APTs), social engineering attacks, and cybercriminal innovation [2]. Advanced Persistent Threats (APTs) focus on targeting valuable network assets across sustained periods since they operate with stealth within system infrastructure. Ransomware attacks, phishing schemes, and data breaches have become much more targeted and challenging to prevent, with cyber attackers tending to target zero-day vulnerabilities and highly complex attack vectors.

15.1.2 Limitations of Traditional Anomaly Detection (Rule-Based/Statistical)

Anomaly detection methods have served as the standard approach to uncover suspicious activities in both cybersecurity and financial systems for a long period. Rule-based systems of detection use pre-established thresholds and rules defined by experts to identify anomalies based on the comparison of network traffic, financial transactions, or system behavior with known patterns of attacks [3]. Though these techniques have worked well for decades, they are more and more unable to stay in step with the complexity and variability of contemporary attacks. Several weaknesses exist in rule-based systems that prevent them from effectively identifying advanced security threats [2]. These systems need regular updating because they fail to adjust to emerging attack methods and fraudulent approaches. Such systems produce numerous false results where they overlook significant threats or mistakenly identify ordinary actions as dubious activities.

15.1.3 Importance of FinTech and Cybersecurity Convergence

Financial institutions have made digital technology their essential operational base, so FinTech convergence with cybersecurity has developed into a major field of interest. The convergence of IT security and digital banking requires joint solutions that effectively handle the multiple security risks associated with digital finance and online banking [9]. The application of graph-based anomaly detection methods in both domains provides an end-to-end solution to real-time threat detection and mitigation [10]. Financial institutions achieve more precise and faster fraud detection systems through the application of GAD. GAD helps analysts detect unusual financial transactions through evaluation of the relationships between entities within the financial system, such as fraudulent transfers [4]. GAD performs network intrusion and advanced threat detection when it models device-user-network traffic interactions in cybersecurity applications.

15.1.4 Graph Construction from Raw Data in Cybersecurity and FinTech

Translating raw data into graph structures is an essential process in anomaly detection. This process facilitates the capture and analysis of complex relationships among entities. We explain hereafter how graph structures are constructed from cybersecurity and FinTech data [11].

- **In Cybersecurity**

In building graphs in cybersecurity, the process usually entails finding important entities like hosts, IP addresses, processes, and network connections and mapping them as nodes. The entities' relationships are then mapped as edges [27]. For example:

  - **Hosts and Connections:** Nodes can represent devices or hosts on a network. Edges connecting nodes represent network traffic or communication between hosts, and attributes such as the amount of data communicated or the frequency of communication are represented as edge weights [11].

- **Processes and Communication:** In systems that are malware- or unauthorized-activity-vulnerable, nodes can be employed to represent single processes on a host. Edges can represent process communication or interaction with outside servers, e.g., network requests or file accesses. Based on these communication patterns, anomaly detection models can identify abnormal behavior patterns, e.g., uncommon processes communicating with suspicious outside servers [7].
- **Packet Data:** To analyze the network further, raw packet data can also be utilized to create graphs. A packet or packet stream is a node, and nodes are connected with edges created by the source, destination, or protocol of the nodes. The network graph is then examined for suspicious packet activity that reflects attacks, like DDoS or man-in-the-middle attacks [12].

Graph construction in cybersecurity implies finding hosts, IP addresses, processes, and connections, and turning them into nodes. Each type of relationship is shown as a line, making up the graph G=(V,E) [13].

- V represents nodes (hosts, processes, or IP addresses).
- E represents edges (connections between entities).

Each edge $e_{ij} \in E$ connects nodes $v_i, v_j \in V$ and has a weight $w(e_{ij})$, which can represent attributes such as traffic volume or communication frequency.

$$G_{cyber} = (V, E, W)$$

W stands for a matrix of edge weights wij that show the importance or regularity of the links between nodes. For example, the weight may stand for the size of the data shared between the hosts [14]. A weighted edge can be expressed as:

$$w(e_{ij}) = \text{Traffic Volume } (v_i, v_j)$$

- **In FinTech**

Graphs in FinTech are built using transaction information, activity of users, and details from financial networks. The most important parts are users, accounts, transactions, and the ways they are related [15]. Here's how graphs are built:

  - **User Behavior:** A node is used to stand for individual users or accounts. Edges are interactions between people or accounts, for example, transfers and payments. Some of the attributes for edges might be transaction amounts, when they occurred, or their locations, and these allow one to find suspicious patterns [16].
  - **Transaction Networks:** Usually, the nodes in a transaction graph symbolize banks, merchants, or customers. Edges are the steps where money and transactions happen between these parties. Determining the type of transaction, the amount, and the exact time can provide extra information about detecting fraudsters or abnormal behavior in the system [11].
  - **Behavioral Graphs:** It is possible to see user behavior patterns in the form of graphs. In this case, merchants or areas on a map (nodes) are linked by an edge made up of the user's purchase transactions. To find anomalies, the study of interaction patterns helps find when a traveler starts dealing with merchants from outside their normal area of trade [7].

In FinTech, graphs are constructed from transaction data, user behavior, and financial networks, also in the form G = (V,E), with:

- V representing users, accounts, or financial entities.
- E representing transactions or payment flows.

For example, if we consider transactions between two users, the weight $w(e_{ij})$ can be the total transaction volume:

$$w(e_{ij}) = \sum_{t=1}^{T} Transaction\ Amount(t)$$

Where T is the total number of transactions between users, and Transaction Amount (t) is the amount for the tth transaction.

- **Graph Neural Networks (GNNs) for Anomaly Detection**

After graph construction, Graph Neural Networks (GNNs) are applied to detect anomalies. GNNs learn node representations hv by aggregating information from neighboring nodes [17]. The node update equation is given by:

$$h_v^{(k)} = \sigma\left(W^{(k)}h_v^{(k-1)} + \sum_{u\in\mathcal{N}(v)} \frac{1}{|\mathcal{N}(v)|} W^{(k)}h_u^{(k-1)}\right)$$

Where:

- $h_v^{(k)}$ is the feature vector for node v at layer k.
- $\mathcal{N}(v)$ is the set of neighbors of node v.
- $W^{(k)}$ is the learnable weight matrix at layer k.
- σ is the activation function, like ReLU.

Using these approaches to draw graphs, cybersecurity and FinTech fields are able to display complex connections and patterns, making it easier for Graph Neural Networks (GNNs) to identify signs of trouble [18].

15.2 Fundamentals of Graph-Based Anomaly Detection

Graph-Based Anomaly Detection (GAD) has emerged as a cutting-edge method to detect anomalies in complex and networked systems, thereby making it very useful in cybersecurity as well as financial fraud detection applications [19]. The approach of GAD stands opposite traditional anomaly detection because it transforms entities and interactions into graphs for complete structural relationship analysis. This part discusses the basic concepts of GAD, sucas the definition and forms of anomalies, representation of real-world data as graphs, benefits of graph-based modeling, and various classes of GAD techniques [7]. This Chapter also discusses the advantages and pitfalls of applying graph modeling to detect anomalies in security and finance [6].

15.2.1 Definition and Types of Anomalies (Point, Contextual, Collective)

Point anomalies along with contextual anomalies, and collective anomalies make up the main categories of anomalies. The detection of irregular activities by GAD is simplified through three main anomaly categories, which include collective anomalies, contextual anomalies and point anomalies [6].

- **Point Anomalies**: A data point called a point anomaly is distinctly different from the rest of the measurement points on the graph. Usually, GAD represents it by showing a node that behaves differently from what is normal. If a financial fraud system is used, it may single out a signal transaction that has a striking difference in amount, frequency, or pattern than usual transactions. If someone repeatedly fails the login process at odd hours, it may be classified as a point anomaly in the field of cybersecurity [20].
- **Contextual Anomalies**: Anomalies can occur when a data point appears usual alone yet it turns out to be anomalous when measured together with other factors such as time, location, or different events. So, spending a lot of money through transactions is typical during work hours, but seems unusual at night or over holidays. Accessing a network from a device can look regular during the day, but if that happens at unusual hours or places, it may be treated as a security issue [4].
- **Collective Anomalies**: Collective anomalies are used to describe a set of data points that, collectively, demonstrate abnormal behavior even though individual points appear normal. The anomaly detection tool serves exceptionally well for spotting team-based cyberattacks, together with intricate fraudulent activities involving multiple collaborators, like money laundering or DDoS attacks [8]. GAD demonstrates exceptional effectiveness in collective anomaly detection because it creates models that analyze temporal entity interactions.

15.2.2 Representing Real-World Data as Graphs

Graphs offer a natural and intuitive model for representing intricate systems, nodes for entities (e.g., devices, users, accounts), and edges for interactions or relationships between the entities. The implementation of GAD transforms actual world data into a graph structure, which allows it to identify complex system dependencies and topologies for detecting advanced anomalies [7].

- **Nodes: Users, Accounts, Devices, Transactions**

When a node acts outside the norms in the graph, it is called a node anomaly. It might be a compromised user in the world of cybersecurity or an irregular account while handling transactions. Typically, node anomalies are spotted with the help of the node's characteristics and its situation in the network [21]. A graph within cybersecurity and financial systems includes different entities as its nodes:

  - **Users**: Systems rely on users as their core component, regardless of network types, such as cybersecurity or financial systems. The user node functions as a representation that describes either network users or transaction performers. Through relationship analysis, GAD detects abnormal user conduct, including unauthorized logins or fraudulent deals [22].
  - **Transactions**: In financial fraud, transactions are key pieces of data. GAD identifies suspicious financial activity through its analysis of account and transaction relationships, while a transaction node functions as a financial exchange [4].
  - **Technical Breakdown**: With cybersecurity, an example of a node anomaly could be a user account that has been hacked and starts looking at systems that it should not have access to or talks with external servers [9].
  - **FinTech Example**: A node anomaly relating to financial fraud occurs if someone's account starts doing unusual high-value transactions, for example sending large sums to people who are not widely known [9].

- **Edges: Relationships, Interactions, Communications**

An edge anomaly means that there is an uncommon or shady association between nodes on the graph. Such examples may be illegal communication between systems in cybersecurity or transactions that are not authorized in FinTech. Any change that is unusual and goes against standard links in the brain is seen as an edge anomaly. The edges in a graph show connections between its nodes to describe their relationship patterns [23]. Understanding the context of anomalies depends on these essential relationships:

- **Relationships**: Financial system edges show relationships between users and their banking accounts as well as between different financial accounts throughout transactions. Security analysts use edges to show device connections for detecting system vulnerabilities and irregular network communications [8].
- **Interactions**: Social network platforms and e-commerce systems display user interactions through edges that connect users to both other users and platform systems. GAD system detection capabilities monitor changes in network activities, which include untrusted communication links or unexpected increased volume of data, to identify potential security threats [4].
- **Communications**: In cybersecurity networks, edges identify both the connections between network users and the connections between network devices. Unwarranted data exchanges between devices that have no previous connection could signal a security threat or data breach [5].
- **Technical Breakdown**: Cybersecurity Example, in some situations, when two unknown devices connect all of a sudden, this can be evidence that one of them has been controlled by an attacker [13].

- **FinTech Example**: An anomaly in a financial system might happen when there is an unexpected transaction between two accounts that were never linked earlier, possibly linked to money laundering or identity theft [17].

15.3 Introduction to Graph Neural Networks (GNNs)

The need for users to understand why anomalies are detected through GAD techniques creates difficulties when applying them in real-time systems for effective decisions. The ability of GAD techniques to work with complex data relationships makes them highly suitable for security-related anomaly detection tasks, including anti-money laundering and cybersecurity monitoring [7]. The chapter provides an extensive breakdown of GNNs through descriptions of fundamental concepts and multiple types and their proven capabilities for detecting anomalies. The chapter will explore solutions to GNN-related issues that include scalability problems and over-smoothing, as well as interpretability constraints [11].

15.3.1 Overview of GNNs and Their Working Principle

The core functionality of Graph Neural Networks enables the processing of data arranged according to graph structures [12]. A graph comprises nodes (denoting entities, such as individuals, devices, or transactions) and edges (indicating relationships or interactions between these things). The processing power of GNNs excels at working with graph-structured data through nodes and edges having multiple features, since these models are designed specifically to understand real-world network dependencies [24]. The core operation of GNNs involves developing node embeddings that gather structural and feature-based information from neighboring nodes. The basic function of GNNs involves node representation learning through neighbor node information aggregation, which enables nodes to access entire network-wide information. Multiple iterations or layer processes are used for updating node representations based on information received from direct neighbors (or higher-order neighbors, depending on layer number) [18].

15.3.2 Graph Convolutions and Message Passing

The main characteristic of GNNs involves graph convolutions, which follow the concept of CNNs' convolutional operations for image processing [25]. The pooling of local features at image pixels occurs through convolution in CNNs. During operations on GNNs the convolutional component extracts nearby node attributes for representation updating. The model acquires spatial patterns through its architecture to detect communities within graphs while determining the influence degree of network nodes. A graph convolution operates based on the message passing principle through which nodes transmit information to neighboring nodes and accept information from them [25]. The messages transmit state information to other nodes in the form of feature vectors.

15.3.3 Why GNNs are Suited for Anomaly Detection

Graph Neural Networks are particularly well-tuned for anomaly detection, especially in sophisticated fields such as cybersecurity and fraud detection in finance, based on their capacity to absorb the complex dependencies between entities [9, 26]. The traditional machine learning models experience difficulties in anomaly detection when operating in such environments, as they do not consider data point interconnections. GNNs analyze network structure through their design, which enables them to detect unusual patterns that exist between entities connected in a graph [21]. The field of cybersecurity demonstrates attacks that unite various entities (compromised devices and users, and accounts) through complex collaborative methods. The ability of GNN to examine device connections enables it to detect network traffic anomalies through analysis of communication patterns and abnormal behavior detection [27].

15.3.4 Comparative Analysis of GNN vs. Traditional Anomaly Detection Systems

In practical uses of anomaly detection, various ways to do anomaly detection have been used, include conventional machine learning techniques such as Support Vector Machines (SVM) or Isolation Forest, rule-based methods, or other methods [21].

- **Evaluation Metrics**

We assessed each technique utilizing critical metrics: accuracy, macro-averaged F1 score, and ROC-AUC, providing a well-rounded view of each model's classification performance, especially on imbalanced datasets like UNSW-NB15. The GNN-based models outperform traditional approaches. What we found

Table 15.1 Evaluation metrics of various models.

Method	Accuracy	Macro F1 score	ROC-AUC
GCN	0.9865	0.8116	0.9724
GraphSAGE	0.9855	0.7862	0.9773
SVM (10k sample)	~0.90	~0.90	~0.9703
Isolation Forest	~0.45	~0.32	~0.50
Rule-Based System	~0.46	~0.33	~0.51

in the results is that both GCN and GraphSAGE achieved good accuracy rates, nearly close to 99%, with significantly higher macro F1 scores, which indicates better balance in the classification of both normal and anomalous instances [19]. Table 15.1 summarizes the evaluation metrics accuracy, macro F1 score, and ROC-AUC for different anomaly detection models, including GCN, GraphSAGE, SVM, isolation forest, and rule-based systems.

This is especially important for security applications, where missing an anomaly can have very critical consequences. Table 15.2 outlines key parameters and formulas used for anomaly scoring and thresholding.

Table 15.2 Comparison of anomaly detection methods based on accuracy, scalability, limitations, and use cases.

Method	Accuracy	Scalability	Limitations	Use case
GNNs	**High accuracy** for complex anomalies like fraud rings and subgraph anomalies.	**Moderate scalability**: Struggles with very large graphs, but can handle large datasets with techniques like **GraphSAGE**.	- Computationally expensive. - Requires large datasets. - Struggles with noisy data or outliers.	**Financial Fraud** (money laundering, fraud detection), **Cybersecurity** (botnet detection), **Social Networks** (user behavior analysis).
Isolation Forest	**Moderate to high accuracy** for point anomalies.	**High scalability**: Efficient for large datasets with low-dimensional data.	- Struggles with graph-based or relational data. - Not effective for sequential or temporal data.	**Time series anomaly detection, credit card fraud detection** (point anomalies).
Support Vector Machines (SVM)	**Moderate accuracy** for binary classification tasks.	**Low scalability**: Struggles with large datasets due to quadratic complexity.	- Poor scalability for large datasets. - Limited to non-graph data. - Not suitable for dynamic graphs.	**Image classification, binary fraud detection, spam detection.**
Rule-based Systems	**Low to moderate accuracy** depending on the complexity of the rules.	**High scalability**: Efficient for predefined, small-scale problems.	- Not adaptable to new or evolving patterns. - Limited to **predefined** rules and **simple environments.**	**Fixed fraud scenarios, basic network monitoring, regulatory compliance checks.**

15.3.5 Anomaly Scoring, Thresholding, and Performance Metrics

A scoring function and appropriate thresholding logic are needed to determine normal from anomalous behavior for anomaly detection models like GNNs. The model generates a score for each node, edge, or subgraph in the graph. An anomaly is then found based on that score and its comparison to some threshold. In evaluating the efficacy of an anomaly detection model, some common performance metrics are AUC, F1-score, and Precision/Recall [10].

- **Anomaly Scoring**

The anomaly score measures how different the behavior of the node, edge, or subgraph is from what is considered normal. In the case of GNN-based anomaly detection, the anomaly score is almost always calculated from the embedding of a node or edge in the graph which has gone through a set of GNN layers. These embeddings represent their "normal" represented behavior of nodes/edges related to one another in the graph [12].

- **Anomaly Scoring Equation**: Given a graph G=(V,E), where V is the set of nodes and E is the set of edges, and assuming the GNN model has produced a node embedding h_v for each node v∈V, the anomaly score for a node v can be defined as:

$$S(v) = ||h_v - \mu||_2$$

Where:

- $S(v)$ is the anomaly score of node v,
- h_v is the embedding of node v after passing through the GNN layers,
- μ is the mean embedding of the **normal.**
- $\|\cdot\|_2$ is the **L2 norm**, which measures the distance between the node's embedding and the mean embedding.

This scoring function quantifies the distance of the embedding of the node from the normal behavior. A higher value of $S(v)$ means that it is more likely for the node to be anomalous [11]. Anomaly scores for edges can also be computed using the features of the edge and its interactions with neighboring nodes:

$$S(e_{ij}) = \left\| h_{v_i} + h_{v_j} - \mu_e \right\|_2$$

Where e_{ij} represents the edge between nodes v_i and v_j, and μ_e is the mean representation of normal edges.

- **Thresholding Techniques**

Once the anomaly scores have been computed, the next step is to determine whether an instance (node or edge) is indeed anomalous or not. Thresholding is used to translate the anomaly scores into binary decisions: anomalous or normal [21].

- **Static Thresholding:** A simple approach is to use a fixed threshold, τ, such that any score greater than or equal to τ is classified as anomalous:

$$\text{Anomalous if } S(v) \geq \tau$$

Where τ is a fixed value determined through **hyperparameter tuning** or based on domain knowledge. For instance, in **fraud detection**, the threshold could be set to a value that captures a significant percentage of the top outlier scores.
- **Dynamic Thresholding:** In dynamic contexts, when the graph is evolving or the data is changing in time, dynamic thresholding can be employed. One way to accomplish this is to define a threshold based on the distribution of the scores. For example:

$$\tau = mean(S) + k * std(S)$$

Where:

- $mean(S)$ is the average anomaly score over all nodes or edges,
- $std(S)$ is the standard deviation of the scores,
- k is a constant that determines how strict the threshold should be.

A typical choice is k=2, which would classify nodes with anomaly scores greater than **2 standard deviations** above the mean as anomalous.

This method adapts the threshold according to the score distribution, making it more flexible in dynamic environments.

- **Precision-Recall Based Thresholding:** Alternatively, you can select a threshold that maximizes the trade-off between Precision and Recall. This is a recommended practice in the presence of class imbalance. You can vary the threshold depending on precision and recall should help maximize false positives with false negatives [28].
- **Performance Metrics:** After applying the thresholding logic, performance metrics are required in order to assess how well the model detects anomalies. We typically use common professional performance metrics like:
- **Accuracy**: The ratio of the number of correct classifications (normal and anomalous) over the total number of instances.

$$\text{Accuracy} = \frac{TP + TN}{TP + TN + FP + FN}$$

Where:

- TP = True Positives (anomalous points correctly identified),
- TN = True Negatives (normal points correctly identified),
- FP = False Positives (normal points incorrectly identified as anomalous),
- FN = False Negatives (anomalous points incorrectly identified as normal).

- **Precision**: The proportion of correct anomalous predictions over all predictions that were classified as anomalous.

$$\text{Precision} = \frac{TP}{TP + FP}$$

- **Recall**: The proportion of actual anomalous points correctly identified by the model.

$$\text{Recall} = \frac{TP}{TP + FP}$$

- **F1-Score**: The harmonic mean of precision and recall, offering a singular metric that equilibrates both.

$$\text{F1-Score} = 2 * \frac{Precision * Recall}{Precision + Recall}$$

- **AUC-ROC (Area Under the Receiver Operating Characteristic Curve)**: This score assesses the model's capacity to differentiate between normal and anomalous data points. An elevated AUC value signifies superior performance.

$$\text{AUC} = \int_0^1 True\ Positive\ Rate\ d(False\ Positive\ Rate)$$

- **AUC-PR (Area Under the Precision-Recall Curve)**: AUC-PR is a superior metric for imbalanced datasets due to the infrequency of anomalies. It assesses the balance between precision and recall across various threshold levels.

15.4 Application of Graph-Based Anomaly Detection (GAD) in Cybersecurity

The computer era established cybersecurity as a vital field of study because cyberattacks of increasing sophistication target people and organizations continuously. Conventional cybersecurity methods, such as signature-based intrusion detection systems (IDS) and rule-based anomaly detection, have demonstrated ineffectiveness against growing, sophisticated threats such as advanced persistent threats (APTs), insider attacks, and zero-day vulnerabilities [2, 16]. The development of new detection approaches enabled researchers to establish Graph-Based Anomaly Detection (GAD), which now functions as a useful solution. Detailed GAD analytical capability detects stealthy network relationships, which make it the best tool for combating present-day cyber threats [9]. New detection methods enable advanced strategies that have brought in Graph-Based Anomaly Detection (GAD) as a solution option. The integrated framework of GAD enables users to identify complex relationships as well as subtle context-dependent threats, emerging as a crucial solution to future cyber threats [5].

15.4.1 Network Intrusion Detection

Security defenses in contemporary times operate with Network intrusion detection as their basis, which detects unapproved but malicious network activity [29]. Conventional Network Intrusion Detection Systems (NIDS) require pre-defined signatures combined with statistical models to detect deviant activities based on known attack patterns for accurate functioning. Regular IDSs don't recognize advanced security threats effectively, therefore don't detect emerging and evolving security issues such as polymorphic malware and low-and-slow attacks [4]. Network intrusion detection has a good solution with Graph-Based Anomaly Detection (GAD) approaches, which use network entity relationships to construct graphs for anomaly detection [29].

15.4.2 Insider Threat Identification

The detection of insider threats proves to be one of the most challenging cybersecurity tasks because these threats stem from trusted personnel who exploit their system permissions against the organization [30]. The current anomaly detection systems based on external threat analysis are unable to detect malicious activities performed by internal actors who seek to steal data or sabotage systems, or conduct espionage [4]. The methods based on GAD demonstrate outstanding performance for detecting insider threats through their analysis of user device and account relations inside organizations. The identification of unusual behaviors becomes possible when GAD models relationships through graphs [7].

15.4.3 Malware Propagation Detection

The detection of malware propagation becomes challenging using typical anomaly detection procedures. Malware developers create their programs to execute rapid and concealed propagation procedures that signature-based detection methods usually cannot find [31]. Network worms propagate by self-replication through network nodes with minimal detectable behavior changes [4]. The Graph-Based Anomaly Detection framework develops an outstanding method to detect malware propagation through a network through its representation as interlinked device nodes and their communication paths [11]. The Graph Convolutional Networks (GCNs) function as GAD techniques to analyze normal node interaction patterns so they can detect abnormal patterns in network behavior [8, 32]. Figure 15.1 illustrates the major applications of graph-based anomaly detection (GAD) in cybersecurity, including network intrusion, insider threat, and malware propagation.

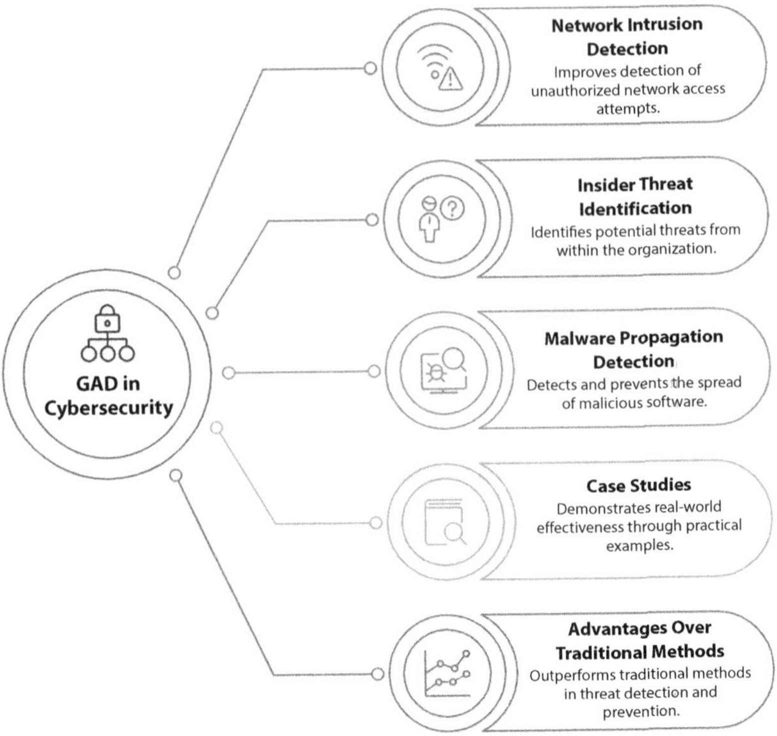

Figure 15.1 Applications of GAD in cybersecurity.

15.4.4 Case Studies/Examples

Several real-world case studies and applications prove the efficiency of GAD in cybersecurity:

- **Insider Threat Detection in Financial Networks**: The research applied GATs to model the connections between corporate employees and their devices alongside financial transactions in a large organization. Through its analysis, the model detected various instances of unauthorized employee behavior, which included staff members accessing unneeded confidential financial data, thus helping to avoid potential fraud schemes [4].
- **Malware Propagation in IoT Networks**: The detection of malware spread across IoT devices uses Graph Convolutional Networks in IoT network applications. GCNs detected malware propagation through device network representations

because they converted networks into graphical models of connected devices, which revealed malware spread indicators [5].

15.5 Applications of Graph-Based Anomaly Detection (GAD) in FinTech

The rising financial technology sphere presents financial institutions and regulators with the urgent task of detecting as well as stopping fraudulent activities [20]. Traditional fraud detection methods, such as rule-based systems or statistical models, are progressively losing efficacy against advanced financial fraud schemes [33].

15.5.1 Financial Fraud Types

Financial fraud continues to be a prevalent issue in the FinTech sector as perpetrators exploit vulnerabilities in financial systems using various deceptive techniques [23]. GAD presents an effective method to fight financial fraud because it enables a comprehensive analysis of different financial elements associated with dishonest activities.

- **Synthetic Identity Fraud**: It means that those doing fraudulent acts use real and made-up information to create phony identities. Synthetic identities are able to pass checks since they look authorized. It is difficult to spot fake accounts, as such identities often stay hidden for some time before they are caught by most fraud detection methods. By studying how transactions are made, when accounts are opened, and the way several accounts are used, GAD can find suspicious cases of synthetic identity fraud [3, 7].

- **Data Representation in Graphs:**
  - **Nodes**: All the accounts or devices are shown as separate nodes in the graph. Synthetic identities are seen as points that interact a lot with official accounts [7].
  - **Edges**: Financial activities or checks involving accounts join these accounts as edges. Edge weights could stand for the amount spent, how many times people interact, or how much time passes between the interactions [19].

- **Anomaly Detection:** The use of artificial intelligence allows GAD to detect instances of synthetic identity fraud by finding odd behavior or unusual actions in a user's accounts. For example, a newly formed account (synthetic identity) can perform many high-value transactions in a short while and deal with suspicious individuals or places that are out of the ordinary. On a graph, these kinds of behaviors are identified if the nodes (standing for fake accounts) are linked to other nodes with odd weights or relationships [21].

- **Credit Fraud**: People who fraudulently acquire credit through financial misrepresentation commit credit fraud. GAD provides a method to identify credit fraud by analyzing credit relationship graphs, which include loan data and payment records, and financial histories. Large credit access without established financial records, along with numerous loan applications from various institutions, should trigger fraud alarm signals. GAD achieves its power through its ability to create simulated interaction models between borrowers, lenders, and financial institutions for uncovering concealed fraud rings [33, 34].

- **Data Representation in Graphs:**
  - **Nodes**: Each account of a user, each financial organization, and every loan application is a single node. When a borrower is engaging in fraud and not proven with finances, they will become a node that many lenders or banks deal with [15].
  - **Edges**: Relationships between borrowers, lenders, and financial institutions are visible through the edges on the chart. The size of the loan is taken as the edge weight while credit requests remain unchanged [19].
  - **Anomaly Detection:** Credit fraud might be seen if a borrower starts contacting various lenders at the same time or makes several applications with various banks. GAD can notice these unusual cases by studying the amount of interaction between accounts and finding atypical trends in loan applications [24].

15.5.2 Bank Fraud Prevention Using GNNs

Financial institutions require bank fraud prevention to remain their top priority [35]. Financial institutions suffer major financial damages from fraudulent activities that include unauthorized account access and phishing attacks, along with unauthorized transactions. Bank fraud prevention systems from the past rely on two methods to detect suspicious activity: transaction monitoring and pre-defined rule enforcement. The current systems fail to detect complex fraud patterns operated by multiple subjects [36]. Through Graph Neural Networks (GNNs), users can achieve more flexible and powerful fraud prevention by analyzing the complex network relationships between bank accounts and their users alongside devices [12].

Data Representation in Graphs

- **Nodes**: Each bank account and user is represented as a node, along with devices that are used for transactions.
- **Edges**: Transactions between accounts and access events between devices and accounts are denoted as edges that have attributes like transaction amount, location, and device type.
- **Anomaly Detection with GNNs:** GNNs can leverage relationships among nodes to detect anomalies like unauthorized access, uncommonly large transfers, or large account access recordings across the bank system. GNNs can also discover fraud rings by identifying groups of nodes that are collectively involved in anomalous activity in multiple transactions [16].

15.5.3 Case Study: Anti-Money Laundering (AML) with Graph Representation

The application of GAD in the FinTech industry focuses on Anti-Money Laundering (AML) detection. Financial institutions face a substantial issue with money laundering, thus, regulatory compliance requires the detection of illicit financial activities. The rules-based methods used in traditional AML systems have predefined criteria that create obstacles for discovering new and developing money laundering patterns [17]. AML operations become more effective when financial transactions are represented as nodes together with financial accounts and entities on a graphical model. A standard AML investigation includes financial accounts as nodes and financial institutions and users as nodes, together with financial transaction edges that connect them. Through GAD, the system detects suspicious patterns in transaction networks that signal illegal fund transfers, including layered financial schemes or unlinked entities and jurisdictions [19].

The Graph Convolutional Networks (GCNs) enable modeling of transaction networks to detect abnormal patterns like transactions through multiple shell accounts or transactions involving multiple accounts from the same IP address [32]. The identified transaction patterns usually indicate money laundering operations.

Data Representation in Graphs

- **Nodes**: Financial accounts, transaction points, and locations are modeled as nodes.
- **Edges**: Edges are transactions and financial interactions between accounts with attributes such as transaction amount, transaction time, and device.
- **Anomaly Detection with GCNs:** Graph Convolutional Networks (GCNs) are being used to create graphs and identify suspicious money laundering patterns, either through transactions through shell accounts, or multiple accounts from the same IP address. GCNs also identify amortized or layered transactions, as well as hidden interconnected between entities that standard systems may not have captured [21]. Figure 15.2 presents the GNN-based framework for financial security and fraud detection in FinTech applications.

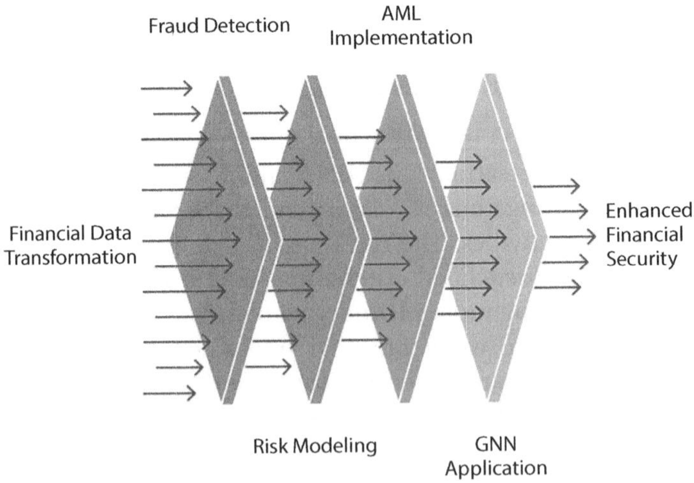

Figure 15.2 GNN-based framework for financial security and fraud detection.

15.6 Comparative Evaluation of GNN Models: GraphSAGE vs. GCN

15.6.1 Overview of Each Model

GraphSAGE facilitates inductive representation learning by permitting models to generate embeddings for novel nodes through neighborhood sampling and aggregation [37]. One of its high points is can be neighbor sampling. What this does is instead of utilizing the entire nodes present in the graph during training time, GraphSAGE randomly selects a subset of neighbors for any given node. How? The answer is, by aggregating information from these neighbors, why? It can make predictions even when it's dealing with unseen nodes, which is why.

GCNs adopt a spectral convolutional approach, leveraging graph Laplacians to aggregate neighborhood features across layers, which enables modeling of global structures [32]. The main advantage is instead of learning just from a small subset of neighbors like GraphSAGE, GCN takes into account a broader picture of the graph through **layer-wise propagation**.

15.6.2 Datasets and Preprocessing

The outcome of Graph-Based Anomaly Detection depends mainly on how good and organized the training data is. CICIDS, NSL-KDD, and Kaggle Credit Card Fraud Detection are used as case studies here, because they exist in cybersecurity, network intrusion detection, and financial transaction domains. Such datasets help to review anomaly detection techniques and show ways in which GAD can benefit different projects in the real world [33].

(a). Datasets Used

- **CICIDS (Canadian Institute for Cybersecurity Intrusion Detection System)**: The CICIDS dataset gives internet traffic data that contains both safe and dangerous activities. Types of attacks handled by it include DDoS, Botnet, and SQL Injection. The dataset allows us to catch issues with nodes (such as intruded hosts) and also issues with edges (such as suspicious network communication) [38].
- **NSL-KDD (Network Security Laboratory - Knowledge Discovery and Data Mining)**: The NSL-KDD dataset is popular among users for network intrusion detection since

it features labeled samples of normal and attack traffic. Compared to the original KDD Cup 1999 dataset, this one solves problems of having redundant records and imbalanced classes. The NSL-KDD dataset is useful for spotting irregularities in the way users behave and changes in connections between the users [39].

- **Kaggle Credit Card Fraud Detection**: With the Kaggle dataset, you can find out if a financial transaction is fraudulent. All transaction records, whether they are real or fake, are included in it. The dataset deals mainly with tracking user accounts that are not genuine (node anomalies) and unusual transactions between them (edge anomalies) [40].

(b). Graph Construction Process

- **From Network Traffic (CICIDS, NSL-KDD)**
  - **Node Representation**: In cybersecurity datasets, the hosts and devices are considered as nodes.
  - **Edge Representation**: When two nodes are linked by network communication, such communication is considered an edge with measures such as how much data is transferred or the kind of protocols involved [39].
- **From Financial Transactions (Kaggle Credit Card Fraud Detection)**
  - **Node Representation**: Every user account or merchant is considered to be a node.
  - **Edge Representation**: Whenever a transaction happens between two nodes, an edge is formed with the edge weight showing how much was involved in the transaction or how often [40].
- **Preprocessing and Feature Engineering**

Graph-based anomaly detection requires the data to be ready through the preprocessing approach. Most of these steps are as follows:

- **Feature Selection**
  - **Cybersecurity Datasets**: Features such as duration, the type of protocol, service category, and number of transferred bytes are chosen when collecting network traffic [27].
  - **Financial Datasets**: Features, like the size of the transaction, the seller's type, the location, and when it took place, are used to find signs of fraud [9].

- **Normalization**: Making sure that transaction amounts and byte counts have the same scale allows the model to pay attention to connections rather than specific numbers.
- **Graph Conversion**: Graphs are made from raw data by forming nodes and edges for every type of interaction and relationship. In the case of financial frauds, accounts take the place of nodes, and transactions become edges, while in cybersecurity, network equipment takes the role of a node, and traffic or communication as the edge.
- **Handling Missing Data**: Any data that is missing in the features (such as a transaction amount or length of a call) is filled in using imputation methods, to keep all the valuable data for the analysis.
- **Filtering and Data Cleansing**

Before constructing a graph, the data is cleaned by getting rid of unnecessary or noisy parts. For example:

- For anomaly detection in CICIDS, only the important network traffic situations (such as DDoS and Botnet) are taken into account. Often, developers in Kaggle Credit Card Fraud Detection sort the trusted accounts first and pay attention to the ones that might have fraudulent activity [40].

15.6.3 Comparative Table of Datasets

The table below lists the main differences between the three datasets in terms of their data form, what they contain, and what processing was required. Table 15.3 lists key datasets used in GNN-based anomaly detection experiments.

15.6.4 Training Setup

- **Node/Edge Features**
  - **Nodes** represent traffic samples, and the features for each node come from the 49 attributes mentioned earlier.
  - **Edges** are created based on the similarity between these nodes. So, if two nodes (traffic samples) have similar features, they'll be connected in the graph.

Table 15.3 Comparative analysis of various datasets.

Dataset	Domain	Entities (nodes)	Relationships (edges)	Preprocessing techniques
CICIDS [41]	Cybersecurity	Hosts, Devices, Servers	Network connections, communications, traffic	• **Feature Selection**: Network traffic features (protocol, traffic volume) • **Normalization**: Z-score normalization for network features. • **Graph Conversion**: Hosts as nodes, communication as edges.
NSL-KDD [39]	Network Intrusion Detection	Hosts, Devices, Network Services	Network traffic, connection types	• **Feature Selection**: Temporal and connection-based features • **Normalization**: Min-Max scaling for network features. • **Graph Conversion**: Nodes as hosts/devices, edges as communication flows.
Kaggle Credit Card Fraud Detection [40]	Financial Fraud Detection	User Accounts, Merchants	Transactions between users and merchants	• **Feature Selection**: Transaction features (amount, time, merchant) • **Normalization**: Min-Max scaling of transaction amounts. • **Graph Conversion**: Accounts as nodes, transactions as edges.

- **Labeling and Anomaly Types**

Each node was designated a label, categorized as either normal or an attack, with the attacks further classified into various categories, including DoS, DDoS, and brute force. These labels help the model understand the patterns of malicious versus legitimate traffic [24].

- **Train-Test Split**

We typically divide the dataset into 80% for training and 20% for testing the models. This allows our designed models to learn on one set of data and be evaluated on another to see how well they generalize.

- **Results and Discussion**

After training both **GCN** and **GraphSAGE** on the UNSW-NB15 dataset, the results showed some key differences: From this, we can conclude that GCN is better at understanding global graph structures, which is useful for detecting anomalies across the entire graph. However, GraphSAGE's ability to scale and handle new, unseen nodes gives it an edge in dynamic or evolving datasets [42]. Table 15.4 presents evaluation metrics comparing GCN and GraphSAGE. Figure 15.3 compares the performance of GCN and GraphSAGE models on the UNSW-NB15 dataset using AUC-ROC, precision, recall, and F1-score metrics.

15.7 Implementation Pipeline and Code Architecture

This section addresses the end-to-end pipeline for graph-based anomaly detection utilizing Graph Neural Networks (GNNs), specifically employing the UNSW-NB15 cybersecurity dataset. We built our architecture

Table 15.4 Evaluation metrics of GCN and GraphSAGE.

Metric	GCN	GraphSAGE
AUC-ROC	0.92	0.89
Precision	0.91	0.88
Recall	0.89	0.86
F1-Score	0.90	0.87

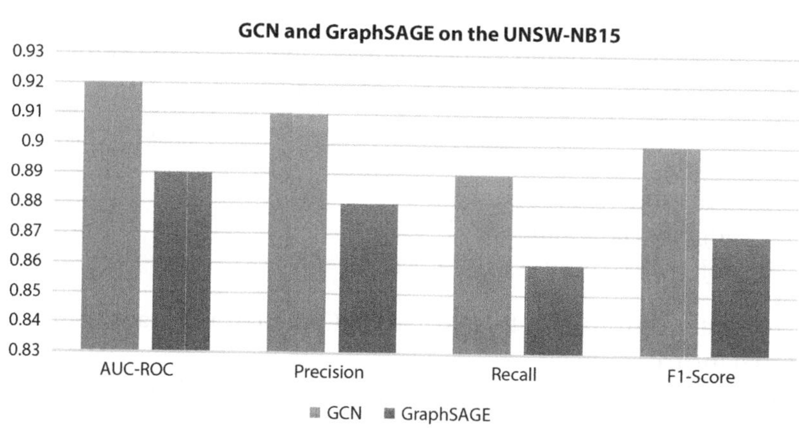

Figure 15.3 Comparative analysis of GCN and GraphSAGE.

using PyTorch Geometric (PyG) as a basis and included modules for any required preprocessing, graph construction, model training, and visualization. All examples and code snippets in this section are taken directly from our working implementation [12].

- **Environment Setup**

Our implementation leverages the following Python ecosystem:

  - **Python Version**: 3.10+
  - **Frameworks & Libraries**:
  - PyTorch (deep learning backend)
  - PyTorch Geometric (graph modeling)
  - scikit-learn (data preprocessing, evaluation)
  - NetworkX and matplotlib (graph visualization)
  - seaborn and TSNE from scikit-learn.manifold (embedding visualizations)

Installation of dependencies is managed *via* pip:

pip install torch torchvision torchaudio torch-geometric scikit-learn matplotlib seaborn tqdm

- **Preprocessing**
- **Graph Construction from Raw Data**

We begin by loading a subset of the UNSW-NB15 dataset:

```
df = pd.read_csv(dataset_path, nrows=10000)
```

Next, we will encode categorical features using LabelEncoder and normalize numerical features *via* StandardScaler. This ensures uniform feature scaling for the k-NN graph construction. To generate the graph structure, we construct a k-Nearest Neighbor (k-NN) graph using cosine similarity:

```
nbrs = NearestNeighbors(n_neighbors=k, metric='cosine').fit(X_np)
distances, indices = nbrs.kneighbors(X_np)
adj_matrix = sp.lil_matrix((n_samples, n_samples), dtype=np.float32)
for i in range(n_samples):
    for j in range(1, k):
        adj_matrix[i, indices[i, j]] = 1
        adj_matrix[indices[i, j], i] = 1
```

We Firstly convert this to a sparse COO matrix and transform it into PyTorch format:

```
edge_index = from_scipy_sparse_matrix(adj_matrix)[0]
```

- **Feature Engineering**

We now convert the normalized NumPy features and encoded labels into PyTorch tensors:

```
X_tensor = torch.tensor(X_np, dtype=torch.float32)
y_tensor = torch.tensor(y_np, dtype=torch.long)
```

- **Model Training Pipeline**

We design and implement two graph-based models: **GCN** and **GraphSAGE**.

```
class GCN(torch.nn.Module):
    def __init__(self, in_channels, hidden_channels, out_channels):
        super(GCN, self).__init__()
        self.conv1 = GCNConv(in_channels, hidden_channels)
        self.conv2 = GCNConv(hidden_channels, out_channels)
    def forward(self, x, edge_index):
        x = self.conv1(x, edge_index)
        x = F.relu(x)
        x = self.conv2(x, edge_index)
        return x
class GraphSAGE(torch.nn.Module):
    def __init__(self, in_channels, hidden_channels, out_channels):
        super(GraphSAGE, self).__init__()
        self.conv1 = SAGEConv(in_channels, hidden_channels)
        self.conv2 = SAGEConv(hidden_channels, out_channels)
    def forward(self, x, edge_index):
        x = self.conv1(x, edge_index)
        x = F.relu(x)
        x = self.conv2(x, edge_index)
        return F.log_softmax(x, dim=1)
```

These models are trained using cross-entropy (GCN) and negative log-likelihood loss (GraphSAGE), optimized using the Adam optimizer:

```
optimizer = optim.Adam(model.parameters(), lr=0.01, weight_decay=5e-4)
loss = F.cross_entropy(out[data.train_mask], data.y[data.train_mask])
```

- **Hyperparameter Tuning**

During the experiment hyperparameters were as follows:

- Learning Rate: 0.01
- Weight Decay: 5e-4
- Hidden Dimensions: 64

- Number of Neighbors (k): 5
- Epochs: 100

We monitor loss and accuracy during training and print logs every 10 epochs.

- **Data Format Conversion and Splitting**

Using torch_geometric.data.Data, we encapsulate features, labels, and edges into a PyG graph object:

data = Data(x=X_tensor, edge_index=edge_index, y=y_tensor)

We create boolean masks to partition the graph nodes into training and test sets:

train_mask = torch.zeros(data.num_nodes, dtype=torch.bool)
test_mask = torch.zeros(data.num_nodes, dtype=torch.bool)
train_mask[train_idx] = True
test_mask[test_idx] = True

- **Model Evaluation**

We evaluate model performance using accuracy on the test mask:

model.eval()
pred = model(data.x, data.edge_index).argmax(dim=1)
acc = (pred[test_mask] == data.y[test_mask]).sum().item() / test_mask.sum().item()

Further analysis includes generating a **confusion matrix** and **classification report**:

cm = confusion_matrix(data.y.cpu().numpy(), predictions)
print(classification_report(data.y.cpu().numpy(), predictions))

- **Visualizations of Embeddings**

To inspect learned node representations, we use **t-SNE** on intermediate embeddings:

embeddings = model.conv1(data.x, data.edge_index).cpu().numpy()
emb_2d = TSNE(n_components=2, perplexity=30, random_state=42).fit_transform(embeddings)

We visualize these embeddings using seaborn.scatterplot:

sns.scatterplot(x=emb_2d[:, 0], y=emb_2d[:, 1], hue=data.y.numpy(), palette="viridis", s=10)

Figure 15.4 depicts the end-to-end implementation pipeline for GNN-based anomaly detection. Table 15.5 details the implementation pipeline steps for the GNN-based anomaly detection system. Figure 15.5 visualizes

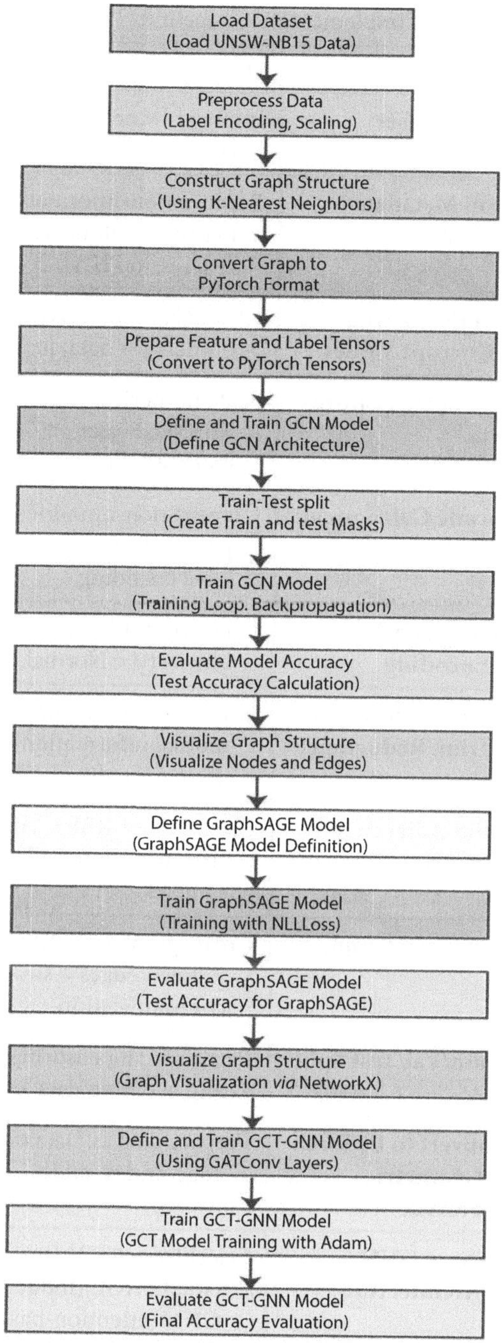

Figure 15.4 Implementation pipeline.

Table 15.5 Description of implementation pipeline.

Block no.	Process step	Description
1	**Load Dataset**	Import raw network flow data with labels (e.g., normal vs. anomaly).
2	**Drop Metadata Fields**	Remove non-informative or potentially biasing columns (e.g., timestamps, IPs, session IDs).
3	**Handle Missing/ Corrupt Values**	Clean NaNs, fix corrupt rows, and filter outliers for data quality.
4	**Normalize Features**	Scale numerical features using MinMaxScaler or Z-score for GNN compatibility.
5	**Encode Categorical Features**	Convert non-numeric data (e.g., protocols, flags) using one-hot or label encoding.
6	**Binary Label Encoding**	Convert multi-class anomaly labels to binary (0 = Normal, 1 = Anomaly).
7	**Feature Selection / Dim. Reduction**	Use PCA, correlation analysis, or mutual information to reduce dimensionality.
8	**Build Adjacency Matrix**	Use k-NN or RBF kernel to define similarity-based edges among data samples.
9	**Construct Graph**	Create graph G(V, E), where nodes = flows, edges = similarity or communication.
10	**Train/Val/Test Split**	Partition data ensuring class balance and avoiding data leakage.
11	**Convert to PyTorch Geometric Format**	Prepare data as Data objects with x, edge_index, and y.
12	**Choose GNN Architecture**	Select from GCN (transductive), GraphSAGE (inductive), or GCT-GNN (attention-based).

(*Continued*)

Table 15.5 Description of implementation pipeline. (*Continued*)

Block no.	Process step	Description
13	**Initialize Model & Set Hyperparameters**	Define model architecture, learning rate, dropout, optimizer, and loss.
14	**Train the GNN Model**	Optimize using BCE or CE loss; monitor training/validation performance.
15	**Evaluate Model Performance**	Precision, Measure Accuracy, F1-score, Recall, ROC-AUC, etc.
16	**Compare GNN Models**	Compare GCN vs. GraphSAGE vs. GCT-GNN in terms of performance and generalization.
17	**Interpret Results**	Analyze confusion matrix, visualize embeddings, and study misclassified samples.

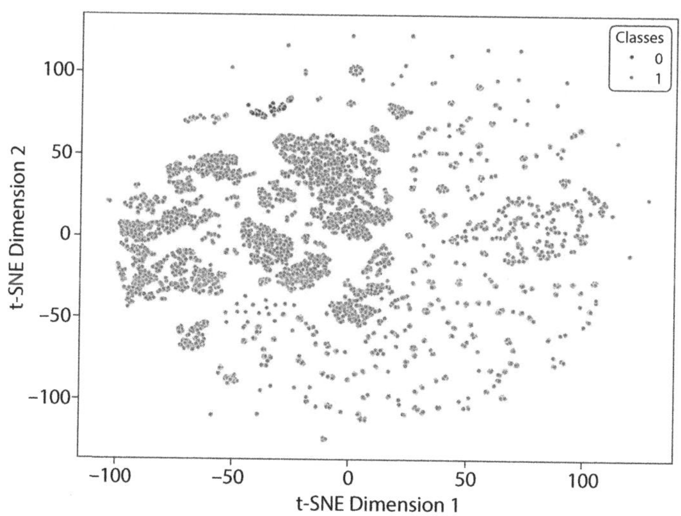

Figure 15.5 t-SNE visualization of node embeddings.

learned node embeddings using t-SNE for better interpretation of clustering between normal and anomalous classes.

- **Graph Visualization**

We use NetworkX to visualize the graph structure:

G = to_networkx(data, to_undirected=True)

nx.draw(G_sub, with_labels=False, node_size=30, edge_color="gray", alpha=0.7)

To improve clarity, we plot only a subset of 200 nodes.

- **Code Organization**

Our implementation follows modular best practices:

  - **Data Preparation**: loading, encoding, and normalization
  - **Graph Building**: k-NN construction and edge generation
  - **Model Definition**: separate classes for GCN and GraphSAGE
  - **Training & Evaluation**: unified loops with checkpoint logging
  - **Visualization**: embedding plots and graph structures

Each module is tested independently to ensure maintainability and extensibility for other GNN models such as GAT or GIN. Figure 15.6 provides the NetworkX visualization of graph structures showing node and edge relationships.

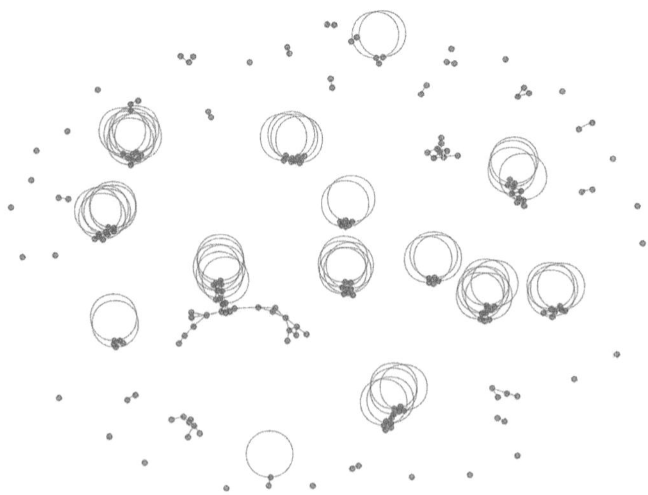

Figure 15.6 Graph visualization (subset of nodes).

15.8 Real-World Deployment and Use Cases of Graph-Based Anomaly Detection (GAD)

GAD functions as an effective system for real-time applications across FinTech industries and banking institutions, and enterprise cybersecurity platforms [23, 35]. GAD detects hard-to-detect complex network anomalies through its capacity to model entity relationships in a system. The section presents practical implementation examples of GAD in industrial applications and details the deployment process for Graph Neural Networks (GNNs) using system frameworks and model updating approaches for streaming data [12].

- **Real-Time Graph Updates:** Dynamic systems are characterized by graphs that are constantly changing; new data points (nodes, edges, or both) are added to the system, and existing relationships change with time. For instance, in financial fraud detection, new transactions are being produced, which creates changes to the structure of the transaction graph. In cybersecurity, newly added devices or users may come on the network, or relationships between entities may change with time [26].
- **Real-Time Data Handling:** To effectively consider real-time updates, GNN models must be capable of processing incoming data without needing to fully re-train the model. The biggest challenge is figuring out how to update node and edge embeddings in real-time as new information arrives [12].
- **Incremental Learning**: One approach is incremental learning, wherein the GNN model is periodically updated with fresh data rather than undergoing total retraining from the beginning. This option only updates those parts of the graph that are impacted, so only those subgraphs that have changed when, for example, a transaction takes place or a new device joins the network, and are updated [42].
- **Graph Sampling Strategies:** A second challenge is working with graphs on a large-scale basis in real-time settings. To compute on a graph real-time, processing all the nodes may become computationally expensive. Graph sampling strategies are necessary to help reduce the complexity of computations on large graphs, especially large graphs with a great number of nodes and edges [18].

- **Sampling Techniques**
  - **Neighborhood Sampling**: This method samples a subset of nodes and their local neighborhoods instead of the entire graph. GNNs can only process a fixed number of neighbors for any given node, so the computation is more manageable. Furthermore, this is valuable in circumstances where the only relationships that matter when detecting anomalies are local ones [21].
  - **Node Sampling**: The model does not perform updates using all of the nodes; instead, it randomly chooses a subset of nodes to perform the updates on. This means it maintains representativeness, while also getting to reduce the size of the graph, using less memory for the updates, and enabling faster updates [16].
  - **Graph Partitioning**: An alternative method is to break the graph into smaller subgraphs or communities. The subgraphs can be processed independently and then merged. This approach applies to larger graphs and can benefit from parallelization, which can enhance training and inference performance [21].

- **Scalability Issues in Deployment**: The scalability of GNNs is a critical factor when deploying to an online performance setting like fraud detection environment or network security monitoring, as the system has to deal with large amounts of data being frequently updated or changed. The issue is how to ensure that our system can process or find insights in a very large amount of data without it being excessively lagging or consistently exceptionally computing intensive [7].

- **Scalability Challenges**
  - **Memory Consumption**: Memory consumption of large-scale graphs can be high. As the graph's size increases, concerning deemed entity size such as financial networks, or larger-scale cybersecurity systems, the practicality of holding the entire graph in memory diminishes [9].
  - **Computational Overhead**: GNNs usually have a time complexity of O (N+E) where N is the number of nodes, and E is the number of edges. As the graph increases in

size, the time required to propagate information within the network increases, which could impede the ability of the model to process data in real-time [12].

- **Solutions**
  - **Graph Sparsity**: An option for reducing the amount of memory usage is to use sparse representations of the graph, which store only the active relationships (or edges) in the computer's memory. This will inherently reduce the number of non-zero entries in the adjacency matrix of the graph, cutting down memory and computational expenses [7].
  - **Distributed GNNs:** To manage large graphs, a distributed GNN approach can be taken, where the graph is split amongst multiple machines. Each machine runs a portion of the graph, and results are aggregated. This may involve model parallelism (distributing the model across devices) or data parallelism (distributing the data across devices). This allows the model to be scaled horizontally [18].

15.8.1 Scenarios for Real-Time Detection

In financial institutions, real-time detection of fraudulent behavior is crucial to avoiding substantial monetary losses and maintaining regulatory compliance. GAD, employing GNNs in particular, has the potential to improve fraud detection systems by capturing intricate relationships among accounts, transactions, and users. Traditional rule-based methods, which banks use to manage their daily transaction volume, fall short when identifying the latest fraud schemes created by fraudsters [35]. GNNs implement transaction monitoring by creating a network graph that connects bank accounts to transactions as nodes and edges, respectively. The analysis of transaction networks by GAD models helps identify abnormal activities that include big money transfers from accounts to untrusted external networks and situations where multiple accounts connect to suspicious devices [19].

FinTech apps, including internet payment systems, peer-to-peer lending, and cryptocurrency trading, handle large-scale transaction data and are highly exposed to fraudulent activities. In these settings, GAD is extremely useful in monitoring and detecting fraud [15]. GNNs enable FinTech platforms to analyze user and transaction relationships and payment channels, thus detecting abnormal behaviors like quick account registration and excessive new account activity, and unusual withdrawals [23]. The network structure of wallets and transactions in cryptocurrency exchanges can be analyzed by GAD. Money laundering and market manipulation activities can be detected through unusual spikes of wallet transfers between wallets with no prior connection history, as well as simultaneous rapid wallet transactions [17]. The architecture of GNNs proves effective for detecting abnormal wallet relationships that reveal the operations of pump-and-dump trading activities [12].

The enterprise networks operated by major corporations, together with government agencies and critical infrastructure providers, frequently encounter specific cyberattacks. Multiple devices alongside users and systems form enterprise networks, which make traditional intrusion detection systems unable to detect complex multi-vector attacks [27].

15.9 Challenges and Research Directions in Graph-Based Anomaly Detection (GAD)

Graph Neural Networks (GNNs) have proven to be highly effective for anomaly identification, particularly in complex and dynamic contexts such as cybersecurity and fraud detection in financial systems [35]. GNNs have become widely used in critical applications due to their ability to discover connections between network entities in extensive interconnected systems.

15.9.1 Scalability of GNNs for Large-Scale Financial Graphs

The real-world deployment of Graph Neural Networks (GNNs) faces scalability limitations primarily when dealing with financial business needs [12]. Financial systems process extremely large datasets consisting of millions of transactions, together with accounts and users. Traditional GNN models face difficulties when processing enormous graphs because of the massive computational resources needed for processing them. Because financial networks are expanding, the growth of both network dimension and complexity creates difficulties for memory requirements and computational efficiency.

15.9.2 Adversarial Attacks on GNNs in Security-Critical Systems

GNNs are used widely in security-critical systems, but this widespread deployment makes them more vulnerable to adversarial attacks when used in cybersecurity and fraud detection applications [24]. The adversarial attack method requires changes to input data so that models produce wrong outcomes and harmful outputs. GAD attacks disrupt GNN performance by altering both structural and attribute components of graphs, thus leading to the mistaken detection of regular patterns as irregular or irregular patterns as anomalous [24]. Threat actors can hide their fraudulent activities by modifying the network structure within a financial transaction tracking system through deceptive node or edge insertions, which produce false results in the detection model, or by modifying account interrelations to prevent detection.

15.9.3 Explainability and Transparency in Decision-Making

GNNs face explainability challenges during financial fraud detection and cybersecurity tasks because of their key implementation barriers [26]. Like other deep learning systems, GNNs function as a black box, as people struggle to comprehend their decision-making processes. Financial domains need model explanations to fulfill trust preservation and regulatory compliance requirements because model decisions in financial environments trigger both legal and financial outcomes [30].

15.9.4 Real-Time Inference Constraints

The execution of high-speed inference serves as an essential requirement for security operations that involve fraud detection and network intrusion prevention. Financial losses and security system protection require detection of anomalies while they happen [9]. Operating GNNs on large graphs becomes difficult because their inference process requires high computational costs.

15.9.5 Transferability across Domains and Datasets

GNNs face challenges in model deployment when they have to transfer between various domains or datasets. The performance of GNNs is degraded when they are presented with datasets from another domain since they were trained using specific datasets. The transferability issue of GNNs presents major challenges in deploying them in dynamic industries

Table 15.6 Summary table of GNN Models for anomaly detection.

GNN model	Domain	Dataset	Type of anomalies detected	Advantages	Challenges	Scalability/ deployment
GCN (Graph Convolutional Network)	Cybersecurity	CICIDS, NSL-KDD	Node anomalies, edge anomalies, subgraph anomalies	Efficient graph convolutions, good for structured graphs	Limited to fixed graphs, scalability issues with large graphs	Can be adapted for dynamic graphs, but computationally intensive
GAT (Graph Attention Network)	Cybersecurity, Financial	CICIDS, Kaggle Credit Card Fraud	Node anomalies, edge anomalies, fraud rings	Uses attention mechanisms to focus on relevant neighbors	High computational cost due to attention mechanism	Scalable with sparse graphs, but slow on large graphs due to attention layers
GraphSAGE (Graph Sample and Aggregation)	Financial Fraud, Cybersecurity	Kaggle Credit Card Fraud, NSL-KDD	Node anomalies, edge anomalies, fraud detection	Scales to large graphs using sampling and aggregation	Loss of information with limited neighborhood sampling	Excellent for large-scale graphs, supports real-time updates through neighborhood sampling

(*Continued*)

Table 15.6 Summary table of GNN Models for anomaly detection. (*Continued*)

GNN model	Domain	Dataset	Type of anomalies detected	Advantages	Challenges	Scalability/ deployment
ChebNet (Chebyshev Graph Convolution)	Cybersecurity	CICIDS	Node and subgraph anomalies	Efficient for spectral graph convolution without requiring full graph Laplacian computation	Computational complexity grows with graph size	Effective for large datasets but still faces scalability issues
Graph Isomorphism Network (GIN)	Financial Fraud, Cybersecurity	Kaggle Credit Card Fraud, NSL-KDD	Node anomalies, transaction fraud, money laundering	Powerful expressive power for graph structure learning	High computational cost, not efficient for real-time processing	Requires efficient hardware for large-scale graphs
Relational Graph Convolution Network (R-GCN)	Financial Fraud, Cybersecurity	Kaggle Credit Card Fraud, CICIDS	Node and edge anomalies in heterogeneous graphs	Handles complex, heterogeneous graph data well	Struggles with very large and dense graphs	Scalable with partitioning, works well with multi-relational data

(*Continued*)

Table 15.6 Summary table of GNN Models for anomaly detection. (*Continued*)

GNN model	Domain	Dataset	Type of anomalies detected	Advantages	Challenges	Scalability/ deployment
GCN+LSTM (Graph Convolutional Network + Long Short-Term Memory)	Financial Fraud, Cybersecurity	Kaggle Credit Card Fraud, CICIDS	Time-series anomalies, transaction behavior anomalies	Combines spatial (graph) and temporal (sequence) features	Complex architecture, training is time-consuming	Can be used for streaming data with temporal dependencies, but computationally expensive

like FinTech and cybersecurity, owing to ongoing fraud type evolution [15]. Table 15.6 summarizes open research challenges and future directions in graph-based anomaly detection.

15.10 Conclusion

GAD utilizes GNNs to position itself as a sophisticated system that detects sophisticated threats in both fraud and cybersecurity contexts. Conventional rule systems cannot keep up with the excellent performance that this anomaly detection method shows when it is in action within the FinTech sector and enterprise cybersecurity detection duties. The GraphSAGE and GCN, and GCT-GNN implementations of GAD models provide security benefits through their ability to understand complex node relationships in user-device-transactions networks for threat detection. GNNs detect suspicious patterns by analyzing sudden large account transfers between disconnected financial accounts and abnormal network device communication. The evaluation results demonstrate that GraphSAGE functions best in dynamic large-scale networks but GCNs perform best on structured graphs and GCT-GNNs achieve higher accuracy through their attention mechanism which concentrates on crucial graph substructures. GNN deployment in production settings faces essential challenges because of scalability issues together with real-time inference restrictions along with adversarial attacks as well as explainability requirements. GNNs must demonstrate high scalability when processing large financial transaction data volumes and real-time inference is essential for preventing fraud as transactions take place. Graph structures face threats from adversarial attacks by attackers who manipulate them to bypass detection systems primarily in applications with security requirements. GNN decision-making processes lack transparency which prevents easy output explanation especially in industries with strict regulations. Looking to the future, developments in self-supervised learning, graph contrastive learning, and federated learning hold the potential to take GAD models to new heights of scalability, flexibility, and privacy. Self-supervised GAD, for instance, has the potential to minimize dependence on labeled data, while federated learning enables joint model training without exposing sensitive information. Cross-modal anomaly detection can further enhance GAD by fusing data from different sources to better equip it with the ability to detect complex threats. GAD, based on GNNs, will thus play a key role in the future of fraud detection and cybersecurity by providing more dynamic and scalable solutions than those currently in use.

AI disclosure statement: The author acknowledges the use of AI-based tools for minor language editing or grammatical correction. However, all intellectual contributions, analyses, and conclusions presented in this chapter are entirely the author's own.

References

1. Fatoki, J.O., The influence of cyber security on financial fraud in the Nigerian banking industry. *Int. J. Sci. Res. Arch.*, 9, 2, 503–515, Aug. 2023, doi: 10.30574/ijsra.2023.9.2.0609.
2. Alshamrani, A., Myneni, S., Chowdhary, A., Huang, D., A Survey on Advanced Persistent Threats: Techniques, Solutions, Challenges, and Research Opportunities. *IEEE Commun. Surv. Tutorials*, 21, 2, 1851–1877, 2019, doi: 10.1109/COMST.2019.2891891.
3. Ahmad, Z., *et al.*, Anomaly Detection Using Deep Neural Network for IoT Architecture. *Appl. Sci.*, 11, 15, 7050, Jul. 2021, doi: 10.3390/app11157050.
4. Azumah, S.W., Elsayed, N., Adewopo, V., Zaghloul, Z.S., Li, C., A Deep LSTM based Approach for Intrusion Detection IoT Devices Network in Smart Home, in: *2021 IEEE 7th World Forum on Internet of Things (WF-IoT)*, New Orleans, LA, USA, Jun. 2021, IEEE, pp. 836–841, doi: 10.1109/WF-IoT51360.2021.9596033.
5. Inuwa, M.M. and Das, R., A comparative analysis of various machine learning methods for anomaly detection in cyber attacks on IoT networks. *Internet Things*, 26, 101162, Jul. 2024, doi: 10.1016/j.iot.2024.101162.
6. Alabadi, M. and Celik, Y., Anomaly Detection for Cyber-Security Based on Convolution Neural Network : A survey, in: *2020 International Congress on Human-Computer Interaction, Optimization and Robotic Applications (HORA)*, Ankara, Turkey, Jun. 2020, IEEE, pp. 1–14, doi: 10.1109/HORA49412.2020.9152899.
7. Ekle, O.A. and Eberle, W., Anomaly Detection in Dynamic Graphs: A Comprehensive Survey. *ACM Trans. Knowl. Discov. Data*, 18, 8, 1–44, Sep. 2024, doi: 10.1145/3669906.
8. Ullah, I. and Mahmoud, Q.H., A Framework for Anomaly Detection in IoT Networks Using Conditional Generative Adversarial Networks. *IEEE Access*, 9, 165907–165931, 2021, doi: 10.1109/ACCESS.2021.3132127.
9. Boorugupalli, K.K., Kulkarni, A.K., Suzana, A., M, D., Ponnusamy, S., Kumar S, S., Cybersecurity Measures in Financial Institutions Protecting Sensitive Data from Emerging Threats and Vulnerabilities. *ITM Web Conf.*, 76, 02002, 2025, doi: 10.1051/itmconf/20257602002.
10. El-Shafeiy, E., Alsabaan, M., Ibrahem, M.I., Elwahsh, H., Real-Time Anomaly Detection for Water Quality Sensor Monitoring Based on Multivariate

Deep Learning Technique. *Sensors*, 23, 20, 8613, Oct. 2023, doi: 10.3390/s23208613.

11. Marfo, W., Tosh, D.K., Moore, S.V., Enhancing Network Anomaly Detection Using Graph Neural Networks, in: *2024 22nd Mediterranean Communication and Computer Networking Conference (MedComNet)*, Nice, France, Jun. 2024, IEEE, pp. 1–10, doi: 10.1109/MedComNet62012.2024.10578278.
12. Khemani, B., Patil, S., Kotecha, K., Tanwar, S., A review of graph neural networks: concepts, architectures, techniques, challenges, datasets, applications, and future directions. *J. Big Data*, 11, 1, 18, Jan. 2024, doi: 10.1186/s40537-023-00876-4.
13. Alshaibi, A., Al-Ani, M., Al-Azzawi, A., Konev, A., Shelupanov, A., The Comparison of Cybersecurity Datasets. *Data*, 7, 2, Jan. 2022, doi: 10.3390/data7020022.
14. Pirani, M. and Sundaram, S., On the Smallest Eigenvalue of Grounded Laplacian Matrices, *arXiv:* arXiv:1406.2271, Jul. 06, 2014, doi: 10.48550/arXiv.1406.2271.
15. Ng, A.W. and Kwok, B.K.B., Emergence of Fintech and cybersecurity in a global financial centre: Strategic approach by a regulator. *JFRC*, 25, 4, 422–434, Nov. 2017, doi: 10.1108/JFRC-01-2017-0013.
16. Jyothsna, V., Rama Prasad, V.V., Munivara Prasad, K., A Review of Anomaly based Intrusion Detection Systems. *IJCA*, 28, 7, 26–35, Aug. 2011, doi: 10.5120/3399-4730.
17. Lo, W.W., Kulatilleke, G.K., Sarhan, M., Layeghy, S., Portmann, M., Inspection-L: self-supervised GNN node embeddings for money laundering detection in bitcoin. *Appl. Intell.*, 53, 16, 19406–19417, Aug. 2023, doi: 10.1007/s10489-023-04504-9.
18. Scarselli, F., Gori, M., Tsoi, A.C., Hagenbuchner, M., Monfardini, G., The Graph Neural Network Model. *IEEE Trans. Neural Netw.*, 20, 1, 61–80, Jan. 2009, doi: 10.1109/TNN.2008.2005605.
19. Bello, O.A., Folorunso, A., EJiofor, O.E., Budale, F.Z., Adebayo, K., Babatunde, O.A., Machine Learning Approaches for Enhancing Fraud Prevention in Financial Transactions, no. 1, 2023.
20. Sina, A., Open AI and its Impact on Fraud Detection in Financial Industry. *JKLST*, 2, 3, 263–281, Sep. 2024, doi: 10.60087/jklst.vol2.n3.p281.
21. Kim, H., Lee, B.S., Shin, W.-Y., Lim, S., Graph Anomaly Detection with Graph Neural Networks: Current Status and Challenges. *IEEE Access*, 10, 111820–111829, 2022, doi: 10.1109/ACCESS.2022.3211306.
22. Moustafa, N., Hu, J., Slay, J., A holistic review of Network Anomaly Detection Systems: A comprehensive survey. *J. Network Comput. Appl.*, 128, 33–55, Feb. 2019, doi: 10.1016/j.jnca.2018.12.006.
23. Saluja, S., Identity theft fraud- major loophole for FinTech industry in India. *JFC*, 31, 1, 146–157, Jan. 2024, doi: 10.1108/JFC-08-2022-0211.
24. Sun, L., *et al.*, Adversarial Attack and Defense on Graph Data: A Survey. *IEEE Trans. Knowl. Data Eng.*, 1–20, 2022, doi: 10.1109/TKDE.2022.3201243.

25. Chen, Q., Xu, J., Koltun, V., Fast Image Processing with Fully-Convolutional Networks, in: *2017 IEEE International Conference on Computer Vision (ICCV)*, Venice, Oct. 2017, IEEE, pp. 2516–2525, doi: 10.1109/ICCV.2017.273.
26. Abdallah, A., Maarof, M.A., Zainal, A., Fraud detection system: A survey. *J. Network Comput. Appl.*, 68, 90–113, Jun. 2016, doi: 10.1016/j.jnca.2016.04.007.
27. Barik, K., Misra, S., Konar, K., Fernandez-Sanz, L., Koyuncu, M., Cybersecurity Deep: Approaches, Attacks Dataset, and Comparative Study. *Appl. Artif. Intell.*, 36, 1, 2055399, Dec. 2022, doi: 10.1080/08839514.2022.2055399.
28. Araújo, S.O., Peres, R.S., Ramalho, J.C., Lidon, F., Barata, J., Machine Learning Applications in Agriculture: Current Trends, Challenges, and Future Perspectives. *Agronomy*, 13, 12, 2976, Dec. 2023, doi: 10.3390/agronomy13122976.
29. Chaabouni, N., Mosbah, M., Zemmari, A., Sauvignac, C., Faruki, P., Network Intrusion Detection for IoT Security Based on Learning Techniques. *IEEE Commun. Surv. Tutorials*, 21, 3, 2671–2701, 2019, doi: 10.1109/COMST.2019.2896380.
30. Yuan, S. and Wu, X., Deep learning for insider threat detection: Review, challenges and opportunities. *Comput. Secur.*, 104, 102221, May 2021, doi: 10.1016/j.cose.2021.102221.
31. Bilot, T., El Madhoun, N., Al Agha, K., Zouaoui, A., A Survey on Malware Detection with Graph Representation Learning. *ACM Comput. Surv.*, 56, 11, 1–36, Nov. 2024, doi: 10.1145/3664649.
32. Kipf, T.N. and Welling, M., Semi-Supervised Classification with Graph Convolutional Networks, *arXiv:* arXiv:1609.02907, Feb. 22, 2017, doi: 10.48550/arXiv.1609.02907.
33. Abdul Salam, M., Fouad, K.M., Elbably, D.L., Elsayed, S.M., Federated learning model for credit card fraud detection with data balancing techniques. *Neural Comput. Applic.*, 36, 11, 6231–6256, Apr. 2024, doi: 10.1007/s00521-023-09410-2.
34. Faisal, N.A., *et al.*, Fraud Detection In Banking Leveraging Ai To Identify And Prevent Fraudulent Activities In Real-Time. *NHJ*, 1, 01, 181–197, Oct. 2024, doi: 10.70008/jmldeds.v1i01.53.
35. Sanusi, Z.M., Rameli, M.N.F., Isa, Y.M., Fraud Schemes in the Banking Institutions: Prevention Measures to Avoid Severe Financial Loss. *Procedia Econ. Finance*, 28, 107–113, 2015, doi: 10.1016/S2212-5671(15)01088-6.
36. Nuthalapati, S.B., AI-Enhanced Detection and Mitigation of Cybersecurity Threats in Digital Banking. *EATP*, 357–368, Feb. 2023, doi: 10.53555/kuey.v29i1.6908.
37. Hamilton, W., Ying, Z., Leskovec, J., Inductive Representation Learning on Large Graphs.
38. Najafi Mohsenabad, H. and Tut, M.A., Optimizing Cybersecurity Attack Detection in Computer Networks: A Comparative Analysis of Bio-Inspired

Optimization Algorithms Using the CSE-CIC-IDS 2018 Dataset. *Appl. Sci.*, 14, 3, 1044, Jan. 2024, doi: 10.3390/app14031044.

39. Thomas, R. and Pavithran, D., A Survey of Intrusion Detection Models based on NSL-KDD Data Set, in: *2018 Fifth HCT Information Technology Trends (ITT)*, Dubai, United Arab Emirates, Nov. 2018, IEEE, pp. 286–291, doi: 10.1109/CTIT.2018.8649498.
40. Jiang, S., Dong, R., Wang, J., Xia, M., Credit Card Fraud Detection Based on Unsupervised Attentional Anomaly Detection Network. *Systems*, 11, 6, 305, Jun. 2023, doi: 10.3390/systems11060305.
41. Leevy, J.L. and Khoshgoftaar, T.M., A survey and analysis of intrusion detection models based on CSE-CIC-IDS2018 Big Data. *J. Big Data*, 7, 1, 104, Dec. 2020, doi: 10.1186/s40537-020-00382-x.
42. Zhang, Q., Chang, J., Meng, G., Xu, S., Xiang, S., Pan, C., Learning graph structure *via* graph convolutional networks. *Pattern Recognit.*, 95, 308–318, Nov. 2019, doi: 10.1016/j.patcog.2019.06.012.

16

Graph Neural Networks in Healthcare

S. Sathyanarayanan

Department of Mathematical and Computational Sciences, Sri Sathya Sai University for Human Excellence, Kalaburagi, Karnataka, India

Abstract

Graph neural networks (GNNs) are a type of deep learning architecture that can process data organized in graphs. A node in the graph represents an entity and an edge explains its relationship with other entities. Traditional neural networks operate on data arranged in grids or sequences. GNNs can handle dynamic and irregular data structures. GNNs share their information with nodes in their immediate vicinity and promote the concept of links to nodes of any size. GNNs can help advance health care. They assist in interpreting molecular topology and interplay. By predicting the interactions between molecules and proteins and identify possible molecules, GNNs might be used to streamline the process of drug discovery. Virtual screening of large databases can be used to identify compounds that have greater potential, thus limiting the trial-and-error process. GNNs may help in diagnosis and treatment planning by modeling the relationships between various features and analyzing medical images. GNNs may also play a significant role in bioinformatics to assist in the analysis of gene regulatory networks and prediction of protein-protein interactions. It may also be applied to reuse drugs in different ways by analyzing compound-target interaction networks. Information siloes can be mitigated with the help of GNNs, because they can merge various data types into a single heterogeneous graph. GNNs are computationally expensive and slow compared with their simpler counterparts. However, this is offset by the lower cost of the experiments, capability to combine multi-modal data, and scalability owing to their effectiveness in the screening of ultra-large libraries of molecules. This chapter provides an overview of the application of GNNs in the healthcare domain and their potential, with illustrations of cases in which their benefits have been demonstrated.

Email: sathyanarayanan.s@sssuhe.ac.in

J. Ramkumar, Sridaran Rajagopal, B. Suchitra and S. Balamurugan (eds.) Graph Neural Networks: Concepts and Applications, (663–700)

***Keywords*:** Graph neural networks, neural networks, healthcare, deep learning, machine learning, electronic health records, relational data

16.1 Introduction

Enhancing health care using artificial intelligence (AI) has made it easier to study and apply medical data to inform medical choices. Studies in this area indicate that traditional machine learning (ML) and deep learning (DL) have significantly improved image classification, disease prediction, and patient grouping based on risk [1–4]. However, these techniques usually treat every observation in healthcare data as an isolated point and are unable to recognize the close connections that are present. Healthcare data are often inconsistent, with diseases spreading through networks and treatment affecting multiple patients at once. Complex networks formed by genes, proteins, and drugs in biomedicine are vital for understanding how diseases occur and for finding new therapies.

Graph neural networks (GNNs) have been developed to address complex relationships between data points. Since GNNs use graphs and nodes for patients or providers, and edges for the ways they relate, they can discover dependencies and structures that standard approaches miss. Consequently, doctors can make more accurate predictions, represent patients better, and understand how diseases develop and their treatment outcomes better.

GNNs are used in healthcare for two reasons: they help with predictive modeling and knowledge discovery by studying the relationships in medical data. Therefore, it is possible to use patient similarity networks, model disease spread, and predict interactions between drugs, all of which are unachievable using standard DL methods. As the amount and details of healthcare data are increasing, it is more important than ever to address how things are related. Utilizing GNNs can help make full use of healthcare data, and as a result, support the delivery of effective, personalized, and fair medical care.

Limitations of Traditional Deep Learning

Conventional DL approaches are limited in healthcare because they represent data as a table and miss the detailed relationships found in the healthcare records. Healthcare data are usually organized in relational databases that use separate tables to show how patients, diseases, treatments, and clinical situations relate to one another. To use conventional DL, researchers typically convert relational data into a regular table by themselves which can lead to significant data loss and demands the manual design of special features by experts [5].

Because this method ignores detailed dependencies and timing, it is unreliable and makes modeling challenging for healthcare applications. For instance, adding features such as the number of reviews or trends over a given period may fail to account for the dynamic connections that a model can uncover if it can use the full connections of the data [5]. Additionally, manually designing features requires considerable domain knowledge, and the work must often be repeated whenever data shifts or other clinical enquiries are introduced, making them less adaptable and scalable.

Traditional DL models pose various other healthcare challenges: they require a large amount of well-labeled data, while high-quality training sets are rarely available in healthcare; they are not easily understandable, so clinical teams may not trust them; and they are prone to problems caused by common distortions and noisy data which are usually seen in medical data [1, 6, 7]. Owing to these concerns, deploying DL in key healthcare areas is difficult because it is essential to explain and validate the actions of the models. In general, traditional DL in healthcare is hindered by its inability to deal with related data structures which leads to data loss, and it faces the problems of limited examples, conformity, and being easily deceived by faulty data [5–7].

The above-mentioned motivations and limitations of DL have led us to explore the use of GNNs to overcome these issues. GNNs are becoming increasingly important because they can easily analyze and learn from complicated relationships in healthcare data. The connections between patients, problems, therapies, and outcomes are captured using patient similarity networks, molecular interaction maps, and electronic health records (EHR). GNNs can reveal information to make more precise predictions regarding diseases, classify risk factors, and provide personalized medicine. As a result, GNNs are now being used for various applications in healthcare, from examining brain connections using medical images to predicting outcomes from a patient's treatment history [8, 9]. We now explore the fundamentals of GNNs.

16.2 Fundamentals of Graph Neural Networks

In graph theory, mathematics is used to explore systems that are composed of objects and their connections. Graph theory describes aspects of a graph.

- **Nodes (or vertices)**: The graph includes individual entities such as patients, proteins, or medical ideas. These are

represented by nodes. For example, each node represents a patient with attributes, such as age, gender, and medical history.
- **Edges (or links)**: Edges represent the relationships between nodes in a network (e.g. actions, likenesses, or shared characteristics). The edges capture relationships such as patient similarity, temporal sequences, and causal interactions.

Graphs can be undirected (edges have no direction) or directed (edges move from one node to another), and may have additional properties, such as edge weights or labels. Graphs are highly flexible and can represent a wide range of complex structures, making them ideal for modeling relational data in healthcare applications. Relational data in healthcare often take the form of relational databases, where information is stored across multiple tables (e.g. a patient table, diagnosis table, and medication table). GNNs excel at modeling the primary-key-to-foreign-key relationships between these tables directly as a graph. GNNs transform isolated medical records into interconnected networks, thereby representing patient health data as graphs. This enables the analysis of complex relationships and patterns among records.

Message Passing

GNNs were built to analyze the data arranged as graphs. GNNs can predict intricate interdependencies in healthcare data owing to their inherent capabilities [7, 10]. The message-passing mechanism is the basis of the functionality of a GNN while working with graph data. Each node in the graph can collect and incorporate data from its close neighbors because of this iterative process, which spreads knowledge across the graph structure.

Process of Message Passing

The message-passing mechanism for each node in each layer of the GNN is developed in two primary steps.

1. **Message Creation and Combination**
   - *Message Generation*: Every node first creates "messages" using the characteristics of its outgoing edges and its own features. The immediate neighbors are then notified of the messages.
   - *Aggregation*: A node compiles all messages it receives from all its inbound neighbors into a single summary. A well-chosen function (such as sum, mean, max, or more intricate neural-network-based aggregators like those in

GraphSAGE) performs this aggregate. The objective is to condense all neighborhood information into a clear picture.

2. **Node Upgrade**
   - Following aggregation, the node merges its current feature representation (also known as its "stored representation") with aggregated neighborhood data.
   - These combined data are then processed by a learnable update function (usually a neural network or nonlinear transformation) to provide a fresh, updated feature representation for the node. The node can now incorporate information about its local neighborhood thanks to this update.

Iterative Propagation and Global Understanding

In the GNN architecture, this message-passing and node-updating procedure is repeated at several levels. Each layer spreads information deeper through the graph.

- **Layer 1**: The representation of a node is modified by its nearest (1-hop) neighbors.
- **Layer 2**: A node implicitly combines information from its 2-hop neighbors (neighbors of its neighbors) owing to the updated representations from layer 1.
- **Subsequent Layers**: Nodes learn about the graph's larger structural background and progressively farther connections as the process progresses. GNNs can identify both direct and indirect dependencies in data because of this iterative information flow.

Graph Construction Approaches

1. *Patient Similarity Networks*: In patient similarity networks, the nodes represent patients, and the edges with weights represent the similarity metrics. One example is the use of the Euclidean distance between embedded clinical features. For example, k-nearest neighbors (kNN) can be applied to graphs of diagnoses, procedures, and medications to link patients with similar treatment histories. This enables GNNs to gather insights from similar patients even when individual data are sparse [11].
2. *Temporal Event Graphs*: In temporal event graphs, nodes represent medical events, such as lab tests and prescriptions

connected chronologically. They can be used to model disease progression and treatment course. An example is a long short-term memory (LSTM)-GNN hybrid that captures both temporal patterns and graph-structured dependencies [12].

3. *Heterogeneous Knowledge Graphs*: Heterogeneous Knowledge Graphs combine multiple entity types such as patients, drugs, and diseases, as well as relationship types such as "prescribed", "causes", etc. An example is the linking of patients to diagnoses (node A) and medications (node B) *via* edges labeled "treated with" [13].

Core GNN Architectures

GNNs have been developed into several core architectures, each with unique mechanisms for collecting and updating information from the graph-structured data. GNN architectures are optimized to suit healthcare data through modifications that address temporal relationships, data sparsity, multi-modal integration, and other challenges. Here, we provide a brief overview of the most widely used state-of-the-art GNN architectures for healthcare system implementations. Figure 16.1 shows the major types of GNNs.

Graph Convolutional Networks (GCN)

A fundamental type of GNN called a graph convolutional network (GCN) extends the concept of convolution from grid-like inputs (such as images) to arbitrary graph topologies. The main concept is to iteratively

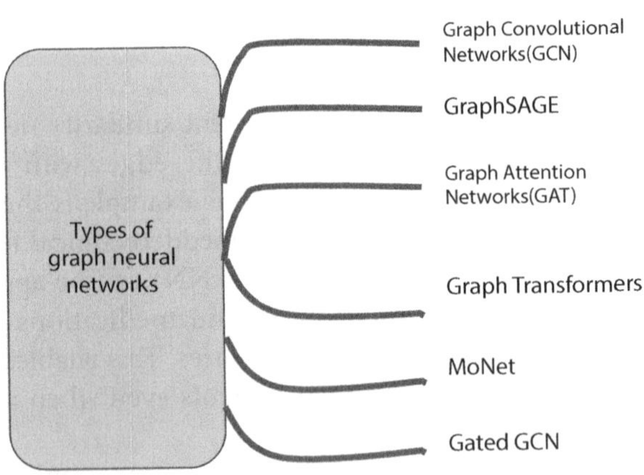

Figure 16.1 Types of graph neural networks.

aggregate and transform data from a node's close neighbors to learn node representations.

A weighted average of a node's neighbors' (and often its own) features is used by the GCN layer to update the node's feature vector. This is followed by a linear transformation and nonlinear activation. Using this approach, nodes can acquire representations that capture both their own characteristics and the structural information from their immediate neighborhood, efficiently "smoothing" the features throughout the graph based on its connections. GCNs are well-known for their effectiveness and capacity to manage graph-based semi-supervised learning tasks.

Temporal Event Handling: GCN is modified with LSTM-like gating mechanisms to process sequential medical events (e.g. lab tests and prescriptions) in heterogeneous graphs. For example, an Event-involved GCN (EGCN) connects patients, diseases, and hospitals to a medical-treatment graph. It uses attention to prioritize relationships like "hospitalization events" while capturing time-series dependencies.

Multi-Modal Fusion: In creating patient similarity graphs, AMA-GCN dynamically weighs phenotypic measures (such as age and imaging traits), increasing the accuracy of Alzheimer's and brain cancer prediction (98.4% on the ABIDE dataset) [14].

Edge Weight Calibration: To minimize noise in scarce EHR data, Edge Weight Calibration modifies patient similarity metrics based on clinical variables. For instance, EGCN uses attention to rank important associations, such as "hospitalization events", in order to mimic medical treatment migration by linking people, illnesses, and hospitals in a heterogeneous network.

GraphSAGE

GraphSAGE stands for Graph SAmple and aggreGatE, a type of GNN. It is designed for learning low-dimensional vector representations (embeddings) of the graph nodes. It uses inductive learning, meaning that it learns from a general rule or function from specific training instances. The aim is to generalize these learned patterns so that they can be used to make predictions on unseen data. This makes it effective for large dynamic graphs that change over time. It learns to generate a node's embeddings by looking at its local neighborhood instead of memorizing an embedding for every single node that has already been observed.

Inductive learning overcomes the limitations of transductive learning used by GCN. GCN models require a complete graph during training. GCNs learn a specific embedding for each node in the training graph. The model must be retrained from scratch if a new node is added to the graph, which is impractical for real-world applications.

The "Sample and Aggregate" Process

For every node, GraphSAGE operates iteratively by performing two steps to generate an embedding, which is typically repeated across several layers. In the first step, GraphSAGE samples a small, fixed number of neighbors instead of looking at all its neighbors. This is done to guarantee the optimality and scalability of the method to perform computation on larger graphs. The model then aggregates the feature representation of the chosen neighbors into a vector through an aggregator function. This embedding represents the combined message of the local neighborhood. The composition of the aggregators varied. Some of the types are:

- *Mean aggregator*: This simple but effective method takes the element-wise mean of the feature vectors of the neighbors.
- *Pooling aggregator*: This method captures the most prominent features from the neighborhood by applying an element-wise function, such as the maximum or minimum, to the neighbors' vectors.
- *LSTM aggregator*: The LSTM aggregator is a powerful and complex aggregator that applies LSTM to a random set of neighbors. This aggregator is computationally expensive, but can learn more intricate relationships.

This single vector, generated after aggregation, is combined with the current feature vector of the target node. The new updated representation of the node for that layer is then produced by passing this combined vector through a neural network with a nonlinear activation function. This procedure is performed layer by layer, enabling every node to collect data from increasingly distant parts of the graph.

Focal Loss: With many healthy or common cases and few cases of the disease of concern, medical databases are frequently inherently skewed. A standard model can attain high accuracy by ignoring uncommon instances. Focal Loss addresses this issue by dynamically reducing the weight of the majority class during the training. This forces the model to focus on harder-to-detect minority classes, improving sensitivity and utility.

GraphSAGE has several advantages. It is scalable because the neighbor sampling technique can be applied to graphs with billions of nodes and edges as the computation required per node is fixed and predictable. It is suitable for real-world applications in which graphs are dynamic, such as social networks, e-commerce, and healthcare systems. A new patient entering a hospital can be analyzed or features for new users can be generated as they register for treatment. Generalization is another important advantage because a trained GraphSAGE model can be applied to a new graph if it has the same types of features as the trained model [7].

Healthcare Adaptions of GraphSAGE: For real-world EHR systems, inductive learning allows the dynamic addition of new patients to pre-trained graphs without retraining [15]. Focus loss is used to prioritize minority classes (e.g., heart failure cases) in datasets that are imbalanced [16]. As an example, GraphSAGE used kNNs to construct similarity graphs of patients based on diagnoses, procedures, and medications to achieve an Area Under the Precision-Recall Curve (AUPRC) comparable to that of Random Forest (0.5476 by the GraphSAGE method compared to 0.5132 with Random Forests) [17]. The GraphSAGE algorithm analyzed the medical histories of patients with similar diagnoses, treatments, and medications, such as shared diagnoses, treatments, and medications. This method matched the accuracy of traditional methods, such as Random Forest, while also revealing how patient relationships affect the outcomes.

Graph Attention Networks (GAT)

Graph Attention Networks (GATs) use an innovative attention mechanism that makes them different from other GNNs. Older GNNs treat all neighboring nodes as equally important or use fixed aggregation schemes. GATs assign different attention weights to each neighbor, giving more importance to its more relevant neighbors when aggregating information and providing a deeper understanding of the local graph structure. This ability also gives GATs inductive capabilities, enabling them to generalize effectively to unseen nodes and graph structures.

The GAT layer first linearly transforms the node features. Next, a tiny neural network is used to compute a raw attention score between each node and its neighbors. A softmax function is then used to normalize these raw scores, yielding attention coefficients that show the importance of each neighbor in relation to the others. Finally, the new feature representation of a node is calculated as the weighted sum of the converted features of its neighbors, where the weights are the attention coefficients. By enabling the model to learn a variety of patterns from the neighborhood, the widely used multi-head attention technique improves GATs even more and produces node representations that are stronger and more resilient [9, 10]. The following are several examples of how GAT has been adapted to healthcare.

- *Patient Similarity Attention*: Patient-GAT uses Dynamic Time Warping (DTW) to measure temporal lab series similarity by combining imputed lab data with demographics and medications. The Dynamic Time Warping (DTW) algorithm is a technique used to determine how comparable two time series are despite differences in duration or speed. DTW "warps" the time axis of one or both sequences to choose the best alignment that minimizes the overall difference between them. This is in contrast to more straightforward techniques, such as Euclidean distance, which demand that sequences have the same length and are precisely aligned point-by-point [18].

- *Interpretable Predictions*: The influential features are highlighted by attention coefficients. An example is specific medications in sarcopenia prediction [18].
- *Hierarchical Embeddings*: HealthGAT improves the accuracy of medical code representations by training them in related tasks. Models benefit from this understanding because it helps them predict whether patients will be readmitted more accurately [19]. Patient-GAT offers superior results for chronic disease prediction compared to baselines, identifying lab trends as key factors [18].

Graph Transformers

Graph Transformers are adapted to the well-performing self-attention mechanisms of conventional transformers. By allowing any node to directly attend to information from every other node in the graph, Graph Transformers improve their ability to capture global context and long-range dependencies, in contrast to conventional GNNs that rely on local message forwarding.

The attention mechanism is modified to include graph topology in the Graph Transformers. This is frequently performed using positional or structural encodings (e.g. those based on graph Laplacian eigenvectors or shortest paths). Graph Transformers consider the significance of every other node, irrespective of their direct connectivity, and are thus able to learn rich node representations while still gaining access to the local graph information. They effectively capture complex graph interactions and adapt well to diverse graph tasks. Recent hybrid models, such as the General, Powerful, Scalable Graph Transformer (GPS), combine GNN layers (such as GCN) with transformer layers to efficiently leverage both the local structure and global context. For example, GPS processes a graph in parallel through both the GNN and Transformer components, and merges the results, achieving scalability and strong performance on large and complex graphs [7].

Other Architectures

There have been additional variants of GNNs such as MoNet (that employs learnable Gaussian filters to aggregate neighbors) and Gated GCN (that introduces gating to better control message passing) which may be uniquely suitable to tasks and data structure [7]. These architectures have been the foundation of modern graph-based ML and are beginning to be deployed to healthcare datasets, where both local interaction and global dependencies are more important for effective modeling and prediction. The different types of GNNs, their core techniques, their pros and cons, and their use in healthcare are given concisely in Table 16.1.

Use in Medical Databases

A significant alteration occurs when this is applied to healthcare databases, and every row of a traditional relational table can be considered a stand-alone node. Importantly, a rich and diverse graph structure is created by transforming the connections between the database's primary and foreign keys into edges. Because a GNN can directly exploit the full relational structure to learn complicated connections and even the temporal

Table 16.1 Comparison of different types of GNNs.

GNN type	Core technique and uniqueness	Features, pros and cons	Healthcare use cases & reference(s)
GCN	Spectral/spatial convolution; efficient local aggregation; can be adapted for temporal/ multimodal data	+ Scalable, semi-supervised, handles heterogeneity - Limited to shallow neighborhoods, ignores edge features	Disease gene prioritization, patient similarity networks [20]
GAT	Attention-based neighbor aggregation; learns importance of each neighbor; interpretable attention weights	+ Interpretable, robust to noise, focuses on relevant neighbors - Computationally intensive, sensitive to hyperparameters	Disease progression from EHR, patient similarity [21]

(Continued)

Table 16.1 Comparison of different types of GNNs. (*Continued*)

GNN type	Core technique and uniqueness	Features, pros and cons	Healthcare use cases & reference(s)
GraphSAGE	Inductive learning with neighbor sampling; supports unseen nodes and dynamic graphs	+ Scalable, real-time updates, preserves feature granularity - Sampling bias, limited global context	Heart failure prediction, patient clustering, predicting drug-drug interactions [17, 22]
Graph Transformer	Global self-attention + GNN; combines local and global context, long-range dependencies	+ Captures long-range dependencies, expressive, interpretable - High memory/cost, complex training	Drug–drug interaction, multimodal fusion [23]
MoNet	Mixture model CNNs; flexible convolution kernels for non-Euclidean domains	+ Flexible for spatial/temporal/heterogeneous data - Complex, less standardized, needs tuning	Brain region analysis in MRI [24]
Gated GCN	Message passing with edge/node gating; controls information flow, models complex dependencies	+ Improved message passing, handles edge features/dynamics - Less common, computationally heavy, needs tuning	Digital twin for patient vital forecasting [25]

characteristics inherent in the data, this method does not require tedious feature engineering or manual data flattening [7, 10]. As a result, GNNs are in a unique position to locate subtle, difficult-to-find links in non-Euclidean healthcare data, which can lead to more accurate predictions, enlightening discoveries, and ultimately, better patient outcomes [9].

16.3 Applications of GNNs in Healthcare

In healthcare, GNNs have demonstrated their usefulness by building and evaluating patient-similarity graphs for predicting diseases, triaging patients, and grouping high-risk patients. All applications in this category place each patient as a node, with edges showing similarities in their diagnoses, medications, procedures, or laboratory outcomes [9, 10]. Owing to this structure, GNNs can detect intricate relationships that regular models cannot capture. The improvement in performance with the adoption of GNN is presented in Table 16.2.

The broad categories of applications of GNNs in healthcare are shown in Figure 16.2.

Table 16.2 Performance comparison.

Application	GNN model	Traditional model	Improvement	Source
CVD Prediction	GAT	Logistic Regression	+21.65% accuracy	[26]
Heart Failure Risk	GraphSAGE	Random Forest	Comparable AUPRC + relational insights	[26]
ED Triage	GCN	LSTM	+18% F1-score	[26]
Breast cancer survival	GCN	CNN	+12% accuracy	[26]

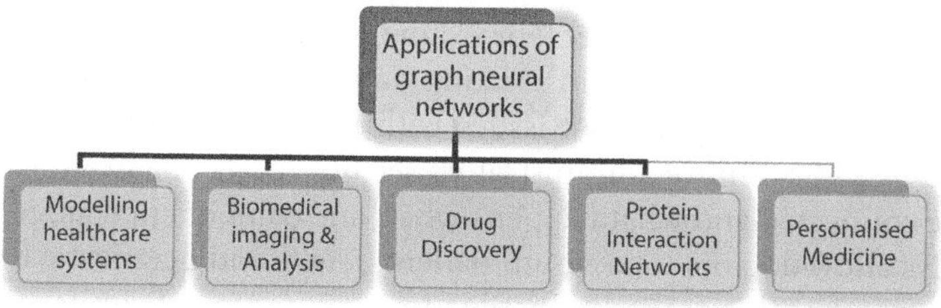

Figure 16.2 Applications of GNNs in healthcare.

Modeling Healthcare Systems Using EHRs: Mapping Relationships Among Patients, Providers, and Institutions

To predict heart failure, researchers built graphs of similar patients from EHR data using codes for disease, procedures, and drugs. Tests on different graphs have revealed that GNN models such as GraphSAGE, Graph Attention Networks (GAT), and graph transformers provide better predictions and are more understandable than standard ML techniques. With graph-based attention, the model is better able to find key associations among patients, and as a result, makes improved and significant predictions. This is why GNNs have also been used in emergency departments for auto triaging, using graphs for patient data, and node classification to assign patients to different risk or priority groups. With a similarity network among patients, GNNs can increase the accuracy of triage and better organize patient treatments.

Similarity graphs for cardiovascular and pulmonary conditions have been used to predict the likelihood of chronic diseases. GNNs can discover hidden patterns and gather useful information from patients with similarities, allowing solid grouping of patients with different types of data. In general, using GNNs with patient-similarity graphs enables the identification of relevant indicators of patient illnesses, predicts diseases sooner, and improves choices in both emergency and ongoing care settings. Using a complete network of EHR data, this approach tends to be more accurate, clearer to understand, and can be adapted more easily than standard DL models [9, 27].

GNNs provide a useful way to model detailed interactions among patients, providers, healthcare institutions, and others. There are many layers in healthcare systems, patients communicate with different providers, and healthcare institutions share and organize care resources. Owing to the GNNs, gaining insight into and improving such complex connections is easier than before. GNNs can demonstrate the network representation of a healthcare system by mapping where patients, doctors, hospitals, and insurers meet or work together. For example, patient-provider-institution network can identify trends in treating patients, identify providers who play a large role and point out issues with patient procedures [28]. This method helps healthcare administrators see how parts of a healthcare system are connected and what happens when one component changes.

GNNs allow doctors to use both patient details and diagnostic findings to draw graphs. This makes it possible to form groups of similar patients and to discover the best way to treat them. They have also helped examine how referrals are made between primary doctors and specialists, showing common patterns and improving how referrals are coordinated. Unlike traditional ML, GNN approaches are more effective for therapy recommendations, referral prediction, and discovery of special clusters and structures in healthcare.

GNNs also support the bringing together of different data, such as EHR, administrative claims, and provider directories, into just one analysis framework. From this perspective, healthcare organizations are better able to predict patient outcomes, see how resources are used, track their performance and take data-informed steps to improve [6]. GNNs stand out in this field by allowing us to study and refine healthcare systems by analyzing how patients, providers, and institutions relate to each other.

Biomedical Imaging and Analysis

GNNs have seen an increase in importance in biomedical imaging and analysis, mainly for tasks such as brain scanning and studying how parts of the brain are connected. Medical images, including MRI, fMRI, and pathology slides can be converted into graphs. Here, the anatomical regions, voxels or image patches are the nodes and the edges show how these areas are related [29, 30].

GNNs are used in brain imaging to portray the brain as a network comprising of different regions. The brain areas are the nodes and edges representing the degrees or equalities in the brain, be it in the brain activity or anatomical routes [10, 31]. Such a design enables GNNs to notice both time and space relationships and perform tasks such as disease diagnosis, abnormality detection, and the exploration of different neurological issues. To give an example, GNNs have learned topological information in fMRI data successfully, which allowed scientists to identify abnormal brain regions and links associated with neurogenerative pathologies [27, 31].

GNNs are used in histopathology to cluster tumors, forecast outcomes and profiling different sections of the tissue, all of which is dependent on matters concerning proximity between adjacent areas of cells and tissues [5, 6]. The GNNs outperforms and is more effective than multilayer perceptrons (MLP) in medical 3D image classification, particularly, when the graph representation is utilized to capture 3D volumetric data [7]. Therefore, a strong biomedical imaging approach provided by GNNs discloses complex natural interrelationships in biomedical data. As a result, these techniques improve the accuracy and transparency of analyses in brain network mapping, disease prediction and image-guided diagnosis, showing their greater importance in computational medicine [6, 10, 29, 30].

Drug Discovery

Because GNNs process graph data, chemists use them in computational drug discovery to represent molecules as graph nodes, and the bonds among those molecules as graph edges. Thanks to this framework, GNNs and, in particular, GCNs can identify molecular characteristics from the structure and chemistry of compounds, leading to accurate prediction of properties including solubility, toxicity and bioactivity [6]. GNNs enable a model to predict several molecular properties simultaneously, thereby improving the drug discovery process and encouraging task-related knowledge transfer. GNNs can represent complex networks, such as drug-drug interactions, drug-target interactions, and drug-disease associations. Some researchers have used these skills to match old medicines with new applications by investigating the complicated patterns in drug networks. Even with this new approach, it remains difficult to access the data, explain the results of the models, and consume a large amount of computer resources. Therefore, researchers continue to find new ways to enable the better use and understanding of GNNs.

Protein Interaction Networks

The interactions of proteins are very important, as learning about these interactions (Protein-Protein Interactions (PPIs)) helps us understand how cells function and what causes disease. GNNs, including GCNs and GATs, have been found to be effective in predicting PPIs by modeling proteins as graphs that link amino acids or residues according to how close or related they are. By integrating both types of data, this method improves and scales PPI predictions better than previously used methods. They make use of both the internal structure and interconnections among proteins, allowing them to understand both how proteins function and where they interact with others [9, 10, 30].

Personalized Medicine

In personalized medicine, GNNs help to make better treatment recommendations by organizing patient data as graphs. All patients become nodes and similarities in clinical, genetic or demographic data form the network's edges [7]. Family physicians who use GNNs can learn from the experiences of similar patients, make more accurate predictions, and design individual treatment plans. Automatic mechanism in advanced GNN frameworks can select the most important features for constructing similarity graphs among patients, refining predictions, and providing assistance for clinical decision-making [9].

Overall, GNNs are essential for finding drugs, analyzing protein interactions, and personalized healthcare, because they help make sense of the difficult relationships found in these fields. As a result of these developments, biomedical innovation is occurring more quickly, while challenges related to how results can be understood, how many datasets can be combined, and the efficiency of computer systems remain [10].

16.4 Case Studies

Case Study 1: Heart Failure Prediction

New research findings have revealed that GNNs perform well in predicting heart failure (HF) by considering the similarity graphs of patients drawn from EHR. In these studies, every patient was represented by a node and the edges connected nodes that had common diagnoses, procedures, or medication records. Prediction of cardiovascular disease (CVD) was performed on Australian insurance claims (2,610 patients) with demographics, diagnoses, and medication data. A modified MAT using a weighted patient network (WPN) was constructed using shared diagnoses and demographics as edges. Critical predictors, such as cholesterol levels and prior heart attack, were identified using the attention mechanism. The accuracy obtained was 93.49%, surpassing those of logistic regression (71.84%) and Random Forest (89.15%). The AUROC for this model was 0.97 compared to the AUROC of 0.74 for the Logistic Regression model and 0.94 for the Random Forests model [26].

Using GraphSAGE, GAT, and GT, a notable approach was used to construct patient similarity graphs using the kNN to predict the likelihood of heart failure. All the models were trained and checked using data from the MIMIC-III database, which is a large EHR resource available to the public. They evaluated performance using common metrics like F1 score, AUROC and AUPRC and then compared them with existing ML techniques such as Random Forests [1]. The technique identified high-risk patient clusters, thus enabling personalized interventions for patients with share comorbidities. The Graph Transformer achieved the best results, reaching an F1 score of 0.5361 and an AUROC value of 0.7925 which were higher than those of the baseline algorithms and all other GNN variants. Using graphs, researchers could interpret which relationships between patients and which clinical features had the most impact on the model's outcomes [31]. These results indicate that GNNs can discover complex nonlinear patterns in EHR data that may not be visible to conventional approaches. Shaped as nodes linked in a similarity graph, HF can be predicted earlier and more accurately and managed better. As a result, management is better and more accurate.

Case Study 2: Discovery of Halacin as an Antibiotic

An example of the discovery of a new drug using GNN is the identification of halacin as an antibiotic by a team of researchers at MIT in 2020. Conventional time-consuming approaches have been used to screen thousands of chemicals in search of their antibiotic activity. In this work, Scientists built a GNN and trained it on a database of antibacterial molecules against E. coli. Consequently, the GNN model determined the molecular and substructures involved in the antibacterial effect. The trained GNN model was used to test the antibacterial activity of a massive library of more than 100 million compounds. The model was accurate in predicting the antibacterial nature of halacin, although it showed a different structure compared to conventional antibiotics. This proves the relevance of GNNs in accelerating the process of finding new drugs [32].

The Insilico Medicine drug Insilico Medicine ISM001-055 is a new candidate drug to treat idiopathic pulmonary fibrosis (IPF) through the creation of models incorporating the GNN and the drug is in clinical trial phase [33].

Case Study 3: Protein-Protein Interaction-Discovery of Drugs for SARS-CoV-2

A famous example is the use of GNNs to forecast the results of SARS-CoV-2 protein-protein interactions (PPI) with human host cells. MGPPI is a multiscale graph convolutional neural network (CNN) that predicts PPI established by the researchers. This model has enabled researchers to develop treatment plans for the COVID-19 pandemic over a short period. In this approach, proteins are treated as graphs, where each amino acid is a node of the graph and the edges are the relationships among them (whether geographical propinquity or chemical bonding). They are associated with features that describe the structural and chemical properties of these nodes and edges. The generated graphs are then captured through a GNN architecture, for example, a GCN. GNN decomposes the entire representation of a protein by studying the interactions and features of the structure.

We can even create a model that will predict the binding association between the human ACE2 receptor and viral spike (S) protein in the case of SARS-CoV-2. GNN identified the essential amino acid residues at the interface that produced this critical interaction. This knowledge is crucial for understanding how viruses infect people and for creating drugs that can stop this interaction and prevent the virus from infecting human cells. Compared to conventional, laborious lab tests, this computer approach greatly speeds up the study process [34, 35].

Case Study 4: Interpretable Cancer Drug Response Prediction using DRPreter

Selecting the appropriate medication for a patient is a key goal of personalized medicine for the treatment of cancer. Cancer is a heterogeneous disease, meaning that a medication that works for one patient with lung cancer may not work for another, even if the two patients have the same type of cancer. The distinct genetic and molecular features of each patient's tumor cause this variation. It is possible to avoid ineffective medications, reduce harmful side effects, and enhance patient outcomes by anticipating a response before therapy administration. This section presents a summary of the innovative methodology. The patient cells are represented as a graph. DRPreter treats a patient's genetic data as a complicated network, rather than a straightforward list of gene mutations or expression levels. Every gene in the cancer cell represents a "node" in the graph. The edges between nodes represent known biological pathways (such as signaling pathways that regulate cell development or death). This is an important step because it directly incorporates expert biological information into the structure of the model. Therefore, the GNN is limited to learning biologically plausible relationships [36].

Drug Graph Representation: The drug molecule is additionally depicted as a graph, where every atom such as oxygen, carbon, and so on, is represented as a node. Chemical bonds are represented by edges that hold the atoms together. This molecular graph was processed using a different GNN to learn a feature representation that encapsulates its structure and chemical characteristics.

Combining Patient and Drug Data with a Transformer: The model represents the relationship between the drug's molecular graph and the patient's biological route graphs using a transformer architecture, which is the same technology that powers strong language models. This enables the model to discover which structural elements of a medication interact with particular biological pathways within the cancer cells.

Significant Findings from the DRPreter Study

- *Superior Predictive Accuracy*: DRPreter outperforms other cutting-edge computational models in properly predicting drug response when tested on the sizable Genomics of Drug Sensitivity in Cancer (GDSC) dataset. The high accuracy achieved was the first step toward its clinical utility.
- *Interpretability and Actionable Insights*: The model excels in personalized treatment owing to its interpretability and actionable insights. Scientists were able to explain the model predictions because the cell line graph was based on

established biological pathways. The Transformer's attention mechanism helped in determining which cancer-related pathways had the biggest impact on the model's choice of medication.

Case Study 5: Automatic Triage in Emergency Departments

The use of automatic triage in EDs helps arrange care for patients in the order and amount according to the severity of their illness. Triage approaches that depend on a doctor's opinion can cause inconsistencies and mistakes, which may affect treatment. Advances in the field recently looked at using AI, particularly GNNs, to improve the errors and time associated with triaging [1, 5]. A GNN-based triage system treats each patient as a node on a graph, and the edges indicate whether the two patients share similar vital signs, symptoms, medical records, and demographics. As the patient similarity network maps out complex connections better than regular tables, the model can detect patterns that other approaches can miss. With preprocessing and lowering the noise level done, the model embeds the graph in a latent space and uses GNNs to assign patients to different risk categories for triage [1, 5, 6].

Recent studies have shown that GNN-based triage methods outperform both the traditional and previous ML methods. The unique part of these systems is that they consider not only patient reports but also simulate similar patient records within the network to improve decision-making when data is scant for a single patient [1, 5]. Because GNN embeddings are learned through the data, information from new patients can be added in real time which helps with triage in extremely busy emergency departments [1]. GNNs in triage support the proper care of patients by managing their orders, reducing mistakes, and maximizing resource usage in emergency departments. A remarkable improvement was achieved in clinical decision support for emergency medicine by uniting specific patient information and relational data [1, 5].

Designing Efficient Healthcare Networks with GNNs

Patients, healthcare providers, hospitals, and insurance organizations are all part of a complex healthcare system. Designing and optimizing such networks using old approaches usually do not fully capture their detailed connections. GNNs solve the problem by treating the healthcare system as a graph, where people or places are the nodes and their links to one another form the edges [5]. Using GNNs, researchers can explore and describe the many connections found in healthcare networks, which helps them to understand the patterns and dependencies necessary for strategic decisions. For example, GNNs may reveal where patients are directed, how the equipment is distributed, and the role of insurance in possible care routes. After reviewing the analyses, these healthcare organizations can use what is learned to build more efficient networks and enhance the standard of care provided to patients [5].

Using GNNs has helped a recent technique sort healthcare claims data and show the relationships among patients, providers, hospitals, and insurers. Network analysis revealed inefficiencies and delays, which encouraged adjustments to the health care network, making care better and less expensive for patients. It has been found through experiments that GNN-based networks work better than standard models and lead to more effective and efficient healthcare [5]. Overall, GNNs help to create healthcare networks that can handle real-world complexities and improve results, providing both monetary savings and better care for patients. Table 16.3 presents the use cases of GNNs in the healthcare domain in concise format.

Table 16.3 Use cases of GNNs in healthcare.

Use case	Description	GNN architectures	Datasets/data sources	References
Patient Similarity Graphs	Modeling patients as nodes with edges based on clinical similarities for disease prediction (e.g., heart failure, triage).	GCN, GraphSAGE, GAT, Graph Transformers	MIMIC-III, EHR	[17, 31, 37]
Healthcare System Modeling	Mapping relationships among patients, providers, and institutions to optimize care delivery.	General GNNs, Hybrid Architectures	Claims data, EHRs	[17, 37]
Biomedical Imaging	Analyzing brain networks (fMRI/MRI) or histopathology slides as graphs for disease diagnosis.	GCN, GAT, Graph Transformers	fMRI/MRI, histopathology datasets	[38–40]

(*Continued*)

Table 16.3 Use cases of GNNs in healthcare. (*Continued*)

Use case	Description	GNN architectures	Datasets/data sources	References
Drug Discovery	Predicting molecular properties or drug-target interactions using molecular graphs.	GCN, GAT, GraphSAGE	ChEMBL, PubChem	[34]
Protein Interaction Networks	Predicting protein-protein interactions (PPIs) to study disease pathways.	GCN, GAT, Hierarchical GNNs	PPI databases (e.g., STRING)	[40]
Personalized Medicine	Tailoring treatments using patient similarity graphs and multimodal data integration.	GAT, Graph Transformers	EHRs, genomic data	[34, 37]
Efficient Healthcare Network Design	Optimizing resource allocation and patient flow by modeling hospitals/providers as graphs.	GNNs with Federated Learning	Claims data, institutional records	[17, 37]
Emergency Department Triage	Classifying patient risk levels using similarity graphs and real-time GNN inference.	GAT, GraphSAGE	EHR, triage records	[31]
Rare Disease Detection	Identifying rare diseases by leveraging patient similarity and multimodal data integration in graphs.	GAT, Multimodal GNNs	EHRs, genetic databases	[34, 37]

(*Continued*)

Table 16.3 Use cases of GNNs in healthcare. (*Continued*)

Use case	Description	GNN architectures	Datasets/data sources	References
Adverse Drug Event Prediction	Detecting and predicting rare adverse drug reactions by modeling drug-patient-event relationships.	GCN, Heterogeneous GNNs	EHRs, pharmacovigilance databases	[34, 37]
Medical Incident Analysis	Mining incident reports to identify patterns and prevent future medical errors [17, 37].	GAT, Graph Transformers	Incident reporting systems	[37, 40]
Clinical Pathway Optimization	Modeling and optimizing patient care pathways by analyzing event sequences and transitions as graphs.	GCN, Temporal GNNs	EHRs, hospital workflow data	[17, 37]
Genomic Data Analysis	Exploring gene-disease relationships and variant effects using biological and patient graphs.	GCN, GAT, Graph Transformers	Genomic databases, EHRs	[34, 40]
Medical Knowledge Graphs	Integrating and querying large-scale biomedical knowledge graphs for decision support and discovery.	GCN, Knowledge GNNs	UMLS, SNOMED-CT, PubMed	[34, 40]

16.5 Technical and Clinical Challenges

Data Heterogeneity and Graph Construction

One of the foremost technical and clinical challenges in the application of GNNs in healthcare is managing the data heterogeneity and constructing meaningful graphs from diverse medical sources. Healthcare data are inherently heterogeneous and encompass EHR, medical images, data from wearable devices, genomic information, and environmental or behavioral records. Each of these modalities comes in different formats – structured, semi-structured, or unstructured-making integration complex and error-prone [1].

Defining nodes and edges in healthcare graphs is a non-trivial process. Choices must be made about what entities to represent as nodes (e.g., patients, clinical events, anatomical regions) and how to define edges (e.g., similarity in clinical features, temporal relationships, physical connections) [1]. These decisions directly influence the quality and interpretability of the resulting graphical models. In patient similarity graphs, connections are sometimes built by common diagnoses, approaches to treatment or demography, but the best way to measure equivalence is often hard to establish and can be different in different uses [7].

Merging multi-modal data into a single graph structure adds additional challenges to the graph construction. To merge and align data coming from separate systems, designers must use advanced preprocessing, extract features and normalize the data [30]. In particular, missing data in clinical documents and diversity in neuroimaging results may weaken the stability and effectiveness of the graph approach [1, 30]. Healthcare graphs must ensure data quality, privacy, and security due to the sensitivity of medical data [1]. Graphs for clinical use must represent reality and protect the privacy of the patients. In brief, the variety found in healthcare data and the complex process of building graphs-including choosing suitable nodes and edges, as well as mixing several types of information-create real difficulties that must be addressed to make the most of GNNs in hospitals [1, 30].

Scalability and Computational Complexity

It is challenging to use GNNs for healthcare applications because of their scalability and difficulty in computing. With the data in healthcare becoming larger and more complex, including large sets of patients, providers or molecules, the graphs used to extract meaning are also growing in size. Large-scale graphs are difficult to process using existing computers and memory which keeps GNNs from being used widely in medical settings [6, 7].

GNNs require many operations to gather information across all layers and neighborhoods within a graph. An increase in both nodes and edges leads to more work being performed during every forward and backward pass during the training. Further issues arise with heterogeneous graphs in healthcare, because nodes and edges can indicate many different elements and ties. Consequently, the models needed are more complex and use more resources [7]. Furthermore, because patients and records are constantly changing, GNNs must easily manage updates and continue to be scalable without requiring retraining.

The mere structure of the graph, and especially the number of edges, directly influences the processing time of different GNN systems. The computational complexity of a typical Graph Convolutional Network (GCN), e.g., is of the order $O\ (L\ |E|\ d)$, where L is the number of layers, L is the number of edges and d the dimension of the node features [20]. The reason behind it is due to the fact that the GCN is applied to each layer on every edge in the graph. Similar dependency is seen in Graph Attention Networks (GATs), which have a complexity of $O(L \cdot (|V|+|E|) \cdot d \cdot K)$, where K is the number of attention heads and |V| is the number of nodes [21]. The fact is that the GATs have to compute attention coefficients of each edge. In graphs as huge as dense patient similarity networks, this limit on the overall number of edges becomes a considerable bottleneck [15].

GraphSAGE addresses this bottleneck by sampling fixed-size neighborhoods per node instead of aggregating all neighbors. This results in an $O(L\ |V| \cdot \prod_{i=1}^{L} S_i\ d)$ complexity, where S_i is the number of neighbors sampled at layer i. Importantly, this complexity is far more scalable for big, dense graphs because it is not dependent on the total number of edges, or ∼E|. To be effective in healthcare settings, this scalability necessitates careful implementation and may come at the expense of a trade-off with model performance [15]. Hence, scalability and efficiency remain important topics and several teams are exploring ways to design strong GNNs for large and continuously developing healthcare datasets [6, 7].

Safety, Bias, and Monitoring

Additionally, the models must be carefully monitored for slight changes, unfair treatment, and unintended problems. If there are changes in the way data are collected or in common clinical practices, GNNs may not work. Monitoring the system and regularly collecting feedback from clinicians are required to preserve safety and results [1, 28]. GNNs have great potential for promoting better diagnosis, patient division, and therapy advice. Nevertheless, these benefits are associated with ethical and practical concerns, mainly bias, safety, and ongoing surveillance.

Safety: The security of GNN-based medical applications should be ensured, as patient care could be impacted directly by inaccurate predictions or recommendations of the system. Regulatory authorities (e.g. the European Medicines Agency (EMA) and the United States) Food and Drug Administration (FDA)) are developing standards that should guide the evaluation and authorization of AI-powered medical technologies with the significant emphasis on validation, transparency, and risk mitigation in the lifecycle of the model development [41, 42]. An example of safety considerations is robust model validation across a variety of patient populations, stress testing in extreme cases, and defining fail-safe procedures to prevent harm in the case of model imperfections or uncertainty.
Bias: Like other ML models, GNNs can inherit training data bias, potentially worsening healthcare inequities [43]. Ethical frameworks and guidelines, as well as the ethical requirements depicted by the European Commission, World Health Organization (WHO), and other organizations, demonstrate the importance of fairness and equity in AI systems. Regular biased audits, the inclusion of representative and diverse sets of data, and bias-reduction strategies are thus essential to deliver equitable outcomes to all populations of patients [44]. For example, a GNN trained on historical ICU data may under-predict sepsis risk in women if the original dataset underrepresented them, leading to delayed intervention [45, 46].
Monitoring: GNN models should be monitored over time to guarantee long-term safety and fairness. Model drift is a phenomenon which may deteriorate performance of the model or create novel bias due to evolving nature of the healthcare settings and shifting populations and methods of patient care or data acquisition [47]. Regulatory guidelines suggest continuous performance analysis, anomaly supervision in real time, or, in case of any issues, rolling back or retraining of the model within the shortest time possible. The dependability and credibility of GNN-based healthcare solutions can be further improved by establishing transparent accountability frameworks and enlisting multidisciplinary monitoring committees.

Interpretability and Explainability of GNN Predictions

Interpretability and explainability are major hurdles for GNNs in healthcare, as health workers depend on understandable and useful results, and rely on the model's suggestions. Both EHR and other medical data can be handled well by GNNs, although understanding their decisions remains difficult [1, 7].

Over the past few years, studies have introduced frameworks designed to facilitate understanding of GNNs. By analyzing the patterns of connections, attention and clinical information within graphs showing patient similarity, researchers can determine the main aspects behind predictions [5]. Paying attention to the principal assets, GAT and Graph Transformers in the models makes it possible to determine which patients and features significantly affect the predicted outcomes. As a result, clinicians can easily relate predictions to specific relationships and medical variables which cannot be obtained using basic black box models [5].

Despite progress, GNN outputs remain less interpretable than other ML models. We must improve how we represent and interpret how the graph shares information, how the roles of nodes and edges matter, and how interactions among subgroups are recognized [5, 7]. Clarifying the explanations of GNN predictions is necessary to earn clinicians trust and make these models usable in real healthcare practices [7, 28].

Handling Class Imbalance and Rare Events in Healthcare Data

Working with class imbalance and rare events is a major problem in healthcare data because the less common class often involves serious outcomes such as rare diseases and adverse responses to drugs or other medical incidents [6, 7, 16]. Usually, in clinical datasets, the class with the most examples is much greater than the rest because few cases are from a minority in practice. If we consider GNNs to be biased toward the most frequent class, their ability to detect and predict rare but significant health outcomes could be poor [6, 7].

The consequences of poor minority class detection are substantial because these rare events often correspond to critical clinical outcomes that require timely intervention [6, 27]. For example, in the context of opioid-related adverse events, even in high-risk populations, the prevalence of such outcomes can be less than 0.5%, making accurate predictions extremely challenging [6]. Similarly, in medical incident reporting, only a small fraction of reports pertains to rare but risky events, such as medication errors arising from drug names that are phonetically or visually similar [5, 7].

Strategies employed to mitigate rare event bias in healthcare GNNs are:

- Applying resampling methodologies (e.g., increasing the representation of the minority class or decreasing the representation of the majority class)
- Use specialized loss functions (such as focal loss or weighted cross-entropy)
- Integration of multi-modal data to enhance rare event detection.
- Implement meta-learning strategies for few-shot strategies

For instance, in heart failure prediction using GNNs, adapting loss functions to account for class imbalance has been shown to improve the identification of minority (positive) cases. Cross-modal learning, which fuses information from diverse data sources, can enrich context and aid rare pattern detection in clinical data. In spite of having made progress, unequal representation among patients in healthcare ML is still not given enough thought and continues to be an issue, so keeping solutions updated and properly tested is essential [6, 9, 10].

16.6 Future Directions

Advances in Graph-Based Interpretability and Explainability

The development of robust, scalable, and actionable explainability methods for GNNs remains a critical area of research in healthcare AI. Current approaches, such as attention mechanisms, provide initial insights but often lack the fidelity required for real-world clinical validation. Future research should focus on the following aspects:

- *Unified Evaluation Frameworks*: Standardized metrics like the AIM framework (Accuracy, Instance-level explanations, Model-level explanations) [1] are needed to assess explainability across diverse healthcare tasks. These frameworks should prioritize clinical relevance, ensuring that explanations align with medical domain knowledge and support actionable decisions.
- *Robustness-Aware Explanations*: According to studies done presently, if adversarial changes are applied to graphs, GNN results and their explanations might change a lot [6]. Future approaches are needed to maintain explanations when noise or attacks are present in the data because robustness is needed for tasks such as healthcare applications in diagnosis or treatment planning.
- *Integration with Multi-Modal Data*: Pairing GNNs with LLMs or vision transformers could cause explanations in graphs to be explained in both text and images [28]. Natural language summaries of subgraph patterns observed by GNNs in patient similarity networks can be automatically prepared using LLMs.
- *Exact Feature Attribution*: Any-order Shapley interactions help identify influential nodes and edges, aiding risk

stratification in complex patient data. This may help clinicians rank risk factors in complicated patient charts [7].

- *Multi-Granular Interpretability*: Converting GNN output into tree structures allows "growing branches" and looks at both single node outcomes and overall graph choices. This is vital for modeling how diseases progress, since features of individual patients and larger patterns among the population are both significant [7].
- *GNNExplainer:* GNNExplainer is designed to explain the predictions of any GNN. It indicates which parts of a graph and which node features are most important for a GNN's specific decision. GNNExplainer provides supporting evidence directly from graph structure and data. GNNExplainer answers the question of which neighbors and which features matter the most. It is model-agnostic, and provides an instance-level explanation. It helps turn the "black box" GNN models into more transparent and trustworthy tools. We hope that this tool will enhance the utilization of GNN based models in the clinical setting. Beyond GNNExplainer, newer techniques like PGExplainer and GraphSVX offer scalable, probabilistic explanations that can be tailored to clinical questions [48].

Integration with Other AI Models (e.g., Transformers, Multi-Modal Learning)

Next generation of healthcare AI will likely combine GNNs with transformers and multi-modal frameworks. This synergy can address the limitations of standalone GNNs while unlocking new capabilities for analyzing heterogeneous and interconnected medical datasets.

1) Hybrid GNN-Transformer Architectures: Transformers excel at processing sequential and contextual data (e.g. clinical notes and time-series EHR), whereas GNNs capture relational structures (e.g. patient networks and molecular interactions). Combining these with Graph Transformers or Co-Transformer-GNN models enables simultaneous learning of local graph structures and global contextual patterns. For example, in longitudinal patient data, transformers can model temporal trends in laboratory results, whereas GNNs can analyze how these trends relate to comorbidities or social determinants in a patient graph.

2) Multi-Modal Graph Learning: Healthcare data include images, text, genomics, and sensor inputs. Multi-modal GNNs can unify these modalities into a single graph where the nodes represent diverse data types (e.g. a patient node connected to their MRI scans, prescriptions, and genetic variants). Cross-modal attention mechanisms can then be used to identify clinically meaningful interactions, such as linking a specific genetic mutation to tumor morphology in imaging. Recent studies on multi-modal graph contrastive learning have shown promise for tasks such as rare disease diagnosis, where aligning features across modalities improves the robustness to missing data.
3) Cross-Modal Knowledge Transfer: Pretrained foundation models (e.g. LLMs and vision transformers) can enhance GNNs by injecting domain-specific knowledge. For instance, fine-tuning a clinical LLM to generate node embeddings for medical concepts (e.g. diagnoses and drugs) allows GNNs to leverage semantic relationships in ontologies, such as SNOMED-CT. Conversely, GNN-derived graph structures can guide transformers to focus on the most relevant tokens or regions in text/image data.
4) Federated GNNs for Distributed Learning: Integrating GNNs with federated learning frameworks enables collaborative model training across institutions, without sharing raw patient data. For example, hospitals can jointly train a GNN on local patient graphs with updates aggregated by using secure protocols. This approach is particularly promising for studying rare diseases, in which data scarcity at individual sites can be mitigated by privacy-preserving graph collaboration.

Challenges and Opportunities

Key challenges include aligning heterogeneous data modalities, managing computational overhead, and ensuring privacy in federated settings. However, the integration of GNNs with transformers and multi-modal systems has the potential to advance personalized medicine, drug discovery, and population health analytics by unifying the healthcare data into a knowledge-rich framework.

If these challenges are handled properly, GNNs in the future could provide explanations that are correct, dependable, and easy for health workers to follow, and might support the use of AI in medical practice.

Real-World Deployment and Clinical Validation
The translation of GNN models from research to real-world clinical practice remains a major future direction and significant challenge. Although numerous studies have demonstrated the technical promise of GNNs for disease prediction, patient risk stratification, and therapy recommendation using EHR, real-world deployment requires rigorous clinical validation, integration with healthcare workflows, and ongoing assessment of safety and utility [1, 7, 28].

Clinical Validation
Experts have discovered that GNNs have surpassed existing ML models in predicting diseases such as heart failure risk, by working from patient similarity graphs and advanced attention features [7]. However, most validations at present are based on analyses of past data such as MIMIC-III. To see GNN models used in healthcare, we need to test and evaluate them using studies and randomized trials in real clinical settings, to check if they produce useful predictions for all kinds of patients [1].

Workflow Integration
GNNs in clinical use must be harmonized with existing clinical systems and fit smoothly into EHR. It means the solutions need to be easy for clinicians to use, respond promptly to questions and present ideas that are easy to follow [7].

Regulatory and Practical Considerations
Medical AI trials require clarity, predictability, and privacy. In deploying their implementation, they ought to focus on their infrastructure, growth in computations, and training of the health staff. GNNs offer clinical promise, but require strong validation and workflow integration to make real impact [1, 7, 28]. The adoption of GNNs in various healthcare scenarios presents a number of challenges and opportunities. To ensure safe and effective integration into clinical routines, both practical and regulatory aspects should be considered, although these models are already being used to conduct clinical decision support activities, drug development, and patient risk assessment.

Regulatory Considerations: AI technology use, specifically GNNs, in the tightly regulated healthcare sector should comply with the existing legal framework. The Food and Drug Administration (FDA) of the United States regulates software as a medical device (SaMD) and AI-based applications and has issued recommendations on validating the use of an AI-based software as well as risk management and transparency throughout the product lifecycle [49]. Likewise, European Commission and the European Medicines Agency (EMA) emphasized on explainability and responsibility, data privacy concerning General Data Privacy Regulation (GDPR), and ethical AI [41, 44, 49]. To meet these standards, developers should demonstrate that GNN-based solutions are reliable, safe, and highly effective in many groups of patients. This involves extensive validation research, accurate reporting of the model development, and the use of the model, as well as post-market surveillance systems to detect and correct potential issues once the model has already been used.

Practical Considerations: Besides regulatory compliance, practical issues must be addressed to ensure the successful implementation of GNNs in healthcare.

- *Data Integrity and Quality*: GNNs rely on high-quality databases that portray real-world clinical scenarios. Missing values, data heterogeneity, and interoperability problems may influence model performance and generalizability.
- *Model Interpretability*: AI systems must become transparent to regulators and clinicians. However, GNNs may be complex and difficult to interpret, and may decrease clinician trust and block their uptake.
- *Workflow Integration*: With intuitive user interfaces and unambiguous decision-support outputs, GNNs must be smoothly incorporated into current clinical processes to provide value.
- *Ongoing Monitoring*: As patient populations and clinical procedures change, GNN systems must be updated regularly to maintain accuracy and dependability [47].

The responsible and efficient application of GNNs in healthcare requires addressing these practical and regulatory issues to guarantee that advancements result in significant enhancements in patient care while adhering to legal and ethical requirements.

16.7 Conclusion

This chapter has shown how Graph Neural Networks (GNNs) represent a paradigm change in healthcare analytics by capturing the relational structure in patient data, moving beyond the analysis of discrete data points.

We summarized the technical contributions of important GNN architectures and demonstrate how their unique designs solve persistent problems in the medical field. For example, inductive learning potential in models like GraphSAGE enables models to analyze new patient data from Electronic Health Records (EHRs) without requiring retraining. Similarly, attention mechanisms in Graph Attention Networks (GATs), as shown in cardiovascular risk prediction, improve model interpretability by highlighting key biomarkers and patient interactions. Despite setting a high-performance standard, GNNs' use in clinical settings depends on resolving a number of significant technical issues that characterize the next stage of research:

- *Advancing Architectures*: Future work should explore causal GNNs to transition from correlation to causation, which is a crucial stage for clinical intervention. Federated learning techniques for GNNs can enhance data privacy by enabling model training across institutions without data sharing. Using methods like GNNExplainer will enhance explainability which will in turn increase physician trust and help in clinical adoption.
- *Clinical Integration and Validation*: Retrospective accuracy measurements such as AUROC are not sufficient to validate a healthcare model. Prospective clinical trials are required to access the impact of GNN-based decision support systems on patient outcomes. Research must ensure GNN systems scale and operate reliably in real-time hospital settings.

GNNs promise to transform healthcare. But their adoption requires scalable, explainable, and privacy-complaint models validated in real-world settings.

AI disclosure statement: The author acknowledges the use of AI-based tools for minor language editing or grammatical correction. However, all intellectual contributions, analyses, and conclusions presented in this chapter are entirely the author's own.

References

1. Advantages and Disadvantages of Deep Learning, VNG Cloud. Accessed: May 15, 2025. [Online]. Available: https://vngcloud.vn/blog/advantages-and-disadvantages-of-deep-learning.

2. Sathyanarayanan, S. and Chitnis, S., A Survey of Machine Learning in Healthcare, in: *Artificial Intelligence Applications for Health Care*, I, M.K. Ahirwal, N.D. Londhe, A. Kumar (Eds.), pp. 1–22, Taylor & Francis Ltd, Boca Raton, 2022, doi: https://doi.org/10.1201/9781003241409-1.
3. Sathyanarayanan, S., Murthy, S., Mallappa, S., Gudada, C., Machine Learning Approach Using HOG and LBP Features of Spectrograms-Based Heart Sounds Analysis for the Detection of Heart Diseases, in: *Proceedings of the 15th International Conference on Soft Computing and Pattern Recognition (SoCPaR 2023)*, vol. 1 - Smart healthcare, A. Bajaj, A. Abraham, O. Castillo (Eds.), Springer, London, Apr. 2025.
4. Sathyanarayanan, S. and Murthy, K.S., Heart Sound Analysis Using SAINet Incorporating CNN and Transfer Learning for Detecting Heart Diseases. *J. Wireless Mob. Netw. Ubiquitous Comput. Dependable Appl. (JoWUA)*, 15, 2 (June), 152–169, 2024, doi: 10.58346/JOWUA.2024.I2.011.
5. Introduction to Relational Deep Learning (RDL), Kumo. Accessed: May 15, 2025. [Online]. Available: https://kumo.ai/research/relational-deep-learning-rdl/.
6. Zohuri, B., Deep Learning Limitations and Flaws. *Mod. Approaches Mater. Sci.*, 2, 3, Jan. 2020, doi: 10.32474/mams.2020.02.000138.
7. Dhar, T., Dey, N., Borra, S., Sherratt, R.S., Challenges of Deep Learning in Medical Image Analysis—Improving Explainability and Trust. *IEEE Trans. Technol. Soc.*, 4, 1, 68–75, Jan. 2023, doi: 10.1109/tts.2023.3234203.
8. The Impact of Deep Learning on Medical Technology, Gener8. Accessed: May 15, 2025. [Online]. Available: https://www.gener8.net/the-impact-of-deep-learning-on-medical-technology/.
9. Shwartz-Ziu-tabular_data_DL is not all u need.
10. Tsuneki, M., Deep learning models in medical image analysis, 2022.
11. Choi, E., Bahadori, M.T., Song, L., Stewart, W.F., Sun, J., GRAM: Graph-based Attention Model for Healthcare Representation Learning. *Proceedings of the ACM SIGKDD International Conference on Knowledge Discovery and Data Mining*, Nov. 2016, vol. Part F129685, pp. 787–795, doi: 10.1145/3097983.3098126.
12. Chen, J., Yin, C., Wang, Y., Zhang, P., Predictive Modeling with Temporal Graphical Representation on Electronic Health Records, Accessed: Jun. 27, 2025. [Online]. Available: https://github.com/The-Real-JerryChen/TRANS.
13. Nicholson, D.N. and Greene, C.S., Constructing knowledge graphs and their biomedical applications, 2020, doi: 10.1016/j.csbj.2020.05.017.
14. Chen, H., *et al.*, AMA-GCN: Adaptive Multi-Layer Aggregation Graph Convolutional Network for Disease Prediction, 2021.
15. Hamilton, W.L., Ying, R., Leskovec, J., Inductive Representation Learning on Large Graphs, Sep. 2018, [Online]. Available: http://arxiv.org/abs/1706.02216.
16. Lin, T.-Y., Goyal, P., Girshick, R., He, K., Dollár, P., Focal Loss for Dense Object Detection, in: *2017 IEEE International Conference on Computer Vision (ICCV)*, Oct. 2017, pp. 2999–3007, doi: 10.1109/ICCV.2017.324.

17. Boll, H.O., Amirahmadi, A., Soliman, A., Byttner, S., Recamonde-Mendoza, M., Graph Neural Networks for Heart Failure Prediction on an EHR-Based Patient Similarity Graph, Nov. 2024, [Online]. Available: http://arxiv.org/abs/2411.19742.
18. Xiao, C., Pham, N., Imel, E., Luo, X., Patient-GAT: Sarcopenia Prediction using Multi-Modal Data Fusion and Weighted Graph Attention Networks, in: *Proceedings of the ACM Symposium on Applied Computing*, Mar. 2023, Association for Computing Machinery, pp. 614–617, doi: 10.1145/3555776.3578731.
19. Piya, F.L., Gupta, M., Beheshti, R., HealthGAT: Node Classifications in Electronic Health Records using Graph Attention Networks, in: *2024 IEEE/ACM Conference on Connected Health: Applications, Systems and Engineering Technologies (CHASE)*, Jun. 2024, pp. 132–141, doi: 10.1109/CHASE60773.2024.00022.
20. Kipf, T.N. and Welling, M., Semi-Supervised Classification with Graph Convolutional Networks, Feb. 2017, [Online]. Available: http://arxiv.org/abs/1609.02907.
21. Veličković, P., Cucurull, G., Casanova, A., Romero, A., Liò, P., Bengio, Y., Graph Attention Networks, Feb. 2018, [Online]. Available: http://arxiv.org/abs/1710.10903.
22. Zhou, B., Ran, B., Chen, L., A GraphSAGE-based model with fingerprints only to predict drug-drug interactions. *Math. Biosci. Eng.*, 21, 2, 2922–2942, 2024, doi: 10.3934/mbe.2024130.
23. Von Berg, J., Ten Dam, M., Van Der Laan, S.W., De Ridder, J., PolarMorphism enables discovery of shared genetic variants across multiple traits from GWAS summary statistics, doi: 10.5281/zenodo.5844193.
24. Monti, F., Boscaini, D., Masci, J., Rodoì, E., Svoboda, J., Bronstein, M.M., Geometric deep learning on graphs and manifolds using mixture model CNNs.
25. Barbiero, P., Viñas Torné, R., Lió, P., Graph Representation Forecasting of Patient's Medical Conditions: Toward a Digital Twin. *Front. Genet.*, 12, Sep. 2021, doi: 10.3389/fgene.2021.652907.
26. Lu, H. and Uddin, S., Disease Prediction Using Graph Machine Learning Based on Electronic Health Data: A Review of Approaches and Trends, MDPI, Apr. 01, 2023, doi: 10.3390/healthcare11071031.
27. Shamshirband, S., Fathi, M., Dehzangi, A., Chronopoulos, A.T., Alinejad-Rokny, H., A review on deep learning approaches in healthcare systems: Taxonomies, challenges, and open issues, Academic Press Inc., Jan. 01, 2021, doi: 10.1016/j.jbi.2020.103627.
28. Miotto, R., Wang, F., Wang, S., Jiang, X., Dudley, J.T., Deep learning for healthcare: Review, opportunities and challenges. *Brief Bioinform.*, 19, 6, 1236–1246, May 2017, doi: 10.1093/bib/bbx044.
29. Battaglia, P.W., *et al.*, Relational inductive biases, deep learning, and graph networks, Jun. 2018, [Online]. Available: http://arxiv.org/abs/1806.01261.

30. O' Mahony, N., *et al.*, Deep Learning vs. Traditional Computer Vision.
31. Defilippo, A., Veltri, P., Lió, P., Guzzi, P.H., Leveraging graph neural networks for supporting automatic triage of patients. *Sci. Rep.*, 14, 1, Dec. 2024, doi: 10.1038/s41598-024-63376-2.
32. Stokes, J.M., *et al.*, A Deep Learning Approach to Antibiotic Discovery Article A Deep Learning Approach to Antibiotic Discovery, doi: 10.1016/j.cell.2020.01.021.
33. Pulmonary Fibrosis Foundation, INS018_055, Pulmonary Fibrosis Foundation.
34. Zhang, X.M., Liang, L., Liu, L., Tang, M.J., Graph Neural Networks and Their Current Applications in Bioinformatics. *Front. Genet.*, 12, Jul. 2021, doi: 10.3389/fgene.2021.690049.
35. Zhao, S., Cui, Z., Zhang, G., Gong, Y., Su, L., MGPPI: multiscale graph neural networks for explainable protein–protein interaction prediction. *Front. Genet.*, 15, 2024, doi: 10.3389/fgene.2024.1440448.
36. Shin, J., Piao, Y., Bang, D., Kim, S., Jo, K., DRPreter: Interpretable Anticancer Drug Response Prediction Using Knowledge-Guided Graph Neural Networks and Transformer. *Int. J. Mol. Sci.*, 23, 22, Nov. 2022, doi: 10.3390/ijms232213919.
37. Hoyt, G., Chatterjee, N., Battaglia, F., Basu, P., Medical Applications of Graph Convolutional Networks Using Electronic Health Records: A Survey, Feb. 2025, [Online]. Available: http://arxiv.org/abs/2502.09781.
38. Meng, X. and Zou, T., Clinical applications of graph neural networks in computational histopathology: A review. *Comput. Biol. Med.*, 164, 107201, Sep. 2023, doi: 10.1016/J.COMPBIOMED.2023.107201.
39. Zhang, R., Zhu, Y., Ge, Z., Mu, H., Qi, D., Ni, H., Transfer Learning for Leaf Small Dataset Using Improved ResNet50 Network with Mixed Activation Functions. *Forests*, 13, 12, 2022, doi: 10.3390/f13122072.
40. Ahmedt-Aristizabal, D., Armin, M.A., Denman, S., Fookes, C., Petersson, L., Graph-based deep learning for medical diagnosis and analysis: Past, present and future, MDPI AG, Jul. 02, 2021, doi: 10.3390/s21144758.
41. US FDA, Artificial Intelligence and Machine Learning in Software as a Medical Device, US FDA.
42. Committee for Medicinal Products for Human Use (CHMP) and Committee for Medicinal Products for Veterinary Use (CVMP), Committee for Medicinal Products for Human Use (CHMP) Committee for Medicinal Products for Veterinary Use (CVMP) Reflection paper on the use of Artificial Intelligence (AI) in the medicinal product lifecycle, Sep. 2024, Accessed: Jun. 20, 2025. [Online]. Available: https://www.ema.europa.eu/en/documents/scientific-guideline/reflection-paper-use-artificial-intelligence-ai-medicinal-product-lifecycle_en.pdf.
43. Ziad, O., Brian, P., Christine, V., Sendhil, M., Dissecting racial bias in an algorithm used to manage the health of populations. *Science (1979)*, 366, 6464, 447–453, Oct. 2019.

44. European Union, Ethics guidelines for trustworthy AI.
45. Wang, H., Li, Y., Naidech, A., Luo, Y., Comparison between machine learning methods for mortality prediction for sepsis patients with different social determinants. *BMC Med. Inform. Decis. Mak.*, 22, 2, 1–13, Dec. 2022, doi: 10.1186/S12911-022-01871-0/TABLES/6.
46. Jiang, W., *et al.*, The influence of gender on the epidemiology of and outcome from sepsis associated acute kidney injury in ICU: A retrospective propensity-matched cohort study. *Eur. J. Med. Res.*, 29, 1, 1–13, Dec. 2024, doi: 10.1186/S40001-024-01651-8/TABLES/5.
47. Ghassemi, M., Oakden-Rayner, L., Beam, A.L., The false hope of current approaches to explainable artificial intelligence in health care, Elsevier Ltd., Nov. 01, 2021, doi: 10.1016/S2589-7500(21)00208-9.
48. Ying, R., Bourgeois, D., You, J., Zitnik, M., Leskovec, J., GNNExplainer: Generating Explanations for Graph Neural Networks. *Adv. Neural Inf. Process Syst.*, 32, Mar. 2019, Accessed: Jun. 28, 2025. [Online]. Available: https://arxiv.org/abs/1903.03894v4.
49. FDA, FDA permits marketing of artificial intelligence-based device to detect certain diabetes-related eye problems, FDA. Accessed: Oct. 01, 2020. [Online]. Available: https://www.fda.gov/news-events/press-announcements/fda-permits-marketing-artificial-intelligence-based-device-detect-certain diabetes-related-eye.

17

Graph Neural Networks for Early Prediction of Osteoporosis Disease

T. Mathankumar and S. Vijayarani*

Department of Computer Science, Bharathiar University, Coimbatore, Tamilnadu, India

Abstract

In contemporary healthcare applications, especially in modern life, data mining is crucial for uncovering latent knowledge in vast medical databases. The technological advancements in the field of healthcare offer numerous benefits. In the health sector, technologies powered by data are utilized for diagnosing, treating, and preventing illnesses. The prediction of future ailments before any symptoms manifest, by understanding patterns present in the medical data, is referred to as disease prediction and is a vital component of medical diagnostics. The discussion in this section deals with the prediction of diseases. More specifically, this chapter deals with diseases of bones. Focusing on osteoporosis, a condition characterized by reduced bone mineral density, resulting in fragile bones. Other types of bone diseases are also considered, including osteomalacia, rickets, Paget's disease, bone infections, etc. Osteoporosis is a bone disease that should be treated immediately. In Artificial Intelligence, the classification of osteoporosis has been accomplished using machine learning, deep learning, and graph neural networks (GNN), aptly due to its interrelated data structure, which assists in enhancing diagnostic precision. GNN models enabled a better understanding of complex relationships inherent within medical data, which leads to better clinical and predictive outcomes.

***Keywords*:** Bone disease, osteoporosis, graph neural networks, graph convolutional network, graph attention networks, graph sample aggregate

**Corresponding author*: vijayarani@buc.edu.in

J. Ramkumar, Sridaran Rajagopal, B. Suchitra and S. Balamurugan (eds.) Graph Neural Networks: Concepts and Applications, (701–740) © 2026 Scrivener Publishing LLC

17.1 Introduction

17.1.1 Healthcare Data Analytics

Healthcare analytics is the collection and analysis of data. Data from the very corners of the health care industry is used to source insights, medical imaging, and clinical notes [11]. When deploying advanced analytics on health care data, the health care provider can enhance operations, improve outcomes of patient care, and access clinicians and their respective clinical decision-making. Data analytics methods involve many paths in terms of analytics, including descriptive analytics, diagnostic analytics, predictive analytics, and prescriptive analytics [12]. Healthcare data analytics uses statistical and computational methods to analyze clinical, operational, and financial data in healthcare organizations. This analysis can provide stockholders of the organization, including healthcare clinicians, administrative decision-makers, and researchers, with 'actionable' insights. While healthcare organizations had relied on the past experiences of clinicians and research recommendations, the advantage of healthcare data analytics has assisted, first and foremost, the healthcare organization to integrate and evaluate a large amount of data from disparate sources; medical records, lab results, patient surveys and questionnaires, and provider surveys, provide value to the healthcare organization. The ability to evaluate disparate data available in healthcare assists healthcare with recognizing patterns, identifying early disease stages, tracking patient development, and streamlining processes [13]. The importance of healthcare data analytics can be implicated in transforming modern healthcare with efficiency, accuracy, and decision-making. Benefits associated with health care analytics are,

- **Better Patient Care:** When able to access real-time comprehensive data about patients, clinicians are able to provide a more accurate diagnosis, and design personalized treatment plans.
- **Better Operational Processes:** Organizations use analytics to recognize inefficient processes within the organization, reduce costs, and best use available resources.
- **Predictive Outcomes:** Predictive analytics enables clinicians to anticipate potential health declines before they require invasive remedial action.
- **Evidence-Based Practices of Care:** Data analytics enhances research used in clinical studies and increased the standard of care, while at the same time promotes evidence-based

practices. For healthcare organizations to provide better, high-quality, personalized, and cost-affordable medical care with the economic and social frailty facing healthcare, use of healthcare data analytics is imperative [14].

17.1.2 AI Applications in Diagnosis

Artificial Intelligence (AI) is revolutionizing the healthcare diagnosis sector with increased accuracy and efficiency of decision-making. AI-based diagnostics include a range of methodologies with machine learning (ML) and deep learning (DL) algorithms at the forefront, followed by others.

- **Medical Imaging:** AI can interpret MRIs, X-rays, CT scans, etc. so the AI identifies anomalies like fractures, tumors, and more sometimes with more accuracy than a human radiologist.
- **Radiology and Oncology:** In radiology and oncology, physicians are diagnosing illness from medical images like X-rays, CT scans, MRIs, and PET scans. Examining images allows physicians to identify fractures, tumors, or infections. Oncologists, however, have to resort to the process of imaging to determine the size, location, and development of a tumor. Thus, this knowledge guarantees correct diagnoses and helps in determining an appropriate course of action that places judgment at the forefront of patient care in these disciplines [15].
- **Pathology:** Pathologists seek out diseases like cancer or tumors in a sample of tissue, which is cut up into fragments and studied clinically with the aid of a microscope. Based on their education and clinical practice, pathologists recognize sick tissues and differentiate types of cancerous tissues, as well as evaluate the tumor's stage or grade. They review multiple biopsies, including breast, prostate, and skin tissues, and their evaluations are very important to reach the correct diagnosis and outline the proper treatment. A single skilled human eye guarantees that the tissue's illness is accurately recognized and provides a logical pathway to be used in treatment.
- **Genomic Medicine:** In genomics, health care professionals analyze an individual's DNA to look for potential genetic variation related to inherited conditions, or to find out how

a patient might respond to multiple different treatments. When used with clinical observations, genetic testing can help specialists identify potential health risks, assist with the diagnosis of genetic conditions, and help tailor treatments based on intrinsic genomic data. For this reason, genome testing enhances person-centered care by relying on human intelligence and provides more accurate care for our patients.

- **Natural Language Processing in Diagnosis:** Healthcare providers consistently read and analyze unstructured clinical text, such as physician notes, discharge summaries, or pathology reports, to extract relevant information about the patient's clinical condition. These clinical narratives often contain only the critical details about symptoms, history, treatment responses, etc. Clinicians utilize their training and judgment for constructing the relevant insights, for proper documentation, for diagnostic support, and for care planning, all of which are grounded on their in-depth human analysis.
- **Artificial Intelligence in Cardiovascular Diagnosis:** In cardiology, physicians read electrocardiograms (ECGs), echocardiograms, and cardiac MRI scans to identify various heart conditions, including arrhythmias, heart failure, and coronary artery disease. Physicians read the patterns and rhythms of cardiac function derived using these tests, which determine the presence or absence of pathological conditions and thereby assist in the decision to treat. The experience and expertise of the physician facilitate early warning sign detection because early medical intervention typically yields improved patient outcomes at the time of onset of disease progression.
- **Artificial Intelligence in Early Disease Detection:** Doctors apply keen observation and clinical experience to detect diseases in an early stage. Through sequentially examining the patient's symptoms, medical history, lab tests, or imaging tests at intervals, physicians are able to identify early indications of diseases such as diabetes, osteoporosis, or cancer from prior to when they are symptomatic. Periodic tests and expert review enable early diagnosis, which enables the patients to start treatment in an early stage, thus considerably enhancing their overall health and quality of life [16].

- **Clinical Decision Support Systems (CDSS):** AI assists in diagnosing disease by comparing medical history, symptoms, and lab test results against a vast database of cases, leading to faster and more accurate diagnosis.
- **Predictive Diagnostics:** AI could predict the occurrence of diseases, including cancer, diabetes, and heart disease, by examining patient records longitudinally, particularly when patterns are analyzed. AI utilizes large data sources, rolling up trends, recognizing patterns, and diagnosing diseases far earlier than traditional means.

17.1.3 Treatment

The utilization of AI together with data analytics is advancing treatment methods through personalized approaches, which are customized for individual patients. The use of AI extends to personalized medicine and robotic surgery while also advancing drug discovery and remote monitoring capabilities.

- **Personalized Medicine:** AI uses genetic data analysis to create tailored medical treatments especially in oncology where drug effectiveness depends on genetic mutations.
- **Robotic Surgery:** The use of AI-powered robots in surgical procedures enables surgeons to perform precise operations while reducing human error and patient recovery times.
- **Drug Discovery:** The expedited progression of novel therapeutic development can be attributed to the implementation of artificial intelligence models in the analysis of biological data, thereby uncovering potential medicaments.
- **Mental Health Treatment:** Mental health interventions offer individuals efficacious strategies to mitigate and manage psychological comorbidities, encompassing a multifaceted approach that integrates pharmacological, psychotherapeutic, lifestyle modifications, and interventional social support networks. Psychotherapy works to transform harmful thought processes and behavioral patterns. Medications help restore proper brain chemical balance and decrease symptom severity. To manage mental health successfully requires implementing lifestyle adjustments through diet plans, physical activity, and good sleep practices. The recovery process

benefits significantly from the social backing provided by family members and friends.

- **Remote Monitoring:** IoMT (internet of medical things) and wearable devices have enabled healthcare providers to track patient health in real time which helps deliver prompt medical interventions. Such advancements enhance treatment effectiveness while concurrently diminishing error-related risks [17].

17.1.4 Predictive Analytics

Predictive analytics relies upon historical data, statistical algorithms, and machine learning techniques to predict health outcomes in the future. This ability will allow healthcare providers the following benefits.

- **Predict disease development:** Predictive models can predict how diseases progress by identifying patterns in patient data which will assist doctors in making informed decisions regarding a treatment plan.
- **Identify at-risk populations:** Predictive analytics in healthcare has the ability to identify people at risk of developing a chronic disease for early intervention and preventive care [18].
- **Reduce hospital readmission:** Predicting patient outcomes and reducing unnecessary readmissions allows hospitals to utilize their resources more effectively.
- **Improve clinical trials:** By identifying which candidates will be most successful in trial, predictive models can improve trial success and lower trial costs.
- The applications of predictive analytics allow healthcare providers to take proactive measures and better patient outcomes [19].

This chapter will be about advancing disease prediction techniques using graph neural networks, incorporating the early prediction of osteoporosis disease. Section 17.2 presents the basics regarding bone diseases and their categories. Section 17.3 is a review of the literature. Section 17.4 is the fundamental of medical imaging and diagnosis, and applications of graph neural networks – an overview. Section 17.5 is the Advancement method for the classification of osteoporosis. Section 17.6 is the result and discussion. Section 17.7 finishes the chapter by offering a quick review of significant takeaways and discussing the more general use of structured data approaches in assessing health conditions.

17.2 Bone Disease

Bone disease is a collection of different disorders that adversely affect bone structure and integrity. They can be metabolic disorders, inherited disorders, or fractures due to injury. Bone disease usually leads to reduced bone strength and thus makes the bone vulnerable to fracture and deformity. Age and hormonal imbalance-related bone diseases such as osteoporosis are increasing. Medical imaging technologies, biomarker discovery, and diagnostic technologies in bone health have greatly added to our understanding and control of these diseases.

Types of Bone Diseases
Bone disease is also classified based on underlying conditions and bone structure effects. Metabolic bone disease invades the process of bone remodeling to cause brittle bone. Inherited bone disease suggests inherited disease conditions, affecting bone development. Trauma and injury-related disease includes fractures of bones and subsequent conditions following injury. Bone infections and tumor conditions such as osteomyelitis or bone cancer. The four most prevalent bone diseases are Paget's disease, Osteomalacia, Rickets, and Osteoporosis.

Osteoporosis: Osteoporosis is a sickness in which bones acquire a fragile and brittle nature due to repeated loss of bone density. Usually developing over time and often goes unnoticed until a fracture happens, commonly in the hip, spine, or wrist. The diseases become increasingly prevalent with age, particularly in women after menopause, when estrogen levels decline. Other causes include poor calcium or vitamin D intake, a sedentary lifestyle, smoking, certain medications, and underlying health issues. Signs of osteoporosis may include back pain, a stooped posture, loss of height, and frequent fractures from minor falls. Treatment involves preserving strength in bones and fracturing resistance. This often involves medication, regular weight-bearing exercise, and a balanced diet with enough calcium and vitamin D. Early screening through bone density tests helps to detect the disease before serious damage occurs [20].
Osteomalacia: Osteomalacia is a medical condition in which bones are softened due to the lack of adequate amounts of vitamin D, calcium, or phosphate, thereby causing weak, deformed, and fractured bones. The main etiologies of this condition are vitamin D deficiency that inhibits the body's ability to absorb calcium and phosphate from food malabsorption syndromes like celiac disease or Crohn's disease that prevent the intake

of necessary nutrients, and renal disease, like chronic kidney disease, that disrupts the regulation of appropriate calcium and phosphate levels. The indication for this disease are bone pain specifically the lower back legs, pelvis, and muscular weakness. To overcome this needs a vitamin D, calcium supplements, dietary modifications, sufficient nutrients, and physiotherapy for muscle strength [21].

Rickets: A bone disorder in children, it is seen on children's bodies and stems from the lack of vitamin D affect their skeletal growth. The major symptoms include skeletal deformities, growth retardation, and bone pain. Major causes are vitamin D deficiency from poor health conditions such as insufficient sunlight or dietary in taking inadequate amounts; dietary deficiencies in calcium or phosphate; and genetic factors that affect metabolism of vitamin D. The most common symptoms of rickets observed in children are bowed legs or arms as well as delayed growth and development. Treatment frequently involves additional vitamin D, calcium and phosphorus along with nutritional counseling to ensure that enough nutrients are taken in. In severe cases, surgical intervention may be necessary to correct bone malformations [22].

Paget's Disease: Paget's bone disease is a long-standing illness that reasons uncharacteristic bone altering, which in turn reasons the bones to develop superior and pathetic. The illness mainly disturbs the pelvis, spine, skull, and legs. The pathogenesis of Paget's infection seems to be multiclausal with genetic issues obtainable in the procedure of augmented danger in persons with a personal past, age existence applicable as the illness is mainly initiated in persons overhead 50 years of age, and perhaps ecological features like viral contaminations like paramyxovirus. Indications contain bone aching, joint aching or stiffness in the unnatural bones, and a developed danger of fractures. Treatment commonly includes medicines such as bisphosphonates to regulator bone breakdown, pain release with anti-inflammatory medications, and surgery to pleasure fractures [23, 24].

Bone Cancer: Bone cancer is an unusual but serious illness that arises when cells in the bones amplify in an uncontrolled fashion. Uncontrolled cell growth can lead to pain, swelling, and fractures in the involved area. Various forms of bone cancer are recognized, and osteosarcoma is the most prevalent form in children and young adults. Treatment is usually surgical removal of the tumor, use of chemotherapy, and, in a few instances, radiation therapy. Early detection of the cancer and intervention are important to enhance the prognosis and effectively manage the disease [25].

Hyperparathyroidism: Parathyroid hormone (PTH) was an abnormality that resulted in excessive secretion of a hormone in Hyperparathyroidism.

It plays a vital role for balancing the calcium levels in a body. If PTH raised abnormally the hormone balance is changed it shows the various health issues, including weakened bones, fractures, and other complications. Hyperparathyroidism split into three categories. The primary hyperparathyroidism commonly arises due to non-cancerous growths in the parathyroid glands. The result of chronic kidney problems, where the kidneys struggle to manage calcium and phosphate levels properly in the secondary hyperparathyroidism. Tertiary hyperparathyroidism happens when secondary hyperparathyroidism continues for a long time, leading to persistent hormone overproduction. Clinical signs of the condition are ostealgia and fractures of the skeleton, nephrolithiasis, malaise, mood disorders, myopathy, and hypercalcemia. To recover from this needs a surgical removal of the hyperactive parathyroid glands, pharmacological therapy such as calcimimetics that function to inhibit PTH release, also calcium and vitamin D supplementation to improve bone density [26].

17.3 Literature Review

Sung Kean Kim [1] has discussed that osteoporosis is a major health issue, particularly for postmenopausal women. The World Health Organization reports that about 30% of them were affected. This condition involves reduced bone mineral density (BMD) and often shows no signs until a fracture happens. The technique used for the diagnosis of osteoporosis is Dual X-ray Absorptiometry (DEXA) which computes BMD in various areas of the human bone. Yet, DEXA has major costs associated with it rendering it impractical for population based screening, and so helps identify individuals to be screened. To help with screening, many screening tools to estimate the risk of osteoporosis have been developed; primarily in postmenopausal women. While user friendly and relatively accurate it may not be reliable enough for a medical decision. The research compared techniques such as Support Vector Machines (SVM), Random Forests (RF), Artificial Neural Networks (ANN) and Logistic Regression (LR) for the prediction and diagnosis of osteoporosis. In each case it achieved results that were better than the standard screening tools. The authors utilized a 10-fold cross-check for all their models. The researchers emphasized medical history, and augmented the results with feature selection. They also identified the problem of not having enough events; the high-risk cases outnumbered the low-risk cases. In general, SVM model performed the best with a beating of OST performance with an area under the curve (AUC) of 0.827. The accuracy improvement of 2.7% will help identify approximately

230,000 individuals at risk. The authors conclude that their methods can help to identify high-risk women in post menopause, which may in turn promote more accurate predictive tool in the healthcare system. The aims by the authors may be seen as leading us away from a traditional screening system and more towards data driven risk prediction [4].

Inui, Atsuyuki [2] had undertaken a research study in osteoporosis on a topical subject that poses a recognized health threat to older people by predisposing people to increased fracture risk. The diagnosis of osteoporosis in adults involved a clinical area of assessment that used bone mineral density (BMD) measurements measured by a dual-energy x-ray absorptiometry (DXA), which diagnosed osteoporosis *via* a T-score of < -2.5, or by way of a fragility fracture. The Fracture Risk Assessment Tool (FRAX) was popularly used to identify higher likelihood estimates of fracture risk in the current time by synthetic BMI measure. The FRAX model was acknowledged, but simplified the measure by removing some risk factors. Inui documented that there was an increased awareness of the machine learning (ML) field from early concepts of artificial intelligence, and it was being utilized in the health care system as it is associated with diagnosis. There have been previous studies that used clustering models for risk, with fracture risk models as in Kruse *et al.* [35] and with Huang *et al.* [32] that used radiomics by ML as a method of increasing prediction in the risk of fracture. Osteoporosis screening and diagnosis are limited by a variety of traditional tools, like OST and FRAX, as developed. Machine learning, offered as a subject and as a possible alternative, could be more positively efficacious. The intention of Inui's study was to find ways to predict low BMD in aged women, without a DXA, that used factors such as age, BMI, blood tests, etc. Different models of prediction were used, including logistic regressions, decision trees, and lightGBM, because lightGBM achieved accuracy, further developing the evidence that ML can be implemented in osteoporosis screening.

Menghan Liu [3] studied osteoporotic fractures in the elderly, particularly in China, where 36% of the elderly population is affected by osteoporosis, leading to frequent fractures and a significant healthcare burden. Despite its high prevalence, diagnosis and treatment rates remain low due to the gradual onset of bone density loss, often discovered late. The research reviewed the FRAX tool, commonly used to assess fracture risk, noting its limitations in recognizing emerging risk factors. Previous studies, such as Atkinson *et al.* [33], used Gradient Boosting Machine (GBM) models, while Chen *et al.* [34] developed a hybrid deep learning model for diabetic patients. Liu's team developed a hybrid model combining Support Vector Machine (SVM) and XGBoost algorithms, achieving over

90% accuracy and precision in predicting fracture risk. The study identified 20 key features related to fracture risk, including smoking, weight loss, and less recognized factors like calcium levels and inflammatory markers. The findings warranted validation in larger datasets and additional investigation of potential contributing risk factors, including heart disease. The research pointed to the prospect of integrating clinical information with cutting-edge tools to transform osteoporosis treatment.

Osteoporosis risk screening has been studied by Bogyeong Suh [6] with the development of an interpretable deep learning model. The paper pointed to the rising prevalence of osteoporosis, most particularly among adults aged 50 and older, of whom there were 10.2 million in 2010, and were expected to rise to 12.3 million in 2020 and to 13.6 million by 2030. This underscores the importance of appropriate preoperative screening. The paper focused on clinical aids such as the Fracture Risk Assessment Tool, which was found to have poor accuracy (area under the curve (AUC) 0.65–0.82) contained in the same special issue and low uptake is ascribed to their inadequacies. The literature demonstrated increasing usage of machine learning in medical diagnostics, but early models suffered from problems such as overfitting and poor validation. The study's novel deep-learning model analyzed complex relationships among various risk factors and included interpretable methods, such as local interpretable model-agnostic explanations, improving clinical understanding of the feature contributions. Using large datasets like NHANES and KNHANES, the model was trained on a significant number of respondents and variables. The results demonstrated that the model significantly outperformed traditional tools, with high AUC values of 0.851 for femoral neck BMD and 0.922 for total femur BMD in NHANES.

Mahsa Ghorbani [7] presented a study addressing the challenge of class imbalance in disease prediction using graph-based methods. The paper focused on the use of graph convolution techniques in medical analysis, particularly for tasks like brain and mammogram analysis. It highlighted the growing use of these methods due to their ability to manage complex data structures. A key concern identified was that many existing models were biased toward majority class samples, often overlooking minority cases, which are crucial for accurate diagnosis. The review categorized imbalance handling methods into two types: data-level approaches, which involve re-sampling techniques, and cost-sensitive approaches, which adjust the importance of each class during training. However, most of these strategies were found to be inadequate in addressing imbalance issues in real medical datasets. The paper also discussed Dual Regularized models, which aimed to manage imbalance through a more complex setup but faced issues with stability, especially when sample sizes were small. In response to these limitations, the proposed RA-GCN model took direct learning of sample weights, a simpler and more stable to learn this problem.

Hejie Cui [8] provided a literature survey of explainable AI (AI), BrainNNExplainer, for brain network analysis in disease prediction models. Interpretable models are essential to understand neural systems biological functions which is critical for early diagnosis of neurological diseases. Research of various models has demonstrated a continuum of complexity of models from shallow (e.g., graph kernels, tensor factorizations) to deep architectures (e.g., convolutional neural networks) increasing in complexity of models. Graph neural networks (GNN) were captured the community's interest as a better representation of complex graph data compared to shallow models.

However, they faced challenges like overfitting and a lack of transparency, limiting their use in healthcare. Several methods were proposed to explain GNN predictions, but most lacked a backbone model tailored to brain networks and provided instance-specific explanations rather than shared insights needed to understand common disease patterns. To address this, BrainNNExplainer was introduced as an interpretable graph framework with a backbone prediction model and an explanation generator identifying significant brain network connections related to diseases. Experiments on real-world datasets demonstrated high prediction accuracy and explanations consistent with neuroscience knowledge.

Heloisa Oss Boll [9] examined the use of patient medical records to predict heart failure by connecting individuals with similar health profiles. The study reviewed earlier work that had used electronic records to better understand and estimate health risks in complex conditions like heart failure. In the past, different systems were used to study health histories, especially those focusing on repeated visits or patterns in medical tests. However, many of these methods treated medical information as separate entries, missing the important links between them. To improve this, the study used a connected approach where patients, illnesses, and treatments were seen as parts of a larger network. This way of organizing data made it possible to find deeper patterns and similarities across patient histories. Earlier research had tested similar methods, but the specific idea of linking patients based on shared medical experiences had not been widely explored. The paper presented a new way to build these connections and compared different models to see which one predicted heart failure more effectively. It also paid attention to how understandable the predictions were, using simple measures to explain how the results were reached.

17.4 Deep Learning in Medical Imaging and Diagnosis

Deep Learning is a Convolutional Neural Network (CNN) that is heavily utilized in medical imaging for diagnosis. They help to analyze large collections of medical images to identify abnormalities, classify conditions, and assist in predicting health outcomes [5, 28]. Figure 17.1 illustrates the applications of deep learning in medical imaging.

17.4.1 Basics of Graph Neural Networks

Graph Neural Networks (GNNs) are specially designed models to handle graph-structured data. Unlike traditional models that assume inputs are in regular formats, such as grids or sequences, GNNs are constructed to process data represented as graphs, where information is encoded in nodes and their connections. Graph Neural Networks are built to work with data structured as graphs, where nodes represent entities and edges represent the relationships between them. They are particularly helpful in areas where information is naturally interconnected, such as social systems, molecular data, or healthcare records involving complex relationships [29]. Figure 17.2 represents the structures of graph neural networks.

A graph is formally defined as G - (V, E), where V is the set of nodes, $E \subseteq V \; x \; V$ is the set of edges that connect pairs of nodes. Each node $u \in V$

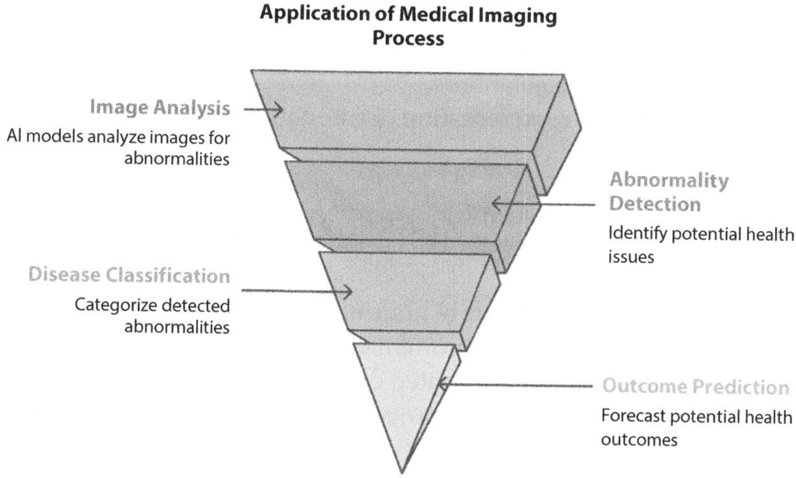

Figure 17.1 Applications of deep learning in medical imaging.

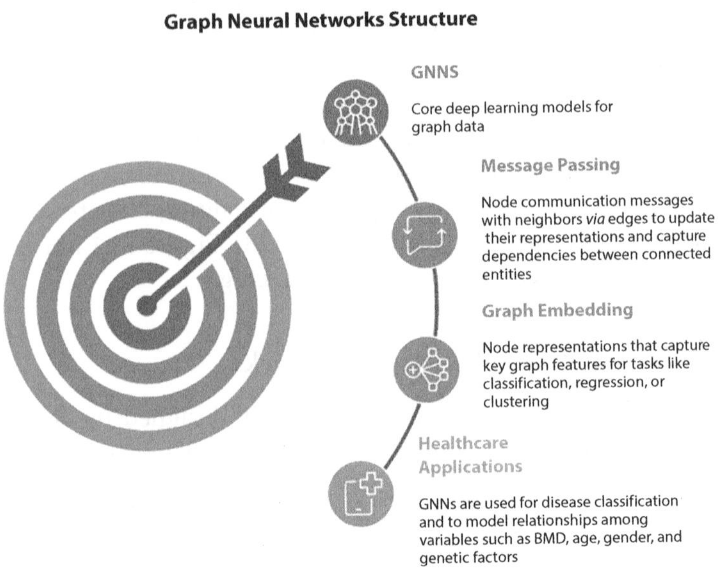

Figure 17.2 Structure of graph neural networks.

may have a feature vector $x_v \in \mathbb{R}^d$, and optionally, edges $(u,v) \in E$ may carry attributes or weights. Graphs can be directed or undirected, and they can be weighted or unweighted, depending on the nature of the connections and the presence of numerical edge values.

A Graph Neural Network is a model that learns representations of nodes, edges, or entire graphs by propagating and transforming feature information across the graph structure. The core mechanism in GNNs is message passing, where each node updates its representation based on its features and those of its neighbors.

At each layer l, the representation of a node v, denoted $h_v^{(l)}$, is updated according to

$$h_v^{(l)} - UPDATE^{(l)} h_v^{(l-1)},\ AGGREGATE^{(i)}\ h_u^{(l-1)} : u \in \mathcal{N}(v)$$

$\mathcal{N}(v)$ denotes the neighbors of node u; AGGREGATE collects information from neighbors. UPDATE combines this information with the node's current state. This process is repeated over multiple layers, allowing nodes to incorporate information from progressively larger neighborhoods.

- **Neighborhoods Aggregation:** Central GNNS, this operation combines features from a node's local neighborhoods to form a new representation.
- **Permutation Invariance:** Since node neighborhoods have no inherent order, the methods used to combine information must be invariant to the order in which neighbors are considered.
- **Parameter Sharing:** GNNs usually transfer the parameters among all nodes and edges. To enable for generalization across various graph structures.
- **Inductive vs. Transductive Learning:** GNNs may be applied in transductive learning scenarios, where the entire graph is known at training time. Inductive learning, where the approach generalizes to invisible nodes or graphs.
- **Layer-Wise Propagation:** Each GNN layer expands the receptive fields of nodes, enabling information to flow throughout the graph.

17.4.2 Applications and Case Studies

The Graph Neural Network (GNN) is phenomenal in the field of healthcare as it helps in understanding and visualizing patient records, multifaceted medical imaging, and even diagnostic results. The GNN-based system already has real-world applications in osteoporosis as well as bone health. In this section, we will explore deeper into the most significant case studies, which, alongside other findings, will be compiled into a table that outlines how useful it is for the patients, doctors, and even the hospitals.

Osteoporosis GNN Screening System Hospitals: These days, diagnostic centers and hospitals are automating a lot of processes to improve efficiency, and GNNs are a great addition that helps automate the scanning process. The current practice of monitoring osteoporosis through DEXA scans or X-ray screenings often relies on subjective readings and does not incorporate a patient's entire health context, which includes both medical history and lifestyle. With GNNs, hospitals can incorporate lifestyle and demographic the radiographic data to flag patients that meet certain criteria. These systems assist medical practitioners in addressing and detecting bone density impairments months, even years, before serious symptoms manifest. GNN-based screening systems help doctors examine the profiles of thousands of patients and understand and identify correlations between early symptoms, such as unexplained minor fractures, which are often associated with later-stage osteoporosis. As a result, patients receive preventive care faster, and resources such as imaging devices are used more efficiently.

Mobile or Wearable Health Apps Predicting Bone Health Risks: Mobile applications and wearable devices provide an important way to monitor health. These systems collect the information on daily activity, step count, fall detection alerts, calcium intake, and other sensor data. GNNs can interpret these inputs in real time to assess ongoing bone health risk and provide actionable feedback. Unlike traditional methods, graph-based models consider how different pieces of information are connected, such as linking reduced movement with patterns seen in earlier medical records. Using a smartphone app, individuals can get tailored health updates and reminders to help them take early action to protect their bone health. These tools are especially helpful for older adults or post-menopausal women, who face a greater risk of losing bone density.

AI in Rural Telemedicine Using Lightweight GNNs: One of the most valuable applications of graph-based methods is in rural healthcare through telemedicine services. With limited access to advanced imaging tools or medical specialists, simplified models can examine basic health data, such as blood tests, physical activity records, and patient-reported symptoms, to assess potential health risks. These systems operate well on basic devices like smartphones or tablets used by community health workers and can send alerts or suggestions remotely to healthcare providers. This setup helps to reduce healthcare inequalities by making expert-level assessments available in underserved areas. Using simplified models ensures that even clinics with limited equipment can support essential health screening. Table 17.1 lists out the benefits of graph-based healthcare solutions [10].

These applications demonstrate that graph-based methods are not confined to academic research. They are actively used in both clinical and community environments. Whether in a city hospital or a rural area, these systems provide effective tools for identifying osteoporosis and supporting bone health. By working with wearable devices, mobile health apps, and remote care services, they help make healthcare more forward-looking, preventative, and tailored to individual needs.

17.4.3 Research Challenges and Future Directions

The association between deep learning technologies and graph-based approaches, in medicine for osteoporosis diagnosis, is a rapidly developing direction. While these technologies can offer substantial benefits, there remain a number of practical and scientific challenges to be overcome before they are fully embraced in the field.

Table 17.1 Benefits of graph-based healthcare solutions for key stakeholders.

Area of benefit	Explanation
Patients	
Early Risk Awareness	GNNs analyze subtle health patterns to detect osteoporosis risk using symptoms before it become serious.
Personalized Advice	Mobile apps offer feedback based on unique health profiles, improving engagement and self-care.
Accessibility	Wearables and telemedicine apps extend healthcare access to rural and elderly population.
Lower Costs	Preventive screening reduces the need for advanced treatments and frequent hospital visits.
Continuous Monitoring	Real-time tracking of activity and nutrition helps patients stay informed about their bone health status
Doctors	
Accurate Decision Support	GNNs synthesize medical imaging and patient records for more precise diagnosis and treatment planning.
Prioritization of High-Risk Cases	Risk stratification enables efficient allocation of time and resources to patients in urgent need.
Seamless Integration	These tools can be embedded into existing hospital systems to enhance workflow without major disruption
Time Efficiency	Automation of screening allows physicians to focus on complex cases rather than routine checks.
Hospitals	
Efficient Use of Resources	GNNs help reduce unnecessary scans or admissions by identifying high-risk patients in advance.
Operational Savings	Decreased readmission rates and optimized use of diagnostic equipment result in cost savings.
Data-Driven Policy Planning	Aggregated insights from GNNs support hospital management in designing better health strategies.
Better Reporting	Enhanced patient outcomes contribute to hospital accreditation and quality improvement programs.

Availability and Quality of Data: Two main issues impacting healthcare AI are trustable and quality data and data availability. There is not a significant number of good quality annotated datasets related to osteoporosis diagnosis especially with respect to medical images (i.e., X-rays, and MRIs) and patient health records. Healthcare datasets often suffer from data imbalance where certain classes (i.e., osteoporosis cases) are underrepresented. This will cause a bias in the outputs of the model. To help balance and augment the training datasets, the consortium used synthetic data generation methods such as Generative Adversarial Networks (GANs) to generate synthetic medical datasets. Another approach, Federated Learning, is an approach to training models of knowledge from models of knowledge coming from large, abundant datasets, where there is no need to share any of the private datasets, thus allowing for greater data reach, while offering privacy advances on the patient data.

Interpretability and Explainability: Deep learning models, including GNNs, are considered black boxes. They may perform well, but it's hard to see how they make decisions on something vital in healthcare. Doctors need to trust and understand AI to use in real-world applications. Explainable AI (XAI) developing techniques to interpret and explain deep learning models in healthcare solutions is crucial. Methods like attention in GAT or explainable layers in GNNs can make the model's decision-making process understandable for clinicians.

Incorporation into Operating Procedures: Incorporation of these systems into everyday clinical work in healthcare is a difficult task. For them to be effective, the models must be sophisticated yet simple, requiring minimal effort on the clinician's part, with clearly interpretable diagnostic outcomes and actionable recommendations. Clinical Validation, along with Standardization, however, are of the utmost importance for the operational efficiency of AI in health care. AI systems need to be validated in actual hospital settings and comply with clinical governance frameworks. They have to be compatible with other systems used in hospitals, and there has to be cross-disciplinary engagement with other system users. The autonomy of clinicians AI has to collaborate deeply is also very important. Contrary to common belief, AI has to augment – enhance, not replace – physician functions to enable better care service planning, decision-making, and patient interaction for improved patient management.

Legal and Sociocultural Ethics as Well as Privacy Issues: The application of AI technologies in the healthcare domain raises significant ethical dilemmas concerning the autonomy of individual patients and the confidentiality of health information. Given that these systems work with

confidential health information, patient records must be protected while keeping the model balanced. Data Privacy and Security protects patient information while obtaining the incomprehensible information from the healthcare data. This can be aided by data-privacy tools like differential privacy and secure multi-party computation. Bias Mitigation lessens the biased views of AI systems to contribute to lessening disparity, while allowing for equitable care to all patients and fairness in data outcomes.
Ethical Implications of AI in Clinical Practice: Deployment of AI models such as GNNs in the clinical setting poses ethical concerns such as patient confidentiality of data, explainability of the model, and accountability of decisions. Explainability of model outputs needs to be open to achieve clinician trust and patient safety. Anonymization of data and transparency of AI-assisted decision-making should be regulated by guidelines to conform to clinical ethics and medical jurisprudence.
Real-Time Processing and Scalability: Deep learning models need powerful computing resources, especially when dealing with big data such as medical images. Their use in real-time clinical environments, where instantaneous yet accurate predictions are required, is highly challenging in terms of scalability. Edge AI and Model Optimization are using lightweight models that are capable in devices such as smartphones and wearables to facilitate processing on the fly. Optimizing models through pruning or quantization helps maintain speed without losing accuracy. For the heavier tasks, Cloud computing offers a way to manage large datasets and keep AI tools scalable for broader healthcare use.

17.5 Objective of the Problem and Methodology

The significant focus of this research is to predict the osteoporosis bone disease severity level using a graph neural network.

17.5.1 Dataset

This study uses four different datasets for osteoporosis classification. These datasets use combined information that includes a range of clinical, demographic, lifestyle, and diagnostic.

- **Osteoporosis Dataset:** Sourced from the ultrasonic bone scanning records, contains 3008 entries and 9 attributes.
- **Knee X-Ray Dataset:** Collected from Mendely repository with 240 records and 8 attributes. It contains clinical records, like joint pain and medical history.

- **Bone Mineral Density Dataset:** Gathered from the Kaggle repository with 169 records focusing on 11 attributes such as medication usage and t-score.
- **Osseous and DXA Exams Dataset:** Used from Zenodo, using QUS and DXA exams, comprising 505 samples and 13, including factors like race and family history of osteoporosis. All these datasets are categorized into three severity osteoporosis levels, they are, normal, mild, and severe. Figure 17.3 illustrates the working architecture of this research.

Architecture Flow of GNN Based Prediction System

Figure 17.4 presents GNN-based end-to-end workflow of osteoporosis prediction. Patient nodes are mapped from domain-specific features (e.g., BMD, age, gender, calcium, vitamin D, activity, family history) after normalization or encoding. Edges are formed on the basis of similarity

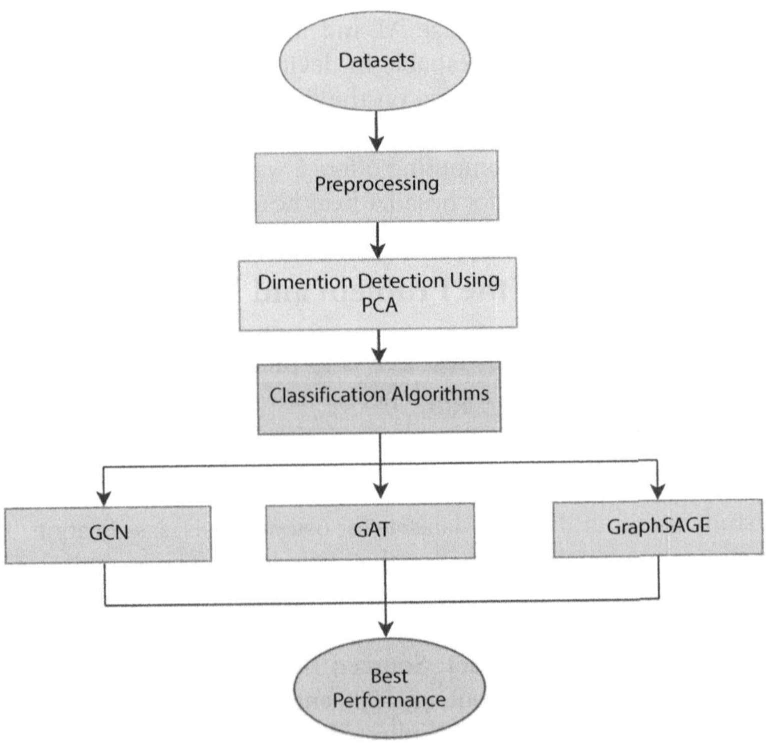

Figure 17.3 Proposed architecture.

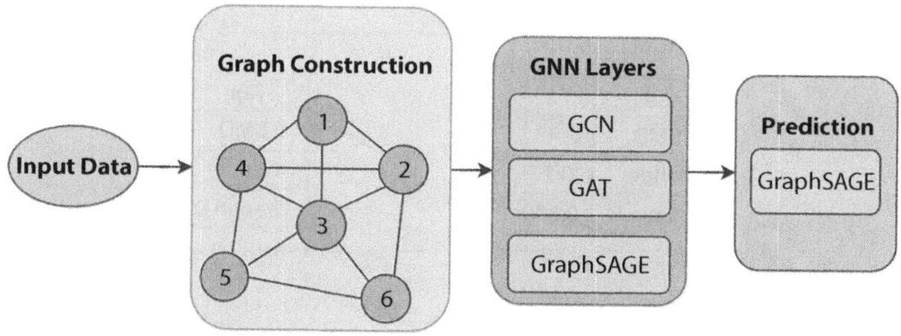

Figure 17.4 Architecture flow GNN-based prediction system.

to construct a patient graph, managed by GNN layers (e.g., GCN, GAT, GraphSAGE) for risk prediction.

17.5.2 Pre-Processing

The important stage in medical data analysis is pre-processing, as it looks to improve the quality of raw data and the quality of deep learning models. It includes some of the activities, such as the attribution of missing values using either record augmentation or deletion of records with missing values, categorical encoding to render them algorithm-ready, and numerical feature normalization to render them data standardization-ready. To reduce the dimensions of the dataset to retain the most beneficial information and remove redundancies with the help of PCA techniques.

Domain-Specific Feature Mapping to Graph Nodes

Figure 17.5 shows graph node features such as BMD, age, gender, calcium, vitamin D, activity, and family history are assigned to graph nodes. Features were chosen based on clinical importance. Categorical features were one-hot encoded and the continuous features were normalized.

Principal Component Analysis (PCA)

Principal component analysis is a data reduction technique with no loss of the most significant information on vast datasets. It transforms the data alienating to new variables or principal components that hold the largest pattern and variation of the original data. This technique is especially applied in health research, like bone disease research, since it enables what matters most in a set of variables, enhancing precision in outcome models and providing greater insight. Table 17.2 has the pseudocode of PCA.

The PCA method begins with a data matrix X in which rows are the samples and columns are the attributes. The data is then centered by

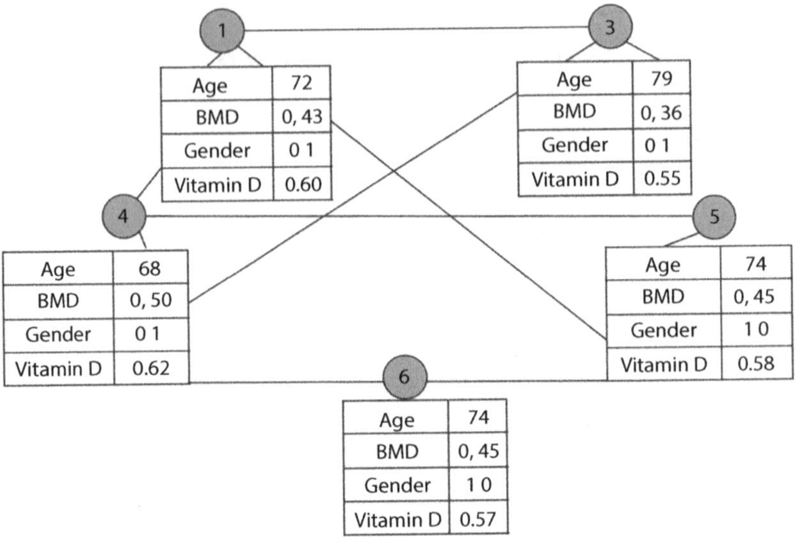

Figure 17.5 Domain-specific feature mapping to graph nodes.

Table 17.2 Pseudocode of PCA.

Step 1:	Start
Step 2:	**Input:** Data matrix X (n_samples x n_features)
Step 3:	Center X by subtracting the mean of every column
Step 4:	Calculate covariance matrix C=1/n) * X^T X
Step 5:	Evaluate eigenvectors and eigenvalues of C
Step 6:	Sort eigenvectors by descending eigenvalues
Step 7:	Select top k eigenvectors to form matrix W
Step 8:	Project data: X_pca =X xW
Step 9:	**Output:** Reduced data X_pca
Step 10:	End

removing the mean of all the attributes to zero-center the data. Then, the covariance matrix C is computed to find the correlations between the attributes. For determining the direction of maximum variance, one computes the eigenvalues and eigenvectors of the covariance matrix C. These eigenvectors are arranged in descending order of their eigenvalues and the first k eigenvectors are utilized for constructing the projection matrix W. Data is projected to new space by the equation $X_{pca} = X \times W$ and produces a lower-dimensional representation with maximum variance.

17.5.3 GNN for Classification

The classification algorithms are employed to predict the data into various classes or tags based on various features. Classification algorithms read the labeled data and detect patterns that are applied on unlabeled data. For instance, in the medical field, classification algorithms can predict whether a patient has a specific condition or not based on symptoms and test results. Typical kinds of algorithms include machine learning models, deep learning neural networks. Each technique develops its logic to distinguish classes of data. The principal aim of classification algorithms is to assist in decision making based on known information to predict a correct response.

Deep learning methods can be applied with Graph Neural Networks (GNNs), now established to be immensely helpful in the diagnosis and classification of osteoporosis. Osteoporosis is marked by lower bone density compared to healthy adults and is coupled with brittle bones and thus have a greater likelihood of fracture. Osteoporosis can be diagnosed by conventional tests, like X-ray imaging and measurement of bone mineral density (BMD); however, utilization of multiple medical imaging datasets with deep learning techniques is most likely to significantly increase the diagnostician's ability to reach a definitive diagnosis. Deep learning is an inexpensive means of managing complex relationships in medical data. For example, the utilization of multiple medical image datasets, along with the patient history and their genetic profile, may provide a better classification of an individual's overall bone health [27]. Focusing on osteoporosis diagnosis, deep learning AI has been used for image classification in medical imaging as part of the classification process, particularly with Convolutional Neural Networks (CNNs) and Recurrent Neural Networks (RNNs), which specifically look for changes in patients' bones associated with osteoporosis. These models are trained to identify patterns in X-ray, CT scan, or MRI images to determine whether a patient has osteoporosis or not. Graph Neural Networks (GNNs) is especially useful in dealing with graph-based data, e.g., the relationship between various medical conditions, patient information, and clinical outcomes. These networks are better suited for cases where conventional deep learning models like CNNs are unable to grasp the full complexity of the relationships among various variables in a patient's health profile.

Graph Construction: Node/Edge Formulation Logic

In this study, every patient record was represented as a graph node whose node attributes were clinical and demographic factors such as age, gender, calcium value, BMD value, comorbidities, and drug history. Edges between patient records were formed on the basis of cosine similarity thresholds (≥ 0.8), which joined patients with highly similar profiles. This enabled the graph structure to retain useful medical associations by representing the clinical similarity, hence improving the graph-based learning performance.

Integrate Medical Terminology and Disease Staging into GNN Modeling
All three clinical phases of osteoporosis (normal, osteopenia or mild, and osteoporosis or severe) were mapped to classification labels relative to WHO-defined thresholds of bone mineral density. These clinical phases were used in the supervised learning of GNN models. In addition, medical terminology concepts, including fragility fractures, DXA T-scores, and calcium metabolism, were mapped as node-level features or label surrogates, back-anchoring GNN learning in disease phases that are relevant to the domain.

Graph Convolutional Network
Graph Convolutional Networks (GCNs) is another form of neural network carefully crafted to process graph-structured data. A lot of traditional models are made to handle data like images or text. The graph-based methods instead aim at learning from the context between connected data points, like users in social networks or proteins in biological systems. They iteratively update the node features using the aggregation of information in neighboring nodes. This makes GCNs capable of capturing local structure information and node attributes, which further allows them to be suitable for tasks such as node classification, link prediction, and graph clustering [27].

Table 17.3 highlights a pseudocode for a Graph Convolutional Network (GCN), and describes main steps involved in deriving node representations from graph-structured data. The procedure starts with an initialization which establishes the graph G = (V, E), node feature matrix X, adjacency matrix A, trainable weight matrix W and activation function f. The major computation is layer by layer. For each layer, for each node v, the neighboring node features are aggregated from the adjacency matrix A. Aggregation refers to the sum of features from each of the adjacent neighbor nodes, thereby obtaining the local graph structure around each node. The original features will be transformed to the new space as follows: the followed by the activation function (ReLU), where is the trainable matrix. We will refer to this new node feature matrix as X, which we will pass to the next layer. This process can be repeated on X as many times as planned, allowing information to be transferred across multiple hops in the graph. Once all the layers process the information, the last node

Table 17.3 Pseudocode of GCN.

Step 1. Start
Step 2. Initialize:
• Graph G = (V, E)
• $X \leftarrow$ Initial node feature matrix
• $A \leftarrow$ Adjacency matrix of G
• $\{W1, W2, \ldots W_L\} \leftarrow$ Weight matrix for each layers
• $f \leftarrow$ ReLU Activation function
Step 3. For every layer l in total_layers:
• For every node v: sum neighbor features:
• Aggregated_features(v) = Sum (A[v, :] * X)
Step 4. Apply the transformation: Transformed_features = Aggregated_features * W
Step 5. Apply the activation function: Updated_features = f(Transformed_features)
Step 6. Upgrade the node feature matrix: X = Updated_features
Step 7. Continue a steps 3 to 6 for every layer
Step 8. Output the last node Characteristics after all layers
Step 9. Use upgraded node features for downstream activities
Step 10. End

features can properly output and used for other purposes appearing in the previously mentioned downstream applications, such as node classification, clustering, or link prediction. The algorithm concludes after producing the learned node embeddings. In this scenario, patients are represented as nodes, with edges defined based on patients' similarities based on their clinical or demographics data. This leads to a more systematic and holistic analysis of disease manifestations in different patient populations. The method also allows discrete data types, such as laboratory results, genetic data, and medical history, to act in unison in the same analytical system.

Graph Attention Networks

Graph Attention Networks are a more sophisticated class of graph neural network GNN type which applies attention on which neighboring nodes are more critical when updating a node's representation. When all neighbors are given the same treatment, GAT pays attention to each neighbor individually and chooses a weight depending on what each neighbor brings to the table. In particular, some graphs might be more complex and features more important than others are more common than people might think. GATs are helpful, for instance, in social networks, recommendation systems, or even biological data analysis [30]. This attention helps the model to filter out a lot of irrelevant connections within a graph, making it easier to use for incomplete medical data. For osteoporosis detection, GATs use clinical indicators such as bone mineral density (BMD) values or some patient histories, but the question is, how much should they take into account for each case? Because this is a case- dependent issue, stronger assertions can be made. These adjustments result in better conclusions despite missing data. Using GATs increases classification precision and encourages innovation because they depend on highlighting patterns and relationships rather than rigid rules, which is critical given the uncertain needs of modern healthcare.

The pseudocode for the Graph Attention Network (GAT) outlined in Table 17.4 explains how attention mechanisms are used to compute node embeddings by focusing on the relative importance of neighboring nodes. The process begins by initializing key components: a graph $G = (V, E)$, the node feature matrix X, the adjacency matrix A, a learnable weight matrix W, attention coefficients α, and an activation function f. In each layer, the model iterates over every node v and calculates attention scores between the node and each of its neighbors u, using a shared attention mechanism

Table 17.4 Pseudocode of GAT.

Step 1: Start
Step 2: Initialize:
- Graph $G = (V, E)$
- Node feature matrix $X \in R\,|V| \times d$
- Adjacency matrix $A \in R\,|V| \times |V|$
- Weight matrix: $W^{(l)}$ for each layer l
- Attention coefficient matrix $\alpha^{(l)}$
- Activation function: ELU: $f \leftarrow ELU\; f(x) = \begin{cases} x & \text{if } x > 0 \\ \alpha(e^x - 1) & \text{if } x \le 0 \end{cases}$

Step 3: For each layer l in total layers:
- For each node $v \in V$:
- For each neighbor $u \in N(v)$:
- Compute attention coefficient: $\alpha_{vu}^{(l)} = \text{Attention}\,(W^{(l)}X_v,\; W^{(l)}X_u)$
- Normalize attention coefficients: $\alpha_{vu}^{(l)} = \text{softmax u}\,(\alpha_{vu}^{(l)})$

Step 4: Combine neighbor characteristics using attention:
$H_v^{(l+1)} = \sum u \in N\,(v)\,\alpha_{vu}^{(l)} \cdot W^{(l)}\, X_u$
Step 5: Apply activation function: $X_v^{(l+1)} = f\,(H_v^{(l+1)})$
Step 6: Update node feature matrix: $X = X^{(l+1)}$
Step 7: Repeat Steps 3–6 for all layers
Step 8: Output the final node embeddings
Step 9: End

that takes their feature vectors as input. These raw attention scores are normalized using the softmax function to ensure a valid distribution over the neighbors. The node then aggregates its neighbors' features, each weighted by the learned attention coefficient αvu, which dynamically emphasizes more relevant neighbors. The aggregated feature is linearly transformed and passed through a non-linear activation function to obtain the updated node representation. This method is done for several layers, gradually capturing higher-order dependencies. Unlike classic methods such as GCNs, GAT uses a learnable and dynamic attention mechanism that adapts to the graph structure and feature space, making it more expressive and powerful. Finally, a good-quality node embedding is generated that can be utilized for downstream graph-based tasks like classification, clustering, or link prediction. The same process is repeated for all of the layers, progressively aggregating the highest-order dependencies. Rather than using static importance weights as previous methods like GCNs, GAT uses a learnable and dynamic attention mechanism that is specifically concerned with the graph structure and feature space, thus making it more expressive and robust. Ultimately, a rich embedding of a node is produced that can be used for downstream graph-based tasks such as classification, clustering or link prediction.

GraphSAGE

GraphSAGE - Graph Sample and Aggregation is a scalable Graph Neural Network architecture. GraphSAGE is one of those graph neural networks that learn node representations by sampling a fixed number of neighbors and performing aggregation over these features. GraphSAGE constructs embeddings from local neighborhoods, thus it is scalable to big graphs. GraphSAGE can handle various methods of aggregation such as mean, LSTM, or pooling so that it can extract useful patterns from the graph structure. This makes GraphSAGE applicable to applications such as node classification, link prediction, and recommendation systems [31]. It creates node embeddings by randomly sampling a fixed size of neighboring nodes and combining their information. It is very effective in managing large healthcare data, with patients as nodes and medical similarities as connections. GraphSAGE facilitates efficient analysis in osteoporosis classification by considering only an informative subset of similar patients. This enables faster and more personalized predictions. its ability to learn patient-specific patterns without excessively overcomplicating computation makes it suitable for medical applications. Because its data-agnostic nature and flexible design, it can support a wide range of data inputs, including clinical, demographic, and diagnostic data, allowing it to fit different aspects of bone disease analysis.

The GraphSAGE algorithm presented in Table 17.5 efficiently learns node embeddings for large-scale graphs through sampling from fixed-size neighbor sets and aggregating their information instead of processing entire neighborhoods. The pseudocode starts by creating the graph structure G = (V, E), the node feature matrix X, an aggregation function which might be mean or LSTM, or pooling, and weight matrices W(l) for each layer, along with a ReLU activation function. For every node v, each network layer selects a random small subset of neighbors N(v) to manage computational complexity. The selected aggregator function processes the neighboring nodes' features to generate a summarized neighborhood representation. A node's current representation receives the aggregated feature through concatenation before undergoing linear transformation with layer-specific weights and subsequent activation by a non-linear function. Normalization of the resulting embedding can be applied optionally to maintain numerical stability. Every layer undergoes the same sequence of steps, which enables nodes to integrate information from both nearby and distant portions of the network. The final output provides a strong low-dimensional representation for each node, which integrates both node features and graph structure to enable scalable inductive learning for tasks including node classification and link prediction.

Table 17.5 Pseudocode of GraphSAGE.

Step 1: Start
Step 2: Initialize
- Graph G = (V, E)
- Node feature matrix X ∈ R^ (|V| × d)
- Aggregator function AGGREGATE (mean, LSTM, pooling)
- The Weight Matrices W^(l) for each layer l
- Activation function: ReLU: $f \leftarrow ReLU\ f(x) = \max(0, x)$

Step 3: For every single layer l in total_layers:
For each node v ∈ V: Sample a fixed-size neighbor set N(v)
Step 4: Combine neighbor Characteristics:
$h_N(v)^{(l)} = AGGREGATE\ (\{h_u^{(l)} | \forall_u \in N(v)\})$
Step 5: Concatenate with the node's characteristics:
$\hat{h}_v^{(l+1)} = CONCAT(h_v^{(l)}, h_N(v)^{(l)})$
Step 6: Apply linear transformation: $h_v^{(l+1)} = W^{(l)} \cdot \hat{h}_v^{(l+1)}$
Step 7: Apply activation function: $h_v^{(l+1)} = f(h_v^{(l+1)})$
Step 8: (Optional) Normalize embedding: $h_v^{(l+1)} = \dfrac{h_v^{(l+1)}}{\|h_v^{(l+1)}\|}$
Step 9: Repeat steps 3–8 for all layers
Step 10: End

17.6 Results and Discussion

Model evaluation and visualization composition fulfill a dynamic role in performance validation of Graph neural network model algorithms. Models get evaluated by measuring their predictive performance through quantitative metrics, which include accuracy, precision, recall, F1-score, and execution time. The evaluation process incorporates visualization methods such as confusion matrices, as well as ROC curves, together with performance comparison charts. These techniques allow users to convert quantitative measures into comprehensible patterns. The evaluation tools give indications of model strengths and provide recommendations for performance enhancement. Real-world applications demand prediction models to produce reliable results that are interpretable to users.

Training and Testing

The database was split into a train and a test, where 70% of the databases were used for training the classifier and 30% of the databases utilized for testing. The dataset is partitioned to make sure the model is trained with enough data, yet also to leave unseen data for model performance evaluation. The dataset is then split into training and test samples to test the model's prediction, and outliers that are detected and treated by statistical functions such as IQR or Z-scores. The pre-processing steps guarantee data is organized, cleaned and prepared for suitable analysis in healthcare solutions.

Accuracy

Accuracy is a basic metric used in evaluating the general performance of a model of classification able to predict outcomes well. When applying to the datasets, accuracy computes the number of positive cases of bone disease as well as negative cases out of predicted cases overall. A high value of accuracy indicates the model's ability to discriminate well between those with osteoporosis and those without osteoporosis, and informs what is best to include in the model. Accuracy is also relevant in class-imbalanced datasets, such as which medical records contain a very few instances of positive. Additionally, accuracy must be provided along with other metrics such as recall, precision, and F1-score, to provide a general balanced and unbiased measure of model performance.

Table 17.6 shows the comparative study that was done to examine the performance of the three graph neural network-based techniques of GCN, GAT, and GraphSAGE. The study examined four different medical datasets. the Osteoporosis Dataset, Knee X-ray Osteoporosis Dataset, and Bone Mineral Density Datasets include the Osseus and DXA Exams Dataset.

Table 17.6 Accuracy performance.

	Algorithms		
Dataset	**GCN**	**GAT**	**GraphSAGE**
Osteoporosis Dataset	95.57%	92.25%	97.45%
Knee X-ray Osteoporosis dataset	79.17%	75.00%	80.56%
Bone Mineral Density dataset	82.35%	76.47%	84.31%
Osseus and DXA Exams Dataset	65.13%	63.16%	69.74%

GraphSAGE was the most accurate of the algorithms, scoring 97.45% on the Osteoporosis Dataset, 80.56% on the Knee X-ray Dataset, 84.31% on the Bone Mineral Density Dataset, and 69.74% on the Osseus and DXA Dataset. The Graph Convolutional Network (GCN) achieved a strong performance with an accuracy of 95.57% on the Osteoporosis Dataset. GAT had slightly lower accuracy across the board, with its best result being 92.25% on the Osteoporosis Dataset. Overall, the GraphSAGE algorithm highest accuracy, GCN provides a strong balance between performance and speed, and GAT remains competitive but slightly behind in terms of predictive consistency. Figure 17.6 illustrates the accuracy performance.

Precision

Precision is the proportion of relevant instances among the retrieved instances. It measures how accurately the model, when it predicts a positive.

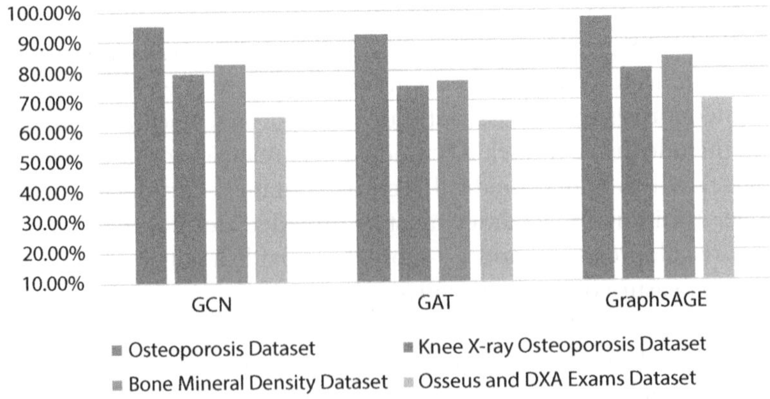

Figure 17.6 Accuracy performance.

As shown in Table 17.7, the Osteoporosis Dataset exhibited the best performance with GraphSAGE attaining 97.42% precision as compared to GCN's 95.53% and GAT's 92.89%. In the Knee X-Ray Osteoporosis Dataset, again, GraphSAGE outperformed with 83.72%, while GCN and GAT trailed at 80.14 and 77.57%. In the Bone Mineral Density Dataset GraphSAGE's 84.33% placed it in the front, just above GCN's 83.89% and GAT's 81.28%. finally, the Osseus and DXA Exams Dataset has GraphSAGE remaining in the lead with 70.29% and GCN and GAT coming afterwards with 67.62% and 66.13%. Overall, most datasets had the highest precision attributed to GraphSAGE, indicating good pattern recognition for important medical data. On the other hand, GCN performed exceptionally on the Bone Mineral Density dataset. Meanwhile, GAT provided consistent performance across all datasets. Figure 17.7 shows the performance of the precision comparison.

Table 17.7 Precision performance.

	Algorithms		
Dataset	**GCN**	**GAT**	**GraphSAGE**
Osteoporosis Dataset	95.53%	92.89%	97.42%
Knee X-ray Osteoporosis dataset	80.14%	77.57%	83.77%
Bone Mineral Density dataset	83.89%	81.28%	84.33%
Osseus and DXA Exams Dataset	67.62%	66.13%	70.29%

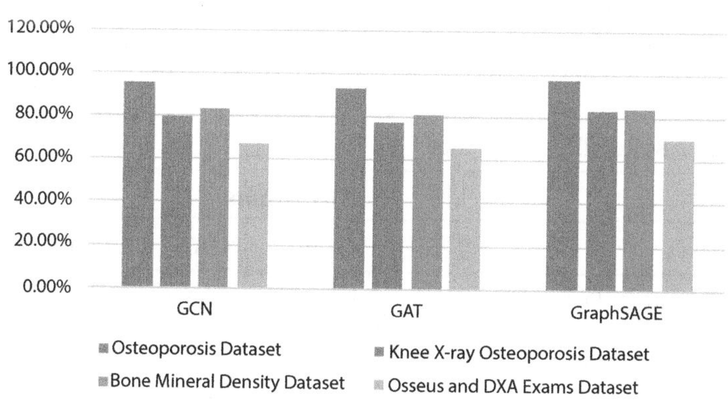

Figure 17.7 Precision comparison.

Recall

Recall is the sensitivity measure of the suggested method. It is the number of returned relevant documents to the total of all relevant documents. Recall calculates the measure of a method to recognize all the potential positive instances.

The recall performance of different algorithms, as shown in Table 17.8, GraphSAGE was the highest out of all approaches in the Recall Performance Evaluation. With the Osteoporosis Dataset, GraphSAGE achieved the highest recall value of 97.45%, surpassing GCN and GAT, which got 95.57% and 92.25% respectively. With a strong recall of 84.31%, GraphSAGE maintained a modest lead over GCN at 82.35%, while GAT trailed at 76.47% in the Bone Mineral Density Dataset. GraphSAGE led the Knee X-ray Osteoporosis Dataset with 80.56%, followed closely by GCN at 79.17% and GAT at 75%. Finally, in the Osseus and DXA Exams Dataset, GraphSAGE once more recorded the highest recall of 69.74% with GCN and GAT following at 65.13% and 63.16% respectively. From the results, it can be noted that GraphSAGE consistently outperformed and provided better recall versatility throughout and pattern recognition across multiple datasets. GCN provided solid results for recall in the Osteoporosis and the Bone Mineral Density datasets and GAT, though consistently in lower numbers, remarkably maintained a good performance throughout all datasets. Figure 17.8 illustrates the comparison of recall.

F1Score

The F1 score is the harmonic mean of precision and recall. It gives a balanced measure between the two metrics, it is most important in situations where the class distribution is not balanced.

The GraphSAGE algorithm has, on all datasets compared, maintained the highest F1-score as illustrated in Table 17.9. In the Osteoporosis Dataset, GraphSAGE also recorded the highest F1-score of 97.43%,

Table 17.8 Recall performance.

	Algorithms		
Dataset	**GCN**	**GAT**	**GraphSAGE**
Osteoporosis Dataset	95.57%	92.25%	97.45%
Knee X-ray Osteoporosis dataset	79.17%	75.00%	80.56%
Bone Mineral Density dataset	82.35%	76.47%	84.31%
Osseus and DXA Exams Dataset	65.13%	63.16%	69.74%

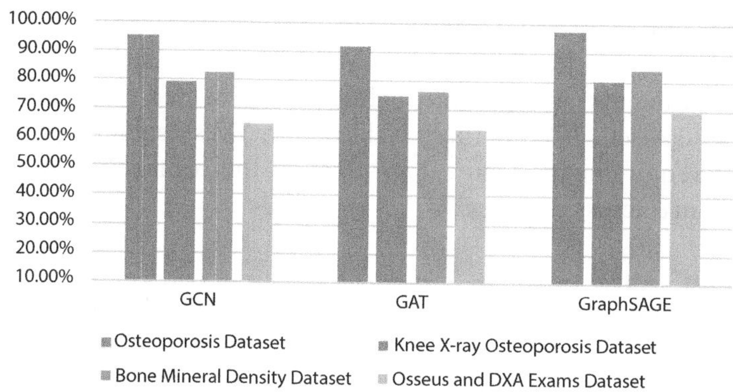

Figure 17.8 Performance of recall comparison.

Table 17.9 F1-score performance.

	Algorithms		
Dataset	**GCN**	**GAT**	**GraphSAGE**
Osteoporosis Dataset	95.51%	90.15%	97.43%
Knee X-ray Osteoporosis dataset	79.28%	70.08%	81.41%
Bone Mineral Density dataset	82.68%	75.85%	84.27%
Osseus and DXA Exams Dataset	65.68%	63.55%	69.90%

beating GCN with 95.51%, and GAT with 90.15%. This means that the algorithm can achieve a strong balance between the precision and recall relating to the classification of the osteoporosis data. in the Bone Mineral Density Dataset, GraphSAGE achieved an 84.27% F1 score, eclipsing GCN at 82.68% and GAT at 75.85%. In the Knee X-ray Osteoporosis Dataset, GraphSAGE also recorded the highest F1-score of 81.41%, beating GCN at 79.28% and GAT at 70.08% in its overall performance. Finally, in the Osseus and DXA Exams Dataset, GraphSAGE was again at the highest F1-score of 69.90%, beating GCN at 65.68% and GAT at 63.55%. This shows that performance is sustained across all datasets. GraphSAGE does well in attaining a balanced performance across datasets, which is useful in areas needing precision and recall, such as analysis and classification of medical data. Thus, it can be stated that GraphSAGE has an advantage in F1-score performance across all datasets suggesting effectiveness of the algorithm for increased complexity in medical data compared to GCN which was good in the Osteoporosis and Bone Mineral Density Datasets, and GAT, though trailing slightly behind, still provided solid performance across all datasets, making them viable alternatives depending on specific use-case requirements. Figure 17.9 presents the comparison of F1-score performance across datasets.

Execution Time

Execution time is the total computational time model training plus prediction time. It is very important because it provides an evaluation of the effectiveness and practical usability of the model in real-time applications.

Table 17.10 outlines the duration requirements for the three algorithms: GCN, GAT, and GraphSAGE across four datasets. The results demonstrate a varying time performance for each algorithm based on the complexity

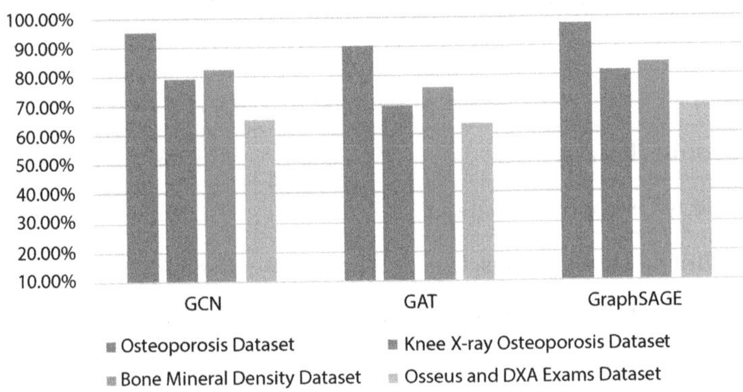

Figure 17.9 Performance of F1-score.

Table 17.10 Execution time performance.

	Algorithms (execution time – in seconds)		
Dataset	**GCN**	**GAT**	**GraphSAGE**
Osteoporosis Dataset	0.0080	0.0142	0.69
Knee X-ray Osteoporosis dataset	0.0020	0.0041	0.03
Bone Mineral Density dataset	0.0024	0.0036	0.01
Osseus and DXA Exams Dataset	0.0098	0.0072	0.05

of the dataset. In the Osteoporosis Dataset, GCN had the fastest execution time at only 0.0080 seconds. The second is to finish the dataset GAT with 0.0142 seconds. GraphSAGE recorded the longest time for the completion of the task, at 0.69 seconds, demonstrating that GraphSAGE is more computationally intensive when evaluating this dataset in comparison to GCN and GAT. GCN also had the fastest execution time for the Knee X-ray Osteoporosis Dataset at 0.0020 seconds, followed closely by GAT with 0.0041 seconds. Although GraphSAGE took longer, at 0.03 seconds, it is still a fast algorithm in comparison to the length of time for the other datasets it evaluated. Again, GCN and GAT were faster algorithms compared to GraphSAGE when evaluating the Bone Mineral Density Dataset. GCN completed the task in 0.0024 seconds, and GAT evaluated the dataset in 0.0036 seconds. Though slower, GraphSAGE tested the dataset in 0.01 seconds, again mirroring the increased complexity and higher computational requirement while testing this dataset. Lastly, time ended with GCN at 0.0098 seconds, GAT at 0.0072 seconds, and GraphSAGE testing this dataset in 0.05 seconds stating that GraphSAGE always tested longer time duration than the other two algorithms in all tested datasets. Table 17.11 summarizes the comprehensive evaluation metrics for all models for four datasets.

Summary of Evaluation Metrics

Table 17.11 Evaluation metrics – different datasets.

Dataset	**Evaluation metrics**	**GCN**	**GAT**	**GraphSAGE**
Osteoporosis Dataset	Accuracy	95.57%	92.25%	97.45%
	Precision	95.53%	92.89%	97.42%
	Recall	95.57%	92.25%	97.45%
	F1score	95.51%	90.15%	97.43%
	Execution Time	0.0080 seconds	0.0142 seconds	0.69 seconds

(*Continued*)

Table 17.11 Evaluation metrics – different datasets. (*Continued*)

Dataset	Evaluation metrics	GCN	GAT	GraphSAGE
Knee X-ray Osteoporosis dataset	Accuracy	79.17%	75.00%	80.56%
	Precision	80.14%	77.57%	83.77%
	Recall	79.17%	75.00%	80.56%
	F1score	79.28%	70.08%	81.41%
	Execution Time	0.0020 seconds	0.0041 seconds	0.03 seconds
Bone Mineral Density dataset	Accuracy	82.35%	76.47%	84.31%
	Precision	83.89%	81.28%	84.33%
	Recall	82.35%	76.47%	84.31%
	F1score	82.68%	75.85%	84.27%
	Execution Time	0.0024 seconds	0.0036 seconds	0.01 seconds
Osseus and DXA Exams Dataset	Accuracy	65.13%	63.16	69.74%
	Precision	67.62%	66.13%	70.29%
	Recall	65.13%	63.16	69.74%
	F1score	65.68%	63.55%	69.90%
	Execution Time	0.0098 seconds	0.0072 seconds	0.05 seconds

17.7 Conclusion

In this study, the use of deep learning methods, specifically Graph Neural Networks (GNNs) such as GCNs, GATs, and GraphSAGE, in conjunction with sophisticated models like GANs, for osteoporosis diagnosis was investigated. The models improved diagnostic performance, reduced human error, and made bone diseases detectable in clinical settings. With different datasets, compared models were assessed based on significant benchmark metrics like Accuracy, Precision, Recall, F1 Score, and Evaluation Time. Among all the models, GraphSAGE gave higher accuracy than

GCN and GAT in all the scenarios with the best accuracy too, at 97.45% in the Osteoporosis Dataset. GCN also performed very well, particularly in the Osteoporosis Dataset. Overall, however, GAT had the lowest accuracy. In addition, GraphSAGE scored highest in precision in most datasets, achieving 97.42% precision in the Osteoporosis Dataset, and GCN closely followed with Bone Mineral Density, with a slightly lower GAT. Similarly, GraphSAGE had the highest recall in every dataset, again the Osteoporosis Dataset with a 97.45% recall, and the Bone Mineral Density dataset with an 84.31 recall, and GCN was good recall here as well in the other datasets and particularly the Osteoporosis and Bone Mineral Density datasets, while overall for recall GAT had the lowest recall. GCN also had the lowest run time by a fair margin, with GAT slightly lower than that, while GraphSAGE performed the slowest since it required more computations. Yet, using GraphSAGE, overall best performance with accuracy, precision, recall, and F1-score would be warranted even on the more challenging Osteoporosis myelography and Knee X-ray Osteoporosis datasets. Overall, in this work, GraphSAGE performed best overall of the three algorithms and thus best option with complicated medical datasets. GCN achieved a recognized faster main time, and although it too did well on the Osteoporosis and Bone Mineral Density datasets, for GAT overall did not achieve the greatest performance overall across the entirety of datasets either. Although these are encouraging directions, there are still significant challenges ahead that include data quality and availability, interpretability of models, ethical issues, inclusion in workflows in clinical practice, and scalability of such models to heterogeneous patient groups. Follow-up studies must place extreme emphasis on explainable AI, privacy-saving methods, and data augmentation methods in surmounting such challenges. Regulatory compliance, and interdisciplinary coordination between healthcare practitioners and AI researchers, are also critical milestones to a successful implementation. All in all, deep learning-based models and GNN-based models have already been applied in diagnosing osteoporosis. Nevertheless, consistent innovation and solid validation remain crucial to leveraging their potential to the fullest. By overcoming current challenges, AI can potentially transform osteoporosis and bone disease diagnosis, treatment, and management, leading ultimately to improved patient care and influencing future healthcare.

AI disclosure statement: The author acknowledges the use of AI-based tools for minor language editing or grammatical correction. However, all intellectual contributions, analyses, and conclusions presented in this chapter are entirely the author's own.

References

1. Lee, J., H., *et al.*, Prediction of osteoporosis using support vector machine and logistic regression models. *J. Bone Metab.*, 20, 2, 97–104, 2013, June.
2. Tsai, T.Y., Wang, Y.C., Chiang, Y.J., Screening for osteoporosis from blood test data in elderly women using a machine learning approach. *Bioengineering*, 10, 3, 277, 2023. https://doi.org/10.3390/bioengineering10030277.
3. Liu, M., Yu, X., Zhang, Y., Huang, Z., Predicting fracture risk for elderly osteoporosis patients by a hybrid machine learning model. *DIGITAL HEALTH*, 10, 2024. 20552076241257456. https://doi.org/10.1177/20552076241257456.
4. Tecle, N., Teferi, T., Esayas, T., Application of artificial intelligence methods on osteoporosis classification with radiographs: A systematic review. *Bioengineering*, 8, 11, 166, 2021. https://doi.org/10.3390/bioengineering8110166.
5. Wang, H., Zhang, M., Li, J., Application of deep convolutional neural networks in the diagnosis of osteoporosis. *Sensors*, 22, 21, 8189, 2022. https://doi.org/10.3390/s22218189.
6. Hsieh, C.Y., Hsu, H.T., Huang, C.C., Interpretable deep-learning approaches for osteoporosis risk screening and individualized feature analysis using large population-based data: Model development and performance evaluation. *J. Med. Internet Res.*, 25, e40179, 2023. https://doi.org/10.2196/40179.
7. Ghorbani, M., Yaghoubi, M., Arabfard, M., RA-GCN: Graph convolutional network for disease prediction problems with imbalanced data. arXiv preprint, arXiv:2103.00221, 2021. https://arxiv.org/abs/2103.00221.
8. Cui, H., Liu, S., Pan, X., Xia, Y., Zhang, A., BrainNNExplainer: An interpretable graph neural network framework for brain network-based disease analysis. arXiv preprint, arXiv:2107.05097, 2021. https://arxiv.org/abs/2107.05097.
9. Boll, H.O., Lin, Y., Li, Z., Graph neural networks for heart failure prediction on an EHR-based patient similarity graph. arXiv preprint, arXiv:2411.19742, 2024. https://arxiv.org/abs/2411.19742.
10. Khemani, B., Patil, S., Kotecha, K., Tanwar, S., A review of graph neural networks: Concepts, architectures, techniques, challenges, datasets, applications, and future directions. *J. Big Data*, 11, 1, 18, 2024. https://doi.org/10.1186/s40537-023-00876-4.
11. Odionu, C.S., *et al.*, Big data analytics in healthcare: A comparative review of USA and global use cases. *Int. J. Multi. Res. Growth Eval.*, 4, 6, 1109–1117, 2023. https://doi.org/10.54660/IJMRGE.2023.4.6.1109-1117.
12. Shaffi, S.M., *et al.*, Transforming healthcare with real-time big data analytics: Opportunities, challenges, and future directions. *Int. J. Multi. Res.*, 7, 1, 2025. JFMR250136459.
13. Alghamdi, A., *et al.*, Healthcare analytics: A comprehensive review. *Engineering Technol. Appl. Sci. Res.*, 11, 1, 6650–6655, 2021.

14. Imran, S., *et al.*, Big data analytics in healthcare — A systematic literature review and roadmap for practical implementation. *IEEE/CAA J. Autom. Sin.*, 8, 1, 1–13, 2021. https://doi.org/10.1109/JAS.2021.1001234.
15. Mir, M.A., Revolutionizing radiology: AI-powered disease detection and diagnosis in healthcare. *Int. J. Innovative Res. Technol.*, 11, 8, 2025. IJIRT 171603.
16. Cabral, B.P., Future use of AI in diagnostic medicine: 2-wave cross-sectional survey study. *J. Med. Internet Res.*, 27, e53892, 2025. https://www.jmir.org/2025/1/e53892.
17. Maurya, N., Artificial intelligence revolution in healthcare: Transforming diagnosis, treatment, and patient care. *Asian J. Adv. Res.*, 7, 1, 241–263, 2024.
18. Kothinti, R.R., *et al.*, AI-powered predictive analytics – Improving preventive healthcare. *Int. J. Creative Res. Thoughts*, 13, 2, 2025. IJCRT2502668ISSN: 2320-2882.
19. Rahim, M.J., *et al.*, Predictive analytics in healthcare: Big data, better decisions. *Int. J. Sci. Res. Mod. Technol.*, 4, 1, 2025. https://doi.org/10.5281/zenodo.14630840.
20. Chin, K.Y. and Ima-Nirwana, S., A mini review on osteoporosis: From biology to pharmacological management of bone loss. *J. Clin. Med.*, 11, 21, 6434, 2022. doi:10.3390/jcm11216434.
21. Arboleya, L., Suárez, L., López, P., Solís, G., Margolles, A., Gueimonde, M., Osteomalacia in adults: A practical insight for clinicians. *J. Clin. Med.*, 12, 7, 2714, 2023. doi:10.3390/jcm12072714.
22. World Health Organization, *Nutritional rickets: A review of disease burden, causes, diagnosis, prevention, and treatment*, WHO, Geneva, 2019, ISBN: 9789241516587. Available from: https://iris.who.int/bitstream/handle/10665/329859/9789241516587-eng.pdf.
23. Ralston, S.H., Corral-Gudino, l., Cooper, C., Francis, R.M., Fraser, W.D., Gennari, L., Guanabens, N., Javaid, M.k., Layfield, R., O'Neill, T.W., *et al.* Diagnosis and Management of Pagets Disease of Bone in Adults: A Clinical Guideline. *J. Bone and Mineral Res.*, 34, 4, 579–604, 2019. https://doi.org/10.1002/jbmr.3657.
24. Banaganapalli, B., Fallatah, I., Alsubhi, F., Awan, Z., Elango, R., Shaik, N.A., Paget's disease: a review of the epidemiology, etiology, genetics, and treatment. *Front. Genet.*, 14, 1131182, 2023. doi: 10.3389/fgene.2023.1131182.
25. Xu, H., Clinical decision-making in bone cancer care management and forecast of ICU needs based on computed tomography. *J. Bone Oncol.*, 49, 100646, 2024. doi:10.1016/j.jbo.2024.100646.
26. Bilezikian, J.P., Khan, A.A., Potts, J.T., Clarke, B.L., Eastell, R., Lewiecki, E.M., *et al.*, Evaluation and management of primary hyperparathyroidism: Summary statement and guidelines from the fifth international workshop. *J. Bone Miner. Res.*, 37, 10, 1969–1999, 2022. doi:10.1002/jbmr.4677.

27. Xu, X., Zhao, X., Wei, M., Li, Z., A comprehensive review of graph convolutional networks: Approaches and applications. *Electron. Res. Arch.*, 31, 7, 4185–4215, 2023. https://doi.org/10.3934/era.2023213.
28. Zhang, H. and Qie, Y., Applying Deep Learning to Medical Imaging: A Review. *Appl. Sci.*, 13, 18, 10521, 2023. https://doi.org/10.3390/app131810521.
29. Wu, Z., Pan, S., Chen, F., Long, G., Zhang, C., Yu, P.S., A comprehensive survey on graph neural networks. *IEEE Trans. Neural Networks Learn. Syst.*, 32, 1, 4–24, 2021. https://doi.org/10.1109/TNNLS.2020.2978386.
30. Vrahatis, A.G., Lazaros, K., Kotsiantis, S., Graph Attention Networks: A Comprehensive Review of Methods and Applications. *Future Internet*, 16, 9, 318, 2024. https://doi.org/10.3390/fi16090318.
31. Zhang, Y., Lei, X., Pan, Y., Wu, F.-X., Drug repositioning with GraphSAGE and clustering constraints based on drug and disease networks. *Front. Pharmacol.*, 13, 872785, 2022. https://doi.org/10.3389/fphar.2022.872785.
32. Huang, N., Zhou, J., Wang, W., Wang, Q., Tang, Y., Sun, Y., Wang, D., Chen, S., Retinol-binding protein 4 is positively associated with bone mineral density in patients with type 2 diabetes and osteopenia or osteoporosis. *Clin. Endocrinol. (Oxford)*, 88, 5, 659–664, 2018, May. doi: 10.1111/cen.13560.
33. Atkinson, E.J., Therneau, T.M., Melton, LJ.3., Camp, J.J., Achenbach, S.J., Amin, S., Khosta, S., Assessing fracture risk using gradient boosting machine (GBM) models. *J. Bone Miner. Res.*, 27, 6, 1397–404, 2012, June. doi: 10.1002/jbmr.1577. PMID: 22367889; PMCID: PMC3408850.
34. Chen, Y., Yang, T., Gao, X., Xu, A., Hybrid deep learning model for risk prediction of fracture in patients with diabetes and osteoporosis. *Front. Med.*, 16, 3, 496–506, 2021. doi: 10.1007/s11684-021-0828-7. Epub 2021 Aug 26. PMID: 34448125.
35. Kruse, C., Eiken, P., Vestergaard, P., Clinical fracture risk evaluated by hierarchical agglomerative clustering. *Osteoporosis Int.*, 28, 819–832, 2017. https://doi.org/10.1007/s00198-016-3828-8.

18

Learning Behavioral Patterns for Autism Prediction with Graph Neural Networks

Deepa B.* and K. S. Jeen Marseline

Department of Information Technology and Cognitive Systems, Sri Krishna Arts and Science College, Coimbatore, India

Abstract

Proper and early identification of Autism spectrum disorder (ASD) is an essential aspect related to effective treatment and improvement of the situation developmental outcomes. The offered work proposes an unusual predictive model of ASD based on Graph Neural Networks (GNNs) on the structured text data such as formal clinical questionnaires, clinician reports, and structured medical records. We will turn structured text data into graph structures where the nodes represent one of the primary attributes or responses and edge represents semantic or relational information between them. GNN architecture is adjusted to allow the extraction of higher-level interactions and concealed connections in structured data to enhance the performance of predictions. To further improve the accuracy, we use advanced techniques of training which include stratified mini-batching, early stopping, and graph-structure specific augmentation list of techniques. Furthermore, we investigate the influence of different loss functions such as focal loss, weighted cross-entropy and label smoothing in order to address the problem of class imbalance as well as reduce this overconfidence of the model. Indeed, experimental findings on the classic ASD datasets demonstrate that our GNN-based framework along with adaptive training and loss setting can substantially increase the classification performance compared to the state-of-the-art machine and deep learning methods. The model that we propose represents an interpretable and scalable ASD screening supported by structured clinical text.

***Keywords*:** ASD, graph neural network, data set, preprocessing, loss function, confusion matrix, accuracy

**Corresponding author*: deepab@skasc.ac.in

J. Ramkumar, Sridaran Rajagopal, B. Suchitra and S. Balamurugan (eds.) Graph Neural Networks: Concepts and Applications, (741–758) © 2026 Scrivener Publishing LLC

18.1 Introduction

A graph model gives a pictorial and intuitive description of the complex interrelationship between objects. It represents an old-fashioned graph structure by using vertex and links to identify the items and relationship between them. A graph model, however, a subsequent action introduces Features and tags for vertex and Links. In this model, vertexes act as objects like persons or things, and link act as the associations of these vertexes with each other, which can be of different types, such as friendships, financial relations, or dependences. Additionally, this model has features and tags. Features are properties that define the vertex and links, like a person's age or the price of a product. Tags are descriptive labels assigned to vertex and links that enable understanding of the structure of the graph. Graph models are used in various domains, such as social media platform, banking sector, and biology. For example, this model can be used within online community to analyze users' connections and find important influencers or communities. In the funding industry, a graph model might be used to display account transactions and detect fraudulent behavior or money laundering. A particularly good aspect of graph models is their ease of visualization and interpretation. By representing intricate relationships in an uncomplicated and intuitive form, graph models enable the discovery of patterns and knowledge that are hard to identify through other means. In Technology aspects, graph models are broadly used in information management systems and data inspection software. Semantic databases are designed to store and manage large sets of data through graph models, thereby allowing for effective querying and analysis of entity relationships. Another important term associated with graph models is conceptual model. Its functions just like a glossary or a group of defined rules, establishing a universal understanding of a certain sector and enabling the exchange and reuse of knowledge systems, along with building recommendation systems.

18.1.1 Types and Application of Graph Models

There exist numerous categories of graph models, each one of them adapted to represent a particular kind of graph in a form of data structure with certain rules and limitations that govern the connections and activity of the nodes and the edges constructed.

- Conceptual Schema Model: Nodes in a conceptual Schema structure represent concepts or entities and links represent the relationships existing between concepts.
- Direct Acyclic Graphs (DAGs): Directed Acyclic Graph (DAG) is the type of graph, whose links are one-way and free of loops. Nodes in a Direct Accessible Graph have a route to specify the flow of data or the flow of power and loops are prohibited, thereby disallowing dependency loops or infinite loops. Other domains that DAGs are used include job scheduling, information processing and source control systems.
- Hypergraphs: It is a generalization of a graph to allow edges to connect nodes several of which simultaneously. Hypergraphs do not have an edge between any two nodes in a plane; there is at least one edge having many nodes, which is depicted by a set of nodes. This type facilitates a more complicated interrelationship. Hypergraphs frequently arise as a representation of a complicated data interconnection structure such as social networks and recommender systems or bioinformatics. They have found application in most of the fields, inclusive of social media analysis and expert systems.
- Knowledge Graphs: A graph of things that are related and the relationship between things defines knowledge in a knowledge graph model. It provides the way to Mold and organize data which makes the reasoning, inference, and query enhancement better. Knowledge graph models are applied to many application scenarios, including search engines, recommendation systems, chatter bot, and question answering systems.
- Network Graphs: A network graph model allows the complex networks to be modeled and analyzed so that the users are able to observe the structures and properties. Network graph models can be used to analyze many types of networks including social, transport, biological and communication networks.
- Property Graphs: The property graph happens to be one of the most popular models of graphs used in graph databases. Property graph – it can be said that property graph supports flexible data structure because vertex and links can have arbitrary key-value properties.

- RDF Graphs: It's a metadata and online ontology standard of expressing information. In resource development framework graph nodes represent concepts or resources and edges bear the relationship among resources. The URIs (Uniform Resources Identifier) describe each of the nodes and the edges. Metadata and other semantic information is frequently modeled by RDF graphs.
- Semantic Networks (Conceptual Map): Semantic networks (also known as Conceptual Graph) are a graph representation form suggesting organizing knowledge in a form easily arguable. It is characterized by an application of edges (or arcs) in the display of relationship between concepts and the use of nodes to represent concepts.
- Spatial Graphs: Spatial graph – a spatial graph is a type of model of a graph used to represent spatial networks. Geographic locations within the framework would be nodes or vertices and edges or links would be the spatial relationships or interactions across these locations. Many classes of networks, including social connections, biological networks, and transportation networks can be studied *via* spatial graph models.

18.1.2 Application of Graph Models

Graph models are the flexible tools that are used to represent and analyze relationships between entities in many fields. They are used in an enormous number of domains, which include computer science, social networks, transport, and others, serving as a powerful tool of knowledge representation and understanding of complex systems.one of the primary applications is in Artificial Intelligence as a model of knowledge representation, building expert systems, and developing recommended systems.

18.1.3 Loss and Training Functions in Graph Models

The training process in graphical forms involves refinement of the model parameters, so as to minimize a loss method, which indicates the difference between predictions and real values. The goal will be finding the parameters, which best match the model forecasts to real values and hence have the most minimal loss. During the training period, their weights are adjusted in order to minimize the loss. Loss in training and the loss in validation is two important measures when determining the effectiveness of the model and its ability to generalize.

18.1.4 Training GNN

Working out the Graph neural network model has several ways in which it can be done

- Forward Propagation: In such propagation the Input data (graph) undergoes the layers of the GNN to generate node embedding.
- Loss Calculation: A loss function is applied which measures how the predicted node embeddings (or graph-level embeddings) compare with the target values that are to be expected (e.g. node labels in classification).
- Backpropagation: It involves the use of the gradient of the loss and is sent backwards in the network thus adjusting the weights.
- Optimization: In an attempt to reduce the loss function, methods of optimization such as stochastic gradient descent are employed to fine-tune the weights.

18.1.5 Loss of Training

The training loss is the calculated difference to which the model is making an assumption on the result as per training set. This parameter is updated, right after every run in forward and backward propagation of the model during the learning periods. As a rule, the loss decays with time as the model becomes to better guess the connections between inputs and outputs. The variance between the predicted labels and the actual ones is measured using a loss function (such as Mean Squared Error, Cross Entropy Loss, etc.).

- Standard Loss Function
- Mean Squared Error
- Cross Entropy Loss node classification
- Binary Cross entropy classification

Mean Squared Error (MSE) is one of the fundamentals in the statistical and machine learning realms, and its role is vital in the process of assessing the accuracy of predictive models. It serves as a parameter to establish the good quality of the model. It is the squared difference between the expected result and actual value on average measured in the data.

18.2 Related Work

Meng Li, Han Liu, Fanyu Kong, Pengju Lv proposed a model known as DTRE (Drug Target Relationship Enhance) specifically crafted to forecast Drugtarget Interactions (DTIs), including prospective drug targets for endometrial cancer [1]. Mengmeng Gaoa, Daokun Zhangb, Yi Chena, Yiwen Zhangc, Zhikang Wangd, Xiaoyu Wangd, Shanshan Lic, Yuming Guoc, Geoffrey I. Webbb, Anh T.N. Nguyene, Lauren Maye, Jiangning Song propose GraphormerDTI, which employs the powerful Graph Transformer neural network to model molecular structure [2]. Seunghwan Jung, Seunghyun Wang, Doheon Lee proposed Cancer GATE, a cancer-driver gene prediction model [3]. Yongjie Xu, Zengjie Yu, Yisheng Li, Yuehan Liu, Ye Lia, Yishan Wang Proposed a model to developed a novel method for detecting ASD with EEG data [4]. S. Rajaprakash, R Sofia, C. Bagath Basha, C. Sunitha Ram I. Ameethbasha, V. Subapriya framed a novel approach based on the fuzzy MSE GCN to detect ASD effectively. In this work, a population graph is generated with a combination of images with phenotypic data [5]. Jiyeon Han, Min Ji Kang, Sanghyuk Lee proposed two GCN models where the first GCN model predict the drug-perturbed gene expression profile and the second GCN model use the prediction result to calculate the drug synergy score [6].

18.3 Proposed Work

Proposed model includes the Cross-entropy loss function and Adam optimization method to do better the model performance to estimate the traits of autism.

18.3.1 Description of the Data Set

The autism screening Dataset acquired on Kaggle and it consists of 6,075 records that should help to detect individuals being at risk of the autism spectrum disorder (ASD). This database includes a set of diagnostic items, demographic data, and signs of medical history, making it a critical tool in a study of the early diagnosis and modeling of ASD. In essence, the data comprises ten binary variables with names of A1-A10. These indicators will probably pertain to the replies based on a regularized autism screening tool with 1 and 0 as answers discharging positive and negative responses to behavioral or developmental questions. All these are the building blocks of the predictive input that we use to assess the level of risk. Other than the screening questions, the data has four other variables;

- Age: A numerical entry which contains the age of the individual being assessed. Even though the majority of records refer to the adolescent range of age (e.g., 15-16 years of age) this is a crucial demographic characteristic. Sex – a dummy variable that expresses the biological sex of the person (m male, f female).
- Jaundice: A dichotomous variable in which the jaundice at birth was characterized as either 'yes' or 'no'. This part of medical history may be considered relevant because of its possible significance in infant neurodevelopment.
- Family_ASD: The other categorical variable that describes the presence of the known family history of ASD that can serve as the genetic risk factor.
- Class: This is the target label in the supervised learning efforts, and the options of the label may include YES or NO to indicate that the individual is or is not characterized as being at risk of ASD.

They are well-formatted, with no missing values, and give a balanced combination of psychological, medical, and demographic variables. It can be used properly by a broad set of machine learning products, especially in the case of classification models that centrally aim at the early detection of autism. The creation of algorithms to predict the risk of ASD and evaluate them, including comparison to those created by other researchers, can be based on this dataset, and the relations between early symptoms, the family and medical histories, and the diagnostic outcomes can be examined.

18.3.2 Used Libraries

In order to build a model, the Libraries below are utilized:

- Pandas: It is used to loading a dataset in form of a tabular column
- PyTorch: It is a deep learning framework to define and train a Graphical Neural network
- torch.nn. functional: It is a usual Neural network operation such as ReLu
- torch geometric: This Library is utilized in constructing and training GNN Model

- GCNConv: It is a graph convolutional layer of PyTorch Geometric which is added.
- sklearn. Preprocessing: It is employed in encoding Categorical data
- learn. model_selection. train_test_split: In this, the data is divided into training and testing.
- sklearn. neighbors. NearestNeighbor:
- NumPy: It is a library dealing with arrays and numerical calculations

18.3.3 Preprocessing

Training the Autism Screening Dataset involves various vital steps to prepare the data to be utilized in machine learning. To begin with, the categorical variables, that is, Sex, Jaundice, Family_ASD, and the target label Class are converted into numerical form using label encoding. An example where the entry of m and f is in the Sex field is translated to 1 and 0 respectively and the two entries of YES and NO are converted to 1 and 0 in the Class field. This step will ensure that data is transformed to fit numerical modeling. The next stage is feature selection where the most relevant columns will be retained such as the ten screening questions (A1 to A10), Age, and the demographical features that have been encoded. These would be input variables, and Class column as output label to make the prediction. The missing values are not present in the dataset; hence, no imputation steps would be needed. Normalization to scale the numerical database such as age that may be represented in a different range is to be performed after the feature selection phase. To determine the same effect of model in training the methodology of Z-score is applied. The second step consisted of the separation of the dataset into the training data and test data with the certain division, and then to test the performance of the model with unseen data. Last adopt the equivalent framework, e.g. Pytorch or TensorFlow.

18.3.4 Splitting Data

The act of splitting data implies breaking down the data set into two categories; training set and testing set. For the recognition of the Features and Labels, the variable X and Y are identified. There is split sklearn to cut the dataset into some portion such as 80 % and 20 percent or 70 percent and 30 percent, and give fixed random state value (e.g. 42), the split will be constant ensuring that the same data points are distributed on training and test data set. It creates a reliable model and enables to offer standard evaluation.

18.3.5 Construction of a KNN Graph

K-Nearest Neighbor KNN: A graphical representation in which nodes represent data points and edges connect a node to its k nearest neighbors according to a similarity or distance measure. It is applied to many machines learning tasks: categorization, anomaly detection, and others, to take advantage of local facts to make predictions or identify anomalies. Construction of a kNN graph involves calculation of distances between points and finding the k-most neighborly points to each point after which edges have to be created connecting a point to all of its k-nearest neighbors. The Figure 18.1 displays a line graph that shows how a K-Nearest Neighbors (KNN) classifier's performance varies as the number of neighbors (K) rises from 1 to 20. At K=1, the accuracy is destitute; it steadily increases and peaks at K=10–11. Following this, the accuracy exhibits tiny ups and downs throughout increasing K values while fluctuating somewhat but remaining largely consistent. Overall, the graph shows that very low or very high K values perform marginally worse than midrange K values, which produce the maximum accuracy.

18.3.5.1 Algorithms Employed in Creation of KNN Graph

Step 1: Preprocessing of data
Step 2: The selection of k values will be up to the neighbors according to each point
Step 3: Compute the distance using the algorithms such as Euclidean, cosine distance or Manhattan distance.
Step 4: Identify the nearest neighbor
Step 5: Create a graph
Step 6: Graph to be stored in the adjacent matrix

This piece of work Find the nearest neighbor using Euclidean distance formula is:

$$d1 = \text{sqrt}\ [\ (x\ 2 - x\ 1)\ 2 + (y\ 2 - y\ 1)\ 2\ ] \tag{18.1}$$

where, One of the points has coordinates (x1, y1). The coordinates of the other point are (x2, y2), d1 is the distance among (x1, y1) and (x2, y2).

18.3.6 GCN Model

GCNs are a zoom in neural network that are shaped to generalize as a node representation in graphs. A graph, which is a data structure, denoted as $G = (V, E)$, is a collection of n nodes also known as vertices and a collection of E edges,

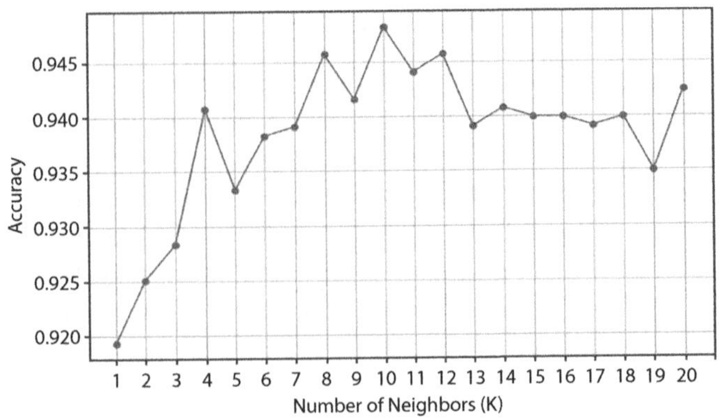

Figure 18.1 KNN accuracy for different K values.

where the edges are used to represent a relationship between the nodes. GCNs use a method known as convolution where the interrelation of the nodes in a graph v (x v) is considered and as a result, it is able to identify more complicated patterns than the common neural networks. Exactly, an adjacency matrix A is a square matrix portraying a graph. The adjacency matrix is a v 2 matrix *A* 1, 1, where the entry *A* 1, 1 consists of a 0 or 1 depending on whether a connection exists between nodes i and j or not in a graph with n nodes. The elements of the matrix are either binary or weighted. The convolutional operations of the GCN are caused by the adjacency matrix as it defines the combination of features of various nodes. This matrix directs the convolution process, since pertinent information is relayed between associated nodes, thus advancing the model that is then able to capture important trends with precision [7].

Autism Screening Dataset, use of spatial features through a per-question adjacency matrix involves designing separate graphs for each separate screening item (A1 to A10). Rather than creating a single global similarity graph, this approach creates a separate adjacency matrix for each question, such that nodes (people) are connected if they had the same response to that particular question. These separate graphs contain localized patterns of behavior tied to distinct characteristics or reactions. Each graph is processed through its respective Graph Convolutional Network (GCN) layer, allowing the model to acquire spatial features that reflect how people interact with each other in terms of specific behaviors. The representations of the features learned through each GCN pathway are then combined—either through concatenation, summation, or attention mechanisms—to generate an all-inclusive embedding that captures the behavioral secrets from all the questions. This composite embedding is then passed through a classifier, typically a fully connected neural layer, to predict if a person is likely to have autism.

The standard propagation rule for GCN layers is

$$H^{(l+1)} = \sigma(\bar{D}^{-\frac{1}{2}} \bar{A} \bar{D}^{-\frac{1}{2}} H^l w^l) \tag{18.2}$$

Where H^l *is layer l feature matrix,*
w^l *is the weight matrix that can be learned in layer l*
The adjacency matrix with self loop is $\bar{A} = A + I$
$\bar{D}$ *is the degree matrix of A is an activation function* $(e.g.\ ReLU)$
The input feature matrix is H ^ 0 = X

18.3.7 GCN Model Architecture

The working procedure for GCN in the Autism dataset involves several steps

Step 1: Data can be represented in the form of node. Each node has individual features and constructs a KNN graph to connect similar individuals.
Step 2: A graph can be constructed using edge index to define a connection between nodes
Step 3: Use of GCN Layers. In GCN two layers are there
Step 4: First layer Collects and modifies data from the neighboring nodes to develop a concealed
Step 5: Interpretation of the node.
Step 6: Add non-linearity by using Relu activation function
Step 7: Second Layer projects the concealed representation into labels of classes.
Step 8: Model learn the concept by minimizing cross entropy loss function
Step 9: Optimize the model by Adem Technique
Step 10: Predict the class Label

The Figure 18.2 depicts a Graph Convolutional Network (GCN) architecture that uses stacked GCN layers to process input characteristics. After entering the main GCN architecture block, the input splits into two consecutive GCN layers. To create the final output label, various GCN layers' outputs are concatenated.

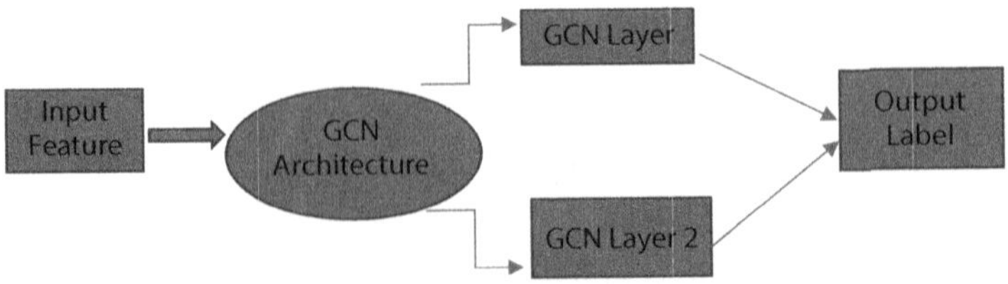

Figure 18.2 GNN architecture.

18.3.8 Cross Entropy Loss Function in Autism

Loss functions are a kind of model that in machine learning enables models to quantify their errors and self-improves. Loss functions are equations (mathematical equations) that measure the discrepancy between the predicted values to the actual values in a machine learning model The measure of discrepancy, what is yielded by the loss functions, also acts as a guide during the optimization progress, as it gives the model some insight on the extent to which it tallies with the data. Hence, the computational process of optimization associated with machine learning algorithms involves the employment of loss function, and parameters adjustment of the model is set to aggregate a desirable outcome: the lower the error, the better the model. The cross-entropy loss estimates the distance between the probability of a certain class assigned by the model and the true class. It punishes the predictions that are far away to the actual label. In the case of binary classification, in cross-entropy loss, one prediction is as follows:

$$\text{Loss} = -[y \cdot \log(\overline{y}) + (1-y) \cdot \log(1-\overline{y})] \tag{18.3}$$

18.3.9 Adam Optimization Technique

The Autism Screening Dataset uses methods of artificial intelligence (neural networks or Graph Convolutional Networks (GCNs)) to classify individuals by such characteristics as responses to assessment questionnaires (A1-A10), age, sex and other aids about the person or behavior. The Adam optimizer is often used when fine-tuning such models to minimize the cross-entropy loss. Adaptive Moment Estimation is a gradient-descended optimization algorithm designed to prompt better and comfortable training.

18.4 Results and Discussion

18.4.1 Performance Matrices

We used five measures to determine the prediction results: Accuracy, Precision, Recall, F1-score, and AUROC. These are obtained by matching the anticipated tags, with the ground truth tags. The mix up matrix was used to carry out these estimations.

A= True Positive, False Positive, True Negative and False Negative

Accuracy = True Positive+ True Negative / A

Precision= True Positive / True positive+False Positive

Recall = True Positive/True positive+False Negative

F1= 2 *Precision*Recall / Precision+Recall

The calculation of AUROC is based on True Positive Rate (TPR) and False Positive Rate (FPR):

TPR=Precision= True Positive /True Positive+False Positive

FPR= False Positive/ False Positive+True Negative The area under the curve of TPR against FPR according to different thresholds is calculated as Area Under the Receiver Operating Characteristic curve (AUROC) and this is displayed [8].

18.4.2 Correlation Matrix

A correlation matrix is graphically represented as a correlation matrix, in which a color scale is used to show the direction of the association between and the strength of the association between a number of variables. A heatmap is an effective way to find patterns and interdependencies of data within a data set as it allows highlighting positive, negative, or zero correlation graphically in a map.

The Figure 18.3 illustrates the connections between several numerical characteristics in the form of correlation matrix. In that Weaker or negative correlations are displayed in lighter blue to dark blue hues, whereas strong positive correlations are indicated by red. The heatmap highlights couples with significant positive or negative dependency, making it easier to see how attributes are related to one another.

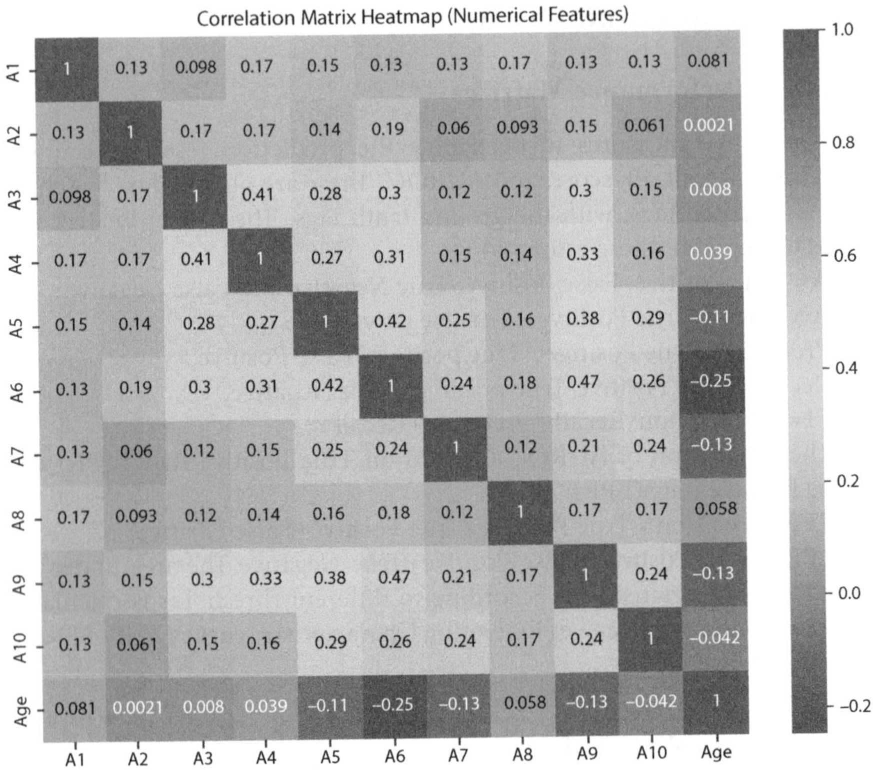

Figure 18.3 Correlation matrix.

18.4.3 Confusion Matrix

The following confusion matrix is to assess how well a classification model, trained on the Autism Screening Dataset, performs. Both the correct and incorrect predictions in the model are visually separated into two categories, namely, True and Predicted label. It exists in two categories. The symbols, YES (indicating the presence of autism) and NO (indicating the absence of autism) are depicted.

The Figure 18.4 displays a confusion matrix comparing true labels with predicted labels from a classification model. The analysis indicates a substantial count of accurately predicted NO cases (801) and accurately predicted YES cases (314). The matrix indicates misclassification also.

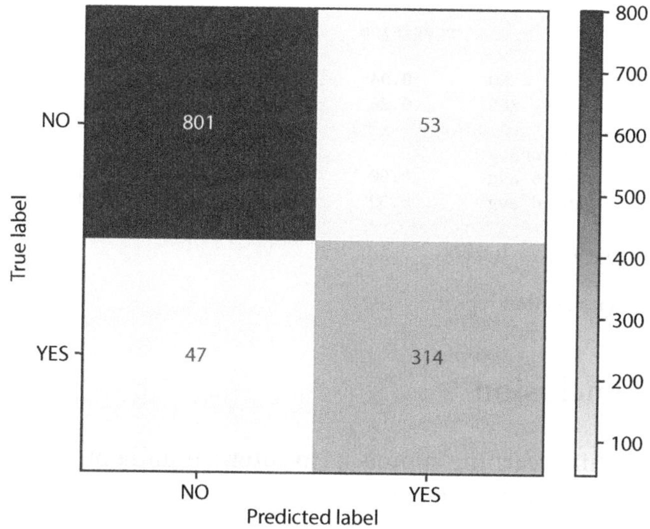

Figure 18.4 Confusion matrix.

18.4.4 Classification Report

The picture contains a report of the classification that gives an overview of the performance of a trained machine learning model on Autism Screening Dataset. The report measures the effectiveness of the model in dividing two classes, namely, the NO (no autism) and YES (autism). The report gives important performance metrics, like, precision, recall, F1-score, and support of each of the classes.

One of the key performance measures, there to estimate the overall performance of a classification model, is accuracy. The model was found to possess an accuracy of 0.9177 in the Autism Screening Dataset which corresponds to 91.77 percent. It indicates that the model was accurate in predicting the condition of people with autism about 92 out of 100 people in the test set.

The Figure 18.5 provides a classification report that summarizes precision, recall, and F1-score for the two classes: NO and YES. The overall accuracy is approximately 91.77%, with macro and weighted averages ranging from 0.90 to 0.92, suggesting a balanced and reliable predictive performance.

```
Classification Report:
              precision    recall  f1-score   support

          NO       0.94      0.94      0.94       854
         YES       0.86      0.87      0.86       361

    accuracy                           0.92      1215
   macro avg       0.90      0.90      0.90      1215
weighted avg       0.92      0.92      0.92      1215

Accuracy: 0.9177
```

Figure 18.5 Classification report.

18.5 Conclusion

This is a machine learning approach (to automate and optimize early identification of the autism spectrum disorder (ASD) through a structured data collection of a behavior screening questionnaire assessment). Through thorough data preprocessing, feature selection (e.g., correlation heatmaps) and with the application of the supervised learning algorithms, a high rate of the classification accuracy (91.77%) was achieved. The effectiveness of the model in separating both the non-autistic and autistic people is proved by performance measures in terms of accuracy (maximum of 0.94 as well as recall similarly as much as 0.87). Evaluation of the confusion matrix also indicated the viability of the model due to the even ratio of true positives and true negatives, and having very minimal false predictions. A mixture of cross-entropy loss combined with Adam used to optimize the training of the model also contributed to enhancing the convergence and training rate of the model. The findings indicate that machine learning could definitely have a potential to be used as an additional clinical decision-making assistant generating fast reliable predictions, all based on behavioral data. Despite the impressive performance of the model it might be interested in future experiments that apply deeper neural networks or other graph learning methods (e.g. GCNs) to learn more expressive relationships (and detect further into more obscured instances). Such systems eventually demonstrate a great potential in supporting early intervention programs and de-stressing the traditional diagnostic procedures.

AI disclosure statement: The author acknowledges the use of AI-based tools for minor language editing or grammatical correction. However, all intellectual contributions, analyses, and conclusions presented in this chapter are entirely the author's own.

References

1. Li, M., Liu, H., Kong, F., Lv, P., DTRE: A model for predicting drug-target interactions of endometrial cancer based on heterogeneous graph. *Future Gener. Comput. Syst.*, 161, 478–486, 2024. https://doi.org/10.1016/j.future.2024.07.012.
2. Gao, M., Zhang, D., Chen, Y., Zhang, Y., Wang, Z., Wang, X., Li, S., Guo, Y., Webb, G.I., Nguyen, A.T.N., May, L., Song, J., GraphormerDTI: A graph transformer-based approach for drug-target interaction prediction. *Comput. Biol. Med.*, 173, 2024. https://doi.org/10.1016/j.compbiomed.2024.108339.
3. Jung, S., Wang, S., Lee, D., CancerGATE: Prediction of cancer-driver genes using graph attention autoencoders. *Comput. Biol. Med.*, 176, 2024. https://doi.org/10.1016/j.compbiomed.2024.108568.
4. Xu, Y., Yu, Z., Li, Y., Liu, Y., Li, Y., Wang, Y., Autism spectrum disorder diagnosis with EEG signals using time series maps of brain functional connectivity and a combined CNN–LSTM model. *Comput. Methods Programs Biomed.*, 250, 2024. https://doi.org/10.1016/j.cmpb.2024.108196.
5. Rajaprakash, S., Basha, C.B., Ram, C.S., Ameethbasha, I., Subapriya, V., Sofia, R., Using convolutional network in graphical model detection of autism disorders with fuzzy inference systems. *Intelligence-Based Med.*, 11, 2025. https://doi.org/10.1016/j.ibmed.2025.100213.
6. Han, J., Kang, M.J., Lee, S., DRSPRING: Graph convolutional network (GCN)-based drug synergy prediction utilizing drug-induced gene expression profile. *Comput. Biol. Med.*, 174, 2024. https://doi.org/10.1016/j.compbiomed.2024.108436.
7. Kourbane, I., Papadakis, P., Andries, M., Optimized assessment of physical rehabilitation exercises using spatiotemporal, sequential graph-convolutional networks. *Comput. Biol. Med.*, 186, 2025. https://doi.org/10.1016/j.compbiomed.2024.109578.
8. Zhou, F., Khushi, M., Brett, J., Uddin, S., Graph neural network-based subgraph analysis for predicting adverse drug events. *Comput. Biol. Med.*, 183, 2024. https://doi.org/10.1016/j.compbiomsed.2024.109282.

Part IV

RESEARCH ANALYTICS, OPTIMIZATION, AND SOFTWARE ASSURANCE

19

Emerging Trends and Collaborative Networks in Indian Graph Neural Networks Research: A Bibliometric Analysis

Muthukrishnan M.[1*], Ghouse Modin Nabeesab Mamdapur[2], Saravanan S.[3] and Parasakthi D.[3]

[1]*Department of Library and Information Science, Dr. N.G.P. Arts and Science College, Coimbatore, Tamil Nadu, India*
[2]*Department of Library and Information Science, Yenepoya (Deemed to be University), Mangaluru, Karnataka, India*
[3]*Department of Commerce, Dr. N.G.P. Arts and Science College, Coimbatore, Tamil Nadu, India*

Abstract
The rise of Graph Neural Networks (GNNs) has significantly impacted artificial intelligence (AI) research, with their ability to model complex relationships in graph-structured data. However, bibliometric analyses of Indian GNNs research have been limited. This study addresses this gap by analyzing 1,148 publications on GNNs from India (2020-2025), sourced from the Scopus database. Using tools such as VOSviewer visualization tool and CiteSpace visualizing tools, we identified key trends, influential works, and the structure of collaborative networks in Indian GNNs research. India's research output grew from 19 papers in 2020 to 536 in 2024, but the citation impact declined, with citations per paper (CPP) dropping from 40.00 in 2020 to 0.45 in 2025. Despite the increase in funded publications (from 5 in 2020 to 100 in 2024), India's global research visibility remains low, with only two highly cited papers (HCPs). The United States emerged as the primary international collaborator, contributing 84 papers and 652 citations. Collaboration analysis highlighted a strong network with India leading in total link strength (TLS: 447), followed by the United States (TLS: 180) and China (TLS: 118). This research emphasizes the need for India to enhance its global influence,

**Corresponding author*: connectmm@gmail.com

J. Ramkumar, Sridaran Rajagopal, B. Suchitra and S. Balamurugan (eds.) Graph Neural Networks: Concepts and Applications, (761–796) © 2026 Scrivener Publishing LLC

particularly by focusing on high-impact journals and strengthening international collaborations. Future directions include increasing funding and exploring interdisciplinary research opportunities to improve the visibility and impact of Indian GNNs research.

Keywords: Graph neural networks, bibliometric analysis, citation impact, international collaboration, Indian research, research productivity, global visibility

19.1 Introduction

Graph Neural Networks (GNNs) become a powerful and general framework for modeling and analyzing graphs, with significant impact on how to handle graph data [1, 2]. This is especially true due to their ability to effectively model complicated relations between different entities using graphs. As a result, these models are now commonly used for applications like link prediction, node classification, and graph generation. India has been particularly active in advancing theoretical constructs and practical implementations within GNNs research [3, 4]. Given the increasing prominence of India in this area, it is important to explore the new patterns and research collaborations in GNNs research from India so as to interpret the future of research from there as well as anticipate possible international partnerships [5]. Therefore, to maximize the effectiveness of the Indian contributions, efforts should be concentrated at domains where GNNs are especially promising. One application is in the field of health care, where GNNs are being employed to represent patient data and improve disease prediction and personalized treatment recommendations [6]. In finance, fraud detection is being improved by uncovering hidden patterns in transaction networks. Monitoring such developments and facilitating partnerships between Indian research agencies and their international counterparts can help foster innovation in this area.

Collaboration will use technological innovation and serve as a mechanism to produce solutions for urgent global problems [7]. Interdisciplinary research and knowledge exchange will thus be key to the development of GNNs technologies [8]. These efforts will have ramifications not only academically but also in an industry thirsty for advanced technologies based on graph knowledge [5]. For anyone looking to be at the forefront of this exciting field, now is a terrific time to get involved with the growing GNNs community in India. Interest in GNNs has sparked a flurry of activity in the research community, with many trying their hand at various aspects of the technology. In the context of GNNs and mobile sensing, only a recent work from Mondal *et al.* [9] has proposed a framework leveraging GNNs for the recognition of human activities based on smartphone sensor data [9]. This project showed how GNNs can be applied to learn patterns in time series data from sensor networks, which is relevant for both health monitoring and activity sensing. Aljohani *et al.* [10] applied deep neural networks using GNNs to quantify the spread of academic research by means of citation networks. The research focused on analyzing the movement of citations in academic papers and proposed that future predictions in the spread of science could be done through models of citation networks through the use of GNNs [10]. It shows a desire to extend the applications of GNNs beyond regular machine learning problems to more complicated issues in the field of scholarly communication. Chaudhary *et al.* [11] tackled the problem of anomaly detection utilizing GNNs and proposed a new method for identifying outliers in large datasets with relational structures. It also contributed to existing research on GNNs for anomaly detection in highly structured and relational domains like fraud detection and network security [11].

Another example is the scientometric analysis by Cheng *et al.* [12], which employed CiteSpace to study how trends in neural networks and GNNs are developing. Their paper also identified research topics and prominent papers that contributed to advances in GNNs, as well as creating a map of the intellectual structure of the field [12]. This study highlights how bibliometric methodologies are increasingly valuable in understanding and tracking trending topics in GNNs studies. Xu *et al.* [13] conducted a scientometric review of event knowledge graphs, which revealed the growing interest in event knowledge representations in general. As discussed in this review, and illustrated in the GNNs models for event prediction and event-based reasoning, GNNs are becoming increasingly relevant for processing dynamic, temporal data about events [13]. Sawhney *et al.* [14] investigated the use of GNNs of algorithmic trading by employing hyperbolic graph learning in order to model the scale free character of financial markets. Their analysis showed GNNs could advance stock movement prediction and risk management, bringing the application of GNNs into finance and economics [14, 15]. This shows how flexible GNNs can be for different high stakes use cases. Saini *et al.* [16], which uses label wise attention in GNNs for extreme classification in GalaXC, is a powerful demonstration of the capabilities of GNNs for large scale, complex classification problems. Their work has important implications of applications on recommendation systems and multi label classification problems [16].

Jiang *et al.* [17] reviewed GNNs for traffic forecasting and highlighted that spatiotemporal models are particularly relevant for traffic prediction. Their research demonstrated the application of GNNs to transportation network modelling and traffic prediction, which has vital implications for smart city development and automated navigation for vehicles [17]. Gupta *et al.* [18] provided a survey on the current state of GNNs, focusing on the challenges and emerging applications of GNNs in a wide array of fields, from materials science, through computer vision, to healthcare [18]. Their work provided a useful introduction to the general landscape of GNNs application and the challenges that persist within the development and use of these models. On top of that, a bibliometric analysis from Da Silva *et al.* [19] provided an overview of the current status of research using GNNs, including insights on such research and application across the globe. This showed the increase of interdisciplinary works and highlighted tendencies within the GNNs literature, which continues to grow [19]. Keramatfar *et al.*, in a similar fashion, also provided a bibliometric overview on GNNs by evaluating the most influential authors, journals, and research topics relevant to GNNs [20]. Though a significant amount of research pertaining to GNNs exists, the specific applications of these theories to Indian pedagogical networks remains unexamined. As discussed earlier, existing studies are largely oriented towards global trends or particular uses and do not pay much attention to the new collaborative research networks developing in India, where the number of publications related to GNNs has rapidly increased in recent years.

19.2 Literature Gaps

The international trends of GNNs have been explored, but a detailed bibliometric study focused exclusively on GNNs in India is still missing. Indian researchers, particularly in global partnerships, are also not fully charted. Also, the India emerging applications of GNNs in important interdisciplinary fields are not very well understood. As the number of publications from India in the field of GNNs continues to grow, it provides a relatively unique opportunity to perform a targeted analysis of trends and collaborations within the field of GNNs research in India.

19.3 Justification for Current Research Topic

This area of research was chosen for the increasing importance of GNNs in both scientific research and practical applications. While India is gradually increasing its share in the GNNs studies research, it becomes important to analyze which trends and collaborations exist within the Indian research ecosystem. Through the bibliometric study conducted in this research, the intent is thus to gather useful information on which research areas are emerging in India, who are the important authors and institutions publishing, and the international collaborations helping to push the field forward.

19.4 Summary of Current Work

The present study seeks to conduct an extensive bibliometric assessment of new developments as well as cooperative relationships in the area of Indian GNNs research. Also, the study will glean important thematic trends as well as major authors and institutions in India by exploring leading peer reviewed journals in the fields of high interest. It will also identify the networks of collaboration involved in the global dissemination of Indian GNNs research. It will better position us to know where GNNs research in India stands and where Indian researchers are in relation to the world in GNN advancements. This research will guide the future of GNNs as well as promote greater international cooperation.

19.5 Materials and Methods

To investigate emerging trends and collaborative patterns in GNNs research originating from India, a bibliometric study was carried out using data sourced from the Scopus database [21], known for its broad and high-quality coverage of peer-reviewed literature across scientific disciplines. Particular emphasis was placed on journal publications appearing in high-impact, peer-reviewed outlets to ensure the reliability and scholarly relevance of the dataset. The data extraction was conducted on March 26, 2025, focusing on the publication window from 2020 through 2025 to reflect the most current advancements in the field. The search strategy employed, the query: "TITLE-ABS-KEY" ("Graph Neural Network*") "AND PUBYEAR" > 2019 "AND PUBYEAR" < 2026 AND ("LIMIT-TO (AFFILCOUNTRY", "India")), ensuring inclusion of studies explicitly cantered on GNNs with at least one author affiliated with an Indian institution. This yielded a dataset of 1,148 publications, of which high-impact journal articles were prioritized in the analysis.

The bibliographic data included essential metadata such as article titles, author details, institutional affiliations, journal names, journal volume numbers, journal issue numbers, articles page numbers, publication years, citation metrics, keywords, funding details, document type and abstracts.

Initial data refinement and descriptive statistical analysis were conducted using Microsoft Excel. Network-based analyses were performed with VOSviewer visualization tool (version 1.6.20) [22–24] and CiteSpace visualization tool (version 6.3. R1 Basic) [25–27], both widely adopted tools for visualizing co-authorship networks, institutional collaborations, and keyword co-occurrence structures.

To examine collaborative intensity, the Total Link Strength (TLS) metric was applied, quantifying the depth and frequency of co-authorship ties [28]. Keyword co-occurrence mapping was used to uncover central research themes and to track shifts in scholarly focus over the years [29, 30]. Additionally, trending topic detection and thematic mapping provided deeper insight into the evolving interests and priorities within India's GNNs research community. Productivity and impact were assessed using standard bibliometric indicators, including publication counts and average citation rates, while relative citation impact was computed to normalize for temporal disparities in citation accrual [31, 32]. This multifaceted approach offers a nuanced understanding of the growth, influence, and collaborative dynamics characterizing India's contributions to high-quality GNNs research [33].

19.6 Publication Overview

In Table 19.1, the volume of publications relating to GNNs in India and around the world demonstrates an increasing trend during the period 2020 to 2025. The analysis also presented year wise details of Global Publication GP, Indian Total Publications ITP, Total Citations TC, Citation per Paper CPP, Funded Papers FP, International Collaboration Papers ICP, Percentage of International Collaboration Papers ICP, Relative Citation Index RCI and Highly Cited Papers HCP. The statistics bring out the contribution made by Indian researchers, in the context of global patterns, especially with respect to high impact, peer reviewed journals. In 2020, world produced 2496 GNNs papers, 19 of which came from India. Total citations for the year were 760, for a CPP of 40.00. Out of these, 5 were funded papers, 5 were papers with international collaborations, which is 26.32% of total Indian publications. Its Relative Citation Index this year is 8.82 and it has 1 Highly Cited Paper (HCP), with more than 100 citations. In the year 2021, there were 10016 publications worldwide, of which the authors from India published 78. The total citations that year was 1,134 and the CPP was 14.54. The funded papers reached 20 and international collaboration papers were

Table 19.1 Year-wise distribution of GNNs research output – global vs. India (2020–2025).

Year	GP	ITP	TC	CPP	FP	ICP	%ICP	RCI	HCP
2020	2496	19	760	40.00	5	5	26.32	8.82	1
2021	10016	78	1134	14.54	20	26	33.33	3.21	
2022	7420	139	1235	8.88	33	36	25.90	1.96	1
2023	4948	266	1376	5.17	68	57	21.43	1.14	
2024	2786	536	651	1.21	100	141	26.31	0.27	
2025	1167	110	49	0.45	16	22	20.00	0.10	
Total	**28833**	**1148**	**5205**	**4.53**	**242**	**287**	**25.00**	**1.00**	**2**

GP = Global Publications; ITP = Indian Total Publications; TC = Total Citations; CPP = Citation per Paper; FP = Funded Papers; ICP = International Collaboration Papers; RCI = Relative Citation Index; HCP = Highly Cited Paper (>100)

26, which was 33.33% of the papers published in India. The RCI for 2021 is 3.21, but, there were no highly cited papers.

Total global publications in 2022 dropped further to 7420 of which India contributed to 139 publications. Total citations for the year were 1,235 with a CPP of 8.88. Funded papers increased to 33, and international collaboration papers increased to 36, also a quarter of Indian publications at 25.90%. Its RCI for 2022 was 1.96, and it had 1 highly cited paper.

In 2023, global publications further declined to 4948, while publications from India increased significantly with n=266. There were 1376 total citations this year for a CPP of 5.17. Funded papers increased to 68, while international collaborative publications slightly decreased to 57, which represents 21.43% of the total publications from India. The RCI for 2023 was 1.14 and highly cited papers are 0. The global numbers in 2024 dropped to 2786 publications, while the number of publications from India increased significantly to 536 publications. Total citations for the year were 651, with a CPP of 1.21. The number of funded papers was 100 and international collaboration papers was 141, 26.31% of the total publications from India. There were no highly cited papers, and the RCI for 2024 is.27. In 2025, total publications dropped to 1167 of which 110 came from Indian researchers. There were 49 total citations this year, with a CPP of 0.45. The number of funded papers decreased to 16 and papers with international collaboration were reduced to 22 which make up 20.00% of the publications from India. The RCI for 2025 was .10, and no highly cited papers.

The total number of publications on GNNs during these six years is 28833 at world level and 1148 from India. Total citations were 5205 with an overall CPP of 4.53. There were 242 papers with funding and 287 papers with international collaboration, while 25.00% of publications in India were internationally collaborative in nature. The overall RCI for the period was 1.00, and two highly cited papers were identified. This increasing trend of contribution from Indian researchers has also been found to be reflected in contributions to GNNs research in peer reviewed journals that have significant impact. Research output from India is on the rise while global output has been trending down in recent years, indicating that India is becoming increasingly important in this area.

This disparity is further highlighted by the differences in citation impact, measured by citations per paper (CPP). Globally, the average CPP is 10.50, while India's CPP stands at 4.53, suggesting a lower visibility and influence of Indian research on the international stage. Moreover, the global landscape boasts 461 Highly Cited Papers (HCPs), while India has contributed only 2 HCPs. This stark contrast underscores the need for India to enhance the global impact and visibility of its research, particularly in high-impact, peer-reviewed journals, to bridge this gap and improve its scientific standing.

Indian research trends show a steady upward trajectory in total publications (TP), rising from 19 in 2020 to 536 in 2024, before slightly declining to 110 in 2025. However, despite this growth, total citations (TC) remain considerably below global standards, peaking at 1,376 in 2023. The citations per paper (CPP) exhibit a declining trend, dropping from 40.00 in 2020 to 0.45 in 2025, suggesting that older publications continue to have a more

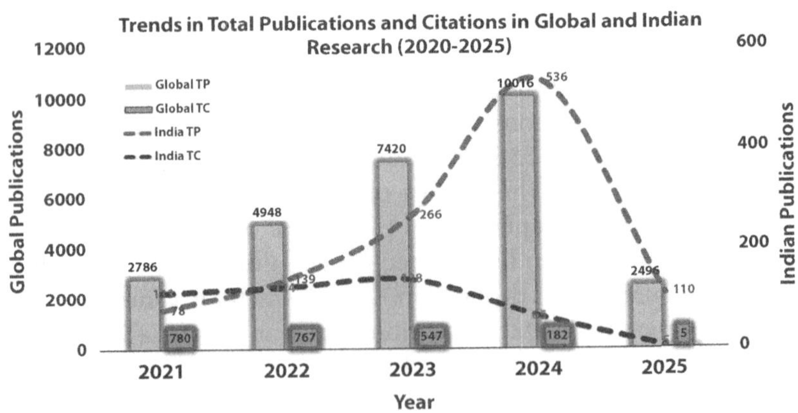

Figure 19.1 Global and Indian research publications on GNNs from 2020 to 2025.

significant impact over time. The increase in funded publications (FP) is notable, rising from 5 in 2020 to 100 in 2024, reflecting a positive growth in research funding. However, this Figure 19.1, indicating potential challenges in maintaining funding levels. Despite the increase in funded publications, the number of Highly Cited Papers (HCPs) from India remains low, with only two HCPs during the period one in 2020 and one in 2022 highlighting the difficulties in achieving substantial global research influence.

Figure 19.1 shows the timeline of GNN studies from 2020-2025. There were more than 10,000 papers published worldwide in 2021, the highest number ever and with a slow decreasing trend since then. Meanwhile, Indian research publications, always at a lower scale, appeared to achieve gradually until a sudden drop in 2025 of research output in 2024. Citation counts present a similar but delayed pattern, since it takes time for research to be acknowledged. Although the absolute number of publications in India with respect to world publications is still low, this shows a growing interest by local researchers on GNNs. Interestingly, the CPP on studies for India is very competitive which shows the quality and relevance of these studies, especially those published in high impact peer-reviewed journals. The increase in collaboration in internationally authored papers and funded projects also shows the relevance of joining India in the global scientific community, which is necessary to develop the field. This is well represented by the time series plot, which depicts Indian publications and citations on trendlines indicating both successes and cause for concern. The drop off in 2025 should be explored further, as it could represent changes in research focus, funding, or publication. On the whole, the share of GNN research in India on the total GNN research output is still relatively new, but the growth trajectory does still remain optimistic until the year of 2024. Additional research infrastructure and international partnerships play key roles in sustaining and increasing India's contributions to this rapidly growing area of research.

19.7 Publication Types Overview

Conference papers dominated the output, accounting for 52.09% (598 papers), with 2,361 citations (CPP: 3.95) and an RCI of 0.87. Articles, comprising 42.06% (483 papers), garnered the most citations (2,684 citations, CPP: 5.56, RCI: 1.23) and had the highest international collaboration rate at 31.26%. Reviews, though fewer in number (18), exhibited a CPP of 5.17 and an RCI of 1.14 (Table 19.2). Other publication types such as book chapters, books, errata, and retracted papers showed lower productivity.

Table 19.2 Distribution of document types in GNNs research output in India.

Documents type	TP	TC	CCP	FP	ICP	%ICP	RCI
Conference paper	598	2361	3.95	91	122	20.40	0.87
Article	483	2684	5.56	144	151	31.26	1.23
Book chapter	42	26	0.62	4	8	19.05	0.14
Review	18	93	5.17	3	2	11.11	1.14
Book	3	0	0.00	-	1	33.33	0.00
Erratum	2	1	0.50	-	1	50.00	0.11
Retracted	2	40	20.00	-	2	100.00	4.41
Total	**1148**	**5205**	**4.53**	**242**	**287**	**25.00**	**1.00**

Surprisingly, retracted works achieved a high CPP in 20.00 and RCI in 4.41, all of which were internationally co-authored.

19.8 Citation Distribution Overview

Citation distribution revealed that 49.74% of the documents remained uncited, though 79 were funded and 111 internationally co-authored. Papers with 1–5 citations (32.23%) had a CPP of 2.19. Higher citation brackets (21–50: CPP 28.20; 51–100: CPP 66.30; and 101–241: CPP 181.00) indicated fewer but significantly impactful studies. These highly cited papers demonstrated greater international engagement, with up to 63.41% co-authorship in the 21–50 citation range (Table 19.3).

19.9 Top International Collaborators

The landscape of GNNs research involving Indian collaborations from 2020 to 2025 demonstrates global engagement and varied international partnerships. India leads the dataset with 1,148 publications and 5,205 citations, averaging 4.53 citations per paper (CPP). India also has the highest Total Link Strength (TLS) of 447, stemming from 39 international links (Table 19.4). The United States is the most prominent international collaborator,

Table 19.3 Distribution of citations across GNNs research publications in India.

Citation slab	TP	TC	CPP	%TP	FP	RCI	ICP	%ICP
Zero Citation	571	0	0.00	49.74	79	0.00	111	19.44
01-05	370	810	2.19	32.23	85	0.48	94	25.41
06-10	71	518	7.30	6.18	24	1.61	19	26.76
11-20	73	1033	14.15	6.36	27	3.12	29	39.73
21-50	41	1156	28.20	3.57	18	6.22	26	63.41
51-100	20	1326	66.30	1.74	9	14.62	8	40.00
101-241	2	362	181.00	0.17	0	39.92	0	0.00
Total	**1148**	**5205**	**4.53**	**100.00**	**242**	**1.00**	**287**	**25.00**

contributing 84 publications, 652 citations (CPP: 7.76), and a TLS of 180, with 35 links. This indicates strong bilateral research efforts. The United Kingdom and China also show notable contributions, with 36 and 33 publications, respectively. China has the highest CPP (15.58) and a TLS of 118, while the UK follows with a CPP of 12.69 and a TLS of 129.

Saudi Arabia's research output includes 36 publications, 194 citations (CPP: 5.39), and a TLS of 93. In Europe, Germany contributed 24 publications with a CPP of 7.42, while France and Italy each contributed 12 publications, with CPPs of 9.00 and 7.92, respectively. Portugal, though with fewer publications (6), also demonstrated substantial engagement, with a CPP of 7.67. Among Asian contributors, Malaysia produced 13 publications (CPP: 5.00), Japan published 9 papers (CPP: 10.00), and South Korea published 10 papers (CPP: 4.20). Taiwan's output of 9 publications yielded the second-highest CPP (11.89). Other emerging contributors include Iraq (17 publications, CPP: 5.82) and the United Arab Emirates (11 publications, CPP: 7.73), reflecting growing regional interest. Switzerland, Canada, and Australia contributed with more moderate outputs, while Norway and Lebanon exhibited targeted contributions with high CPPs of 5.50 and 5.67, respectively. This overview illustrates the wide-ranging international collaboration in GNNs research, highlighting the significant engagement of countries across continents and the varying citation impacts of these collaborations. The data reflects a robust exchange of knowledge,

Table 19.4 Top 22 countries with more than six publications in collaboration with Indian GNNs research (2020–2025).

Country	TP	TC	CCP	TLS	Links
India	1148	5205	4.53	447	39
United States	84	652	7.76	180	35
Saudi Arabia	36	194	5.39	93	29
United Kingdom	36	457	12.69	129	37
China	33	514	15.58	118	35
Germany	24	178	7.42	102	35
Iraq	17	99	5.82	30	13
Canada	15	42	2.80	50	19
Malaysia	13	65	5.00	69	32
France	12	108	9.00	61	32
Italy	12	95	7.92	84	33
Russian Federation	12	65	5.42	20	7
Australia	11	46	4.18	65	32
United Arab Emirates	11	85	7.73	33	18
South Korea	10	42	4.20	68	33
Spain	10	60	6.00	42	29
Japan	9	90	10.00	40	17
Switzerland	9	26	2.89	80	34
Taiwan	9	107	11.89	83	34
Lebanon	6	34	5.67	35	24
Norway	6	33	5.50	9	4
Portugal	6	46	7.67	51	31

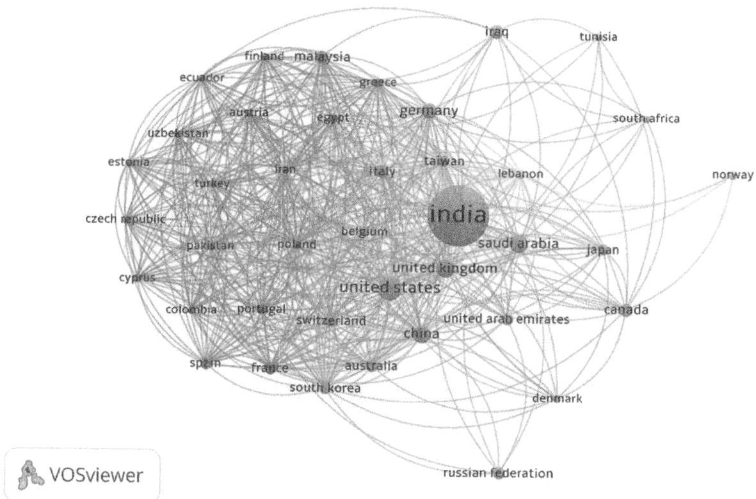

Figure 19.2 International collaboration map of GNNs research output.

with particular emphasis on high-impact contributions from countries like China, the United States, and the United Kingdom.

As depicted in Figure 19.2, a map was generated to analyse the distribution and collaboration patterns among countries involved in Graph Neural Network research. The study included 80 countries, each contributing a minimum of three documents and at least ten citations. Out of these, 41 countries met the threshold criteria for inclusion in the co-authorship and citation analysis. For further analysis, the top 40 countries—those with the highest Total Link Strength (TLS), with a minimum TLS of 40—were selected. These countries were subsequently grouped into five clusters, indicating collaborative groupings, with a total of 540 links and a cumulative TLS of 1,281. The range of TLS among these countries varied notably. Norway, with a TLS of 9, represented the lowest link strength, while India, with a TLS of 447, exhibited the highest. Each of the 40 countries maintained at least one collaborative link, with the maximum number of 39 links observed in countries such as India, while Norway had the least, with only 4 links.

19.10 Top Contributing Indian Authors

The most prolific Indian authors in GNNs research between 2020 and 2025 reveal a combination of publication, volume and scholarly impact. Among the top contributors by productivity, Sayan Ranu (IIT Delhi) leads with 15 publications, accumulating 141 citations, averaging 9.40 citations per paper (CPP) (Table 19.5). He also recorded six funded papers and an international collaboration rate of 40.00%, yielding a Relative Citation

Table 19.5 Leading Indian authors in GNNs research (2020–2025): productivity and impact.

Name	Affiliations with respect to the Indian authors	TP	TC	CPP	FP	ICP	%ICP	RCI
Top 6 authors publication productivity								
Ranu, Sayan	Indian Institute of Technology Delhi	15	141	9.40	6	6	40.00	2.07
Jadeja, Mahipal	Malaviya National Institute of Technology Jaipur	8	19	2.38	0	3	37.50	0.52
Patgiri, Ripon	National Institute of Technology Silchar	7	78	11.14	0	0	0.00	2.46
Krishnan, N.M. Anoop	Indian Institute of Technology Delhi	7	67	9.57	3	1	14.29	2.11
Bandyopadhyay, Sanghamitra	Indian Statistical Institute Kolkata	7	7	1.00	2	0	0.00	0.22
Arya, Arti	PES University Bangalore	7	2	0.29	0	0	0.00	0.06
Top 6 impactful authors								
Sarkar, Ram	Jadavpur University Kolkata	6	108	18.00	2	3	50.00	3.97
Waikhom Lilapati	National Institute of Technology Silchar	6	75	12.50	0	0	0.00	2.76
Chepuri, Sundeep Prabhakar	Indian Institute of Science Bangalore	6	40	6.67	4	2	33.33	1.47
Bandyopadhyay Sambaran	Indian Institute of Science, Bangalore	5	29	5.80	3	0	0.00	1.28
Chandra, Joydeep	Indian Institute of Technology Patna	6	29	4.83	4	3	50.00	1.07
Mausam	Indian Institute of Technology Delhi	6	23	3.83	4	2	33.33	0.85

Index (RCI) of 2.07. Other productive contributors include Mahipal Jadeja (MNIT Jaipur) with 8 papers (CPP: 2.38; RCI: 0.52) and Ripon Patgiri (NIT Silchar), who, despite no international collaborations or funding, garnered 78 citations across 7 papers (CPP: 11.14; RCI: 2.46). N.M. Anoop Krishnan (IIT Delhi) also made a notable impact with 7 papers, a CPP of 9.57, and an RCI of 2.11. In terms of impact, Ram Sarkar (Jadavpur University) stands out with an RCI of 3.97, based on 6 papers and 108 citations (CPP: 18.00), with 50.00% international collaboration. Similarly, Waikhom Lilapati (NIT Silchar) achieved a CPP of 12.50 and an RCI of 2.76 from 6 publications. Notably, Sundeep P. Chepuri and Joydeep Chandra also combined quality with collaboration, each contributing significantly through both funded research and international partnerships.

19.11 Top Contributing Foreign Authors

The table provides insights into the contributions of the top 6 foreign authors in GNNs research associated with Indian publications between 2020 and 2025 (Table 19.6). These authors, primarily from Southern Federal University in Russia and the University of Illinois in the United States, have significantly influenced the field. Among them, Bozhenyuk,

Table 19.6 Top 6 foreign authors in GNNs research output in India (2020–2025).

Name	AAFA	TP	TC	CPP	FP	ICP	%ICP	RCI	TLS
Bozhenyuk, Alexander	SFU	8	37	4.63	8	8	100.00	1.02	33
Sharma, Amit	SFU	8	37	4.63	8	8	100.00	1.02	34
Tselykh, Alexe	SFU	8	37	4.63	8	8	100.00	1.02	33
Gavrilenko, Vadim	SFU	6	32	5.33	6	6	100.00	1.18	28
Medya, Sourav	UL	5	62	12.40	2	5	100.00	2.73	5
Nikashina, Polina	SFU	5	28	5.60	5	5	100.00	1.24	24

AAFA=Affiliations associated with Foreign Authors, SFU= Southern Federal University, Taganrog, Russian Federation; UL=University of Illinois, Chicago, United States

Alexander, Sharma, Amit, and Tselykh, Alexe, all from Southern Federal University, Taganrog, Russian Federation, have each published 8 papers (TP) with 37 citations (TC). Their Citations per Paper (CPP) stand at 4.63, indicating a moderate scholarly impact. All three authors share an International Collaboration Percentage (ICP) of 100%, which reflects their full international engagement in research. Their Relative Citation Index (RCI) is 1.02, suggesting their scholarly impact is well balanced compared to other researchers in the field. Gavrilenko, Vadim, also from Southern Federal University, has published 6 papers (TP) and garnered 32 citations (TC), resulting in a CPP of 5.33. His CPP is higher than that of Bozhenyuk, Sharma, and Tselykh, but his ICP remains at 100%, indicating that he is completely engaged in international collaborations. His RCI of 1.18 indicates that he has a marginally greater impact than the other Russian scholars affiliated with the same university, reflecting his slightly higher contribution to the field. Medya, Sourav from the University of Illinois, Chicago, United States star CPP of 12.40. Although he has published only 5 papers, the total TC of Medya is 62, which positions him in terms of research impact among the top researchers in the GNNs field. His ICP is also 100%, and his RCI of 2.73 further highlight his pre-eminence and high impact of the international GNNs research. Finally, Nikashina, Polina, another researcher from Southern Federal University has published 5 papers (TP) and has 28 citations (TC) resulting in CPP of 5.60. Like the remaining foreign authors affiliated with this university, her ICP is 100%, and her RCI of 1.24 shows that she has a moderate level of research impact.

Figure 19.3 presents the analysis of author productivity and collaboration patterns through metrics including total papers, citations, and co-authorship link strength. VOSviewer was used to analyse co-authorship networks among 5,824 researchers, applying thresholds of ≥3 documents and ≥1 citation per author, resulting in 137 qualifying authors. The analysis revealed three collaborative clusters with total link strength of 756, where node size represents publication output and edge thickness indicates collaboration strength. The findings highlight authors with the strongest collaborative ties, offering insights into research collaboration patterns. The network visualization illustrates knowledge dissemination hubs, showing how thresholds and link strength metrics objectively define research connectivity in this domain. The analysis reveals Ranu, Sayan as the most productive scholar with 15 publications and 141 citations, demonstrating significant research impact. A cluster of Russian researchers from Southern Federal University shows particularly strong collaborative ties, with Bozhenyuk, Alexander and Tselykh, Alexey each contributing 8 papers with 37 citations and demonstrating high co-authorship link strengths of 33.

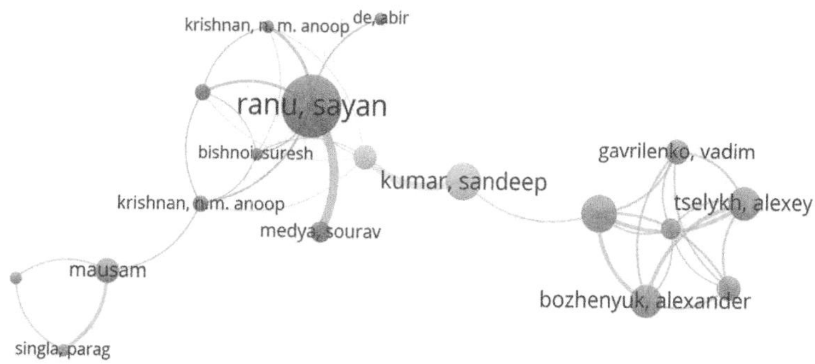

Figure 19.3 Co-authorship network visualization of GNNs.

Their colleague Gavrilenko, Vadim maintains slightly lower but still substantial metrics (6 papers, 32 citations, link strength 28), while Nikashina, Polina contributes 5 papers with 28 citations and a link strength of 24. Sharma, Amit emerges as the most connected researcher with a co-authorship link strength of 34 despite having fewer papers (9) and citations (38). Other notable contributors include Singh, Pradeep Kumar (6 papers, link strength 27), Bandyopadhyay, Sanghamitra (7 papers, link strength 10), and Patgiri, Ripon (7 papers, 78 citations, link strength 6).

19.12 Top Indian Research Institutions

The Table 19.7 presents a detailed analysis of the top 12 affiliations within the Indian academic landscape, focusing on key metrics. The top 6 most productive affiliations include the Indian Institute of Technology, Delhi, leads with 44 publications, receiving 383 citations and CPP of 8.70. Particularly, out of those 44 outputs, 16 were publications with an external funding and the same number of these are internationally collaborating, meaning that its international collaboration output percentage is 36.36%, with an RCI of 1.92. 5 papers were funded, international collaboration was present in 16%; RCI is 0.39 The K L deemed to be university, Vijayawada placed next with 36 publications, 64 citations and CPP at 1.78. The Saveetha Institute of Medical and Technical Sciences, Chennai has 31 papers with 45 citations (CPP=1.45), 3 of them funded, 13% of its output coming from international collaborations, and has an RCI of 0.32. Next, PES University, Bangalore published 30 publications with 71 citations (CPP=2.37), with 3 funded papers and papers developed by 2 international collaborations (6.67%), with an RCI of 0.52. Indian Institute of Technology, Kharagpur had a total of 29 publications, 130 citations with CPP of 4.48, 10 papers under funding and 7 in collaboration with foreign countries (24.14%), obtaining an RCI of 0.99. Last but not least, the Indian Institute of Science, Bangalore had 28 publications with 436 citations and an excellent CPP of 15.57. Of these publications, 15 with funding and 10 with international collaboration (35.71%), presenting an RCI of 3.43.

Table 19.7 Top 12 affiliations within the Indian scenario with the most productive and impactful affiliations.

Top 12 affiliations within the Indian scenario	TP	TC	CPP	FP	ICP	%ICP	RCI
The top 6 most productive affiliations							
Indian Institute of Technology, Delhi	44	383	8.70	16	16	36.36	1.92
K L Deemed to be University, Vijayawada	36	64	1.78	5	16	44.44	0.39
Saveetha Institute of Medical and Technical Sciences, Chennai	31	45	1.45	3	13	41.94	0.32
PES University, Bangalore	30	71	2.37	3	2	6.67	0.52
Indian Institute of Technology, Kharagpur	29	130	4.48	10	7	24.14	0.99
Indian Institute of Science, Bangalore	28	436	15.57	15	10	35.71	3.43
Top 6 most impactful affiliations							
Indraprastha Institute of Information Technology, Delhi	12	237	19.75	6	4	33.33	4.36
Manipal Institute of Technology, Manipal	10	195	19.50	1	2	20.00	4.30
Indian Institute of Technology, Hyderabad	18	288	19.00	5	6	33.33	3.53
Indian Institute of Technology, Patna	14	168	12.00	7	3	21.43	2.65
R.V. College of Engineering, Bangalore	10	106	10.60	0	1	10.00	2.34
International Institute of Information Technology, Hyderabad	14	143	10.21	5	4	28.57	2.25

But, in terms of impact, the Indraprastha Institute of Information Technology, Delhi is the most impactful affiliation with 12 publications, 237 citations, and a CPP of 19.75. For its part, this institution sustains 6 funded by it and 4 international collaborative publications, which represents 33.33% with RCI of 4.36. Also, Manipal Institute of Technology, Manipal had a CPP of 19.50 from 10 publications with 195 citations, but, from these only 1 was funded and 2 (20.00%) were internationally collaborative with an RCI of 4.30. The Indian Institute of Technology, Hyderabad published 18 papers with 288 citations and a CPP of 19.00; it had 5 funded papers, 6 (33.33%) international collaborations and its RCI was 3.53. The Indian Institute of Technology, Patna contributed 14 documents, and received 168 citations (CPP=12.00); it is worth noticing that, from the total of its records, 7 papers were funded and 3 documents were internationally collaborated papers (21.43%), the RCI value for this university was 2.65. Plus, R.V. College of Engineering, Bangalore published 10 papers which accumulated 106 citations, with a CPP of 10.60, while no funded papers appeared, also exhibiting a low international collaboration rate of 10.00%, detecting an RCI of 2.34. And lastly the International Institute of Information Technology, Hyderabad has 14 publications and 143 citations (CPP= 10.21) with 5 funded publications and 4 international collaborations (28.57%) and an RCI of 2.25. This shows the different kinds of profiles that Indian institutions have in GNNs research. While some, IIT Delhi and IISc Bangalore being two, present higher total productivity, others, particularly from the impact-oriented group such as IIIT Delhi and Manipal Institute of Technology prefer higher impact factor per paper. These institutions have made significant progress in their field by publishing scholarly papers in top ranked academic journals.

19.13 Top Global Research Institutions

Table 19.8 shows the statistics that reflect the global scenario of GNNs research from 2020 to 2025 regarding productivity and impact, which reveals a clear stratified research. Among the most 6 prolific affiliations, University of Illinois at Chicago, USA, presented 8 publications and 86 citations, with a CPP of 10.75. On the same note, Southern Federal University, Russia published 8 papers but received just 37 citations (CPP=4.63), although it also funded 8 papers, denoting that there is a strategy to focus on funding research. The University of Wisconsin Madison USA contributed 7 publications with 24 citations CPP=3.43, 4 papers funded and has an RCI of 0.76. In this scenario, Microsoft Research, USA was remarkable

Table 19.8 Top 12 affiliations within the global scenario with the most productive and impactful affiliations.

Top 12 affiliations within a global scenario	TP	TC	CPP	FP	RCI
The top 6 most productive affiliations					
University of Illinois at Chicago, USA	8	86	10.75	4	2.37
Southern Federal University, Russia	8	37	4.63	8	1.02
University of Wisconsin-Madison, USA	7	24	3.43	4	0.76
Microsoft Research, USA	6	144	24.00	2	5.29
CNRS Centre National de la Recherche Scientifique, France	6	74	12.33	5	2.72
Lebanese American University, Lebanon	6	34	5.67	4	1.25
Top 6 most impactful affiliations					
Institute for Particle Physics Phenomenology, UK	3	111	37.00	3	8.16
Columbia University, USA	5	103	20.60	4	4.54
Harvard University, USA	4	58	14.50	3	3.20
University of California, Santa Barbara, USA	5	53	10.60	4	2.34
University of Cyprus, Cyprus	4	45	11.25	4	2.48
Google LLC, USA	4	41	10.25	1	2.26

with a CPP of 24.00 from 6 papers with 144 citations and 2 funded papers, reflecting the impact of their research. On top of that, CNRS Centre National de la Recherche Scientifique, France, presented 6 publications with 74 total citations (CPP= 12.33) and 5 funded papers, while Lebanese American University, Lebanon, recorded 6 publications that resulted in 34 citations (CPP= 5.67) and 4 funded papers, resulting in an RCI of 1.25.

For the most influential affiliations, the leader is Institute for Particle Physics Phenomenology, UK with a CPP of 37.00 from 3 publications, 111 total citations, 3 funded and an RCI of 8.16, clearly the most influential in the discipline. Columbia University, USA had 5 publications, a total of 103 citations (CPP=20.60), 4 funded papers, and an RCI of 4.54. Also, from the USA, Harvard University had 4 papers and 58 citations (CPP=14.50), as well as being involved in 3 funded articles and having an RCI of 3.20. The University of California Santa Barbara, USA, was also highly productive with 5 publications, 53 citations (CPP=10.60), 4 funded papers and an RCI=2.34. Also, in the University of Cyprus, Cyprus, 4 publications were associated to 45 citations (CPP=11.25), with 4 papers on funding and an RCI of 2.48. Finally, Google LLC, USA was responsible of 4 publications with a total of 41 citations (CPP=10.25), from which 1 was funded, with an RCI of 2.26.

Together, these indicators provide a picture of divergent research priorities in different parts of the world. Microsoft Research institute from United States establishes significant impact through its high citation levels in each published paper while the Institute for Particle Physics Phenomenology from United Kingdom establishes its research influence through its outstanding RCI. This is more evidence of how academic research on Graph Neural Networks is spread out in different world excellence institutions and that, in all likelihood, leading breakthroughs continue to be made in high impact refereed journals.

The author group collaboration network presented in Figure 19.4, shows connections between 261 institutions in the five years, from 2020 to 2025, where the origin of the ties is the year of collaboration, represented by the color of the link. It contains 341 links and visualizes collaborative networks of these institutions in research. Among them, Indian Institute of Technology (IIT) is the most dominating and central, with the maximum number of citations (50), degree (19), centrality (0.39), and sigma values (2.19), making it the key player in establishing connections across research fields. The other institutions that present high values in burst strength, along with high citation numbers are the Koneru Lakshmaiah Education Foundation and Indian Institute of Science, both with 27 citations. The Indian Institute of Science shows the highest burst of citation with a burst strength of 4.81. Degree and betweenness centrality, two examples of network-based measures, also position institutions like Saveetha School of Engineering and Technical University of Munich as central nodes within this academic network. This visualization tool is helpful in exploring the structure and dynamics of these academic alliances, created using CiteSpace. It reveals the interconnectedness of the top institutions and will thus provide clues as to the trends in knowledge flows and the development of research networks within fields.

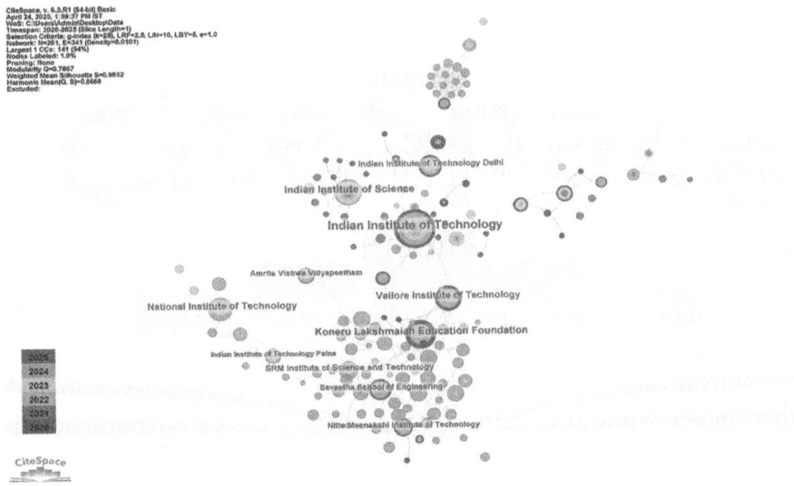

Figure 19.4 Key institutional collaboration network in GNNs research.

19.14 Top Keyword Metrics and Research Focus Areas

A detailed summary of keyword statistics for the top fifty keyword occurrences is presented in Table 19.9, which contains specific data on a selection of keywords as they relate to their Occurrences, Total Links to Search, External Links, and cluster designation. Some other significant keywords include deep learning with 263 occurrences and 975 Total Links to Search, convolutional neural networks, with 238 occurrences and 958 Total Links to Search, and graph neural networks at a massive 928 occurrences and 2529 Total Links to Search. The table includes entries on machine learning, artificial intelligence, neural networks, and applications including computer vision and natural language processing. The following details, such keyword analysis of a keyword, providing some understanding of keywords that need perhaps more explanation and/or are more prevalent or useful in terms of the academic and research fields concerned, shedding some light on keywords that may be of importance, focus, and/or interest to the field of study.

According to the keywords discussed, this resulted in a dataset of 1148 documents available in CSV format, focused on graph neural networks for the years 2020 to 2025. For this study, all VOS viewer Visualization analyses were based on keywords and initiated with the choice of Co-occurrence. All keywords are also available as units of analysis. A set threshold level

Table 19.9 Keyword analysis metrics for top 50 keyword occurrences.

Keyword	Occurrence	TLS	Links	Cluster	Keyword	Occurrence	TLS	Links	Cluster
adversarial machine learning	43	199	33	2	graphic methods	112	423	44	1
anomaly detection	35	111	31	2	image enhancement	26	100	29	3
artificial intelligence	26	97	29	3	image segmentation	28	92	39	3
artificial neural network	46	135	37	3	internet of things	26	95	30	2
benchmarking	33	103	36	3	knowledge graph	43	166	39	1
complex networks	41	156	38	1	learning algorithms	26	129	34	2
computer vision	26	94	27	4	learning models	22	104	33	2
contrastive learning	50	228	40	2	learning systems	89	382	47	3
convolutional networks	91	425	46	4	link prediction	36	150	37	1
convolutional neural networks	238	958	49	4	machine learning	114	370	39	3

(Continued)

Table 19.9 Keyword analysis metrics for top 50 keyword occurrences. (*Continued*)

Keyword	Occurrence	TLS	Links	Cluster	Keyword	Occurrence	TLS	Links	Cluster
data mining	25	98	36	1	multilayer neural networks	34	148	42	3
deep learning	263	975	49	3	natural language processing	23	102	30	1
deep neural networks	92	387	47	2	natural languages	24	122	32	1
deep reinforcement learning	21	107	27	2	network architecture	48	215	40	4
diagnosis	42	152	32	3	network security	47	139	32	2
embeddings	65	279	42	1	network theory (graphs)	40	179	36	2
federated learning	31	136	34	2	network-based	49	139	39	1
forecasting	111	386	43	1	neural network models	44	169	37	4
graph algorithms	44	182	40	2	neural-networks	98	379	48	2
graph attention network	26	89	30	4	recurrent neural networks	47	186	43	2

(*Continued*)

Table 19.9 Keyword analysis metrics for top 50 keyword occurrences. (*Continued*)

Keyword	Occurrence	TLS	Links	Cluster	Keyword	Occurrence	TLS	Links	Cluster
graph convolutional network	51	252	38	4	reinforcement learning	43	155	38	2
graph embeddings	48	220	39	1	semantics	50	222	44	1
graph neural networks	**928**	**2529**	**49**	**1**	sentiment analysis	23	78	26	1
graph structures	54	248	37	1	social media	26	128	31	1
graph theory	99	407	47	1	social networking (online)	39	169	31	1

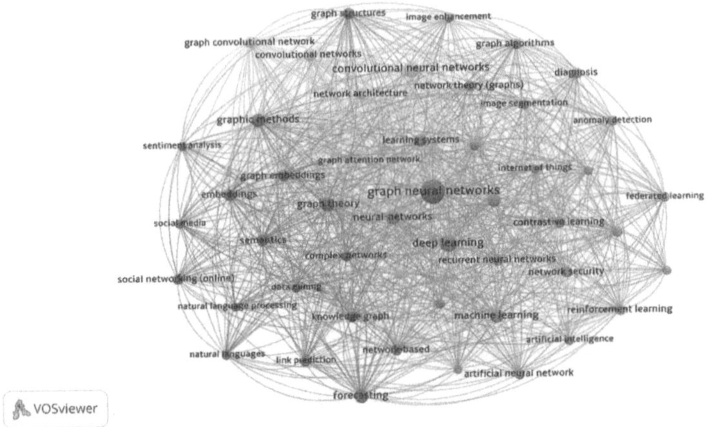

Figure 19.5 Top 50 most frequent keyword co-occurrences in GNNs research.

requires minimum occurrences in order for the keywords to be considered; out of 7975 keywords, 93 pass this threshold. The next step is to sum the strength of the co-occurrence links for each keyword with other keywords in order to get the strongest linked keywords. The 50 keywords with the highest occurrence frequency in this analysis are shown in four clusters with 938 links among them, appearing with a TLS of 6747. The objective of this careful process is to attempt to provide useful information about the relevance and relationship between keywords in GNNs over the time frame in question. Figure 19.5 offers a visual representation and example of the clustering and relationships among the keywords, which demonstrates the interconnectedness of these keywords in the field of study.

19.15 Top Influential Journals

Table 19.10 showcases the top 12 journals with key metrics, including the most productive and impactful journals in the field. Among the most productive journals, IEEE Access leads to 18 publications, 57 citations, and a Citation per Paper (CPP) of 3.17, showcasing a high international collaboration rate of 38.89%. Multimedia Tools and Applications follows closely with 18 publications, 68 citations, and a CPP of 3.78, along with an international collaboration rate of 16.67%. Biomedical Signal Processing and Control contributes 10 publications with 31 citations and a CPP of 3.10, maintaining a modest international collaboration percentage of 10%.

Table 19.10 Top 12 journal with key metrics, including the most productive and impactful journals.

Top 12 journals	TP	TC	CPP	FP	ICP	%ICP	ES	NES	AIS	IF	II
The top 6 most productive journals											
IEEE Access	18	57	3.17	2	7	38.89	0.36	78.33	0.70	3.70	0.60
Multimedia Tools and Applications	18	68	3.78	4	3	16.67	0.03	7.46	0.44	2.90	0.50
Biomedical Signal Processing and Control	10	31	3.10	3	1	10.00	0.02	4.43	0.81	4.90	0.90
International Journal of Communication Systems	9	18	2.00	0	1	11.11	0.00	0.88	0.29	1.70	0.50
Applied Intelligence	8	67	8.38	2	1	12.50	0.02	4.29	0.70	3.90	0.40
International Journal of Information Technology Singapore	8	56	7.00	3	0	0.00	N/A	N/A	N/A	N/A	N/A
Top 6 most impactful journals											
Scientific Reports	6	161	26.83	2	4	66.67	0.91	199.36	1.06	4.30	0.80
Information Sciences	2	126	63.00	1	1	50.00	0.06	13.59	1.39	N/A	N/A
Journal of High Energy Physics	3	120	40.00	3	3	100.00	0.09	21.53	0.99	4.60	1.70

(Continued)

Table 19.10 Top 12 journal with key metrics, including the most productive and impactful journals. (*Continued*)

Top 12 journals	**TP**	**TC**	**CPP**	**FP**	**ICP**	**%ICP**	**ES**	**NES**	**AIS**	**IF**	**II**
Journal of Big Data	1	86	86.00	1	0	0.00	0.01	2.59	2.55	12.40	1.40
IEEE Transactions on Industrial Informatics	2	84	42.00	2	2	100.00	0.08	17.72	2.75	11.40	3.30
Entropy	1	56	56.00	1	1	100.00	0.03	6.37	0.49	2.20	0.60

ES=Eigen factor Score; NES=Normalized Eigen factor Score; AIS=Article influence Score; IF=5 Year Impact Factor; II=Immediacy Index

International Journal of Communication Systems has 9 publications, 18 citations, and a CPP of 2.00, with an international collaboration rate of 11.11%. Applied Intelligence records for 8 publications, 67 citations, and a higher CPP of 8.38, with 12.50% international collaboration. The International Journal of Information Technology Singapore also has 8 publications, 56 citations, but no international collaboration.

In terms of impact, Scientific Reports stands out with 161 citations, a CPP of 26.83, and an international collaboration rate of 66.67%. Information Sciences follows with 126 citations and a CPP of 63.00, with 50.00% international collaboration. Journal of High Energy Physics achieves a CPP of 40.00 with 120 citations and 100% international collaboration. The Journal of Big Data has 86 citations and a CPP of 86.00, while IEEE Transactions on Industrial Informatics and Entropy both exhibit 100% international collaboration with high impact scores.

19.16 Discussion

This bibliometric study presents a comprehensive overview of the publication trends and patterns in the research of Indian Graph Neural Networks GNNs between the years 2020 and 2025. This is despite an increasing trend in the number of publications, such as between the years 2020 and 2024, and a relatively low citation impact, which may point out to the conclusions that although research output is increasing, it has still not translated into a major global impact in terms of academic citations. The fall of annual CPP after 2020 represents the disparity the volume and international resonance of research publications. A relevant point here is that, while there seem to be a lot of GNNs papers coming from India, they are not being published in leading journals to the same extent as GNNs papers from abroad. This is especially clear in citation dynamics. International cooperation seems to have had a positive effect in this regard, as detected by Total Link Strength (TLS) values of Indian researchers. Yet in these collaborations, the citation impact is still lower than it should be, indicating that higher quality partnerships leading to publishing in high impact journals are still needed.

IIT Delhi is also one of the major contributors to publications and citations in this area. Though even contributions from this institution have failed to escape the larger problem of international visibility, which points to symptomatic problems within the academic ecosystem in India. The increasing trend of papers with funding from 2020 to 2024 indicates the popularity of GNNs, but the falling number of funded papers in 2025 indicates difficulties in securing continued funding for research, which may limit its long-term development. More than a half of the publications are conference papers, as is common in Indian GNNs studies.

Conference proceedings are faster to publish but lower in status than peer reviewed journal publications, which are what matter for serious publication records. This highlights the necessity for researchers in India to focus on articles in journals as they receive more citations and have a greater tendency for international collaboration and raise the profile of research more in terms of international visibility. The analysis also points to a lack of topic diversity. Terms such as deep learning and graph neural networks are prominent in the Indian research landscape, but there is scope for more interdisciplinary work and innovative applications. In order to keep with international trends in research as well as not becoming too narrowly focused on certain topics, Indian researchers should try to expand the nature of their research.

A very low number of HCPs also highlights the disconnect between GNNs research in India and the world. In this regard, only two HCPs were identified in India during this period, while the U.S., China, and the UK have a much larger number of HCPs. This indicates that there is an opportunity for Indian researchers to work on increasing the quality and impact of their research in that the volume of Indian GNNs research has already seen a growth trend and is essential for Indian researchers to aim at achieving better quality work that can translate to impact in terms of citation and be internationally appreciated. In order for India to improve its positioning in the GNNs global research community, this will necessitate strategic collaborative efforts with international organizations, publishing in mission relevant journals, and an increasingly broader repertoire of research interests.

19.17 Strengths and Limitations of the Study

A bibliometric study of Indian GNNs research for the period 2020 to 2025, it offers an overview of publication and citation trends, as well as collaborative networks. Its use of Scopus, a widely accepted database, is a significant strength in terms of the reliability and quality of the data. Focusing on peer reviewed journal articles and conference papers, the study documents select important contributions to the developing area of GNNs in AI and machine learning research. The analysis incorporates state of the art bibliometric techniques using VOSviewer and CiteSpace to analyze patterns of co-authorship, institutional collaboration and keyword co-occurrence to provide a deep dive into the intellectual structure of GNNs research. The time frame of 2020 to 2025 is particularly relevant to the fast pace of technological development in GNNs. On top of that, by examining GNNs in the context of India, the research contributes new knowledge on regionally specific GNNs phenomena, addressing a noted gap in the literature.

But the study has its limitations. Though extensive, Scopus does not include all publications, such as preprints and grey literature. Citation based metrics, such as Citations per Paper, are also subject to distortions such as self-citations or the age of publications. Informal collaboration and the role of funding for research quality are also not considered in this study and should be explored further.

19.18 Findings and Prospects for Future Research

This research offers an extensive bibliometric study on GNNs in India spanning from 2020 until 2025. Its added value is in bibliometric analyses with VOSviewer visualization tool, CiteSpace visualization tool, and Microsoft Excel that map publishing trends, discover influential publications, and explore patterns of institutional and international collaboration. This indicates an increasing research output from India, but also highlights the citation impact gap leading to a lesser global visibility. This analysis provides important information about collaboration, finding that the U.S., China, and the U.K. are important partners. It also highlights who the major players are within India in terms of influential researchers and institutions. The insights obtained here can be useful in shaping the policies for enhancing collaborations between academia and industry, whether nationally or globally. The analysis also highlights the need for more funding and institutional support in order to improve the position of India in high impact research. As GNNs studies in India increase, our inability to address a gap in citation impact will continue to be problematic. Continuing research should emphasize both quality and quantity with an aim to place publications in prestigious journals. The inclusion of interdisciplinary studies and recent areas of research, such as dynamic graph networks, provides additional pathways for research. To continue expanding and to make a sustained academic impact on the global AI research scene, India will need to bolster its institutions around international partnerships and sustainable funding.

19.19 Conclusion

The present bibliometric study offers an overall picture of GNNs research in India during the period 2020-2025, showing a marked increase in the volume of research output. Nevertheless, the citational impact of Indian GNNs studies is still comparatively low, in a sense that strategic measures could be taken to improve its global presence. Collaborations internationally, especially with the US, China and the UK are important in creating collaborative research networks. But if the goal is also to maximize academic impact this has to be a priority to publish in high- quality and high-impact peer- reviewed journals. Lastly, this study also articulates the need to develop collaborative and research strategies to enable India to create a niche GNNs research community as part of the larger global community. In order to enhance its position globally, India needs to work towards publishing in high impact journals, developing interdisciplinary collaborations, and obtaining long term funding to make a meaningful impact in the future of AI research.

AI disclosure statement: The author acknowledges the use of AI-based tools for minor language editing or grammatical correction. However, all intellectual contributions, analyses, and conclusions presented in this chapter are entirely the author's own.

References

1. Corso, G., Stark, H., Jegelka, S., Jaakkola, T., Barzilay, R., Graph neural networks. *Nat. Rev. Methods Primers*, 4, 1, 17, Mar. 2024, doi: 10.1038/s43586-024-00294-7.
2. Fan, W., *et al.*, Graph Neural Networks for Social Recommendation, in: *The World Wide Web Conference*, San Francisco CA USA, May 2019, ACM, pp. 417–426, doi: 10.1145/3308558.3313488.
3. Zhou, J., *et al.*, Graph neural networks: A review of methods and applications. *AI Open*, 1, 57–81, 2020, doi: 10.1016/j.aiopen.2021.01.001.
4. Zhang, C., Song, D., Huang, C., Swami, A., Chawla, N.V., Heterogeneous Graph Neural Network, in: *Proceedings of the 25th ACM SIGKDD International Conference on Knowledge Discovery & Data Mining*, Anchorage AK USA, Jul. 2019, ACM, pp. 793–803, doi: 10.1145/3292500.3330961.
5. Xu, K., Hu, W., Leskovec, J., Jegelka, S., How Powerful are Graph Neural Networks?, arXiv: arXiv:1810.00826, Feb. 22, 2019, doi: 10.48550/arXiv.1810.00826.

6. Scarselli, F., Gori, M., Tsoi, A.C., Hagenbuchner, M., Monfardini, G., The Graph Neural Network Model. *IEEE Trans. Neural Netw.*, 20, 1, 61–80, Jan. 2009, doi: 10.1109/TNN.2008.2005605.
7. Kuwahara, T., Irino, H., Suzuki, K.-I., Global Collaboration Initiatives Revolutionizing Research and Development of Network Technologies. *NTT Tech. Rev.*, 16, 6, 41–47, Jun. 2018, doi: 10.53829/ntr201806fa7.
8. Scherer, R.C., Interdisciplinary Research Collaboration. *Voice Speech Rev.*, 4, 1, 329–336, Jan. 2005, doi: 10.1080/23268263.2005.10739490.
9. Mondal, R., Mukherjee, D., Singh, P.K., Bhateja, V., Sarkar, R., A New Framework for Smartphone Sensor-Based Human Activity Recognition Using Graph Neural Network. *IEEE Sens. J.*, 21, 10, 11461–11468, May 2021, doi: 10.1109/JSEN.2020.3015726.
10. Aljohani, N.R., Fayoumi, A., Hassan, S.-U., A Novel Deep Neural Network-Based Approach to Measure Scholarly Research Dissemination Using Citations Network. *Appl. Sci.*, 11, 22, 10970, Nov. 2021, doi: 10.3390/app112210970.
11. Chaudhary, A., Mittal, H., Arora, A., Anomaly Detection using Graph Neural Networks, in: *2019 International Conference on Machine Learning, Big Data, Cloud and Parallel Computing (COMITCon)*, Faridabad, India, Feb. 2019, IEEE, pp. 346–350, doi: 10.1109/COMITCon.2019.8862186.
12. Cheng, S., *et al.*, Cartography and Neural Networks: A Scientometric Analysis Based on CiteSpace. *ISPRS Int. J. Geo-Inf.*, 13, 6, 178, May 2024, doi: 10.3390/ijgi13060178.
13. Xu, S., Liu, S., Jing, C., Li, S., Event Knowledge Graph: A Review Based on Scientometric Analysis. *Appl. Sci.*, 13, 22, 12338, Nov. 2023, doi: 10.3390/app132212338.
14. Sawhney, R., Agarwal, S., Wadhwa, A., Shah, R., Exploring the Scale-Free Nature of Stock Markets: Hyperbolic Graph Learning for Algorithmic Trading, in: *Proceedings of the Web Conference 2021*, Ljubljana Slovenia, Apr. 2021, ACM, pp. 11–22, doi: 10.1145/3442381.3450095.
15. Sawhney, R., Agarwal, S., Wadhwa, A., Shah, R.R., Spatiotemporal Hypergraph Convolution Network for Stock Movement Forecasting, in: *2020 IEEE International Conference on Data Mining (ICDM)*, Sorrento, Italy, Nov. 2020, IEEE, pp. 482–491, doi: 10.1109/ICDM50108.2020.00057.
16. Saini, D., *et al.*, GalaXC: Graph Neural Networks with Labelwise Attention for Extreme Classification, in: *Proceedings of the Web Conference 2021*, Ljubljana Slovenia, Apr. 2021, ACM, pp. 3733–3744, doi: 10.1145/3442381.3449937.
17. Jiang, W., Luo, J., He, M., Gu, W., Graph Neural Network for Traffic Forecasting: The Research Progress. *ISPRS Int. J. Geo-Inf.*, 12, 3, 100, Feb. 2023, doi: 10.3390/ijgi12030100.
18. Gupta, A., Matta, P., Pant, B., Graph neural network: Current state of art, challenges and applications. *Mater. Today Proc.*, 46, 10927–10932, 2021, doi: 10.1016/j.matpr.2021.01.950.

19. Da Silva, A.M.B., Ferreira, N.C.D.S., Braga, L.A.M., Mota, F.B., Maricato, V., Alves, L.A., Graph Neural Networks: A Bibliometric Mapping of the Research Landscape and Applications. *Information*, 15, 10, 626, Oct. 2024, doi: 10.3390/info15100626.
20. Keramatfar, A., Rafiee, M., Amirkhani, H., Graph Neural Networks: A bibliometrics overview. *Mach. Learn. Appl.*, 10, 100401, Dec. 2022, doi: 10.1016/j.mlwa.2022.100401.
21. Zhang, L., Banihashemi, S., Zhang, Y., Chen, S., The confluence of project and innovation management: A scientometric analysis of emerging trends and research frontiers. *Proj. Leadersh. Soc.*, 6, 100181, Dec. 2025, doi: 10.1016/j.plas.2025.100181.
22. Van Eck, N.J., Waltman, L., Dekker, R., Van Den Berg, J., A comparison of two techniques for bibliometric mapping: Multidimensional scaling and VOS. *J. Am. Soc. Inf. Sci. Technol.*, 61, 12, 2405–2416, Dec. 2010, doi: 10.1002/asi.21421.
23. Waltman, L., Van Eck, N.J., Noyons, E.C.M., A unified approach to mapping and clustering of bibliometric networks. *J. Informetr.*, 4, 4, 629–635, Oct. 2010, doi: 10.1016/j.joi.2010.07.002.
24. Van Eck, N.J. and Waltman, L., VOS: A New Method for Visualizing Similarities Between Objects, in: *Advances in Data Analysis*, R. Decker and H.-J. Lenz (Eds.), Studies in Classification, Data Analysis, and Knowledge Organization, pp. 299–306, Springer Berlin Heidelberg, Berlin, Heidelberg, 2007, doi: 10.1007/978-3-540-70981-7_34.
25. Chen, C., CiteSpace II: Detecting and visualizing emerging trends and transient patterns in scientific literature. *J. Am. Soc. Inf. Sci. Technol.*, 57, 3, 359–377, Feb. 2006, doi: 10.1002/asi.20317.
26. Sabe, M., *et al.*, Half a century of research on antipsychotics and schizophrenia: A scientometric study of hotspots, nodes, bursts, and trends. *Neurosci. Biobehav. Rev.*, 136, 104608, May 2022, doi: 10.1016/j.neubiorev.2022.104608.
27. Chen, C., Searching for intellectual turning points: Progressive knowledge domain visualization. *Proc. Natl. Acad. Sci.*, 101, suppl_1, 5303–5310, Apr. 2004, doi: 10.1073/pnas.0307513100.
28. Zeng, M., Wang, X., Chen, W., Worldwide research landscape of artificial intelligence in lung disease: A scientometric study. *Heliyon*, 10, 10, e31129, May 2024, doi: 10.1016/j.heliyon.2024.e31129.
29. Yang, R., Yuan, Q., Zhang, W., Cai, H., Wu, Y., Application of Artificial Intelligence in rehabilitation science: A scientometric investigation Utilizing Citespace. *SLAS Technol.*, 29, 4, 100162, Aug. 2024, doi: 10.1016/j.slast.2024.100162.
30. Zhao, M., *et al.*, Influence of glutamine metabolism on diabetes Development:A scientometric review. *Heliyon*, 10, 4, e25258, Feb. 2024, doi: 10.1016/j.heliyon.2024.e25258.
31. Muthukrishnan, M., R., S.K., Mamdapur, G.M.N., K. R., S., A Focused Bibliometric Analysis of Blockchain Research in India (2014-2023): A

Decade of Transformation, in: *Advances in Library and Information Science*, K. Senthilkumar, R. Jagajeevan, G. Krishnakumar (Eds.), pp. 33–50, IGI Global, 2024, doi: 10.4018/979-8-3693-7783-3.ch002.

32. Dhawan, D.S.M., Gupta, G.B.M., Mamdapur, G.M.N., Walke, R., Bansal, M., Bibliometrics Research in India: A Scientometric Assessment of High-Cited Publications During 1994-2023. *J. Data Sci. Informetr. Cit. Stud.*, 3, 3, 275–293, Dec. 2024, doi: 10.5530/jcitation.3.3.29.
33. Mamdapur, G.M.N., Gupta, B.M., Ahemad, M.N., Vaish, A., Mapping High-Impact Publications in Diabetes and Sarcopenia Research: A Bibliometric Analysis (2003-2024).

20

Graph Neural Networks for Optimization: A New Paradigm in Complex Problem Solving

Spelmen Vimalraj Santhanam[1]* and Vignesh Ramamoorthy H.[2]

[1]*Department of Information Technology, Sri Krishna Arts and Science College, Coimbatore, India*
[2]*Amity Institute of Technology, Amity University, Bangalore, India*

Abstract

Optimization is a fundamental problem in many areas like logistics, network design, resource allocation, and even machine learning. Heuristics and linear programming often fail on high dimensional and intricate structured data. A more promising alternative are Graph Neural Networks (GNNs) because they utilize the structure of the data which is in the form of graphs. We discuss the enhancement of optimization GNNs offer by learning the intricacies of graph-structured problems. In contrast to traditional techniques that depend on fixed rules and features, GNNs depend on nodes, edges, and graph topology to retrieve pertinent data. This has overall improved accuracy and adaptability for a wider range of applications including but not limited to routing, supply chains, chip layout, and social networks. This chapter focuses on key GNN models used in optimization: Graph Convolutional Networks (GCNs), Graph Attention Networks (GATs), and Message Passing Neural Networks (MPNNs). These architectures enable better decision-making by capturing spatial and relational dependencies. Performance on problems such as the TSP, vehicle routing, and scheduling is enhanced with GNNs and reinforcement learning. Like any other approach, optimization based on GNN has its own problems to solve. Some of these problems are scaling to bigger graphs, their computational requirements, and the ability to generalize to new problem instances. Newer approaches like self-supervised training, transfer learning, and hybrid GNN-optimization frameworks attempt to solve these problems. This chapter illustrates how GNNs connect graph theory with machine learning

**Corresponding author*: spelmenvimalrajs@skasc.ac.in

J. Ramkumar, Sridaran Rajagopal, B. Suchitra and S. Balamurugan (eds.) Graph Neural Networks: Concepts and Applications, (797–822) © 2026 Scrivener Publishing LLC

to develop advanced and adaptive optimization tools. We showcase the deepening influence of GNNs across sectors with case studies from different domains and describe areas for further investigation.

***Keywords*:** Perceptron, graph, deep learning, optimization, heuristic algorithms

20.1 Introduction

Optimization is one of the central tasks in domains like computation and applied mathematics applied over various problems like engineering, science, economics, and artificial intelligence. The classical optimization specifically focuses on finding efficient solutions to the given problem from a set of feasible solutions derived by algorithms. The applications of optimization range from network routing, supply chain logistics, portfolio optimization, manufacturing, and scheduling. In the past era of research, optimization techniques were dominated by methods like linear programming (LP), nonlinear programming, dynamic programming, and combinatorial algorithms [1]. The metaheuristic approaches—namely genetic algorithms, particle swarm optimization, and ant colony optimization—have been adopted to solve complex problems, especially in non-deterministic environments [2]. Despite their power, these algorithms also have limitations. They often depend on handcrafted features and non-deterministic heuristics, making them less accurate for real-world data where the data is noisy, high-dimensional, and non-Euclidean in nature. In general, the data in optimization problems is represented as graph structures made up of nodes and edges, where nodes represent entities and edges represent relationships—for instance, power grids, road networks, social media graphs, telecommunications, infrastructure, and citation networks. Conventional ML models like convolutional neural networks, multilayer perceptrons, and recurrent neural networks are designed to work with regular data and often struggle with graph data [5]. This limitation led researchers to invent a new method called Graph Neural Networks (GNNs).

Generally, GNNs are designed to infer from graph-based data by aggregating information across interconnected nodes. GNNs are capable of learning node embeddings and vector representations that analyze features of independent nodes and their neighborhoods. This is achieved *via* the message-passing framework, where each node collects information from its neighbors and updates its state based on that information. Repetitive updating across multiple layers enables the model to capture both local and global structures, making GNNs well-suited for complex relational data [5].

GNNs are particularly effective for optimization problems due to their ability to learn from relational structures, the configurations of components, and constraints encoded in the graph. For instance:

- Transportation and logistics nodes are locations and edges are roads, where the GNN predicts best routes.
- Protein-protein interactions and gene regulatory networks in bioinformatics are graphically represented; therefore, GNNs can recognize crucial nodes or paths that are associated with disease [5].
- In telecommunication, network topology is needed to enable the highest transmission of signal and bandwidth usage [23].
- Transaction networks aid finance in fraud detection or asset portfolio optimization by relationship modeling [5].

The key characteristic of GNNs is their inductive bias: they generalize over graphs by capturing connectivity patterns and not storing individual layouts. This makes GNNs highly transferable to out-of-sample data and scalable across domains. Compared to conventional models, which use input data in flat feature vectors form, GNNs learn dynamically from graph structure, leading to improved generalization and efficiency. GNNs are one of the strongest players for the combinatorial optimization problem solution—a domain traditionally dominated by exhaustive search, greedy methods, and heuristic-based solvers. Knapsack Problem, Traveling Salesman Problem (TSP), Graph Coloring, and Minimum Vertex Cover problems have employed GNN architecture to solve efficiently. These applications prove that GNNs can offer near-optimal solutions at lower computational cost, particularly when implemented in combination with pointer networks or reinforcement learning.

For instance, Bello *et al.* suggested a model in which GNNs, combined with reinforcement learning, are used to tackle the TSP. The neural combinatorial optimization model performs better than conventional methods in most cases and learns to adjust to different problem instances under dynamic environments [4]. Another innovation is hybrid GNN-Optimization models in which GNNs precondition conventional solvers—for example, inferring likely binding constraints in a Mixed-Integer Linear Program (MILP), enabling efficient solving [19]. GNNs have even been coupled with metaheuristics like Ant Colony, Bee Colony, or Genetic Algorithms to enhance decision-making in each iteration [6]. The use of GNNs in industrial-scale problems is gaining attention. For instance,

Google has applied GNNs in chip layout design, optimizing routing and placement [13]. Governments and companies use GNN-based models to predict recommended routes, traffic flows, and optimize public transportation in urban mobility systems. In energy systems, GNNs help dynamically balance loads across distributed networks, improving resilience and minimizing losses [18]. Despite these advantages, GNNs also face challenges. Scalability remains a primary concern, as message passing becomes expensive for large graphs with many nodes and edges. Another limitation is over-smoothing—a phenomenon where repeated aggregation causes node representations to become indistinguishable, limiting GNN depth [5]. Interoperability remains an open research area, especially in critical domains like finance and healthcare.

Recent advances aim to address these issues. GraphSAGE introduces sampling to limit neighborhood size, reducing computational cost [11]. Graph pooling and hierarchical GNNs reduce graph complexity while preserving essential structure. Explainable GNNs and attention mechanisms improve model transparency [8]. As the field evolves, integration with self-supervised, transfer learning, and meta-learning is expected to enhance GNNs' predictive capabilities in optimization. Models trained on specific problem classes can adapt to new domains with minimal retraining. For unlabeled data, contrastive learning, graph augmentation, and learned graph rewiring show promising performance gains [5]. Briefly, GNNs represent a paradigm shift in optimization. They leverage graph-structured data and relational reasoning to provide good alternatives to traditional optimization processes. Flexible, versatile, and learnable, GNNs are evolving into pillar tools for scalable, strong, and smart optimizers that industries require.

20.2 Background and Motivation

Optimization is one of the central topics of study of operations research, computer science, and applied mathematics. In order to resolve an enormous collection of decision-making problems, the traditional methods such as linear programming, dynamic programming, integer programming, and metaheuristic methods have provided robust frameworks. The traditional methods were applied in the case of classic problems such as network design, logistics, scheduling, and resource allocation [1, 2]. With the growing complexity in areas such as logistics, bioinformatics, energy networks, and transportation, traditional optimization methods are challenged to a larger extent. The models generally handle non-Euclidean information, independent variables, and non-regular relations that are optimally represented as graphs. Traditional optimization methods handle the information as flat-structured and based on engineered features, ignoring the underlying dependencies—especially in the case of dynamic, structured information.

Researchers began building graph models to encode complex relationships between things in the early 2000s. Graphs naturally encode relational data with things (e.g., proteins, genes, or sites) as nodes and interactions or dependencies (e.g., traffic, biochemical relationships, or social relationships) as edges. Graph-theoretic minimum spanning tree, shortest path, and maximum flow algorithms ensued from these advances [3]. Although robust, the algorithms tend to be complicated, rigid, and coarse-grained in their ability to handle new data distributions. These limitations gave rise to the shift towards flexible, data-driven graph problem solutions and stimulated the evolution of Graph Neural Networks (GNNs). Some early versions of GNNs emerged in the mid-2000s to learn node representations using recursive aggregation of neighborhoods [5]. Such models were inspired by the belief that a node representation has to be a function of its own and its neighbors' features. Khalil, E.B *et al.* cast this intuition into a model that updated node representations in graph topology language and utilized fixed-point iterations to converge [6]. The GNN landscape gained huge traction with the emergence of Graph Convolutional Networks (GCNs) by Kipf and Welling [7]. These models used spectral filters on the graph domain with a first-order approximation of the graph Laplacian, rendering the possibility of training neural networks on even gargantuan graphs feasible. Graph Attention Networks (GATs) followed suit, with attention mechanisms dynamically weighing the relative importance of the neighbors in aggregation time [8]. The success of GCNs and GATs spurred a research explosion in richer GNN architectures, including GraphSAGE, Message Passing Neural Networks (MPNNs), and Graph Isomorphic Networks (GINs) [5, 9, 12]. These models enhanced expressiveness and scalability and enabled the GNNs to solve a wide range of problems. The most essential point is that these models offer unique frameworks to combine node features, global graph features, and edge features—ingredients critical to solving graph-based optimization problems. While initial uses of GNNs focused on classification (e.g., graph and node classification) and link prediction, scientists soon recognized their capability to solve optimization tasks—more precisely, combinatorial optimization. Traveling Salesman Problem (TSP), Minimum Vertex Cover, and Vehicle Routing Problem (VRP) are structured decision problems. Traditional solutions are typically brute-force search or problem-specific heuristics, which are flexible and scalable. To solve these tasks, scientists began using GNNs to learn heuristics end-to-end from examples. Khalil *et al.* [6] were the first to demonstrate that one could employ GNNs as solvers by learning from optimization problem examples. Their NeuroSAT system used a GNN to determine satisfiability in Boolean formulas—resulting in learned models as approximations to NP-hard problems. Similarly, Bello *et al.* [4] applied GNN-based architectures combined with reinforcement learning to TSP, achieving near-optimal results on benchmark datasets. The key motivations for using GNNs in optimization include reducing dependence on manual feature engineering, generalizing across problem instances, and incorporating structural knowledge into learning methods. This is why industrial applications increasingly rely on GNNs for high-dimensional, dynamic, and time-critical decision-making. GNNs are now used in:

- Power grid optimization, modeling energy flows and predicting failures [18, 23];
- Drug discovery, optimizing molecular structures for binding affinity and toxicity in bioinformatics [5];
- Smart cities, optimizing traffic routing, energy distribution, and public services based on dynamic network topologies [22].

The rise of data-rich environments further necessitates GNN-based optimization. With growing numbers of IoT devices, real-time telemetry, and high-resolution sensors, organizations now possess massive graph-structured datasets. Traditional metaheuristics struggle to process such data efficiently, whereas GNNs can learn directly from it, uncovering patterns missed by rule-based or statistical methods. Additionally, GNNs align well with self-supervised and transfer learning trends, enabling pre-training on large graph corpora and fine-tuning for specific tasks—reducing the need for labeled data and improving model efficiency. Academic toolkits and libraries such as PyTorch Geometric and DGL (Deep Graph Library) have made it easier to develop and deploy GNN models for optimization. Benchmarks like OGB (Open Graph Benchmark) and SATCOMP enhance focus on performance, interpretability, and reproducibility. Despite these advances, challenges remain—particularly over-smoothing, scalability, and interoperability—which continue to limit industrial adoption. Nonetheless, the drive to bridge the gap between data-driven learning and algorithmic optimization fuels ongoing innovation. The convergence of GNNs and optimization represents not just a technological evolution, but a fundamental paradigm shift in how we formulate, solve, and deploy intelligent systems.

20.3 Fundamentals of Graph Neural Networks

Graph Neural Network (GNN) is one of the learning architecture of deep learning specifically designed for graph-structured data. A graph can be defined as G=(V,E), where V represents the set of nodes or vertices and E represents the set of edges that represents the relationships or interactions among the nodes. Many real world problems in nature aligned with graphs; for instance, transportation systems, social networks, knowledge graphs and molecular structures. The traditional neural networks work with Euclidean data such as images or sequences where GNN operates on the non-Euclidean data. GNN allows the nodes to aggregate the features from the neighborhood which is called as message passing is the key innovation.

20.3.1 Message Passing Mechanism

As we know the GNN is the iterative message passing framework, which comprises of two key steps per layer. The first aggregation is where each node collects the information that is message from its neighbor. The second one is the node updates its representation using the aggregated message. The update rule at the k-th layer is as follows in Equation 20.1.

$$h_v^{(k)} = UPDATE^{(k)}\left(h_v^{(k-1)}, AGGREGATE^{(k)}\left(\left\{h_u^{(k-1)} | u \in N(v)\right\}\right)\right) \tag{20.1}$$

In the above equation $h_v^{(k)}$ represents the node v at layer k, $N(v)$ represents the set of neighboring nodes to v. AGGREGATE combines the neighbor features like sum, mean and max. UPDATE can be a neural network or a simple activation function like ReLU, Softmax [5].

20.3.2 Core GNN Architectures

Many researchers have proposed various GNN variants, each of them modifying the aggregation and update functions in its own way to achieve the improved scalability, performance and expressiveness.

Graph Convolutional Networks (GCN)
Kpif and Welling *et al.* [7] proposed the GCN based on the spectral graph theory. GCN applies the filters to generalize the concept of convolution over a nodes neighbor. The update equation is as follows in the Equation 20.2.

$$H^{(l+1)} = \sigma(\widehat{D}^{-\frac{1}{2}}\widehat{A}\widehat{D}^{-\frac{1}{2}}H^{(l)}W^{(l)}) \tag{20.2}$$

In the above equation $\widehat{A} = A + I$ represents the adjacency matrix with self-loops and $\widehat{D}$ represents the degree matrix, $H^{(l)}$ represents the node representation at layer l, and $W^{(l)}$ represents the learnable weight matrix and σ represents the activation function. This model allows the nodes to incorporate features from the neighbors which are nearby in an effective manner.

Graph Attention Network (GAT)
Velickovic *et al.* [8] invented the GAT, which improves the GCN by introducing the attention mechanism to dynamically weigh the neighbor contributions. Instead of uniformly aggregating all neighbors, GAT Assign the importance score by using the following Equation (20.3).

$$a_{ij} = Softmax_j(LeaakyReLU(a^T[Wh_i||Wh_j])) \tag{20.3}$$

The above equation allows the nodes to specifically focus on the most relevant neighbors during the message passing and improve the learning in graphs with noisy or heterogeneous connections.

GraphSAGE
Hamilton *et al.* [11] invented the GraphSAGE that supports inductive learning by learning how to aggregate the neighborhood information over the transductively training on a fixed graph. The neighborhood size was controlled by equipping the random sampling technique and supports different kind of aggregation functions (mean, LSTM, Pooling).

Message Passing Neural Networks (MPNN)
Gilmer *et al.* [9] proposed MPNN that is a general framework encompassing most GNN models. MPNN operates in two phases, the first one is message computation by equipping the Equation 20.4. Second one is state update by equipping the Equation 20.5.

$$m_{uv}^{(k)} = M(h_u^{(k)}, h_v^{(k)}, e_{uv} \tag{20.4}$$

$$h_v^{(k+1)} = U(h_v^{(k)}, \sum m_{uv}^{(k)} \tag{20.5}$$

Graph Isomorphism Networks (GIN)
Xu *et al.* [12] proposed the concept called GIN which aims to match the expressive power of the Weisfeiler –Lehman test. This method uses the multi-layer perceptron to increase discriminative capacity by using the Equation 20.6.

$$h_v^{(k)} = MLP^{(k)}((1+\epsilon).h_v^{(k-1)} + \sum_{u\epsilon N(v)} h_v^{(k-1)}) \tag{20.6}$$

The core architectures have been found breakthroughs in various domains like chemistry to predicting the molecular properties of atom-bond graph using MPNN; in Social networks, the GATs are used to identify the communities or the influential users; in recommender systems, the GraphSAGE is used for like prediction in user-item graphs; and in cyber security it is used to identify the fraud and anomalies in transaction graphs.

20.4 GNNs in Combinatorial Optimization

Combinatorial optimization entails problem-solving in which the space of solutions is discrete and usually exponentially large. Such problems include well-known NP-hard problems like the Traveling Salesman Problem (TSP), Knapsack Problem, Graph Coloring, Max-Cut, and Vehicle Routing Problem (VRP) [3]. Conventional solutions depend on greedy heuristics, branch-and-bound, integer programming, or approximation algorithms [1, 3]. These however are not able to scale or cope with complex, high-dimensional data in real time, particularly in dynamic settings [2]. The application of Graph Neural Networks (GNNs) to combinatorial optimization problems is a big step towards finding scalable, adaptive, and learnable optimization techniques [5]. GNNs provide the capacity to learn heuristics from experience, generalize over instances of the problem, and leverage the problem space structure using the inductive bias inherent in graphs [5, 6].

Why Use GNNs for Combinatorial Problems?

Numerous combinatorial problems lend themselves to being expressed naturally as graphs. Take the case of the TSP, in which cities are vertices and edges are distances between them. The solution structure involves picking a subset of edges to complete the shortest tour possible [3]. Classic solvers such as Concorde or Gurobi use search or relaxation approaches [1]. Such approaches are computationally costly for large-scale, real-time applications [2].

As opposed to that, GNNs can directly learn to approximate optimal solutions by learning to map graph inputs to solution outputs [5]. This method allows for tremendous speed-ups at inference time and is thus appropriate for online decision-making applications like dynamic routing, scheduling, and logistics [13, 14].

Learning-Based Approaches

Supervised Learning

In supervised learning methods, GNNs are learned using instances of problems and their optimal (or near-optimal) solutions in problem instance datasets. For example, in vertex cover or graph coloring, one may predict the nodes or edges that are in the solution set. The loss function would usually calculate the discrepancy between the predicted set and the actual set. While supervised models can achieve high accuracy, they depend heavily on high-quality labels—often generated by slow solvers [6].

Reinforcement Learning (RL)

Reinforcement Learning (RL) provides a label-free paradigm, enabling GNNs to acquire optimization tactics through exploration of an environment. In RL-based environments, the GNN is used as a policy network, which directs search for viable or near-optimal solutions. Rewards are provided in terms of solution quality (e.g., reduced path, reduced set, reduced conflicts). One pioneering example is the research by Bello *et al.* [4], where a GNN was augmented with RL to optimize the TSP. The model recursively chooses cities to explore, learning to optimize the tour distance as a reward function. This method beat legacy heuristics and showed generalization to graphs larger than what it was trained on [26].

Highlighted GNN Models for Optimization

NeuroSAT (Gasse *et al.*, 2019) [19] – Used GNNs to solve SAT instances through learning satisfiability patterns between clauses and variables. The model gave rise to more extensive applications in symbolic reasoning and discrete decision-making.

GNN-TSP (Kool *et al.*) [13] – Proposed an attention-based GNN that learned through policy gradient RL for TSP and VRP. The model was able to generalize across unseen examples and different sizes at a slight loss in performance.

CombOptNet (Gasse *et al.*, 2019) [19] – A system for combining GNNs with classical optimization solvers, with the GNN directing the decision-making of the solver (e.g., which variables to optimize).

SP-GNN (Dai *et al.*) [6] – Showed that GNNs are able to learn to efficiently approximate shortest path problem solutions better than Dijkstra's algorithm with limited computation budgets.

Performance and Generalization

One significant strength of GNNs in combinatorial optimization is that they generalize across instances. After training, a GNN can make an inference about a solution for a new graph instance immediately without having to solve it again from the beginning [6, 13]. This is vastly different from classical optimization solvers, which frequently begin anew for each instance. GNN-based solvers also scale more efficiently with graph size through sampling methods [5]. They can adjust to structural changes in graphs (e.g., added nodes or edges) [5]. GNNs discover patterns in solution spaces, reducing search inefficiencies [6]. But GNNs come with trade-offs as well: solution quality relies on the training distribution, absence of formal optimality guarantees, and difficulty with hard constraint encoding [5, 6].

20.5 GNNs Combined with Reinforcement Learning

The combination of Graph Neural Networks (GNNs) with Reinforcement Learning (RL) paradigms is a strong method for addressing deep decision-making problems on graph-structured data. Although GNNs offer strong mechanisms to pull and distribute representations in structured worlds, RL provides a paradigm for sequential optimization from trial-and-error learning [4, 5]. The synergy allows models to make intelligent decisions in dynamic and partially observable worlds like vehicle routing, task scheduling, network optimization, and resource allocation [14, 21, 22].

Why Use GNNs with RL?

Numerous real-world problems formulated as Markov Decision Processes (MDPs) or Partially Observable MDPs (POMDPs) are naturally graph-structured. Some examples: Vehicle Routing Problems (VRP): Nodes are cities and depots; edges are travel routes. Job-Shop Scheduling: Nodes are tasks; edges are dependencies or precedence constraints. Telecommunication Networks: Nodes are routers and servers; edges are data channels. Smart Grids: Nodes are distribution nodes; edges are transmission lines.

In these applications, the agent not only has to reason about the properties of individual nodes but also the relationships between nodes. Classical RL methods—e.g., Deep Q-Networks (DQN) or Policy Gradient methods—usually treat the input as unstructured feature vectors or images, thus less well suited for graph-structured state spaces. GNNs overcome this by injecting the relational inductive bias into the agent's policy or value function, which improves generalization and efficiency.

Hybridize reinforcement learning (RL) with graph neural networks (GNNs) to address combinatorial optimization challenges. The graph structure with its constituent node and edge features such as (location, demand, visited) captures the state representation. A GNN encoder of the type GAT processes the state to produce context-aware embedding for each of the nodes. A policy network, which operates on the context-aware embedding, issues the next action (selecting the next city in TSP for example) based on learned priorities. The reward function is shaped based on the task and is designed to promote valid and efficient solutions. In TSP, for example, the agent incurs a negative reward that mimics the proportionality of the tour length, while in VRP or CSP, additional penalties or bonuses are employed to encourage learning focused on compliance with constraints and solution quality.

Comparative Advantages

- **Graph Awareness**: GNN-RL agents perform better than flat vector-based agents in applications where topology and connectivity are important [13, 14].
- **Generalization**: GNN-RL agents learn on one graph size or topology and tend to generalize to larger or unknown graphs [21].
- **Constraint Handling**: Structural constraints (e.g., visiting connected nodes) can be learned or encoded in the GNN architecture [22].
- **Scalability**: Through message passing and graph sampling, GNN-RL approaches can scale more economically to larger problem instances compared to traditional RL [18, 23].

How GNN Topologies Support Optimization

- **Message Passing Dynamics**: Designed to transmit task-specific constraints (e.g., limits, intervals, possible, plausible).
- **Edge-Aware Learning**: Notably applicable in MPNN and GAT models where edge elements (e.g., value, fit) influence the choices made.
- **Permutation Invariance**: Important for issues with a combinatorial nature, dealt with by GIN and MPNN.
- **Neural Heuristics**: Various GNNs are trained to replicate or augment the functions of greedy algorithms and other heuristics found in classical problem solvers. The Table 20.1 briefs the comparison of different graph neural network based methods with with its pros and cons.

Table 20.1 Comparison of the graph neural network-based methods.

Architecture	Aggregation strategy	Scalability	Inductive capability	Optimization strengths/ notes
GCN (Graph Convolutional Network)	Mean of neighbor features (normalized Laplacian)	Limited; full-graph computation required	Transductive (originally)	Efficient for small/ medium graphs; simple and widely used
GAT (Graph Attention Network)	Attention-weighted sum over neighbors	Moderate; per-node attention adds overhead	Inductive	Learns neighbor importance dynamically; better for heterogeneous graphs
GraphSAGE (Graph Sample and Aggregate)	Mean/LSTM/ pooling (sampled neighborhood)	Scalable; uses neighborhood sampling	Inductive	Efficient for large-scale graphs; generalizes to unseen nodes
MPNN (Message Passing Neural Network)	Learnable message + update functions over edges	Depends on implementation	Inductive	Highly expressive; handles edge features and custom rules well
GIN (Graph Isomorphism Network)	Sum aggregation with MLP	Limited scalability; full-graph access needed	Transductive (typically)	Strong discriminative power; theoretically powerful (matches WL test)

20.6 Hybrid GNN-Optimization Models

Although Graph Neural Networks (GNNs) are versatile in learning representations from structured data, they tend to falter in completely substituting traditional optimization algorithms, particularly for exact solution problems, hard constraint satisfaction, or long-term consistency [3, 19]. On the other hand, traditional optimization methods—such as Mixed-Integer Linear Programming (MILP), heuristic search, and metaheuristics—are typically constrained in flexibility, scalability, and responsiveness to novel problem instances [1, 2]. The integration of GNNs with conventional or heuristic optimization approaches—so-called hybrid GNN-optimization models—is trying to leverage the strengths of each approach. These hybrid approaches intend to take advantage of the expressive learning power of GNNs yet try to benefit from the mathematical guarantees and structured problem-solving of optimization theory [6, 19]. This part discusses the architecture, motivations, applications, and challenges of hybrid GNN-optimization frameworks.

20.6.1 Why Hybrid Models?

Though GNNs can effectively model intricate interactions, their solution might not obey problem-specific restrictions like feasibility, integrality, or optimal objective value optimization [19]. Hybrid models pursue:

Employ GNNs for learning or estimating heuristics to diminish search space [6].

Facilitate verification and improvement by optimization solvers of GNN-recommended solutions [19].

Incorporate optimization solvers within GNNs (or vice versa), creating end-to-end learnable systems [19, 23].

Enhance performance on big-scale or real-time issues where only pure GNNs or optimization solvers would not be sufficient [6].

Hybrid Model Architectures

Hybrid models can be generally classified into the following architectures:

GNN as Preprocessing or Heuristic Generator

In this configuration, the GNN predicts variables, priorities, or heuristics, which are fed into a classical solver. Forecasting node or edge importance scores in Max-Cut, which determines the first solution of a greedy solver is the great example for this [6].

GNN Inside Optimization Loops

GNNs can be applied inside iterative solvers like metaheuristics or local search. Applying GNNs as value estimators in Genetic Algorithms for fitness approximation. Directing local search directions in Tabu Search or Simulated Annealing. Substituting hand-engineered scoring functions in Branch-and-Bound with GNN predictions [6, 19].

Optimization-Inspired GNNs

Here, optimization principles influence the GNN architecture design or training objectives. Rolling out optimization algorithms such as Iterative Soft Thresholding (IST) as GNN layers (e.g., Learnable Proximal GNNs). Synthesizing GNN training loss to embody optimization objectives (e.g., penalties on constraint violations). Case Study: During graph-based matrix factorization, optimization loss functions are incorporated into a GNN's training process to maintain spectral properties and matrix rank [19].

Differentiable Optimization Solvers in GNNs

Current progress supports integration of differentiable solvers (e.g., cvxpylayers, OptNet) into a GNN's graph, making end-to-end backpropagation possible across optimization layers. Resource allocation problems with GNNs embedded with differentiable QP solvers. Joint training of GNN policies and supply chain differentiable routing layers. This architecture supports training GNNs to not only produce good solutions but learn how to solve constrained problems formally [23].

20.6.2 Hybrid GNN-Optimization Applications

Supply Chain Optimization

GNNs can represent a logistics network structure—nodes for warehouses, edges for shipping routes. A GNN predicts demand and costs, and a linear optimizer does multi-commodity flow optimization.

Bioinformatics

Protein docking and drug discovery tend to involve searching an enormous configuration space. GNNs make predictions of candidate conformations or matching scores, with stochastic optimization or Monte Carlo refinement following.

Smart Infrastructure Planning

Urban planning issues are multi-objective optimization—cost, accessibility, efficiency. GNNs can learn from past designs to seed solvers, cutting time-to-convergence in design tasks such as utility network design. In smart grids, hybrid models can employ GNNs to predict probability of faults or load flows, which are passed to convex optimization models scheduling repair crews or re-routing power.

20.6.3 Performance Benefits

- Feature Pure GNN Classical Solver Hybrid Model
- Scalability High Medium High
- Interpretability Low High Medium
- Constraint Satisfaction Weak Strong Strong (*via* refinement)
- Adaptability/Learning Strong Weak Strong
- Solution Guarantee, No Yes Conditional

20.6.4 Hybrid GNN-Optimization Model Challenges

Even with potential, there are some challenges that remain:

- **Integration Complexity**: Merging neural network back-propagation with non-differentiable or GPU-unfriendly solvers may not be easy.
- **Training Stability**: Non-convex loss surfaces and solver inconsistencies can make learning unstable.
- **Scalability of Solvers**: Solvers can be bottlenecks even when GNN inference is efficient.
- **Lack of End-to-End Benchmarks**: There are not many standardized tasks and datasets for assessing full hybrid pipelines.

20.7 Applications in Real-World Scenarios

The real-world applicability of Graph Neural Networks (GNNs) to optimize problems has attracted more and more attention from a wide range of industries and research areas. The special property of GNNs to learn relational dependencies and propagate contextual information renders them especially applicable in situations where the underlying data is naturally represented as graphs. This section briefly describes significant areas where GNNs, especially when combined with optimization methods, have been utilized successfully to solve hard real-world problems. In transportation and urban mobility, GNNs have played a crucial role in creating intelligent traffic management and routing systems. Road networks are graph-structured naturally, with intersections forming nodes and roads forming edges. Standard routing optimization algorithms, like Dijkstra's or A*, are optimal but restrictive in dynamic environments where traffic patterns shift in real-time. GNNs overcome this limitation by learning traffic behaviors from historical and real-time data, hence allowing adaptive routing strategies. For example, in smart city infrastructure, models based on GNNs are trained to forecast traffic flow and congestion and use this information to provide optimal traffic light control and public transport scheduling. These models perform better than heuristic-based models in terms of both scalability and responsiveness, particularly when integrated with reinforcement learning frameworks for real-time decision-making [23].

In logistics and supply chain, GNNs have proven promising applications in scheduling deliveries, optimizing distribution of inventories, and allocation of resources. Supply chains are graphically equivalent to directed graphs with warehouses, distribution facilities, or stores being represented as nodes, and the edges signifying transportation links. GNNs are well-suited to model spatial and temporal dependencies among these nodes so that they can learn sophisticated logistics policies considering lead times, demand uncertainty, as well as capacity constraints. By combining GNNs with conventional optimization solvers, researchers have developed hybrid models that learn initial heuristics and then refine them with MILP or metaheuristics. This has resulted in shorter delivery times, improved cost savings, and increased resistance to disruptions such as demand surges or transit delays. Energy networks are another domain of application of GNNs to optimization of interest. The electrical power grid, perhaps the world's most advanced engineered system, whose operation is critically reliant on graph-based models, is one such area. Substations, transformers, and load centers can be represented with nodes, transmission lines with edges. GNNs have been used in load forecasting, fault detection, and optimal power flow problems. Specifically, GNNs facilitate decentralized control approaches through local and global topological features' learning, enabling better predictions of grid behavior under fluctuating loads. Hybrid GNN-optimization models have been employed in developing algorithms for demand response, energy storage allocation, and reconfiguration of power networks in real time. These models are essential to the current shift toward renewable energy systems that are prone to intermittent power from solar and wind power [19, 29].

Bioinformatics and computational biology have also adopted GNNs for numerous optimization issues. Biological networks are inherently graph-modeled, such as protein-protein interaction networks, gene regulatory networks, and metabolic pathways. GNNs were used to predict protein functions, simulate genetic mutations, and recognize disease-causing substructures. Drug discovery is one of the major uses where GNNs are utilized to represent molecular graphs—atoms as nodes and chemical bonds as edges—to predict the bioactivity of compounds. Optimization is utilized to recognize the optimal compound structures by traversing the enormous chemical space in an efficient manner. By combining GNNs with evolutionary algorithms or Bayesian optimization, scientists have created generative models that can generate new drug candidates with wanted properties like solubility, toxicity, and binding affinity. In financial networks, GNNs are revolutionizing risk evaluation, fraud detection, and credit scoring. Financial objects like individuals, firms, or accounts are related through transactional relations that create complex graph structures. GNNs are able to identify deviant patterns in these networks that might suggest fraud activity, i.e., money laundering or identity theft. With the incorporation of optimization methods, GNNs can also be applied to portfolio optimization, where the aim is to achieve maximum return while minimizing risk under market constraints. Encoding market correlations and dependencies into a financial graph, GNN-augmented optimizers can more accurately predict systemic risks and dynamically rebalance asset allocations [25, 27].

Cybersecurity and threat detection are another area in which GNN-based optimization frameworks have proven effective. Computer networks such as the Internet or intranets of organizations can be graphically represented with devices as nodes and communication links as edges. GNNs are utilized to identify abnormal network traffic, trace malware spreading routes, and rank vulnerabilities in terms of patching. Optimization is applied to distribute scarce security resources like firewalls or monitoring agents on the network. GNNs guide this distribution by learning what nodes or subgraphs are most significant to the security stance of the network. Additionally, GNNs have been applied in attack graph analysis, wherein they represent the possible paths an attacker would follow to attack a system. Optimization methods are employed thereafter to secure these paths through placing countermeasures effectively. Both recommendation systems and e-commerce model user-item interactions as bipartite graphs using GNNs. Nodes are used to denote users and items, and edges are used to encode interaction events like clicks, purchases, or ratings. GNNs can learn item similarities and user preferences by message passing, which allows for personalized recommendation generation. Optimization methods are incorporated to solve allocation problems, including choosing the most profitable subset of items to show under constraints like screen space, budget, or promotion policies. These hybrid frameworks integrate the representational power of GNNs with the strength of constrained optimization, yielding better user engagement and monetization tactics. Industrial manufacturing systems, especially for Industry 4.0, also gain from GNN-optimization solutions. New-generation production lines consist of networked machines, sensors, and workstations that form cyber-physical systems. These are inherently graph-structured, dynamic, and exhibit complex behaviors. GNNs can represent machine-to-machine interactions and foresee bottlenecks or failures. Optimization layers subsequently modify scheduling, maintenance scheduling, or inventory management according to these forecasts. For instance, for predictive maintenance, GNNs detect equipment that is likely to fail, and optimization decides on the maintenance schedule that saves time and money [9, 18, 26].

Finally, for the environmental and geospatial sciences, GNNs are applied for optimizing sensor locations, tracking ecological networks, and forecasting climatic patterns. Environmental processes like the spread of pollutants, wildlife movement, or forest fires are spatially and temporally located, and can be represented using graphs. GNNs are able to forecast the trajectory of such processes over time as optimization techniques decide where monitor sensors are to be placed or how relief material is to be distributed in case of disasters. Such applications become inevitable in a bid to manage worldwide phenomena like climate change, loss of biodiversity, and the management of natural disasters [21].

In total, the combination of GNNs and optimization has made it possible to develop smart, scalable, and knowledge-aware systems for real-world use. The applications demonstrate the effectiveness of GNN-based systems in filling the gap between decision-based optimization and learning from data. With the scalability, expressiveness, and compatibility of GNNs growing, their applications in major industries will become more, and smart systems and automation will be achieved.

20.8 Challenges and Limitations

Practical application of Graph Neural Networks (GNNs) in optimization problems has attracted significant attention from multiple industries and research communities. The inherent capability of GNNs to learn contextual information propagation and relational dependencies renders them the best fit to problems with underlying data graph-represented inherently [1]. This section provides a brief overview of the pioneering areas wherein

GNNs, especially when integrated with optimization, have been deployed to tackle complex real-world problems effectively. In traffic and urban mobility planning, GNNs played a pivotal role to create intelligent traffic routing and control schemes [2]. Road networks are naturally graph-structured, where roads are edges and intersections are nodes. Routing optimization algorithms such as Dijkstra's or A* become static in real-time dynamic traffic patterns [3]. GNNs circumvent this by learning traffic patterns from real-time and historical data, hence providing adaptive routing schemes [4]. For example, in smart city infrastructure, GNN-based models are learned to predict traffic flow and congestion and utilize the same to send optimal traffic light control and public transport timetables [5]. Such models become responsive and scalable in terms of responsiveness and scalability in comparison to heuristic-based models, especially when integrated with reinforcement learning architect for real-time decision-making [6]. In supply chain and logistics, GNNs have been explored with potential application in scheduling delivery, optimization of distribution of inventories, and resource allocation [7]. Supply chains are graphically isomorphic to directed graphs where nodes are warehouses, distribution nodes, or stores and the edges are transportation links.

GNNs are specifically created to learn spatial and temporal dependencies between such nodes [8] to learn complex policies for lead time logistics, demand uncertainty, and capacity [8]. Researchers have come up with hybrid models by combining GNNs with traditional optimization solvers that learn initial heuristics and refine them through MILP or metaheuristics [9]. This has made it possible to reduce delivery times, increase cost savings, and make the system more robust to disruptions such as demand shocks or transportation delays [10]. Energy grids are another significant field of application of GNNs for optimization [11]. Electricity power grid is one of the world's most sophisticated man-made systems, and its control is highly dependent on graph-based models. Sub-stations, transformers, and load centers can be modeled as nodes, and transmission lines can be modeled as edges [12]. GNNs have been used in application such as load forecasting, fault detection, and optimal power flow [13]. GNNs, in fact, enable decentralized control strategies based on learning of global and local topological characteristics, which offers better predication of the grid dynamics under different fluctuations of the different loads [14]. Hybrid models of GNN-optimization have been used in the algorithm design of demand response, energy storage allocation, and real-time reconfiguration of power networks [15]. These models are at the forefront of the shift to renewable energy systems based on intermittent power from solar and wind power [16]. Bioinformatics and computational biology also utilized

GNNs as an optimization problem-solving technique [17]. Biological networks naturally have graph-models, e.g., protein-protein interaction networks, gene regulatory networks, and metabolic pathways [18]. GNNs were used in protein function prediction, simulation of genetic mutations, and detection of disease-causing substructures [19]. One of the most significant applications is drug discovery, where GNNs are used to model the molecular graphs—atoms as nodes and chemical bonds as edges—to predict the bioactivity of compounds [20]. Optimization is used to find the best compound structures by searching the enormous chemical space efficiently [6]. Using GNNs and evolutionary algorithms or Bayesian optimization, scientists have developed generative models that can generate new drug candidates with targeted properties such as solubility, toxicity, and binding affinity [21].

In financial networks, GNNs are revolutionizing risk evaluation, fraud detection, and credit scoring [22]. Financial objects like individuals, firms, or accounts are related through transactional relations that create complex graph structures. GNNs are able to identify deviant patterns in these networks that might suggest fraud activity, i.e., money laundering or identity theft [23]. With the incorporation of optimization methods, GNNs can also be applied to portfolio optimization, where the aim is to achieve maximum return while minimizing risk under market constraints [24]. Encoding market correlations and dependencies into a financial graph, GNN-augmented optimizers can more accurately predict systemic risks and dynamically rebalance asset allocations [25]. Cybersecurity and threat detection are another area in which GNN-based optimization frameworks have proven effective [26]. Computer networks such as the Internet or intranets of organizations can be graphically represented with devices as nodes and communication links as edges. GNNs are utilized to identify abnormal network traffic, trace malware spreading routes, and rank vulnerabilities in terms of patching [27]. Optimization is applied to distribute scarce security resources like firewalls or monitoring agents on the network. GNNs guide this distribution by learning what nodes or subgraphs are most significant to the security stance of the network. Additionally, GNNs have been applied in attack graph analysis, wherein they represent the possible paths an attacker would follow to attack a system. Optimization methods are employed thereafter to secure these paths through placing countermeasures effectively [28].

For recommendation systems and e-commerce, GNNs are utilized to represent user-item interactions as bipartite graphs [29]. The nodes model the items and users, and the interaction event such as click, purchase, or rating are the edges. The GNNs learn item similarity and user preference through message aggregation and produce personalized recommendations. The allocation problems, like the most valuable set of items to present under limitations of available screen estate, budget, or promotion policy, are optimized with optimization techniques. The hybrid models combine the representational power of GNNs and the ability of constrained optimization to produce more optimal plans for user interaction and monetization.

20.9 Future Research Directions

Graph Neural Networks (GNNs) have been demonstrating outstanding ability in addressing complex optimization challenges with the assistance of the graph-encoded data representation [1]. Despite unprecedented progress in architecture, training, and their combination with classic optimization methods, there remains an enormous void of open challenges and unexplored opportunities [2]. The future research wave in this direction will be driven by the intersection of theoretical advancement, algorithmic innovation space, domain-specific tuning, and real-world deployment demands [3].

This sub-section discusses some of the novel research areas that will define the next generation of GNN-based optimizers. Developing scalable GNN architectures in a bid to cope with large graphs is one of the key research areas of the future [4]. Real-world applications in transportation, telecommunications, finance, and biology often involve graphs with millions or even billions of nodes and edges [5]. Current GNN models struggle with such scales due to the computational overhead of message passing and the memory requirements for storing intermediate representations [6]. While sampling techniques and hierarchical pooling strategies have been proposed, they often sacrifice representational fidelity or are problem-specific [7]. A promising area is the construction of asynchronous and distributed GNNs, which allow parallel computation over graph partitions [8]. Models in this direction would have to tackle communication bottlenecks, consistency constraints, and information propagation between partitions so that they converge and are accurate [9]. Another important direction is the improvement of the theoretical comprehension of the expressivity of GNNs and how it connects to combinatorial optimization [10]. Although advancements such as Graph Isomorphism Networks (GINs) and higher-order models have augmented the capacity of GNNs to differentiate graph structures, a fundamental disconnect continues to exist between GNNs and traditional algorithmic reasoning [11]. As research evolves, it may entail the creation of logic-aided GNNs or models that engage discrete mathematics and algebraic topology more explicitly with the learning process [12]. Such endeavors seek to bridge statistical and symbolic reasoning, allowing GNNs to simulate or approximate algorithmic logic and execute optimally on tasks with strict combinatorial constraints [13].

One area that is receiving growing attention is the combination of GNNs with differentiable optimization layers [14]. Such architectures enable neural models to produce parameters that are used as inputs to an optimization solver integrated into the computational graph [15]. Methods such as OptNet, cvxpylayers, and differentiable mixed-integer solvers are initial attempts in this direction [16]. Modern methods are still narrow in scope and usually assume convexity or relaxation [17]. Non-convex, combinatorial, and constraint programming solvers that are differentiable or can be well-approximated are potential avenues to investigate in future studies [18]. This would allow for end-to-end training of models that learn not just to represent graph data but also optimize decisions in a principled and constrained way [19].

Enhancing the generalization power of GNNs on heterogeneous and dynamic graphs is another critical area [20]. Real graphs in the world are seldom static; they expand, contract, and evolve over time [6]. In addition, various examples of a problem can have very different node degree distributions, densities, or community structures [21]. Current GNN models tend to degrade in performance when tested on out-of-distribution graphs [22]. Solving this will involve breakthroughs in domain adaptation, graph meta-learning, and graph augmentation [23]. Self-supervised pretraining of large-scale graph corpora, similar to NLP and computer vision transfer learning, also has the potential to yield general-purpose GNN encoders that may be fine-tuned for optimization tasks of specific interest using minimal labeled data [24].

The prospects of multi-agent and hierarchical GNN architectures also merit further investigation [25]. Decentralized decision-making by agents or subsystems is present in most optimization problems [26]. For instance, in supply networks, various facilities can have partial knowledge while having to globally coordinate [27]. GNNs may be implemented to capture local interactions, and hierarchical representations allow information to be aggregated at several scales of granularity [28]. Multi-level GNNs that trade off local decision efficiency with the accuracy of global coordination could be a direction for future research [29]. These architectures may be especially beneficial in use cases such as smart grids, distributed robotics, and logistics at large scale.

20.10 Conclusion

Graph Neural Networks (GNNs) have shaken optimization with their capacity to learn from graph-structured data. In contrast to machine learning models or solvers relying on handcrafted features and heuristics, GNNs can potentially learn both topological relationships and node/edge features, making GNNs the best choice for structured, complicated problems. As networks such as traffic networks, supply chains, or even communication systems become increasingly connected, the demand for graph-based modeling keeps growing. In this chapter, we understood how the GNNs learn structural patterns *via* message passing, enabling generalization across various graph instances, a desirable feature for optimization problems with context-sensitive interdependent decisions. The use of GNNs combined with reinforcement learning (RL) has also shown good results for sequential decision problems, e.g., Traveling Salesman Problem (TSP), Vehicle Routing Problem (VRP), and scheduling. In RL, the GNNs are employed as policy networks, constructing adaptive policies that are superior and generalized compared to conventional heuristics. The optimization obtained using hybrid solutions, which combine GNNs with conventional approaches such as Mixed-Integer Programming or metaheuristics, has a good trade-off between learned adaptability and guaranteed feasibility. Such systems have already been rolled out in several industries such as logistics, energy, healthcare, and cybersecurity. GNNs are not without problems. The size of a graph puts a computation time limit which is a challenge to efficiency. Other challenges include deep GNN over-smoothing, generalization to unseen graph types, and lack of interpretability which is extremely harmful in finance and healthcare. Advances in these areas will need to center on architecture, training methods, and more socially conscious usage. Nevertheless, GNNs have tremendous potential for smart, adaptive, and responsible optimization for the real world.

AI disclosure statement: The author acknowledges the use of AI-based tools for minor language editing or grammatical correction. However, all intellectual contributions, analyses, and conclusions presented in this chapter are entirely the author's own.

References

1. Boyd, S. and Vandenberghe, L., *Convex optimization*, Cambridge University Press, England, 2004.

2. Yang, X.S., *Nature-inspired metaheuristic algorithms*, Luniver Press, United Kingdom, 2010.
3. Papadimitriou, C.H. and Steiglitz, K., *Combinatorial optimization: Algorithms and complexity*, Dover Publications, New York, 1998.
4. Bello, I., Pham, H., Le, Q.V., Norouzi, M., Bengio, S., Neural combinatorial optimization with reinforcement learning. arXiv preprint arXiv:1611.09940, 2016.
5. Wu, Z., Pan, S., Chen, F., Long, G., Zhang, C., Yu, P.S., A comprehensive survey on graph neural networks. *IEEE Trans. Neural Networks Learn. Syst.*, 32, 1, 4–24, 2021.
6. Khalil, E.B., Dai, H., Zhang, Y., Dilkina, B., Song, L., Learning combinatorial optimization algorithms over graphs, in: *Advances in Neural Information Processing Systems (NeurIPS)*, 2017.
7. Kipf, T.N. and Welling, M., Semi-supervised classification with graph convolutional networks, in: *International Conference on Learning Representations (ICLR)*, 2017.
8. Veličković, P., Cucurull, G., Casanova, A., Romero, A., Liò, P., Bengio, Y., Graph attention networks, in: *International Conference on Learning Representations (ICLR)*, 2018.
9. Gilmer, J., Schoenholz, S.S., Riley, P.F., Vinyals, O., Dahl, G.E., Neural message passing for quantum chemistry, in: *Proceedings of the 34th International Conference on Machine Learning (ICML)*, 2017.
10. Ying, R., He, R., Chen, K., Eksombatchai, P., Hamilton, W.L., Leskovec, J., Graph convolutional neural networks for web-scale recommender systems, in: *Proceedings of the 24th ACM SIGKDD International Conference on Knowledge Discovery & Data Mining (KDD)*, 2018.
11. Hamilton, W., Ying, R., Leskovec, J., Inductive representation learning on large graphs, in: *Advances in Neural Information Processing Systems (NeurIPS)*, 2017.
12. Xu, K., Hu, W., Leskovec, J., Jegelka, S., How powerful are graph neural networks?, in: *International Conference on Learning Representations (ICLR)*, 2019.
13. Kool, W., Van Hoof, H., Welling, M., Attention, learn to solve routing problems!, in: *International Conference on Learning Representations (ICLR)*, 2019.
14. Nazari, M., Oroojlooy, A., Snyder, L., Takác, M., Reinforcement learning for solving the vehicle routing problem, in: *Advances in Neural Information Processing Systems (NeurIPS)*, 2018.
15. Joshi, C.K., Laurent, T., Bresson, X., Learning TSP requires rethinking generalization, in: *International Conference on Learning Representations (ICLR)*, 2020.
16. Cai, H., Zheng, V.W., Chang, K.C., A comprehensive survey of graph neural networks. *IEEE Trans. Neural Networks Learn. Syst.*, 32, 1, 4–24, 2020.
17. Wan, Y., Wu, J., Lu, Q., Zhang, Y., Hyperparameter optimization *via* graph neural networks. *J. Mach. Learn. Res.*, 23, 56, 1–26, 2022.

18. Chen, Z., Liu, Z., Qiu, J., Power flow prediction in smart grids with graph neural networks. *IEEE Trans. Smart Grid*, 12, 5, 4004–4013, 2021.
19. Gasse, M., Chételat, D., Ferroni, N., Charlin, L., Lodi, A., Exact combinatorial optimization with graph convolutional neural networks, in: *Advances in Neural Information Processing Systems (NeurIPS)*, 2019.
20. Nowak, A., Villar, S., Bandeira, A., Bruna, J., A note on learning algorithms for quadratic assignment with graph neural networks arXiv preprint arXiv:1706.07450, 2017.
21. Zhang, H., Song, Y., Hu, X., Deep learning for scheduling: A survey. arXiv preprint arXiv:2004.13335, 2020.
22. Sun, H., Wang, Y., Li, Z., Graph neural networks for smart transportation: A survey, in: *IEEE Transactions on Intelligent Transportation Systems*, 2021.
23. Zhou, Q., Yang, H., Liu, J., Power grid optimization using graph neural networks: Opportunities and challenges. *IEEE Access*, 10, 19820–19830, 2022.
24. Dwivedi, V.P., Joshi, C.K., Laurent, T., Bengio, Y., Bresson, X., Benchmarking graph neural networks. arXiv preprint arXiv:2003.00982, 2020.
25. Vinyals, O., Blundell, C., Lillicrap, T., Wierstra, D., Kavukcuoglu, K., Matching networks for one shot learning, in: *Advances in Neural Information Processing Systems (NeurIPS)*, 2016.
26. Bello, I., Pham, H., Le, Q.V., Norouzi, M., Bengio, S., Neural combinatorial optimization with reinforcement learning, in: *International Conference on Learning Representations (ICLR)*, 2017.
27. Li, Y., Yu, R., Shahabi, C., Liu, Y., Diffusion convolutional recurrent neural network: Data-driven traffic forecasting, in: *International Conference on Learning Representations (ICLR)*, 2018.
28. Fey, M. and Lenssen, J.E., Fast graph representation learning with PyTorch Geometric. arXiv preprint arXiv:1903.02428, 2019.
29. Bronstein, M.M., Bruna, J., LeCun, Y., Szlam, A., Vandergheynst, P., Geometric deep learning: Going beyond Euclidean data. *IEEE Signal Process Mag.*, 34, 4, 18–42, 2017.

21

Quality Assurance of Software Incorporating Graph Neural Networks for Defect Prediction

Medhunhashini D. R.[1]*, K. S. Jeen Marseline[1] and B. Varun[2]

[1]Department of IT & Cognitive Systems, Sri Krishna Arts and Science College, Coimbatore, Tamilnadu, India
[2]Department of Mechanical Engineering, Sri Ramakrishna Institute of Technology, Coimbatore, Tamilnadu, India

Abstract
This chapter explores GNNs as a powerful for defect prediction as they aggregate information from nearby nodes and create embeddings that capture local and global structures and these embeddings can reveal the code structure patterns which expose modules to defects. New studies show that GNNs can additionally improve defect localization, bug detection, and also software quality assurance. By learning from historical software repositories, GNNs can be able to have better accuracy in defect detection and less effort in testing. GNNs can generalize across a great plurality of software projects without the need for any significant amount of feature engineering in traditional models, being transferable in nature and flexible. Although some of the challenges still remain such as availability of large labeled datasets, model interpretability and handling with computational complexity in large scale systems. To tackle these challenges, active research is going on the hybrid methodologies combining GNNs with the traditional methods. Future research work is to improve the scalability, efficiency, and theoretical foundation of GNNs, possible advances in meta-learning, self-supervised learning and integrate domain knowledge for defect prediction and software quality assurance applications.

***Keywords*:** Software, quality assurance, defect prediction, machine learning, graph neural networks

**Corresponding author*: medhunhashinidr@skasc.ac.in

J. Ramkumar, Sridaran Rajagopal, B. Suchitra and S. Balamurugan (eds.) Graph Neural Networks: Concepts and Applications, (823–858) © 2026 Scrivener Publishing LLC

21.1 Introduction to Software Quality Assurance (SQA)

Considering that most of these days' private and professional activities are based on software systems such as healthcare, finance, transportation, and space exploration, it safe to say that quality of software and hence on the quality of the test processes is vital. Software Quality Assurance (SQA) is a top-level process that contains several practices, standards, and tools to verify that a software product meets preconditions and functions as designed in practical setups. Low quality of the software affects not only the monetary cost but also user and organizations confidence, safety and their reputation. With scale and complexity of software systems growing exponentially the need for robust and scalable quality assurance practices grows proportionally. Static (such as code review) and dynamic (such as testing, formal verification and metrics analysis) methods are covered by SQA. However, because these methods are reactive, they require post development testing to find and fix defects [1]. This is a very expensive and ineffective process, especially once defects are found late in the software development process. In response accordingly, there has been a growing trend in predictive and intelligent early detecting of defects in the development process in order to minimize rework and increase reliability.

21.1.1 Role of Defect Prediction in SQA

Proactive Software Quality Assurance (SQA) relies on Defect prediction as a key component. It uses history data, metrics, and machine learning to predict which software modules are going to have the defects. This ability enables development teams to focus test effort and code inspections, optimize resource management to better focus effort where it is most needed (in high risk areas). In improving the quality assurance efficacy, defect prediction effectiveness greatly reduces development costs and shortens time-to-market. Conventionally, defect prediction models were based on a set of manually engineered features from software including code complexity, churn, and developer activity measures. Decision trees, Support vector machines (SVM) and Random forests are very commonly used in machine learning setting to predict likelihood of having defects through these features. Although these models have been encouraging, they are usually not capable of learning the intricate structural dependencies in software code (e.g., relationship among functions, control flow or call hierarchies) that constrain their predictivity [2].

21.1.2 Rise of Graph Neural Networks (GNNs) in Software Analytics

In recent years, Graph Neural Networks (GNNs) have brought increased opportunities for modeling and understanding structured data. GNNs are deep learning model that specially specialized to act on graph inputs. This is especially useful for the domain of software because the form of software code already aligns with graph representations like Abstract Syntax Trees (ASTs), Control Flow Graphs (CFGs), Call Graphs and Dependency Graphs. GNNs are able to discover sophisticated syntactic and semantic structures in code that traditional machine learning models simply cannot recognize, because of their ability to represent software graphs. This allows them to represent interrelationship and interdependency between pieces of code and this allows for more accurate defect prediction. Being able to learn representations from the graph architecture itself, GNNs also spare us from laborious manual feature engineering [3]. In recent years GNNs have been used heavily in software engineering tasks like vulnerability detection, code summarization, clone detection, bug localization, etc. Amongst these applications, defect prediction is one of the more promising applications where GNNs can make difference early in fault detection of the faulty modules to maintain the software quality assurance processes.

21.1.3 Objectives and Contributions

In this chapter we take a look at ways in which Graph Neural Networks can be employed within software quality assurance workflows with special focus on the problem of defect prediction. The primary objectives of the chapter are:

- To **explain the role and significance of defect prediction** in enhancing software quality assurance.
- To **introduce the theoretical foundations and working principles of GNNs**, especially as they apply to structured software data.
- To **illustrate how software can be modeled as graphs** and how these representations feed into GNN-based models.
- To **review and evaluate the effectiveness of GNNs for defect prediction** through comparative studies and case-based examples.

- To **highlight the quality assurance challenges and considerations** unique to GNN-based systems, including reproducibility, interpretability, and generalizability.
- To **propose a comprehensive framework** for incorporating GNN-based defect prediction into real-world software development and quality assurance pipelines.

The chapter seeks to make the following key contributions achieving these objectives as:

- A detailed survey and synthesis of existing research on GNN-based defect prediction.
- A methodological guide for practitioners on modeling code as graphs and applying GNNs.
- An empirical evaluation demonstrating the superiority (or limitations) of GNNs compared to traditional models.
- Insights into integrating GNN-based predictions into software development lifecycles and QA practices.
- Discussion of open challenges, including dataset quality, model explainability, and scalability.

21.2 Background and Related Work

This section outlines the background and related research in the area of software defect prediction with emphasis on traditional and machine learning based methods. Conventional tools tend to use hand-designed metrics and statistical models that are not adaptable to complicated code structures. Machine learning: Decision trees, support vector machines and other machine learning methods were used to learn patterns in the historical data, thereby giving better predictive accuracy. Graph Neural Networks (GNNs) have recently been attracting attention as a way to represent source code as a graph, incorporating both syntactic and semantic edges. Experiments indicate that GNN-based models can effectively outperform non-structural information-based traditional approaches to predict defects more accurately due to the structural information present in software programs.

21.2.1 Traditional Defect Prediction Methods

Software defect prediction has long relied on machine learning and statistical approaches applied to software metrics such as code complexity, size, and churn. These metrics are typically extracted at the file, class, or function level and serve as the features for predictive modeling. Some of the most widely used traditional algorithms in defect prediction include **Support Vector Machines (SVM), Random Forests (RF), Logistic Regression (LR), and Naïve Bayes and Decision Trees** [4].

- **Support Vector Machines (SVM)** are robust classifiers that work by constructing hyperplanes in high-dimensional space to separate defective from non-defective modules. SVMs are particularly effective for binary classification tasks and can handle imbalanced datasets well when properly tuned. They do a good job of determining the best separating hyperplane, but they don't do very well with large data sets or really complex and nonlinear relationships, unless optimized well [5].
- **Random Forests (RF)** is an ensemble method that while training builds a large ensemble of decision trees and outputs the class predicted by most of them. However because of their high accuracy, their ability to handle high dimensional datasets and their robustness in the case of noisy data, Random Forests are favored. Because these are less prone to overfitting than single decision trees, they are considered fit for defect prediction in most software systems.
- **Logistic Regression (LR)** is a powerful, but simple method to perform binary classification of outcomes. This is done as LR provides a probabilistic model which makes the model outputs more interpretable. Despite this, LR has poor performance in nonlinear feature interaction situations which is commonplace in complex software systems where code component interactions are beyond additive relationships [6].
- **A lightweight model like naïve Bayes and Decision Tree** gives a quick training and convenience. On the other hand, the assumption of independence of features in Naïve Bayes does not hold true in software defect prediction hence poor performance. Decision Trees are also light with the drawback of missing the more complex data patterns in the data leading to overfit or underfit in data sets.

These traditional algorithms have certain well-liked benefits, but they do not work so well with challenging, high dimensional software data where the feature relationships can be challenging. Thus, in these scenarios Graph Neural Networks (GNN) are a good alternative as they use the structure of these relationships already present in code.

21.2.2 Overview of Machine Learning in Software Quality Assurance

SQA has completely developed from being rule based and manual to data driven due to the advent of machine learning in SQA. Machine Learning (ML) in Software Quality Assurance (SQA) has seen tremendous applications and has enhanced the process of automating and optimizing a number of aspects of the software development and maintenance. The three areas from the above in which machine learning models, including traditional algorithms such as SVMs and Random Forests and more complex models such as Graph Neural Networks (GNNs), have been commonly employed are:

- **Defect Prediction and Localization:** They learn from historical datasets (past defects and other software metrics) that inform them which areas may be the bad ones. Localization means as precisely localize the position of defect in the codebase so the developers can focus in one small piece of the overall code base [7].
- **Code Smells and Anti-Patterns Detection:** Code smells are code patterns with specific properties that indicate something is wrong with the codes behind them such as a problem in poor design or maintainability. The smells are detected by using machine learning which in turn enables developers to deal with possible complications before they turn into issues. In a similar manner, learning from labeled training sets allows for the detection of anti-patterns (habitual sub optimal practices or design flaws), making possible software quality enhancements.

Machine learning can also be employed to automatically generate suitable test cases from the source code and a past defect data that help first test the most important parts of the software. Test case prioritization is maximization of the testing process by concentrating the efforts in the regions that have higher probability of containing defects, thereby maximization of testing efficiency by minimization of regression test time [8]. Machine learning models can be able to serve the purpose of effort estimation associated with different activities carries out during a software development lifecycle, for instance, fixing bugs or implementing new features. Using historical data these models make more accurate predictions that allow more efficient resource planning. Better project management can result from this, as resources are focused in those areas of highest need.

Machine learning is then used to detect anomalies in software logs which can be signals of reduced performance, security issues, or unusual behavior. These models are able to pick up on strange patterns without having to be told which means possible problems are identified quicker. In large systems, checking logs manually is at once time consuming and error prone and is precisely where anomaly detection is most helpful. In most software packages, Machine Learning (ML) has proven useful as a beneficial tool for optimizing the quality and dependability of software systems, automating what used to be manually executed procedures, and helping facilitate the greater efficiency and productivity of software creation [9]. The implementation of ML within Software Quality Assurance (SQA) for defect prediction purposes commonly entails a set of required procedures:

1. **Steps to Data Collection:** Collect data from various data sources like version control system (Git), issue tracker (JIRA) and the software codebase. Historical commits, changes, and defect history can be had from data mined from version control systems. The data on reported defects, priorities and resolution time of these defects are presented by issue trackers such as JIRA and the codebase allows to gather data of the structure and other metrics of the software modules [10].
2. After collecting the data, the **second task is feature engineering** which means, identify the meaningful features that can be used to train the machine learning model. Some of these features might be lines of code (LOC), cyclomatic complexity, developer interaction or software metrics such as Halstead complexity, code churn or test coverage. These are the sign of complexity and quality of different parts of code base and can also point towards possible defects. Similarly, developer activity (commit rate and patterns) can also uncover underlying issues, as some coding patterns have high defect rates [11].

3. **Model Training**: That is done through utilizing various learning methods according to the issue they're dealing with:
   - Using labeled data, **Supervised Learning** treats each example (e.g. code module, commit) and assigns a defect label (defective, non-defective). Common methods for this are Support Vector Machines (SVM), Random Forests and Neural Networks [12].
   - Clustering can be used (**unsupervised learning**) in case of scarce labeled data to group similar chunks of the code or similar commits and then identifying patterns or abnormality in the code that might be prone to errors.
   - **Semi-supervised learning** in case of very little of the data being labeled, semi supervised learning uses both the labeled and unlabeled data to get better performance out of the model. Especially when it is expensive or painful to achieve labeled defect data, Semi supervised learning is very powerful [13].
4. **Evaluation**: After training the model, its performance is evaluated by several metrics on some metrics to see how accurately it can predict defects:
   - **Accuracy:** It is the overall correctness of the model (the proportion of correct prediction to total predictions).
   - It shows the proportion of true positives (defective modules correctly identified) to all positives.
   - **Recall**: The ratio of actual positives that are correctly identified as such to the number that actually exist in the population (number of actual defective modules).
   - This is known as **F1 Score** which combines precision and recall scores to give a balanced measure of how good the model is, particularly when there is a class imbalance (defective vs. non-defective).
   - **AUC-ROC** (Area Under the Receiver Operating Characteristic Curve) – how well the model discriminates between positive and negative classes. Higher AUC means better model performance.

These steps combine together to provide an end-to-end ML based defect prediction framework for software teams to use to predict when bugs may occur early during software development and focus their efforts in specific parts of the code which are more likely to have defects. However most of the existing techniques suffer from the drawback that they view source code as flat vectors of features ignoring the relational structure present in software artifacts. Graph based approaches become relevant here.

21.2.3 Fundamentals of Graph Neural Networks

Graph Neural Networks (GNNs) are a class of deep learning models that operate directly on graph-structured data. In contrast to traditional neural networks that assume grid-like input (e.g., images, sequences), GNNs can learn representations from nodes, edges, and their topological relationships. The general idea is to **propagate and aggregate information** across the nodes of a graph over several layers. This enables GNNs to capture both **local (node-level)** and **global (graph-level)** structural information [14].

Key types of GNN architectures include:

- **Graph Convolutional Networks (GCNs):** Extend the concept of convolution to graphs by aggregating features from neighboring nodes.
- **Graph Attention Networks (GATs):** Introduce attention mechanisms to learn the relative importance of neighbor nodes.
- **GraphSAGE:** Samples and aggregates information from a node's neighborhood in an inductive setting, enabling generalization to unseen graphs.
- **Message Passing Neural Networks (MPNNs):** A general framework where nodes update their states through message-passing functions [15].

Training GNNs involves minimizing a task-specific loss (e.g., classification loss for defect prediction) using backpropagation. Unlike traditional ML models, GNNs inherently exploit the structure of software data, making them highly suitable for applications in software engineering.

21.2.4 Existing Applications of GNNs in Software Engineering

GNNs have found increasing utility in various software engineering tasks where code can be naturally represented as graphs. Some notable applications include:

- **Code Representation Learning:** ASTs and CFGs are used to represent the syntax and flow of programs. GNNs can learn latent representations that preserve this structure.

- **Vulnerability Detection:** GNNs can detect security vulnerabilities in code by analyzing control and data flow between functions.
- **Code Clone Detection:** By comparing graph embeddings of different code segments, GNNs can identify functionally similar or duplicate code.
- **Bug Localization:** GNNs allow to trace fault-prone paths in call graphs or execution traces [16].
- **Requirement-to-Code Traceability:** Mapping between them using graph similarity techniques is possible by representing requirements and source code as graphs.

GNNs exploit software dependency graphs or function call graphs that identify relationships between software code components in the domain of defect prediction. Recently, GNN-based models are shown to outperform traditional classifiers using structural information effectively.

For example, DeepWalk and node2vec as two techniques have successfully been used to generate embeddings from software graphs which are then fed into classifiers to do defect prediction. These algorithms instead represent the structural features of the code as low dimensional vector embeddings, so as to enable machine learning models to more easily learn code relationships. Likewise, by learning from the past commit graphs, Graph Convolutional Networks (GCNs) have been applied to Abstract Syntax Trees (ASTs) to detect potential defect patterns. They leverage the graph structure of the code and capture the hierarchical relationships and dependencies between different parts of the code, to identify sections of the codebase that are more likely to contain defects. It shows the ways GNNs can improve defect prediction by taking advantage of software's structural and historical aspects.

21.2.5 Limitations in Current Research

While Graph Neural Network (GNN) has large potential in defect prediction, several challenges have prohibited such GNN from being widely used in real world software quality assurance. These challenges include:

1. **Scalability:** However, one of the main difficulties of GNNs to the defect prediction problem is the computational cost which tends to be very high when the software system is large scale. The graphs have millions of nodes and edges due to software systems with millions lines of code having

complex interdependencies. Such large graphs would be expensive and time consuming to train GNN models over. Due to the I/O and computation costs of operating graphs on such nodes, optimizing the GNN architectures themselves for scalability (e.g., *via* graph sampling and/or mini-batch training) is important to make these models feasible for large software systems [17].

2. **Lack of Labeled Data:** GNN models are for supervised learning tasks that require them to learn from labeled examples. In defect prediction, particularly, it is often difficult to obtain labeled data (that is to say, to identify which modules are defective or non-defective). In many real-world software projects the data is either scarce, imbalanced, or noisy. Not all developers explicitly mark defects and the defects they label may be partial. This may restrict the performance of GNN models, especially when the training data is not a correct representation of the whole range of defect patterns in software [18].
3. **Graph Construction Complexity:** The process of constructing graphs from source code is a complex process which entails parsing an AST or Control Flow Graph (CFG) or call graph. This step is non-trivial and can therefore spuriously introduce errors and inconsistencies in the resulting graph. For example, if there was a syntactic or semantic error with a parser, the parsing process would produce incorrect graph structures which in turn could lower the quality of the defect prediction. Furthermore which parts of the code should be represented as nodes and which should be represented as edges can be subjective and may differ among projects and software domains.
4. **Overfitting and Generalizability:** Like other ML models, GNNs are prone to overfitting which is even more exaggerated on small or highly specific datasets. If you train a model on defects from one project, the trained model can be poorly applied to a different project (party), a different language, or even a different module within the same system; however, the elements we propose to be extracted from the code can serve as independent features, rather than taking the form of dependent variables. There is a fundamental challenge in making GNNs generalize across different software systems, programming languages and defect types. A possible way to

handle this issue is developing techniques to enhance generalizability, e.g., transfer learning or domain adaptation [19].

5. **Interpretability:** Since GNNs hide the underlying data (called black-box model), the reasoning behind their predictions are difficult to understand or explain. Since interpretability of a GNN model is one of the most important issues in software development where knowing what makes a module defective (predicted by a GNN) is critical, this shortcoming would lead to the limited adoption of GNN based models. However, without clear explanations around how the model deduced the provided prediction, developers need to trust such predictions which hinders the broad acceptance of GNNs in the software development process.
6. **Evaluation Inconsistencies:** Evaluation of GNN based defect prediction methods is also another challenge. Different datasets, evaluation metrics and validation protocols are used in many studies and it is hard to benchmark methods or compare results between studies. Evaluation practices are inconsistent which prevents proper evaluation of the performance of GNN-based models and make reproducing results a difficult task. Standardizing datasets, evaluation metrics, and validation protocols would better ground comparisons between various GNN approaches.

It demonstrates the challenge of how to apply GNNs to defect prediction in software systems. GNNs hold the promise of significantly improving the accuracy of defect prediction, but there are several challenges to be faced, including scalability, data labeling, graph construction, overfitting, interpretability and evaluation practices still to be addressed so that these advances in practice could be deployed in the real world.

21.3 Graph Representations of Software Systems

Source code is structured and hierarchical which makes it readily represented as a graph. Both syntactic and semantic relationships in software can be captured by graphs and graphs are a powerful abstraction of these relationships, being fundamental for accurate defect prediction using Graph Neural Networks (GNNs) [20]. In this section, several graph representations of source code, the kind of nodes and edges involved, how real world software can be represented as a graph and the preprocessing steps for preparation of the graph data for use in GNN based models are explored.

21.3.1 Modeling Source Code As Graphs

Depending on what kind of relationships (and information) one is interested in capturing, source code can be modeled in a variety of graph forms. There are three most common representations.

21.3.1.1 Abstract Syntax Trees (ASTs)

The AST is a tree based representation which shows the syntactic structure of the original source code. A node represents a construct in the programming language (for example a loop, conditional and declaration) [21].

- **Use Case in Defect Prediction:** ASTs help capture coding patterns, nested structures, and syntax errors that may correlate with defects. Example as given with int sum(int a,int b){return a+b;}
- This function would be represented as an AST with nodes for function definition, parameters, return statement, and binary operator.

21.3.1.2 Control Flow Graphs (CFGs)

The possible execution paths in a program are represented by CFGs. Nodes usually represent basic blocks (consecutive code fragments), while edges represent control flow e.g. in case of conditions, loops or jumps [22].

Use Case: Works well for finding logical complexity, unreachable code, or faulty branching which are all indicative of defects.

21.3.1.3 Call Graphs (CGs)

A function to function invocation relationship is demonstrated in call graphs which chart across a code base. Functions we call nodes, calls between them are edges.

Use Case: Faulty interactions between modules often leads to Defective behavior. GNNs learn cross module dependency patterns with call graphs.

21.3.2 Node and Edge Types in Software Graphs

In converting source code to graph structure, the right node and edge types should be defined so as to capture meaningful relationships.

21.3.2.1 Node Types

In software systems represented using graph-based representations, the choice of proper node types is important to properly capture the structural and semantic aspects of code. Nodes at the lowest level might represent source code syntax elements like keywords, operators, and literals, for example when used in Abstract Syntax Trees (ASTs) that also reflect the grammatical structure of the original source code. At higher levels of the abstraction hierarchy, nodes represent statements or expressions like assignment, loops, or function calls which are required to accurately model the control and data flow in a program. At higher levels of abstraction nodes may represent functions, classes or whole files, offering a modular view of software components [23]. Nodes in Control Flow Graphs (CFGs) are basic blocks (linear sequences of code that have one entry and one exit point) which are analyzed solely in terms of the execution paths that they enable. In large codebases, graphs can be used at the system level – with the packages or modules shown as nodes representing inter module dependencies. This hierarchical flexibility enables graph based models to function at different granularity level supporting both fine grained and coarse grained defect prediction tasks.

As well as the structural roles, graph nodes carry attributes or features which can bring semantic and statistical context to learning models. For example, attributes commonly used are code metrics like Lines of Code (LOC), cyclomatic complexity and Halstead metrics to understand the size and complexity of code elements. Some nodes may also carry frequency related features, e.g., how many times a function or class is invoked, modified or referenced throughout the code base. It indicates the operational importance of a function or class. Additionally, the token embeddings used in modern approaches are taken from pretrained language models such as CodeBERT or word2vec which use large scale code corpus to learn the contextual meaning of code elements. This better enables Graph Neural Networks (GNNs) to learn more discriminative representations over enriched node features which can better identify potential defects over a myriad of diverse software structures.

21.3.2.2 Edge Types

Edges or the relationships in software graphs define the interaction or the relationships between the nodes, making it the structural backbone holding together the different code elements. This includes syntactic links, for example, the parent child relationships as evidenced in Abstract Syntax Trees (ASTs) which shows the hierarchical organization of the code constructs. Conditional branches, loops and execution paths which determine the flow of a program, are captured as control dependencies (more or less in the form of Control Flow Graphs-CFGs). Caller and callee functions are put into call dependencies, showing how components cooperate *via* method invocations. They contain data dependencies, too which represent shared use of variables or parameters between code units and implies coupling and state change. On a higher more abstract level, object oriented features like inheritance, function overriding and interface implementation are captured by semantic relationships like inheritance relationships, function overriding relationships or interface implementation relationships. The different edge types let Graph Neural Networks (GNN) learn complex dependencies in software systems which leads to more depth and accuracy in defect prediction tasks [24].

Within software graphs, edges can then be characterized as being directed or not, by their type and by their weight, an extra layer in the representation of relationships. The relations may be directed, meaning that the relationship is represented by a clear source to destination flow (e.g. in function calls or control flows) or undirected (e.g. when two variables are accessing each other's value). Furthermore, they can be typed or untyped, allowing typed edges to discern between different types of relationships (e.g., control versus data dependency) and permitting models like Relational Graph Neural Networks (R-GNNs) to handle heterogeneous interactions. Edges can also be weighted, indicating the strength or importance of a relationship (or likelihood) like, e.g. frequency of method invocation, execution cost, or static analysis confidence scores. Such edge properties enable a more nuanced, more expressive program behavior representation which is essential for enabling Graph Neural Networks to learn and generalize defect prediction patterns [25].

Recent developments in GNNs such as Relational GCNs (R-GCN) enable them to support heterogeneous graphs with multiple node and edge types which makes them very effective to apply in software modeling.

21.3.3 Case Study: Representing Modules from a Real-World Software Project

As an example of how to use software modules as graphs, we use a subset of the JDT (Java Development Tools) from the Eclipse project. JDT is heavily used in software defect prediction research and it has modules for parsing and taking care of Java code.

Step 1: Module Selection
Let us consider a utility class, ASTParserHelper which parses Java syntax trees. There are many functions inside this class to analyze, traverse, or transform the AST.

Step 2: AST Construction
Java parsers such as Eclipse JDT Core can use each method in a class to parse and transform to an Abstract Syntax Tree (AST) that represents its structural composition. Individual nodes in these ASTs stand for syntactic elements in the code. As an example, a node might stand for a variable declaration, capturing how local variables or fields are introduced. Function or procedure calls are as represented by method invocation nodes which are as important in understanding interaction patterns. Branching nodes represent control structures, like if statements or loop structures (e.g., for, while) reflecting the conditional and iterative logic of the program. Additionally, exception handling behavior (which could model the robustness and fault tolerance of a program) is modeled using try-catch blocks. The outputs of the above system from step 2 identifies a fine-grained syntactic representation of the code which serves as an input to graph based models for detailed code analysis and defect prediction.

Step 3: Call Graph Generation
Static analysis tools like Soot or WALA are used to extract a call graph from the entire project. Now with each method on the ASTParserHelper becoming a node and if it calls another method (within the same class or not), we draw a directed edge [26].

Step 4: Feature Encoding
A variety of features of each node in the graph (e.g. a method or function) would give structural and behavioral context to learning algorithms. It is possible that these features are (1) the number of lines in the method which is correlated to the size and its possible complexity, (2) the number of method invocations (the level of coupling with other components) and

(3) the depth in the call hierarchy (the depth where the method is located in the call graph). The other important factor is in cyclomatic complexity which measures the number of independent execution paths and is a good indication of code complexity and defect proneness. Further such features are added to each node and edge which from the resulting graphs makes them an informative input to Graph Neural Networks (GNNs) for learning these defective or non-defective code component patterns [27].

21.3.4 Preprocessing and Graph Construction Techniques

In order for a software graph to be used effectively in a Graph Neural Network (GNN) for a defect prediction task, it needs to undergo a number of key preprocessing steps to guarantee the consistency and quality of the data and to make sure that the generated graph is structured in the way that can be used with the machine learning pipelines. Code parsing and extraction starts with tools like ANTLR, tree-sitter or Eclipse JDT to get to Abstract Syntax Trees (ASTs) or, static analysis frameworks like Soot, WALA to get control flow graphs and call graphs. Relevant features and metadata (file locations, developer history) is also extracted during this stage. Graph normalization then cleanses identifiers (e.g. variable names) to remove project specific noise, prunes noisy nodes (e.g. import statements) and cuts large graphs to manageable subgraphs with strategies such as sliding windows. Initial node representations are constructed in the feature engineering phase by (e.g.) one-hot encoding or by embedding models like DeepWalk, node2vec or pretrained tokenizers which are then normalized across the feature scales. With meaningful structure and attributes on the graph, the graph needs to be prepared for computation by storing the graph in a nice way and serializing the graphs in formats like JSON, XML, GraphML or using nice data structures like adjacency matrix or edge list. Often the libraries such as DGL, PyTorch Geometric or NetworkX are used to deal with large scale graphs. Thirdly, label assignment is very important in the case of supervised learning, where nodes or subgraphs are given ground truth labels (e.g., 1 for defective and 0 for clean). In order to offset the problem of class imbalance which are common in defect dataset, techniques like SMOTE (Synthetic Minority Over-sampling Technique) or cost sensitive learning are employed. These preprocessing stages work together so that the resulting graph data are robust, informative and ready for effective training in GNN based defect prediction models.

21.4 GNN Architectures for Defect Prediction

In this section we introduce different GNN architectures that have been investigated on software defect prediction. GNNs have shown success in learning structural and semantic relationships in code, represented software components as graphs, including Abstract Syntax Trees (ASTs), control flow graphs, and dependency graphs. Such architectures as Graph Convolutional Networks (GCN), Graph Attention Networks (GAT), GraphSAGE, and Gated Graph Neural Networks (GGNN) have demonstrated good potential. The variation of these models lies in the way they aggregate and pass information between nodes, and attention mechanisms or recurrent units can improve their capabilities in modeling complex code behaviors. GNNs are quite flexible, which makes them very applicable in enhancing defect prediction.

21.4.1 Popular GNN Models: GCN, GAT, GraphSAGE

A variety of architectures falling under the category of Graph Neural Networks (GNNs) have recently become popular to learn on graph structured data. The Graph Convolutional Network (GCN) is one of the foundational GNN architectures among the most widely adopted models. A GCN performs graph convolutions on nodes of a graph and hence propagates features to a node's neighboring nodes by aggregating input features by spectral operations. Using this approach, GCNs are able to view the structure locally, allowing the network to infer local patterns and is useful for tasks such as node classification. The second popular model is the Graph Attention Network (GAT) which brings attention mechanisms to the graph convolution process. Unlike other graph attention models such as Graph Attention Network (GAT) that treats all neighbors uniformly, GAT gave learnable attention scores to neighbors, enabling the model to pay more attention to more relevant nodes in the graph. In particular GATs are effective in networks where there is much variance in the importance of neighboring nodes. The other side of the coin is GraphSAGE (Graph Sample and Aggregation) which is intended to address the case of large graphs in which only a subset of neighbors is selected for training [28]. To efficiently learn from graph structures, it also uses different aggregation functions such as mean, max and LSTM based aggregation (for aggregating the node features), even in an inductive setting when the training has not seen new nodes or graphs. These and other models (e.g. Relational Graph Convolutional Networks (R-GCN), Message Passing Neural Networks

(MPNNs)) for learning graph representations each utilize distinct techniques for working with different varieties of graph data.

21.4.2 Adapting GNNs to Code-Based Graphs

Defect prediction in software engineering uses GNNs that process source code by converting it into a graph structure. Typical ways to show programs include using Abstract Syntax Trees (ASTs), Control Flow Graphs (CFGs) and call graphs. ASTs use nodes to show things like expressions, method declarations and variables, while edges capture how these items are related in the code (for example which items act as the parent of other items). Basically, a node in CFG represents a segment of code and an edge represents an instance where one node comes after another because of a conditional or loop. Call graphs pay attention to how different functions or methods are related through calling. Nodes and edges in the program graph are given special features like cyclomatic complexity, the number of method invocations and holders of tokens. They let GNNs identify both the way code is organized and the way it works when executed. Besides, given that many software systems are not only large but also complex, GNNs must handle difficulties related to the sparseness, many types and size of the graphs which might result in the GNNs making different aggregation and neighborhood sampling choices for better learning.

21.4.3 Model Training and Optimization

Carefully considering several aspects is important when training GNNs for defect prediction. The initial step is to feed labeled graphs (such as examples of defects and non-defects) into the model. GNNs are trained to give each node or subgraph a binary label telling us whether or not the code could have defects. Optimization with Stochastic Gradient Descent (SGD) or variants such as Adam or RMSprop is applied during training to bring the loss down to a minimum, usually the cross-entropy loss for binary classification. Based on the unbalanced structure of defect prediction data, using dropout, early stopping and L2 regularization can prevent overfitting and so improve generalization [29]. It is also possible to use batch normalization and residual connections to ensure that training doesn't fluctuate and that convergence happens faster, especially in deep graph neural networks (GNNs). When learning is complete, the system is assessed on a validation dataset, so it is clear whether it will recognize defects in code that was not included in its training.

21.4.4 Comparison with Baseline ML/DL Models

When using GNNs to predict product defects, their results are compared to those obtained using existing machine learning (ML) and deep learning (DL) models to better understand their advantages. Defect prediction often relies on traditional models called Support Vector Machines (SVM) and Random Forests (RF). Most of these models make use of lines of code (LOC) and other handcrafted features like cyclomatic complexity and Halstead metrics to predict results. Neither do these techniques usually reveal complex connections between codes nor use all the valuable code structures available. Alternatively, feedforward neural networks (FNNs) and convolutional neural networks (CNNs) are capable of recognizing more complex features in raw data, but they fail to effectively represent the network structure of software systems. However, GNNs are particularly effective because they explicitly define the relationships between nodes which boosts their ability to understand code. Using data in experiments, it has often been found that GNNs usually perform better than other models if learning from how the graph is organized is key to the problem. If you can model interactions between parts of a system, you can often make better predictions, especially for things traditional methods might not notice.

21.4.5 Hyperparameter Tuning Approaches

Finding the right values for hyperparameters greatly helps in improving how GNNs work for defect prediction. Some main hyperparameters of GNNs are the depth of the network, the size of each hidden layer, the chosen learning rate, the use of dropout and the number of examples in a single batch. The kind of aggregation function you use (e.g., mean, max or LSTM-based) can greatly influence how well the model captures diverse interactions between the nodes. To find the combination of hyperparameters that gives the model the highest predictive accuracy, you need to test different settings very carefully. People often use grid search which checks every hyperparameter combination from a pre-made set and random search, where the algorithm picks hyperparameters from a wide range. If computational resources are limited, Bayesian optimization and Genetic Algorithms are good ways to look for the best hyperparameter values. One can also stop training early using early stopping when the model starts to do less well on a validation set, helping to avoid overfitting and saving resources. If hyperparameters are properly chosen, GNNs can run smoothly and give more accurate defect predictions [30].

21.5 Quality Assurance Considerations in GNN-Based Systems

This section talks about the important quality assurance (QA) aspects to keep in mind when implementing GNN-based systems in software defect prediction. On contrary to traditional models, GNNs act on complicated graph representations, and thus interpretability, scalability, and reliability are essential considerations. The quality of data being important, the proper construction of graphs out of the source code should ensure that no noisy data or poorly- parsed data will result in a lower-quality performance. The explainability of the models also matters as the developers should have trust and understanding of predictions. Also, GNNs need to be cautiously validated in order to prevent overfitting because of their high capacity. Resilience to changing codebases, and project generalization are essential to effective QA in the long term.

21.5.1 Dataset Quality and Reproducibility

Dataset quality is very important in any machine learning job, especially when working with Graph Neural Networks (GNNs) to predict software issues. Such systems rely on great datasets that precisely capture both the structure and the ways software projects behave. It is very important that the dataset consists of enough examples of both correctly functioning code and its defective counterparts. Make sure the constructs covered by the model such as loops, conditionals, function calls and exception handling, are true to how they are used in real programming tasks. The issue of reproducibility also exists, as collecting, preparing and converting source code into graphs should be done the same way for all experiments. Making sure the graphs are reproducible by using standard parsers, static analysis tools and graph building methods. Also, it is necessary to record the way each dataset is prepared, especially the methods used to extract node and edge features and to assign labels, to ensure future experiments can be done with trust. Taking care of data leakage (when information from the tested data impacts learning) and class imbalance (where there are fewer defects in the samples) plays a big role in keeping the data accurate and the model fair.

21.5.2 Model Interpretability and Explainability

People involved in software development need to understand how GNNs make their predictions which is why interpretability is important. Since

GNNs move data over the nodes and edges of a graph, they are more complicated than traditional machine learning models and harder to interpret. Even though model-driven software quality assurance automates much of the process, developers and QAs need to review and understand the reason behind each code being classified as potentially defective. When a human can explain the model's reasons for a decision, it helps form trust and guarantees the results make sense for operations. For instance, with the help of graph attention mechanisms (for example, GAT), the model can point out the most vital areas to look for mistakes in the code. You can also use visual tools to follow the path of the data and see which characteristics are most important for the outcome. Researchers are investigating LIME (Local Interpretable Model-Agnostic Explanations) and SHAP (Shapley Additive Explanations) methods to help explain better what GNNs are predicting. Giving these explanations is needed to ensure transparency and to help debug and develop the model, because developers can realize which parts may be going wrong.

21.5.3 Fairness and Bias in Defect Prediction

It is very important to consider fairness and bias in any machine learning process and this is most true for defect prediction using GNNs. Concerns may arise if a defect prediction model is built using data that highlights some defects, coding styles or developers while ignoring others, since this could cause the model to repeat the same biases and therefore reinforce them. If, for example, a dataset has a higher number of defects from some types of applications or coding practices, the predicted defect rate for those could be inflated while the model may fail to notice many defects in other applications. As a result, the quality assurance might turn biased and single out certain types of code or engineers. Such biases need to be handled by paying attention to the diversity of the training data, covering programs from a wide range of styles, frameworks and developers. Regularly checking the model's accuracy for fairness is necessary, mainly when it is used to recognize defects in critical security software or in software developed by teams with diverse backgrounds. Applying different learning techniques or putting fairness rules during model optimization can lower the effects of these biases. Also, monitoring fairness metrics such as demographic parity or equal opportunity should be part of the regular evaluation process so that defect predictions do not harm certain categories of people or kinds of software disproportionately.

21.5.4 Generalizability across Projects and Domains

Being able to generalize the outcome of a GNN-based defect prediction model is also very important for practical use. Their predictions should be accurate no matter what the differences are in programming languages, libraries, frameworks or coding styles among the chosen software projects. A main obstacle is to stop the model from giving wrong answers because it is tailored too much to just one project or domain. The model can get stuck learning the exact features in the training data and not handle other projects outside of them. Enhancing generalizability means exploiting techniques such as transfer learning or domain adaptation. When a model is trained using one dataset, but then uses training data from another, it uses its past experience to handle the new data. By using domain adaptation, models avoid being influenced too much by the features of a single project or programming language. In addition, evaluating the model on more than one project ensures it performs similarly in different settings and does not get confused by those project-specific attributes. Testing the model with many types of projects allows developers to make sure it can handle various code and defect situations.

21.5.5 Integration with Continuous QA Pipelines

For deployment in production, it is important to consider integrating the GNN models with continuous quality assurance (QA) pipelines. Today, developers primarily apply agile methodologies and DevOps, keeping software development going through continuous integration (CI) and continuous deployment (CD). When GNNs are used in defect prediction models, they can give instant feedback while code is written. As new code is added by developers, the system automatically constructs a graph from the changes, feeds that into the trained GNN and soon gives out results for defects. It helps teams discover defects in development which saves money by reducing the need to fix them and improves the whole software. Also, by regularly retraining the model with additional data, the model can adjust to new programming mistakes and practices. Because of this process, the defect prediction system is trustworthy and speaks to needs as time goes on. Connecting with other technologies in the CI/CD pipeline, for example, static analysis tools and automated testing tools, makes it possible to check code quality more completely and efficiently.

Key quality assurance challenges and opportunities in using GNN based defect prediction systems are captured by these considerations. This work demonstrates a reliable and effective defect prediction tool in modern software development settings, however standardized development process (e.g., ensuring dataset quality, improving model interpretability, addressing fairness and bias in the criteria for defining defectiveness, ensuring generalizability and interconnecting the system with continuously integral QA pipelines).

21.6 Experimental Setup and Evaluation

The section defines the experimental setup and the evaluation strategy in terms of which it is planned to determine the effectiveness of GNN-based models in software defect prediction. The benchmark datasets include NASA MDP, JDT, or Code4Bench on which the experiments are carried out and modules are defective or non-defective. The source code is converted to graph structures (e.g. ASTs or dependency graphs) and fed as input to different GNN architectures such as GCN, GAT, and GGNN. Stratified k-fold cross-validation is applied to train and test models. The measures used in evaluation are Accuracy, Precision, Recall, F1-score, and AUC to ascertain that the evaluations focus on all aspects of predictive performance and generalization.

21.6.1 Dataset

To evaluate GNN based defect prediction models, it necessitates to use high quality, real world datasets that cover a wide variety of software characteristics. The commonly used software defect prediction dataset is the JDT (Java Development Tools) dataset composed of software modules from Java based projects. Because it has both defective and non-defective instances, the JDT dataset is widely used for the purposes of binary classification tasks, due to the availability of its rich set of features. The PROMISE repository is also an important dataset and it includes multiple datasets derived from different software systems (i.e. from NASA's software repositories). The PROMISE datasets are used widely for evaluating defect prediction models, contain detailed feature sets such as lines of code (LOC), complexity metrics, and other software metrics required for defect prediction models. The second valuable dataset is Code4Bench which has been designed specifically for evaluating machine learning models on software engineering tasks. Code4Bench offers a standardized benchmark for defect prediction by supplying code fragments together with the information if they are defective or clean so that systematic model comparisons are possible. For the study of generalizability, these datasets provide diverse software artifacts from small scale modules to large software systems and thus well suited for training and evaluating GNN based models [31].

In addition, diverse datasets from various domains and programming languages are used to improve understanding on how the GNN models generalize across software systems as well as different coding styles, to prevent the findings of having strong dependencies on a single dataset or the code type.

21.6.2 Evaluation Metrics (Precision, Recall, F1, AUC, MCC)

Since GNN based defect prediction is an emerging practice, a variety of performance metrics are needed to evaluate the effectiveness of GNN based defect prediction models. For the reason that defect prediction is usually placed as a binary classification problem, the main evaluation metrics are:

- **Precision**: This metric measures the proportion of true positive defect predictions among all instances predicted as defective. Precision is critical when the cost of false positives (incorrectly classifying non-defective code as defective) is high, as in the case of identifying software defects during the development process.
- **Recall**: Recall measures the proportion of actual defective instances that are correctly identified by the model. This metric is essential when the goal is to minimize the risk of missing actual defects, even at the cost of producing some false positives.
- **F1-Score**: The F1-score is the harmonic mean of precision and recall. It is a good metric for evaluating the balance between precision and recall, particularly when the class distribution is imbalanced (i.e., fewer defective instances).
- **AUC (Area Under the ROC Curve)**: The metric AUC is used to judge how well the model is able to make a distinction between positive and negative classes at different thresholds. In this case, A high AUC means our model can differentiate between defective and non-defective code, very effectively.
- **MCC (Matthews Correlation Coefficient)**: MCC is a less biased measure since it takes in account the true positives, true negatives, false positives and false negatives. When working with highly imbalanced datasets it is particularly useful and gives a more holistic view of model performance.

They are each important in measuring model performance of defect prediction models because they reveal the magnitude of different factors of a model's behavior (i.e. model accuracy, capacity to avoid false negative predictions and overall robustness).

21.6.3 Baseline Comparisons

Compared to traditional machine learning models such as Support Vector Machines (SVM), Random Forest (RF) and Logistic Regression (LR), comparing against deep learning models like Fully Connected Neural Nets (FNNs) is required to benchmark the performance of GNNs for defect prediction models. These are usually used as baseline methods and are used for software defect prediction. SVM is very popular classifier in defect prediction domain as it is able to handle high dimensional data and find the best hyperplanes for classification. Used widely for defect prediction tasks, it uses radial basis function (RBF) kernels for nonlinear decision boundaries. Random Forests (RF) are ensemble methods by averaging multiple tree predictions, resistant to overfitting and often improve over other classifiers on imbalanced datasets due to their ability to detect nonlinear interactions in the data [32].

A simpler baseline method that is still popularly used in defect prediction is Logistic Regression (LR) because it is interpretative and can describe linear relationships between features and the target variable. Although simpler than GNN, Fully Connected Neural Networks (FNN) have been used to predict defects by learning complex combinations of features. While GNNs can explicitly model graph structure, however they lack the ability to explicitly model such graph structures which they depend on.

Such comparisons will highlight advantages of graph based approaches as opposed to these models, in particular when the graph structure of the software system greatly influences the ability to predict defects.

21.6.4 Cross-Validation Strategies

A very important way of how to check the generalizability of the defect prediction model is cross validation. The most commonly used type of cross validation is k-fold cross validation where the dataset is split into k subsets (folds). However, k-1 folds are trained and the last fold is validated. Each fold is used as validation set, once and the process is repeated k times. It also helps the model to be deterred on various parts of the dataset so that we get a better estimate of its performance and also reduce the chances of overfitting. Also, it is common practice in the defect prediction tasks to employ stratified k-fold cross-validation so that each fold possesses a representative share of instances containing defect and others without defect. This is very important when dataset is imbalanced, because this can protect the model from being biased towards major class [33].

It can also be used in leave-one-out cross validation (LOO-CV), in cases where the dataset is small. In LOO-CV, the training data consists of all instances but one which is the test set. Then, this process is repeated for each instance such that the model is evaluated on (trains on) every data point. Cross validation strategies give valuable insights to the performance of model and its capability to generalize on unseen data.

21.6.5 Ablation Studies and Statistical Validation

Ablation studies are necessary to understand the role of each model component to its overall performance. With respect to GNNs based defect prediction, it is possible that ablation studies may involve getting rid of or changing away some components of the graph construction process, for instance:

These can be excluded, for instance, if some node or edge features (e.g., complexity metrics or function call dependencies) are not available or are irrelevant.

- Reducing the complexity of GNN architectures (e.g., using GCN instead of GAT).
- Aggregation can be changed (e.g., from mean aggregation to LSTM aggregation).
- By systematically varying these changes and testing the model performance, researchers are better able to understand the component level role and if need be optimize the model.

It is also important to use statistical validation methods (e.g. t-tests or ANOVA) to confirm that the observed differences in model performance are statistically significant [34]. Statistical tests enable the sorting of multiple models or configurations by knowing whether the observed differences in performance metrics are real (real) or just chance (random variation). Adding this level of validation to the evaluation process validates reported results and makes sure they are meaningful and reliable.

21.7 Challenges and Future Directions

This section points out the most crucial future directions and challenges of using GNNs to predict software defects. The complexity of graph construction is one of the major challenges since it is not a trivial task to extract accurate and consistent representations out of a variety of programming languages and code styles. Another issue is scalability, particularly in using GNNs on large-scale systems or in real-time applications. Interpretability is also low, and developers would find it difficult to have confidence in and take actions based on predictions. Future directions: Cross-project generalization Cross-project generalization is an under-explored direction that is important to future progress. Explainability and CI/CD integration of GNNs are also avenues of future progress.

21.7.1 Scalability and Computational Costs

Applying Graph Neural Networks (GNN) on a software defect prediction problem poses one of the major scalability and computational cost associated with the large software systems. The size of the graphs grows along with the size of the codebase and it leads to increased memory requirements as well as an increase in complexity of computations needed to process these graphs. Very large graphs, millions of node to millions of edges, can be generated from software projects, particularly where there are millions of log lines of code (LOC). The nature of GNNs is to use a iterative message passing across nodes and edges and can be computationally expensive.

The graph structure of software itself can be quite complex and interconnected and thus this problem is exacerbated. GNNs are effective at learning these relationships however the time and space complexity of processing these graphs become a bottleneck. Finally, to resolve this issue, multiple approximation techniques and graph sparsification methods can be experimented with to keep the graph size as small as possible, yet preserving the main structural information. Moreover, the large scale computation that is required for training GNNs on massive software datasets can be handled by the means of distributed computing and parallel processing. Scaling up GNNs to efficiently process large graphs could rely on optimizing GNN architectures to consume graph inputs more efficiently, combined with subgraph learning techniques that minimize the loss of accuracy through focusing on learning only relevant parts of large graphs [35].

21.7.2 Evolution of Software and Incremental Learning

There is also another challenge with using GNN based defect prediction: the evolution of software. Frequent updates, bug fixes and addition of new features keep changing the software systems. That is, a defect prediction model trained on one version of the software may become outdated when the software evolves. To ensure that the model will stay effective for the long term in a production environment it is essential to be able to keep the model updated without retraining it from scratch.

An interesting direction is to use incremental (aka online) learning to update the learned model as more data arrives (e.g., new code versions or defect labels). In such a case, the GNN can be trained so that it is able to learn from the new graphs without having to retrain the whole model. Agile development environments, where the software is in constant change, make this especially important. The model can incrementally learn to adapt to changes in coding styles, defect patterns and software architecture. Still, there is still active research in making GNNs support incremental learning, including how to cope with catastrophic forgetting where the model can stop predicting well on previous code versions as it learns on new data.

21.7.3 Explainable GNNs for Developer Trust

Explainability is a critical concern for the adoption of GNNs in defect prediction. For developers to trust and act upon the predictions made by a GNN model, they need to understand why certain predictions were made. This is particularly important in software quality assurance (QA) processes, where developers must justify the identification of defects in specific code sections.

Explainable AI (XAI) methods for GNNs are still an evolving field, and while GNNs inherently offer some interpretability (e.g., through attention mechanisms in models like GAT), providing clear and actionable explanations remains a challenge. Research into explainable GNNs focuses on providing insights into which parts of the code (i.e., which nodes and edges in the graph) contribute most to the defect prediction. Techniques like attention-based mechanisms or graph saliency maps can be useful to highlight the most important parts of the graph that influenced the decision.

Local explainability techniques such as LIME or SHAP can be adapted for graph data to offer developers clear, understandable explanations of individual defect predictions. Such techniques can indicate certain structures of code, functions or lines of code that are most likely to be defective which can help to raise developer trust and improve the user friendliness of defect prediction systems. To accept and integrate GNNs into real world development workflow, it will be critical to ensure that GNN based defect prediction models do not become "black boxes", rather they do provide clear reasoning for their predictions [36].

21.7.4 Transfer Learning Across Projects

One promising avenue to improving generalization of GNN-based defect prediction models is transfer learning. Defect prediction models are trained in practice for a particular software project and it restricts their capability of achieving good performance on other projects possessing dissimilar codebases, structures and features. This is the problem that transfer learning attempts to tackle by transferring knowledge from one project (or domain) to another. For instance, a defect prediction model which is learnt on a Java projects codebase can be used on a Python or C++ project, even though the syntax and structure is completely different [37].

Fine tuning a pre trained model to a new dataset is called the transfer learning. When little labeled data is available for a new project, this is especially useful, because the model can learn from preexisting knowledge from other projects. Transfer learning in the context of GNNs is to share embeddings or graph representations in related projects. Yet a difficulty exists in adapting GNNs across various projects as various project characteristics are extremely different. Transfer learning may not be very effective at all, requiring domain adaptation techniques or cross domain graph representations instead. Models also need to be analyzed to determine if they suffer from negative transfer, that is transferring knowledge from one project to another worsens its performance instead.

21.7.5 Opportunities for Hybrid Models (GNN + LLMs)

A second exciting avenue for future research is to develop the theory and tools to make hybrid models that combine the features of GNNs with other state of the art machine learning algorithms. A case in point is to combine Large Language Models (LLMs; e.g., GPT, CodeBERT) and GNNs for defect prediction. What GNNs are good at, is learning from the structure relationships in code; LLMs, on the other hand, are good at understanding the semantic content of code (detecting potential logic errors, supporting commonly used coding patterns or understanding natural language comments).

Hybrid models combine graph-based learning capabilities of GNNs with the language modeling abilities of language learning models (LLMs) to learn from both structural features (i.e., the graph of the code) and the semantic features (i.e., the textual content of the code) and make more accurate and comprehensive defect predictions. For instance, LLMs can give insights around the use of functions (e.g. the use of locale, the usage of the variable name 'key'), documentation (e.g. function/class/module) or any differences between function archives, whereas GNNs can look at the syntactic structure of the code, the relationships between functions, classes and modules, etc. Similarly, other hybrid approaches of combining GNN with reinforcement learning can be for automated bug fixing and combining GNN with CNN for graph image processing tasks translating graph to grid representation (e.g. adjacency matrices). These new hybrid models could provide the means for fusing different (hybrid) forms of data and (hybrid) forms of learning to attack the problem of defect prediction in a complex manner.

21.8 Conclusion

This chapter discusses the use of graph neural networks (GNNs) in software quality assurance (QA) within the context of defect prediction. The results indicate that GNN prove to be a promising technique for discovering and predicting defects in software systems by exploiting well the structural relationships among source code. However, the reliance on manual feature engineering and the limited capability of encoding complex code structure limit the feasibility of traditional machine learning techniques, e.g. SVMs and Random Forests, for defect prediction. In contrast to this scenario, GNNs can learn very complicated relationships between different parts of code (function calls, control flows and data dependencies) automatically and thus prove suitable to this problem.

As GNNs view software as graphs, nodes characterize code components and edges capture the interconnections between nodes; these GNNs are expected to achieve better defect prediction accuracy by considering both local and global patterns in the software. Quality assurance (QA) is an important component in the production of reliable and robust software which comes down to writing error free code. Although capable to a point, traditional defect prediction methods are limited in their ability to uncover these sorts of complex, context bound problems which arise from complex software architectures. Traditional defect prediction techniques are extended by GNN based defect prediction models which account for structural code context as well as semantic understanding of code level interactions within the software.

It utilizes GNNs in software defect prediction as a transformative means of automated QA, whereby models can continually and on the fly learn and adapt to new data as code changes. In addition, GNNs provide the ability to scale defect prediction costs for complicated and large-scale codebases which becomes an increasingly important issue as software systems become ever larger and more complex. GNNs promise impressive results, but their utilization in actual software development is still at its infancy, shifting from manual and rule-based QA techniques to data driven and predictive models will no longer only improve the accuracy of defect prediction, but also reduce the cost of manual inspection and testing. There are several remaining challenges, especially with regard to scalability, explainability and generalization to other projects and other domains. Training GNNs on large scale software systems can still be quite costly computationally and there is still further research required to optimize these models for real time, real-world, defect prediction. Furthermore, GNNs need to be made more interpretable for both developers to trust the outputs of these models and to know how to act upon them.

Incremental learning, transfer learning and hybrid models with the combination of GNN with other machine learning approaches show the bright future of GNN based defect prediction. Therefore, such models will likely become more flexible and provide more personalized and context specific predictions of defects in different types of software systems. Finally, because they are new and disruptive technology will be coming out soon, GNNs might be even more powerful at understanding underlying structure and semantics of software and could be integrated with large language models (LLMs). Current strengths and limitations of GNN based systems are outlined and their potential to revolutionize how defects are detected, predicted and address in software through their enabling of GNNs as the technical drivers for future software systems which can utilizing technology as it matures of the defects maturing and overcoming current limitations are discussed. This will integrate these models into continuous QA pipeline will improve software reliability, improving the quality of the software developed at faster rates with lesser defects in it and less time to market.

AI disclosure statement: The author acknowledges the use of AI-based tools for minor language editing or grammatical correction. However, all intellectual contributions, analyses, and conclusions presented in this chapter are entirely the author's own.

References

1. Jose, S., Zemouri, R., Nguyen, K.T.P., Medjaher, K., Lévesque, M., Tahan, A., Prognostics of complex machinery with sparse multilabel multimodal run-to-failure data: A graph neural network approach. *Adv. Eng. Inf.*, 65, 103361, 2025. doi: https://doi.org/10.1016/j.aei.2025.103361.
2. Gao, H., Guo, X.-W., Li, G., Li, C., Yang, C., GCPNet: An interpretable Generic Crystal Pattern graph neural Network for predicting material properties. *Neural Networks*, 188, 107466, 2025. doi: https://doi.org/10.1016/j.neunet.2025.107466.
3. An, D., *et al.*, Reinforcement learning-based secure training for adversarial defense in graph neural networks. *Neurocomputing*, 630, 129704, 2025. doi: https://doi.org/10.1016/j.neucom.2025.129704.
4. Qian, J. and Tong, G., Metapath-guided graph neural networks for financial fraud detection. *Comput. Electr. Eng.*, 126, 110428, 2025. doi: https://doi.org/10.1016/j.compeleceng.2025.110428.
5. Maryam, Rehman, M.U., to Chong, K., Tayara, H., Drugs inhibition prediction in P-gp enzyme: a comparative study of machine learning and graph neural network. *Comput. Toxicol.*, 34, 100344, 2025. doi: https://doi.org/10.1016/j.comtox.2025.100344.
6. Chen, J., Shi, L., Chen, B., RSEA-MVGNN: Multi-view graph neural network with reliable structural enhancement and aggregation. *Inf. Fusion*, 121, 103143, 2025. doi: https://doi.org/10.1016/j.inffus.2025.103143.
7. Yang, F., Zhong, F., Wei, X., Zeng, G., A Fine-Grained Defect Prediction Method Based on Drift-Immune Graph Neural Networks. *Comput. Mater. Contin.*, 82, 2, 3563–3590, 2025. doi: https://doi.org/10.32604/cmc.2024.057697.
8. Zhang, M., Jia, J., Capretz, L.F., Hou, X., Tan, H., Graph neural network-based long method and blob code smell detection. *Sci. Comput. Program.*, 243, 103284, 2025. doi: https://doi.org/10.1016/j.scico.2025.103284.
9. Liu, C., Liu, T., Yang, K., Hong, K., Chen, X., Differentiable simulation and optimization of particle-fluid flows using graph neural networks. *Powder Technol.*, 462, 121143, 2025. doi: https://doi.org/10.1016/j.powtec.2025.121143.
10. Xu, L., Peng, J., Jiang, X., Chen, E., Luo, B., Graph neural network based on graph kernel: A survey. *Pattern Recognit.*, 161, 111307, 2025. doi: https://doi.org/10.1016/j.patcog.2024.111307.
11. Liu, J., Ai, J., Lu, M., Wang, J., Shi, H., Semantic feature learning for software defect prediction from source code and external knowledge. *J. Syst. Softw.*, 204, 111753, 2023. doi: https://doi.org/10.1016/j.jss.2023.111753.
12. Mehta, S., Kumar, L., Misra, S., Patnaik, K.S., Singh, V., Nested deep learning with learned network embeddings for software defect prediction. *Appl. Soft Comput.*, 174, 113057, 2025. doi: https://doi.org/10.1016/j.asoc.2025.113057.

13. Zhang, J., Ding, S., Zhang, J., Guo, L., Ding, L., Multi-channel set polynomial based label regularized graph neural networks against extreme data scarcity. *Pattern Recognit.*, 166, 111754, 2025. doi: https://doi.org/10.1016/j.patcog.2025.111754.
14. Wang, X., Wang, H., Peng, M., Interpretability study of a typical fault diagnosis model for nuclear power plant primary circuit based on a graph neural network. *Reliab. Eng. Syst. Saf.*, 261, 111151, 2025. doi: https://doi.org/10.1016/j.ress.2025.111151.
15. Wang, Z., *et al.*, BlastGraphNet: An Intelligent Computational Method for the Precise and Rapid Prediction of Blast Loads on Complex 3D Buildings Using Graph Neural Networks. *Engineering*, 2025. doi: https://doi.org/10.1016/j.eng.2025.03.007.
16. Giray, G., Bennin, K.E., Köksal, O., Babur, O., Tekinerdogan, B., On the use of deep learning in software defect prediction. *J. Syst. Softw.*, 195, 111537, 2023. doi: https://doi.org/10.1016/j.jss.2022.111537.
17. Zhang, H., *et al.*, Combining physics-informed graph neural network and finite difference for solving forward and inverse spatiotemporal PDEs. *Comput. Phys. Commun.*, 308, 109462, 2025. doi: https://doi.org/10.1016/j.cpc.2024.109462.
18. Yang, H., Fang, J., Wu, J., Li, D., Wang, Y., Zheng, Z., Soft label enhanced graph neural network under heterophily. *Knowledge-Based Syst.*, 309, 112861, 2025. doi: https://doi.org/10.1016/j.knosys.2024.112861.
19. Qiu, S., He, J., Wang, Y., E, B., A Feature Selection Method for Software Defect Prediction Based on Improved Beluga Whale Optimization Algorithm. *Comput. Mater. Contin.*, 83, 3, 4879–4898, 2025. doi: https://doi.org/10.32604/cmc.2025.061532.
20. Lin, A., Wang, H., He, W., Owyong, S., Chen, H.T., Nguyen, T.H., Bayesian enhanced graph neural networks: Refining design spaces for hollow concrete components with optimum mechanical performance. *Eng. Struct.*, 327, 119628, 2025. doi: https://doi.org/10.1016/j.engstruct.2025.119628.
21. Sharma, S., Ratmele, A., Seth, A.D., Multiclass vulnerability and clone detection in Ethereum smart contracts using Block-wise Abstract Syntax Tree based Federated Graph Neural Networks. *Comput. Electr. Eng.*, 123, 110220, 2025. doi: https://doi.org/10.1016/j.compeleceng.2025.110220.
22. Prasad, E.S., Evangelin Sonia, S.V., Suresh, K.N., Shivapanchakshari, T.G., Active and Reactive Power Control in Three-Phase Grid-Connected Electric Vehicles using Zebra Optimization Algorithm and Multimodal Adaptive Spatio-Temporal Graph Neural Network. *Renew. Energy Focus*, 54, 100715, 2025. doi: https://doi.org/10.1016/j.ref.2025.100715.
23. Li, H., Xu, J., Yin, L., Wang, Q., Jiang, Y., Liu, J., Metapath-free adversarial attacks against heterogeneous graph neural networks. *Inf. Sci. (Ny)*, 713, 122143, 2025. doi: https://doi.org/10.1016/j.ins.2025.122143.
24. Khakhar, P. and Dubey, R.K., Chapter 5, in: *The integrity of machine learning algorithms against software defect prediction*, R. Pandey, S.K. Khatri, N. kumar

Singh, and P. B. T.-A. I. and M. L. for E. C. Verma (Eds.), pp. 65–74, Academic Press, London, 2022, doi: https://doi.org/10.1016/B978-0-12-824054-0.00027-7.

25. Liu, X., Meng, Y., Chen, F., Qiao, D., Wu, F., GCTGNN: A forecasting method for time series based on graph neural networks and graph clustering. *Neurocomputing*, 626, 129544, 2025. doi: https://doi.org/10.1016/j.neucom.2025.129544.
26. Fei, Q., Hu, H., Yin, G., Sun, Z., A Software Defect Prediction Method Using a Multivariate Heterogeneous Hybrid Deep Learning Algorithm. *Comput. Mater. Contin.*, 82, 2, 3251–3279, 2025. doi: https://doi.org/10.32604/cmc.2024.058931.
27. Akpaku, E., Chen, J., Ahmed, M., Agbenyegah, F.K., Brown-Acquaye, W.L., RAGN: Detecting unknown malicious network traffic using a robust adaptive graph neural network. *Comput. Networks*, 262, 111184, 2025. doi: https://doi.org/10.1016/j.comnet.2025.111184.
28. Zhang, D., Roy, N., Wang, R., Frost, J.D., Predicting tornado-induced building damage: A comparative study of tree-based models and graph neural networks. *Int. J. Disaster Risk Reduct.*, 123, 105525, 2025. doi: https://doi.org/10.1016/j.ijdrr.2025.105525.
29. Mishra, A. and Sharma, A., Deep learning based continuous integration and continuous delivery software defect prediction with effective optimization strategy. *Knowledge-Based Syst.*, 296, 111835, 2024. doi: https://doi.org/10.1016/j.knosys.2024.111835.
30. Lu, M., Liu, Z., Wang, P., Wang, H., Yang, D., A Bayesian graph structure inference neural network based on adaptive connection sampling. *Appl. Soft Comput.*, 175, 113018, 2025. doi: https://doi.org/10.1016/j.asoc.2025.113018.
31. Zain, Z.M., Sakri, S., Ismail, N.H.A., Application of Deep Learning in Software Defect Prediction: Systematic Literature Review and Meta-analysis. *Inf. Softw. Technol.*, 158, 107175, 2023. doi: https://doi.org/10.1016/j.infsof.2023.107175.
32. Mohammad, F., Al-Ahmadi, S., Al-Muhtadi, J., SESDP: A Sentiment Analysis-Driven Approach for Enhancing Software Product Security by Identifying Defects through Social Media Reviews. *Comput. Mater. Contin.*, 83, 1, 1327–1345, 2025. doi: https://doi.org/10.32604/cmc.2025.060228.
33. Daza, A., Software defect prediction based on a multiclassifier with hyperparameters: Future work. *Results Eng.*, 25, 104123, 2025. doi: https://doi.org/10.1016/j.rineng.2025.104123.
34. Ning, X., Li, H., Wang, Q., Yang, F., Zhang, Y., Wang, G., Two-stream neural network for the prediction of multiple defects in continuous casting. *Appl. Soft Comput.*, 175, 113116, 2025. doi: https://doi.org/10.1016/j.asoc.2025.113116.
35. Azzeh, M., Elsheikh, Y., Nassif, A.B., Angelis, L., Examining the performance of kernel methods for software defect prediction based on support vector machine. *Sci. Comput. Program.*, 226, 102916, 2023. doi: https://doi.org/10.1016/j.scico.2022.102916.

36. Rankovic, N., Rankovic, D., Ivanovic, M., Kaljevic, J., Interpretable software estimation with graph neural networks and orthogonal array tuning method. *Inf. Process. Manag.*, 61, 5, 103778, 2024. doi: https://doi.org/10.1016/j.ipm.2024.103778.
37. Chen, Y., Li, L., Ma, Z., Wen, X., Pang, J., Yuan, W., Improving weak magnetic detection of ferromagnetic material defects diagnostics *via* transfer learning-enhanced residual networks. *Eng. Appl. Artif. Intell.*, 150, 110647, 2025. doi: https://doi.org/10.1016/j.engappai.2025.110647.

22

Future Research Directions in Graph Neural Networks

Valarmathi Viswanathan

Department of Information Technology, Sri Krishna Arts and Science College, Coimbatore, Tamilnadu, India

Abstract

This chapter explains how Graph Neural Networks (GNNs) have made non-Euclidean machine learning both better and more useful by exploring social networks, recommendation systems, biological informatics, traffic prediction and knowledge graphs. Because of graph structures and relational biases, GNNs can efficiently represent both structured and semi structured data. At present, current GNN models have major difficulties when it comes to their scalability, how expressive they are and their underlying theory. At the end, we offer new research goals that get past the flaws and point towards the future for GNNs. Many (if not all) current GNNs are expressive, but most are Weisfeiler–Lehman test restricted which limits their abilities. Improving the expressiveness of syntax is a main research focus without making it less efficient. Also, the issue of overprocessing and overcompressing by deep GNNs means they cannot properly preserve features of each node and still learn distant dependencies. We need to resolve these problems to build better and more reliable architectures. Having the ability to process large and changing graphs is very important as well. These types of solutions, such as sampling and distributed training, can harm accuracy or completeness. We also urgently need models that can handle more data so that handling real life billion-edge graphs becomes feasible. Other promising areas are working with graphs that change over time, self-supervised methods and using little or no training data. It is possible for computer science to work with other fields by using methods from causal inference, probabilistic reasoning and quantum computing. Ease of explanation, fairness and robustness are all extra requirements for GNNs in important cases. All in all, the chapter gives a detailed overview of the major achievements of GNNs.

Email: valarskasc1@gmail.com

J. Ramkumar, Sridaran Rajagopal, B. Suchitra and S. Balamurugan (eds.) Graph Neural Networks: Concepts and Applications, (859–904) © 2026 Scrivener Publishing LLC

Keywords: Graph neural networks, oversmoothing, oversquashing, Weisfeiler-Lehman test, billion-edge graphs

22.1 Introduction

Graph Neural Networks (GNNs) are a family of deep neural networks trained and optimized with graphs. Whereas the traditional neural networks operate on normal Euclidean data (images, sequences, etc.), GNNs are proven to capture the complicated patterns and interconnectivities between objects in the form of nodes and edges of a graph. This makes them very applicable to areas where data are originally relational, e.g., social networks, biological systems, traffic networks, and knowledge graphs. Core GNNs learn expressive node, edge or graph-level representations by cumulatively combining and transforming information about the neighbors of a node. This mechanism, which is commonly called the message passing, enables the network to capture both the local and global patterns of the structural layout within the graph.

There exist numerous architectures of GNNs: Graph Convolutional Networks (GCNs), Graph Attention Networks (GATs), GraphSAGE, Graph Isomorphism Networks (GINs), etc., each of them possessing different properties and suited to different tasks. GNNs have come to state-of-the-art performance over the last few years across a large number of applications, node classification, link prediction, graph classification, and clustering. Their flexibility, scalability, and capability of modeling irregular data mean that GNNs are quickly being recognized as a core technology in the study of AI and, through this, are driving innovations in science, engineering, and practical systems.

22.1.1 Background and Evolution of Graph Neural Networks

The field of Graph Neural Networks (GNNs) marks an improvement in developing deep learning models for data structured as graphs (non-Euclidean). Neural networks trained with grids such as images or sequences, are limited to that type of data, while GNNs can learn both the structure and the data of graphs directly. GNNs are now important once more in social network analysis, recommendation, computational biology, chemistry, traffic prediction and knowledge graphs [1]. Early 2000s research on recursive networks and graph representation was the starting point for GNN. True advancements began when Mozilla introduced their ICBM machine and then this field really took off when Kipf and Welling published their work on Graph Convolutional Networks (GCNs) in 2016. With GCN we can perform spectral filtering on graphs and this set the stage for a host of Graph Attention Networks, GraphSAGE, Graph Isomorphism Networks (GINs) and other variants. These designs (i) made GNNs more capable and (ii) easily connected to pre-existing GNNs, helping to expand their use on real problems. Among the elements seen in GNNs today are all of these: attention mechanisms, hierarchical pooling, message passing and transformer structures. Even though these advantages have sparked new ideas for GNNs [2], it is still difficult for most of these models to identify global characteristics of graphs because they collect information from just the nearby sections of a graph [3]. Besides, over-smoothing and squashing of residuals for auto regression and the difficulties with deep graph neural networks encourage scientists to invent new solutions and techniques.

22.1.2 Role of GNNs in Modern AI Applications

Right now, GNNs are most often used to apply AI to data that has a specific organization. GNNs in this area can show the trend of information spreading, create user groups and identify common patterns in user behavior. Beyond bipartite and heterogeneous graphs, they make it simple to look at user-item interactions in different ways. In science, GNNs predict what molecules can do, find drugs that may work well, and model the interaction between proteins. Modeling different aspects of road networks and current events on them lets them predict traffic and monitor transport systems by applying GNNs. Besides, GNNs have been used for reasoning in knowledge graphs, recognizing the roles of words in sentences and classifying documents within the field of natural language processing (NLP) [4]. GNNs work even better when used with reinforcement learning, generative models and causal inference. For example, in multi-agent situations, GNNs assist everybody by promoting team spirit and guiding each person toward a better strategy using their connections. They can work with various kinds of data and accommodate changes, which is important for genuine AI applications.

22.1.3 Research Gaps and Opportunities

Many existing GNNs rely on being equivalent to the Weisfeiler-Lehman test which keeps them from seeing all the differences among graphs. There exists a research space to make GNNs capable of detecting single motifs or subgraphs which can help them circumvent these limits [5]. If GNNs are made too complex, their node embeddings look almost the same (over-smoothing) and fail to notice relationships beyond a few connections (over-squashing). By using new architecture designs, residual connections and common regularization approaches, the issues can be solved. Bringing Data to Extreme Scales: Most approaches cannot be applied to graphs that consist of billions of elements. Sampling, use of distributed computing and managing graph sparsity all present important opportunities for further improvement.

Majority of GNN models run with the assumption of a fixed and unchanging graph structure, but this is not realistic in most cases. To process changing data well, scientists must focus on temporal graph learning, modeling events and memory-augmented GNNs. An important issue is that labeled data for graph supervised learning is hard to get and costly. Acquiring and labeling data can be time-consuming, but new techniques allow us to reduce this need. There is important work to be done in law, education, agriculture and climate science with GNNs. Researchers could in the future design GNNs to suit the structural and semantic traits of specific domains. Since GNNs play an increasing role in important uses, ensuring fairness and easy understanding is especially significant. It means creating models that can stand up to attacks and also help prevent bias from spreading. It is the aim of this chapter to focus on these drawbacks in the following sections, preparing for the next phase of GNN research.

22.2 Core Research Challenges in GNNs

The scalability of Graph Neural Networks (GNNs) remains a serious challenge when such models are deployed to modern, large-scale industrial areas. The applications of today (e.g. social media analysis, molecular simulations, recommendation system) require systems that can handle graphs containing millions or even billions of nodes and edges [6]. Despite the immense conceptual abilities of GNNs, they are often subjected to performance drawbacks in those circumstances, owing to their memory-efficiency and extensive computational requirements.

To overcome such limitations, the conventional mitigation techniques, including GraphSAGE, neighborhood sampling, and mini-batched training have been introduced. However, they tend to come with tradeoffs, neighborhood sampling within example compromises structural information, GraphSAGE and (most forms of) mini-batch training limit model expressivity. The search of scalable and, at the same time, accurate GNN architectures is, therefore, a burgeoning research agenda.

Researchers ought to thus to focus their attention to designing memory efficient and light weight architectures. The sparsity-awareness of matrix operations, low-rank model of the adjacency matrices, and selective message passing approaches have been demonstrated to be able to significantly invoke the computational load. Furthermore, scalability can also be enhanced in the form of on-the-fly neighborhood sampling and parameter-efficient learning that do not compromise performance.

Similar development of value in other aspects provides new directions of scale breakthrough. The distributed GNN models in which computations are distributed to several nodes assist in parallel training and allow efficient use of resources. Federated GNNs, in their turn, offer a framework in which nodes are in a heterogeneous setting (e.g. are mobile devices or different institutional domains) and can cooperate, making such frameworks very appealing to sensitive application needs in the areas of finance and healthcare.

Additional practices are graphs compression and models quantization. Edge pruning, coarsening, and subgraph clustering compression methods reduce the size of the graph, but maintain prominent topological information, reducing graph memory footprint. Deployment on resource-limited (or edge) platforms can be achieved by quantization which sacrifices precision of weights and activations without significantly sacrificing overall performance.

The conclusion to this is that the deployment of GNNs at scale demands a concerting effort that combines algorithmic optimizations and system-level optimizations. These activities not only expand the potential area of applications of GNNs but also allow applying them to industry and science, which can be characterized by a high level of impact.

22.2.1 Noisy, Sparse, and Incomplete Data Learning

Real-world graphs are often polluted with noise and under sampled and in-fully observed and thus present sophisticated challenges to the data-driven methods [7]. Unlike well-curated datasets typically used in academic benchmarks, industrial and practical graph data are frequently noisy, sparse, and incomplete. Social networks may have missing links or unreliable relationships, recommendation systems often suffer from cold-start and data sparsity issues, and biological networks frequently contain uncertain or approximate associations. These imperfections significantly affect the performance and generalizability of Graph Neural Networks (GNNs), making this a critical area for future research.

22.2.1.1 Challenges in Noisy and Sparse Graph Learning

GNNs rely heavily on the assumption that graph structure and node features contain meaningful, high-quality information. However, in noisy graphs, spurious edges or incorrect node attributes can mislead the message-passing process, propagating errors across the network and reducing model accuracy. Similarly, sparse graphs with very few connections or low feature richness undermine the effectiveness of aggregation mechanisms and limit the model's ability to generalize.

These challenges are especially pronounced in low-resource domains, long-tail applications, and new users/items in recommender systems, where information is naturally limited. Furthermore, incomplete graphs, where critical nodes, edges, or attributes are missing, introduce uncertainty and reduce the expressive power of GNNs, particularly in downstream tasks like node classification or link prediction.

22.2.1.2 Future Research Directions

The common use of graph neural networks (GNNs) is currently limited by the amount of noisy, sparse and incomplete data sets. Future scholars are therefore advised to advance the study in the following strategic directions in order to mollify the above limitations:

1. **Stable GNN Architectures**: It is always vital to develop GNNs, which have inbuilt noise resistance. Robust aggregation (e.g., median based or attention-weighted schemes), adversarially training, and noise- injection regularization are some of the techniques that help to avoid significant performance degradation in the presence of corrupted data.
2. **Graph Preprocessing and Denoising**: Pretreatments that can align the neighbors (graph smoothing), renormalize the edges (edge correction) or remove noise before GNN training can improve the quality of the model. The detection and reduction of high-frequency noise can be achieved with the aid of methods borrowed out of the fields of graph signal processing or spectral filtering.
3. **Missing Edges/Features**: Incomplete data require the use of algorithms that can work in such a partial observable environment. The reconstruction of missing structural or feature values may be addressed using graph autoencoders, variational techniques, and generative adversarial networks

(GANs), but the explicit consideration of input-graph uncertainty is a new promising direction of research.

4. **Self-Supervised and Contrastive Learning**: In some settings where labeled data is limited, self-supervised learning (SSL) and contrastive learning have proven to be quite effective. The techniques carry out useful representations at node-level or graph-level by getting advantage of inherent structural similarities and lessen the reliance on dense supervision and flawless connectivity of graphs.
5. **Bayesian GNNs and Probabilistic GNNs**: Introduction of probabilistic reasoning into GNNs allows uncertainty modeling of data as well as making predictions. Bayesian GNNs, or any version with uncertainty quantification, are well suited to address this ambiguity and do so in providing calibrated results an aspect that is of critical importance in safety-sensitive applications.
6. **Meta-Learning and Transfer Learning**: To mitigate the problem of limited data, meta-learning could allow GNNs to quickly transfer between tasks with little or no supervision, and transfer learning performed on high-data regions to low-data or incomplete environments could allow generalizing.

In short, noisy, sparse and incomplete data is a real barrier on the way of the widespread application of GNNs. The mitigation process requires multi-layered research programmatic with strong modeling probabilistic reasoning, generative reconstruction, noisy data preprocessing and self-supervised contrastive learning [8]. Develop requests in these direction will not only make GNNs very strong in perfect conditions, but also robust, changing, and reliable in realistic settings marked by uncertainty and imperfection.

22.2.2 Dynamic, Streaming and Temporal Graph Processing

The structure of graphs is usually dynamic and continuously changes with time in most real life scenarios. Examples are in social networks where there is daily ebb and flow in the interactions of the users, financial transaction, etc. The networks with the dynamic growth of edges (transactions), or the IoT communication graphs [9]. In which a mobile dynamic interconnection and disconnection of devices is possible. The original Graph Neural Networks (GNNs), Static graph inputs oriented models find it difficult in adapting to such temporally dynamic data. Due to this, study toward dynamic, streaming and temporal graph processing has become an important trend in facilitating GNNs to work in dynamically changing environments. In the real world, many of these graphs are not fixed but develop with time due to dynamic aspects, the character of actions, affairs and affairs. This lead to two broad categories of time-sensitive graph structures: streaming graphs and temporal graphs. Awareness of these paradigms is important to facilitate a development of Graph Neural Networks (GNNs) with useful time-sensitive behavior modeling. Perform real-time forecasting, and are capable of running in scale.

22.2.2.1 Understanding Temporal Graphs and Streaming Graphs

A graph whose existence of edges and even nodes is conditional upon time is called a temporal graph. At certain moment, each edge can be labeled with a timestamp heralding the time when a relationship or a contact happened. Through these graphs, the structures can be captured besides the structural information. Temporal change and so the algorithms of learning are able to identify pattern like the recency and frequency of something and sequencing of relationships [10]. As an example, in social networks, the order of the messages that is discussed has been transferred by means of a series of messages. Interactions amongst users may reflect the changing social relationships; and trades in the financial system could be used as a trace changing behaviors with time. Task areas involving use of temporal graph include temporal link prediction. Dynamic community detection, anomaly detection, and influence propagation that all need a subtle interpretation of time-coded relations. Streaming graphs, on the other hand, are a more operational view of graph data, in that streaming graphs is sent one node at a time, one edge at a time as opposed to being presented as a whole beforehand [11].

Unlike temporal graphs, which may be processed in time-stamped batches or intervals, streaming graphs demand real-time or near-real-time updates to both the graph structure and the model's internal state. In this paradigm, the learning system must handle frequent and potentially unbounded changes, such as new users joining a network, new relationships forming, or existing connections disappearing. Applications in cybersecurity (e.g., monitoring network traffic for intrusions), recommendation systems (e.g., updating preferences in real-time), and IoT sensor networks rely heavily on streaming graph processing. Table 22.1 shows the comparison of static, temporal, and streaming graphs with use cases.

In both temporal and streaming contexts, the assumption of a fixed and fully known graph topology, which underpins most traditional GNN training paradigms, becomes invalid and often computationally infeasible.

Table 22.1 Comparison of static, temporal, and streaming graphs with use cases.

Aspect	Static graphs	Temporal graphs	Streaming graphs
Definition	Graphs with fixed topology; nodes and edges remain unchanged over time.	Graphs with time-stamped edges (and possibly nodes), capturing evolution over time.	Continuously evolving graphs where data (edges/nodes) arrives as a stream.
Graph Structure	Static and complete during training and inference.	Structure changes over time; time is embedded in the data.	Structure is incrementally updated; graph is never fully known at once.
Data Access	Full graph available upfront.	Data available in historical sequences or time windows.	Data arrives continuously; system reacts to new updates in real time.
Model Update Strategy	Trained once on fixed data; updated periodically with retraining.	Often updated in batches or time-aware iterations.	Requires incremental updates and low-latency processing.
Key Challenges	Scalability, overfitting.	Temporal dependency modeling, long-term memory.	Real-time processing, memory efficiency, catastrophic forgetting.

(*Continued*)

Table 22.1 Comparison of static, temporal, and streaming graphs with use cases. (*Continued*)

Aspect	Static graphs	Temporal graphs	Streaming graphs
GNN Architectures Used	GCN, GAT, GraphSAGE.	TGAT, DyRep, EvolveGCN, Temporal GAT.	TGN, JODIE, real-time event-based GNNs.
Primary Use Cases	- Citation networks- Molecular property prediction- Knowledge graphs.	- Social network evolution- Influence propagation- User behavior analysis.	- Intrusion detection- Real-time fraud detection- Dynamic recommendation systems.

Conventional batch-processing GNNs suffer from performance degradation due to their inability to update representations incrementally and their reliance on outdated or incomplete context [12]. Moreover, maintaining global synchronization and consistency of embeddings becomes impractical when the graph is constantly evolving. This requires fundamental change in the design of GNN architecture that goes beyond awareness of time to elastic, lightweight models that learn in real time.

Thus, both the temporal and streaming graphs belong not only to a richer representation of the data but also they belong to an entirely different model of computing. They provoke the researchers to reconsider everything: the aspect of data availability, training schedule, designing that stores memory, and development of representation learning. Overcoming them will be the key to unlocking a new world of possibilities in applications such as finance and cybersecurity, healthcare and autonomous systems, and so on, by extending GNNs into real-time, dynamic, high-frequency domains. Table 22.2 provides information regarding the static graph datasets.

Highlights:

- Static Graphs are ideal for domains where the structure is relatively stable, such as citation networks or chemical molecules.
- Temporal Graphs are useful when historical evolution matters, such as social interactions or transaction timelines.

Table 22.2 Static graph datasets.

Dataset name	Domain	Description	Common tasks
Cora	Citation networks	Scientific publications as nodes; edges represent citations	Node classification
PubMed	Biomedical citations	Articles on diabetes, with features and citation links	Node classification
PROTEINS	Bioinformatics	Graphs of proteins with nodes as amino acids	Graph classification
ENZYMES	Biochemistry	Graphs representing enzyme structures	Graph classification
OGBN-arXiv	Large-scale citation	Papers from arXiv; OGB benchmark with node features and labels	Node classification (OGB dataset)

- Streaming Graphs are essential for real-time systems where immediate action is needed, such as anomaly detection in financial or IoT networks.

22.2.2.2 *Challenges in Dynamic Graph Learning*

Graph Neural Networks (GNNs) have demonstrated powerful capabilities on static graphs; however, many real-world applications involve graphs that evolve over time. These dynamic or streaming graphs arise in domains such as social networks, financial transaction systems, cybersecurity, and recommendation engines [13]. Learning over such temporal structures presents unique and complex challenges, primarily because graph topology, node attributes, and edge relationships are constantly in flux. Capturing these changes in a meaningful and efficient way is central to the development of robust dynamic GNN architectures. Table 22.3 provides temporal graph datasets information.

One of the primary challenges is temporal dependency modeling. Unlike static graphs, where the structure is fixed, dynamic graphs require the model to capture time-sensitive patterns such as the frequency, recency, and duration of interactions [14]. An example of such anomalous use cases is the detection

Table 22.3 Temporal graph datasets.

Dataset name	Domain	Description	Common tasks
Reddit (TGB)	Social media	User-post interactions with timestamps (used in Temporal Graph Benchmark)	Temporal link prediction
Wikipedia	Collaborative editing	Users editing pages over time, forming dynamic bipartite graphs	Temporal link prediction
MOOC	Online education	Student-content interactions over time	Node classification, link prediction
Enron Emails	Communication	Email exchanges with time annotations	Temporal community detection
DBLP Temporal	Academic collaboration	Author collaboration network over time	Dynamic graph modeling

of fraud where a series of high valued transactions may show a burst activity, whereas in case of social networks, the very recent activities are weighed more than the older activities. It is an open research question how to design architectures that can efficiently code these temporal patterns and maintain the dynamics of the relationships developed with the various time-points.

Real-time and low-latency processing is another point that is important. In many applications, e.g. use as intrusion detection systems or recommendation engines, the predictions need to be made virtually immediately as new nodes or edges come in. This requires GNNs with the ability to conduct incremental updates on embeddings, predictions without re-training a model. Current batch-based training methods are inadequate under such constraints and are required specialized models with event driven or streaming mechanisms to stay responsive. Table 22.4 shows information regarding streaming graph datasets.

Dynamically learning graphs also generate enormous storage and computational bottlenecks. Since the size of graphs are increasing with time (due to the amassing of interactions/nodes and attributes), models need

Table 22.4 Streaming graph datasets.

Dataset name	Domain	Description	Common tasks
UCI Messages	Online forum	Time-stamped messages between users forming a dynamic graph	Node classification, anomaly detection
Elliptic	Blockchain	Bitcoin transactions with temporal labels for illicit vs. licit activities	Streaming fraud detection
Yahoo Finance Stream	Financial networks	Continuous trade records of assets over time	Streaming event prediction
Twitch (Social Evolution)	Social media	Follow relationships and user activities in Twitch streamers	Edge classification, temporal link prediction
GraphStream Twitter	Social networking	Real-time tweet stream converted into follower/ mention graphs	Real-time community detection

to be able to deal with large graph scale, without essentially bogging down memory and computational capacity [15]. This difficulty is further aggravated in the need to maintain temporal histories because the storage of several versions of the graph or time-stamped edge features becomes prohibitively expensive. To that end, developing memory-efficient models having approximate or compressed representations is essential to scalability.

Lastly, the use of dynamic graphs does have the problem of catastrophic forgetting. With changing structure and characteristics, models will be more likely to accommodate newer modifications and most times at the cost of what had been previously learned. This is especially awkward in systems where one is interested in long-term monitoring (e.g. network security systems or systems to analyze customer behavior) as trends created in the past contain useful predictive information. To capture such need, a combination of memory processes, time-selective attention mechanisms or replay processes that maintains long-term connections but are flexible at the same time is needed. Table 22.5 summarizes key challenges in dynamic graph learning.

Table 22.5 Summary of key challenges in dynamic graph learning.

Challenge	Description	Examples	Research focus
Temporal Dependency Modeling	Capturing recency, frequency, and order of interactions	Fraud detection, social activity prediction	Time-aware message passing, temporal graph encoding
Real-Time Processing	Fast updates and predictions in streaming environments	Intrusion detection, live recommendations	Event-based GNNs, continual learning, lightweight updates
Storage and Computation	Efficient handling of continuously growing graphs with historical data retention	IoT networks, financial transaction graphs	Memory optimization, dynamic graph compression
Catastrophic Forgetting	Loss of previously learned patterns as graphs evolve	User modeling, security monitoring	Long-term memory, attention-based history retention

These challenges are the range of issues that are crucial to their possible solution to enable GNNs to enter real-world applications. Dynamics are not incidental but they are the focus of the problem [16]. Further studies will have to weigh in between computational efficiency; structural complexity, temporal faithfulness and, at the same time, learned representations do not become obsolete. This is because of the need to develop dynamisms in GNNs that can be agile. Resilient will provide a massive expansion of the functionality and depth of graph-based deep learning.

22.2.2.3 *Emerging Research Directions*

To overcome the above difficulties, it is investigated some attractive research directions:

- Temporal GNN Architectures: Time-temporal-based architecture, e.g. TGAT (Temporal Graph Attention Network), TGN and DyRep and (Temporal Graph Network) are intentionally developed to work with temporal graphs. By integrating memory components and time encoding components. These models could be a topic of future work in order to make them more interpretable and efficient and project them to multi-scale time patterns.
- Incremental and Online Learning: The sessions covering the full GNN are to be repeated when new information comes instead of retraining the whole GNN. Future studies need to rely on the creation of increments methods of learning that permit the insertions of updates in real-time.
- Adaptive learning: Online GNN might achieve adaptive learning in settings where graph structure and labels change fast.
- Graph Memory Networks: Adding some peripheral memory modules to store past communications or temporal motifs may enable models to recall proper recent events and minimize the loss of information in the process time. Time efficient architecture design is one of the essential directions.
- Continuous-Time Representation Learning: Several dynamic GNNs sample time to snapshots, which could miss fine changes on time. Dynamic GNNs in continuous time space, which may be instantiated *via* time-dependent message passing and neural ODEs, are more accurate and temporal (their abilities).
- Edge/Event-Level Prediction: Beyond the node-level delay perceiving, temporal GNNs ought to be trained on a task where the knowledge of the impending interaction/the time of edge formation could be predicted, or the evolution of communities could be modeled over time. Such tasks demand subtle modeling of temporal dependencies and may have as a helpful side-effect the introduction of recurrent or transformer-based sequence encoders.

- Streaming Graph Compression and Sampling: In the case of streaming there can be optimal sampling and summarizing that can be applied to reduce memory consumption and processing overheads. Methods which retain important temporal elements and remove noise will gain greater significance.

Dynamic, streaming, and temporal graph processing is an indispensable area of future research in GNNs, especially as applications demand adaptability to real-world, ever-evolving data [17]. Progress in this domain will require a fusion of temporal modeling, memory-efficient computation, and real-time learning mechanisms. By advancing methods that can not only understand static topology but also learn from the fluid nature of real-world interactions, GNNs can be extended to a wide range of high-impact, time-sensitive applications from cybersecurity and fraud detection to recommendation engines and real-time analytics.

22.2.3 Graph Construction and Topology Learning

The success of Graph Neural Networks (GNNs) critically depends on the quality and fidelity of the input graph structure. Most existing GNN models assume a predefined and static graph as input, where node connections (edges) are either observed or constructed using domain knowledge or heuristics. However, in many real-world scenarios, the underlying graph structure may be incomplete, inaccurate, or entirely absent. This makes graph construction and topology learning a pivotal and under-explored research frontier in GNN development.

22.2.3.1 The Challenge of Graph Quality

In domains such as recommendation systems, healthcare analytics, cybersecurity, and NLP, relationships among entities (nodes) are often implicit rather than explicitly defined. Relying on a noisy or improperly constructed graph can degrade the effectiveness of GNN models [18]. For instance, missing edges may hinder information propagation, while spurious connections can introduce noise and mislead the learning process. This necessitates developing principled approaches to learn the graph topology jointly with model training or to refine existing graphs during the learning process.

22.2.3.2 Future Research Directions in Graph Construction and Topology Learning

To address these limitations, several emerging research paths are gaining traction:

a. **Data-Driven Graph Construction**: Rather than depending on static rule-based or heuristic methods, data-driven approaches can infer graph structure directly from raw data, such as text, images, tabular features, or time-series signals. Techniques such as similarity-based methods (e.g., k-nearest neighbors), co-occurrence statistics, or learned metrics *via* neural attention mechanisms can automatically determine edge connectivity. Future work may focus on learning adaptive, context-sensitive edge weights that vary across tasks or domains.
b. **Joint Learning of Graph Structure and Node Representations**: Recent advances propose jointly optimizing both the graph topology and node embeddings during training. These methods iteratively update the graph structure based on feedback from downstream tasks (e.g., classification, link prediction). Such structure learning GNNs adaptively modify the adjacency matrix, enhancing model flexibility and robustness [19]. This paradigm shift opens new possibilities for GNNs to operate in unstructured environments.
c. **Differentiable Graph Structure Learning**: Incorporating graph construction into an end-to-end trainable pipeline is a promising research direction. Differentiable frameworks allow the optimization of edge existence and weights using gradient-based techniques. Continuous relaxation (examples) Gumbel-Softmax. It has demonstrated that adjacency matrices, and bilevel optimization have some promise in learning graph structuring without losing training efficiency.
d. **Implicit Graphs and Latent Spaces Based Learning**: A graph may be latent implied, as opposed to being directly observed in some applications. For instance: A latent interaction graph may take the form of the user preferences in recommendation systems. Future a research problem should be on the development of models that could be

used to discover and exploit these latent, index embedding structures through latent variable models, graph variational autoencoders (VAEs) or graphical models, probabilistic.

e. **Dynamic and Task-Specific Topology Learning**: Graph topology is not always fixed over time or work tasks. The dynamic models can be worked on, change the structure of the graph according to demands of tasks, or changes in time, may be of great value, enhance performance and generalization. As an example, building the various topologies of when it does classification, regression or detect anomalies on same dataset, this is a promising frontier.

f. **Domain Constrained Topology Learning**: In risk-sensitive fields of application such as healthcare or finance, domain-specific constraints (e.g., causality, regulatory logic) shall be observed when constructing the graph. Integrating prior knowledge or limitations into topology learning perhaps through rule based regularization symbolic learning provides a chance to develop more trustworthy and interpretable GNNs.

Drawing graphs and topology acquisition are primary elements of GNN performance, especially in areas where the graph is poor, incomplete or absent [20]. In the future work, it is necessary to focus on pursue the excellence of adaptive, differentiable, and domain-aware graph construction techniques. Such breakthroughs will enable GNNs to acquire meaningful relational configurations. Improving their capability of generalization in complex, unstructured or changing data sets. As that this subject develops in the future, we expect that topology learning will become a key module of the next-generation GNN frameworks, and hence increasing their appropriateness and dependability in the diversification of fields.

22.3 Emerging Application Domains for Graph Neural Networks

Graph Neural Networks (GNNs) are changing the face of artificial intelligence by which they have extended structured, relational information to deep learning. The situation is different in established health disasters that have been reversed considerably. Computer vision, natural language processing and bioinformatics, GNNs are finding their way to a broad range of emerging, and cross-disciplinary applications [21]. The forces behind these new applications are driven by the discovering the extent to which numerous real world problems within the industries to sciences to the society can be inherently solved. That are represented as graphs, and in which the entities engaged in their interaction occupy complex, non-Euclidean spaces. This body part examines ten of those rising areas in which GNNs are proving supremely promising.

The industrial automation and smart manufacturing use digital twins as virtual representation of physical ones and underpinning communications, as times and places of production become ever more systemized [22]. Such are systems consisting of dynamic networks, interacting machines, tools and sensors, which can be modeled adequately, using GNNs. GNNs are able to use these well-organized graphs of machine interaction to educate themselves. Predictive maintenance, optimizing of processes and detection of faults thereby augmenting productivity and increasing the rate of Industry 4.0 downtimes.

The functional and structural connectivity of the brain can be put in terms of graphs in the neuroscience field. The neuroimaging data like, fMRI and EEG. It is possible to analyze these brain graphs with GNNs. Discover tendencies regarding neurological disorders, thought processes, or even mental conditions. They are also able to vary between the patients and scales because of their capability of processing graphs with different topology. Contributing to precision neuroscience and brain-computer interfaces due to its contribution of new instruments.

The legal field that has always been dependent on literal interpretations and cross references by hand is also challenged, which have been transformed through GNNs. Interlinked rich graphs are of legal documents, cases, statutes and arguments. GNNs can model the citation network, the structure of arguments and the semantic relationship, judgment prediction, legal document classification and legal question answering. This encourages the establishment of the AI-assisted legal systems that can overcome the complicated judicial system.

The other promising front is education. Within intelligent tutoring systems, the students engage with material, learners, and evaluations in such a manner that they construct graphical structures of groupings of learners and content interactions. GNNs can Model to track learners progress, recommend items in form of content and learning gaps these interactions. They have the capacity of capturing the linkages between concepts and behaviors of learners improves the versatility of the digital learning environments.

Within the context of space science and astronomy, the scientists handle spatial data on stellar bodies Galaxies, stars, and planetary systems, were some of these. These objects create naturally occurring graphs as per the following, Gravitational and geographical relations. Celestial bodies GNNs are being used to classify interstellar encounters, and substantially speed up the analysis of telescope and simulation data, thus helping to astrophysical automated discovery.

Supply chain and the logistics business include very graph-like problems too multi-level supply-chain relations among suppliers, manufacturers, and distributors. GNNs will be useful in modeling such networks in real time to assess risk, optimize the inventory and the supply chain the ability to persevere in cases of uncertain situations like geopolitical instability or pandemic. They enable relational thinking that cannot be well served by traditional ML models.

One more area of high applicability is fake news classification and social content control networks. GNNs are capable of capturing spread of information, user credibility and social influence across platforms. GNNs are able to find misinformation by examining user-post interactions graphs, coordinated inauthentic behavior, bot activity, and campaigns, and scalable solutions to the digital issue platform governance [23].

GNNs are transforming the field of quantum chemistry and materials science with regards to the way by which molecules and materials are constructed. Molecular graphs are formed by atoms and bonds that can be trained on learning quantum properties by GNNs model chemical reactions, forecast chemical reactivity, and establish new compounds with the desired properties. This allows computing the problem dramatically cheaper than with classical quantum mechanical calculations and is helping to hasten advancements in areas such as drug design, battery development and photonic materials.

The graphs structures form naturally in blockchain and decentralized finance systems as well. Wallets, transactions, and smart contracts can be formulated in a form of node and edges. GNNs are more and more applied to detection of fraud, network forensics, and the analysis of vulnerabilities of smart contracts in decentralized systems. They enable transparency whereas their capacity to simulate both time-based and multi-relational transaction graphs improves transparency, believe in the Web3 ecosystems. Lastly, in digital humanities and cultural analytics, GNNs are massively used to analyze literature corpus, historical records and culture artifacts. The books and writers are interlinked with citation thematic similarities, graphs or historical references. Narrative structure analysis can be assisted with GNNs, author attribution, historical network reconstruction, and provide quantitative means of answering new questions, particularly humanities research.

The wider consideration of new fields of GNN use emphasizes the expanding application of the GNN to areas other than stated above. Contemporary data science has graph-based learning. Many spheres place more and more emphasis on being relational. GNNs will also be valuable to install in all fields since thanks to their variance, they will have an intelligence, interpretable, and context-sensitive solutions which go far beyond conventional AI methods.

22.3.1 Future Research Issues Caused by Emerging Uses of GNNs

Since Graph Neural Networks (GNNs) are growing to encompass new and cross-cutting fields, they are faced with novel set of problems which extend beyond the conventional graph learning issues. These issues are so entrenched in the issues of concept-specific facts, computational limitations real-time needs, interpretability needs, the desire to be trustworthy, and ethical standard-compliant. It is vital to work over the issues raised to guarantee the safe and efficient deployment of GNNs high-impact settings like care, law, industry and safety. How to come up with meaningful graphs in fields where the endeavor to do so is very basic is one of the most fundamental situations. Structure of relations is not easily accessible is shifting in nature. For example, in digital twins or smart manufacturing, interactions between machine components may change over time. Similarly, in education systems, learner-concept relationships evolve based on engagement and learning progress. There is a need for automated, adaptive, and context-aware graph construction algorithms that can learn the optimal topology from raw multimodal data sources. Furthermore, domain-informed representation learning that integrates physical, semantic, or symbolic constraints remains underexplored. Table 22.6 summarizes the emerging domains for GNNs.

a) Scalability to Large and Heterogeneous Graphs: Emerging applications, especially in blockchain analytics, social media platforms, and scientific simulations, involve massive-scale graphs with millions or even billions of nodes and edges [24]. These graphs are often heterogeneous, containing multiple types of nodes and relationships. Traditional GNN models struggle with such scale due to memory bottlenecks, message-passing inefficiencies, and loss of local-global balance.

Table 22.6 Summary of emerging domains for GNNs.

Domain	Graph representation	Primary applications	GNN benefits
Smart Manufacturing	Machines, sensors, and workflows as interaction graphs	Predictive maintenance, anomaly detection, optimization	Models physical system dependencies, real-time learning
Neuroscience	Neuronal regions and their connections (brain graphs)	Disorder diagnosis, cognitive mapping, BCI support	Adapts across individuals, learns complex connectivity
Legal Reasoning	Cases, statutes, citations, and arguments as legal graphs	Judgment prediction, legal search, document classification	Supports structured reasoning, interpretability
Educational Systems	Students, courses, concepts, and assessments as interaction graphs	Personalized learning, dropout prediction	Learns adaptive pathways, supports learner modeling
Astronomy	Celestial objects with spatial/gravitational relationships	Object classification, wave detection, simulation analysis	Handles sparse spatial data, enables faster inference
Supply Chain and Logistics	Multi-tier networks of suppliers and products	Bottleneck prediction, resilience modeling, inventory planning	Captures cascading dependencies, supports real-time ops
Social Media and Fake News	User-content interaction and propagation networks	Fake news detection, influence analysis, bot identification	Detects structural anomalies, tracks misinformation
Quantum Chemistry and Materials	Atoms and molecules as interaction graphs	Property prediction, molecular generation, materials discovery	Respects symmetry, models quantum interactions
Blockchain and Crypto Analytics	Wallets, transactions, and contracts as transaction graphs	Fraud detection, smart contract analysis, network tracing	Models temporal behavior, ensures traceability
Cultural Analytics	Authors, texts, and events as knowledge and citation graphs	Authorship analysis, thematic mapping, historical reconstruction	Supports semantic reasoning, cross-domain insights

Future research must focus on scalable GNN architectures, graph sampling, distributed training, and edge computing techniques tailored to real-time, resource-constrained environments.

b) Spatio-Temporal and Dynamic Graph Modeling: Many applications in IoT, traffic prediction, neuroscience, and supply chain analytics require GNNs to reason over graphs that evolve over time, sometimes rapidly and unpredictably. Current temporal GNNs often rely on static snapshots or limited time windows, missing long-term dependencies and fine-grained event transitions. Future work should address the development of continuous-time GNNs, event-driven message propagation, and temporal attention mechanisms that can model asynchronous and non-uniform dynamics in real-world graphs.

c) Interpretability and Explainability in Critical Domains: In safety-critical applications such as healthcare, law, or finance, GNNs must provide transparent and explainable outputs. Current models often act as black boxes, which hinders their acceptance in high-stakes environments where users need to understand the "why" behind a decision. Research into *post-hoc* explanation tools, intrinsically interpretable GNNs, and visualization of graph reasoning paths is essential to improve user trust and enable model debugging and validation by domain experts.

d) Privacy, Security, and Ethical Challenges: The implication of GNNs in such spheres as education, healthcare, and social networks causes a lot of concern regarding the privacy of data, abuse, and equity. As an example, malicious users may use the graph pattern to deduce sensitive interpersonal relationships or identity features. On the same note, biased graph the biasness inherent in construction (e.g. discriminatory practices in hiring networks or legal networks) would trigger the discrimination. The research should deal with privacy-preserving methods of GNN training, e.g. federated learning and differential privacy, etc. bias detection, bias reduction and bias auditing mechanisms on graph-based AI.

e) Generalization and Transferability Across Domains: Compared to other domains that have a grid structure (e.g. images or text), graphs are highly heterogeneous in structure and size semantics. This causes the problem in generalization of tasks and domains to GNNs to be particularly difficult. As an example, a system that has been trained to spot fraud in a financial net may fail to identify frauds in another net. Transfer easily into a different structure having other patterns. Developing meta-learning, domain adaptation, and task-agnostic GNNs is crucial to make graph models more robust, reusable, and accessible in real-world settings.

f) Integration with Symbolic and Physics-Based Models: In domains such as scientific discovery, smart manufacturing, and chemistry, purely data-driven models may overlook fundamental domain laws or constraints. There is a growing demand for hybrid GNN architectures that combine symbolic reasoning, first-principles physics models, or knowledge graphs with learned representations. Such integration not only improves model accuracy and physical plausibility but also offers interpretable insights aligned with scientific or engineering principles.

g) Real-Time Inference and Deployment on Edge Devices: In IoT, robotics, and embedded systems, real-time decision-making is essential. However, the computational complexity of many GNN variants makes them unsuitable for edge deployment. Future GNN research should prioritize model compression, quantization, hardware-aware design, and approximate inference techniques to facilitate deployment on devices with limited power and latency constraints.

As GNNs enter diverse real-world domains, solving these challenges will be pivotal for building intelligent, ethical, and resilient graph-based systems. Progress in these areas will not only extend the applicability of GNNs but also bridge the gap between theoretical advances and practical impact, ensuring that future GNN models are as responsible as they are powerful.

22.4 Graph Neural Networks in Healthcare and Bioinformatics

Healthcare and bioinformatics are inherently complex and data-intensive fields, characterized by heterogeneous, high-dimensional, and interrelated data sources. These domains include electronic health records (EHRs), protein-protein interaction networks, genomic data, disease ontologies, and patient similarity graphs—all of which naturally exhibit graph-like structures. Traditional machine learning techniques often struggle to effectively capture and leverage the intricate dependencies present in such data. On the contrary, Graph Neural Networks (GNNs) have cultivated as a potent solution, a model that can model these lush relational patterns. GNNs are better than learning using irregular both ordered and unstructured data inputs, and providing interfaced and interconnected data explanatory forethoughts a prerequisite to any life or death health care application [25]. Table 22.7 lists cross-domain research challenges in emerging GNN applications.

Table 22.7 Cross-domain research challenges in emerging GNN applications.

Challenge	Application examples	Research needs
Adaptive Graph Construction	Education, Digital Twins	Automated, dynamic, and semantic-aware graph creation
Scalability and Heterogeneity	Blockchain, Social Networks, Supply Chains	Distributed GNNs, graph sampling, heterogeneity handling
Spatio-Temporal Dynamics	IoT, Traffic, Neuroscience	Continuous-time GNNs, temporal attention, real-time updating
Interpretability and Trust	Law, Healthcare, Finance	Graph explanations, reasoning path tracing, user-centric interpretability
Privacy and Fairness	Social Media, Education, Finance	Federated GNNs, bias detection, fairness-preserving learning
Transfer Learning and Adaptation	Finance, Culture, Legal Systems	Domain-invariant GNNs, graph meta-learning
Physics-Integrated GNNs	Chemistry, Scientific Simulation, Manufacturing	Hybrid GNNs, symbolic fusion, physics-informed learning
Edge Deployment	Robotics, IoT, Real-time Monitoring	Lightweight GNNs, quantized inference, edge-optimized architectures

22.4.1 Motivation for GNNs in Healthcare and Bioinformatics

The necessity to implement GNNs in these areas derives from the multi-dimensional interdependence concept, the character of biomedical data. A combination of clinical outcomes of a single patient can be affected by the several following reasons: physiological measures, concurrent diseases, hereditary changes, lifestyle, and therapy responses. In bioinformatics, too, biological phenomena like gene regulatory networks, drugs and their interactions are taught. The interactions between targets, and molecular pathways are graphical in nature. Unlike conventional techniques that have the tendency of making this interconnected richness into being flat or simplistic, GNNs retain such a topology and contextual data in information data. They can model complex dependencies among biomedical entities, learn meaningful patterns from small and noisy datasets, integrate multi-modal inputs, and deliver clinically interpretable predictions making them particularly suitable for precision medicine and genomic research.

22.4.2 Applications of GNNs in Healthcare and Bioinformatics

GNNs have demonstrated immense potential across a diverse array of applications in healthcare and life sciences. Table 22.8 summarizes the key use cases.

22.4.3 Future Research Directions

Despite the transformative capabilities of GNNs in healthcare and bioinformatics, several challenges and promising directions remain:

1. **Integration of Multi-Modal and Multi-Scale Data**: Future GNN architectures must be designed to integrate diverse data types—ranging from genomic sequences and radiology scans to clinical narratives—while bridging multiple biological scales (e.g., molecular to organ-level).
2. **Dynamic and Temporal Graph Modeling**: As healthcare is inherently time-dependent, GNNs need to evolve to handle dynamic graphs that capture temporal changes in patient conditions or molecular interactions, enhancing predictions of longitudinal outcomes.
3. **Interpretability and Clinical Trust**: Clinicians require models that not only predict outcomes but also explain them. Mechanisms such as attention-based interpretability,

Table 22.8 Summary of key use cases.

Application area	Description
Drug Discovery and Drug-Target Interaction Prediction	Molecules are modeled as graphs (atoms as nodes, bonds as edges), enabling GNNs to learn rich molecular representations. These representations facilitate drug property prediction, virtual screening, and identification of potential drug-drug interactions.
Disease-Gene Association and Gene Function Prediction	By operating on gene co-expression and protein-protein interaction (PPI) networks, GNNs can infer disease associations and predict unknown gene functions, thereby advancing genomic research.
Electronic Health Records (EHR) Modeling	GNNs model temporal graphs where medical events (e.g., diagnoses, treatments, lab results) form interconnected nodes, improving the prediction of patient outcomes such as readmission, disease progression, or mortality.
Medical Imaging and Radiomics	Medical images (e.g., MRI, CT scans) can be converted into graphs where regions or lesions are nodes. GNNs can then analyze spatial relationships, improving diagnosis accuracy (e.g., tumor classification) and treatment planning.
Patient Similarity Networks and Personalized Medicine	Patients are represented as nodes in a similarity graph based on genetic, clinical, or lifestyle data. GNNs extract patterns that enable tailored diagnostics and personalized treatment recommendations.
Biomedical Knowledge Graph Reasoning	Biomedical literature and structured databases can be represented as knowledge graphs linking drugs, diseases, genes, symptoms, and pathways. GNNs are used for link prediction, hypothesis generation, and discovery of novel biomedical insights.

subgraph visualization, or causal pathways within GNNs are crucial for transparent clinical decision-making.

4. **Privacy-Preserving and Federated Learning**: Given the sensitive nature of healthcare data, there is a strong need to develop federated GNN frameworks that can learn from

distributed data silos without transferring patient-level data, thus ensuring compliance with privacy regulations.

5. **Scalability and Efficiency**: Biomedical data can scale to population-level health records or entire interactomes. Efficient GNN architectures, possibly through graph sampling, pruning, or compression, are essential to ensure scalable deployment in real-world systems.
6. **Benchmarking and Standardization**: The community would benefit from the development of standardized, domain-specific benchmarks and datasets to evaluate GNN performance consistently. It will promote reproducibility, speed up innovation, and enable the comparisons of models in a convenient way.

Graph Neural Networks have already provided an opportunity to explore unexplored spaces in healthcare and bioinformatics and provide models that capture the actual depth and richness of biological systems and patient care pathways. GNNs allow more precise, reliable, and comprehensible results, all the way at the grapheme/bimolecular levels of drug discovery to the whole-system level of health projections, through the use of graph structure implicit in biomedical data [26]. Yet, to achieve their full potential, directions looking ahead will center on issues such as interpretability, data privacy and scaling architectures to the size and diversity levels of real-world healthcare data. As more studies are realized at the border between graph theory, deep learning, and clinical science, GNNs will play a central role in the coming world of biomedical informatics and precision healthcare.

22.5 Graph Neural Networks in Cybersecurity

The cybersecurity field is rapidly developing and problematic in the sense that it is faced with dynamic complicated high-dimensional tasks, interconnection between data sources, including log files network traffic, user patterns of behavior, system logs, etc. threat intelligence. Such data types are strictly structured and interdepended. They are perfect candidates of graph based modeling. Conventional machine learning techniques tend to fail deficient in recording these subtle relationships because they are based on fixed-format, tabular, or sequential inputs. On the other hand, Graph Neural Networks (GNNs) are learned with a specific purpose to learn graph-related tasks [27]. Based on structured non-Euclidean data, thus they are perfectly suited to the contemporary cybersecurity approach challenges. Representing something as a user, device, file, process, and a network connection etc. GNNs can learn contextual dependencies by the way of representing graphs as nodes and their connection as edges, abnormalities and detect threat trends with the depth of knowledge that is impossible to attain otherwise conventional methods.

22.5.1 Why GNNs for Cybersecurity?

There is also an implicit relation reasoning in cybersecurity, wherein in identifying threat there is foreseeability on the basis of relation, on the interaction between various entities in the course of time. Activities such as detection of insider threats, surveillance of lateral the propagation of a malware or the structure of a network, or a study of an attack is by extension a graph problems.

This area offers clear benefits by use of GNNs:

- Multi-hop dependency capture: Because of the dependency modeling property, the GNNs can approximate the indirect-relationships (e.g., across) hidden chains of threats may be unveiled through multiple hosts or users).
- Adaptability to dynamic network topologies: Modern networks are constantly evolving; GNNs can adapt to changes in structure over time.
- Contextual behavioral understanding: GNNs leverage the local and global graph structure to distinguish between legitimate and malicious behavior.
- Scalability: GNNs can scale to large enterprise-level networks, which include millions of nodes and edges.

22.5.2 Applications of GNNs in Cybersecurity

GNNs are increasingly being used across multiple key cybersecurity tasks. The Table 22.9 below summarizes these applications and their relevance:

22.5.3 Future Research Directions

While the utility of GNNs in cybersecurity is evident, several challenges and promising directions guide future research in this space:

Table 22.9 Various applications of GNN in cybersecurity.

Application area	Description
Intrusion and Anomaly Detection	Network entities (e.g., devices or servers) and their communications form dynamic graphs. GNNs can learn typical patterns of interaction and identify outliers suggestive of intrusions.
Malware Detection and Classification	Graphs derived from code structure (API calls, control-flow, binary relationships) are used to differentiate between malicious and benign software through learned structural representations.
Phishing and Fraud Detection	Transactional and behavioral relationships (emails, messages, clicks) can be modeled as graphs to identify suspicious clusters or patterns consistent with phishing or fraudulent activities.
User and Entity Behavior Analytics (UEBA)	User activity graphs help detect insider threats by modeling access behaviors and identifying abnormal privilege escalations or data exfiltration attempts.
Botnet and APT Detection	Botnets often form structural communication patterns. GNNs capture these recurring motifs in network flow graphs, aiding early detection of coordinated and stealthy attacks.
Threat Intelligence Graphs	Entities like vulnerabilities, threat actors, exploits, and indicators of compromise (IOCs) can be connected in knowledge graphs. GNNs are used for real-time reasoning, linking unseen threats to known malicious campaigns.

a. **Dynamic and Streaming Graph Learning:** Cyber graphs evolve continuously with each network event or threat. Developing online or incremental GNNs that learn in real time is critical to maintaining relevance in changing environments.
b. **Temporal and Spatio-Temporal GNNs:** Security events are time-sensitive. Incorporating temporal context into GNNs (e.g., time-aware message passing or recurrent graph

models) can help detect sequences like slow-moving attacks or delayed privilege escalations.

c. **Robustness Against Adversarial Attacks:** GNNs, like other ML models, are susceptible to poisoning and evasion attacks. Ensuring the robustness of GNN-based security systems through adversarial training or graph certification is a growing area of research.

d. **Explainability and Trustworthiness:** Analysts need interpretable models. GNNs must evolve to support explanations *via* subgraph highlights, node importance scores, or interpretable embedding spaces, increasing confidence in security decisions.

e. **Privacy-Preserving and Federated GNNs:** Much of security data is distributed and sensitive. Federated GNN learning allows training across multiple institutions without sharing raw data, promoting collaboration in threat intelligence without compromising confidentiality.

f. **Benchmarks and Standard Evaluation Protocols:** Progress in GNN-based cybersecurity is hindered by the lack of standardized datasets and evaluation metrics. Creating open, domain-relevant benchmarks is crucial for meaningful model assessment and reproducibility.

Graph Neural Networks are transforming the landscape of cybersecurity by enabling intelligent, adaptive, and scalable threat detection systems. Their strength lies in their ability to model the complex interdependencies that exist between cyber entities and their interactions, allowing for the detection of subtle and sophisticated attack patterns. From real-time intrusion detection to proactive threat intelligence, GNNs are making it possible to develop systems that not only respond to threats but anticipate them [28]. Nonetheless, to establish GNNs as cornerstones of operational security systems, current research is needed to establish answers to urgent questions of model robustness, interpretability, and scalability, and privacy. GNNs may be used to develop resilient, next-generation and context-aware cyber-security solutions that are currently not available. Moreover, cyber threats can continue to gain sophistication and scale, and GNNs may present a viable solution.

22.6 Graph Neural Networks in Smart Cities and IoT

The future of urban development is smart cities and in a smart city, everything is researched, planned and developed to be intelligent. The Internet of Things (IoT) links citizens in a complex way. These are cities that result in enormous the volumes of dynamic and heterogeneous data of different sources such as traffic sensors, energy meters, surveillance tools, phones and social networks. One of the major problems is how to receiving, analyzing and using this very complicated, interdependent data to make real-time decisions, effectively proactive urban services, resource management, and proactive urban services. Graph Neural Networks (GNNs) come into presence [29]. Effective solution through simulating interaction between such interdependent elements of a city. GNNs allow systems to record patterns in space, time, semantics covering the digital fabric of a city, which makes them well suited to aid the smart cities to become intelligent, sustainable and adaptive ecosystems.

22.6.1 Why GNNs for Smart Cities and IoT?

Traditional AI models often view data as isolated or sequential, ignoring the critical spatial-temporal and relational dependencies that naturally exist in urban and IoT systems. In contrast, GNNs are inherently structured to process graph-based data, where:

- Nodes represent key entities such as buildings, vehicles, sensors, people, or service hubs.
- Edges capture spatial proximity, data communication, physical connections, or interactions.

The relational structure of smart city systems enables GNNs to:

- Perform context-aware localized learning, tailoring predictions based on neighboring entities.
- Enable spatial and temporal pattern recognition, crucial for mobility, climate, or energy use modeling.
- Scale across expansive infrastructure networks, such as smart grids or transportation systems.
- Support coordinated decision-making among distributed and diverse IoT devices.

22.6.2 Applications of GNNs in Smart Cities and IoT

The flexibility of GNNs allows them to be applied across a wide range of smart city domains. The Table 22.10 below summarizes key use cases:

Table 22.10 Key use cases of GNN in smart cities and IoT.

Application area	Description
Traffic Prediction and Optimization	GNNs model urban roads as spatial graphs to predict traffic flow, optimize routing, and control signals in real time, helping reduce congestion and emissions.
Energy Grid Monitoring and Demand Forecasting	Smart energy systems are represented as power flow graphs. GNNs forecast energy demand, detect anomalies, and enable load balancing, enhancing efficiency and stability.
Environmental and Air Quality Monitoring	Networks of environmental sensors are captured as spatial graphs. GNNs help predict pollution levels, identify sensor faults, and guide air quality interventions.
Public Safety and Surveillance	Surveillance cameras and sensors form monitoring graphs. GNNs model human movement and crowd dynamics to detect suspicious activity and assist emergency response.
Smart Building Automation and Maintenance	IoT-enabled buildings produce structured interaction data among HVAC, lighting, and security systems. GNNs optimize energy consumption and detect system faults.
Waste and Resource Management	Smart bins and resource facilities create dynamic service graphs. GNNs support optimal waste collection routes and resource distribution planning.
Citizen Mobility and Social Dynamics	GNNs help model mobility graphs involving people, vehicles, and locations to analyze commuting behavior and enhance smart transportation and ride-sharing systems.

22.6.3 Future Research Directions

Despite their promise, applying GNNs to smart cities and IoT introduces domain-specific challenges and opportunities for innovation:

a. **Spatio-Temporal GNNs (ST-GNNs):** Urban systems involve spatial and temporal dependencies e.g., traffic congestion varies by hour and location. Efficient ST-GNN models are needed to jointly learn from both dimensions for real-time city analytics.
b. **Federated and Edge GNNs for IoT:** Many IoT devices have limited compute and storage capabilities. Developing lightweight GNN architectures that can run at the edge or participate in federated learning is crucial for distributed intelligence across devices.
c. **Dynamic Graph Learning:** Smart cities are constantly evolving. Roads close, sensors fail, and new infrastructure is added. GNNs must support incremental learning over dynamic graphs to remain accurate and responsive in real time.
d. **Heterogeneous and Multi-Modal Graphs:** Urban data comes from diverse modalities text (tweets), vision (CCTV), geospatial (GPS), and structured logs. Designing heterogeneous GNNs that can integrate multiple data types and relationship types is a pressing need.
e. **Interpretability and Trust:** Urban applications like traffic control or public safety require decisions that are understandable and actionable. GNNs must provide interpretable insights such as node influence, subgraph attribution, or decision rationale.
f. **Privacy, Security, and Resilience:** As smart city systems become cyber-physical, data privacy, system robustness, and resilience to cyber-attacks are essential. Future GNN research should explore secure learning protocols, differential privacy, and resilient architectures.

Graph Neural Networks offer a transformative approach to building intelligent, adaptive, and connected smart city ecosystems. By capturing the relational structure of urban environments, GNNs enable breakthroughs in traffic management, environmental monitoring, energy optimization, surveillance, and citizen services. Their ability to model real-time, dynamic, and heterogeneous data makes them a foundational technology in the era of IoT-driven urbanization. Nevertheless, to become widespread, it needs to solve these issues regarding real-time learning, computational efficiency, heterogeneity, security, and interpretability. With urban places being more digitized and interconnected, GNNs will take the lead in bringing sustainable, secure, and human-oriented cities in the future.

22.7 Graph Neural Networks in Scientific Computing and Discovery

Scientific computing is experiencing a paradigm shift towards the data-driven approaches, be complementary to conventional modeling and simulation. Institutions like physics or chemistry, biology and engineering entail systems of interacting entities a and in this case, the entities are atoms, molecules, points in space meshes or particle in dynamic simulations which can naturally be represented by a graph. They are expressed in form of graphs. The Graph Neural Networks (GNNs) provide a reliable paradigm to represent. These kinds of ordered systems and their talents in encoding physical constraints, principally relationships. and generalize across different configurations that they are especially suitable to new ones. scientific exploration and high end computing operations.

22.7.1 Why GNNs for Scientific Computing?

Scientific information tends to be relational, spatial, and temporal, which is why graph structures have potential applications of such data modeling. Although classic numerical methods, such as finite element methods and molecular dynamics simulations, are correct, they are costly to run and might not scale well in large systems or complex systems. The essential benefits of GNNs arise:

Inductive Generalization: GNNs are able to generalize in mesh resolutions, topologies, geometries, and topologies and are therefore appropriate to use to simulate new or novel topologies and so on configurations.
Physical Consistency: It is possible to use GNNs to capture essential physical properties like symmetries, conservation laws, and equivariance, to make sure models obey known scientific principles.
Accelerated Simulation: Simulation results may be estimated using surrogate-modeling by GNN on the order of magnitude faster, so they quickly fit

into scientific workflows in terms of design and optimization. The Hybrid Modeling GNNs are also able to integrate with models that are symbolic or physics-based and merge them strengths of first-principles methods combined with data-based learning.

22.7.2 Applications of GNNs in Scientific Discovery

GNNs are nowadays quickly implemented into other fields related to sciences. Major applications can be summarized as shown in the Table 22.11 below:

Table 22.11 Major applications of GNN in scientific discovery.

Application area	Description
Physics-Informed Modeling	GNNs simulate complex physical systems like fluids, deformable solids, or electromagnetic fields by modeling interactions governed by PDEs or particle-based dynamics.
Molecular and Materials Science	Molecules and materials are represented as atom-bond graphs. GNNs predict properties (e.g., toxicity, stability), simulate quantum interactions, and guide new material design.
Climate and Environmental Modeling	GNNs capture spatial and temporal dependencies in climate data (from sensors or satellites) to support tasks like weather prediction and environmental forecasting.
Biological Networks and Systems Biology	GNNs analyze biological graphs such as gene regulatory networks, protein-protein interactions, and metabolic pathways for functional analysis and intervention prediction.
Combinatorial Optimization and Theoretical Computation	GNNs approximate solutions for NP-hard problems like SAT solving, graph coloring, or molecular docking, outperforming traditional heuristics by learning patterns from instances.
Scientific Knowledge Graphs and Literature Mining	GNNs extract and reason over scientific knowledge graphs derived from research literature, aiding hypothesis generation and discovery across disciplines.

22.7.3 Future Research Directions

While GNNs show immense promise in scientific computing, the field is still emerging and presents several key research opportunities:

a. **Physics-Informed GNN Architectures:** Scientific models must be physically grounded. Future GNNs need to encode domain knowledge, such as symmetry constraints, conservation laws, and invariances. Research into equivariant GNNs, constrained message-passing, and differentiable solvers is central to this goal.
b. **Data Efficiency and Sim-to-Real Transfer:** In many domains, high-quality labeled data is scarce. Developing GNNs that support few-shot, self-supervised, or unsupervised learning will improve their usability. Techniques for **sim-to-real transfer** can help models trained on simulations adapt to real-world data.
c. **Uncertainty Quantification and Trustworthy AI:** Scientific use cases demand reliable, calibrated predictions with quantifiable uncertainty. Integration of Bayesian GNNs, ensemble models, or probabilistic message-passing is vital for risk-aware scientific decision-making.
d. **Integration with Traditional Simulators:** Hybrid frameworks that embed GNNs within classical simulators can leverage domain knowledge while accelerating computation. These physics-ML hybrids can support tasks such as design exploration, multi-objective optimization, and control.
e. **Scaling to Massive Scientific Graphs:** Many scientific datasets, such as those from genomics or climate simulations, involve graphs with billions of nodes and edges. Scalable GNN architectures using graph sparsification, sampling, or distributed training are essential for inference at scale.
f. **Automated Scientific Discovery:** The long-term vision is the development of AI systems that act as collaborative scientific agents: proposing hypotheses, designing experiments, and autonomously uncovering new principles. GNNs will take essential part in such systems by using structured reasoning and symbolic generalization.

Graph Neural Networks are turning the world of science computing and discovery on its head by allow learning in a structured data that is efficient. Whether it's modeling fluid flow, estimating the physical values of chemicals, learning about life or requiring faster combinatorial computing, GNNs give rise to a data driven perspective in which to generate and make novel discoveries science. With the field maturing, combining GNN with domain-specific knowledge, supplementing them with, is bound to be a positive step in the field [30]. The interpretability aspect, and making them reliable will be crucial in the functioning of their potentials. Ultimately, GNNs are not only a computational tool but a paradigm shift in the manner through which science is practiced in tearing down the wall between data, theory and discovery.

22.8 Case Studies in Graph Neural Networks (GNNs)

Different case studies have also shown that graph neural networks (GNNs) can give a big impact in different real-world scenarios as heterogeneous relational structures can be modeled and meaningful patterns identified in graph-structured data. GNNs have been put to successful use in healthcare and bioinformatics to predict disease-gene relationships, model similarity graphs on patients, and analyzing electronic health records to conduct early diagnosis and optimize the treatment process. In cybersecurity, the potential applications of GNNs are featured in case studies through the application of GNNs in intrusion detection, malware classification, and threat intelligence graphs analysis to facilitate proactive defense mechanisms. The application of Smart cities displays, how GNNs facilitate traffic deposit, quality of air assessment, and energy network optimization by leveling the city infrastructure as a spatio-temporal graph. GNNs have been applied to scientific computing, used to simulate physical phenomena, make predictions of chemical properties in molecular graphs and speed up material discovery [31]. There is more, financial systems and social networks employ GNNs to detect fraud, recommendation systems, and influence propagation. The combination of these different case studies highlights the range of, and flexibility of, GNN in solving more complex problems within the context of structured, dynamic and multi-modal data.

22.8.1 Smashing Insights About the Use of Graph Neural Networks in Smart Manufacturing through the Digital Twin Modeling

With the rise of digital environments in the contemporary industrial setting, smart manufacturing systems have taken on the digital fold twins virtual models of physical machines, processes or whole section of a production system. These twins are constantly filtered with live information provided by IoT sensors, physical processes that are able to simulate, monitor and optimize the occurrence in factories digitally. The complexity bottom-line the digital twin is an eligible option due to the number of interactions between machine-to-machine, machine-to-tools, and machine-to-control systems it considers to use graph-based models. Operational, failure and redundancy models can all be represented as nodes of a particular physical entity. The relationship between mechanical, or data-driven relationship can be modeled as edges within a dynamic graph.

Problem Statement

One of the top companies in the electronics manufacture sector intended to drive out unscheduled shutdowns within its surface-mount technology (SMT) assembly line by anticipating possible malfunctions of the machines in real-time. Traditional ML models failed to capture dependencies across machines and processes, leading to inaccurate or delayed predictions. The goal was to build a graph-based failure prediction model that accounts for interdependencies in the workflow, machine health indicators, and operational context.

GNN-Based Solution

To address this, the research team proposed a Graph Neural Network framework built upon the digital twin of the production line. The SMT line was represented as a graph $G = (V, E)$, where each node in V corresponds to a machine station (e.g., solder paste printer, pick-and-place machine, reflow oven), and edges in E encode data flows and process dependencies.

Each node was enriched with real-time sensor attributes such as:

- Temperature fluctuations
- Vibration levels
- Error logs and alarms
- Operational cycles
- Component placement accuracy

Edges were annotated with data transfer rates, production order information, and shared workload constraints. The team implemented a spatio-temporal GNN architecture with attention mechanisms to capture dynamic changes in both the node states and edge interactions.

The GNN model was trained on 3 months of sensor and production log data, including failure events. It was benchmarked against baseline models including LSTM (for temporal modeling) and XGBoost (for feature-based classification). The dataset was split into sequences representing complete production shifts. Table 22.12 shows the performance evaluation of various GNN Models.

Key evaluation metrics included:

- Prediction Accuracy
- Precision/Recall of Failure Events
- Lead Time Before Failure
- False Positive Rate

Results

The GNN-based system significantly outperformed baseline models:

The graph-based approach not only improved prediction accuracy but also provided interpretability by highlighting which parts of the production graph contributed to the predicted failure. For instance, abnormal heat in one node and a sharp drop in data flow through an adjacent node were flagged as critical early warning indicators.

Impact and Deployment

This model was incorporated in the already existing digital twin dashboard of the company. The operators of plants now be in a position to obtain pro-active cautions when the GNN indicates that a high likelihood of failures would occur in the near future production cycle. This has caused:

Table 22.12 Performance evaluation.

Model	Accuracy	Precision	Recall	Avg. lead time before failure
XGBoost	78.2%	70.4%	65.9%	~20 minutes
LSTM	81.6%	74.1%	68.7%	~25 minutes
Spatio-Temporal GNN	89.4%	83.7%	80.5%	~42 minutes

- A 38% reduction in unplanned downtime
- Improved maintenance scheduling efficiency
- Enhanced operator trust due to interpretability of failure causes

The GNN model has been retrained quarterly to adapt to changes in machine configurations, product types, and sensor drift—showcasing the model's adaptability.

22.8.2 Graph Neural Networks for Brain Connectivity Analysis in Neurological Disorder Classification

Hard to comprehend is the way of functioning and malfunctioning of the brain, something imminent in the neuroscience. The recent developments in neuroimaging, e.g. functional Magnetic Resonance Imaging (fMRI) and Electroencephalography (EEG) enable scientists to monitor sequences of neural activities, the connection between the various parts of the brain. Such connectivity patterns are natural presented in the form of graphs with brain areas corresponding to the nodes and functional or structural links between them. They are joined together at edges. Graph Neural Networks (GNNs) have been developed over the past years effective instrument of analyzing such brain graphs and in particular neurological disorder. One of the leads to it being diagnosed, including Alzheimer's Disease, Autism Spectrum Disorder (ASD), and epilepsy.

Problem Statement

A multi-effective neuroscience working team aimed at achieving better accuracy in early identification of Autism Spectrum. The images of disorder (ASD) will be based on resting state fMRI data. The conventional approaches of machine learning with Linear connectivity measures or handcrafted features had not been very successful since they cost high dimensionality, noise and non-linear dependencies of brain networks. This was to design indicative features of brain-connectivity structures that may be learned using graph-based models maintaining topological and functional relations.

GNN-Based Solution

The authors individually plotted the graphs of functional connectivity of each subject. ABIDE (Autism Brain Imaging Data Exchange) db. Every brain graph was made of nodes representing regions of interest (ROIs) based on brain atlases (e.g. AAL or Harvard-Oxford) and edges were determined using Pearson coefficients of time series of fMRI indications of every area.

They used Graph Attention Network (GAT) to learn node features in an adaptively-attention way weights, concentrating on maximally informative functional interactions. The last way of graph-level embedding was obtained with the help of the readout functions (global average and max pooling), and a fully linked layer binary distinction (ASD vs. Control). To enhance generalization, the model was trained with **data augmentation strategies** such as graph masking and random edge perturbation, simulating variability in real-world fMRI measurements. They also compared performance with baseline models including Support Vector Machines (SVMs), CNNs on connectivity matrices, and vanilla GCNs.

Training and Evaluation

The model was trained on a stratified 5-fold cross-validation split of the ABIDE dataset, ensuring demographic balance. Evaluation metrics included:

- Classification Accuracy
- Sensitivity (True Positive Rate)
- Specificity (True Negative Rate)
- AUC-ROC (Area Under Curve – Receiver Operating Characteristic)
- Interpretability of attended brain regions

Results

The GAT-based GNN model outperformed baseline models across all metrics:

The attention weights learned by the GNN highlighted ROIs within the default mode network (DMN) and salience network, both of which have been previously implicated in ASD. These insights aligned with existing neurobiological findings, lending interpretability and scientific credibility to the model. Table 22.13 displays the results using various performance evaluation metrics of GNN models.

Table 22.13 Performance evaluation metrics.

Model	Accuracy	Sensitivity	Specificity	AUC-ROC
SVM	72.1%	68.4%	75.2%	0.76
CNN (Matrix Input)	74.5%	69.3%	78.9%	0.79
GCN	77.8%	73.6%	80.1%	0.82
GAT (Proposed)	82.3%	80.4%	84.5%	0.87

Impact and Implications

The successful deployment of GNNs in this context showed their potential to serve as clinical decision support tools. By leveraging relational patterns in brain connectivity, GNNs provided early, accurate, and interpretable predictions of ASD essential for initiating timely interventions. Furthermore, the modularity of the graph model allows for future extensions to multi-modal data fusion, such as combining fMRI with genetic or behavioral data.

This case study illustrates that GNNs can bridge the gap between neural data complexity and clinically actionable insights, offering a scalable framework for precision neuroscience. Their use could revolutionize how we diagnose, monitor, and even personalize treatment for a range of neurological and psychiatric conditions.

22.9 Conclusion

Graph Neural Networks (GNNs) have emerged as one of the most transformative advances in modern deep learning, uniquely positioned to model data with inherent relational, structural, and non-Euclidean characteristics [32]. Their power lies in their ability to perform deep representation learning over graphs, enabling breakthroughs across a wide range of domains—from computer vision and natural language processing to cybersecurity, healthcare, finance, smart cities, and scientific discovery. Despite their success, the field of GNNs is still evolving. Numerous challenges persist in terms of scalability, interpretability, dynamic graph modeling, learning from noisy or incomplete data, and efficient graph construction. Additionally, as GNNs are applied to increasingly sensitive and high-impact areas, concerns around privacy, robustness, ethical use, and regulatory compliance become critical to address.

Future research must focus on developing Graph Neural Network (GNN) architectures that are capable of addressing the growing demands of real-world applications. These architectures need to be scalable to efficiently process massive graphs containing billions of nodes and edges, enabling their deployment in domains such as scientific computing, social networks, and urban infrastructure. They must also be adaptive to dynamic and streaming data, supporting real-time learning and continuous updates in environments where graph structures evolve rapidly, such as cybersecurity and IoT [33]. Furthermore, GNNs should be generalizable across diverse domains and graph types, including heterogeneous, temporal, and multi-modal graphs, ensuring flexibility and applicability in complex scenarios. Enhancing interpretability is equally crucial, as human-understandable explanations are essential for building trust and facilitating decision-making in high-stakes areas like healthcare and finance. Lastly, GNNs must be designed to be robust and privacy-aware, incorporating mechanisms to resist adversarial manipulation and protect sensitive information, especially in critical systems where data confidentiality and security are paramount. Furthermore, integrating GNNs with federated learning, physics-informed models, knowledge graphs, and large language models (LLMs) represents a rich frontier for creating more intelligent, explainable, and hybrid AI systems.

As researchers and practitioners, we stand at the convergence of theory, application, and innovation. GNNs offer not only a toolkit for solving complex data problems but also a new lens through which we can understand and shape the interconnected world around us. By addressing current limitations and pursuing principled, interdisciplinary research, the next generation of Graph Neural Networks will play a foundational role in powering intelligent systems and scientific breakthroughs in the years to come.

AI disclosure statement: The author acknowledges the use of AI-based tools for minor language editing or grammatical correction. However, all intellectual contributions, analyses, and conclusions presented in this chapter are entirely the author's own.

References

1. Wu, Z., Pan, S., Chen, F., Long, G., Zhang, C., Philip, S.Y., A comprehensive survey on graph neural networks. *IEEE Trans. Neural Networks Learn. Syst.*, 32, 1, 4–24, 2020.
2. Zhou, J., Cui, G., Zhang, Z., Yang, C., Liu, Z., Wang, L., Sun, M., Graph neural networks: A review of methods and applications. *AI Open*, 1, 57–81, 2020.
3. Hamilton, W.L., Ying, Z., Leskovec, J., Inductive representation learning on large graphs, in: *NeurIPS*, 2017.
4. Zitnik, M. and Leskovec, J., Predicting multicellular function through multi-layer tissue networks. *Bioinformatics*, 33, 14, i190–i198, 2017.
5. Gao, K., Nguyen, D.D., Chen, J., Wei, G.W., A review of GNNs and their applications in bioinformatics. *Briefings Bioinf.*, 23, 2, bbac113, 2022.
6. Sun, M., Zhao, S., Gilvary, C., Elemento, O., Zhou, J., Wang, F., Graph convolutional networks for computational drug development and discovery. *Briefings Bioinf.*, 22, 1, 155–171, 2019.
7. Zhang, Z., Cui, P., Zhu, W., Deep learning on graphs: A survey. *IEEE Trans. Knowl. Data Eng.*, 34, 1, 249–270, 2020.

8. Tong, Z., Wu, J., Liu, J., Liu, L., Graph-based anomaly detection and description: A survey. *ACM Comput. Surv. (CSUR)*, 54, 2, 1–37, 2020.
9. Zhang, Y., Wang, Y., Tang, J., Wang, Y., Yu, P.S., GCN-GAN: A non-linear temporal link prediction model for weighted dynamic networks, in: *CIKM*, 2020.
10. Han, K., Wang, Y., Li, Z., Zhang, Y., A survey on adversarial attacks and defenses in graph data, in: *IEEE Transactions on Knowledge and Data Engineering*, 2022.
11. Deng, Y., Bao, F., Kong, Y., Ren, Z., Dai, Q., Deep direct reinforcement learning for financial signal representation and trading. *IEEE Trans. Neural Networks Learn. Syst.*, 28, 3, 653–664, 2016.
12. Feng, F., He, X., Wang, X., Luo, C., Liu, Y., Chua, T.S., Temporal relational ranking for stock prediction. *ACM Trans. Inf. Syst.*, 38, 2, 1–39, 2019.
13. Wu, L., Li, X., Kong, L., A survey on graph neural networks and their applications in financial risk analysis. *Big Data Min. Anal.*, 5, 3, 207–223, 2022.
14. Yu, B., Yin, H., Zhu, Z., Spatio-temporal graph convolutional networks: A deep learning framework for traffic forecasting, in: *IJCAI*, 2018.
15. Li, Y., Yu, R., Shahabi, C., Liu, Y., Diffusion convolutional recurrent neural network: Data-driven traffic forecasting, in: *ICLR*, 2018.
16. Wang, Z. and Liu, X., A review of GNN applications in smart cities. *ACM Comput. Surv. (CSUR)*, 54, 9, 1–35, 2021.
17. Jin, X., Cao, J., Zhang, Y., GNNs for community detection: Advances, applications, and opportunities. *IEEE Trans. Comput. Social Syst.*, 8, 4, 936–948, 2021.
18. Nguyen, G., Nguyen, T., Nguyen, D., Phung, D., Dao, B., Venkatesh, S., Fake news detection using deep learning: A survey. *arXiv preprint arXiv:2007.12679*, 2020.
19. Wu, L., Wu, Z., Huang, Y., A survey of recommendation systems based on GNNs. *IEEE Access*, 9, 173127–173143, 2021.
20. Sanchez-Gonzalez, A., Godwin, J., Pfaff, T., Ying, R., Leskovec, J., Battaglia, P., Learning to simulate complex physics with graph networks, in: *ICML*, 2020.
21. Battaglia, P.W., Pascanu, R., Lai, M., Rezende, D.J., Kavukcuoglu, K., Interaction networks for learning about objects, relations and physics, in: *NeurIPS*, 2016.
22. Gilmer, J., Schoenholz, S.S., Riley, P.F., Vinyals, O., Dahl, G.E., Neural message passing for quantum chemistry, in: *ICML*, 2017.
23. Dey, R., A graph-based framework for policy modeling and compliance verification. *J. Inf. Syst.*, 36, 2, 123–137, 2022.
24. Yao, L., Chen, Y., Huang, Q., Zhang, J., Bi, J., GNNs in education: Modeling learning outcomes from student interaction graphs. *Comput. Educ.*, 182, 104483, 2022.
25. Liu, L., Zhang, X., Wang, C., GNN applications in digital humanities: A survey. *Digital Scholarship Humanit.*, 36, 3, 623–639, 2021.

26. Dai, H., Chen, X., Tian, Y., Song, L., Towards interpretable GNNs. *IEEE Trans. Neural Networks Learn. Syst.*, 32, 8, 3420–3433, 2021.
27. Bica, I., Ghassemi, M., Schwab, P., GNNs for medicine: Fairness, interpretability, and robustness, in: *NeurIPS Workshop on Graph Representation Learning and Beyond*, 2021.
28. Ma, X., Chen, L., Cui, P., Yu, P.S., A comprehensive survey on adversarial attacks and defenses in graph data. *IEEE Trans. Neural Networks Learn. Syst.*, 32, 8, 3445–3467, 2020.
29. Ying, R., Cai, T., Luo, C., Zheng, S., Ke, G., He, D., Liu, T.Y., Do transformers really perform badly for graph representation?, in: *NeurIPS*, 2021.
30. Liu, Y., Xia, Z., Zhang, X., Yu, P., Ding, Y., Continual learning on graphs: A framework and challenges. *arXiv preprint arXiv:2110.06687*, 2021.
31. Li, Y., Li, Y., Wang, H., Spiking GNNs for neuromorphic computing. *Nat. Mach. Intell.*, 2, 12, 678–687, 2020.
32. Ma, T., Chen, J., Xiao, C., Hardware-aware graph neural networks: A survey. *IEEE Trans. Artif. Intell.*, 3, 1, 1–19, 2022.
33. Zhang, X., Wang, C., Wang, C., Towards universal GNNs: Challenges and opportunities, in: *IEEE Transactions on Neural Networks and Learning Systems*, 2023.

Index

Also of Interest

Check out these published and forthcoming titles in the "Leading-Edge Breakthroughs in Artificial Intelligence" series from Scrivener Publishing

Affective Artificial Intelligence
Fundamentals, Challenges, and Applications
Published 2026. ISBN 978-1-394-42272-2

Platform Engineering
Concepts, Challenges and Applications
Edited by K. Suresh Kumar, S. Sundaresan, R. Prithiviraj, T. Ananth Kumar and S. Balamurugan
Published 2026. ISBN 978-1-394-39588-0

Responsible AI
Principles and Practices
Edited by Manish Kumar, Nitigya Sambyal, Leena Priyadarshini Singh, Ramasamy V. and S. Balamurugan
Published 2026. ISBN 978-1-394-35544-0

Digital Immune System
Principles and Practices
Edited by Sujata Priyambada Dash, Vaibhav Mishra, Bijeta Shaw, Sandeep Kumar Panda, and S. Balamurugan
Published 2025. ISBN 978-1-394-38375-7

Adaptive AI
Fundamentals, Challenges, and Applications
Edited by P. Pavan Kumar, G. Suresh Kumar, Ajay Kumar Jena, Sandeep Kumar Panda and S. Balamurugan
Published 2025. ISBN 978-1-394-38904-9

AI Trust, Risk and Security Management
Framework, Principles, and Practices
Edited by R. Karthick Manoj, S. Senthilnathan, S. Arunmozhi Selvi, T. Ananth Kumar, S. Balamurugan
Published 2025. ISBN 978-1-394-39299-5

Forthcoming 2026
Neuro-Symbolic AI: Concepts and Applications
Decision Intelligence: Concepts, Applications, and Case Studies
Human-Centered AI: Frameworks and Applications
Composite AI: Fundamentals, Challenges, and Applications
Visual AI: Fundamentals, Challenges, and Applications
Causal AI: Fundamentals and Applications
Embodied AI: Concepts, Challenges, and Applications
Agentic AI: Concepts, Applications, and Case Studies
Data-Centric AI: Concepts, Applications and Case Studies
Sovereign AI: Frameworks, Principles, and Practices
Accelerated AI: Energy Efficient Computing Systems for Large Language Models (LLMs)
Actionable AI: Powered by Large Action Models (LAMs)
AI-Augmented Software Engineering: Principles and Practices
Ambient Invisible Intelligence: Fundamentals, Challenges, and Applications
Artificial Cognition: Technology, Applications, and Challenges
Artificial General Intelligence: Principles and Practices
Autonomous AI Agents: Foundations and Applications
Principles of Confidential Computing
Disinformation Security for Generative AI Applications: Concepts, Methodologies, and Use Cases
Embedded AI: Concepts, Challenges, and Applications
Graph Neural Networks (GNNs): Concepts and Applications
Industry Cloud Platforms: Fundamentals, Applications, and Case Studies
Multi-Modal Generative AI: Frameworks, Principles, and Practices
Objective-Driven AI: Concepts, Applications, and Case Studies
Physical AI for Robotics and Automation: Concepts, Techniques, and Applications
Predictive AI: Concepts, Applications, and Case Studies
Prescriptive AI: Concepts, Applications, and Case Studies

Shadow AI: Concepts, Applications, and Security
Small Language Models (SLMs) for Compact AI: Architectures, Principles, and Applications

www.scrivenerpublishing.com

Printed and bound by CPI Group (UK) Ltd, Croydon, CR0 4YY

21/09/2026

14965806-0001